THE NORMAL DISTRIBUTION

The best known and most widely used of all probability models, the normal family of distributions, is the foundation for almost all classical statistical procedures. It is commonly referred to as the bell-shaped curve. The normal distribution is used in a wide variety of situations to describe natural phenomena that have measurements clustering around a central value with deviations above and below being equally likely and occurring with decreasing frequency as the deviations increase in absolute value.

The normal distribution was developed or discovered independently by three different men at different times. In 1733, Abraham DeMoivre (1667–1754) found it to be the limiting form for binomial distributions. In 1778, Marquis Pierre Simon De LaPlace (1749–1827) derived it as the distribution of errors of astronomical observations. In the early 1800s, Carl Friedrich Gauss (1777–1855) derived it for a third time, also as the distribution of errors of astronomical observations. Due to the latter development by Gauss, the normal distribution is known in physics and engineering as the Gaussian distribution.

The first application of the normal to other areas appears to be due to Lambert A. Quetelet (1796–1874), a Belgian astronomer who had a strong interest in the social sciences. He discovered that the normal distribution provided a good fit to distributions of anthropological measurements. After Quetelet, several empirical studies in a variety of areas demonstrated that the normal provided an excellent fit for the distributions of sums of individual measurements. Over time, phenomena that could be described by the Gaussian distribution began to be considered as normal, and anything that was non-Gaussian was deemed abnormal. Thus, the term "normal" became associated with this family of distributions.

Business Statistics

GARY E. MEEK
HOWARD L. TAYLOR
KENNETH A. DUNNING
KEITH A. KLAFEHN

The University of Akron

Allyn and Bacon, Inc.

Boston / London / Sydney / Toronto

To Margo
Betty
Nan
Muriel
May we find the time to get to know each other again.

Editorial-production service: Technical Texts, Inc.
Text designer: Sylvia Dovner
Cover designer: Richard Hannus
Production administrator: Lorraine Perrotta
Manufacturing buyer: William J. Alberti

Library of Congress Cataloging-in-Publication Data

Business statistics.

Includes index.
1. Commercial statistics. 2. Statistics.
I. Meek, Gary E.
HF1017.B87 1987 519.5'024'658 86-14142
ISBN 0-205-10281-6

Printed in the United States of America.
10 9 8 7 6 5 4 3 91 90 89 88

CONTENTS

* This section is optional.

PREFACE

INTENDED AUDIENCE

As the title *Business Statistics* implies, this text is intended for use primarily in a business curriculum. Because many of the examples and applications relate to allied disciplines such as public administration, the text can easily be used for similar courses in other curricula. The book is designed for use in a two-term sequence (semesters or quarters), but it may be used for a one-term course by selectively choosing the chapters or topics to be covered.

The mathematical background required for the majority of the material in the text is the equivalent of high school algebra. Some sections do require a knowledge of introductory calculus, and a calculus review in the appendix provides the necessary background. The sections that utilize calculus can be omitted without loss of continuity if the instructor so desires.

Throughout the text, there is an emphasis on intuitive development rather than mathematical derivation. In addition, the assumptions inherent in the use of each procedure are identified and stressed. We firmly believe that if the user knows and understands the assumptions of a procedure, he or she is much less likely to misuse it.

SPECIAL FEATURES

There are a number of features that separate this book from the majority of texts on the market today. One of these features is the placement of descriptive statistics after the probability chapters. The rationale for this placement is the belief that topics should be introduced at the time they are to be used. It is traditional to place the descriptive material in the first few chapters, but since this material is not utilized until several chapters later, it must be reviewed before continuing. One of the authors argued for the traditional placement of descriptive statistics, but he agreed to try the current positioning in the classroom on a trial basis. Now he is convinced of the advantages of this approach and is the most vociferous opponent of the traditional organization.

A second feature that we believe to be unique to a text at this level is the coverage of the test for lack of fit of the linear model in regression analysis. To our knowledge, all other comparable texts assume the linear model with no mention of the fact that this assumption can easily be tested. For situations where insufficient data are available to permit a formal test of the linearity assumption, we have introduced residual plots to provide a simple but informal means of checking the regression assumptions.

The coverage of distribution-free or nonparametric statistical procedures is more extensive than any other text at this level. Except for the chi-square test of a single variance, we have provided a distribution-free alternative for each classical procedure. Included are one- and two-sample procedures for location for both paired data and independent samples, a procedure for testing variances from two independent samples, the runs test for randomness, a rank correlation procedure, and ranking alternatives to analysis of variance.

The presentation of analysis of variance and experimental design is extensive with coverage through a two-factor design with replication. Particular attention is given to the concepts of fixed effects and random effects. It is stressed that the test statistics and *a posteriori* analysis that are presented assume a fixed effects model. The reader is cautioned to evaluate carefully the types of effects present for his or her situation and not to use the test statistics in this text if any effects are random. In addition, the role and the importance of Cochran's theorem in the development of all analysis of variance procedures is highlighted.

A complete chapter on quality control procedures is included. The chapter covers not only the classical acceptance sampling techniques but also the modern statistical process control concepts necessary for total quality control.

Various chapters contain topics that often are not included in other books. Among these topics are the hypergeometric and negative exponential distributions in Chapter 4; stem-and-leaf plots and the coefficient of variation in Chapter 5; the interrelationship of the Z, χ^2, t, and F distributions in Chapter 6; an emphasis on the fact that confidence intervals and hypothesis tests are probability statements in Chapters 7 and 8; and multicollinearity and indicator variables in Chapter 11.

ORGANIZATION OF MATERIAL

Our primary concern in the organization of the material throughout the text has been the introduction of material and concepts at the place where they are to be used. That is the rationale for placing descriptive statistics after probability and probability distributions. Special topics such as nonparametric statistics, index numbers, Bayesian analysis, and quality control are placed at the end so that the user with limited time may emphasize the classical approach and may omit these chapters without a loss of continuity.

Each chapter begins with a brief scenario of a practical problem. The scenario highlights one or more of the concepts introduced in the chapter. In addition, each chapter concludes with a tabular summary highlighting the concepts and formulas of the chapter. Throughout the text, a total of 268 examples are worked out in detail to aid the student in mastering the concepts and formulas.

Many of the 971 exercises in the text follow the individual sections within the chapters. These provide the student with an immediate opportunity to utilize the concepts of the section. Answers to the end-of-section exercises are given in Appendix C.

The majority of exercises are in supplementary sets at the ends of the chapters. These exercises have been randomly positioned so that the student must consider the chapter as a whole in working them. We believe that this randomization aids the student in synthesizing the material since a problem cannot be identified by its positioning. In addition, a few of the supplementary exercises are marked with asterisks. A single asterisk identifies an exercise that either is a minor proof or requires a better understanding and synthesis of the material. A double asterisk indicates an exercise that is a major variation and requires a thorough understanding of concepts.

SUPPLEMENTS

Available with the text are a study guide, an instructor's manual, a test bank, and a computer diskette. The study guide is keyed to the text and includes additional examples and exercises. The instructor's manual provides detailed solutions to all exercises in the text. The test bank provides both problems and multiple choice questions for each chapter. The computer diskette contains actual data bases as well as statistical programs. The programs are designed for use as teaching tools rather than sophisticated statistical packages. We encourage the reader to become familiar with the various commercial packages that are available, such as SAS, SPSS, ADEPT, MINITAB, and BMD. We have included actual printouts from the SAS and MINITAB packages. The use of these two is not an endorsement of them but is merely to represent what is available.

ACKNOWLEDGMENTS

We are indebted to the various reviewers who made many helpful comments and suggestions to greatly improve the manuscript. Every effort has been made to incorporate their suggestions. Among these individuals are

Shirley Dowdy, West Virginia University
Jack Hayya, The Pennsylvania State University
Ravinda Nath, Memphis State University
James E. Willis, Louisiana State University
Jack Yurkiewicz, Hofstra University

We also thank the staffs of Allyn and Bacon and Technical Texts for their continual support, encouragement, and never-ending attention to detail.

Most importantly, we thank the many students who served as test subjects in the development of this text and the typists who spent hundreds of hours laboring over our handwriting. Special thanks go to Debra Catanzarite, Julie Sweet, Barb Lucas, Margo Meek, Mary Lou Ondack, and Gail Wild, whose expertise in typing and word processing is unparalleled.

1 Introduction

1.1 MEANING AND NATURE OF STATISTICS

Henry Morrison is in charge of the records for the 143,769 credit customers of the Enterprise Oil Corporation. The accounts of credit customers represent an important part of the firm's assets. As a result, Henry is continually receiving requests for information about the accounts for various decision-making purposes. Some of the requests come from officials of the company, but many are from outside auditors who must confirm balances or determine the quality of the data in terms of an error rate.

The following questions are typical of those asked of Henry: What is the average unpaid balance of the accounts? What proportion of the accounts is more than sixty days past due and what is the dollar amount of such accounts? Has there been a significant change from a year ago in the proportion of accounts more than sixty days past due? What proportion of the accounts is believed to have incorrect balances? Is there a relationship between account size and whether there was a complaint regarding the balance?

Henry Morrison realizes that many questions can be answered for all the company's accounts through use of the firm's computer system. He also knows that the outside auditors require information from only a sample of accounts to provide satisfactory answers to certain questions of interest. Procedures that are useful for making managerial decisions and that require only sample information fall in the area of science known as "statistics."

Nearly everyone has heard of statistics and correctly associates the word with numbers or figures. But the manner of the association differs widely from person to person. Some people think of "vital statistics" and believe that a statistician's main function is to report numbers of births or deaths. Others picture the statistician arranging the numbers in a table or displaying them graphically. And others have the impression that statistics involves more complicated computations, such as adding the numbers or summarizing them in some way, or somehow describing the data by a single number.

The activities mentioned above are a part of *descriptive statistics*—the handling of data when no interpretation beyond the data at hand is intended. Obviously, this constitutes one portion of statistical methodology. However, it is not the most important part, and it is not the part we emphasize.

When we use statistics, we are primarily concerned with *analysis of data* in such a way that the reliability of conclusions we draw can be assessed by means of probability. For example, from a national telephone poll of 1200 people, we might conclude that 43% of all citizens approve the president's economic policy and that there is a probability of .95 that the value of 43% is within 2.8% of the true percentage. Thus, we place our emphasis on what is known as *inductive statistics*—making inferences from sample data to the population from which the sample was drawn. In the preceding statement, we have used two words—population and sample—that need to be defined. (Probability is covered in detail in Chapter 2.)

DEFINITION 1.1 A **population** is the total collection or set of values for which we desire information.

Examples of populations would be prices of all stocks listed on the New York Stock Exchange, amounts of all the invoices of the Halifax Corporation, incomes of all residents of Russellville, and diameters of all ball bearings produced by the Hartrack Company. Note that populations do not necessarily refer to human beings.

Sometimes the word *universe* is used to refer to the collection of elements, and the word *population* is used to refer to the collection of values for some characteristic of the elements in the universe. Distinguishing between a universe and a population in this way may be important when there is more than one property of interest. For example, a universe could consist of a set of stones. One population of interest might be the weights of the stones. A second population might be the volumes. Both populations pertain to the same universe.

Even though it would be useful in certain instances to distinguish a population from a universe in the manner just suggested, we do not do so in this text. Instead, we refer only to populations. It should always be clear from the context whether reference is being made to the set of elements or to the set of values for a characteristic of interest.

DEFINITION 1.2 A **sample** is any subset or portion of a population.

Examples of samples would be items selected from a continuous production line, a selection of accounts receivable ledgers of any corporation, and selected companies whose stocks are listed on the New York Exchange.

DEFINITION 1.3 A sample is **random** if all possible samples of the same size have an equal chance of occurring.

Definition 1.3 is appropriate for what is usually called *simple random sampling*. A more precise definition of a random sample is given in Section 5.2. In many places throughout the text, the word *random* is omitted, but we still have in mind a random sample. In effect, then, *sample* and *random sample* are used interchangeably in this text.

The following definition of statistical methods incorporates the preceding definitions and, to a large extent, indicates the approach we would like to emphasize in this text.

DEFINITION 1.4 **Statistical methods** are procedures for drawing conclusions about populations utilizing information provided by random samples.

In Chapter 5 we discuss the rationale for using sample data as an alternative to examining all items in a population.

1.2 IMPORTANCE OF STATISTICS

Statistics has been an important subject of study for many years, and as time passes, it is becoming even more important for several good reasons. First, as our economy grows and business becomes more complex, the volume of quantitative data available for analysis keeps growing. Second, more and more administrators are willing to apply statistical tools to quantitative data to aid them in their decisions. Third, mathematical statisticians are continually deriving new methods that eventually find their way into practice. Fourth, the impact of computers is so great that many statistical problems that were difficult or impossible to solve in the past can now be solved rather easily.

Quantitative data abound in all types of sciences, disciplines, and activities. Regardless of their source, properly collected data can be analyzed by statistical methods. Thus, statistics is used, for example, in engineering, education, biology, sociology, business, and in nearly every field of study. However, in our illustrations, we draw, for the most part, from the area of business. We now cite some examples to illustrate statistical problems in business.

1. A company packages baby food in small jars. The intended net weight is 4 ounces. The process is not operating properly if jars are considerably overfilled or underfilled. Considering the risks of a wrong decision, what deviation is tolerable before the process is judged to be out of control?

2. A supermarket chain has records for the past ten years on expenditures for newspaper advertising and volume of sales. There appears to be a strong relationship between the two variables—that is, the more that is spent on advertising, the larger the sales. Can the nature of this relationship be determined so that sales for next year could be predicted if advertising expenditures are set at a certain amount?

3. A company is planning a large-scale survey in which mailed questionnaires will be used. A test mailing using three different mailing approaches is made. The number of responses for each approach is recorded. The company wonders whether the observed differences can be attributed to chance or to a real difference among the approaches.

4. Data are available for the number of items produced in a certain time period by four workers, each using three machines. Is there a significant difference in amounts of production between either the workers or the machines?

These examples illustrate only a few of the possible applications of statistics to business problems. Statistics has become a part of nearly all decision making in the business environment.

For most business problems that we wish to analyze, a fundamental knowledge of algebra will be sufficient mathematical preparation. However, in a few instances, some elementary calculus is required. Students who do not have previous exposure to calculus can benefit from carefully working through the fundamental calculus concepts given in Appendix A.

1.3 SOURCES OF BUSINESS DATA

A primary source of business data is the information routinely recorded by an organization in its day-to-day operation, such as sales records, payroll information, and manufacturing or production data. Most firms have a large mass of internal data of this type.

External data are also frequently used in the analysis of business problems. The data come largely from information published by government agencies or specialized research organizations. A familiar example is the Consumer Price Index, published regularly by the Bureau of Labor Statistics. The *Federal Reserve Bulletin* contains many statistics in the area of finance. We could, of course, cite many other examples.

If the information desired is not available either internally or externally, it may be necessary to conduct a special investigation to obtain it. We now consider briefly the two types of statistical investigations that generate data for analysis: the *experiment* and the *survey*.

The experiment is usually associated with physical sciences, but it is not necessarily restricted to those areas. The distinguishing characteristic of experiments is that they are usually performed under controlled conditions. The yield of a chemical process, for example, can be observed for various levels of pressure and temperature controlled by the experimenter.

In most surveys, it is not possible to have the control that exists in an experiment. Surveys are usually associated with business and the social sciences, where extraneous factors, often uncontrollable, are most troublesome. Many surveys are concerned with human populations where exact duplication of conditions is impossible.

Let us consider an example to illustrate the difference between surveys and experiments. Marketing research workers are frequently interested in studying the effect of various merchandising practices upon the volume of consumer purchases of specific products. If, in advance of obtaining the data, no effort is made to control any of the large number of other variables that also affect retail sales, then the study is classed as a survey. If the effect of other variables is eliminated at all, it must be done by statistical analysis after the data are obtained. In an experiment, it may be possible to control an extraneous variable that is considered likely to affect the results. The effect would be to eliminate any influence of such a variable on sales volume.

To be specific, let us consider two merchandising practices, A and B, and two store sizes—large and small. If the practices are already in effect in the stores and the investigator observes things as they are, the study is classed as a survey. The sales data could be affected by store size as well as the type of practice. It might be possible to remove the effect of store size by a technique such as regression analysis (see Chapter 10). However, an arrangement that controls store size could easily be achieved if it is possible to assign practices to stores. In our example, if half the stores of each size receive one merchandising method, and half receive the other method, then differences observed in sales for the two methods should not be affected by store size because that variable was controlled to be the same for both methods. Such a study could be classed as an experiment.

In summary, both surveys and experiments are useful methods for obtaining data. The experiment is usually considered the more powerful method. However, much of statistical analysis, especially in business and economics, is based on survey data.

1.4 LEVELS OF MEASUREMENT

For most of the statistical procedures presented in the text, a certain level of measurement must exist for the sample data in order for the particular procedure to be applicable.

DEFINITION 1.5 The **level of measurement** associated with a set of observations is the amount of quantitative information contained in the sample data. There are four levels of measurement: nominal, ordinal, interval, and ratio.

To analyze sample data, we need to be familiar with the four measurement levels. The simplest or weakest level of measurement is the *nominal* scale. Objects in the sample are simply classified into categories. That is, each item may be assigned a number, but the numerical value is actually little more than a name—hence, the term "nominal." An example of nominal data is the returns from a political preference survey where each person was asked for a yes or no response on a bond issue. There is no inherent quantitative difference between a yes and a no; thus, with most nominal data, the variable used is the frequency or number of observations belonging to each category.

The next step up in the hierarchy of levels of measurement is the *ordinal* scale. As the name implies, data satisfying this level can be ordered or ranked in some way. The ranking need not be accomplished using actual numerical values; it can take the form of preferences. For example, in comparing the flavors of various margarines, a person may indicate which is least preferable and then order them from the least to the most preferable, assigning 1 to the least preferable and n to the most preferable. We adhere to this ranking procedure in the text, but the order may be reversed without affecting the procedures. Though numerical values may designate the observations, they serve only to identify the relative positions of the observations when rated from "worst" to "best." With strictly ordinal data, it is meaningless to compare two observations in terms of the distance between them.

The next highest level of measurement that we consider is the *interval* scale. For data to satisfy an interval scale, not only are the observations able to be ranked, but also an exact difference between any two observations can be obtained and is meaningful. The interval level of measurement does not assume a natural origin—that is, a natural zero—but it is necessary that a one-unit change on the scale correspond to a one-unit change in the object being measured. An example of the interval scale is temperature, since the selection of an origin is arbitrary, and 80° is not twice as hot as 40°. We find in future chapters that many test procedures assume at least an interval level of measurement.

The fourth and strongest type of measurement level is called the *ratio* scale. A ratio scale has all the properties of an interval scale, but it also has a natural origin. With measurements obtained on a ratio scale, it is possible to make a statement such as "x_1 has twice the value of x_2," where x_1 and x_2 represent specific values. Length and weight are examples of measurements on a ratio scale. In this text, none of the procedures presented require a level of measurement stronger than the interval level.

Note that data satisfying one level of measurement also meet the condition of any less sophisticated level. For example, observations on an interval scale can obviously be ranked and therefore satisfy the requirements of the ordinal scale. It is critical that we be familiar with the *manner* in which numerical data are obtained. In many situations, if only the numbers are considered, the data may appear to be interval; whereas, if their manner of origin is considered, the scale of measurement may be ordinal at best and possibly no better than nominal.

To illustrate this point, consider a sample of values: 3.2, 7.7, 6.0, 6.3, 5.4, 2.5, 8.1, 4.9, 9.1. At first glance, the data appear to be of an interval level, but let us consider how they were obtained. The nine values correspond to ratings of this text by nine randomly selected students. They were instructed to rate the text on a continuous scale from one to ten as to its readability when compared to other quantitatively oriented texts with which they were familiar. Obviously, it is ridiculous to attach any significance to the difference of 5.9 that occurs between the ratings of the first and ninth students. At best, we could order the values—that is, rank the ratings—and possibly we should do no more than categorize them in terms of ratings greater than 5.0 versus ratings less than or equal to 5.0.

EXERCISES

1.1 Indicate the level of measurement associated with each of the following: **(a)** classification of production workers as fast, average, and slow; **(b)** numerals assigned to a team of baseball players; **(c)** weights obtained from a well-adjusted scale; **(d)** measurements of temperature obtained from a Celsius thermometer; **(e)** identification of police officers by rank; **(f)** identification of service stations by brand of gasoline sold; and **(g)** IQ measurements.

1.5 COMPUTERS AND STATISTICS

Modern digital computers can perform highly complex computations many times faster and more accurately than the human brain. Hence, the usefulness of the computer in business statistics, where quantitative data abound, is obvious.

Time saved by the computer can be devoted by the statistician to the analysis of other problems that might not otherwise be considered. The statistician is further aided by much available software—that is, prepared programs already available for frequently occurring problems. Software for standard statistical techniques is readily available for most computer systems. Some of the statistical packages that are available are MINITAB, BMD (Biomedical Programs), SAS (Statistical Analysis System), SPSS (Statistical Package for the Social Sciences), and ADEPT (Advanced Data Enquiry Package—Time-shared). New programs need be written only for less common situations. In this text, we have provided sample output from the SAS and MINITAB packages. These printouts are for illustrative and comparative purposes only and are not an endorsement.

Writing computer programs is not discussed in this book. This fact does not mean that the use of computers is discouraged. In fact, the opposite is true: Students should use the computer whenever possible. Using a computer does not remove the necessity for understanding basic statistical principles. The computer can do only what it is instructed to do. In this text, our attention will be devoted to covering basic principles of statistics. However, in nearly every chapter, there is opportunity for students to use the computer with packages or statistical software that may be available at their computer centers.

1.6 SET OPERATIONS

Familiarity with set theory will help in our understanding the basic concepts of probability in the next chapter. In this section, we consider some of the fundamental concepts and operations associated with sets.

For our purposes, the following definition of a set will suffice.

DEFINITION 1.6 A **set** is a collection of distinct objects having some attribute(s) in common. ■

Examples of sets are the department managers of the Jones Company, the paved streets of Laredo, the stocks in Abner's portfolio, and the positive integers. A set may be either finite or infinite. The distinct objects in a set are called *elements*.

We frequently need to distinguish between all the elements of interest to us and a portion of them.

DEFINITION 1.7 A **universal set** is the total set of elements of interest. ■

The universal set might be the set of positive integers, for example.

DEFINITION 1.8 A **subset** is a portion of the universal set defined in some unambiguous way. ■

All the elements of a subset are contained in the universal set. If the universal set is the set of positive integers, then the first four positive integers represent a subset. Elements of a set or subset are usually separated by commas and enclosed in braces. Thus, the subset consisting of the first four positive integers is written

$$\{1,2,3,4\}$$

Sometimes a set with no elements in it is useful.

DEFINITION 1.9 An **empty,** or **null, set** is the subset of the universal set that contains no elements. ■

If the universal set is the accountants employed by the Strasberg Company, then an example of an empty set is the set of accountants having 100 or more years of service with the company. The symbol $\emptyset$ is often used to represent the empty set. By definition, the null set is a subset of every set.

Universal sets and subsets can be represented by *Venn diagrams*. The usual representation of the universal set is as a rectangle with subsets shown as circles inside the rectangle. In Figure 1.1, A and B are both subsets of the

FIGURE 1.1 Mutually exclusive subsets A and B

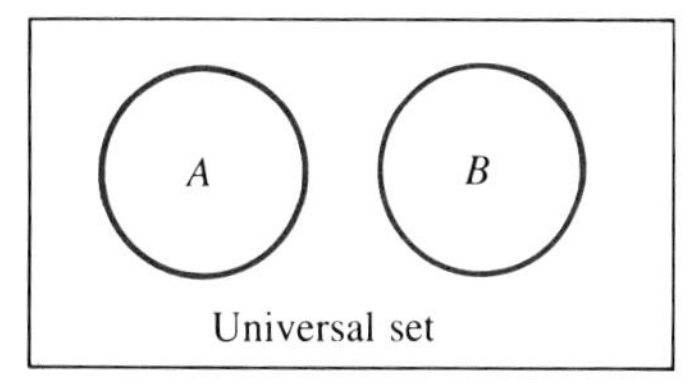

universal set. We note that A and B have no elements in common. In such an instance, A and B are said to be *mutually exclusive* or *disjoint*.

Several operations with sets are of interest to us. Let the universal set be $\{1,2,3,4,5,6\}$ and $A = \{1,2,3,4\}$ and $B = \{3,4,5\}$ be two subsets.

DEFINITION 1.10 The **union** of two sets is the set of elements that are elements of either set or that are elements of both sets.

For example, the union of A and B, denoted by $A \cup B$, is $\{1,2,3,4,5\}$.

DEFINITION 1.11 The **intersection** of two sets is the set consisting of the elements common to both sets.

For example, the intersection of A and B, denoted $A \cap B$, is $\{3,4\}$. In Chapter 2, AB is also used to represent an intersection.

DEFINITION 1.12 The **complement** of a set contains all elements that are in the universal set but not in the set itself.

The complement of A, symbolized as A', is $\{5,6\}$; the complement of B is $B' = \{1,2,6\}$.

The set operations of union, intersection, and complement may be conveniently represented by using Venn diagrams. In Figure 1.2, the shaded area

FIGURE 1.2 $A \cup B$, the union of sets A and B

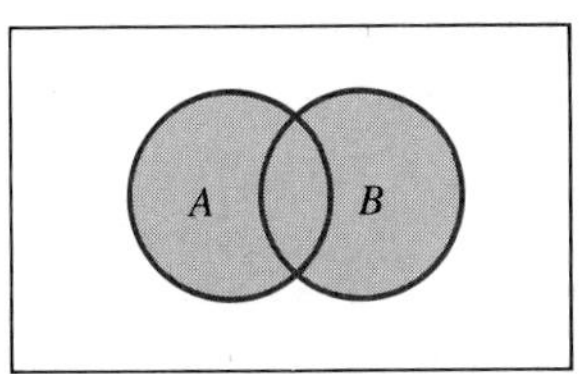

FIGURE 1.3 $A \cap B$, the intersection of sets A and B

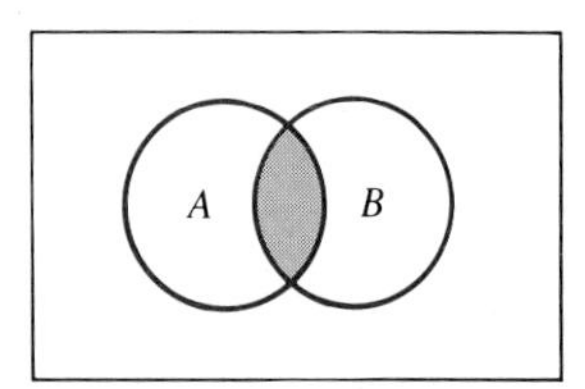

FIGURE 1.4 A', the complement of the set A

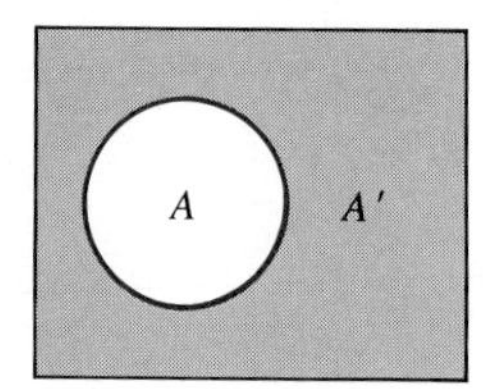

represents $A \cup B$, the union of A and B. In Figure 1.3, the shaded area represents the intersection of A and B. In Figure 1.4, the shaded area corresponds to A', the complement of the set A.

EXERCISES

1.2 Consider the universal set consisting of the employees of the Buskirk Company:

{Able, Baker, Cromwell, Dennis, Ernst, Fitz}

Suppose the set of employees in the sales department is given by A = {Baker, Dennis, Ernst}. Suppose the set of female employees is given by B = {Able, Dennis}.

- **(a)** What are the elements in the set $A \cup B$?
- **(b)** What is the set given by $A \cap B$?
- **(c)** What employees are in A'?
- **(d)** Who is in B'?
- **(e)** Who is in $A' \cap B'$?
- **(f)** Who are the male employees not in the sales department?
- **(g)** Determine the elements in $A' \cup B'$. For which set is this the complement?
- **(h)** For (a) through (e), characterize each set in words—for example, female employees in the sales department.

1.3 Are $\emptyset$ and $\{0\}$ the same? If not, explain the difference.

1.4 Give an example of a finite set.

1.5 Give an example of an infinite set.

1.6 Consider the set $A = \{d,e,f\}$ and the set $B = \{e,f,g,h\}$.

- **(a)** What are the elements in $A \cup B$?
- **(b)** What elements are in $A \cap B$?
- **(c)** Is it possible to identify the elements in A'?

1.7 SUMMATION NOTATION

We close this chapter with a discussion of summation notation that assumes no prior knowledge of the subject. This important topic is used throughout the text, and our experience indicates that even students already familiar with the subject will benefit from a review.

Throughout the text many letters from the Greek alphabet are used. For reference, all the Greek letters that appear in the text are given in Table 1.1.

The capital Greek letter sigma (Σ), which represents summation, has tremendous importance in statistics. Students must understand and be able to manipulate expressions that use this symbol.

The summation symbol is used in conjunction with variables to indicate the sum of a number of terms. In this text, we represent a variable of interest with a capital letter, such as X, Y, or Z. Particular values of the variable are represented with lowercase letters, such as x, y, or z. Usually, the value has a subscript that refers to the location of the value in the sequence of data.

The following example illustrates this notation. Let

$$x_1 = 3 \qquad x_2 = -2 \qquad \text{and} \qquad x_3 = 5$$

We want a method to express the sum of these three variables. That is,

$$x_1 + x_2 + x_3 = 3 + (-2) + 5 = 6$$

We use

$$\sum_{i=1}^{3} x_i = x_1 + x_2 + x_3 = 6$$

In general, the formula is

$$\sum_{i=1}^{n} x_i = x_1 + x_2 + x_3 + \cdots + x_n$$

The value of the variable is designated by x. The left side of the equation is read as "the summation of x_i, i going from 1 to n." Below the summation sign appears "$i = 1$," which identifies the beginning term in the sum. That is, in the general term to be added, replace i by 1 to obtain the first term in the sum. In x_i, replace i by 1 to obtain x_1. The second term is found by replacing i with 2. This

TABLE 1.1 Greek letters used in this text

α	alpha	ν	nu
β	beta	π	pi
ε	epsilon	ρ	rho
θ	theta	Σ	sigma (capital)
λ	lambda	σ	sigma (lowercase)
μ	mu	χ	chi

process is continued until the term is reached where i is replaced by the symbol appearing above Σ. In this case, it is n. Therefore, the last term in the sum is x_n.

This procedure can be followed for general terms that are more complicated than x_i. Of course, we must observe the laws of exponents, treat parenthetical expressions properly, and, in general, follow the laws of algebra.

Before proceeding to a worked-out example, let us consider a symbol such as a, c, or k without subscripts. Such a symbol indicates a *constant* rather than a *variable*.

EXAMPLE 1.1 Write the following expression as a summation.

$$(x_1^2 - a) + (x_2^2 - a) + (x_3^2 - a) + (x_4^2 - a)$$

Solution: The symbol a is a constant that is subtracted from the square of the value of the variable x_i. The general term is thus $x_i^2 - a$. There are four terms in the sum. Therefore, the answer is

$$\sum_{i=1}^{4} (x_i^2 - a)$$

EXAMPLE 1.2 Use summation symbolism and rewrite the following.

$$(2y_1 - 2)^2 + (2y_2 - 2)^2 + (2y_3 - 2)^2$$

Solution: Here, y is a variable with the number 2 as a constant multiplier. In each term, the number 2 is also to be subtracted from twice the y value, and the whole quantity is to be squared. There are three terms in the sum. The answer is

$$\sum_{i=1}^{3} (2y_i - 2)^2$$

EXAMPLE 1.3 If $n = 3$, write the following expression in full without using summation signs.

$$\frac{\sum_{j=1}^{n} (y_j - \mu)^2}{n}$$

Solution: The variable in this problem is y, and the subscript is j. The symbol μ designates a constant. The general term in the sum is $(y_j - \mu)^2$, and the entire sum is divided by 3. The answer is

$$\frac{(y_1 - \mu)^2 + (y_2 - \mu)^2 + (y_3 - \mu)^2}{3}$$

EXAMPLE 1.4 Write out the following without using summation signs

$$\sum_{i=1}^{3} 2x_i - 2$$

Solution: There are three terms in the sum. The variable x is multiplied by the constant 2 in each term. It might seem that 2 is also to be subtracted in each term, but such is not the case. For this to be true, the expression would need to be written as

$$\sum_{i=1}^{3} (2x_i - 2)$$

In the original expression, the number 2 is to be subtracted only once. Therefore, the answer is

$$2x_1 + 2x_2 + 2x_3 - 2$$

There are certain rules or properties relating to summations that facilitate certain expressions. We state these rules without proofs.

RULE 1.1

$$\sum_{i=1}^{n} (x_i + y_i) = \sum_{i=1}^{n} x_i + \sum_{i=1}^{n} y_i \quad \textbf{(1.1)}$$

RULE 1.2

$$\sum_{i=1}^{n} kx_i = k \sum_{i=1}^{n} x_i \quad \textbf{(1.2)}$$

RULE 1.3

$$\sum_{i=1}^{n} k = nk \quad \textbf{(1.3)}$$

EXAMPLE 1.5 Show that Rules 1.1, 1.2, and 1.3 are valid for the following data: $n = 2$, $k = 5$, $x_1 = 3$, $x_2 = 1$, $y_1 = -1$, $y_2 = 4$.

Solution: For Rule 1.1,

$$\sum_{i=1}^{2} (x_i + y_i) \stackrel{?}{=} \sum_{i=1}^{2} x_i + \sum_{i=1}^{2} y_i$$
$$(x_1 + y_1) + (x_2 + y_2) \stackrel{?}{=} (x_1 + x_2) + (y_1 + y_2)$$
$$(3 - 1) + (1 + 4) \stackrel{?}{=} (3 + 1) + (-1 + 4)$$
$$2 + 5 \stackrel{?}{=} 4 + 3$$
$$7 = 7$$

For Rule 1.2,

$$\sum_{i=1}^{2} 5x_i \stackrel{?}{=} 5 \sum_{i=1}^{2} x_i$$
$$5x_1 + 5x_2 \stackrel{?}{=} 5(x_1 + x_2)$$
$$(5)(3) + (5)(1) \stackrel{?}{=} 5(3 + 1)$$
$$15 + 5 \stackrel{?}{=} (5)(4)$$
$$20 = 20$$

For Rule 1.3,

$$\sum_{i=1}^{2} 5 \stackrel{?}{=} (2)(5)$$
$$5 + 5 \stackrel{?}{=} (2)(5)$$
$$10 = 10$$

We note that the last example is not a proof because it is a special case.

EXERCISES

1.7 In each of the following expressions, write out the individual terms without using summation signs. Do not simplify the algebra.

(a) $\dfrac{\sum_{j=1}^{4} x_j^2}{\sum_{j=1}^{4} y_j}$

(b) $\dfrac{\sum_{i=1}^{4} (x_i^2 - a)^2}{c}$

(c) $\sum_{i=1}^{3} x_i y_i - 3\left(\sum_{i=1}^{3} x_i\right)\left(\sum_{i=1}^{3} y_i\right)$

(d) $\sum_{i=2}^{4} 4x_i y_i$

1.8 If $x_1 = 1$, $x_2 = 3$, and $x_3 = 4$, calculate

$$\sum_{i=1}^{3} (x_i + 2)^2$$

1.9 If $x_1 = 1$, $x_2 = 0$, $x_3 = 4$, $y_1 = 1$, $y_2 = 1$, and $y_3 = 1$, calculate

$$2\sum_{i=1}^{3} x_i^2 - \sum_{i=1}^{3} \frac{x_i}{y_i}$$

1.10 Write each of the following using a single summation sign whenever possible.

(a) $4(x_1^2 - y_1) + 4(x_2^2 - y_2) + 4(x_3^2 - y_3) + 4(x_4^2 - y_4)$

(b) $2(y_1^2 - a) + 2(y_2^2 - a)$

(c) $\dfrac{ax_1y_1 + ax_2y_2 + ax_3y_3}{z_1 + z_2 + z_3}$

1.8 SUMMARY

In this chapter, we considered the meaning and nature of statistics, illustrated its importance by means of several practical examples from business, and observed that the text emphasizes making inferences from sample data to the population from which the sample was drawn. We observed that data used in the analysis of business problems may come from internal sources in an organization; external sources, such as published government statistics; or from an experiment or a survey.

We noted the importance of using the computer, although no computer programs are given in the book. We considered the four levels of data measurement (nominal, ordinal, interval, and ratio) in considerable detail. We reviewed fundamentals of set theory. We concluded the chapter with a discussion of summation notation which is encountered in many of the expressions given in forthcoming chapters. The main topics contained in this chapter are summarized in Table 1.2.

SUPPLEMENTARY EXERCISES

1.11 What concept of statistics will be emphasized in this book?

1.12 Distinguish between population and sample. Give examples.

1.13 Could a given set of observations be a population in one problem and a sample in another?

1.14 Give a business problem in which statistical methods would be used.

1.15 Give examples of internal and external data.

1.16 Distinguish between a survey and an experiment.

1.17 Identify the level of measurement associated with each of the following: **(a)** the UPI ratings of major college football teams; **(b)** a political preference survey on presidential candidates; **(c)** the failure times of individual light bulbs; **(d)** items being identified as defective or nondefective; **(e)** the letter grades assigned to students in this class; and **(f)** the fill weights of randomly selected boxes of cereal.

1.18 How has the computer affected the statistician?

1.19 Write each of the following using summation notation.

(a) $(2t_1x_1 - h) + (2t_2x_2 - h) + (2t_3x_3 - h)$

(b) $\dfrac{x_1y_1}{z_1} + \dfrac{x_2y_2}{z_2} + \dfrac{x_3y_3}{z_3}$

(c) $3(x_1^2y_1^2 + x_2^2y_2^2 + x_3^2y_3^2) - (x_1y_1 + x_2y_2 + x_3y_3)^2$

(d) $\dfrac{x_1f_1 + x_2f_2 + x_3f_3 + x_4f_4}{f_1 + f_2 + f_3 + f_4}$

1.20 If $x_1 = 1$, $x_2 = 0$, $x_3 = 4$, $y_1 = 1$, $y_2 = 2$, and $y_3 = 1$, calculate

$$\sum_{i=1}^{3} x_i^2 + \left(\sum_{i=1}^{3} x_iy_i\right)^2$$

TABLE 1.2 Summary table

General Concepts	Analysis of quantitative data is essential for business decisions. Emphasis is on inductive statistics—the making of inferences from sample data to the population from which the sample was drawn. Importance of the computer is recognized, and its use encouraged.
Sources of Data	Internal, such as production or sales data External, such as government statistics Survey or experiment
Levels of Measurement	*Nominal:* Classification of objects into categories *Ordinal:* Measurement with data ranked or ordered *Interval:* Measurement linearly related to its true magnitude with an arbitrary origin assumed *Ratio:* Measurement proportional to its true magnitude with a natural origin
Set Theory	Terminology *Universal set:* Entire set of particular interest *Elements:* Distinct objects of a set *Subset:* A portion of the universal set *Empty set:* Set containing no elements of the universal set Operations (assuming A and B are subsets of the universal set) *Union:* Set containing all elements in either A or B; symbolized by $A \cup B$ *Intersection:* Set containing all elements in both A and B; symbolized by $A \cap B$. *Complement:* Set containing all elements not in A; symbolized by A'
Summation Notation	Symbols Σ: Represents the sum of terms x_i: x represents a variable with the subscript i referring to the position in a sequence k: Represents a constant Rules $\sum_{i=1}^{n} (x_i + y_i) = \sum_{i=1}^{n} x_i + \sum_{i=1}^{n} y_i$ $\sum_{i=1}^{n} kx_i = k \sum_{i=1}^{n} x_i$ $\sum_{i=1}^{n} k = nk$

1.21 Write out the individual terms indicated in each of the following.

(a) $c \sum_{i=1}^{3} (x_i - z_i)^2$

(b) $\sum_{i=1}^{4} a(bx_i - cy_i)^2$

(c) $\sum_{i=1}^{n} \frac{\log x_i}{n}$

1.22 Consider the universal set $\{1,2,3,4,5,6\}$ and subsets $A = \{1,2,5\}$ and $B = \{2,4,6\}$. Indicate what elements are in each of the following sets: **(a)** A', **(b)** B', **(c)** $A \cap B$, and **(d)** $A \cup B$.

1.23 Prove Equation 1.1.

1.24 Prove Equation 1.2.

1.25 Prove Equation 1.3.

1.26 If $\bar{x} = \left(\sum_{i=1}^{n} x_i\right)\Big/n$, show that

$$\frac{\sum_{i=1}^{n} (x_i - \bar{x})^2}{n - 1} = \frac{\sum_{i=1}^{n} x_i^2 - \frac{\left(\sum_{i=1}^{n} x_i\right)^2}{n}}{n - 1}$$

1.27 Write out the individual terms for each of the following.

(a) $3 \sum_{i=1}^{5} (y_i^2 - b)$

(b) $\sum_{j=3}^{4} (2y_j - z_j)$

(c) $\sum_{i=-4}^{4} x_i$

1.28 Given that $x_1 = 2$, $x_2 = 4$, $x_3 = 3$, $y_1 = 1$, $y_2 = 5$, $y_3 = 3$, and $n = 3$, evaluate each of the following.

(a) $\sum_{i=1}^{2} x_i^2$

(b) $\sum_{i=1}^{n} y_i^2$

(c) $\left(\sum_{i=1}^{n} y_i\right)^2$

(d) $\left(\sum_{i=2}^{3} x_i\right)^2$

(e) $n \sum_{i=1}^{n} (y_i^2 - 2) + \left(\sum_{i=1}^{2} x_i\right)^3$

(f) $\left(\sum_{i=1}^{n} x_i y_i\right)^2$

(g) $\sum_{i=1}^{n} (x_i y_i)^2$

(h) $\left[\left(\sum_{i=1}^{n} x_i\right)\left(\sum_{i=1}^{n} y_i\right)\right]^2$

(i) $$\frac{n \sum_{i=1}^{n} x_i y_i - \left(\sum_{i=1}^{n} x_i\right)\left(\sum_{i=1}^{n} y_i\right)}{\sqrt{n \sum_{i=1}^{n} x_i^2 - \left(\sum_{i=1}^{n} x_i\right)^2}\sqrt{n \sum_{i=1}^{n} y_i^2 - \left(\sum_{i=1}^{n} y_i\right)^2}}$$

1.29 If $\bar{y} = \frac{1}{n} \sum_{i=1}^{n} y_i$, show that

$$\sum_{i=1}^{n} (y_i - \bar{y}) = 0$$

1.30 For the data of Exercise 1.28, show that

$$\sum_{i=1}^{n} \frac{x_i}{y_i} \neq \frac{\sum_{i=1}^{n} x_i}{\sum_{i=1}^{n} y_i}$$

1.31 If the universal set consists of $\{a,b,c,d,e,f,g,h\}$ and $A = \{b,d,f,g\}$ and $B = \{c,d,g,h\}$, list the elements in each of the following: **(a)** $A \cup B$, **(b)** $A \cap B$, **(c)** A', **(d)** B', **(e)** $(A \cup B)'$, **(f)** $(A \cap B)'$, **(g)** $A' \cup B'$, **(h)** $A' \cap B'$, **(i)** $A \cup B'$, **(j)** $A \cap B'$, **(k)** $A' \cup B$, **(l)** $A' \cap B$, and **(m)** $(A' \cap B') \cap (A \cap B')$.

1.32 If $\sum_{i=1}^{5} x_i^2 = 51$ and $\sum_{i=1}^{5} x_i = 14$, determine

$$\sum_{i=1}^{5} (x_i - 4)^2$$

1.33 If $x_1 = 4$, $x_2 = 2$, $x_3 = 0$, $y_1 = 3$, $y_2 = 1$, and $y_3 = 5$, find the value for

$$\sum_{i=1}^{3} x_i^2 - 2 \sum_{i=1}^{3} x_i y_i + \left[\sum_{i=1}^{3} (x_i + y_i)\right]^2$$

1.34 Use summation notation to express the following.

$$\frac{2(f_1 x_1^3 + f_2 x_2^3 + \cdots + f_{20} x_{20}^3)}{3y_1 + 3y_2 + \cdots + 3y_{20}}$$

1.35 Consider the data of Exercise 1.20 and verify that

$$\sum_{i=1}^{n} \sqrt{x_i} \neq \sqrt{\sum_{i=1}^{n} x_i}$$

1.36 Mr. Starr, an amateur meteorologist, recorded the maximum daily temperature in degrees Celsius during the past week as 2, 4, −1, 0, 6, −3, −1. If temperatures are represented by x_i, calculate

(a) $\bar{x} = \dfrac{\sum_{i=1}^{n} x_i}{n}$

(b) $s = \sqrt{\sum_{i=1}^{n} \dfrac{(x_i - \bar{x})^2}{n-1}}$

1.37 If $a = 5$, $b = 2$, $x_1 = 4$, $x_2 = 3$, and $x_3 = 1$, evaluate

$$a \sum_{i=1}^{3} x_i - \frac{1}{b}\left(\sum_{i=1}^{3} x_i\right)^2$$

1.38 If $\sum_{i=1}^{n} (x_i + 2)^2 = 100$ and $\sum_{i=1}^{n} (4x_i + 4) = 44$, find

$$\sum_{i=1}^{n} x_i^2$$

1.39 Write out the sum represented by

$$\sum_{i=1}^{4} af_i x_i^2$$

1.40 Consider the data of Exercise 1.20 and verify that

$$\sum_{i=1}^{n} x_i^2 \neq \left(\sum_{i=1}^{n} x_i\right)^2$$

1.41 Consider the universal set {1,2,3,4,5} and subsets $A = \{1,2,3\}$, $B = \{2,3,4\}$, and $C = \{3,4,5\}$. What are the elements in **(a)** A', **(b)** $A \cap B \cap C$, **(c)** $A \cup B$, and **(d)** $((A \cap B) \cup C)'$?

1.42 If $x_1 = 3$, $x_2 = 1$, $x_3 = 4$, $a = 5$, and $b = 2$, evaluate

$$\sum_{i=1}^{3} a(x_i - b)^2$$

1.43 If $m = \left(\sum_{i=1}^{3} x_i\right) \Big/ 3$, show that

$$\sum_{i=1}^{3} x_i^2 - \sum_{i=1}^{3} (x_i - 4m)(x_i + m) = 21m^2$$

1.44 Write the following using a summation sign and appropriate limits.

$$(x_1 - m)^2 + (x_2 - m)^2 + (x_3 - m)^2$$

1.45 Smith, Case, and Sumner, certified public accountants, have obtained the following monthly information on cost (in dollars) of electrical power for one of their clients:

300	240	280	310	340	350
360	320	300	260	290	290

Corresponding observations for the total number of hours (in thousands) that machines requiring electrical power were in use are

9	8	9	10	11	13
12	11	10	8	9	10

(a) Let y_i represent power costs and calculate

$$\sum_{i=1}^{n} y_i$$

(b) Let x_i represent machine hours and calculate

$$\sum_{i=1}^{n} x_i$$

(c) Calculate

$$\sum_{i=1}^{n} x_i y_i$$

(d) Calculate

$$\sum_{i=1}^{n} x_i^2$$

(e) Determine

$$b = \frac{\sum_{i=1}^{n} x_i y_i - \frac{\left(\sum_{i=1}^{n} x_i\right)\left(\sum_{i=1}^{n} y_i\right)}{n}}{\sum_{i=1}^{n} x_i^2 - \frac{\left(\sum_{i=1}^{n} x_i\right)^2}{n}}$$

(f) Determine

$$a = \frac{\sum_{i=1}^{n} y_i}{n} - b\,\frac{\sum_{i=1}^{n} x_i}{n}$$

1.46 Use the data of Exercise 1.20 and calculate

$$\sum_{i=1}^{n} \sqrt{\frac{1}{y_i}}$$

1.47 If $x_1 = 5$, $x_2 = 2$, and $x_3 = 6$, determine

$$\left(\sum_{i=1}^{3} x_i\right)^2$$

2 Probabilities for Discrete Events

2.1 INTRODUCTION

The World Motor Car Division of LeAjax, Inc., produces both sports cars and performance sedans for the international market. The marketing department is constantly making surveys to determine the types and characteristics of new cars that might sell well, while the engineering department is providing new engine and performance designs that they hope will be incorporated into new models. Project teams take the results of these two departments and, after considerable study and engineering, develop new models or enhancements for current models. The numerous project teams produce a number of competing designs each year, not all of which will be accepted for implementation by the top corporate management.

Management insists that each project be assigned a probability for success, before the project is presented to them for a decision. Vic O'Riley, a financial analyst, has supervised the assignment of probabilities for four years. Currently, he is working on three projects, all of which are competing designs for a new mid-engine sports car. Top management must decide whether or not to produce this type of car, and if so, which of the three designs. The quality of their decision will be partly determined by the accuracy of the probabilities that O'Riley assigns to each project.

O'Riley can take one of three approaches to probability assignment: subjective, relative frequency, or objective. He might also combine a couple of these methods through the use of Bayes' formula.

Probability concepts are important for studying decision making under conditions of uncertainty and for the study of statistics in general. This chapter presents the terms and concepts needed for understanding and calculating probabilities. Following chapters will cover the calculation and use of probability distributions.

2.2 BASIC TERMINOLOGY

The concepts of probability are usually presented using the notions of a sample space and events. In this section, we consider some of the basic terminology and definitions, with most examples, for simplicity, based on the flip of a coin.

Sample Space

All probability concepts start with the determination of the set of all possible outcomes for an experiment.

DEFINITION 2.1 A **sample space,** denoted S, is the set of all possible outcomes.

Therefore, if we were to flip a coin, the population of possible outcomes includes a head and a tail, denoted H and T. Therefore,

$$S = \{H, T\}$$

where the items in braces are all the possible outcomes in the sample space S. If we were to flip a coin twice, the sample space would show all possible outcomes for the first and second flip in that order.

$$S = \{HH, HT, TH, TT\}$$

Notice that we must consider the order in which the faces turned up—that is, HT is not the same outcome as TH.

A tree diagram can be used as a technique for observing these outcomes. The first branching on the tree shows the possible outcomes on the first flip. The second branching shows the possible outcomes on the second flip. The tree is shown in Figure 2.1.

If we were to flip two coins at one time, the sample space would be

$$S = \{HH, HT, TH, TT\}$$

which is the same as flipping one coin twice. The HT would represent a head on coin number one and a tail on coin number two, whereas TH represents a tail on coin one and a head on coin two.

In this chapter, we consider sample spaces with both a finite number of

FIGURE 2.1 Tree diagram for two coin flips

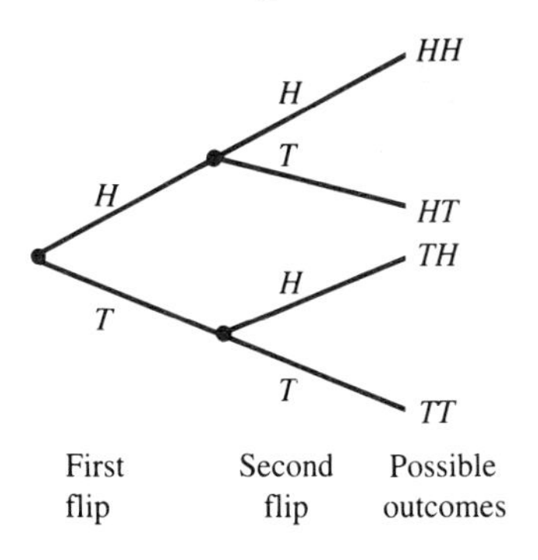

events and a countable number of events. It is possible, however, to encounter problems that have an infinite number of events. It is not feasible to talk about enumerating them, although they can be described. For example,

$$A = \{X: 0 < X < 1\}$$

Events

It is common to speak of the probability of a particular event's occurrence.

DEFINITION 2.2 An **event** is any subset of a sample space.

Definition 2.2 permits an event to include one or more of the possible outcomes or possibly none of the outcomes. If a single outcome is of interest, we have a simple event.

DEFINITION 2.3 A **simple event** is any subset of the sample space that consists of a single outcome.

If more than one outcome is of interest, we have a compound event.

DEFINITION 2.4 A **compound event** is any subset of the sample space that consists of two or more simple events.

When a coin is flipped twice, we have, as noted earlier, the sample space

$$S = \{HH, HT, TH, TT\}$$

each element (outcome) of which is a simple event. We could denote these simple events with the notation E_i; then

$$S = \{E_1, E_2, E_3, E_4\}$$

We may wish to find the probability that in two flips of the coin we obtain a head and then a tail in that order, which is the probability of the simple event E_2. We may then wish to find the probability that in two flips of a coin we obtain at least one head. Here we have the probability of a compound event, which we will call A, because it is made up of three simple events.

$$A = \{HH, HT, TH\} = \{E_1, E_2, E_3\}$$

Note that a compound event can be decomposed into simple events.

EXAMPLE 2.1 Each Friday and Saturday a student works as a used-car salesman. On some days, he sells 2 or more cars (T), on some days he sells 1 car (O), and on some days he sells no cars (N).

a. Enumerate the sample space using a tree diagram, and identify the simple events.
b. Event A consists of simple events where one or more cars are sold on only 1 of the 2 days. List the simple events of event A.
c. Event B is the event that he will sell 1 or more cars at least 1 day this weekend. Enumerate the simple events.
d. Event C is the event that he will sell no cars either day. Enumerate the simple events.
e. Which of the events A, B, C are simple? Which are compound?

Solution:

a. The tree diagram is shown in Figure 2.2. The simple events are

$$\begin{aligned} S &= \{TT, TO, TN, OT, OO, ON, NT, NO, NN\} \\ &= \{E_1, E_2, E_3, E_4, E_5, E_6, E_7, E_8, E_9\} \end{aligned}$$

b. $A = \{TN, ON, NT, NO\} = \{E_3, E_6, E_7, E_8\}$

c. $$\begin{aligned} B &= \{TT, TO, TN, OT, OO, ON, NT, NO\} \\ &= \{E_1, E_2, E_3, E_4, E_5, E_6, E_7, E_8\} \end{aligned}$$

d. $C = \{NN\} = \{E_9\}$

e. Event C is a simple event. Events A and B are compound events.

Mutually Exclusive Events

Two different compound events may include one or more of the same simple events from a given sample space.

DEFINITION 2.5 Two events are said to be **mutually exclusive** if they cannot occur simultaneously or jointly.

Therefore, if two compound events have any simple events in common, they are not mutually exclusive.

FIGURE 2.2 Tree diagram for Example 2.1

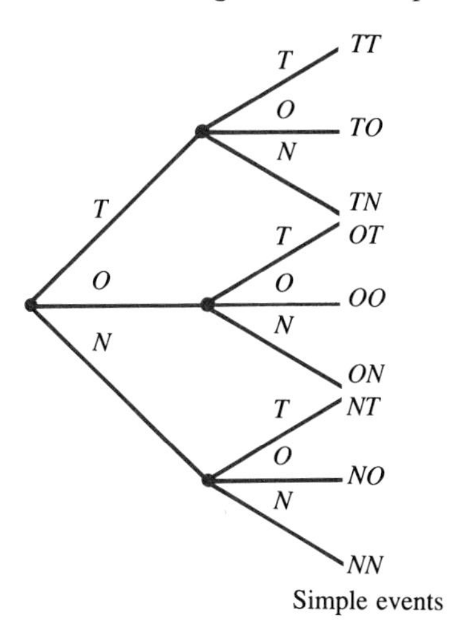

Let us again consider flipping two coins. The sample space is

$$S = \{HH, HT, TH, TT\} = \{E_1, E_2, E_3, E_4\}$$

Event A is the compound event that we have at least one head.

$$A = \{HH, HT, TH\} = \{E_1, E_2, E_3\}$$

Event B is the compound event that we have at least one tail.

$$B = \{HT, TH, TT\} = \{E_2, E_3, E_4\},$$

We can see that both events A and B contain the simple events E_2 and E_3. Events A and B are therefore not mutually exclusive.

Now consider event C, which is the event that we have two coins the same, and event D, in which both coins differ.

$$C = \{HH, TT\} = \{E_1, E_4\}$$
$$D = \{HT, TH\} = \{E_2, E_3\}$$

Events C and D are mutually exclusive because they do not contain any elements in common. All simple events are mutually exclusive because they cannot occur simultaneously.

A Venn diagram is often used to illustrate sets and subsets. In Figure 2.3, Venn diagrams are used to depict the mutually exclusive and nonmutually exclusive events for the coin toss example.

FIGURE 2.3 Venn diagrams for the coin toss example

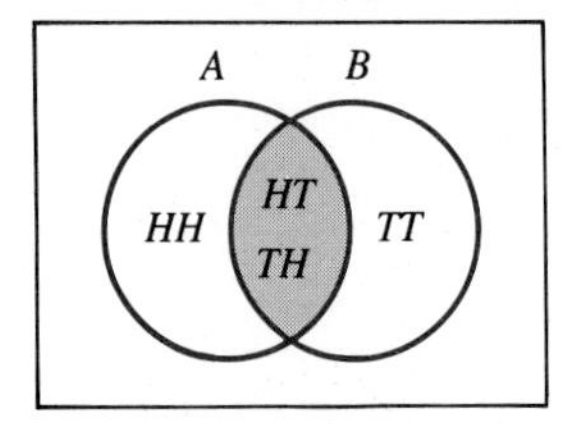

(a) Events *A* and *B* not mutually exclusive

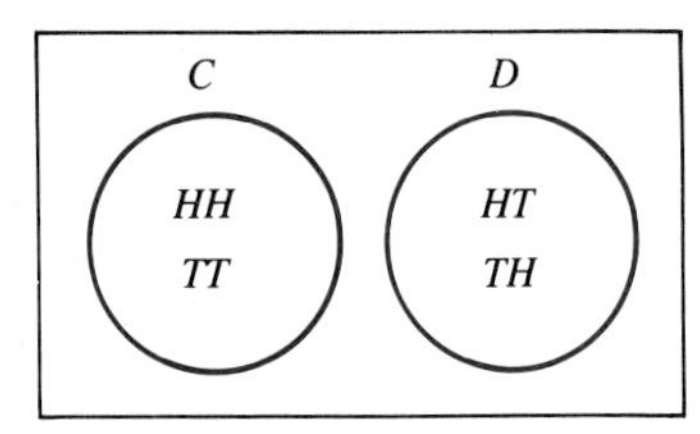

(b) Events *C* and *D* mutually exclusive

EXAMPLE 2.2 Using events *A, B,* and *C* of Example 2.1, answer the following.

a. Are events *A* and *B* mutually exclusive?
b. Are events *B* and *C* mutually exclusive?

Solution: We illustrate the solutions with Venn diagrams in Figure 2.4.

a. Events *A* and *B* are not mutually exclusive because they have simple events in common.
b. Events *B* and *C* are mutually exclusive because they have no simple events in common.

Collectively Exhaustive Events

A collection of events may be collectively exhaustive.

FIGURE 2.4 Venn diagrams for Example 2.2

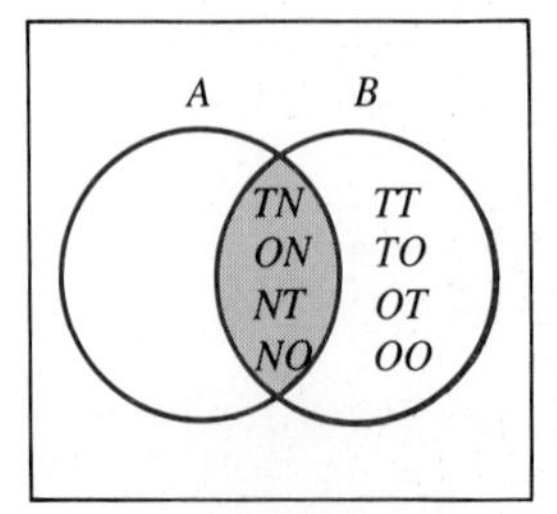

(a) Events *A* and *B* not mutually exclusive

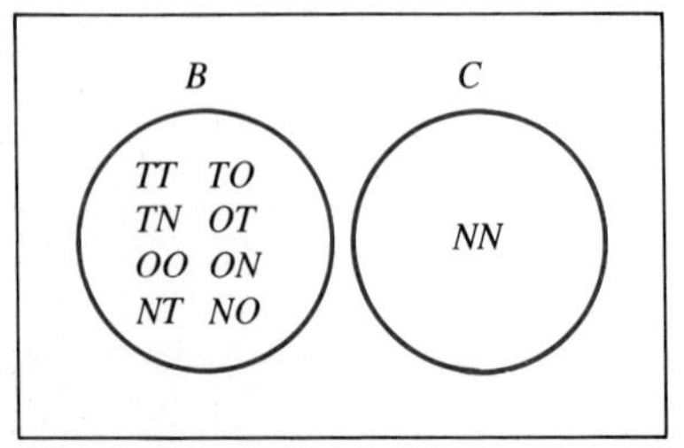

(b) Events *B* and *C* mutuaily exclusive

DEFINITION 2.6 A set of events B_i, $i = 1, \ldots, k$, is said to be **collectively exhaustive** with respect to A if it includes all the simple outcomes—that is, if the union of the B_i equals A.

For example, in Example 2.1, the sample space is

$$S = \{TT, TO, TN, OT, OO, ON, NT, NO, NN\}$$

Since the union of all these simple events includes all the possible outcomes, the events are collectively exhaustive with respect to the sample space S. In the example, we found that these events are also mutually exclusive. It is, therefore, possible for events to be both mutually exclusive and collectively exhaustive.

Other combinations are also possible. In the same example are events A and B.

$$A = \{TN, ON, NT, NO\}$$
$$B = \{TT, TO, TN, OT, OO, ON, NT, NO\}$$

Events A and B are not collectively exhaustive with respect to S because their union does not include all the simple events: NN is missing. Events A and B are not mutually exclusive.

EXERCISES

2.1 A local bakery sells 3 kinds of donuts: glazed (G), sugared (S), and powdered (P). Some days the bakery sells more glazed; some days more sugared; and some days more powdered. The bakery cannot predict which donuts will sell the best on a given day.

- **(a)** What is the sample space for most sales for the next 2 days?
- **(b)** List the elements of event A, namely, that during 1 or more of the 2 days, glazed donuts will sell best.
- **(c)** List the elements of event B, namely, that glazed donuts will not sell best in either of the two days.
- **(d)** List the elements of event C, namely, that the same donut will sell best both days.
- **(e)** What type of events are A, B, and C?
- **(f)** Are events A and B mutually exclusive?
- **(g)** Are events B and C mutually exclusive?

2.2 Draw a Venn diagram showing events B and C.

2.3 Draw a Venn diagram showing events A, B, and C.

2.3 CONCEPTS OF PROBABILITY

We are often faced with situations in which the outcomes are uncertain. Such a situation could be considered a random experiment.

DEFINITION 2.7 A **random experiment** is any process of observation for which the outcome is not certain.

If we throw a coin into the air, we know it will fall a few moments later. This event is not a random experiment because chance is not involved in the process. Only one outcome is possible: The coin will fall, and that is certain.

On the other hand, if we are flipping the coin to see which face appears, two outcomes exist (heads or tails), and the outcome is not certain. The flipping process can be considered a random experiment in which each outcome has a probability associated with its occurrence.

DEFINITION 2.8 **Probability** is a measure of the likelihood of the occurrence of an event.

The measurement is numerical and represents the chance that a particular outcome will occur in a random experiment. Comments such as, "The chance is 60%," or "Chances are 2 to 1" are typical probability statements.

The notation used to indicate a probability is the capital letter P followed by the event in parentheses. The probability of event A is written $P(A)$. An alternate notation is $Pr(A)$.

Probability Axioms

All probability statements must satisfy the following three probability axioms.

AXIOM 2.1 The probability of any event A must be a value between zero and unity inclusive.

$$0 \leq P(A) \leq 1$$

Values outside this interval are impossible.

AXIOM 2.2 If B_i, $i = 1, \ldots, k$, is a set of mutually exclusive and collectively exhaustive events in the sample space S, then the sum of the probabilities of these events is unity—that is,

$$\sum_{i=1}^{k} P(B_i) = 1$$

where k is the number of events in S.

AXIOM 2.3 If A is the union of a set of mutually exclusive events, A_i, $i = 1, \ldots, k$, then

$$P(A) = \sum_{i=1}^{k} P(A_i)$$

where k is the number of events.

All probabilities must be values no smaller than 0 nor larger than 1, because any event is a subset or a portion of the total. The portion cannot be greater than the total, so it cannot be greater than 1. Likewise, it is impossible to have a negative portion, which would be a value less than 0.

Consider the coin flipping of the previous section and assume the outcomes are equally likely. Then each outcome has a probability of .25. We list all the possible outcomes.

Outcome	*Probability*	
HH	.25	
HT	.25	.75 (HH, HT, TH)
TH	.25	
TT	.25	
	1	

We note that each outcome is mutually exclusive, which is true of all simple events in a sample space. If we have a head on coin 1 and tail on coin 2, then we cannot simultaneously have a tail on coin 1 and a head on coin 2. Finally, observe that the sum of all the possible outcomes is unity.

To illustrate the third axiom, we find the probability of at least one head in the flipping of two coins. This compound event is composed of three simple events, *HH*, *HT*, and *TH*, all of which are mutually exclusive, as noted above. Therefore, the probability is the sum of the individual probabilities of the simple events.

$$.25 + .25 + .25 = .75$$

EXAMPLE 2.3 A certain sports car is available in silver, red, or orange. Sales records indicate that the probability of each color's selection by a buyer is .40, .35, and .25, respectively. Show that these probabilities obey the probability axioms.

Solution

a. Each of the three probabilities is between 0 and 1. Thus, Axiom 2.1 is satisfied.

b. The three cars represent all possible outcomes, and the sum of their probabilities is unity.

$$.40 + .35 + .25 = 1$$

Axiom 2.2 is satisfied.

c. The events are all simple events and are mutually exclusive. If we have a compound event (A)—for example, the probability that the next car sold was silver or red—then the probability of A is

$$P(A) = P(S) + P(R) = .40 + .35 = .75$$

Axiom 2.3 is satisfied.

Assignment of Probabilities

Up to this point, we have defined what a probability is and the rules or axioms it must satisfy, but we have not yet determined how to calculate a probability or how to assign a value to a probability statement. We can do so by hunch, by experiment, and by knowledge of the sample space.

1. *Subjective probabilities* are simply guesses or hunches that we may have concerning the outcome. They may be based on experience but are not calculated from any historical data. A new-product manager may ''feel'' that a product about to be introduced has a 75% chance of success. This is a ''seat of the pants'' guess, and although it may be an educated guess, it was not calculated from any real data.

2. *Relative frequency probabilities* are generally the result of experimentation, such as flipping a coin a number of times to determine the percentage of the flips in which heads occur. Similarly, in business circumstances, using historical data, we can calculate the percentage of the time that an event of interest has occurred. Equation 2.1 represents this relationship.

$$P(A) = \frac{n(A)}{N} \tag{2.1}$$

where: A = the event of interest
$n(A)$ = the number of times the event has occurred
N = the number of times the experiment has been repeated

EXAMPLE 2.4 Not being able to tell a good job candidate from a poor one, the president of a computer software firm has randomly selected graduates of the computer science program at a local university. Experience has shown that of the past 42 graduates hired by the firm, 34 have performed competently. Another graduate is about to be randomly hired. What is the probability that this graduate will be competent?

Solution: Letting C be the event that the graduate is competent and using Equation 2.1, we find that the probability is

$$P(C) = \frac{n(C)}{N} = \frac{34}{42} = .81$$

There is an 81% chance that this one will turn out to be competent.

3. *Objective probabilities* are calculated from a knowledge of the number of possible outcomes of a sample space, $n(S)$, and the number of those outcomes that are of interest, $n(A)$. This approach may only be used if each outcome in the sample space has an equally likely chance of being selected. Equation 2.2 represents this relationship.

$$P(A) = \frac{n(A)}{n(S)} \tag{2.2}$$

To find the probability of A, we first count the number of items in the sample space $n(S)$, and then count how many items satisfy the subset of interest $n(A)$. Dividing the latter by the former yields the probability of A.

EXAMPLE 2.5 A tire installer has just completed the installation of a set of 4 new tires on a car when he learns that 2 of them are defective and were marked as such inside the tire. What is the probability that the first tire he picks at random to remove will be a defective tire?

Solution: We know 4 tires were installed, so $n(S) = 4$. We also know 2 of the tires are defective. Thus, $n(D) = 2$, and the probability that the first tire selected is defective can be found by using Equation 2.2.

$$P(D) = \frac{n(D)}{n(S)} = \frac{2}{4} = .5$$

The installer has a 50–50 chance of finding a defective tire on the first try.

EXAMPLE 2.6 A company plans to randomly assign 4 junior executives, A, B, C, and D, to 4 adjacent new offices along a hall. What is the probability that the order of assignment will be $BDAC$?

Solution: We first need to enumerate all the possible orders in which the junior executives can be assigned to offices.

ABCD	*BACD*	*CABD*	*DABC*
ABDC	*BADC*	*CADB*	*DACB*
ACBD	*BCAD*	*CBAD*	*DBAC*
ACDB	*BCDA*	*CBDA*	*DBCA*
ADBC	*BDAC*	*CDAB*	*DCAB*
ADCB	*BDCA*	*CDBA*	*DCBA*

The sample space contains 24 possible outcomes of which each is equally likely to occur. Only one of these is the order of interest, $BDAC$. Using Equation 2.2, we obtain

$$P(BDAC) = \frac{n(BDAC)}{n(S)} = \frac{1}{24} = .0417$$

Although three methods of probability assignments have been presented, one could argue that there are actually only two types: subjective and relative frequency. One might also argue that the relative frequency method is the only method by which probability could be considered to be objectively measured. Since the objective method makes subjective assumptions, such as the coins are fair and both sides are equally likely to occur, it could also be considered a subjective method.

EXERCISES

2.4 A student conjectures that on the next exam the probability of earning an A, B, or C is .20, .60, and .20, respectively. (The professor does not give D's or F's.) Show that these probabilities obey the probability axioms.

2.5 Another student judges that the probability of getting at least an A, at least a B, or at least a C is .10, .70, and .20, respectively. Do these probabilities obey the axioms? Why?

2.6 Are the probabilities mentioned in the preceding exercises objective, subjective, or relative frequency?

2.7 If the owner of the bakery in Exercise 2.1 thinks that each type of donut (glazed, powdered, and sugared) has an equal chance of selling best each day, what is the probability that on the next 2 days the glazed sells best each day?

2.8 If you guessed at the answer to Exercise 2.6, what is the probability that you got it correct? What type of probability did you just compute?

2.9 Flip a coin 5 times and record your results.

- **(a)** Calculate the probability of obtaining a head in the flip of a coin.
- **(b)** What type of probability is this?
- **(c)** Did you just prove that the probability of a head is not $\frac{1}{2}$? Why?

2.4 METHODS OF COUNTING

It is not always as easy to enumerate all the possible outcomes as it was in Example 2.6. When the number of items involved is large, it is useful to be able to count mathematically. Depending on the type of problem, different counting methods may be appropriate: the $m \cdot n$ rule, permutations, or combinations. At times, more than one method could be used in a problem.

The $m \cdot n$ Rule

The $m \cdot n$ rule is the most fundamental counting technique. The number of possible outcomes is the product of the number of ways that each event of a number of individual events can happen.

RULE 2.1 The $\boldsymbol{m \cdot n}$ **rule** states that if we let n_i be the number of ways that the ith event can occur, then N, the total number of possible outcomes for the combined event, is

$$N = (n_1)(n_2) \cdots (n_k) = \prod_{i=1}^{k} n_i \tag{2.3}$$

The n_i's can fluctuate in magnitude, be constant, decrease by one each time, or be computed by other counting formulas.

EXAMPLE 2.7 A systems designer has 3 spaces on a computer card available for punching in a product code. Under consideration is the use of alphabetic characters for each product. The company has 20,000 different products. Will a three-letter code allow each product to be uniquely identified?

Solution: If we tried to enumerate in this problem, we would soon run out of patience.

AAA, AAB, AAC, AAD, AAE, . . .
ABA, ABB, ABC, ABD, ABE, . . .
.
.
.
ZZA, ZZB, ZZC, ZZD, ZZE, . . .

Note that the same letter, such as A, can be used in all three positions. What was used in the first position does not limit or influence what is used in the second position. Using the $m \cdot n$ rule, we find the

$$\text{Number of possible codes} = (26)(26)(26) = 17{,}576$$

This system is not adequate to represent all 20,000 products. If we consider using numbers or letters in each position, it is possible to obtain, with 10 numbers plus 26 letters possible,

$$(36)(36)(36) = 46{,}656$$

possible outcomes. This total is adequate.

The object of counting is usually to obtain the numerator and/or the denominator for the calculation of an objective probability. Example 2.8 illustrates this use.

EXAMPLE 2.8 Refer to the solution of Example 2.7 and assume that either numbers or letters are used in each position. What is the probability of randomly selecting A2F (or any other specified code) as the code for a product?

Solution: The probability is calculated as

$$P(\text{A2F}) = \frac{n(\text{A2F})}{n(S)} = \frac{1}{46{,}656}$$

There are many variations on this type of problem, usually as the result of constraints or restrictions. For example, we may want to eliminate all three-letter codes that suggest vulgar words. If there were five such possibilities, they would be subtracted from the 46,656 possible outcomes, leaving 46,651 possible outcomes.

Throughout this book, we encounter problems that involve only two possible outcomes at each stage. We need to be able to count the number of possible outcomes, using the $m \cdot n$ rule.

EXAMPLE 2.9 A machine making ball bearings occasionally turns out defective items. If the next 5 ball bearings coming from the machine are observed, how many different orders of good (G) and defective (D) items are possible?

Solution: It is a good practice to enumerate a few possibilities to get a perspective for ordered sequences. Here are some.

D G G G G
G G G D G
G D D D D
G D G D D

Notice that each item could be good or defective. So there are two possible outcomes for the first, two for the second, and so forth. Using the $m \cdot n$ rule, we can calculate the total number of possibilities.

$$2 \cdot 2 \cdot 2 \cdot 2 \cdot 2 = 2^5 = 32$$

Permutations

There are numerous situations in which the items being selected are obtained without replacement from a finite set. Whenever this is the case the $m \cdot n$ rule can be expressed in two special forms. The first of these is used when the order of the items selected is important.

If we had four items, A, B, C, and D, and we wished to select two, we could enumerate in order to count the number of different arrangements.

AB	BA	CA	DA
AC	BC	CB	DB
AD	BD	CD	DC

There are 12 possible orders. We note that AB and BA are different orders and are counted as being different. We also have a mathematical approach for obtaining the same results. Again, the $m \cdot n$ rule can be applied.

$$\begin{array}{c}\text{Number of}\\ \text{possible}\\ \text{outcomes}\end{array} = \begin{pmatrix}\text{Number of ways}\\ \text{of selecting}\\ \text{first item}\end{pmatrix}\begin{pmatrix}\text{Number of ways}\\ \text{of selecting}\\ \text{second item}\end{pmatrix}\cdots$$
$$= (4)(3) = 12.$$

In this case, we notice that once we have selected an A for the first position, we have only B, C, and D to choose from for the second position. We cannot use A again because we are simply arranging a given set of items. Therefore, we have four items to select from for the first position, but only three for the second.

DEFINITION 2.9 Each distinct ordering or arrangement of x items selected without replacement from a set of n different items is called a **permutation.** The number of distinct permutations of n items taken x at a time is given by

$$P_x^n = \frac{n!}{(n-x)!} \quad \textbf{(2.4)}$$

where: $n! = n(n-1)(n-2)\cdots 1$ (n factorial)

$0! \equiv 1$

n = total number of items available

x = number of items selected

We observe in the following example that Equation 2.4 gives exactly the same results as the $m \cdot n$ rule.

EXAMPLE 2.10 A hotel flies foreign flags from its 4 flagpoles each day. If it has 10 flags from which to select, how many different arrangements of flags are possible?

Solution: This example can be recognized as a permutation problem for two reasons.

a. The word "arrangements" implies that order is important.
b. When 1 flag is chosen for 1 position, fewer flags are available for the next position.

We could, of course, try to enumerate the different selections and their order, but the mathematical approach is usually preferred.

$$n = 10$$
$$x = 4$$
$$P_4^{10} = \frac{10!}{(10-4)!} = \frac{10!}{6!} = \frac{10 \cdot 9 \cdot 8 \cdot 7 \cdot 6 \cdot 5 \cdot 4 \cdot 3 \cdot 2 \cdot 1}{6 \cdot 5 \cdot 4 \cdot 3 \cdot 2 \cdot 1}$$
$$= 10 \cdot 9 \cdot 8 \cdot 7 = 5040$$

If we use the $m \cdot n$ rule, we get

$$10 \cdot 9 \cdot 8 \cdot 7 = 5040$$

We are not always selecting a few items out of a group. Sometimes we are ordering *all* the items in the group, which gives us a special case of Equation 2.4.

$$P_n^n = \frac{n!}{(n-n)!} = \frac{n!}{0!} = \frac{n!}{1} = n!$$

EXAMPLE 2.11 A store wishes to display 5 carpet samples along the wall. How many arrangements are possible?

Solution: Since all 5 are to be displayed, $n = 5$, $x = 5$, so

$$P_5^5 = 5! = 120$$

Combinations

Other situations exist for which the order of the selection of items is not important and where the two permutations *AB* and *BA* would be duplicates and counted as the same event. In this case, we use the second special form of the $m \cdot n$ rule where the order of the items selected is not important.

DEFINITION 2.10 Each set or grouping of x items selected without replacement from a finite set of n different items without regard to order is called a **combination.** The number of combinations of n items taken x at a time is given by

$$C_x^n = \frac{n!}{x!(n-x)!} \tag{2.5}$$

Refer to the introduction to permutations and the four items *A, B, C, D*. How many different groupings of two items can be obtained from these four items if order is not considered? Using enumeration, we find that six combinations exist.

AB	BC
AC	BD
AD	CD

Since order is not important, *AB* is the same combination as *BA*, and we could have written it either way but not both ways. We can see that a combination results from eliminating all other orders of the same pairs of items from the permutation list we had earlier. For the special case of the permutation formula, we have shown that for each selection of n items there are $n!$ permutations. Since we are now considering x items, there are $x!$ permutations. In this problem, there are 2! or 2 permutations for each possible pair of items. Therefore, to obtain the number of combinations, we divide the number of permutations by 2. This implies that there are twice as many permutations as combinations for the illustration used here. In general, by dividing the permutation formula by $x!$, we eliminate the different orderings and obtain the number of combinations.

$$C_x^n = \frac{P_x^n}{x!} = \frac{\dfrac{n!}{(n-x)!}}{x!} = \frac{n!}{x!(n-x)!}$$

A traditional problem to illustrate combinations is the selection of persons to be on a committee.

EXAMPLE 2.12 How many different committees of 4 vice-presidents can be selected from the 7 vice-presidents of a firm?

Solution: To identify that this example is a combination problem, we notice that the order of selection is not germane to the makeup of the committee. Thus, with $n = 7$ and $x = 4$ and using Equation 2.5, we have

$$C_4^7 = \frac{7!}{4!(7-4)!} = \frac{7!}{4!3!} = \frac{7 \cdot 6 \cdot 5}{3 \cdot 2 \cdot 1} = 35$$

We can observe that the 4! in the denominator canceled with the $4 \cdot 3 \cdot 2 \cdot 1$ in the numerator. The 4! also indicates that in selecting 4 items from 7, there are 4! or 24 times as many permutations as there are combinations.

Again, we can observe the use of counting in a probability problem, using the next example.

EXAMPLE 2.13 If the committee of 4 in the previous example is selected at random, what is the probability that vice-presidents *A, B, D,* and *F* will be on the committee?

Solution: The answer can be determined as an objective probability. The denominator $n(S)$ represents the total number of committees of 4 vice-presidents that can be selected from 7 vice-presidents where order is not important. The numerator indicates that only one committee can have that particular grouping.

$$P(A, B, D, F) = \frac{n(A, B, D, F)}{n(S)} = \frac{1}{C_4^7} = \frac{1}{35}$$

Another notation commonly used for combinations is

$$\binom{n}{x}$$

Therefore, C_4^7 could also be written as $\binom{7}{4}$.

Permutation of Like Objects

In some situations, we are ordering or arranging a number of items, some of which are identical—for example, two *A*'s, one *B*, and one *C*. The order *AABC* appears to be the same as *AABC* even though the two *A*'s were actually switched. Because we cannot distinguish the difference, we do not wish to count these as two different orderings. Hence, we need to calculate the permutation of like objects. The computations are performed using Equation 2.6.

$$P_{n_1,\ldots,n_k}^n = \frac{n!}{n_1!n_2! \cdots n_k!} \qquad \textbf{(2.6)}$$

where the denominator contains an n_i for each of the k different items to be arranged.

EXAMPLE 2.14 In how many different ways can we assign the positions of president, vice-president (2 available), and manager (4 available) for a new subsidiary to 7 job applicants? Assume all applicants are equally qualified, so assignment of the positions is made randomly.

Solution: This problem is a permutation because the order of assignment is important, but because we have like objects in the vice-president and manager category, we need to use Equation 2.6 to obtain the results. Here, $k = 3$.

$$P^{7}_{1,2,4} = \frac{7!}{1!2!4!} = 105$$

In the case where we have the permutation of like objects with $k = 2$, we calculate

$$P^{n}_{n_1,n_2} = \frac{n!}{n_1!n_2!} = \frac{n!}{n_1!(n - n_1)!} = C^{n}_{n_1} \quad \text{or} \quad \binom{n}{n_1}$$

which is identical to the equation for a combination of n things taken n_1 at a time. A combination may be considered as a permutation of like objects. In this case, the two equations give the same results, and combination notation is often used in this situation for a permutation of like objects.

EXAMPLE 2.15 As valves come off the assembly line, a quality control clerk checks for defectives. In how many orders might the next 5 come off the machine in which there are 2 defective (D) and 3 good (G)?

Solution: One possible order is $GDGGD$. Since the interchange of the 2 D's or 3 G's would not make the arrangement look any different, the answer can be obtained as a permutation of like objects.

$$P^{5}_{2,3} = \frac{5!}{2!3!} = \frac{5 \cdot 4}{2 \cdot 1} = 10$$

This same result can be found by using Equation 2.5, the combination formula, since it is the special case of only two possible outcomes.

$$C^{5}_{2} = \frac{5!}{2!3!} \quad \text{or} \quad C^{5}_{3} = \frac{5!}{3!2!}$$

Compare this example to Example 2.9 and observe the difference. Here we are specifying the number of defective items in the sample of five. In Example 2.9, the number of defective items could range from 0 to 5. Since the number was not specified, all possibilities were investigated.

EXERCISES

2.10 A furniture store offers 1 style of sofa, 1 style of chair, and 1 style of drapes for sale. The store has 20 different patterns of material that can be used to cover a sofa or a chair or for draperies. The items may all use the same pattern or different patterns. If a customer buys 1 chair, 1 sofa, and 1 pair of drapes, in how many different ways can this purchase be made?

2.11 Some furniture arrives at the store with defects due to manufacture. Others are considered perfect. A sofa, chair, and set of drapes have just arrived. Assume that defects are equally likely in each.

- **(a)** How many different conditions can they have?
- **(b)** What is the probability that all 3 are perfect?

2.12 A shoe store has room for 4 shoes in a window display and has 6 shoes from which to pick. If the 4 shoes are selected and displayed randomly, how many different arrangements are possible?

2.13 A certain manufacturing plant has a flagpole on which it raises 4 flags together every day. The American flag must be on top, but the other 3 can be in any order. How many arrangements are possible?

2.14 A local television talk show has 3 fresh plants delivered by a local florist each day for promotional consideration. If 7 different plants are on hand on a particular day, in how many ways can 3 be selected for delivery?

2.15 Of the next 10 electronic components passing through the testing machine, 2 will test as defective, 5 will be passed by the machine, and 3 will be given the yellow light, which means the machine is uncertain and they must be tested by hand. In how many ways can these items come through the machine? (One possibility is *DPPDYYPPYP*.)

2.5 JOINT, MARGINAL, AND CONDITIONAL PROBABILITIES

So far we have discussed only the concepts of simple probability—that is, the probability of a single event. More complex probability situations, involving *joint, marginal,* and *conditional* probabilities, are quite common.

Joint Probability

DEFINITION 2.11 A **joint probability** is the probability that two or more events will happen simultaneously or sequentially.

A joint probability is indicated when the word "and" appears in the statement of a probability, signifying the intersection of two sets. Two different event relationships lead to joint probabilities: events occurring at the same time, and events happening in some specified order. We will use the notation $P(AB)$ to indicate the joint probability of A *and* B. Other common notations are $P(A,B)$ and $P(A \cap B)$. In the simultaneous case, $P(AB) = P(BA)$ since both events happen at the same time.

EXAMPLE 2.16 Which of the following statements indicate joint probabilities?

a. Find the probability that in flipping a coin, three heads in a row will occur.

b. Find the probability that in flipping a coin, a head or a tail will result.
c. Find the probability that in flipping two coins at once, a head and tail will result.

Solution:

a. This problem is a sequential type of joint probability that can be read "a head, and then a head, and then a head." The "ands" point to the fact that it is a joint probability. Notationally we have $P(HHH)$.
b. The word "and" does not appear in the question, so a joint probability is not appropriate.
c. Obtaining a head and tail implies a joint probability of the simultaneous type, but it can happen in two ways.

Coin 1	*Coin 2*
H	T
T	H

The probability statement of a head and a tail becomes $P(HT)$ or $P(TH)$, where the two outcomes are mutually exclusive and, according to Axiom 2.3, require the addition of the two joint probabilities.

For convenience we will cover only the simultaneous case in this section. Sequential situations will be covered in Section 2.6, using the multiplication rule.

EXAMPLE 2.17 A synthetic yarn factory employs 50 people. Each worker is classified according to his or her age group and job level. Table 2.1 shows the results. Find the probability that a person selected at random from the 50 will be both a production worker (A_1) and in the 30–49 age group (B_2).

Solution: The data presented in Table 2.1 are the result of *cross-classification*, which means that an individual in any of the 9 cells is simultaneously included in

TABLE 2.1 Tabulation of the number of workers by certain job levels and age groups

	Job Level	*Age Group* B_1 *18–29*	B_2 *30–49*	B_3 *50–65*	
A_1	*Production workers*	12	10	8	30
A_2	*Clerks/supervisors*	3	6	4	13
A_3	*Managers/executives*	1	2	4	7
		16	18	16	50

TABLE 2.2 Joint probability table using data from Table 2.1.

	Job Level	*Age Group* B_1 18–29	B_2 30–49	B_3 50–65	
A_1	*Production workers*	.24	.20	.16	.60
A_2	*Clerks/supervisors*	.06	.12	.08	.26
A_3	*Managers/executives*	.02	.04	.08	.14
		.32	.36	.32	1.00

2 different sets. Thus, there are 10 individuals that are both production workers (A_1) and in the age group 30–49 (B_2) simultaneously. If we assume that all individuals have an equal chance of being selected, we can calculate the required joint probability as an objective probability.

$$P(A_1B_2) = \frac{n(A_1B_2)}{n(S)} = \frac{10}{50} = .2$$

In many applications, it is useful to create a joint probability table where the joint probabilities are all computed and displayed in their corresponding cells. The row totals and column totals are also converted into probabilities. Table 2.2 shows the joint probability table based on the data in Table 2.1. Each value in Table 2.2 was derived by dividing each cell, row, and column total by 50, the grand total.

Marginal Probability

DEFINITION 2.12 Given a sample space S and any event A, the probability that A occurs—denoted $P(A)$—is called a **marginal probability.**

When we are dealing with discrete events, which is the case in this chapter, a marginal probability is the sum of all relevant joint probabilities, often displayed in a joint probability table. The name ''marginal'' is used because this sum is a value on the margin of the joint probability table, or it can be calculated using objective probabilities on values from the margin of a data table, as in Example 2.18.

EXAMPLE 2.18 Using Table 2.1, find the probability that 1 person selected at random from the factory will be a production worker.

Solution: We observe that 30 of the 50 workers are production workers (A_1); hence,

$$P(A_1) = \frac{30}{50} = .6$$

The same result could be found directly in Table 2.2, the joint probability table, by using the sum of joint probabilities in row A_1.

$$P(A_1) = P(A_1B_1) + P(A_1B_2) + P(A_1B_3)$$
$$= \frac{12}{50} + \frac{10}{50} + \frac{8}{50} = \frac{30}{50} = .6$$

We note that it is permissible to add the joint probabilities because the events are mutually exclusive and Axiom 2.3 can be used.

Conditional Probability

DEFINITION 2.13 A **conditional probability** is the likelihood that an event will occur, given that some other event is already known.

The notation used is $P(A|B)$, which is read as "the probability of A *given* B." The $|B$ implies that B is already known to have occurred or is a prerequisite and that the set of possible outcomes has been reduced to just those in B.

EXAMPLE 2.19 Using the data in Table 2.1, which is repeated in Table 2.3, find the probability that a person selected at random from the factory, who is observed to be a production worker, is from the 30–49 age group.

TABLE 2.3 Illustration of conditional subset

		Age Group			
		B_1	B_2	B_3	
	Job Level	*18–29*	*30–49*	*50–65*	
A_1	*Production workers*	12	10	8	30
A_2	*Clerks/supervisors*	3	6	4	13
A_3	*Managers/executives*	1	2	4	7
		16	18	16	50

Solution: Because it is given that the person is a production worker, our probability is now based just on those 30 people, of which 10 are in the required age group.

$$P(B_2|A_1) = \frac{n(A_1B_2)}{n(A_1)} = \frac{10}{30} = \frac{1}{3}$$

Relationship of Joint, Marginal, and Conditional Probabilities

The calculation of the conditional probability in the previous example was accomplished as an objective probability. However, as can be seen in Equation 2.7, there is a relationship between joint, marginal, and conditional probabilities.

$$P(B|A) = \frac{P(AB)}{P(A)} \qquad P(A) \neq 0 \tag{2.7}$$

Equation 2.7 states that a conditional probability is equal to a joint probability divided by a marginal probability. We note that the event used for the marginal probability is also the given event in the conditional probability. These events are pointed out in the equation below.

$$P(B|A) = \frac{P(AB)}{P(A)}$$

EXAMPLE 2.20 Rework Example 2.19, using Equation 2.7 to calculate the conditional probability.

Solution:

$$P(B_2|A_1) = \frac{P(A_1B_2)}{P(A_1)} = \frac{10/50}{30/50} = \frac{10}{30} = \frac{1}{3}$$

The conditional probability equation can be solved algebraically to find either the joint probability or the marginal probability (assuming the denominators are not zero).

Conditional $$P(B|A) = \frac{P(AB)}{P(A)} \tag{2.8}$$

Marginal $$P(A) = \frac{P(AB)}{P(B|A)} \tag{2.9}$$

Joint $$P(AB) = P(A) \cdot P(B|A) \tag{2.10}$$

EXAMPLE 2.21 The probability that a tax return will show that a refund is required is .40. The probability that a return will show that both a refund is required *and* the return contains errors in arithmetic is .10. Find the probability that examiners, randomly selecting a tax return with a refund required, will find that an arithmetic error was made.

Solution: The first step is to determine what is given and to assign the appropriate notation.

$P(R) = .40$ (refund required)

$P(RE) = .10$ (refund required and contains errors in arithmetic)

Find

$P(E|R)$ (errors in arithmetic, given a refund is required)

Since a conditional probability is required to obtain the answer, we can use Equation 2.8.

$$P(E|R) = \frac{P(RE)}{P(R)} = \frac{.10}{.40} = .25$$

EXAMPLE 2.22 The probability that a car is getting poor mileage relative to what it can obtain when in perfect working order is .50. If a car is getting poor mileage, the probability that the distributor vacuum advance is defective is .10. What is the probability that the next car brought into a service station will have poor mileage and a defective vacuum advance?

Solution: The first step is to determine what is given and to assign the appropriate notation.

$P(M) = .50$ (poor mileage)

$P(V|M) = .10$ (defective vacuum advance given poor mileage)

Find

$P(MV)$ (poor mileage and defective vacuum advance)

Since a joint probability is required to obtain the answer, we can use Equation 2.10.

$$P(MV) = P(M) \cdot P(V|M) = (.50)(.10) = .05$$

EXERCISES

2.16 A marketing research group has classified 80 people according to their answers on a questionnaire about a new product. See Table 2.4. Suppose 1 respondent is selected at random.

(a) What is the probability that the person likes the product and is in the \$10,000 to \$20,000 income bracket?

TABLE 2.4 Tabulations for attitude and income

Attitude	*Income ($1000)*		
	< 10	*10–20*	*> 20*
Like	2	10	15
Neutral	3	7	5
Dislike	20	13	5

(b) What is the probability that the person makes less than $10,000?

(c) What is the probability that the person makes over $20,000 if he or she is neutral about the product?

2.17 According to a doctor's records, 30% of her patients pay for services as they receive them; 20% make an income of less than $8000 a year and pay for services as they receive them. What is the probability that the patient who just paid for services has an income of less than $8000 a year?

2.18 At a florist shop, 50% of the plants have received fertilizer and water. A few have received only water, and some have received only fertilizer. If a plant is wet, the probability that it received fertilizer is .7. What percentage of the plants received water?

2.6 MULTIPLICATION OF PROBABILITIES

When we wish to determine the probability that two events will happen simultaneously or sequentially, we are asking to find the probability that A and B will occur. The word "and" can be translated to mean a joint probability; it can also be interpreted to mean that multiplication of probabilities is required. In the previous section Equation 2.10 was derived for the purpose of computing joint probabilities. We restate that equation as the multiplication rule.

RULE 2.2 The **multiplication rule** states that the probability of the occurrence of events A and B in that given order is equal to the probability of A occurring times the probability that B occurs, given that A has already occurred.

$$P(AB) = P(A) \cdot P(B|A) \tag{2.11}$$

This is the formula for the multiplication of probabilities. It says that the probability of A and B in that given order is equal to the probability of A occurring times the probability that B occurs, given that A has already occurred.

Earlier, when discussing joint probabilities for *simultaneous* events, we stated that

$$P(AB) = P(BA)$$

This equality is true for sequential events, as will be illustrated in Example 2.23. Using the multiplication rule, we can state that

$$P(AB) = P(A) \cdot P(B|A)$$

$$P(BA) = P(B) \cdot P(A|B)$$

Therefore

$$P(A) \cdot P(B|A) = P(B) \cdot P(A|B)$$

EXAMPLE 2.23 Two consumers are to be selected at random from a group of 5 (called A, B, C, D, and E) to perform a commercial. The first consumer to be selected will be the interviewer, and the second consumer will be the respondent. What is the probability that C will interview E? What is the probability that E will interview C?

Solution: To obtain the event, C must have been selected first and E second, so we are looking for the probability of C and E.

$$P(CE) = P(C) \cdot P(E|C) = \frac{1}{5} \cdot \frac{1}{4} = \frac{1}{20}$$

Now, reversing the order, we obtain

$$P(EC) = P(E) \cdot P(C|E) = \frac{1}{5} \cdot \frac{1}{4} = \frac{1}{20}$$

It is obvious that $P(CE) = P(EC)$.

In this example, all five people had an equal chance of being selected first. However, in selecting the second consumer, only four people were available because C had been previously chosen. We would say that the second choice was not independent of the first, since the same person could not be used twice.

Statistical Independence

DEFINITION 2.14 Two events are **statistically independent** if the occurrence of one does not affect the probability of the occurrence of the other, thereby satisfying the following conditions:

$$P(A|B) = P(A) \qquad P(B|A) = P(B) \tag{2.12}$$

Thus, in the calculation of a joint probability, for two independent events, the conditional probability term can be replaced by a marginal probability. For example,

$$P(AB) = P(A) \cdot P(B|A)$$

becomes

$$P(AB) = P(A) \cdot P(B) \tag{2.13}$$

where

$$P(B|A) = P(B)$$

when A and B are independent. Equation 2.13 is the multiplication rule for independent events.

EXAMPLE 2.24 From Table 2.5, determine whether events A and C are independent.

Solution:

$$P(A) = \frac{10}{20} = \frac{1}{2}$$

$$P(A|C) = \frac{3}{6} = \frac{1}{2}$$

Therefore,

$$P(A) = P(A|C)$$

and events A and C are independent. It can also be demonstrated that in this example

$$P(C) = P(C|A)$$

And it can be shown that events B and D are not independent.

TABLE 2.5 Data table for Example 2.24

Events	*A*	*B*	
C	3	3	6
D	4	5	9
E	3	2	5
	10	10	20

We should note that there is a difference between statistically independent events and mutually exclusive events. For mutually exclusive events, $P(A|B) = 0$, whereas for statistically independent events $P(A|B) = P(A)$, which is zero only if A is null. For example, assume that one student is to be randomly selected to be a representative on the Dean's Advisory Committee. Let F be the event that the student is a female and M be the event that the student is a male. The group of candidates consists of six females and four males. Now

$$P(F) = .6 \quad \text{and} \quad P(F|M) = 0$$

since F and M are mutually exclusive—that is, no students are both male and female at the same time. If M and F were statistically independent, then $P(F|M)$ still would be .6.

It is possible to artificially create independence by replacing the selected items. If, for example, in a pile of five wrenches, two were of metric size, the probability of selecting a metric wrench at random would be $\frac{2}{5}$. If a metric wrench were selected and not replaced, the probability of selecting a metric wrench on the next try would be $\frac{1}{4}$. The first selection influences the probability of the second, and hence the two events are not independent. Suppose, however, that the wrench had been replaced after selection. The probability of selecting a metric wrench on the second try would again be $\frac{2}{5}$. Since the first selection did not influence the probability of the second, the events are independent. Hence, sampling with replacement insures independence, whereas sampling without replacement does not.

Extension of the Multiplication Rule

Not all joint probabilities comprise only two events. Many more events can be involved. For example, we may wish to find the probability of A, B, C, and D occurring in that order. Computation of the probability simply involves a revision of the multiplication rule to include more conditional probability terms.

$$P(ABCD) = P(A) \cdot P(B|A) \cdot P(C|A,B) \cdot P(D|A,B,C) \qquad \textbf{(2.14)}$$

The commas separating the multiple events indicate that they all occur. If events A, B, C, and D are mutually independent, the formula becomes

$$P(ABCD) = P(A) \cdot P(B) \cdot P(C) \cdot P(D) \qquad \textbf{(2.15)}$$

The next two examples illustrate these extensions.

EXAMPLE 2.25 Two security guards, A and B, have been working alternate Saturdays but have now come up with a new system. They decided to determine by flipping a coin who would work each Saturday. What is the probability that under this new system, A would have to work four Saturdays in a row?

Solution: Since the probability is $\frac{1}{2}$ that either will work, and since the outcome of each coin flip is independent of the previous week's flip, we get, using an expanded version of the multiplication rule for independent events,

$$P(AAAA) = P(A) \cdot P(A) \cdot P(A) \cdot P(A)$$
$$= \frac{1}{2} \cdot \frac{1}{2} \cdot \frac{1}{2} \cdot \frac{1}{2} = \frac{1}{16}$$

As is often true in probability problems, more than one method exists for finding a solution. We consider the next sample, in which the events are not independent.

EXAMPLE 2.26 If the colors of carpet to be displayed in a store window are green, orange, blue, yellow, and red, and if the order of the display is picked at random, what is the probability that the order will be red, orange, yellow, green, and blue?

Solution: Because a specific order was given for the outcome, the multiplication rule is useful. We will use the first letter of each color to represent the color.

$$P(ROYGB) = P(R) \cdot P(O|R) \cdot P(Y|RO) \cdot P(G|ROY) \cdot P(B|ROYG)$$
$$= \frac{1}{5} \cdot \frac{1}{4} \cdot \frac{1}{3} \cdot \frac{1}{2} \cdot \frac{1}{1} = \frac{1}{120}$$

It is also possible to obtain a solution by using permutations, which we discussed in Section 2.4.

$$P(ROYGB) = \frac{n(ROYGB)}{n(S)} = \frac{1}{P_5^5} = \frac{1}{5!} = \frac{1}{120}$$

EXERCISES

2.19 Among 10 job applicants sitting in the waiting room, 4 are qualified for the job. If 2 applicants are selected at random, what is the probability that the first will be qualified and the second will not?

2.20 There are 5 typists in the executive typing pool at Entertainment Unlimited, Inc., only one of whom is qualified. They are assigned randomly each day to 5 executives.

- **(a)** What is the probability that the qualified secretary will be assigned to a given executive 2 days in a row?
- **(b)** What is the probability that the qualified secretary will be assigned to a given executive the second day if that secretary was assigned to that executive the first day?
- **(c)** Comment on independence in this problem.

2.21 What is the probability that the qualified secretary in Exercise 2.20 will be assigned to the executive 4 days in a row?

2.22 Each year the 8-member board of trustees of a university randomly selects 3 of its members to serve as a finance committee. The first one selected is chairman, the second is secretary, and the third does the work. If the members are named A, B, C, D, E, F, G, H for the selection process, what is the probability that C is chairman, F is secretary, and B does all the work?

2.7 ADDITION OF PROBABILITIES

Just as we had a rule for the multiplication of probabilities, we also have one for the addition of probabilities.

RULE 2.3 The **addition rule** states that the probability that event A or event B or both events may occur is given by the addition of the probabilities of each event minus the joint probability of the two events.

$$P(A \cup B) = P(A) + P(B) - P(AB) \tag{2.16}$$

The ∪ means *union,* a term from set theory, and the last term represents the probability of the two events occurring together. Whenever the word "or" is encountered in probability statements, it indicates the possible use of the addition formula.

Figure 2.5 shows a Venn diagram that illustrates the addition rule. The circle labeled A represents event A, and the circle labeled B represents event B. If the probabilities corresponding to events A and B are added, we can see that the probability associated with the intersection of A and B is included twice, once with A and once with B. Therefore, the probability of AB must be subtracted from the table to eliminate the dual inclusion.

EXAMPLE 2.27 Weather forecasters indicate that the probability that any given day in this area will be windy is $\frac{2}{5}$, and the probability that any day will be cloudy is $\frac{3}{5}$. The probability that a day will be both cloudy and windy is $\frac{1}{5}$. What is the probability that a day selected at random will be cloudy or windy?

FIGURE 2.5 Venn diagram illustrating the addition rule

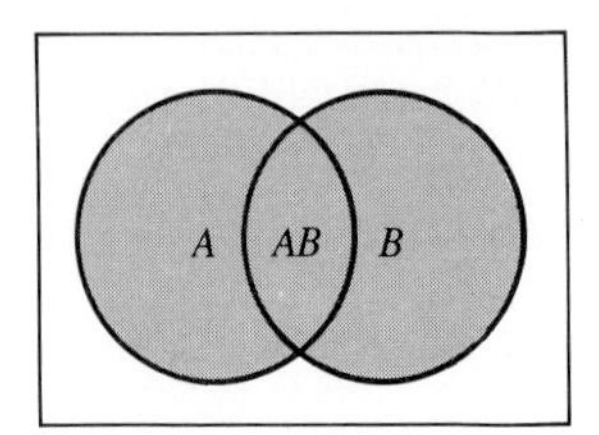

Solution: Whether stated or not, "or both" is also implied in the question. The word "or" indicates the need for the addition rule, Equation 2.16.

$$P(W) = \frac{2}{5} \qquad \text{(windy)}$$

$$P(C) = \frac{3}{5} \qquad \text{(cloudy)}$$

$$P(CW) = \frac{1}{5} \qquad \text{(cloudy and windy)}$$

$$P(C \cup W) = P(C) + P(W) - P(CW) = \frac{3}{5} + \frac{2}{5} - \frac{1}{5} = \frac{4}{5}$$

A Venn diagram for the example is given in Figure 2.6. When we state the probability of a cloudy day as $\frac{3}{5}$, $\frac{1}{5}$ is included in the intersection and the remaining $\frac{2}{5}$ is in the remainder of the set. The $\frac{1}{5}$ in the intersection is also part of the probability for windy days. Since we have added the intersection into the total twice, we must remove it once, using the last term of the equation.

If the two events are mutually exclusive, there is no intersection, and the equation is reduced to

$$P(A \cup B) = P(A) + P(B) \qquad \textbf{(2.17)}$$

EXAMPLE 2.28 When a well is sunk, there is a probability of $\frac{5}{10}$ that it will hit water, $\frac{2}{10}$ that it will hit oil, and $\frac{3}{10}$ that it will hit nothing. What is the probability that the next well will hit either water or oil?

Solution: From the Venn diagram in Figure 2.7, we see that the events are mutually exclusive.

$$P(W \cup O) = P(W) + P(O) = \frac{5}{10} + \frac{2}{10} = \frac{7}{10}$$

FIGURE 2.6 Venn diagram for Example 2.27

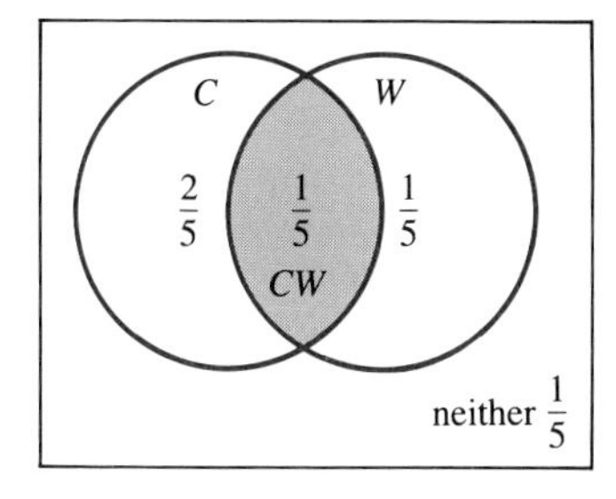

FIGURE 2.7 Venn diagram for Example 2.28

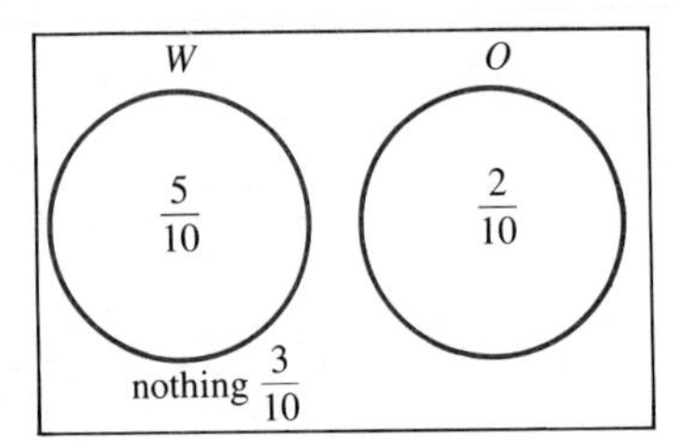

As is indicated in the next example, we can work with complementary probabilities. In doing so, we are using the conditions of Axiom 2.2 given earlier that

$$\sum_{i=1}^{n} P(E_i) = 1$$

EXAMPLE 2.29 To continue Example 2.28, we calculate the probability that water or oil will not be encountered when a well is sunk.

Solution:

$$P(W \cup O)' = 1 - P(W \cup O) = 1 - \frac{7}{10} = \frac{3}{10}$$

The prime notation indicates the negation (or complement) of the set. The result can be observed on the Venn diagram of Figure 2.7.

EXERCISES

2.23 A vase shop orders a certain vase from the Orient. Although the order cannot specify color, only 2 colors, white and rust, are available. The colors are shipped randomly, but there is a 60% chance that a vase will be white. There is also a 40% chance that it will be damaged in shipment. There is only a 10% chance that it will be a damaged white vase. The next vase arrives.

- **(a)** What is the probability that it is white or damaged?
- **(b)** What is the probability that it is rust?
- **(c)** What is the probability that it is not white or not damaged?

2.24 Draw a Venn diagram for Exercise 2.23.

2.25 New cars coming off the assembly line will fail to start with a probability of .2, will start with difficulty with a probability of .4, and will start with ease with a probability of .4.

(a) What is the probability that a car will either start with ease or will start with difficulty?
(b) What is the probability that it will at least start?
(c) Compare parts (a) and (b).

2.8 BAYES' RULE

In previous problems, when asked to compute a conditional probability, we expected to be given a marginal probability and a joint probability, and we would substitute them into Equation 2.7, which is

$$P(B|A) = \frac{P(AB)}{P(A)}$$

Some problems that request that we compute a conditional probability do not provide the information desired; instead, they give us other conditional probabilities and marginal probabilities for data. For purposes of this discussion, assume that we are given two marginal probabilities, $P(B_1)$ and $P(B_2)$, and two conditional probabilities, $P(A|B_1)$ and $P(A|B_2)$, and that we are asked to find the conditional probability $P(B_1|A)$. It does not immediately appear that Equation 2.7 is of any use. However, after reviewing three principles that we learned earlier, it becomes obvious that Equation 2.7 can be used to find the required conditional probability.

We begin by restating Equation 2.7 in terms of B_1 and A,

$$P(B_1|A) = \frac{P(AB_1)}{P(A)}$$

and recognizing that neither the required joint probability, $P(B_1A)$, nor the marginal probability, $P(A)$, is given.

The first principle is that a joint probability can be found by using the multiplication rule. We would expect that

$$P(AB_1) = P(A) \cdot P(B_1|A)$$

but since we are ultimately solving for $P(B_1|A)$, we cannot use it to find $P(AB_1)$.

Second, recall that for joint probabilities, the order of the events does not influence the product. Hence,

$$P(AB_1) = P(B_1A)$$

If we make this substitution, then

$$P(B_1|A) = \frac{P(B_1A)}{P(A)}$$

Using the multiplication rule, we get

$$P(B_1A) = P(B_1) \cdot P(A|B_1)$$

TABLE 2.6 Joint probability table

	A	A'	
B_1	$P(B_1A)$	$P(B_1A')$	$P(B_1)$
B_2	$P(B_2A)$	$P(B_2A')$	$P(B_2)$
	$P(A)$	$P(A')$	

Both $P(B_1)$ and $P(A|B_1)$ are given, so we are able to compute the numerator. However, the denominator, $P(A)$, is still a difficulty since this also was not given.

Third, we can use the rule of partitioning.

RULE 2.4 The **rule of partitioning** states that a marginal probability is the sum of joint probabilities.

$$P(A) = \sum_{i=1}^{n} P(B_iA) \tag{2.18}$$

Table 2.6 illustrates partitioning.

If we partition $P(A)$, we obtain

$$P(A) = P(B_1A) + P(B_2A)$$

We are now able to substitute into our original equation.

$$P(B_1|A) = \frac{P(B_1A)}{P(A)} = \frac{P(B_1A)}{P(B_1A) + P(B_2A)}$$

Now, applying the multiplication rule for each joint probability, we obtain

$$P(B_1|A) = \frac{P(B_1) \cdot P(A|B_1)}{P(B_1) \cdot P(A|B_1) + P(B_2) \cdot P(A|B_2)}$$

We can also find $P(B_2|A)$.

$$P(B_2|A) = \frac{P(B_2) \cdot P(A|B_2)}{P(B_1) \cdot P(A|B_1) + P(B_2) \cdot P(A|B_2)}$$

Finally, generalizing from the above discussion, we have the equation for Bayes' rule.

RULE 2.5 **Bayes' rule** states that a conditional probability may be calculated as a joint probability divided by the sum of all possible corresponding joint probabilities, where each joint probability is determined by the multiplication rule.

$$P(B_j|A) = \frac{P(B_j) \cdot P(A|B_j)}{\sum_{i=1}^{n} P(B_i) \cdot P(A|B_i)} \tag{2.19}$$

EXAMPLE 2.30 A certain supermarket chain makes its own sausage. Each Saturday, an employee cooks the sausage in the store and gives out free samples. From observation, it has been determined that 60% of the persons passing the display will take a sample. The probability that they will purchase some sausage after tasting the free sample is .2, and the probability that a person who has not tasted the sample will purchase the sausage is .1.

When a customer purchases some sausage, what is the probability that the customer tasted the sample? What is the probability that the customer did not taste the sample?

Solution: If we let T represent tasting a sample and B represent making a purchase, then, in this example we are asked to find $P(T|B)$ and $P(T'|B)$, which would be equivalent to $P(B_1|A)$ and $P(B_2|A)$ in our general formula. We are given the following information.

$P(T) = .6$ (tasted)

$P(B|T) = .2$ (purchased given tasted)

$P(B|T') = .1$ (purchased given not tasted)

We note that, while not given, we can find $P(T')$ as the complement of $P(T)$.

$$P(T') = 1 - P(T) = 1 - .6 = .4$$

Using Bayes' rule, we find

$$P(T|B) = \frac{P(T) \cdot P(B|T)}{P(T) \cdot P(B|T) + P(T') \cdot P(B|T')}$$

$$= \frac{(.6)(.2)}{(.6)(.2) + (.4)(.1)} = \frac{.12}{.12 + .04} = \frac{.12}{.16} = .75$$

$$P(T'|B) = \frac{P(T') \cdot P(B|T')}{P(T) \cdot P(B|T) + P(T') \cdot P(B|T')}$$

$$= \frac{(.4)(.1)}{(.6)(.2) + (.4)(.1)}$$

$$= \frac{.04}{.12 + .04} = \frac{.04}{.16} = .25$$

We observe that

$$P(T|B) + P(T'|B) = 1$$

since given that a person purchases the sausage, those are the only possible outcomes.

Many problems have more than two possible outcomes. However, the same general procedure is used.

EXAMPLE 2.31 Expensive pottery is shipped from the Orient to an oriental vase store. The probability that a vase will be broken, chipped, or perfect upon arrival is .1, .2, and .7, respectively. Vases are shipped from Korea and other oriental countries. Experience shows that if a vase is broken, the probability that it is from Korea is .02; if a vase is chipped, the probability that it is from Korea is .25; and if a vase is perfect, the probability that it is from Korea is .86. When a package containing a vase arrives from Korea, what is the probability that the vase is chipped?

Solution: Since we know that there are three possible outcomes, let B_1 = broken, B_2 = chipped, B_3 = perfect. Then,

$$P(B_2|K) = \frac{P(B_2) \cdot P(K|B_2)}{P(B_1) \cdot P(K|B_1) + P(B_2) \cdot P(K|B_2) + P(B_3) \cdot P(K|B_3)}$$

$$= \frac{(.2)(.25)}{(.1)(.02) + (.2)(.25) + (.7)(.86)}$$

$$= \frac{.05}{.002 + .05 + .602} = \frac{.05}{.654} = .076$$

Bayes' probabilities are used extensively in an area of statistics known as *Bayesian decision making,* a topic that is covered in Chapter 16 of this book. The use of Bayes' rule in Bayesian decision making is generally done in a tabular format rather than by formula. In a three-outcome problem, the table values are as follows.

Marginal	***Conditional***	***Joint***	***Conditional***
$P(B_1)$	$P(A\|B_1)$	$P(B_1) \cdot P(A\|B_1)$	$\frac{P(B_1) \cdot P(A\|B_1)}{P(A)}$
$P(B_2)$	$P(A\|B_2)$	$P(B_2) \cdot P(A\|B_2)$	$\frac{P(B_2) \cdot P(A\|B_2)}{P(A)}$
$P(B_3)$	$P(A\|B_3)$	$P(B_3) \cdot P(A\|B_3)$	$\frac{P(B_3) \cdot P(A\|B_3)}{P(A)}$
1.0		$P(A)$	1.0

We note the use of the rule of partitioning, where

$$P(A) = \sum_{i=1}^{n} P(B_iA)$$

The entries in the table coincide with Bayes' rule.

EXAMPLE 2.32 Work the previous example in table form.

Solution: Following that format, the table is as follows.

		Marginal	*Conditional*	*Joint*	*Conditional*
B_1	*Broken*	.1	.02	.002	.003
B_2	*Chipped*	.2	.25	.050	.076 $P(B_2 \mid K)$
B_3	*Perfect*	.7	.86	.602	.921
		1.0		.654	1.000

We also have the answers for a broken vase and a perfect vase as well as the required answer for a chipped vase.

EXERCISES

2.26 People who go through a home for sale are initially attracted either because they saw an ad in the paper or because they saw a sign on the lawn while passing by. Sixty percent are attracted by the sign. If people came through because of the sign, there is a 20% chance they will purchase the house. If they came because of the ad, there is a 10% chance they will purchase. The home has just sold. What is the probability that the buyers were attracted by the sign?

2.27 Three workers are of unequal skill in using a certain gear-making machine. Worker 1 produces 25% of the output, worker 2 produces 40%, and worker 3 produces 35%. Worker 1 has a 20% chance of producing a defective gear; workers 2 and 3 have 15% and 10% chances, respectively. If a gear is found to be defective, what is the probability that it came from worker 3?

2.9 SUMMARY

This chapter is an introduction to the subject of probability. Table 2.7 summarizes the topics discussed. Probability is the measure of the likelihood that a particular event will occur. Probability has a value between 0 and 1 inclusive. It may be assigned subjectively, objectively, or experimentally.

Objective probabilities require a knowledge of the number of outcomes in the sample space and require that all outcomes are equally likely. Counting methods exist to calculate the number of such events. Included are the $m \cdot n$ rule, combinations, and permutations.

Key words or concepts that aid in determining the required probability include:

—*and*—joint probabilities, multiplication rule
—*or*—addition rule
—*given*—conditional probability

TABLE 2.7 Summary table

Probability Axioms	$0 \le P(A) \le 1$ $\sum_{i=1}^{k} P(B_i) = 1$ $P(A) = \sum_{i=1}^{k} P(A_i)$, k = number of simple events in A
Assignment of Probabilities	Subjective—hunches or guesses Relative frequency—experimental: $P(A) = \frac{n(A)}{N}$ Objective—sample space known and all outcomes are equally likely: $P(A) = \frac{n(A)}{n(S)}$
Methods of Counting	$m \cdot n$ rule gives number of possible outcomes for a combined event: $N = (n_1)(n_2) \cdots (n_k) = \prod_{i=1}^{k} n_i$ Permutations are an arrangement of a given number of items (order is important): $P_x^n = \frac{n!}{(n-x)!}$ (arrangements of *some* items) $P_n^n = n!$ (arrangements of *all* items) $P_{n_1 \cdots n_k}^n = \frac{n!}{n_1! n_2! \cdots n_k!}$ (arrangements of *like* objects) Combinations are groups from a given number of items (order is not important): $C_x^n = \frac{n!}{x!\,(n-x)!}$
Rules	Addition $P(A \cup B) = P(A) + P(B) - P(AB)$ $P(A \cup B) = P(A) + P(B)$ (if A and B are mutually exclusive) Multiplication $P(AB) = P(A) \cdot P(B \mid A)$ $P(AB) = P(A) \cdot P(B)$ (if A and B are independent)

Rules	Independence $P(A \mid B) = P(A)$ $P(B \mid A) = P(B)$ Bayes' $P(B_j \mid A) = \dfrac{P(B_j) \cdot P(A \mid B_j)}{\sum_{i=1}^{n} P(B_i) \cdot P(A \mid B_i)}$
Probabilities	Conditional $P(B \mid A) = \dfrac{P(AB)}{P(A)}$ Marginal $P(A) = \dfrac{P(AB)}{P(B \mid A)}$ Joint $P(AB) = P(A) \cdot P(B \mid A)$

We examined a relationship between joint, marginal, and conditional probabilities and used it to derive the multiplication rule. A special form of the equation exists when we have statistically independent events. This also leads to the development of a test for independence.

We also considered the addition rule along with a special form of the equation to be used when the events are mutually exclusive. Venn diagrams are helpful for analyzing problems involving the addition of probabilities.

Finally, we derived Bayes' rule and used it to solve problems where marginal and conditional probabilities were the given data and conditional probabilities were required for the solution.

SUPPLEMENTARY EXERCISES

2.28 How many different license plates can be created if it is possible to have any letter in the first position, any letter but O, I, Q in the second, any number in the third and fourth positions, and any number or letter (except O, I, Q) in the fifth?

2.29 In a fashion show, there is time to show only 6 out of 10 dresses in a new collection. In how many different ways can these 6 be chosen?

2.30 The owner of a camera shop knows from experience that 10% of all cameras manufactured by the Tabayashi Company prove to be defective. Four cameras have been recently purchased.

(a) Find the probability that the first or second or both cameras examined prove to be defective.

(b) Find the probability that none are defective.

(c) Find the probability that at least one is defective.

2.31 The probability that a store selected at random from a telephone book is located downtown is $\frac{5}{9}$. The probability that a store is both downtown and less than five years old is $\frac{17}{90}$. What is the probability that a store is less than five years old if it is downtown?

2.32 An attitude survey in the 4 sections of the United States asked randomly selected groups of voters whether they were in favor of or against the president's proposed tax changes. The numbers in the various categories are presented below.

Preference	*NE*	*SE*	*NW*	*SW*
For	50	30	30	25
Against	50	70	20	35

Suppose an individual response is selected at random from this group.

(a) What is the probability that the respondent is either from the SE or against the proposal or both?

(b) What is the probability that the respondent is from the NW or SW if the respondent is for the proposal?

(c) What is the probability that the respondent is against the proposal given that the respondent is from the NW?

(d) What is the probability that the respondent is from the NE and is for the proposal?

2.33 The Acme Company has 40 female and 60 male employees, and 2 employees are selected at random.

(a) What is the probability that both will be male?

(b) What is the probability that there will be one of each sex?

2.34 If a ship has 10 signal flags of different colors, what is the maximum number of different 3-flag signals that can be created?

2.35 The owner of a hot dog concession at an amusement park provides 6 different condiments for the hot dogs. A customer can request that none, some, or all 6 be put on a hot dog. The owner wishes to advertise the number of different ways a hot dog can be purchased. How many ways are there?

2.36 Given that $P(A) = .4$, $P(B') = .5$, and $P(AB') = .3$, find **(a)** $P(A \cup B)$ and **(b)** $P(A'|B')$.

2.37 What is the probability of getting a king and an ace in two draws from a regular deck of cards where the first card is not replaced before the second is drawn?

2.38 In how many ways can 12 jurors be selected from 14 people eligible for jury duty?

2.39 The Altman Manufacturing Company receives a particular subassembly from 2 suppliers, Brown Company and Cramden, Inc., in the proportions $\frac{1}{4}$ and $\frac{3}{4}$, respectively. Experience indicates that a subassembly from Brown will be defective with a probability of 0.05 and that one from Cramden will be defective with a probability of 0.01. A single item has been drawn from inventory and found to be defective. What is the probability it was manufactured by the Brown Company?

2.40 Five new executive offices located in a row along one side of a new building are to be randomly assigned to 5 junior executives. In how many different ways can these offices be assigned?

2.41 Records show that 70% of all full-time students work at least 10 hours per week, 20% withdraw from at least one course, and 10% withdraw from at least one course while working at least 10 hours per week. What is the probability that a full-time student, selected at random, neither works at least 10 hours nor withdraws from a course?

2.42 In a certain grocery store, the probability that a loaf of bread picked at random is stale is .3. Two shopping trips are made.

- **(a)** What is the probability that the selections, in order, will be fresh and stale?
- **(b)** What is the probability they will be stale and stale?

2.43 A certain machine turns out good parts and defective parts. If we sample a sequence of 10 items from the machine, we observe a sequence such as *GGGDGGGGDG*. How many different sequences are possible for a sample of 10 items?

2.44 A manager wishes to select 3 employees to form a special committee. How many different committees are possible if 8 employees are available for selection?

2.45 In a population of workers, it is known that $\frac{1}{4}$ are college graduates. Among college graduates, $\frac{1}{20}$ are unemployed. Among other workers, $\frac{1}{5}$ are unemployed. A worker is chosen at random and found to be unemployed. What is the probability that the worker is a college graduate?

2.46 A men's toiletry company is examining the relation between the brand of aftershave and the method of shaving used by 200 men. The results follow:

	Brand of Aftershave		
Shaving Method	***Brutal (B)***	***New Spice (N)***	***Green Velvet (G)***
Electric razor (E)	70	30	20
Safety razor (S)	50	30	0

If one person is selected at random, determine **(a)** $P(B)$, **(b)** $P(B|E)$, **(c)** $P(S|G)$, and **(d)** $P(EN)$.

2.47 The following probabilities exist at a Fourth of July picnic.

No tables available	.8
No tables available and hot day	.5
Hot day and concession out of drinks	.4
Concession out of drinks given a hot day	.8

What is the probability that someone going on a picnic will encounter no tables or a hot day or both?

2.48 The probability that an electrical component will work is .9. Two such components are wired so that a machine will run if either component works.

- **(a)** If the components are independent, what is the probability that both work?
- **(b)** What is the probability that the machine will run?

2.49 Stock market analysts have stated that the probabilities of the market rising, staying the same, and dropping are, respectively, .3, .5, and .2. It is also believed that the tax surcharge will pass with a probability of .2 if the market rises, with a probability of .4 if the market stays the same, and with a probability of .6 if the market drops. If the surcharge just passed, what is the probability that the market went up?

2.50 A box contains 15 pieces of paper. Four of the sheets are carbon paper numbered from 1 to 4, six are watermarked bond numbered from 1 to 6, and the remaining sheets are onionskin numbered from 1 to 5. One sheet is randomly selected from the box.

(a) What is the probability that it is carbon paper?
(b) What is the probability that it is either onionskin or numbered 3?
(c) What is the probability that it is watermarked bond if it is numbered 1?
(d) What is the probability that it is numbered either 2 or 5?
(e) What is the probability that it is carbon paper and numbered 4?

2.51 If the NCAA has applications from Los Angeles, Dallas, Miami, and Boston for hosting its intercollegiate tennis championships in 1990 and 1991, in how many ways can the NCAA select the tournament sites? Assume the site for 1991 must be different from that for 1990.

2.52 Winter weather bureau records for the city of Buffalo indicate that it snows 30% of the days, 50% are cloudy, and 10% are both snowy and cloudy.

(a) What percentage of the days are either snowy or cloudy?
(b) What percentage are neither snowy nor cloudy?

2.53 The Sunshine and Snow Ice Cream Company has stores in 3 states. The 80 stores are categorized below according to the state in which they are located and the number of years they have been in operation.

	State		
Years in Operation	***New York***	***Ohio***	***Pennsylvania***
Under 2	8	16	2
2–5	5	22	5
Over 5	8	9	5

One of the 80 stores is selected at random.

(a) Find the probability that the store is located in Ohio.
(b) Find the probability that the store has been in operation at least two years.
(c) Given that the store is in New York, find the probability that it has been in operation five years or less.
(d) Find the probability that the store has been in operation for two to five years and is located in Pennsylvania.
(e) Find the probability that the store is located outside Ohio.

2.54 In how many ways can three persons be selected to be flight attendants on a particular flight if there are six persons from which to select?

2.55 There are 10 professors in a given department. Six belong to professional organization A, 4 belong to organization B, and 1 is not a member of either organization. If 1 professor is selected at random, what is the probability that he belongs to both organizations?

2.56 Given that two dice have been rolled twice with a 7 occurring each time, what is the probability that on the next throw a 7 will occur again?

2.57 If 2 sheets are drawn at random in Exercise 2.50, what is the probability of getting 2 pieces of carbon paper if the first sheet is not replaced before the second is drawn?

2.58 Quarters (25¢) are enclosed at random in 20% of questionnaire mailings. The probability that a questionnaire will be answered if a quarter is enclosed is .6, and the probability that it will be answered if no quarter is enclosed is .3. If a given questionnaire is returned, what is the probability that a quarter had been enclosed with it?

2.59 Among the 8 nominees for the board of directors of a grocery cooperative are 4 men and 4 women. Four persons are to be elected. In how many ways can 2 men and 2 women be elected?

2.60 An accountant randomly selects inventory records in an audit of a clothing store. Each item of clothing has a separate card for its inventory record. Some cards agree with the actual physical count of the inventory, and some are not correct. The accountant selects 3 cards at random and checks the number on the card against an actual count.

(a) What is the sample space?

(b) Let event A be that one of the cards will not agree with the physical count. List the elements.

(c) Let event B be that all of the cards will agree with the physical count. List the elements.

(d) Let event C be that two or more of the cards will agree with the physical count. List the elements.

(e) Are events A and B mutually exclusive? Draw a Venn diagram.

(f) Are events B and C mutually exclusive? Draw a Venn diagram.

(g) Are events A and C mutually exclusive? Draw a Venn diagram.

2.61 A certain magazine puts on what it calls a "sweepstakes" in which all prizes "must be given away," and "you cannot win until you enter," and "you may have already won." They claim that the following probabilities hold.

1st prize	.0000001
2nd prize	.0001
3rd prize	.01
4th prize	.39
5th prize	.64

What comments might the Federal Trade Commission make if they apply the probability axioms?

2.62 A company sends out 3 checks in the local mail for \$100, \$250, and \$175. Two days later, the company discovers it only had \$370 in the bank to cover the checks. It calls the bank and learns that two of the checks have been cashed.

(a) What is the sample space?

(b) Let event A be that both checks cleared. What are the elements?

(c) What is the probability that both checks cleared?

2.63 A car manufacturer has replaced its car door locks with a series of 5 buttons that must be pushed in a certain order to open the door. If 5 buttons must be pushed, how many entry codes are possible?

2.64 The car manufacturer in Exercise 2.63 is thinking that security can be improved if a 10-digit keypad, similar to a calculator, is installed instead of the 5 buttons, and if a 5-digit numeric code must be entered to open the car.

(a) How many codes are possible?

(b) Will this improve security over the 5-button system employed in Exercise 2.63?

(c) If the answer to (b) is yes, by what factor will security be improved?

2.65 A chief executive has a round conference table with 6 chairs. At a conference, she selects a chair at random to sit in, and then the 5 vice-presidents sit down. How many different arrangements around the table are possible?

2.66 Programmers in a computer center have observed that all kinds of errors can show up on programs that they run: syntax errors (incorrect use of the computer language), logic errors (bad thinking), keypunching errors, and job control errors (incorrect control cards). A program may have none, some, or all of these errors. In how many ways can their program end up with errors?

2.67 If all the errors in Exercise 2.66 are equally likely, what is the probability that a program with an error will have only keypunching errors?

2.68 A certain savings and loan institution issues certificates of deposit with a one-year maturity. To some of its CD customers, the bank sends letters reminding them of the expiration date of the CDs and suggesting that renewal be made. To others, they do not. Seventy percent of their customers do not renew. Of those who renew, 30% received letters. Of those who do not renew, 50% received letters. Given that a customer just received a letter about renewal, what is the probability that the customer will renew?

2.69 In the previous exercise, given that a customer does not receive a letter, what is the probability that the customer will not renew?

2.70 An auditor is about to select a sample of 25 accounts from 100 available. A nervous employee has stolen from 2 of these accounts. What is the probability of the employee's being caught?

2.71 A government motor pool has 3 Fords and 7 Chevrolets. If a department requests four cars for the day, in how many ways can it get **(a)** 2 Fords and 2 Chevrolets, **(b)** 3 Fords and 1 Chevrolet, and **(c)** 4 Chevrolets?

2.72 An IRS clerk has the job of displaying in a display case 3 forms and 3 booklets out of their collection of 10 forms and 8 booklets. The 3 forms must be next to each other, and the 3 booklets must also be next to each other. How many different orders are possible in the display?

2.73 A newly formed department in a firm has been granted the right to select at random one clerk from each of 4 existing departments in the firm. Departments A, B, C, and D are currently composed of 5, 3, 7, and 4 clerks, respectively. In how many ways can the four clerks be selected?

2.74 A car dealer has 2 identical sedans, 2 identical trucks, 5 identical vans, and 1 recreational vehicle. In how many different ways can he arrange these vehicles down a row in his lot?

2.75 The probability that it will rain or snow under given conditions is .85, the probability of rain is .35, and the probability of snow is .55. What is the probability that it will both rain and snow at the same time?

2.76 A builder cannot continue a certain job until the doors and windows arrive, each from different suppliers. The builder figures that the probability that the doors will arrive is .6, the probability that they both will arrive is .2, and the probability that one or the other will arrive is .6. What is the probability that the windows will arrive?

2.77 Three contractors always bid for city sidewalk construction in a certain city. The probability of winning the contract, based on past experience, is .32, .41, .27 for firms A, B, and C, respectively. Two contracts are to be awarded.

(a) What is the probability that A wins the first contract?

(b) What is the probability that B or C wins the first contract?

(c) What is the probability that A wins the first contract and C wins the second?
(d) What is the probability that A or B wins the first contract and B or C wins the second?
(e) What is the probability that A wins at least one of the contracts?
(f) What is the probability that B wins neither contract?

2.78 Sales of homes over the past month in Superville are typical for the town and are shown below.

	House Style		
Lot Size	*Cape*	*2-Story Colonial*	*2-Story Tudor*
Small	34	24	2
Medium	20	40	10
Large	12	26	22

Calculate the probability that a sale selected at random is for **(a)** a small lot, **(b)** either a large lot or a two-story tudor, **(c)** a cape or a medium lot, **(d)** a two-story tudor, given it is on a large lot, and **(e)** a small lot or medium lot, given it is a cape.

2.79 An insurance actuary has determined that in automobile accidents, the probability that a driver is drunk and on a suspended license is .1, that the driver is sober is .6, that the driver is either drunk or driving with a suspended license is .6. What is the probability that a driver's license is not suspended?

2.80 The owner of a small business, in starting a new financial advisory service, sent out letters informing potential clients that she would like to meet with them to explain her services. In 60% of the letters, she enclosed a booklet about the benefits of financial services. She discovered that if she did not send a booklet, the probability that she could get a meeting was $\frac{1}{4}$. If she did send a booklet, the probability she could not get a meeting was $\frac{2}{3}$. For all the letters she sent out, what is the probability that she could get a meeting?

2.81 The probability that an employee will pass a test for promotion is .7. The probability that an employee will pass the same exam on a second try is .8. What is the probability that an employee will fail to be promoted in two tries on the exam?

2.82 In the previous exercise, are the results on the exam independent? Support your answer numerically.

2.83 The personnel department gives a true–false exam with 25 questions on it to all prospective employees. If the person being tested guessed at every question, what is the probability of all questions being answered correctly?

2.84 A banker has 12 requests for loans of equal size but only enough money to grant 9 of them. He decides to grant the loans randomly.
(a) In how many different ways can the loans be granted?
(b) What is the probability of receiving a loan?

3 Random Variables, Probability Distributions, and Expected Values

3.1 INTRODUCTION

The Burton Jacobs Company is setting up a new production process to manufacture and package liquid bleach. The company is interested in the quality of the items produced. Carla Deming, BJ's manager of product quality, is faced with answering the following questions. First, what characteristic(s) should she identify to monitor the quality of the output? Second, should she merely have the number of defectives counted, or should she actually have some characteristic, such as the concentration of sodium hypochlorite or the fill content, measured? After she has determined the characteristic(s) and type of measurement(s) to use, Carla must answer a third question: What values or sets of values should be associated with poor quality, and with what frequency do they occur?

To answer the first two questions, Carla must define a random variable. To answer the latter question, she must determine the probability distribution of the values for the random variable—that is, all possible values and the frequencies with which they are expected to occur.

In Chapter 2, we studied the concept of a random experiment and its resultant sample space or set of simple outcomes. (Recall the coin tossing experiment. The basic outcomes of this experiment are head and tail, and, since the coin is honest, a probability of $\frac{1}{2}$ is assigned to each outcome.) We also examined and used the basic rules for calculating probabilities for compound events. A problem that is not solved by this approach, however, is the requirement of numerical values for making statistical inferences and drawing conclusions. As we saw, many experiments do not have numbers for outcomes. For example, if Carla decides to monitor the fill content and merely classifies each bottle as underfilled (U), correctly filled (C), or overfilled (O), her outcome from observing a sample of three bottles would be a sequence of the letters U, C, and O. In this chapter, we consider procedures for assigning numbers to experimental outcomes and for associating probabilities with these numerical outcomes. We also look at the basic formulas for calculating values to describe the resulting sample space.

3.2 RANDOM VARIABLES

We first consider assigning numerical values to experimental outcomes by introducing the concept of a random variable.

DEFINITION 3.1 A **random variable** X is a measurable function relating the sample space S to some subset of the real line.

Essentially, a random variable is nothing more than a function that assigns numbers to the basic experimental outcomes. Consider once more the coin toss, which had basic outcomes of head and tail. Numerous functions could be used to assign numbers to the outcomes: For example, $X(H) = 22$, $X(T) = -7$ is one possibility, though not a very logical one, as we will see. The basic requirement is that X be a function, which implies that one or more outcomes may be assigned to the same number, but only one number may be assigned to any simple outcome (Figure 3.1). The unacceptable relationship of two numbers for the same outcome is due to the use of one domain element for two range elements.

FIGURE 3.1 An illustration of acceptable and unacceptable definitions of random variables

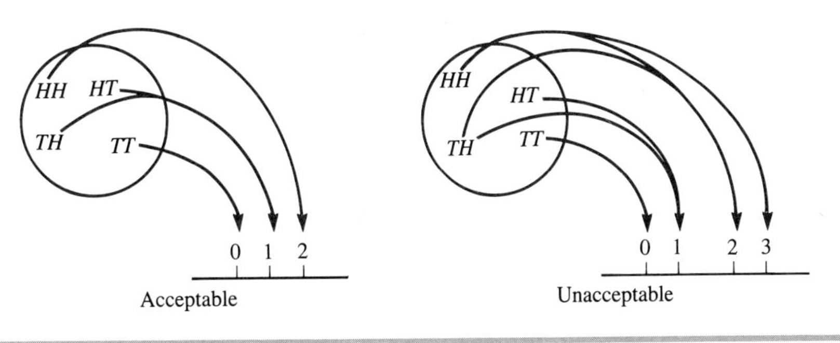

The random variable should also assign values according to the purpose at hand. For the coin experiment, it makes sense to let $X(T) = 0$ and $X(H) = 1$, in which case the variable X can be defined verbally and logically as the number of heads occurring. On the other hand, we could let $X(H) = \$5$ and $X(T) = -\$5$ if a gambling situation involving a $5 bet is to be represented.

The logical association becomes even more important in larger experiments. For example, consider items coming off a production line. The items are either defective or good. If one takes a sample of 10 items, the actual outcomes observed are sequences of defective and good items—for example, GGGDGGGGGG, GGGGGGGGGG, DGGGGGGGGG, and so on. (Note that by the $m \cdot n$ rule, there are $2^{10} = 1024$ possible sequences.) Since one is generally interested in the number of defectives in the sample rather than the order of occurrence, the logical random variable here would be the number of defectives in the sample. Thus, the variable could assume the 11 possible values of 0, 1, 2, 3, 4, . . . , 9, 10, of which the first and third sample sequences would be assigned the value 1, and the second would be assigned the value 0.

If the original experiment already results in numerical values, then the random variable used is generally the identity function. That is, each numerical outcome for X is simply defined to be the original value. For example, suppose three students were selected from the class and weighed, giving the following outcomes in pounds: 161, 112, 174. Then the random variable would simply be defined as $X(161) = 161$, $X(112) = 112$, $X(174) = 174$. Clearly, X has assumed the same values as were originally observed.

EXERCISE

3.1 Which of the following would be considered random variables: **(a)** the number of dots on the upturned face of a die, **(b)** the number of points scored in a football game, **(c)** the day of the week on which July 1 falls, **(d)** the thickness of an arbitrary piece of sheet metal, **(e)** the distance from Cleveland to Akron, and **(f)** the number of blemishes on a piece of wood paneling coming off a production line?

3.3 DISCRETE AND CONTINUOUS RANDOM VARIABLES

Generally, random variables are classified as discrete or continuous. It is important to learn to distinguish between these types because different mathematical techniques are necessary for each type.

DEFINITION 3.2 A random variable is said to be **discrete** if it can assume only a finite or, at most, a countably infinite number of values.

If the number of possible values in the original sample space is finite, then there is no difficulty in identifying the variable; it is discrete. The difficulty arises when the number of possible values is infinite. In this case, perhaps the easiest way to distinguish between discrete and continuous random variables is to ask, Can the values be counted starting with the smallest, then the next smallest, and so on? That is, Is there a one-to-one correspondence with the positive integers? If so, then the set of outcomes is countably infinite, and hence the variable is discrete. Examples of discrete variables are those defined in coin-tossing experiments and in the production-line experiment given earlier.

DEFINITION 3.3 A random variable is said to be **continuous** if it can assume any value in some interval or set of intervals.

In the continuous case, the set of possible outcomes is always infinite—in fact, uncountably so. Hence, it is impossible to list the sample space by individual values in any form, because although it is possible to determine the smallest possible value, it is impossible to find the next smallest. An example of a continuous random variable is the weight of a randomly selected student. Even though the weight might be recorded to the nearest pound, say 165, because of the inexactness of the measuring instrument, it could be any value between 164 pounds and 166 pounds exclusive.

In practice, it is generally not too difficult to distinguish between discrete and continuous random variables. Discrete random variables are often counting variables—for example, the number of defectives, or the number of people in a sample favoring a particular political candidate. Continuous variables, on the other hand, are generally measuring variables—for example, the weight of an individual or the length of a bolt.

EXERCISE

3.2 In Exercise 3.1, identify which random variables are discrete.

3.4 DISTRIBUTIONS OF RANDOM VARIABLES

Since random variables are assignments of numbers to the outcomes of a random experiment, the values of the random variable also have certain probabilities of occurrence.

DEFINITION 3.4 The **probability distribution** of a random variable is an identification of the possible values that the variable may assume and of the probabilities associated with those values.

The possible ways of specifying the probability distribution vary with the type of random variable being considered. If the variable has a finite number of outcomes, it is sometimes possible to give the probability distribution in tabular form—that is, as a *relative frequency table*. For example, suppose that

50% of the residents in a given area are believed to be in favor of a zoning proposal. If the belief is true and three residents are selected at random and asked their opinions, then eight sequences of occurrence are possible. Let F designate a resident in favor of the proposal, and N a resident not in favor of it. The random variable denoting the number of residents in the sample who favor the proposal has the following distribution.

	Distribution of X	
Original Sample Space	***X***	***p(x)***
NNN	0	1/8
FNN, NFN, NNF	1	3/8
FFN, FNF, NFF	2	3/8
FFF	3	1/8

In this table, X denotes the random variable in general, x a specific value of the variable, and $p(x)$ the probability that X equals x—that is, $p(x) = P(X = x)$.

In most cases, the number of outcomes, even though it may be finite, is too large for the tabular method of identification to be feasible. Generally, then, the representation of the distribution will take the form of a mathematical model or function. The preceding tabular representation can also be identified with such a mathematical function. The function would be

$$p(x) = C_x^3 \left(\frac{1}{2}\right)^x \left(\frac{1}{2}\right)^{3-x} \qquad x = 0,\ 1,\ 2,\ 3$$

In this function, $(\frac{1}{2})^x(\frac{1}{2})^{3-x}$ is the probability that x successes and $3 - x$ failures occur in a particular sequence, and C_x^3 is the number of different sequences that are possible. Note that this function can be applied only in situations that are analogous to tossing an honest coin.

Regardless of the method of definition used, the probability distribution must obey the basic rules of probability. In the discrete case, as in the zoning example, these conditions are

$$p(x) \geq 0 \qquad \text{for all } x \tag{3.1}$$

$$\sum_{\text{all } x} p(x) = 1 \tag{3.2}$$

where the summation is over all possible values for X. The function $p(x)$ assigns nonzero probabilities to specific values for X—that is, it assigns small masses of probability to the possible values. Thus, it is generally known as a *probability mass function* (pmf). This concept is shown graphically in Figure 3.2(a), where the random variable for the zoning proposal responses is used. The probabilities are represented as small masses at a height equal to the actual probability—hence the name "probability mass function."

Often, a bar graph or histogram is used to demonstrate the concept of probability as an area. The area in each bar is equal to the probability associated with the center point of the bar. See Figure 3.2(b).

FIGURE 3.2 Graphical representation of $p(x)$ and corresponding histogram for the number of favorable responses to the zoning proposal survey

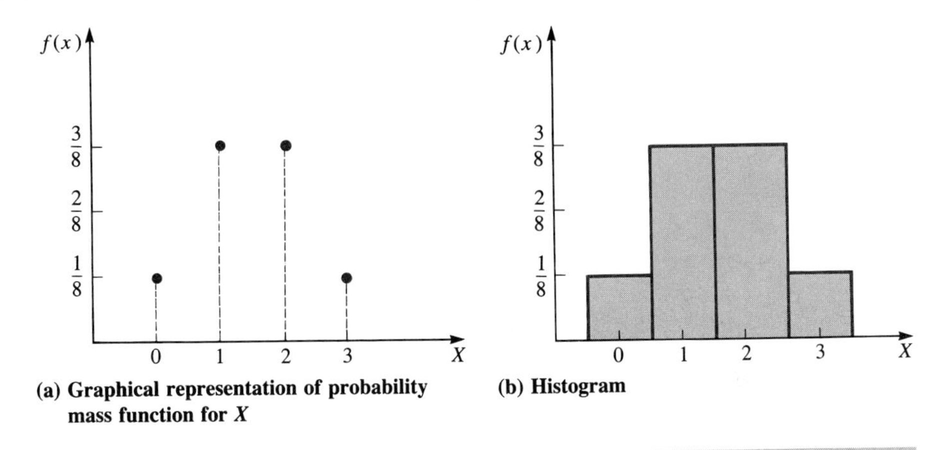

Equations 3.1 and 3.2 relate directly to the probability axioms of Chapter 2. Equation 3.1 simply states that one cannot assign a negative probability, and Equation 3.2 indicates that the total sample space must have a probability of 1.

For continuous random variables, it is impossible to use the tabular format. Generally, the distributions are given by mathematical functions, but occasionally a graphical representation is used. In standard notation for functions, $f(x)$ denotes the *probability density function* (pdf) for a continuous random variable; the probability mass function $p(x)$ is reserved for discrete variables. The probability rules indicated for $p(x)$ are also basic to continuous distributions. For the pdf, these conditions are

$$f(x) \geq 0 \qquad \text{all } x \tag{3.3}$$

$$\int_{-\infty}^{+\infty} f(x)\, dx = 1 \tag{3.4}$$

where $\int_{-\infty}^{+\infty}$ is the integral over the set of real numbers. Equation 3.3 says simply that the pdf must be a nonnegative function, which is logical since probabilities themselves cannot be negative. For continuous variables, events are defined as intervals of values, and probabilities correspond to areas; thus, Equation 3.4 states that the total area between the function $f(x)$ and the x-axis must equal 1. The function $f(x)$ gives the height of the curve at any value of x. Intuitively, the difference between the pmf and the pdf is the difference between placing globules of paint at specific points, as indicated in Figure 3.2(a), and spreading those globules over a surface, or the tops of the bars as indicated in Figure 3.2(b). The pmf identifies where the globule is to be placed; the pdf identifies the surface over which the globules are to be spread. For example, a pdf can be defined as follows to represent the surface depicted in Figure 3.2(b).

$$f(x) = \begin{cases} \frac{1}{8} & -.5 \le X < .5 \\ \frac{3}{8} & .5 \le X < 2.5 \\ \frac{1}{8} & 2.5 \le X < 3.5 \\ 0 & \text{elsewhere} \end{cases}$$

In this case, the pdf might be viewed as a continuous approximation to the original discrete pmf. Note that $p(0)$ now corresponds to the area under $f(x)$ between $-.5$ and $.5$. The following examples further illustrate these concepts.

EXAMPLE 3.1 Let us consider the function $f(x)$ and determine whether it can be used as a probability density function.

$$f(x) = \begin{cases} \frac{1}{3} & -3 \le X \le +3 \\ 0 & \text{elsewhere} \end{cases}$$

Solution: To determine whether $f(x)$ is a valid pdf, we must check Equations 3.3 and 3.4. Since $f(x)$ is always either $\frac{1}{3}$ or 0, Equation 3.3 is satisfied. For Equation 3.4, we look at the graph of $f(x)$ given in Figure 3.3. The base of the rectangle is 6, so the area is $6 \times \frac{1}{3} = 2$. Since the total area is 2 rather than 1, Equation 3.4 is *not* satisfied. Hence, $f(x)$ is not a valid pdf.

EXAMPLE 3.2 We define the function of Example 3.1 so that it becomes a valid pdf for $0 \le X \le 3$, and we find the probability that a randomly selected value for X will be between 1.5 and 2.1.

FIGURE 3.3 Graphical representation of $f(x)$ for Example 3.1

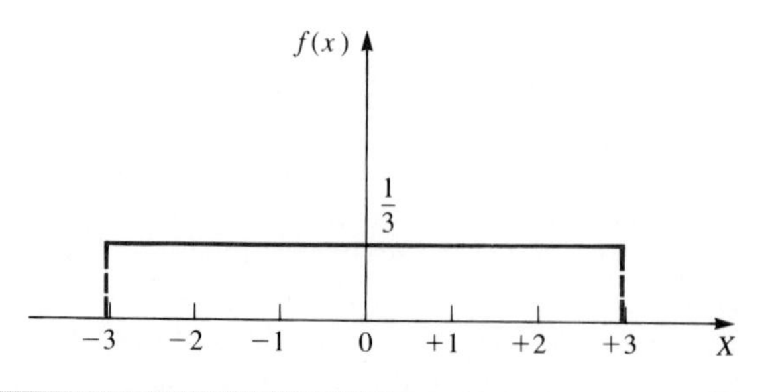

FIGURE 3.4 Graphical representation of pdf for Example 3.2

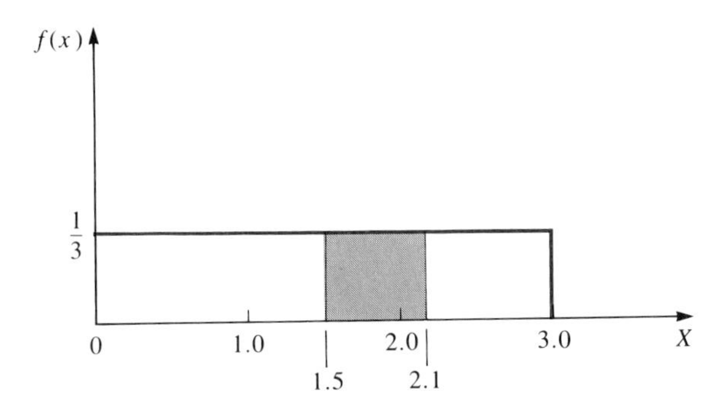

Solution: In this case,

$$f(x) = \begin{cases} \frac{1}{3} & 0 \leq X \leq 3 \\ 0 & \text{elsewhere} \end{cases}$$

The graph is given in Figure 3.4. It is easy to verify that Equations 3.3 and 3.4 are now satisfied. To find the required probability, we note that the probability corresponds to the area under $f(x)$ between 1.5 and 2.1. That is, the proportion of times that a value for X falls between 1.5 and 2.1 is identical to the proportion of the total area between 1.5 and 2.1. Thus,

$$P(1.5 \leq X \leq 2.1) = \text{base} \times \text{height} = (2.1 - 1.5) \times \frac{1}{3} = .2$$

One of the major differences between discrete and continuous random variables is the interpretation of the probability statements. With a discrete variable, it is reasonable to consider the probability of a single outcome. For example, in tossing an honest coin three times, the statement $P(2 \text{ heads}) = P(X = 2)$ has meaning—in fact, it equals $\frac{3}{8}$. But, in the continuous case, probability is interpreted as an area. Thus, to speak of the probability of the occurrence of a specific number is generally meaningless because the area corresponding to the specific point would be that of a line and lines have no area; hence the probability would be zero. Another way of seeing this is to observe that for a specific value there would be only one point out of infinitely many. Thus, in the continuous case, the probabilities considered are those of intervals or sets of intervals—that is, statements such as $P(a \leq X \leq b)$, which is read as the probability that the observed value is between the numbers a and b inclusive (see Figure 3.5). The following are illustrations of continuous probability distributions.

FIGURE 3.5 Graphical representation of probability for a continuous random variable

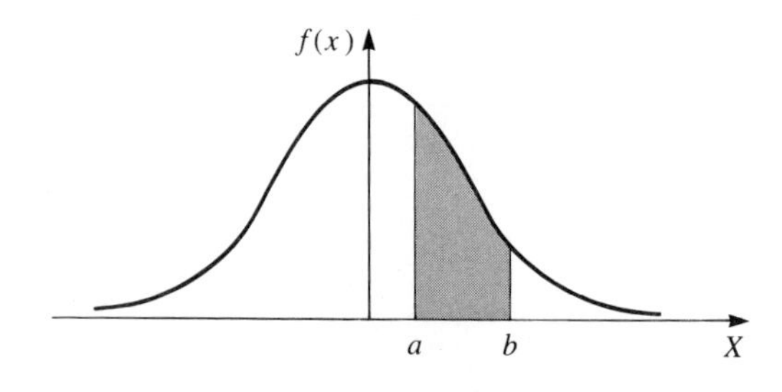

Uniform or Rectangular Distribution

Because the probability is uniformly distributed over some interval, the uniform distribution is probably the simplest of the continuous distributions. Consider the standard or unit uniform distribution, which is defined to have nonnegative probability on the interval 0 to 1 and 0 probability elsewhere.

DEFINITION 3.5 The function describing the **standard uniform distribution** is

$$f(x) = \begin{cases} 1 & 0 \leq X \leq 1 \\ 0 & \text{elsewhere} \end{cases}$$

This distribution also may be described graphically as shown in Figure 3.6. Thus, $P(\frac{1}{4} \leq X \leq \frac{3}{4}) = \frac{1}{2}$, which is the area under the curve between $\frac{1}{4}$ and $\frac{3}{4}$ and is the shaded portion in Figure 3.6. Because the graph describes a rectangle, another name for the distribution is rectangular distribution.

DEFINITION 3.6 The general form for a **rectangular distribution** defined on any interval a to b is

$$f(x) = \begin{cases} \dfrac{1}{b - a} & \text{for } a \leq X \leq b \\ 0 & \text{elsewhere} \end{cases}$$

Triangular Distribution

DEFINITION 3.7 For a **triangular distribution,** the pdf is a steadily increasing or decreasing function over some interval a to b (or both over different subintervals) and has the shape of a triangle over that interval.

FIGURE 3.6 Standard uniform distribution

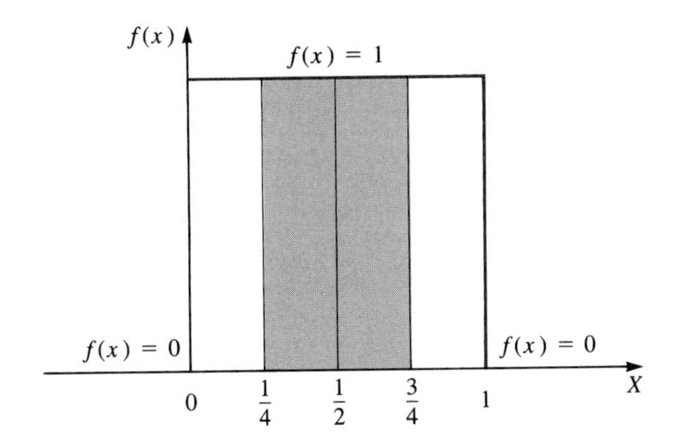

Specifically consider the interval 0 to 2. Then one probability distribution is

$$f(x) = \begin{cases} \frac{x}{2} & 0 \le X \le 2 \\ 0 & \text{elsewhere} \end{cases}$$

The graph is shown in Figure 3.7. Thus, the probability of an observation's being between 0 and 1 would be represented as the area under the curve between 0 and 1 (see the shaded portion in Figure 3.7). The probability associated with this area is $P(0 \le X \le 1) = \frac{1}{4}$.

FIGURE 3.7 A triangular distribution

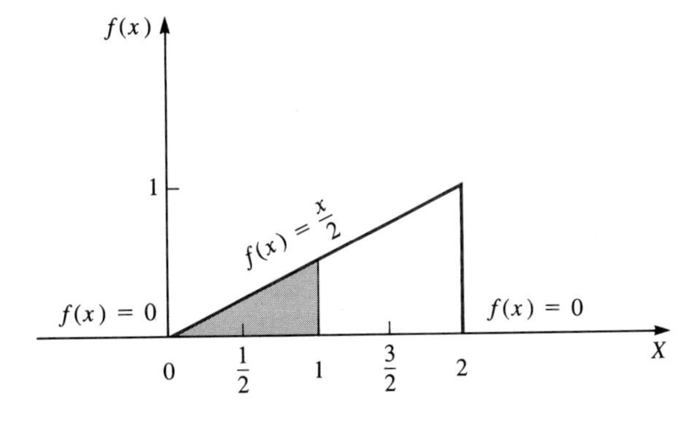

EXAMPLE 3.3 The life or time to failure of a particular type of equipment is a random variable with a probability density function as given below. The value .08 represents the average decrease per year in life expectancy for equipment of this type.

$$f(x) = \begin{cases} .4 - .08x & 0 \le X \le 5 \\ 0 & \text{elsewhere} \end{cases}$$

Determine the probability that a piece of this equipment (a) will fail in the first year and (b) will last at least 3 years.

Solution: Note that the maximum equipment life is 5 years and that $f(x)$ has a graphical representation as shown in Figure 3.8.

a. $$P(X \le 1) = \int_{-\infty}^{1} f(x)\,dx = \int_{-\infty}^{0} f(x)\,dx + \int_{0}^{1} f(x)\,dx$$
$$= \int_{0}^{1} (.4 - .08x)\,dx \qquad [\text{since } f(x) = 0 \text{ for } -\infty < X < 0]$$
$$= \int_{0}^{1} .4\,dx - \int_{0}^{1} .08x\,dx = .4x\Big|_{0}^{1} - (.08)\frac{x^2}{2}\Big|_{0}^{1} = .4 - .08\left(\frac{1}{2}\right)$$
$$= .4 - .04 = .36$$

That is, there is a 36% chance it will fail in the first year.

b. $$P(X \ge 3) = \int_{3}^{\infty} f(x)\,dx = \int_{3}^{5} f(x)\,dx + \int_{5}^{\infty} f(x)\,dx$$
$$= \int_{3}^{5} (.4 - .08x)\,dx + 0$$
$$= .4x\Big|_{3}^{5} - .08\left(\frac{x^2}{2}\right)\Big|_{3}^{5} = .80 - .64 = .16$$

or a 16% chance of lasting at least 3 years.

FIGURE 3.8 Graphical representation of pdf for Example 3.3

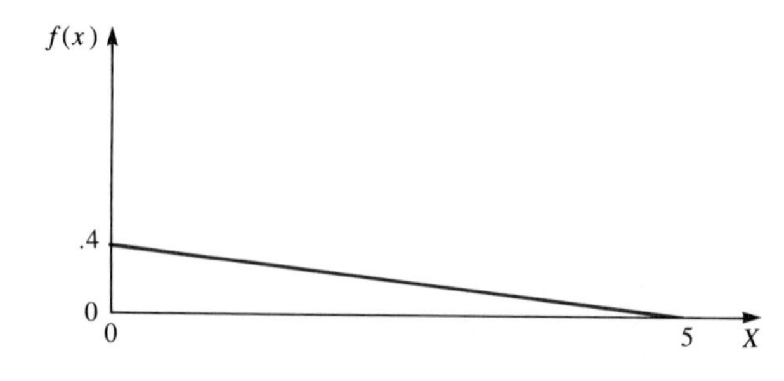

Cumulative Distribution Functions

For most of the probability distributions that are considered later, tables are available. With discrete random variables, these tables may give probabilities for specific values, or they may give the probability corresponding to that value or a smaller one. If the random variable of interest is continuous, then the table must give the probability corresponding to an interval of values. Usually the interval to which the probability will refer will be $(-\infty, x)$—that is, the probability will represent the statement $P(X \leq x)$. This type of probability is known as a "less than or equal to" cumulative probability. The name arises because the result is the cumulation of the probability for all values less than or equal to x.

DEFINITION 3.8 The nondecreasing function

$$F(x) = P(X \leq x)$$

that represents this particular type of probability is called the **cumulative distribution function** and is defined over the total real line. For discrete random variables, it is a step function because it increases only at points that have positive probabilities; for continuous random variables, it describes a curve having a shape similar to an elongated or drawn-out S.

As an example, consider the uniform distribution on the interval $[a,b]$. In this case, the probability density function is

$$f(x) = \begin{cases} \dfrac{1}{b-a} & a \leq X \leq b \\ 0 & \text{elsewhere} \end{cases}$$

By integrating the function from $-\infty$ to x, we obtain the cumulative distribution function

$$F(x) = \begin{cases} 0 & X < a \\ \dfrac{x-a}{b-a} & a \leq X \leq b \\ 1 & X > b \end{cases}$$

The cumulative distribution function is illustrated in Figure 3.9.

The cumulative distribution function as defined represents less than or equal to probability statements. Occasionally, we encounter cumulative functions that represent greater than or equal to probability statements. When using a cumulative table, we must be certain that we know which way it cumulates.

FIGURE 3.9 Graphical representation of the cumulative distribution function for a uniform distribution

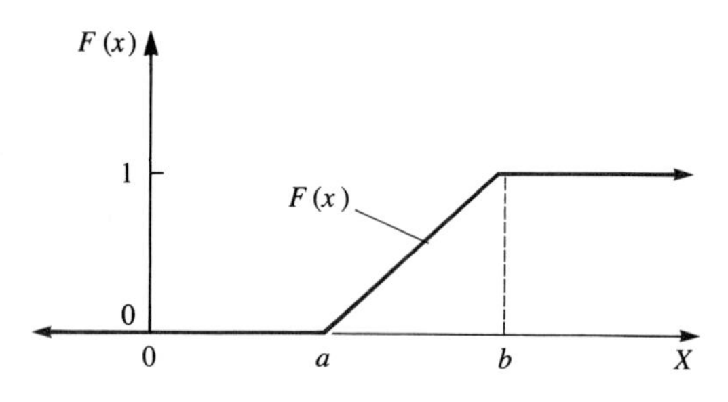

EXERCISES

3.3 A small company has 10 employees—6 men and 4 women. From the personnel files, 3 cards are drawn at random. Construct the probability distribution of the random variable X, the number of women drawn.

3.4 A baker knows that 90% of all wedding cakes she bakes will be good. She has orders for 2 such cakes for next week. Calculate the probability distribution of Y, the number of good wedding cakes resulting from the next two attempts. What is the probability that Y will equal two?

3.5 The distribution of the number of trials, N, required to obtain a given number of successes, r, is given by

$$p(n) = \begin{cases} C_{r-1}^{n-1}\, \pi^r (1-\pi)^{n-r} & n = r, r+1, \ldots \\ 0 & \text{elsewhere} \end{cases}$$

where π = probability that a single trial will result in a success. Refer to Exercise 3.4 and identify r and π. What is the probability that the baker must make exactly 4 cakes to fill her orders for next week? What is the probability that she must make at most 3 cakes?

3.6 Refer to the uniform distribution and let $a = 2.0$, $b = 5.0$. Find the probability that a value will be between 3.0 and 4.0. Find the probability that it will be at most 4.5.

3.7 For the triangular distribution given by

$$f(x) = \begin{cases} \dfrac{2x}{9} & 0 \le X \le 3 \\ 0 & \text{elsewhere} \end{cases}$$

determine **(a)** $P(X \le 1.5)$ and **(b)** $P(.5 \le X \le 2.5)$.

3.8 Determine the probabilities in Example 3.3 without using calculus.

3.5 PARAMETERS OF DISTRIBUTIONS

As we saw in the uniform distribution and in the coin-tossing example, a particular type of mathematical function may be used to describe the distribution of many different variables. For example, the name "uniform" can be associated with an infinite number of distributions all having pdf's of the form $1/(b - a)$. Hence, "uniform" is really a family name and identifies a family of distributions all indexed by the same pdf, just as Smith is the name of a family group. When an individual of the Smith family is of interest, he or she must be further identified by more information, such as first and middle names. The same thing is true of distributions. Simply stating that a variable has a uniform distribution is not enough. To calculate probabilities, one must also know which member of the uniform family is to be used.

DEFINITION 3.9 A **parameter** of a distribution is a numerical value that is necessary for complete identification of the probability distribution.

Thus, parameters are similar to an individual's given names in the sense that they identify which member of the family is being considered. All parameters must be known before probabilities can be calculated. In the uniform distribution, the parameters are a and b. Once the uniform family is specified and values for a and b are given, then any probability can be calculated, and the distribution is said to be completely specified. In the standard uniform distribution, the values were $a = 0$, $b = 1$. Standard notation for a uniform random variable is $X \sim U(a,b)$, where the symbol $\sim$ means "has the distribution," U identifies the uniform family or the pdf, $1/(b-a)$, and a and b are the parameters of that family where $a < b$. For the standard uniform distribution, we use $X \sim U(0,1)$.

To see that this is also true for discrete distributions, recall that in a coin-tossing experiment, the function describing the probability distribution is, in general,

$$p(x) = \begin{cases} C_x^n \, \pi^x \, (1 - \pi)^{n-x} & x = 0, 1, 2, \ldots, n \quad 0 < \pi < 1 \\ 0 & \text{elsewhere} \end{cases}$$

where n is the number of trials (tosses) and π is the probability of a success (head) on a single trial. In our situation, the coin is fair, which implies that $\pi = \frac{1}{2}$; if we toss it three times, $n = 3$.

In Chapter 4, we will study other commonly used distributions such as the binomial, Poisson, normal, and others. These names identify various families of distributions and the corresponding mathematical functions to be used in evaluating specific probability statements. Each family has its own parameters for identifying specific members.

3.6 EXPECTED VALUES

In working with random variables, it is helpful to know certain values that aid in describing the distribution. The most commonly used values are those that identify the location or physical center of the distribution and the dispersion or the way the values of the variable are spread out about the center.

DEFINITION 3.10a Given a discrete random variable, X, with probability function $p(x)$, the **expected value of X,** also called the **mean** and denoted $E[X]$ or μ_X, is defined to be the weighted average of the values that X may assume, where the weights are the corresponding probabilities. That is,

$$\mu_X = E[X] = \frac{\sum_{\text{all } x} xp(x)}{\sum_{\text{all } x} p(x)} = \sum_{\text{all } x} xp(x) \tag{3.5}$$

since $\Sigma p(x) = 1$.

DEFINITION 3.10b For a continuous random variable, $p(x)$ is replaced by the probability density function and Σ is replaced by $\int$ to obtain the **expected value of X,** or **mean.** That is,

$$\mu_X = E[X] = \int_{\text{all } x} xf(x)\,dx \tag{3.6}$$

In succeeding formulas involving discrete random variables, we will omit the denominator, $\Sigma p(x)$, because it equals 1.0. Intuitively, the expected value is just the long-run average of what one would observe and is generally denoted by the Greek letter μ (pronounced mu).

EXAMPLE 3.4 In a class of 10 students, the grades are to be assigned randomly so that 1 A, 2 B's, 4 C's, 2 D's, and 1 F are given. Define the random variable X such that $X(\text{A}) = 4$, $X(\text{B}) = 3$, $X(\text{C}) = 2$, $X(\text{D}) = 1$, $X(\text{F}) = 0$. That is, X is the number of quality points associated with the letter grade. Then the tabular representation of the distribution is

	Letter Grade				
	F	***D***	***C***	***B***	***A***
X	0	1	2	3	4
p(x)	$\frac{1}{10}$	$\frac{2}{10}$	$\frac{4}{10}$	$\frac{2}{10}$	$\frac{1}{10}$

What is the grade that can be expected by a randomly selected student? In other words, what is the expected value or mean of the distribution?

Solution: Using Equation 3.5, we determine the mean as follows:

$$\mu_X = E[X] = \sum_{x=0}^{4} xp(x)$$

$$= 0\left(\frac{1}{10}\right) + 1\left(\frac{2}{10}\right) + 2\left(\frac{4}{10}\right) + 3\left(\frac{2}{10}\right) + 4\left(\frac{1}{10}\right) = \frac{20}{10} = 2$$

Thus, the average or expected grade is C.

EXAMPLE 3.5 Consider a game in which 3 fair coins are tossed simultaneously. If all heads or all tails occur, a player will win \$10; for all other outcomes, the player must pay \$3. What is the player's expected winnings per play?

Solution: Let A be the amount of money gained or lost; then its possible values are \$10 and −\$3. The probability of 3 heads (or of 3 tails) is $\frac{1}{8}$. Thus, the probability of anything else is $\frac{6}{8}$, and the distribution of X becomes

X	$p(x)$
-3	$\frac{3}{4}$
$+10$	$\frac{1}{4}$

The expected winnings per play are then given by

$$\mu_X = E[X] = \sum_{\text{all } x} xp(x) = -3\left(\frac{3}{4}\right) + 10\left(\frac{1}{4}\right)$$

$$= \frac{1}{4} \text{ or } \$.25$$

Equation 3.5 can be extended to give the expected value of any function of a random variable. For example, if the payoffs in Example 3.5 are doubled, we have a new random variable, Y, that can be expressed as a function of X, namely, $Y = 2X$.

DEFINITION 3.11 Let X be a discrete random variable with probability function $p(x)$, and let $g(x)$ be any function of X. Then the **expected value of $g(X)$** is

$$E[g(X)] = \sum_{\text{all } x} g(x)p(x) \tag{3.7}$$

In words, this equation says that to evaluate the function $g(X)$ at each value of x, multiply the result by the probability of that x and add all the products. A specific $g(X)$ of interest is $g(X) = (X - \mu_X)^2$, whose expected value yields the variance of the distribution.

DEFINITION 3.12 Let X be a discrete random variable with probability function $p(x)$. Then the **variance of X,** denoted σ_X^2 (sigma squared), is defined to be

$$\sigma_X^2 = E[(X - \mu_X)^2] = \sum_{\text{all } x} (x - \mu_X)^2 p(x) \tag{3.8}$$

that is, it is the expected or average squared deviation from the mean.

DEFINITION 3.13 The **standard deviation of *X*** is then defined to be the square root of the variance, or the standard deviation of a discrete random variable X is

$$\sigma_X = \sqrt{E[(X - \mu_X)^2]} = \sqrt{\sum_{\text{all } x} (x - \mu_X)^2 p(x)} \quad \textbf{(3.9)}$$

In a practical sense, the standard deviation is the more useful value because its units are the same as the units of the random variable, whereas the variance is in units squared. However, the variance must be determined to obtain the standard deviation.

EXAMPLE 3.6 Consider the grade allocation of Example 3.4 and determine the variance and standard deviation of the variable X.

Solution: The distribution of X was

X	***p(x)***
0	$\frac{1}{10}$
1	$\frac{2}{10}$
2	$\frac{4}{10}$
3	$\frac{2}{10}$
4	$\frac{1}{10}$

and the mean was found to be 2. Hence the variance, as given by Equation 3.8, is

$$\begin{aligned}\sigma_X^2 &= E[(X - \mu_X)^2] = \sum_{x=0}^{4} (x - 2)^2 p(x) \\ &= (0 - 2)^2\left(\frac{1}{10}\right) + (1 - 2)^2\left(\frac{2}{10}\right) + (2 - 2)^2\left(\frac{4}{10}\right) + (3 - 2)^2\left(\frac{2}{10}\right) \\ &\quad + (4 - 2)^2\left(\frac{1}{10}\right) \\ &= \frac{4}{10} + \frac{2}{10} + 0 + \frac{2}{10} + \frac{4}{10} = \frac{12}{10} = 1.2\end{aligned}$$

Then the standard deviation is just the square root of the variance—that is,

$$\sigma_X = \sqrt{\sigma_X^2} = \sqrt{1.2} = 1.095$$

Another way to calculate the variance is by simply expanding the square in Equation 3.8 and using the algebraic properties of Σ from Chapter 1. That is,

$$\sigma_X^2 = \sum_{\text{all } x} (x - \mu_X)^2 p(x) = \sum_{\text{all } x} (x^2 - 2x\mu_X + \mu_X^2) p(x)$$

$$= \sum_{\text{all } x} x^2 p(x) - 2\mu_X \sum_{\text{all } x} x p(x) + \mu_X^2 \sum_{\text{all } x} p(x)$$

$$= \sum_{\text{all } x} x^2 p(x) - 2\mu_X(\mu_X) + \mu_X^2 = \sum_{\text{all } x} x^2 p(x) - 2\mu_X^2 + \mu_X^2$$

$$= \sum_{\text{all } x} x^2 p(x) - \mu_X^2 = E[X^2] - \mu_X^2 \qquad \textbf{(3.10)}$$

which is the expected value of X^2 minus the square of the expected value of X.

EXAMPLE 3.7 Using Equation 3.10, recalculate the variance of X in Example 3.4.

Solution: We first calculate $E[X^2]$.

$$E[X^2] = \sum_{x=0}^{4} x^2 p(x)$$

$$= 0^2\left(\frac{1}{10}\right) + 1^2\left(\frac{2}{10}\right) + 2^2\left(\frac{4}{10}\right) + 3^2\left(\frac{2}{10}\right) + 4^2\left(\frac{1}{10}\right)$$

$$= \frac{52}{10} = 5.2$$

Then

$$\sigma_X^2 = E[X^2] - \mu_X^2 = 5.2 - (2.0)^2 = 5.2 - 4.0 = 1.2$$

and

$$\sigma_X = \sqrt{1.2} = 1.095$$

The previous definitions as well as the following properties also hold for continuous random variables and can be obtained by substituting an integral for the summation sign. The limits of integration are $-\infty$ to $+\infty$. The corresponding equations for the mean and the variance of a continuous random variable are

$$\mu_X = E[X] = \int_{-\infty}^{\infty} x f(x)\, dx \qquad \textbf{(3.11)}$$

and

$$\sigma_X^2 = E[(X - \mu_X)^2] = \int_{-\infty}^{\infty} (x - \mu_X)^2 f(x)\, dx \qquad \textbf{(3.12a)}$$

or

$$\sigma_X^2 = E[X^2] - \mu_X^2 = \int_{-\infty}^{\infty} x^2 f(x)\, dx - \mu_X^2 \qquad \textbf{(3.12b)}$$

The following properties of expectations can be derived in the same manner as Equation 3.10.

$$E[c] = c \tag{3.13}$$

$$E[aX + b] = aE[X] + b \tag{3.14}$$

$$E[aX \pm bY] = aE[X] \pm bE[Y] \tag{3.15}$$

$$\text{Var}(aX \pm b) = a^2 \text{ Var}(X) = a^2\sigma_X^2 \tag{3.16}$$

where a, b, c are constants and X and Y are random variables having probability functions $p(x)$ and $p(y)$, respectively. Furthermore, if X and Y are independent random variables, then knowing a value for one variable, say Y, does not alter the probability function for the other—that is, $P(X|Y = y) = p(x)$.[1] Then

$$\text{Var}(X \pm Y) = \text{Var}(X) + \text{Var}(Y) = \sigma_X^2 + \sigma_Y^2 \tag{3.17}$$

The results in Equations 3.13 through 3.17 are valid for both discrete and continuous random variables.

EXAMPLE 3.8 In the coin-tossing game (Example 3.5), define a new variable, W, such that $W = 2X + 1$, in which case the payoffs become −\$5 and \$21. Thus, corresponding to Equation 3.14, $a = 2$ and $b = 1$. Find the mean and the standard deviation of W.

Solution: The probability distribution of W is

W	$p(w)$
−5	$\frac{3}{4}$
21	$\frac{1}{4}$

and

$$E[W] = \mu_W = -5\left(\frac{3}{4}\right) + 21\left(\frac{1}{4}\right) = \frac{6}{4} = 1.50$$

or

$$E[W] = aE[X] + b = 2\mu_X + 1 = 2(.25) + 1 = 1.50$$

which corresponds to Equation 3.14.

From Equation 3.10, the variances of X and W are, respectively,

$$\sigma_X^2 = 9\left(\frac{3}{4}\right) + 100\left(\frac{1}{4}\right) - (.25)^2 = \frac{507}{16} = 31.69$$

[1] Formally, two discrete random variables, X and Y, are said to be *independent* if and only if the joint probability function $p(x,y) = P(X = x \text{ and } Y = y) = P(X = x)\,P(Y = y) = p(x)p(y)$.

and

$$\sigma_W^2 = 25\left(\frac{3}{4}\right) + 441\left(\frac{1}{4}\right) - (1.50)^2 = \frac{2028}{16} = 126.75$$

or

$$\sigma_W^2 = a^2\sigma_X^2 = (2)^2\left(\frac{507}{16}\right) = 126.75$$

which corresponds to Equation 3.16.

EXAMPLE 3.9 Suppose a real-estate investor is considering investing in two pieces of property, one in southern Ohio, say X, and one in northern Ohio, say Y. The possible returns on investment for property X are (in \$1000), -5, 0, $+10$, with corresponding probabilities of .2, .5, .3, respectively; for Y, they are (in \$1000), -10, 0, 100, with respective probabilities of .3, .5, .2. The means and variances of X and Y are $\mu_X = 2$, $\sigma_X^2 = 31$, $\mu_Y = 17$, $\sigma_Y^2 = 1741$. (Verify.) The total return on the two investments is then given by $V = X + Y$, where a, b are both 1 in Equation 3.15. Find the expected total return and the variance of the total return.

Solution: Assuming independence and using

$$P(V = v) = P(X = x, Y = y) = p(x)p(y)$$

we see that the probability distribution of the total return, V, is as shown in Table 3.1.

Note: $P(V = 0) = P((X = 0, Y = 0) \text{ or } (X = 10, Y = -10))$.

Then

$$\mu_V = E[V] = \sum_{\text{all } v} vp(v) = 19$$

TABLE 3.1 Summary of calculations for Example 3.9

$x + y$	$p(x + y)$	v	$p(v)$	$vp(v)$	$v^2p(v)$
$-5 + (-10)$	.2(.3)	-15	.06	-0.90	13.5
$0 + (-10)$	.5(.3)	-10	.15	-1.50	15.0
$-5 + 0$	.2(.5)	-5	.10	-0.50	2.5
$10 + (-10)$	.3(.3)	0	.34	0.00	0.0
or $0 + 0$	.5(.5)				
$10 + 0$	.3(.5)	10	.15	1.50	15.0
$-5 + 100$	.2(.2)	95	.04	3.80	361.0
$0 + 100$	.5(.2)	100	.10	10.00	1000.0
$10 + 100$	.3(.2)	110	.06	6.60	726.0

or, by Equation 3.15,

$$\mu_V = E[V] = E[X + Y] = \mu_X + \mu_Y = 2 + 17 = 19$$

For the variance,

$$\sigma_V^2 = \text{Var}(V) = E[V^2] - \mu_V^2 = \sum_{\text{all } v} v^2 p(v) - \mu_V^2$$

$$= 2133 - 361 = 1772$$

or, more quickly, by Equation 3.17,

$$\sigma_V^2 = \text{Var}(X + Y) = \sigma_X^2 + \sigma_Y^2 = 31 + 1741 = 1772$$

EXAMPLE 3.10 Determine the average (expected) life and the standard deviation of the life of the equipment in Example 3.3.

Solution: The distribution of the life of that piece of equipment is given by

$$f(x) = \begin{cases} .4 - .08x & 0 \le X \le 5 \\ 0 & \text{elsewhere} \end{cases}$$

Thus, its expected life is

$$\mu_X = E[X] = \int_{-\infty}^{\infty} xf(x)\,dx = \int_0^5 x(.4 - .08x)\,dx$$

$$= \int_0^5 (.4x - .08x^2)\,dx = \int_0^5 .4x\,dx - \int_0^5 .08x^2\,dx$$

$$= .4\left(\frac{x^2}{2}\right)\Big|_0^5 - .08\left(\frac{x^3}{3}\right)\Big|_0^5 = 5 - 3\frac{1}{3} = 1\frac{2}{3} \text{ or } 1.667 \text{ years}$$

To find the standard deviation, we first obtain the variance by using Equation 3.12b.

$$\sigma_X^2 = E[X^2] - \mu^2 = \int_{-\infty}^{\infty} x^2 f(x)\,dx - \mu^2$$

$$= \int_0^5 x^2\,(.4 - .08x)\,dx - (1.667)^2$$

$$= \left\{\int_0^5 .4x^2\,dx - \int_0^5 .08x^3\,dx\right\} - (1.667)^2$$

$$= \left\{.4\left(\frac{x^3}{3}\right)\Big|_0^5 - .08\left(\frac{x^4}{4}\right)\Big|_0^5\right\} - (1.667)^2$$

$$= \{16.667 - 12.5\} - 2.778 = 1.389 \text{ (years)}^2$$

Thus, the standard deviation is

$$\sqrt{\sigma_X^2} = \sqrt{1.389} = 1.179 \text{ years}$$

EXAMPLE 3.11 Suppose two pieces of the equipment in Example 3.3 are purchased. Using Equations 3.15 and 3.17, determine the mean and standard deviation of the sum of their lives.

Solution: To determine the mean, we simply add $E[X_1]$ and $E[X_2]$, where X_1 represents the life of the first piece and X_2, the life of the second.

$$E[X_1 + X_2] = E[X_1] + E[X_2] = 1.667 + 1.667$$
$$= 3.334 \text{ years}$$

since the two pieces are assumed to have the same distribution.

For the standard deviation, we again must find the variance first. The determination of the variance requires the additional assumption of independence. If the two pieces of equipment are not interconnected, this assumption is reasonable. Thus,

$$\text{Var}(X_1 + X_2) = \text{Var}(X_1) + \text{Var}(X_2) = \sigma_{X_1}^2 + \sigma_{X_2}^2$$
$$= 1.389 + 1.389 = 2.778$$

and

$$\sigma_{X_1+X_2} = \sqrt{2.778} = 1.667 \text{ years}$$

If the variables were not independent, it would be necessary to find the joint distribution of X_1 and X_2 and to perform a double integration to obtain the variance of $X_1 + X_2$.

EXERCISES

3.9 Refer to Exercise 3.3. Calculate the expected number of women in the sample.

3.10 Refer to Exercise 3.3. Calculate the variance of the number of women in the sample by both Equation 3.8 and the alternative form in Equation 3.10.

3.11 Refer to Exercise 3.4. Calculate the mean and the variance of the number of good cakes from 2 attempts.

3.12 Refer to Exercise 3.6. Calculate the mean and standard deviation of X.

3.13 Find the mean, variance, and standard deviation for the triangular distribution of Exercise 3.7.

3.14 The Cleveland Browns have been averaging 24 points per game with a standard deviation of 7, and the Cleveland Cavaliers have been averaging 98 points per game with a standard deviation of 9. They are both playing on a particular Sunday. Calculate the mean and the variance of the total number of points scored by the two teams.

3.7 TSCHEBYCHEFF'S INEQUALITY

The reasons for considering the mean or average value of a variable are rather obvious, whereas the standard deviation may not be as clear. Consider a process that makes bolts that are to be 5.00 inches long on the average. Admittedly, due to uncontrollable variability in the process, some bolts will be slightly off in length. Assume the allowable deviations are ±.05 inch. If the only thing considered in checking the bolts is the mean length, it would be possible, though not probable, for a machine to be making half of the bolts 4.90 inches long and half 5.10 inches long. In this extreme case, the mean length would still be 5.00 inches, but none of the bolts would be any good. The standard deviation gives a means for detecting this type of situation through the following inequality. The inequality assumes only values for μ_X and σ_X and is true for *any* distribution. Note that if a specific form for the pdf or the probability distribution is known, then the percentage within will increase due to the additional information.

TSCHEBYCHEFF'S THEOREM Let X be a random variable with mean μ_X and standard deviation σ_X. Then the percentage of values that fall within K standard deviations of the mean will be at least $[1 - (1/K^2)]$ (100%) for $K \geq 1$. That is,

$$P(|X - \mu_X| \leq K\sigma_X) \geq 1 - \frac{1}{K^2} \tag{3.18}$$

Conversely, the probability of a value falling outside of K standard deviations is at most $1/K^2$.

EXAMPLE 3.12 Consider again the bolt manufacturing process. When in control, the process makes bolts with an average length of 5.00 inches and a standard deviation of .025 inch. The specifications on the length are ± .05 inch. What percentage of the bolts will meet specifications?

Solution: Here $K = .05/\sigma_X = .05/.025 = 2$. Then the percentage meeting specifications—that is, between 4.95 and 5.05 inches—will be at least

$$1 - \frac{1}{K^2} = 1 - \left(\frac{1}{2}\right)^2 = 1 - \frac{1}{4} = \frac{3}{4} \quad \text{or } 75\%$$

In the previous situation, no bolts were within specifications. If σ were calculated, it would be found to be .1, which is definitely too large; in other words, the process is out of control.

Tschebycheff's theorem is also called *Tschebycheff's inequality* and can be applied for any random variable, discrete or continuous, with any probability distribution. If a specific distribution or family can be determined, then tighter bounds on the probability can be obtained. Tschebycheff's inequality gives the minimum percentage of values within K standard deviations of the mean and is the pessimistic point of view. You will find it instructive to compare bounds obtained with Tschebycheff's inequality with bounds that can be determined from some of the specific distributions in Chapter 4. In particular, compare Tschebycheff's results with those arising under normal distributions.

EXERCISES

3.15 Refer to Exercise 3.14. What is the minimum probability that on any Sunday the Browns will score between 10 and 38 points? Within what point range can the Cavaliers be expected to score at least 75% of the time?

3.16 A machine produces items with a mean of 25 and a standard deviation of 2. What percentage of the items in the population will be between 20 and 30? Within what interval (symmetric about the mean) will 94% of the population lie?

3.8 SUMMARY

In this chapter, we considered the basic concepts of a *random variable* and its probability distribution. A random variable is merely a function that assigns numerical values to the outcomes of a random experiment. There are two basic types of random variables: *discrete* and *continuous*. In practice, discrete random variables generally will be counting-type variables, whereas continuous random variables normally will be of the measuring type. In determining the probability distribution for a random variable, it must be remembered that probabilities cannot be negative, and the total probability must be unity. Additionally, the association of probabilities with the values of the variable should be in direct correspondence with the original outcomes of the random experiment.

After the random variable has been defined and its probability distribution determined, it is desirable to describe the distribution by determining its center and a measure of its dispersion. Hence, we reviewed the concept of expected values and defined the *mean* to be the expected value of the variable. In a physical sense, the mean identifies the center of gravity of the distribution. As measures of dispersion, we use the *variance,* which is the average squared deviation from the mean, and its square root, the *standard deviation*. Properties of the expected-value operator also were given for linear combinations of random variables. In particular, the mean of a linear combination of random variables is the linear combination of the means, and the variance of a sum or difference of *independent* random variables is the sum of the variances.

Tschebycheff's inequality can be used to determine the minimum percentage of a population that will be within K standard deviations of the mean for values of K greater than unity. Tschebycheff's inequality makes no assumptions about the specific type of distribution that applies. Thus, it is completely general, and its primary importance at this level is to illustrate how the percentage increases if we can identify the type of distribution as well as its mean and standard deviation. The basic concepts and equations of this chapter are summarized in Table 3.2.

TABLE 3.2 Summary table

Terms	*Random variable*: Associates numerical values with the outcomes of a random experiment *Discrete random variable*: Generally counts occurrences *Continuous random variable*: Generally measures some characteristic of an item *Parameter*: Is required to completely identify a probability distribution
Probability Distribution for a Random Variable	Discrete random variables $p(x) \geq 0 \quad \text{all } x$ $\sum_{\text{all } x} p(x) = 1$ Continuous random variables $f(x) \geq 0 \quad \text{all } x$ $\int_{-\infty}^{+\infty} f(x)\,dx = 1$ (probability is interpreted as an area)
Expected Values and Variances	Random variables Mean $= \mu_X = E[X]$ Variance $= \sigma_X^2 = E[(X - \mu_X)^2]$ Standard deviation $= \sigma_X = \sqrt{\sigma_X^2}$ Linear combinations of random variables $E[aX + b] = aE[X] + b$ $\text{Var}(aX + b) = a^2\,\text{Var}(X)$ $E[aX \pm bY] = aE[X] \pm bE[Y]$ $\text{Var}(X \pm Y) = \text{Var}(X) + \text{Var}(Y)$ if X, Y are independent
Tschebycheff's Inequality	$P(\lvert X - \mu_X \rvert \leq K\sigma_X) \geq (1 - 1/K^2) \quad \text{for } K \geq 1$ or $[1 - (1/K^2)]$ (100%) of any population will be within $K\sigma_X$ of μ_X for $K \geq 1$.

SUPPLEMENTARY EXERCISES

3.17 A certain lottery has possible payoffs of \$0, \$25, \$100, \$500, \$1,000, and \$10,000. Its proponents have advertised that the probability distribution of payoffs to date is .60, .25, .10, .08, .05, .03, respectively. Could they be prosecuted for false advertising?

3.18 The following table gives the probability distribution, $p(x)$, of the number of customers, X, arriving in a bank during an arbitrary 4-minute period.

X	$p(x)$
0	.14
1	.29
2	.25
3	.18
4	.10
5	.04

Find the expected number of customers arriving in the next 4-minute period.

***3.19** Prove Equation 3.14.

***3.20** Prove Equation 3.15.

***3.21** A certain probability distribution is given by the pdf

$$f(x) = \begin{cases} +\sqrt{r^2 - x^2} & -r \leq X \leq +r \\ 0 & \text{elsewhere} \end{cases}$$

Determine the value of r that will satisfy the conditions for a probability density function. (*Hint:* Graphically, this function describes a semicircle.)

***3.22** Refer to Exercise 3.5. Calculate the expected number of attempts required of the baker.

3.23 A government contractor is working on 3 independent projects, X, Y, Z. Expected revenues on the projects are \$500,000, \$200,000, and \$100,000 with variances of 50,000, 30,000, and 10,000 $(\$)^2$, respectively.

(a) What is the contractor's total expected revenue from these projects?

***(b)** What is the standard deviation of the total revenue?

3.24 Among 8 nominees for the board of directors of a grocery cooperative are 4 men and 4 women. If 4 persons are to be elected, determine the probability distribution for the number of men elected if it is assumed that the election is a random experiment with all groups of 4 being equally likely.

3.25 In a grocery store, the probability that a randomly selected loaf of bread is more than 1 day old is .3. If you purchase 4 loaves at random, what is the probability distribution for the number of loaves purchased that are more than 1 day old?

3.26 A quality control plan used in a destructive testing situation calls for sampling items at random from the process and testing them to determine whether they are defective or nondefective. The plan calls for testing the items one at a time until either 2 defectives are found, in which case it is decided that the process is out of control, or 5 items are tested (with less than 2 being defective), indicating that the process is in control. The probability of an item's being defective is .1.

(a) Determine the probability distribution for the total number of items tested.

(b) Determine the expected number of items to be tested each time.

3.27 Novelties, Inc., produces items that would generally be classified as fads. They have found that the probability distribution that appears to describe the market lives of their products generally shows a steady increase for the first 1.5 years after introduction and a

---

* An asterisk indicates a higher level of difficulty.

steady decrease for the next 1.5 years. No product has had a demand for more than 3 years. The actual distribution is as follows, with M representing the market life in years.

$$f(m) = \begin{cases} \frac{20}{45} m & 0 \le M \le 1.5 \\ \frac{4}{3} - \frac{20}{45} m & 1.5 \le M \le 3 \\ 0 & \text{elsewhere} \end{cases}$$

(a) Graph the distribution.
(b) Determine the probability that a product will have a market life in excess of 1 year.
(c) Determine the probability that a product will phase out in less than 2 years.
(d) Verify that the expected market life of a product is 1.5 years.
***(e)** Determine the standard deviation of product market life.

3.28 Refer to Exercise 3.27. The average market life of a product for Novelties, Inc., is 1.5 years, and the standard deviation of the market lives is .6124 year.
(a) Using Tschebycheff's inequality, find the minimum percentage of products with market lives between 0.5 and 2.5 years.
(b) Using Tschebycheff's inequality, find the bounds on market life such that 88.89% of all products are included.
(c) What is the actual percentage that is within the bounds of part (b) for this distribution?
***(d)** For this distribution, what is the actual percentage that is within the interval specified in part (a)?

3.29 The expected number of customers arriving in the bank in Exercise 3.18 was found to be 1.93, and the standard deviation can be shown to be 1.336.
(a) Using Tschebycheff's inequality, find the minimum percentage of all 4-minute periods that will have between 0 and 3.86 customers arriving. Using the actual distribution given in Exercise 3.18, find this percentage.
(b) Use Tschebycheff's inequality to find bounds on the number of arriving customers such that at least 75% of all 4-minute periods will be included. What percentage of periods will be in the interval for the actual distribution?
(c) What is the maximum percentage of 4-minute periods that will have less than .041 or more than 3.819 customers arriving, according to Tschebycheff's inequality?

3.30 A particular investment has a .4 probability of yielding a 10% gain, .3 of no gain, .2 of a 10% loss, and .1 of a 20% loss. You are to invest $2000 in this opportunity.
(a) Construct the probability distribution for X, your net change in value.
(b) Calculate your expected monetary return on investment.
(c) Calculate the standard deviation of your return.

3.31 Assume that scores on a particular exam have a mean of 70 and a standard deviation of 10. Assume that the distribution is symmetric, but that nothing else is known about it. If it is desired that the upper 10% receive A's, what should be the minimum score for an A?

3.32 Determine whether or not the following is a probability distribution. Justify your answer.

X	-7	-2	0	5	13
$p(x)$	.20	.15	.30	.25	.15

3.33 A product has an average demand of 10 units per week and a standard deviation of 3 units per week. The accounting office has determined that from a cost standpoint it is best to produce 8 weeks' supply at a time. The demand is assumed to be independent from one week to the next.

- **(a)** What is the size of a production run?
- **(b)** What is the standard deviation of demand over the reorder period? (The reorder period is the time from one production run to the next.)
- **(c)** If the distribution of demand is symmetric, how much safety stock should the company carry in inventory to ensure that a stockout will occur with a probability of less than $\frac{1}{18}$? (Safety stock is inventory carried in excess of expected demand.)

3.34 St. Finian's Church is planning a carnival to raise money for its charity programs. Mrs. Random, who is in charge of games of chance, has asked you, St. Finian's accountant, to advise her on the amount to charge per play for the various games. In one game, the player draws a card from a standard, well-shuffled bridge deck (52 cards). If the card drawn is the jack of hearts, jack of diamonds, or jack of clubs, the player receives $6.50. If the card is a spade, the player receives $2.50. How much should Mrs. Random charge a person for each play of this game so that the expected value of the game is zero?

3.35 Refer to Exercise 3.34. Since the purpose of the carnival is to raise money, Mrs. Random would like to realize an average profit of 15¢ per play. How much should she charge in this case?

3.36 As accountant and financial adviser to Mr. Stoneman, you have been asked for advice on which of two alternatives to choose. One is to sell an investment now for $12,000. The other is to hold the investment for one month, after which he can sell it for $6000 with a probability of .4, $10,000 with a probability of .2, $15,000 with a probability of .3, and $34,000 with a probability of .1. Based on expected value, what do you advise him to do?

3.37 In financial analysis, the standard deviation of the return is often used as a measure of the risk of an investment. Calculate the standard deviation of the payoffs for the second alternative in Exercise 3.36. Which alternative is more risky?

3.38 A certain brand of rubber bands loses elasticity with age, and all rubber bands will break under normal usage within 2 years. Let X represent the time from production to when a rubber band breaks under normal usage. Assume the density function for X is given by

$$f(x) = \begin{cases} 1 - .5x & 0 \le X \le 2.0 \\ 0 & \text{otherwise} \end{cases}$$

- **(a)** What is the probability that the next rubber band produced will last at least 18 months under normal usage before breaking?
- **(b)** What is the probability that it will break some time between 6 months and 1 year under normal usage?

***3.39** What is the average life of a rubber band under normal usage? (Refer to Exercise 3.38.)

***3.40** Refer to Exercise 3.38 and 3.39 and calculate the standard deviation of lives for these rubber bands under normal usage.

3.41 Molson Products, Inc., has kept records of the weekly demand for part #827114 over the past 3 years. The following frequency distribution summarizes this demand.

No. of Parts Required	*Proportion of Weeks Occurring*
10	.10
11	.15
12	.22
13	.27
14	.13
15	.09
16	.04

(a) Suppose Molson can obtain delivery of this part only on Monday mornings. How many parts should be on hand Monday morning (after delivery) to have no more than a 15% chance of running out during the week?

(b) If Molson has storage space for only 12 parts, what is the probability that they will be out of stock during any given week?

3.42 Refer to Exercise 3.41 and determine the average weekly demand for Molson's part #827114.

3.43 Refer to Exercise 3.41 and determine the standard deviation of weekly demand for part #827114.

3.44 When Molson runs out of part #827114, it costs them $180 per part per week. Based on the information in Exercise 3.41, what is Molson's expected weekly out-of-stock cost for this part?

3.45 Refer to Exercises 3.41 and 3.44. Determine the standard deviation of Molson's weekly out-of-stock costs for part #827114.

3.46 Refer to Exercise 3.41 and assume that Molson has expanded the available storage space for part #827114 to handle a maximum of 16 units. The accounting department has determined that it costs Molson $50 per week for each part that is not used during that week. If Molson starts each week with 12 parts, what is their expected holding cost per week?

***3.47** Refer to Exercises 3.44 and 3.46. Calculate the number of parts that Molson should have on hand at the beginning of each week to minimize the total holding and out-of-stock costs per week for part #827114.

3.48 Molson Products, Inc. also has data on part #006969. The distribution of demand for this part is summarized in the following table.

No. of Parts Used	*Proportion of Weeks Occurring*
3	.22
4	.36
5	.18
6	.15
7	.09

Determine the mean and standard deviation of the weekly demand for part #006969.

3.49 If out-of-stock costs for part #006969 in Exercise 3.48 are $500 per week per part and Molson has storage space for only four of these parts, what is Molson's expected weekly out-of-stock cost for this part?

***3.50** See Exercise 3.48. If Molson had sufficient storage space for part #006969, with how many parts should they start the week to minimize total out-of-stock and storage costs if storage costs run $85 per week for each part not used that week?

***3.51** Refer to Exercises 3.47 and 3.50. Determine the combined expected weekly out-of-stock and storage costs for parts #827114 and #006969 if the minimizing values are used to start each week.

***3.52** Did you need to assume anything that was not stated to answer Exercise 3.51? If so, what?

***3.53** Refer to Exercise 3.47. Determine the standard deviation of the weekly out-of-stock and storage costs for part #827114 if the optimum number on hand is used.

***3.54** Refer to Exercises 3.50 and 3.53. Calculate the corresponding standard deviation for part #006969.

***3.55** Refer to Exercises 3.53 and 3.54. Determine the standard deviation of the total weekly out-of-stock and storage costs for parts #827114 and #006969 combined, assuming the optimum on-hand values are used to start each week.

***3.56** Did you need to assume anything that was not stated to answer Exercise 3.55? If so, what?

4 Specific Probability Distributions

4.1 INTRODUCTION

For the past three years, Sandy Burgwardt has been in charge of quality control for Hazlett Electronics. Each day, he randomly selects a sample of 25 electronic units produced at each work location to determine whether the quality of the output is satisfactory. On some days, there have been no defective units; on some days, 2; and on others, as many as 8. Burgwardt judges that he can expect an electronic unit to be defective about 5% of the time, but then he does not want to stop the machinery needlessly to make adjustments. Procedurally, he has stopped the machine only when there seemed to be a pattern of an unusually large number of defectives in each daily sample and, particularly, if a trend appeared. Knowing the probable occurrence of defectives in the daily sample when each electronic unit has a 5% probability of being defective would be helpful for establishing a standard that will tell him precisely when the machine should be stopped for adjustment.

In Chapter 3, we saw that if we identified all of the possible outcomes that a random variable could assume and the probability associated with each outcome, we would have a probability distribution. Further, if the outcomes are finite or countably infinite, we have a discrete probability distribution. If the outcomes are uncountably infinite, we have a continuous probability distribution. In this chapter, we will investigate specific probability distributions. Being aware of the behavior of a random variable with a specific probability distribution would enable Sandy Burgwardt to compare the actual occurrence of defective items with the chance occurrence of defectives as depicted by a probability distribution.

As we proceed through the chapter, further examples of probability distributions are given. Each example is associated with a specific distribution. The main emphasis in the chapter is on the families of distributions. Included in the discussion are the discrete probability distributions known as the hypergeometric, the binomial, and the Poisson. For the continuous probability distributions, we discuss the exponential and the normal. For each of these distributions, we will be concerned with the parameters that distinguish a particular distribution within its family. With these distinguishing characteristics, we can then find specific probabilities for various outcomes.

4.2 THE HYPERGEOMETRIC DISTRIBUTION

Very often businesses are faced with situations where the selection of an item or the occurrence of an event cannot be duplicated. In other words, the item or event cannot be returned to the pool of items or events to be selected again. In effect, items are drawn without replacement, which is illustrated in the following example.

EXAMPLE 4.1 A company has returned 10 parts that it had purchased and indicated that 4 were defective. However, the company failed to tag the defective parts. If an inspector randomly selects 3 of the parts, what is the probability that 2 will be defective?

Solution: In viewing the problem, we can see that after a part is selected from the set of 10, it will not be put back into the pool to be selected again. Thus, the probability of getting a defective or not getting a defective on the second selection is altered.

Initially, let us view this problem from the standpoint of statistically dependent events. One possible outcome for 2 defectives in 3 selections would be

$$\text{DEFECTIVE}_1, \text{DEFECTIVE}_2, \text{NOT DEFECTIVE}_3$$

or

$$D_1, D_2, D_3'$$

In probabilistic form for the individual outcomes, we have

$$P(\text{the sequence above}) = P(D_1) \times P(D_2|D_1) \times P(D_3'|D_1, D_2)$$

Substituting the individual probabilities gives

$$P(D_1, D_2, D_3') = \frac{4}{10} \times \frac{3}{9} \times \frac{6}{8} = \frac{1}{10}$$

However, we also could have had 2 other outcome configurations that satisfy the same requirements. These are

$$D_1, D_2', D_3 \qquad \text{and} \qquad D_1', D_2, D_3$$

The probability associated with each of these configurations is $\frac{1}{10}$, and the probability of 2 defectives in 3 selections is

$$P(\text{2 defectives in 3 selections}) = \frac{1}{10} + \frac{1}{10} + \frac{1}{10} = \frac{3}{10}$$

Since the outcome sets are mutually exclusive, we must add the individual probabilities to obtain the result. ■

We, of course, do not want to go through this procedure each time a similar situation is encountered. The probability can be calculated more directly if we utilize the hypergeometric distribution. This distribution has as its base the general probability concept of

$$p(x) = \frac{n(x)}{n(s)} = \frac{\text{number of successes}}{\text{number of possible outcomes}}$$

For our example, the denominator, which is all possible outcomes, is the total number of ways that three parts can be selected from the ten parts we have available. This will be the combination C_3^{10} or 120. The numerator, which is the number of successes, is determined in a similar manner. We must be careful that we satisfy the description of success. Success in our case is selecting exactly two defective parts and one nondefective part. Thus, we have the combination C_2^4, or six possible configurations where two defectives could be selected. In addition, to complete the total of three parts requires that one of them must be nondefective. A nondefective can be selected as the combination C_1^6, or six possible ways. The resulting numerator is then an application of the $m \cdot n$ rule, and ultimately we have

$$P(\text{2 defectives in 3 selections}) = \frac{C_2^4 C_1^6}{C_3^{10}} = \frac{36}{120} = \frac{3}{10}$$

which is, of course, the same answer we determined before. This calculation is an example of the hypergeometric distribution.

DEFINITION 4.1 If a random experiment consists of sampling without replacement from a finite population where the probability of success changes from trial to trial, the probability of the resulting number of successes $p(x)$ is determined from the **hypergeometric distribution.** The general form of the family of hypergeometric distributions is

$$p(x) = \begin{cases} \dfrac{C_x^R \, C_{n-x}^{N-R}}{C_n^N} & x = 0,1,2, \ldots, \min(n,R) \\ & n < N \\ & R < N \\ 0 & \text{elsewhere} \end{cases} \quad \textbf{(4.1)}$$

where: N = size of the population of interest
R = size of the subset that represents successes
n = size of the subset to be selected as a sample

The value of x in Equation 4.1 is limited to the minimum of n or R. Accordingly, the maximum value of x will be predicated on whichever parameter (n or R) is smaller. Thus, if $n = 4$ and $R = 3$, x will have a maximum value of 3; if $n = 4$ and $R = 5$, x will have a maximum value of 4.

EXAMPLE 4.2 The director of training for Ann Arbor Plastics has 8 persons going through the training program. The vice-president of personnel has requested that 3 of the individuals be selected to meet with her as a committee to discuss their current training. The director of training has been informed by her trainers that 2 of the individuals are unhappy about the program, but nobody has been named specifically. The director randomly selects the 3 persons.

a. What is the probability that no disgruntled person will be on the committee?

b. What is the probability that no less than 1 disgruntled person will be on the committee?

Solution: Before calculating each of the probabilities, we should determine the parameters of this hypergeometric distribution. For this example, we have $N = 8$, $R = 2$, and $n = 3$.

a. The probability requested is

$$P(X = 0) = \frac{C_0^2 \, C_3^6}{C_3^8} = \frac{20}{56}$$

b. We must realize that to have "no less than 1" means the committee could contain either 1 or 2 disgruntled persons. Therefore, we have

$$P(X \geq 1) = \frac{C_1^2 \, C_2^6}{C_3^8} + \frac{C_2^2 \, C_1^6}{C_3^8} = \frac{30}{56} + \frac{6}{56} = \frac{36}{56}$$

This answer could have been obtained in another manner. If we recognize that $\Sigma_{\text{all } x} \, p(x) = 1$, then by subtracting from 1 the probability that none of the disgruntled persons are on the committee, we are left with the probability that the committee may have 1 or 2 such persons.

$$P(X \geq 1) = 1 - P(X = 0) = 1 - \frac{20}{56} = \frac{36}{56}$$

In the family of hypergeometric distributions, the parameters pinpoint the exact distribution with which we are concerned. Suppose we had the following parameters: $N = 7$, $n = 4$, and $R = 4$. The outcomes for x could then be 1, 2, 3, and 4, and we would have a probability distribution that would look like Figure 4.1. It should be obvious that $P(X = 0) = 0$, because at least one item must come from the success group. Thus, we have another indication that the parameters make each distribution unique.

Figure 4.1b graphically depicts the distribution of the probability of X and is called a *histogram*. Each rectangle represents an area or a probability for a given value of x. This area is the product of the height of the bar, the probable occurrence of x, and the width of the bar, a unit interval. The unit interval can be expressed as the difference between the upper boundary and the lower boundary for a given value of x—for example, if $X = 2$, the boundaries are 1.5 and 2.5 and the unit interval is $2.5 - 1.5 = 1.0$. Thus, each area coincides with the probability of X.

The distribution shown in Figure 4.1 can be further described by its mean and variance, which will be unique for a given set of parameters. These values can be calculated as expected values, as determined in Chapter 3. For any set of data, these parameters are

$$\mu_X = E[X] = \sum_{\text{all } x} xp(x) \tag{4.2}$$

FIGURE 4.1 The hypergeometric probability distribution for $N = 7$, $n = 4$, and $R = 4$

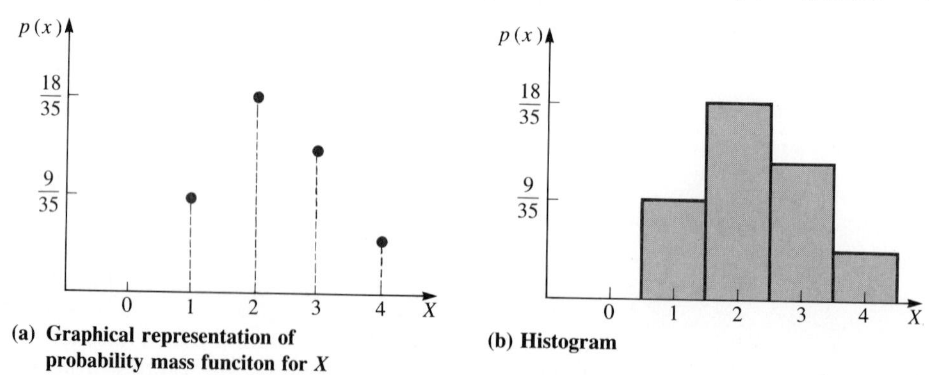

(a) Graphical representation of probability mass funciton for X

(b) Histogram

and

$$\sigma_X^2 = E[(X - \mu_X)^2] = \sum_{\text{all } x} (X - \mu_X)^2 p(x) \tag{4.3a}$$

In Chapter 3, the following alternative was derived for the calculation of the variance.

$$\sigma_X^2 = E[X^2] - (E[X])^2 = \sum_{\text{all } x} x^2 p(x) - \mu_X^2 \tag{4.3b}$$

For the distribution in Figure 4.1 and utilizing Equations 4.2 and 4.3b, we obtain the specific parameters

$$\mu_X = E[X] = 1\left(\frac{4}{35}\right) + 2\left(\frac{18}{35}\right) + 3\left(\frac{12}{35}\right) + 4\left(\frac{1}{35}\right) = \frac{80}{35}$$

$$= 2.2857$$

and

$$\sigma_X^2 = \left[1\left(\frac{4}{35}\right) + 4\left(\frac{18}{35}\right) + 9\left(\frac{12}{35}\right) + 16\left(\frac{1}{35}\right)\right] - (2.2857)^2$$

$$= 5.7143 - 5.2245 = 0.4898$$

From the variance, we can easily obtain the standard deviation, which is a more usable piece of information.

$$\sigma_X = \sqrt{\sigma_X^2} = \sqrt{.4898} = .6999$$

We have another way to calculate the mean and standard deviation using a more direct approach.

FORMULA 4.1 For the hypergeometric distribution, the **mean** is

$$\mu_X = \frac{nR}{N} \tag{4.4}$$

For the hypergeometric distribution, the **standard deviation** is

$$\sigma_X = \sqrt{\frac{nR(N - R)(N - n)}{N^2(N - 1)}} \tag{4.5}$$

Equation 4.4 may be written as

$$\mu_X = n\left(\frac{R}{N}\right) \tag{4.4a}$$

which represents the number of observations multiplied by the initial proportion of successes. Equation 4.5 could be written as

$$\sigma_X = \sqrt{\left[\frac{Rn(N-R)}{N^2}\right]\left[\frac{N-n}{N-1}\right]} \qquad \textbf{(4.5a)}$$

where the second term in this equation is known as the *finite population correction factor*. This correction factor will be discussed further in Chapter 6. Utilizing Equations 4.4 and 4.5 and the distribution of Figure 4.1, we obtain the parameters with some differences due to rounding.

$$\mu_X = \frac{4(4)}{7} = 2.2857$$

and

$$\sigma_X = \sqrt{\frac{4(4)(7-4)(7-4)}{7^2(7-1)}} = \sqrt{.4898} = .6998$$

which are the same values we obtained from the expected value approach.

Before leaving the hypergeometric distribution, it is appropriate to consider the case of multiple types of successes. The hypergeometric distribution can be extended to the case of several types of success.

DEFINITION 4.2 The general form for the **extended hypergeometric distribution** involving multiple categories is

$$p(x_1, x_2, \ldots, x_m) = \begin{cases} \dfrac{C_{x_1}^{R_1} C_{x_2}^{R_2} \cdots C_{x_m}^{R_m}}{C_n^N} & \begin{array}{l} \Sigma x_i = n \\ \Sigma R_i = N \\ n < N \\ 0 \leq x_i \leq \min(n, R_i) \\ \quad \text{for all } i \end{array} \\ 0 & \text{elsewhere} \end{cases} \qquad \textbf{(4.6)}$$

EXAMPLE 4.3 A company fleet of 12 trucks consists of 4 trucks that are 1 year old, 3 trucks that are 2 years old, and the rest are from 3 to 6 years old. Four trucks are to be randomly selected for a job. What is the probability that the 4 trucks will consist of 1 that is 1 year old, 1 that is 2 years old, and 2 that are 3 or more years old.

Solution: The parameters for this example are $N = 12$, $n = 4$, $R_1 = 4$, $R_2 = 3$, and $R_3 = 5$. The variables, X_i, take on the values 1, 1, and 2, and the probability is

$$P(X_1 = 1,\ X_2 = 1,\ X_3 = 2) = \frac{C_1^4 C_1^3 C_2^5}{C_4^{12}} = \frac{(4)(3)(10)}{495} = \frac{8}{33}$$

As the populations of interest increase in number, the calculations for specific probabilities become more tedious, but they could, nevertheless, be computed utilizing the general equation and, if necessary, logarithms of combinatorials.

EXERCISES

4.1 The Alcorn Basket Company desires to purchase a new kind of wicker for the manufacture of a new product line. The purchasing manager has received bids from 6 companies. Of the 6 companies, 4 have supplied Alcorn before. Ultimately, 2 bids will be selected.

- **(a)** What is the probability that both will be from new firms?
- **(b)** What is the probability that at least one will be from a previous supplier?
- **(c)** What is the average number of previous suppliers selected?

4.2 A firm has hired 15 new college graduates, including 7 engineers, 4 business majors, 2 liberal arts majors, and 2 social sciences majors. Of this group, 5 were selected at random.

- **(a)** What is the probability the 5 individuals selected were engineers?
- **(b)** What is the probability that 2 engineers, 2 business majors, and 1 from either of the other areas were individuals that were selected?
- **(c)** What is the mean number of engineers selected? What is the standard deviation?
- **(d)** What is the probability that no engineers were selected?

4.3 A city council is composed of 12 members. Further information about these persons is shown in the following table.

	Republican	*Democrat*
Male	4	4
Female	3	1

A subcommittee of the council is to be selected and will include 5 persons.

- **(a)** What is the probability that the committee will have all Republicans?
- **(b)** What is the probability that the committee will contain at least 3 men?
- **(c)** What is the probability that the committee will contain 2 male Democrats, 2 female Republicans, and anyone else?
- **(d)** What is the mean number of Democrats on the committee? What is the standard deviation?
- **(e)** Plot the distribution of Republicans on the committee.

4.4 Edgerton Products markets gift packages containing sausage and cheese. The company has 6 kinds of sausage and 5 kinds of cheese. Each package contains 4 items that are selected randomly.

- **(a)** What is the probability that a package contains all sausage?
- **(b)** What is the probability that a package contains at least 2 kinds of cheese?
- **(c)** What is the mean number of cheeses in a package?

4.3 THE BINOMIAL DISTRIBUTION

The simplest random variable is one that has only one value. However, such a random variable would have little interest beyond what value it assumed. More useful is a random variable that may assume either of two possible outcomes. Such a variable could be used to describe any experiment that can be classified as resulting in either a "success" or a "failure." As indicated in Chapter 3, the actual numerical values that the variable assigns to the experimental outcomes are irrelevant, except for interpretive purposes. Thus, for simplicity, we assume that the random variable assigns the value 1 to a "success" and 0 to a "failure," with probabilities π and $1 - \pi$, respectively. We observed in Chapter 3 that a variable of this type, which is generally known as a *Bernoulli random variable,* has a probability distribution given by

$$p(x) = \begin{cases} \pi^x (1 - \pi)^{1-x} & x = 0, 1 \\ & 0 < \pi < 1 \\ 0 & \text{elsewhere} \end{cases}$$

The binomial distribution arises as a simple extension of the Bernoulli random variable. Suppose the experiment in the previous paragraph is performed a fixed number of times, $n > 1$, and further assume that the trials are independent of each other and that the probability of a success remains constant from trial to trial. Then, if we define the random variable X to be the total number of successes from the n trials, X may assume any integer between 0 and n inclusive. To complete the probability distribution of X, we need to determine how probabilities will be assigned for each value.

First, let us consider the two extremes, 0 and n. For $p(0) = P(X = 0)$, we note that the only way in which X will equal 0 is if no successes are observed, or, equivalently, if all trials result in failures. Thus, it follows that

$$P(X = 0) = P(F\ F\ F \cdots F)$$

where $F\ F\ F \cdots F$ denotes the event that the first result was a failure, and the second a failure, and so forth. Since we assumed the trials were independent, we have from the multiplication rule that

$$\begin{aligned} P(F\ F\ F \cdots F) &= P(F)P(F)P(F) \cdots P(F) \\ &= (1 - \pi)(1 - \pi)(1 - \pi) \cdots (1 - \pi) \\ &= (1 - \pi)^n \end{aligned}$$

where $1 - \pi = P(\text{failure})$. Similarly, for X to equal n, every trial must result in a success. Hence, it follows that

$$\begin{aligned} p(n) &= P(X = n) = P(S\ S\ S \cdots S) \\ &= P(S)P(S)P(S) \cdots P(S) = \pi \cdot \pi \cdot \pi \cdots \pi = \pi^n \end{aligned}$$

since $\pi = P(\text{success})$ and the trials are independent.

For any values of X between 0 and n, say x, we note that $X = x$ only if x successes are observed. But if we observe x successes in n trials, then we must also observe $n - x$ failures. That is,

$$p(x) = P(X = x) = P(x \text{ successes and } n - x \text{ failures})$$

One way in which this can occur is for the first x trials to be successes and for the next $n - x$ to be failures. That is, if we obtain the sequence

$$\underbrace{S\,S \cdots S}_{x} \quad \underbrace{F\,F \cdots F}_{n-x}$$

Using the multiplication rule, the probability of this sequence is, under our assumptions,

$$\begin{aligned} P(S\,S \cdots S\,F\,F \cdots F) &= P(S)P(S) \cdots P(S)P(F)P(F) \cdots P(F) \\ &= \underbrace{\pi \cdot \pi \cdots \pi}_{x} \ \underbrace{(1-\pi)(1-\pi) \cdots (1-\pi)}_{n-x} \\ &= \pi^x (1-\pi)^{n-x} \end{aligned}$$

But, the occurrence of successes followed by failures is only one sequence that yields x successes and $n - x$ failures. Another possibility is to observe failures on the first $n - x$ trials and successes on the last x trials—that is, obtaining the sequence

$$\underbrace{F\,F \cdots F}_{n-x} \quad \underbrace{S\,S \cdots S}_{x}$$

The occurrence of failures followed by successes also has the probability of occurrence of

$$\begin{aligned} P(F\,F \cdots F\,S\,S \cdots S) &= P(F)P(F) \cdots P(F)P(S)P(S) \cdots P(S) \\ &= (1-\pi)(1-\pi) \cdots (1-\pi)\,\pi \cdot \pi \cdots \pi \\ &= (1-\pi)^{n-x}\,\pi^x = \pi^x(1-\pi)^{n-x} \end{aligned}$$

In fact, any sequence containing exactly x successes and $n - x$ failures will have a probability of occurring of $\pi^x(1 - \pi)^{n-x}$. The next question is, How many such sequences are there?

To determine the number of sequences having x successes, we note that the order of S's (or F's) within themselves is irrelevant since interchanging any two S's (or F's) leaves the sequence unchanged. It is merely a permutation of like objects. Thus, with only two outcomes, the number of different sequences that can be constructed is simply the number of combinations of n things taken x at a time. That is, there are

$$C_x^n = \frac{n!}{x!(n-x)!}$$

ways in which one can observe x successes. Thus, X will equal x if any of these sequences occur and

$$p(x) = P(X = x) = P(\text{1st sequence or 2nd sequence} \ldots)$$

Using the addition rule and the fact that two different sequences cannot be observed at the same time—that is, the sequences are mutually exclusive—we have

$$\begin{aligned} p(x) &= P(\text{1st sequence}) + P(\text{2nd sequence}) + \cdots \\ &= C_x^n\, \pi^x\, (1 - \pi)^{n-x} \qquad x = 1, 2, \ldots, n - 1 \end{aligned}$$

since each sequence has probability $\pi^x(1 - \pi)^{n-x}$ and there are C_x^n of them. Noting that $C_0^n = C_n^n = 1$, we can extend the limits on x to include the extremes 0 and n.

In deriving the preceding formulas, we made some specific assumptions about the experiment or situation considered. These assumptions serve to identify when the general binomial model is applicable.

DEFINITION 4.3 If a random experiment consists of making n independent trials from an infinite population where the probability of success π is constant from trial to trial, the probability of the resulting number of successes $p(x)$ is determined from the **binomial distribution.** The general form of the probability function for the family of binomial distributions is

$$p(x) = \begin{cases} C_x^n \pi^x (1 - \pi)^{n-x} & x = 0, 1, 2, \ldots, n \\ & 0 < \pi < 1 \\ 0 & \text{elsewhere} \end{cases} \tag{4.7}$$

where: n = number of trials to be undertaken

π = probability of success on a single trial

The family of binomial distributions is still applicable if the population is finite and sampling is with replacement. Note that sampling with replacement not only provides independence of trials but also essentially makes the population infinite.

The parameters must be specified to determine which member of the family is applicable. For example, if we said we were going to toss a coin and asked you to determine the probability of obtaining three heads, you could identify the situation as binomial but you could not calculate the actual probability. To calculate $p(3)$, you need to know how many times the coin will be tossed (n) and the probability of obtaining a head on one toss (π).

In any business organization, there are numerous situations where the outcome of a particular event has only two possibilities. For instance, a go–no-go inspection procedure, a sale–no sale for a product, or the absence–nonabsence of an employee. These examples do not necessarily satisfy the other assumptions of a binomial distribution, but they illustrate the two-event outcome that is imperative to this distribution.

EXAMPLE 4.4 Hazlett Electronics produces an item that has a known defective rate of .10. Periodically through the day, samples of 10 are selected from the large output of a production run. For one of these samples of 10, 3 defective items are found. What is the probability of this occurring by chance?

Solution: Initially, we see that the hypergeometric distribution is not appropriate for this problem because the assumptions of a finite population and the probability of success changing from trial to trial are not met. The known defective rate is assumed to be constant from trial to trial, and each mutually exclusive outcome will be either defective, a success, or nondefective, a failure. The samples of 10 are independent trials, and the variable of interest counts defective items. Thus, the binomial distribution has been specified, and the two parameters are $n = 10$ and $\pi = .10$. Thus, the probability of three defectives is

$$P(X = 3) = C_3^{10}(.10)^3(.90)^7 = 120(.001)(.4783) = .0574$$

In attempting to evaluate this figure in a decision-making sense, we are saying that there is just under a 6% likelihood of finding three defectives in ten trials strictly by chance, where the probable occurrence of a single defective is .10. Thus, since three defectives have occurred, we may want to stop the production line and adjust our equipment. This decision-making aspect is discussed in greater detail under the subject of hypothesis testing, the topic of Chapter 8.

EXAMPLE 4.5 Acme Trucking Company has purchased a lot of 20 new tires that have been made especially for them. Each tire has been guaranteed for 40,000 miles under normal wear, and there is only a 5% chance that the tire will not last 40,000 miles. What is the prospect of 4 tires wearing out before reaching 40,000 miles?

Solution: This problem can be easily solved by using Equation 4.7. The parameters involved are $n = 20$, $\pi = .05$. Therefore,

$$P(X = 4) = C_4^{20}(.05)^4(.95)^{16} = 4845\ (.00000625)(.4401) = .0133$$

The calculations involved make this problem difficult, but utilization of Appendix Table B.1 enables us to determine the answer easily. A portion of Appendix Table B.1 is shown in Table 4.1, and the number calculated in Example 4.5 is found at the intersection of the proper row and column and is shown highlighted.

To use Appendix Table B.1 we need only know n and π. After locating the appropriate n and π for a given problem, the probability for a given x can be located at the intersection of the specific column, headed by the chosen π, and the specific row representing the desired value of x by reading the values of x on the left side from the top down.

Should the value of π exceed .50, each column of probabilities serves as values for the complement of the probability that heads the column, and values for x are found in the right-hand column and read from the bottom up. For example, if $n = 20$ and $\pi = .90$ and we desire the probability that $x = 17$, we can

TABLE 4.1 Selected probabilities for the binomial distribution, $n = 20$

	π					
x	.01	.05	.10	...	.50	x
0	.8179	.3585	.1216	...	.0000	20
1	.1652	.3774	.2702	...	.0000	19
2	.0159	.1887	.2852	...	.0002	18
3	.0010	.0596	.1901	...	.0011	17
4	.0000	.0133	.0898	...	.0046	16
5	.0000	.0022	.0319	...	.0148	15
.	.	.	.	...	.	.
.	.	.	.	...	.	.
.	.	.	.	...	.	.
19	.0000	.0000	.0000	...	.0000	1
20	.0000	.0000	.0000	...	.0000	0

find this on the abbreviated portion of Appendix Table B.1 shown in Table 4.1 by using the column headed .10, the complement of .90, and finding the x value in the far right column, reading from the bottom toward the top. Thus, $P(x = 17|n = 20, \pi = .90) = .1901$.

EXAMPLE 4.6 What is the probability that Acme may have 3 or less tires wear out before reaching 40,000 miles?

Solution: The phrase "3 or less" means that Acme could experience 3, 2, 1, or 0 tire failures. Thus,

$$P(X \leq 3) = P(X = 0) + P(X = 1) + P(X = 2) + P(X = 3)$$

From Appendix Table B.1, for $n = 20$ and $\pi = .05$, this equation becomes

$$P(X \leq 3) = .3585 + .3774 + .1887 + .0596 = .9842$$

However, this result could be obtained directly by consulting Appendix Table B.2, which compiles the cumulative probabilities for the binomial distribution. Appendix Table B.2 is a less than or equal to ($\leq$) cumulative table with x values in the left-hand column. Thus, using this table in the same manner as Appendix Table B.1, we can find the value for the probability $X \leq 3$, given $n = 20$ and $\pi = .05$, at the intersection of the appropriate column ($\pi = .05$) and the appropriate row ($X = 3$). The value is .9841, which is slightly different due to rounding.

EXAMPLE 4.7 What is the probability that Acme may have as many as 5 but no less than 2 tires that fail prior to 40,000 miles?

Solution: The answer to this question can be found from either Appendix Table B.1 or Appendix Table B.2. In either case, we desire to find

$$\begin{aligned} P(2 \le X \le 5) &= P(X = 2) + P(X = 3) + P(X = 4) + P(X = 5) \\ &= .1887 + .0596 + .0133 + .0022 \qquad \text{(from Appendix Table B.1)} \\ &= .2638 \end{aligned}$$

We could also write this as

$$\begin{aligned} P(2 \le X \le 5) &= P(X \le 5) - P(X \le 1) \\ &= .9997 - .7358 \qquad \text{(from Appendix Table B.2)} \\ &= .2639 \end{aligned}$$

The first value from Appendix Table B.2, .9997, represents the accumulation of the probabilities $X = 0$ to $X = 5$. The second value, .7358, represents the sum of the probabilities for $X = 0$ and $X = 1$. Thus, the difference represents the probability for $X = 2$ through $X = 5$, which is what we want.

The probabilities in the binomial distribution that are determined for specific values of X could be used in a histogram. Probability could be thought of as the relative frequency for all values of X. The shape of the distribution is greatly influenced by the parameters. Figure 4.2 illustrates four relative frequency distributions with different parameters.

Of the four distributions shown in Figure 4.2, only (c), with $\pi = .50$, is symmetrical. This figure shows that if $\pi > .50$ (d) or $\pi < .50$ (a), the shape of the relative frequency distribution becomes skewed left and right, respectively. The probabilities used to plot illustration (d) can be found in Appendix Table B.1 in the column where $n = 15$ and $\pi = .20$, the complement of $\pi = .80$. However, the figures must be read from the bottom of the column to the top—that is, using the x values on the right-hand side of the table.

The identity of a particular binomial distribution within the family of binomial distributions is specified by giving values to the parameters of n and π. We can further describe binomial distributions by determining their means and standard deviations.

The expected value concept could be applied, but a more direct approach is provided for the calculation of the mean and standard deviation of binomial random variables.

FORMULA 4.2 For a binomial distribution, the **mean** is

$$\mu_X = n\pi \tag{4.8}$$

For a binomial distribution, the **standard deviation** is

$$\sigma_X = \sqrt{n\pi(1 - \pi)} \tag{4.9}$$

FIGURE 4.2 Relative frequency distribution for all values of x, where $n = 15$ for selected values of π

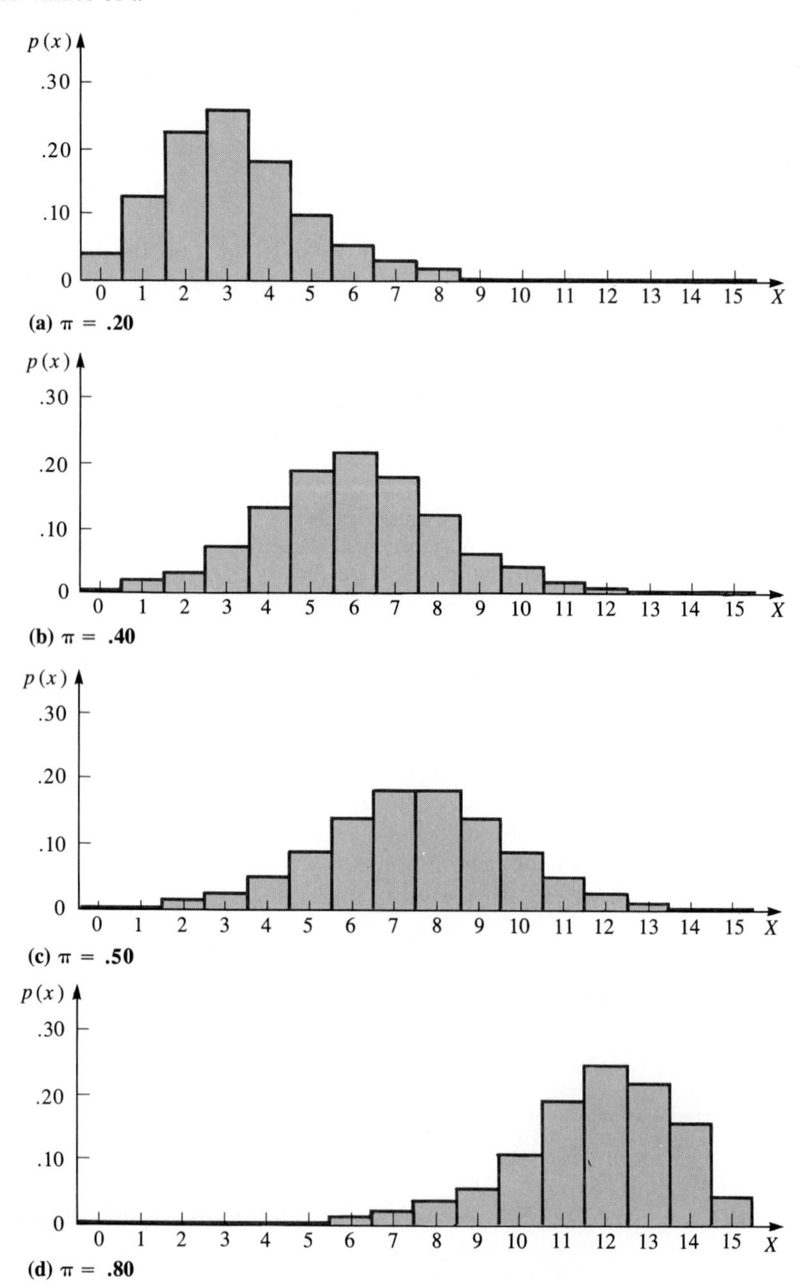

EXAMPLE 4.8 In a large company, it was known that 40% of the workers were involved in some form of a car pool. If 150 persons were randomly selected and questioned regarding how they got to work, what is the mean number expected to be in a car pool? What is the standard deviation?

Solution: Utilizing Equations 4.8 and 4.9, we have

$$\mu_X = n\pi$$
$$= 150(.40) = 60$$

and

$$\sigma_X = \sqrt{n\pi(1 - \pi)}$$
$$= \sqrt{150(.40)(.60)}$$
$$= \sqrt{36} = 6$$

Therefore, we would expect 60 of the 150 persons to be involved in a car pool. The standard deviation of persons involved is 6.

DEFINITION 4.4 The binomial distribution that deals with two possible outcomes is a special case of the **multinomial distribution,** which is not restricted to two outcomes. The general form of the multinomial distribution is

$$p(x_1, x_2, \ldots, x_m) = \begin{cases} \left(\dfrac{n!}{x_1!\, x_2! \cdots x_m!}\right) \pi_1^{x_1}\pi_2^{x_2} \cdots \pi_m^{x_m} & \Sigma x_i = n \\ & \Sigma\pi_i = 1 \\ & \pi_i > 0 \\ 0 & \text{elsewhere} \end{cases} \quad \textbf{(4.10)}$$

In this general form, π_i represents the probability of occurrence of the ith category on a single trial.

EXAMPLE 4.9 Baltic Enterprises has determined that its sales invoices fall into 4 classes. These classes, together with the probable occurrence of each class, are shown in the following table.

Dollar Volume	*Probability*
Less than \$500	.20
\$500 but less than \$1000	.30
\$1000 but less than \$3000	.40
\$3000 and over	.10

If 12 invoices were randomly selected from all invoices, what is the probability that there would be 2, 4, 3, and 3 invoices, respectively, from the 4 classes?

Solution: The probability can be found by using Equation 4.10.

$$P(X_1 = 2, X_2 = 4, X_3 = 3, X_4 = 3)$$
$$= \frac{12!}{2!4!3!3!}(.20)^2(.30)^4(.40)^3(.10)^3$$
$$= 277200(.04)(.0081)(.064)(.001) = .00575$$

This probability indicates that there is less than a 1% chance of obtaining the outcome specified in the problem.

EXAMPLE 4.10 In reference to Baltic Enterprises, what is the probability of obtaining 0 invoices less than \$500, 4 invoices in the \$500 to \$1000 range, 6 invoices in the \$1000 to \$3000 range, and 2 invoices over \$3000 when 12 invoices are randomly drawn from all the invoices?

Solution: Considering this question, we might expect that the probability increases significantly from the previous example because the larger occurrences are in the range of the higher probabilities. However, we must realize that we are working with statistically independent trials that, when combined, may reduce the probability. The question is again answered by using Equation 4.10.

$$P(X_1 = 0, X_2 = 4, X_3 = 6, X_4 = 2)$$
$$= \frac{12!}{0!4!6!2!}(.20)^0(.30)^4(.40)^6(.10)^2$$
$$= 13860(1)(.0081)(.004096)(.01) = .004598$$

Probabilities relative to binomial distribution problems are easily determined using Appendix Tables B.1 and B.2. To provide a table that would give multinomial probabilities would be totally impractical due to the large number of values that n could take as well as the many combinations of the numbers of π's and x's.

EXERCISES

4.5 An inspector for Balboa Ball Bearings selects samples of 25 to test for out-of-round ball bearings, where a defective has a probable occurrence of .01.
- **(a)** What is the probability of having 4 defective ball bearings?
- **(b)** What is the probability of a perfect sample?
- **(c)** What would you expect the mean number of defectives to be?

4.6 A company is test marketing a new product. Through a private testing firm, 20 homes are selected and the product is introduced for a period of 5 days. A consumer is expected to act favorably or unfavorably to the product with equal probability.
- **(a)** What is the probability that exactly half of the households will favor the product?
- **(b)** What is the probability that at least 6 but no more than 12 households will favor the product?
- **(c)** What is the probability that 14 or more households will not favor the product?
- **(d)** Construct a relative frequency histogram of the distribution of households that favor the product.
- **(e)** Determine the mean and the standard deviation of households favoring the product.

4.7 A sparkplug is guaranteed to last 1 year with a probability of 80%. In a box of 8 plugs, what is the probability that half of them will not last the year?

4.8 Seventy percent of all college graduates complete their program in the prescribed time period.
- **(a)** If the records of 6 graduates are checked, what is the probability that 2 of them did not graduate in the prescribed time?
- **(b)** What is the probability of more than 5 but less than 9 graduating in the prescribed time, when 10 persons are checked?
- **(c)** What is the mean number of individuals graduating in the prescribed time, when 180 records are checked? What is the standard deviation?
- **(d)** If 70% complete their program on time, 20% take an additional year, and 10% take even longer, what is the probability that out of 6 alumni 2 would be in each category?
- **(e)** Using the information from (d), find the probability that when 8 alumni are questioned, 4 completed their program on time, 3 took one additional year, and 1 took even longer.

4.4 THE POISSON DISTRIBUTION

In our discussion of the binomial distribution, we were concerned with the number of successful and nonsuccessful outcomes in a finite number of trials. However, there are outcomes that are not part of a finite number of trials but occur at random in a continuous interval. In such situations, the variable of interest is the number of occurrences of the outcome in a specified "interval." For example, we might be concerned with the number of flaws that may be present on a square inch of a deep-draw panel or the arrival of mechanics at a tool crib in an hour. Variables of this type can best be described by the Poisson distribution.

DEFINITION 4.5 If the variable of interest is the number of occurrences of the outcomes in a specified interval, the probable occurrence of x is determined from the **Poisson distribution.** The general form of the family of Poisson distributions is

$$p(x) = \begin{cases} \dfrac{e^{-\mu}\mu^x}{x!} & \mu > 0 \\ & x = 0,1,2,3\ldots \\ 0 & \text{elsewhere} \end{cases} \tag{4.11}$$

where: μ = mean rate of occurrence of the variable X in a given interval

In addition, other elements of Equation 4.11 are

x = desired number of occurrences for a given interval

e = Euler's constant, approximately 2.71828

The use of the Poisson distribution is predicated on the following five assumptions that the variable is expected to meet.

1. The interval in question must be able to be divided into many smaller subintervals.
2. The probability of two or more occurrences in a sufficiently small subinterval is small enough to be nonexistent.
3. The probability of an occurrence in a subinterval remains uniform throughout the interval.
4. The probability of occurrence in one subinterval does not affect the probability of occurrence in any other nonoverlapping subinterval.
5. The expected number of occurrences in any interval is directly proportional to the size of the interval.

For example, the interval could be time, and the subinterval could be an hour, half-hour, or minute, or the interval could be area, and the subinterval could be a square foot, a page, or a sheet. A third possibility for the interval is distance, and the subinterval could be a foot, a yard, or a mile. Lastly, consideration could be given to volume, and the interval could be a liter or a cubic centimeter.

EXAMPLE 4.11 Mechanics at the Munson Machine Shop arrive for service at the tool crib at the rate of 4 per 15-minute period. What is the probability that there will be 6 arrivals in a 15-minute period? The assumptions appropriate to the Poisson distribution are presumed to hold.

Solution: The parameter associated with this distribution is $\mu = 4.0$ per 15-minute period. Thus, the probability of 6 arrivals is

$$P(X = 6) = \frac{e^{-4}4^6}{6!} = \frac{(.0183)(4096)}{720} = .1042$$

In calculating the probability in Example 4.11, e^{-4} was determined from Appendix Table B.5 and the other two values were calculated directly. For larger values of μ and x, the solution becomes more difficult to determine. However, knowing the value of μ, we can obtain the appropriate probability for x by using Appendix Table B.3. A portion of the table is given in Table 4.2. The probability calculated in Example 4.11 is the highlighted value at the intersection of the appropriate column, $\mu = 4.0$, and the appropriate row, $x = 6$.

In general, finding a probability in Appendix Table B.3 is accomplished by finding the desired value of μ and selecting the desired x value in the left-hand column. The intersection of the selected row and column represents the probability.

EXAMPLE 4.12 Refer to the Munson Machine Shop situation in Example 4.11.

a. Find the probability of 2 arrivals in a 15-minute period.
b. Find the probability of 2 arrivals in a 30-minute period.

TABLE 4.2 Selected probabilities for the Poisson distribution

	μ				
x	*3.1*	*3.2*	. . .	*3.9*	*4.0*
0	.0450	.0408	. . .	.0202	.0183
1	.1397	.1304	. . .	.0789	.0733
2	.2165	.2087	. . .	.1539	.1465
3	.2237	.2226	. . .	.2001	.1954
4	.1733	.1781	. . .	.1951	.1954
5	.1075	.1140	. . .	.1522	.1563
6	.0555	.0608	. . .	.0989	.1042
7	.0246	.0278	. . .	.0551	.0595
.	.	.	. . .	.	.
.	.	.	. . .	.	.
.	.	.	. . .	.	.
15	.0000	.0000	. . .	.0000	.0000

Solution:

a. This problem is relative to the same time period as in Example 4.11, so $\mu = 4.0$, and from Appendix Table B.3, we find

$$P(X = 2) = \frac{e^{-4}4^2}{2!} = .1465$$

This value can be seen in the abbreviated Poisson table in Table 4.2.

b. The time interval of interest differs from Example 4.11. Thus, it is necessary to change the rate of occurrence to coincide with the time interval of the question. From the last assumption, we know that the rate of occurrence is proportional to the size of the interval. Therefore, $\mu = 8.0$ in a 30-minute period would be equivalent to $\mu = 4.0$ in a 15-minute period. Utilizing $\mu = 8.0$, we can now solve for the probability of 2 arrivals in 30 minutes.

$$P(X = 2) = \frac{e^{-8}8^2}{2!} = .0107$$

This value was also found directly by using Appendix Table B.3. ■

EXAMPLE 4.13 ■ Outgoing trucks of the Roadrunner Trucking Company arrive at the weighing station at the rate of 3 per hour. What is the probability that 2 or less would arrive in a 1-hour period?

Solution: The variable X can take on values of 0, 1, or 2. Therefore, using Equation 4.11, we have

$$P(X \leq 2) = \sum_{x=0}^{2} \frac{e^{-3}3^x}{x!}$$
$$= \frac{e^{-3}3^0}{0!} + \frac{e^{-3}3^1}{1!} + \frac{e^{-3}3^2}{2!}$$

and from Appendix Table B.3, using $\mu = 3.0$, the values are

$$.0498 + .1494 + .2240 = .4232$$

■

Computing probabilities of this type could be accomplished more directly by using Appendix Table B.4, which is the table for cumulative probabilities of the Poisson distribution. The probability computed in the example can be read directly from Appendix Table B.4 with $\mu = 3.0$ and $x = 2$.

EXAMPLE 4.14 ■ Referring to the Roadrunner Trucking Company situation in Example 4.13, find the probability that at least 3 but no more than 5 trucks arrive in a 30-minute period.

FIGURE 4.3 Poisson probability distribution

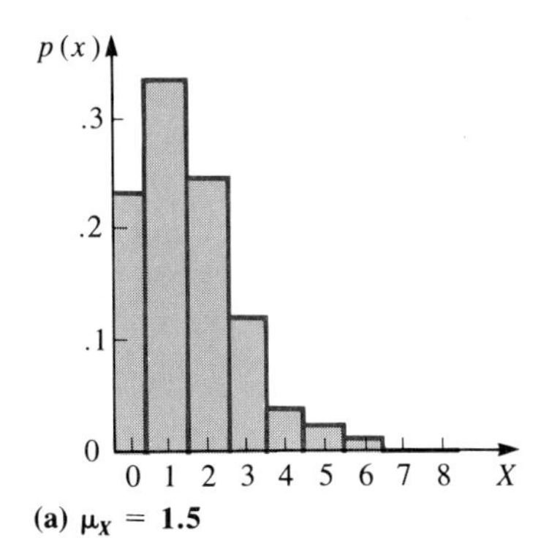

(a) $\mu_X = 1.5$

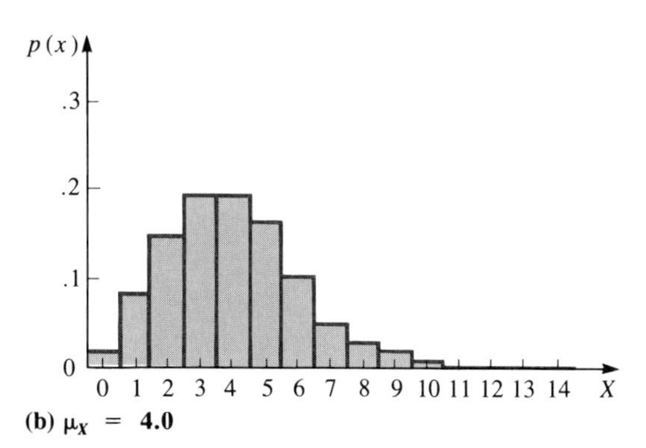

(b) $\mu_X = 4.0$

Solution: Utilizing the equivalent value for the mean arrival rate of $\mu = 1.5$ per 30-minute period, we can find the probability in Appendix Table B.4.

$$P(3 \leq X \leq 5) = P(X \leq 5) - P(X \leq 2) = .9955 - .8088 = .1867$$

Since a Poisson distribution is one of the family of Poisson distributions, the parameters associated with an individual distribution will dictate the appearance of the relative frequency distribution. In Figure 4.3, two distributions are shown where the value of μ has been changed. Observing the distributions in Figure 4.3, we see they are skewed to the right. This results from the fact that X cannot be less than zero but can be any positive integer value from zero to infinity. The Poisson distribution for any parameter μ can be further described by the mean and the standard deviation.

FORMULA 4.3 For the Poisson distribution, the **mean** is

$$\mu_X = \mu \tag{4.12}$$

For the Poisson distribution, the **standard deviation** is

$$\sigma_X = \sqrt{\mu} \tag{4.13}$$

EXAMPLE 4.15 Filled pallets at the end of an assembly line are removed to storage at the rate of 4 per hour. What is the mean removal rate? What is the standard deviation? What percentage of the Poisson distribution is within 2 standard deviations of the mean?

Solution: The answer to the first two questions can be determined by using Equations 4.12 and 4.13. Thus, the mean is

$$\mu_X = 4$$

and the standard deviation is

$$\sigma_X = \sqrt{\mu} = \sqrt{4} = 2$$

For the third question, we are looking for the probability that x would fall in the range $\mu - 2\sigma$ to $\mu + 2\sigma$. Substituting the values for μ and σ, we have a range of 0 to 8. Therefore, the percentage of the Poisson distribution to be in that range can be determined as

$$P(X \leq 8) = \sum_{x=0}^{8} \frac{e^{-4}4^x}{x!} = .9786$$

which is found directly in Appendix Table B.4.

EXERCISES

4.9 At a major trucking terminal, tractor trailers arrive at the rate of 1 every 10 minutes.
- **(a)** What is the probability of 3 arriving in 10 minutes? 10 in 10 minutes?
- **(b)** What is the probability that at least 4 but no more than 8 arrive between 10:00 A.M. and 11:00 A.M.?
- **(c)** What is the mean number of arrivals in a half-hour period?

4.10 A weaver has found that for a certain kind of cloth, her looms produce minor flaws at a rate of 4.8 per 100 square feet of material.
- **(a)** What is the probability that more than 7 flaws would appear in 100 square feet of cloth?
- **(b)** What is the probability that more than 1 but less than 5 flaws would appear in 50 square feet of cloth?
- **(c)** If 200 square feet of cloth were examined, is it probable that no flaws would be present? Is it probable that more than 15 flaws would be present?

4.11 A city administrator has found that during a 6-month period, he averaged 4 telephone calls per working day (8 hours) from citizens of the community.
- **(a)** What is the probability of receiving 8 calls in a day?
- **(b)** What is the probability of all 4 calls coming in the morning (9:00 A.M. to 12:00 noon)?
- **(c)** What is the probability of at least 10 but no more than 25 calls in a week?
- **(d)** What is the expected number of calls in a year (52 weeks)? What is the standard deviation?

4.12 A small town has an average of 3.8 potholes per city block.
- **(a)** What is the probability of 8 potholes in 2 city blocks?
- **(b)** What is the expected number of potholes in 8 city blocks? What is the standard deviation?
- **(c)** What is the probability of at least 14 but no more than 20 potholes in a 5-block distance?

4.5 THE POISSON APPROXIMATION OF THE BINOMIAL

To obtain the probabilities for a binomial distribution, we often are able to use Appendix Tables B.1 and B.2. However, if π is extremely small or extremely large, or for certain values of n or π, we may not be able to use these tables. Under some of these circumstances, we may be able to use the Poisson distribution to approximate the binomial distribution. For the binomial distribution, the mean is $n\pi$, and for the Poisson distribution the mean is μ.

FORMULA 4.4 We can use the **Poisson approximation of the binomial** when the conditions that $n\pi < 5$ or $n(1 - \pi) < 5$ are met. The probable occurrence of x can then be determined using the family of Poisson distributions as given in Equation 4.11.

$$p(x) = \frac{e^{-\mu}\mu^x}{x!}$$

where: $\mu = n\pi$

Thus, the use of the Poisson approximation for the binomial is appropriate when n is large and either π or $1 - \pi$ decreases.

EXAMPLE 4.16 The Tankers Brewery has found that, due to improper positioning, 1% of all bottles going through the capping process are not sealed properly. If a sample of 470 bottles is randomly selected, what is the probability that 4 are not properly sealed?

Solution: This example satisfies the conditions for the binomial distribution. That is, a bottle is sealed or not sealed, and there are $n = 470$ trials and $\pi = .01$, which is the probability of success on a single trial. Further, the trials are independent. Thus, the probability of 4 bottles not properly sealed is

$$P(X = 4) = C_4^{470}\,(.01)^4(.99)^{466}$$

Appendix Table B.1 cannot be used, and a hand calculation is out of the question. Noting that the product of $n\pi = 4.7$ is less than 5, we can approximate the solution with the Poisson distribution. The mean $\mu = 470 \times .01 = 4.7$ is the Poisson parameter, and using Appendix Table B.3, we find

$$P(X = 4) = \frac{e^{-4.7}4.7^4}{4!} = .1849$$

Comparing our answer to the value of the binomial generated on a computer, which was .1856, we have a good approximation.

Since we may not always have access to a computer, the Poisson distribution provides a good approximation of the binomial distribution when we satisfy the condition $n\pi < 5$ or $n(1 - \pi) < 5$.

EXERCISES

4.13 Fatalities in auto accidents occur .5 times out of every 100 accidents. If a total of 800 accidents is investigated, what is the probability of exactly 4 fatalities? of more than 4 fatalities?

4.14 A company prides itself in meeting shipping deadlines in every 96 out of 100 orders even though some of the orders may not be totally filled. In a 2-day period, 75 orders were processed. What is the probability that **(a)** 3 orders were not shipped on time, **(b)** at most 4 orders were not shipped on time, and **(c)** 7 or more orders were not shipped on time?

4.6 THE EXPONENTIAL DISTRIBUTION

Beginning with this section, we move from discrete probability distributions to continuous distributions. We consider first a distribution that is often used to model situations where the random variable of interest is the time, distance, or volume between occurrences. Examples of such are the distance between potholes on a highway, the length of extruded plastic between bubble defects, the volume of liquid passing through a valve before it fails to operate, or the time between the arrival of buses at a bus stop. We discuss the exponential distribution in relation to the Poisson distribution. If the Poisson distribution represents the arrival of orders at a warehouse facility at a mean rate of μ, then the exponential distribution can be thought of as the time between arrivals. In this example, the exponential distribution is a function of time and is a continuous probability distribution.

DEFINITION 4.6 The probability density function of the **exponential distribution** can be expressed as

$$f(t) = \begin{cases} \lambda e^{-\lambda t} & \lambda > 0 \\ & t \geq 0 \\ 0 & \text{elsewhere} \end{cases} \tag{4.14}$$

where: λ = rate of occurrence per standard unit

The parameter λ distinguishes a particular exponential distribution from the others in the family of exponential distributions. Substituting values of t into Equation 4.14 gives the height of the curve at that point. To determine the probability that t or more amount of time will elapse between arrivals, we must integrate this function for the time period in question.

$$P(T \geq t) = \int_t^\infty \lambda e^{-\lambda t}\, dt \tag{4.15}$$

FIGURE 4.4 $P(T \geq t)$ for an exponential distribution

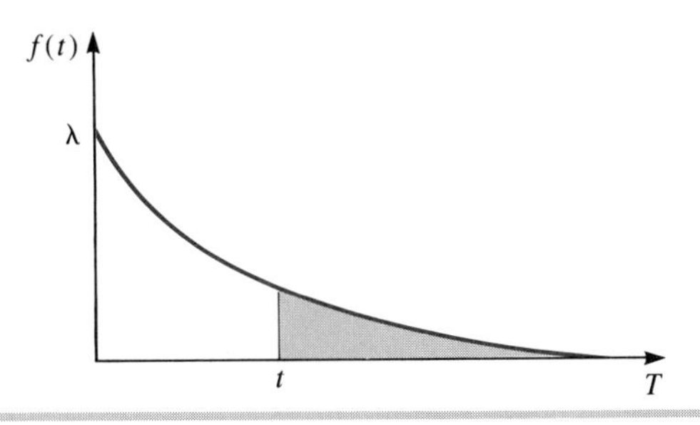

Generally, whenever we are concerned with a given t as a lower limit and ∞ as the upper limit, we could write the probability statement as

$$P(T \geq t) = e^{-\lambda t} \tag{4.16}$$

Graphically this probability can be depicted as the shaded area in the curve shown in Figure 4.4. We might want to know the probability of T being at most t—that is, zero is the lower limit and t is the upper limit. In this case, the probability could be expressed as

$$P(T \leq t) = 1 - e^{-\lambda t} \tag{4.17}$$

Since the total area under the curve is unity, subtracting the area for t or more from 1 provides the area that is desired. This area is shown in Figure 4.5.

If we utilize Equations 4.16 and 4.17 where appropriate, the need to integrate will arise only when we desire to find the probability of a time interval of a different type, say between 2 and 5 hours. However, due to the simplicity of the exponential function, the evaluation of the integral $\int_a^b$ becomes simply $(-e^{-b}) -$

FIGURE 4.5 $P(T \leq t)$ for an exponential distribution

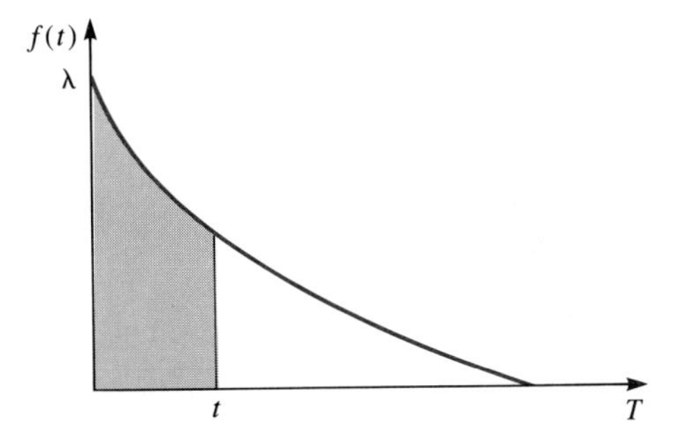

FIGURE 4.6 $P(t_1 \leq T \leq t_2)$ for an exponential distribution

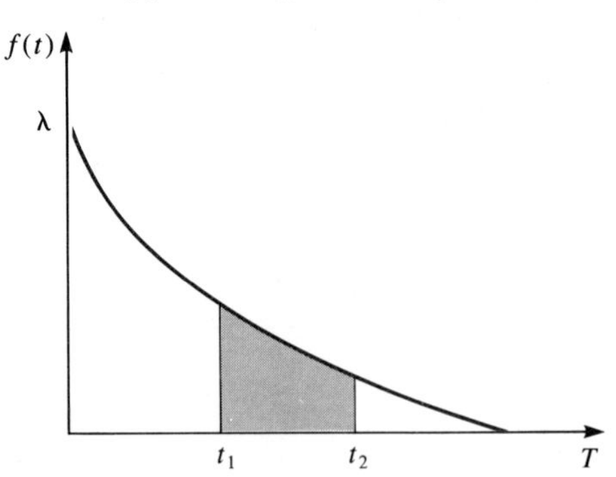

$(-e^{-a})$. Thus, the probability that T falls between time periods t_1 and t_2 can be determined directly as

$$P(t_1 \leq T \leq t_2) = e^{-\lambda t_1} - e^{-\lambda t_2} \quad \textbf{(4.18)}$$

The area under the curve depicted by this probability statement is shown in Figure 4.6. In all cases, $e^{-\lambda t}$ can be found in Appendix Table B.5.

In Chapter 3 we determined the probable occurrence of x for both the uniform and triangular distributions—two continuous distributions—by integration. Even though the exponential distribution is a continuous distribution, we solve for the probable occurrences directly. The probability statements discussed above, together with the integral form and direct determination, are summarized below.

$$P(T \geq t) = \int_t^{\infty} \lambda e^{-\lambda t}\, dt = e^{-\lambda t}$$

$$P(T \geq t) = \int_0^{t} \lambda e^{-\lambda t}\, dt = 1 - e^{-\lambda t}$$

$$P(t_1 \leq T \leq t_2) = \int_{t_1}^{t_2} \lambda e^{-\lambda t}\, dt = e^{-\lambda t_1} - e^{-\lambda t_2}$$

EXAMPLE 4.17 Parkinson Printers has observed breakdowns on its high-speed press at the rate of 2 every 4 hours. What is the probability that no more than 5 hours elapse between breakdowns?

Solution: Since the question calls for a time period expressed in hours, it is necessary to standardize λ on a per hour basis. Thus, λ will become .5 breakdowns per hour, and the probability of no more than 5 hours elapsing between breakdowns is determined by using Equation 4.17.

$$P(T \leq 5) = 1 - e^{-.5(5)} = 1 - e^{-2.5} = 1 - .0821 = .9179$$

The value of $e^{-\lambda t}$ is found by using Appendix Table B.5. It should also be pointed out that the probability of t time elapsing can also be determined by changing the limits of integration to coincide with the given conditions of λ. For Example 4.17, this means the 5 hours could be expressed as 1.25 units of 4 hours, and the solution would be as follows.

$$P(T \leq 1.25) = 1 - e^{-2(1.25)} = 1 - e^{-2.5} = 1 - .0821 = .9179$$

This result is the same as before. Generally, altering λ to standard units is a more reasonable approach.

EXAMPLE 4.18 Referring to the Parkinson Printers' situation in Example 4.17, find the probability that a breakdown occurs between 4 and 8 hours inclusive.

Solution: Using the standardized value of $\lambda = .5$ and Equation 4.18, we can calculate the probability as

$$P(4 \leq T \leq 8) = e^{-.5(4)} - e^{-.5(8)} = e^{-2} - e^{-4} = .1353 - .0183 = .1170$$

The shape of the probability density function will be dictated by the value of λ. In Figure 4.7, two probability density functions are plotted by using Equation 4.14 and selected values of t. We can see that in both cases, when Equation 4.14 is evaluated at $t = 0$, we simply obtain the value of λ. The

FIGURE 4.7 The probability density function for $f(t) = \lambda e^{-\lambda t}$

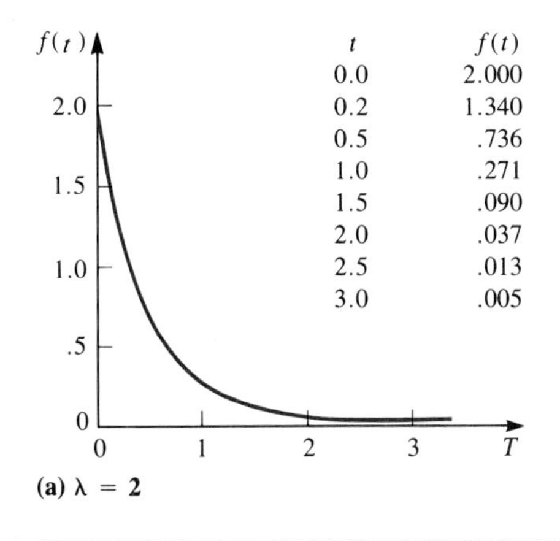

(a) $\lambda = 2$

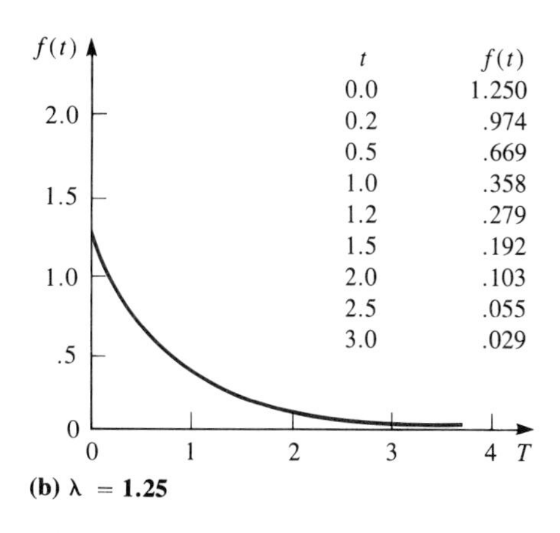

(b) $\lambda = 1.25$

interpretation of the graphic display is the probability distribution of the time between arrivals or, essentially, the service time. Since we have a probability distribution, we can further describe it by finding the mean and standard deviation.

FORMULA 4.5 For an exponential distribution, the **mean** is

$$\mu_T = \frac{1}{\lambda} \tag{4.19}$$

For an exponential distribution, the **standard deviation** is

$$\sigma_T = \sqrt{\frac{1}{\lambda^2}} \tag{4.20}$$

EXAMPLE 4.19 Referring again to Parkinson Printers, find the expected time between breakdowns. What is the standard deviation?

Solution: Utilizing Equations 4.19 and 4.20, the mean is

$$\mu_T = \frac{1}{.5} = 2$$

and the standard deviation is

$$\sigma_T = \sqrt{\frac{1}{(.5)^2}} = \sqrt{4} = 2$$

Therefore, we can say that the mean time between breakdowns is 2 hours with a standard deviation of 2 hours.

EXERCISES

4.15 The arrival of constituents to a senatorial office in Washington is at the mean rate of 4 every 2 days.

- **(a)** What is the probability that at most 3 days elapse between arrivals?
- **(b)** What is the probability that more than 1 day elapses between arrivals?
- **(c)** What is the mean time between arrivals? What is the standard deviation?

4.16 Winsome, a wholesale grocer, receives orders at the mean rate of 5 every hour.

- **(a)** What is the probability that at least a half hour elapses between arrivals of orders?
- **(b)** What is the probability that 12 minutes to 48 minutes elapse between arrivals?

4.17 On the average, 2 ships arrive at the docking facilities at a small gulf port in a 4-hour period.

(a) What is the probability that the time between two arrivals will be at least an hour but no more than 2 hours?
(b) What is the probability that no more than 2 hours elapse between arrivals?

4.7 THE NORMAL DISTRIBUTION

The second family of continuous distributions to be considered is the normal family. It is probably the best known and most widely used of all probability models. The model was actually developed or discovered independently by three different men at different times. In 1733, DeMoivre found it as the limiting form for the binomial distribution. In 1778, Laplace discovered it a second time, and in the early 1800s Gauss developed it a third time. Both Laplace and Gauss derived it as the distribution of errors of astronomical observations. Due to its development by Gauss, the normal distribution is known in physics and engineering as the Gaussian distribution.

Empirically, the Belgian astronomer Quételet, who had a strong interest in the social sciences, discovered that the normal distribution also provided a good fit to distributions of anthropological measurements. Many other empirical studies involved distributions of sums of individual measurements. The yield of an acre of grain is the sum of the yields of the individual plants. A community's consumption of electricity is the sum of the consumptions of the individual users. The yield of a stock portfolio is the sum of the yields of the individual stocks. A person's height is the sum of the lengths of the individual bones in the body. Note that all of these represent applications of the *central limit theorem,* which is presented in Chapter 6. Over time, phenomena that could be described by the Gaussian distribution began to be considered as normal, and anything that was non-Gaussian was deemed abnormal. Thus, the term "normal" became associated with this family of distributions.

Today the normal family is used, at least as a good approximation, to describe situations where measurements cluster around some central value, with deviations above and below that value being equally likely and having decreasing frequency as the deviations increase in absolute value. The graphical representation of this is a bell-shaped curve. Many phenomena in the business world exhibit this pattern. Two examples are the portfolio yield and electricity consumption stated earlier. Other examples are the yields of a polymerization process, scores on a standardized placement exam, starting salaries of accounting graduates, and monthly sales of Krunchy Granola, to name a few.

DEFINITION 4.7 The probability density function for the **normal distribution** is

$$f(x) = \frac{1}{\sigma_X \sqrt{2\pi}} e^{-(x-\mu_X)^2/2\sigma_X^2} \qquad \begin{array}{c} -\infty < x < \infty \\ -\infty < \mu_X < \infty \\ 0 < \sigma_X < \infty \end{array} \tag{4.21}$$

where: μ_X = mean of the distribution

σ_X = standard deviation of the distribution

Two other items appearing in the function are e, Euler's constant, and π, approximately 3.14159. An evaluation of the function at any point x will determine the height of the curve at that point.

The normal distribution has several attributes that further characterize it.

1. It is symmetrical around the mean.
2. It asymptotically approaches 0 at $-\infty$ and $+\infty$.
3. It is a bell-shaped curve.

In Figure 4.8, a normal curve has been partially constructed for the parameters $\mu_X = 25$ and $\sigma_X = 2$, and the function has been evaluated for specific values of x. If we desire to find the probability that x would fall between 25 and 27 in Figure 4.8, it is necessary to integrate the function between those two values.

$$P(25 \leq X \leq 27) = \int_{25}^{27} \frac{1}{2\sqrt{2\pi}}\, e^{-(x-25)^2/2(2)^2}\, dx$$

The actual integration of this function is impossible over finite ranges, but a very accurate answer can be obtained by using numerical approximation techniques. However, it becomes a needless task if we utilize the standardized normal curve. The standardized normal curve is established with the mean equal to zero and the standard deviation equal to unity. These parameters are used to generate, by using numerical approximations, the probabilities shown in Appendix Table B.6. The table is designed to provide the area from the mean out to some point z, the number of standard deviations away from the mean. Thus, to evaluate a statement such as $P(0 \leq Z \leq 1.32)$, we need only locate the value for $z = 1.32$ in Appendix Table B.6. The area under the curve represents the probability of obtaining a value for Z between the stated limits. A portion of the table is

FIGURE 4.8 Normal curve for selected values of x for the parameters $\mu_X = 25$ and $\sigma_X = 2$

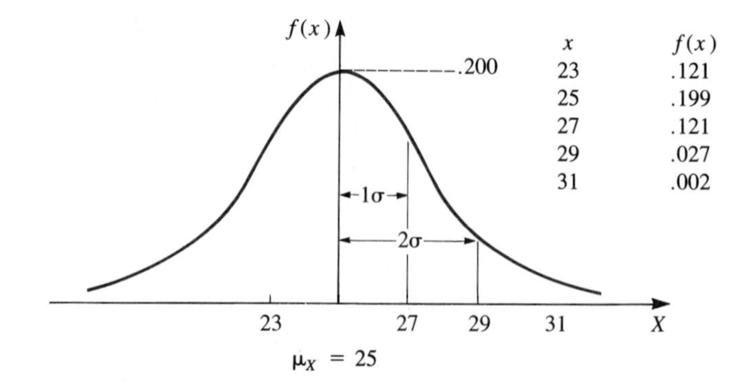

TABLE 4.3 Selected probabilities for the normal distribution

Z	*.00*	*.01*	*.02*	*.03*	...	*.09*
0.0	.0000	.0040	.0080	.0120	...	.0359
0.1	.0398	.0438	.0478	.0517	...	.0753
.	.	.	.	.	...	.
.	.	.	.	.	...	.
.	.	.	.	.	...	.
1.2	.3849	.3869	.3888	.3907	...	.4015
1.3	.4032	.4049	.4066	.4082	...	.4177
1.4	.4192	.4207	.4222	.4236	...	.4319
.	.	.	.	.		
.	.	.	.	.		
.	.	.	.	.		
3.9	.5000	.5000	.5000	.5000	...	.5000

shown in Table 4.3, with the probability value highlighted at the intersection of the appropriate column (the hundredths digit of the z value) and the appropriate row (the unit and tenths digit of the z value).

Any normal random variable can be converted to a standard normal random variable through the use of the Z transformation. The Z transformation,

$$Z = \frac{X - \mu_X}{\sigma_X} \tag{4.22}$$

enables us to represent any x value in terms of its distance from the mean in standard deviations. If X is a normally distributed variable with mean μ_X and standard deviation σ_X, then the variable obtained from this transformation is also normally distributed. This distribution of Z reduces any normally distributed variable to the standardized normal distribution with a mean $\mu = 0$ and a standard deviation $\sigma = 1$. It is shown in Figure 4.9.

The areas shown in Figure 4.9 are found in Appendix Table B.6 and represent the probable occurrence of Z for the probability statements $P(-1 \leq Z \leq 1)$, $P(-2 \leq Z \leq 2)$, and $P(-3 \leq Z \leq 3)$. In essence, these statements and their respective probabilities represent the range of one, two, or three standard deviations on either side of the mean for any normally distributed random variable. The value of the standard normal distribution is that it can be used to find probabilities associated with any normally distributed variable with a mean μ_X and a standard deviation σ_X. The value of the probability is obtained by using Appendix Table B.6. To use this table, we need only know the value of Z. As indicated in the portion given in Table 4.3, the table gives the area under the standardized normal curve from the mean to the value of Z. Thus,

$$P(-1.20 \leq Z \leq 1.29) = .7864$$

FIGURE 4.9 Standardized normal curve

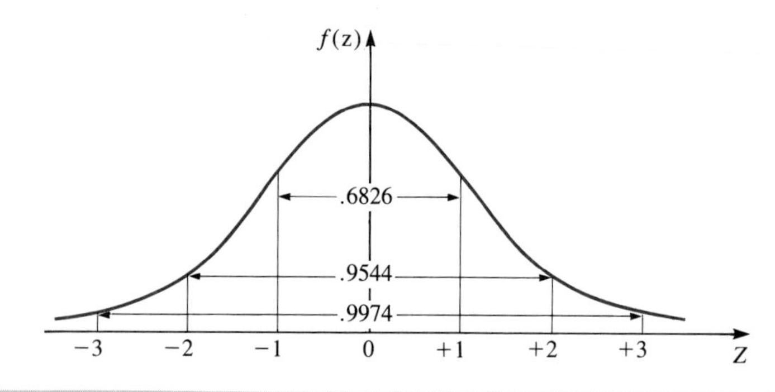

Using Appendix Table B.6, we find the tabular value for $z = 1.29$ to be .4015. The negative value associated with the $z = -1.20$ denotes that the area is on the left of the mean and is equivalent to the tabular value associated with $z = 1.20$—that is, .3849. Therefore, the total area is the sum of these two tabular values, .7864.

$$P(Z \leq 1.43) = .9236$$

From Appendix Table B.6, the tabular value for $z = 1.43$ is .4236. Since the probability statement implies that all the area to the left of the mean is to be included, the total probability is .4236 + .5000 = .9236.

Figure 4.10 shows two normal curves. For each curve, an area of size .6826 is indicated. As can be seen, regardless of the height or breadth of the

FIGURE 4.10 Two normal curves for given μ_X and σ_X values

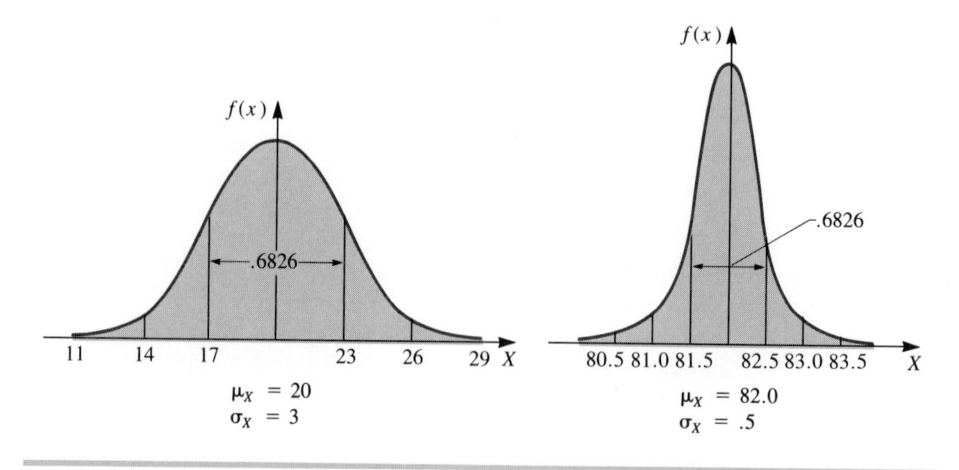

FIGURE 4.11 Graphical representation of $P(X < 32.75)$ for Example 4.20

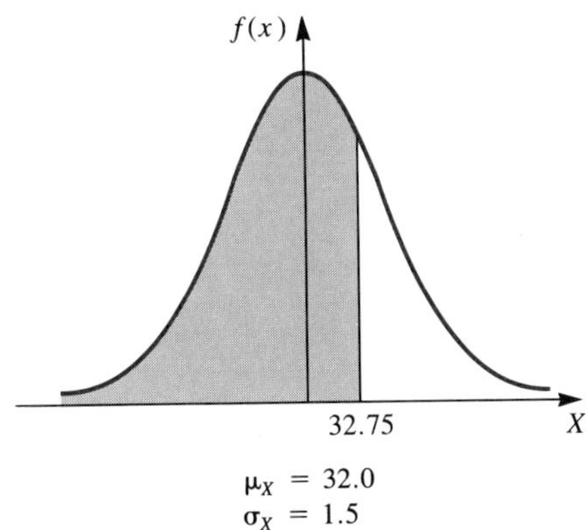

normal distribution, the probability that the variable is within ±1 standard deviation of the mean is the same.

EXAMPLE 4.20 Boxes of laundry detergent are filled automatically, and the weights are approximated by a normal distribution with a mean of 32 ounces and a standard deviation of 1.5 ounces. What is the probability of obtaining a box that weighs less than 32.75 ounces?

Solution: Any time we are attempting to find an area under the normal curve, it is important to draw a picture so that we are aware of exactly what area we are trying to evaluate. Therefore, to answer the question in Example 4.20, we draw the picture as shown in Figure 4.11 and use Equation 4.22.

$$z = \frac{32.75 - 32}{1.50} = \frac{0.75}{1.50} = 0.5$$

From Appendix Table B.6

$$P(0 \leq Z \leq .5) = .1915$$

Thus,

$$P(X < 32.75) = .1915 + .5 = .6915$$

In solving this probability, note that the size of an area on each side of the mean was found and the results were added together. Since we know that the total area under the curve is 1.0, the area to the left of the mean has to be .5.

EXAMPLE 4.21 What is the probability of having a box of laundry detergent that weighs less than 31.1 ounces?

FIGURE 4.12 Graphical representation of $P(X \leq 31.1)$ for Example 4.21

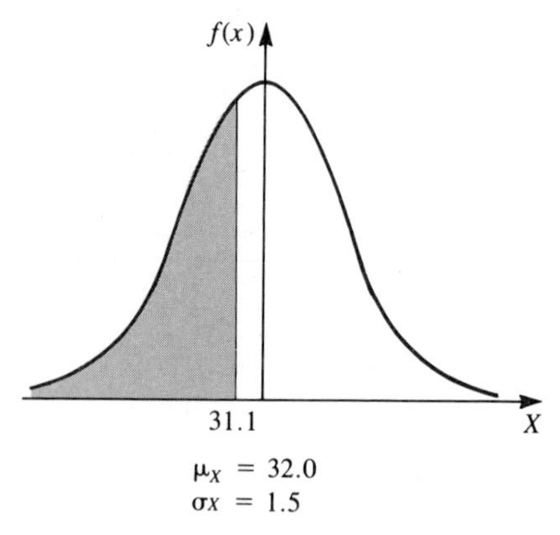

Solution: Again, we want to specify the area in question by drawing a picture and shading in the desired area, as shown in Figure 4.12.

$$z = \frac{31.1 - 32}{1.5} = \frac{-0.9}{1.5} = -.6$$

From Appendix Table B.6

$$P(-.6 \leq Z \leq 0) = .2257$$

Therefore,

$$P(X \leq 31.1) = .5000 - .2257 = .2743$$

Example 4.21 illustrates two important points. The first is the determination of areas specified by a negative z. We notice that Appendix Table B.6 has no negative values. However, as indicated previously, the standardized normal curve is symmetrical: The area under the curve between 0 and .6 is the same as the area from 0 to $-.6$. Therefore, the minus sign indicates that we are concerned with an area that is located to the left of the mean. The second point to notice is that the value found in the table represents the area from 31.1 to 32.0, so it is necessary to subtract this from .5000 to find the desired area.

EXAMPLE 4.22 What is the probability of finding a box that weighs between 32.6 ounces and 33.2 ounces inclusive? Refer to Example 4.20.

Solution: In Figure 4.13, we can see that we are trying to find a segment of area that is on one side of the distribution. We are unable to find this area directly. In this case, it is necessary to find two areas, $P(32 \leq X \leq 33.2)$ and $P(32 \leq X \leq$

FIGURE 4.13 Graphical representation of $P(32.6 < X < 33.2)$ for Example 4.22

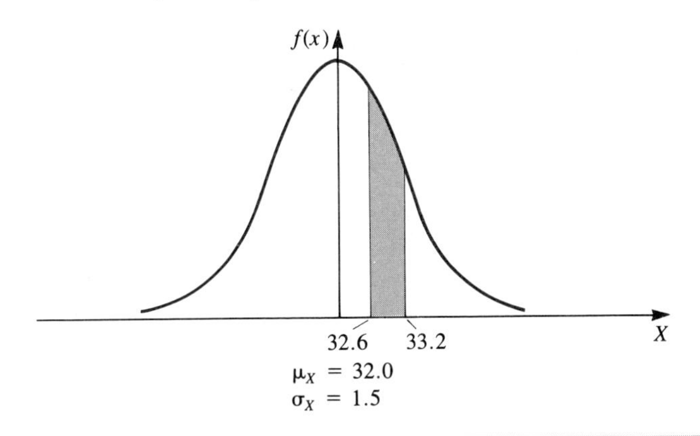

32.6), and subtract the two.

$$z = \frac{33.2 - 32.0}{1.5} = \frac{1.2}{1.5} = .8$$

$$P(0 \leq Z \leq .8) = .2881$$

$$z = \frac{32.6 - 32.0}{1.5} = \frac{0.6}{1.5} = .4$$

$$P(0 \leq Z \leq .4) = .1554$$

Therefore,

$$P(32.6 \leq X \leq 33.2) = .2881 - .1554 = .1327$$

Graphically, the probability found in Example 4.22 is depicted in Figure 4.14. It should be emphasized that we *are only able to add and subtract areas and not values for Z.*

FIGURE 4.14 Graphical representation of $P(32.6 \leq X \leq 33.2)$

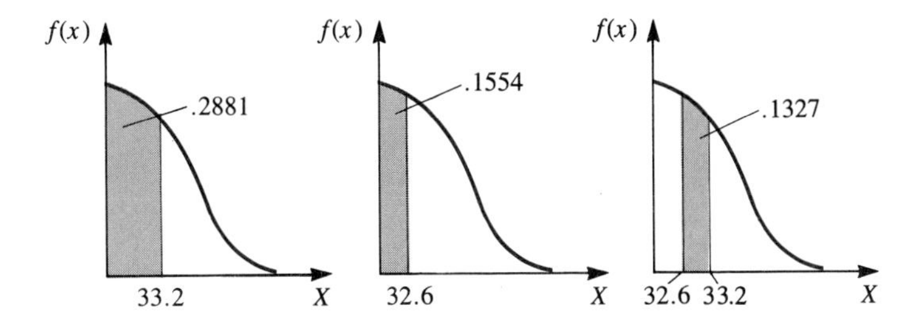

EXAMPLE 4.23 What is the minimum weight a box of laundry detergent could be and remain in the top 40% of all boxes filled? Refer to Example 4.20.

Solution: This example is an illustration of working in reverse. In this case, we are given the probability, and we want to find the value of Z as depicted in Figure 4.15. We are not given a value for Z directly, but it can be found if we find the proper area in Appendix Table B.6 and determine what value of Z is related to that area. The proper area for Example 4.23 is also shown in Figure 4.15 as .1000, since Appendix Table B.6 is designed to provide areas from the mean out to a given X value. In consulting Appendix Table B.6, we are unable to find the exact value of .1000. However, as we can see in the following portion of the table, .1000 falls between .0987 and .1026.

Z	.00	. . .	.05	.06	. . .
.	.		.	.	
.	.		.	.	
.	.		.	.	
0.2	.0793	. . .	.0987	.1026	. . .

Since it is closer to .0987, we will choose $z = .25$ as the one to use in our equation. Solving for x indicates

$$.25 = \frac{x - 32}{1.5}$$

$$.25(1.5) = x - 32$$

$$x = 32 + .375 = 32.375$$

Thus, we see that it is necessary to have a box of laundry detergent weighing 32.375 ounces or more to be in the top 40% of all boxes.

FIGURE 4.15 Graphical representation of the specified and desired probabilities for Example 4.23

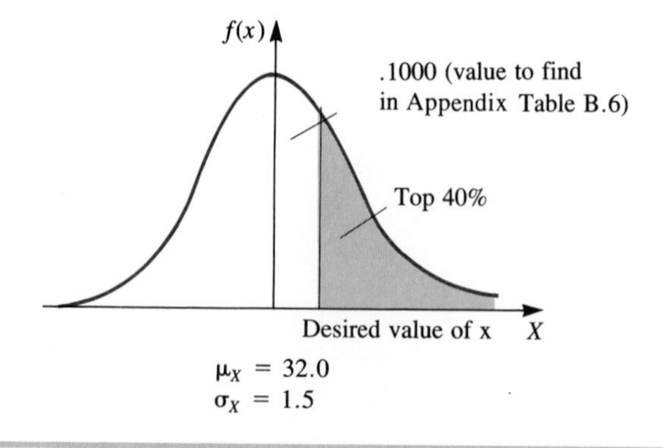

EXERCISES

4.18 Determine the probability for each of the following statements.

(a) $P(0 \leq Z \leq 1.47)$

(b) $P(Z \geq -1.38)$

(c) $P(-.78 \leq Z \leq 1.06)$

(d) $P(Z \leq -.43)$

4.19 For each of the following, determine the value of z_0.

(a) $P(-z_o \leq Z \leq 0) = .4370$

(b) $P(Z \geq z_0) = .1190$

(c) $P(-z_0 \leq Z \leq z_0) = .9970$

4.20 An experiment was performed to determine the time necessary for the human eye to fully dilate. The mean time for the experimental group was 22.1 milliseconds (ms) with a standard deviation of 3.1 ms. Times are assumed to be normally distributed.

(a) What is the probability that an eye would take longer than 25.3 ms to dilate?

(b) What is the probability that an eye would take at least 19 ms but no more than 24 ms to dilate?

(c) Would you feel confident in saying that after 32 ms all eyes are fully dilated?

4.21 A standard rotary mower is set to cut the grass at a mean height of 2 inches with a standard deviation of 0.4 inch. The heights of the grasses are assumed to be normally distributed.

(a) Is it conceivable that grass could be found to be shorter than 1 inch?

(b) What are the maximum and minimum heights of grasses such that they are included in the middle 70% of the distribution of all grasses?

4.22 An automatic measuring device is designed to permit a mean flow of 5 grams with a standard deviation of 0.2 gram. Flow quantities are deemed to be normally distributed.

(a) What is the probability of finding a flow quantity less than 4.5 grams?

(b) What is the probability that a flow quantity falls between 4.65 and 5.65 grams?

4.8 THE NORMAL APPROXIMATION OF THE BINOMIAL

In Section 4.5, the Poisson distribution was used to approximate the binomial distribution for the condition $n\pi < 5$ or $n(1 - \pi) < 5$. However, as these two products become equal to or greater than 5, an alternative approximation is appropriate. In this case, the normal distribution provides a more accurate approximation.

When the normal distribution is used as an approximation of the binomial, we are approximating a discrete probability distribution with a continuous probability distribution. As a result, it is necessary to correct for continuity. In Figure 4.16, a normal curve is superimposed on a binomial distribution. If we want to find the probability that X is equal to a single value, using a normal approximation, we must solve for the area on the normal curve that would be approximately equivalent to the rectangular area in the binomial distribution. Thus, to find the probability $P(X = 4)$, it is necessary to find the probability

FIGURE 4.16 The normal distribution superimposed on the binomial distribution for $n = 14$ and $\pi = .5$

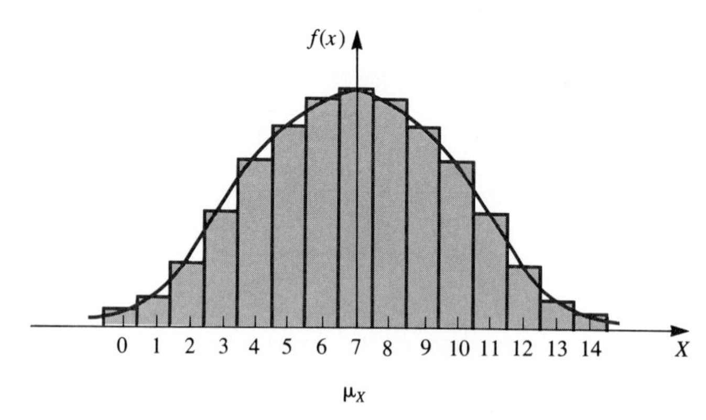

$P(3.5 < X < 4.5)$ in the normal distribution. The obvious addition and subtraction of .5 with respect to the original value of 4 is an example of the continuity correction factor (*CCF*). The *CCF* is used when a continuous probability distribution is used to approximate a discrete probability distribution. The $\pm .5$ with respect to the value of X will improve the approximation. The addition or subtraction of the .5 will be predicated on whether we desire to include or exclude the value in the calculation of Z. Corrected probability statements are shown next for given probabilities of X.

Original Statement	*Statement Corrected for Continuity*
$P(X = x)$	$P(x - .5 \leq X \leq x + .5)$
$P(X \leq x)$	$P(X \leq x + .5)$
$P(X \geq x)$	$P(X \geq x - .5)$
$P(X < x)$	$P(X < x - .5)$
$P(X > x)$	$P(X > x + .5)$

FORMULA 4.6 The mean, $\mu_X = n\pi$, and the standard deviation, $\sigma_X = \sqrt{n\pi(1 - \pi)}$, of the binomial distribution serve as the corresponding parameters for the **normal approximation of the binomial.** Under the conditions that $n\pi \geq 5$ and $n(1 - \pi) \geq 5$, the probable occurrence of X can be determined using the Z transformation

$$Z = \frac{(X \pm .5) - \mu_X}{\sigma_X} \tag{4.23}$$

where: $\mu_X = n\pi$

$\sigma_X = \sqrt{n\pi(1 - \pi)}$

$\pm.5$ = the correction for continuity

EXAMPLE 4.24 Sentinal Sales, Inc., a mail-order house, has experienced daily sales greater than \$100,000 30% of the time. What is the probability that during the next 40 days, at least half of the days have sales that exceed \$100,000?

Solution: This example is a binomial distribution with $n = 40$ and $\pi = .3$. Thus, the probability that at least half the days have sales exceeding \$100,000 is

$$P(X \geq 20) = \sum_{x=20}^{40} C_x^{40}\ .3^x\ .7^{40-x}$$

We are unable to determine this probability from our tables, so the normal approximation can be employed, since $n\pi = 12$, and $n(1 - \pi) = 28$. The standard deviation is $\sqrt{n\pi(1 - \pi)} = 2.90$. Use of the normal distribution suggests we draw a picture, as shown in Figure 4.17, to denote the area under consideration. The solution for Example 4.24 follows.

$$z = \frac{(20 - .5) - 12}{2.90} = \frac{7.50}{2.90} = 2.59$$

$$P(0 \leq Z \leq 2.59) = .4952$$

Thus,

$$P(X \geq 20) = .5000 - .4952 = .0048$$

The *CCF* is reflected in the calculation $(20 - .5)$ used for x in Equation 4.23. As our answer indicates, the probability of daily sales exceeding \$100,000 at least half the days is slightly less than half of a percent.

FIGURE 4.17 Graphical representation of $P(X \geq 20)$ for Example 4.24

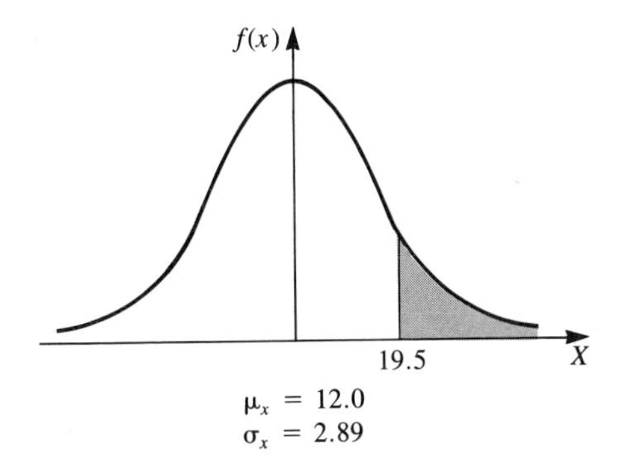

FIGURE 4.18 Graphical representation of $P(X = 8)$ for Example 4.25

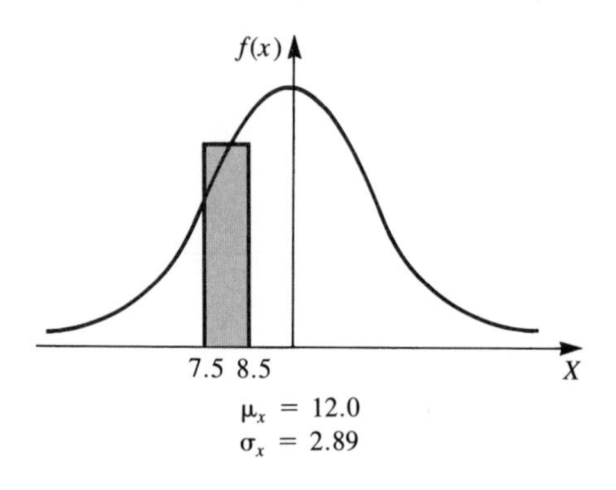

EXAMPLE 4.25 Referring to the Sentinal Sales situation in Example 4.24, find the probability that exactly 8 of the days would have sales exceeding $100,000.

Solution: As illustrated in Figure 4.18, in utilizing the normal approximation for Example 4.25, it is necessary to find the area of the segment that represents exactly 8 days.

$$z = \frac{(8 - .5) - 12}{2.90} = \frac{-4.50}{2.90} = -1.55$$

$$P(-1.55 \leq Z \leq 0) = .4394$$

$$z = \frac{(8 + .5) - 12}{2.90} = \frac{-3.50}{2.90} = -1.21$$

$$P(-1.21 \leq Z \leq 0) = .3869$$

Therefore,

$$P(X = 8) = .4394 - .3869 = .0525$$

We can conclude that there is a 5.25% chance of having exactly 8 days when daily sales will exceed $100,000.

EXERCISES

4.23 Anchor Manufacturing has found that 60% of their orders represent repeat business. In the next 150 orders, what is the probability that 100 or more will be repeat orders?

4.24 Underwood Consultants submits bids on jobs and generally receives the contract 30% of the time. If Underwood presently has 24 bids outstanding, what is the probability of obtaining exactly 9 bids? less than 5 bids?

4.9 SUMMARY

In this chapter, we have seen several distributions that enable us to determine the probable occurrence of events in real-life situations. The first three distributions, shown in Table 4.4a, are discrete probability distributions for which only whole-number outcomes are possible. The hypergeometric distribution is characterized by a changing probability from trial to trial, because successive observations are made without replacement, making the trials dependent. This is in contrast to the binomial distribution, where the probability remains constant from trial to trial and the trials are independent. The third discrete distribution is the Poisson, which models the occurrence of events in an interval. In addition, the binomial and the hypergeometric distributions can be extended to multiple outcomes. These extensions are shown in Table 4.4b.

TABLE 4.4 Summary table

(a) Discrete Probability Distributions for p(x)

Hypergeometric Distribution	$p(x) = \begin{cases} \dfrac{C_x^R \, C_{n-x}^{N-R}}{C_n^N} & x = 0,1,2, \ldots, \min(n,R) \\ & n < N \\ & R < N \\ 0 & \text{elsewhere} \end{cases}$ Parameters: R, n, N Mean: $\mu_X = \dfrac{nR}{N}$ Standard deviation: $\sigma_X = \sqrt{\dfrac{nR(N-R)(N-n)}{N^2(N-1)}}$
Binomial Distribution	$p(x) = \begin{cases} C_x^n \, \pi^x (1-\pi)^{n-x} & x = 0, 1, \ldots, n \\ & 0 < \pi < 1 \\ 0 & \text{elsewhere} \end{cases}$ Parameters: n, π Mean: $\mu_X = n\pi$ Standard deviation: $\sigma_X = \sqrt{n\pi(1-\pi)}$
Poisson Distribution	$p(x) = \begin{cases} \dfrac{e^{-\mu}\mu^x}{x!} & x = 0, 1, 2, \ldots \\ & \mu > 0 \\ 0 & \text{elsewhere} \end{cases}$ Parameter: μ Mean: $\mu_X = \mu$ Standard deviation: $\sigma_X = \sqrt{\mu}$

TABLE 4.4 (Continued)

(b) Discrete Probability Distributions for $p(x_1, x_2, \ldots, x_m)$

Extended Hypergeometric Distribution	$p(x_1, x_2, \ldots, x_m) = \begin{cases} \dfrac{C_{x_1}^{R_1} C_{x_2}^{R_2} \cdots C_{x_m}^{R_m}}{C_n^N} & \Sigma x_i = n,\ \Sigma R_i = N,\ n < N,\ 0 \le x_i \le \min(n, R_i) \text{ for all } i \\ 0 & \text{elsewhere} \end{cases}$
Multinomial Distribution	$p(x_1, x_2, \ldots, x_m) = \begin{cases} \left(\dfrac{n!}{x_1!x_2! \cdots x_m!}\right) & \pi_1^{x_1} \pi_2^{x_2} \cdots \pi_m^{x_m},\ \Sigma x_i = n,\ \Sigma \pi_i = 1,\ \pi_i > 0 \\ 0 & \text{elsewhere} \end{cases}$

(c) Continuous Distributions

Exponential Distribution	Function: $f(t) = \begin{cases} \lambda e^{-\lambda t} & \lambda > 0,\ t \ge 0 \\ 0 & \text{elsewhere} \end{cases}$ Parameter: λ Probability: $P(T \le t) = \int_0^t \lambda e^{-\lambda t}\, dt$ Mean: $\mu_T = 1/\lambda$ Standard deviation: $\sigma_T = 1/\lambda$
Normal Distribution	Function: $f(x) = \dfrac{1}{\sigma_X \sqrt{2\pi}} e^{-(x-\mu_X)^2/2\sigma_X^2}$ $-\infty < x < \infty$, $-\infty < \mu_X < \infty$, $0 < \sigma_X < \infty$ Parameters: μ_X, σ_X Probability: $P(X \le x) = P(Z \le z)$ *where:* $z = \dfrac{x - \mu_X}{\sigma_X}$ Mean: μ_X Standard deviation: σ_X

(d) Approximations of the Binomial Distribution

Poisson Approximation	$p(x) = \dfrac{e^{-\mu}\mu^x}{x!}$ $n\pi < 5$ or $n(1 - \pi) < 5$ Parameter: $\mu = n\pi$
Normal Approximation	$P(X = x) = P(z_1 \le Z \le z_2)$ $n\pi \ge 5$ and $n(1 - \pi) \ge 5$ *where:* $z = \dfrac{(x \pm .5) - \mu_X}{\sigma_X}$ Parameters: $\mu_X = n\pi$ $\sigma_X = \sqrt{n\pi(1 - \pi)}$

Two continuous distributions were introduced in this chapter: the exponential and the normal. These two distributions are shown in Table 4.4c. For the exponential distribution, the determination of a probability is by integration. However, the introduction of the integrated forms allows a more direct solution to the determination of the probability. For the normal distribution, the probability is also determined by integration by using numerical approximations, but the use of the Z transformation allows the direct use of the Z table.

We looked at two approximation procedures, both of them for the binomial distribution. Whether we use the Poisson approximation or the normal approximation is predicated on the value of $n\pi$ and $n(1 - \pi)$. In Table 4.4d these approximations and the specific conditions are shown.

SUPPLEMENTARY EXERCISES

4.25 What distinguishes the hypergeometric distribution involving two outcomes from the binomial?

4.26 Under what circumstances is the binomial distribution symmetric? Is the hypergeometric ever symmetric? Consider the hypergeometric distribution with $N = 6$, $R = 3$, and $n = 2$.

4.27 When the population is finite, why is it necessary that sampling be with replacement to use the binomial?

4.28 In using the multinomial, what restriction must we place on the outcomes?

4.29 What happens to the binomial distribution when n becomes large (> 100) for values of π near .5?

***4.30** Use the definition of expected value to show that the mean of the binomial is $n\pi$.

***4.31** Is the binomial a good approximation of the hypergeometric as n increases?

4.32 Why must we impose a restriction when the Poisson is used as an approximation of the binomial?

4.33 Can two normal distributions have equal means and unequal standard deviations? Can the reverse situation occur?

4.34 Describe the standardized normal curve and indicate how it is used in statistics.

4.35 A fleet of 15 company cars consists of 5 Chevrolets, 5 Fords, 3 Chryslers, and 2 Volkswagens. If 6 cars are randomly selected from the fleet, determine each of the following probabilities of the 6 cars being composed of **(a)** 3 Chevrolets and 3 anything else, **(b)** 3 Chevrolets and 3 Fords, **(c)** at least 2 Fords, **(d)** only Fords and Chevrolets, and **(e)** 1 Ford, 1 Chevrolet, 2 Chryslers, and 2 Volkswagens.

4.36 A complement of middle managers from Toni Hanawak Associates consists of 6 "old guard" and 5 "young rebels." What is the probability that in the selection of 4 persons to form a committee, it would consist of **(a)** 2 old guards and 2 young rebels, **(b)** at most 2 young rebels, or **(c)** only old guard? **(d)** What is the mean number of young rebels that would be expected on the committee? What is the standard deviation?

---

* An asterisk indicates a higher level of difficulty.

4.37 A company promotion list consists of nine persons that have become eligible for promotion. Included in the group are 4 women and 5 men. Two of the women and 3 of the men have college degrees. Four persons will be promoted.

- **(a)** What is the probability that all will be women?
- **(b)** What is the probability that at least 2 will be women?
- **(c)** What is the probability that all will be college graduates?
- **(d)** What is the probability that 2 women with college degrees and at least one man with a college degree will be promoted?
- **(e)** What is the mean number of degree holders to be promoted? the mean number of men? the mean number of women?
- **(f)** Draw the probability distribution for the number of college graduates to be promoted.

4.38 For one of several reasons, 15% of the parts produced by a machine are defective. A sample of 20 parts is randomly selected from a production run.

- **(a)** What is the probability that 3 parts are defective?
- **(b)** What is the probability that no more than 3 parts are defective?
- **(c)** What is the probability that at least 3 but no more than 6 parts are defective?
- **(d)** What is the probability that 5 or more parts are defective?

4.39 What is the mean number of defective parts for Exercise 4.38? What is the standard deviation?

4.40 Use the information of Exercise 4.38 to determine how many defectives could be found if we desire to have the probability of such an occurrence to be approximately 5%.

4.41 Although a company sells a product, it produces only 60% of all the units sold. The remainder are produced by a job shop under a trade contract.

- **(a)** If you purchased 10 items, what is the probability that they are all produced by the selling company?
- **(b)** What is the mean number produced outside if a lot of 50 is purchased? What is the standard deviation?
- **(c)** If 2 lots of 20 are purchased, what is the probability of obtaining 20 items produced by the selling company broken down as 10 in each lot? as 12 from the first lot and 8 from the second?
- **(d)** Obtain an answer to part (c) if the number from each lot is not designated. How would you justify your answer?

4.42 A standard bridge hand contains 13 cards dealt from a regular deck of 52. Set up the following probabilities, but do not do the computations. **(a)** 5 hearts, **(b)** 6 clubs and 4 spades, **(c)** 4 spades or 4 diamonds, **(d)** 7 red cards, **(e)** 7 face cards (K, Q, and J), **(f)** 4 aces, 4 face cards, and 3 hearts that are not an ace or a face card, and **(g)** 3 spades, 5 diamonds, the ace and king of hearts, and 3 clubs.

4.43 In writing a computer program, students usually fail to achieve a successful run the first time. It is found that logic errors account for 40% of the problems, keypunching errors for 30%, omission of a symbol for 20%, and incorrect card order for 10%. What is the probability in 10 randomly selected programs that did not run successfully that 4 had logic errors, 2 had keypunching errors, 2 had omitted a symbol, and 2 had cards out of order?

4.44 It has been determined that 40% of all junior-high students read below their grade level. Seven students are randomly selected.

(a) What is the probability that only 2 are below average?
(b) What is the probability that no more than 2 are below average?
(c) What is the probability that 3 or more are below average?

4.45 Prime-time television programs are divided as follows.

Movies	.30
Sports	.15
Documentaries	.10
Situation comedies	.20
All others	.25

If 10 different times were randomly selected, what is the probability of obtaining 2 shows in each category?

4.46 Trucks arrive at the loading dock of Conical Supply at the rate of 8 trucks per hour. Arrivals are assumed to follow a Poisson distribution.
(a) What is the probability that 4 trucks arrive in an hour?
(b) What is the probability that 4 trucks arrive in 39 minutes?
(c) What is the probability that 6 or less trucks arrive in 45 minutes?
(d) What is the probability that 20 or more arrive in a 2-hour period?
(e) What is the probability that 3 to 5 inclusive arrive in a 15-minute period?

4.47 In a recent survey, only 7% of the persons passing through a toll booth were observed to be wearing a shoulder harness. For the next 50 cars, calculate the probabilities that **(a)** only 4 people are wearing a shoulder harness, and **(b)** 5 or more people are wearing a shoulder harness.

4.48 What is the average number of errors a typist could have per page if we desired to have no more than 3 errors with a probability of approximately 98%?

4.49 A fraternity has 50 members, 40% of whom are seniors and the rest are equally divided between juniors and sophomores. If 6 persons were randomly selected from the total membership, what is the probability of obtaining two from each class?

4.50 A dressmaker has found that sewing machines break down at the rate of 1 every 2 hours and are Poisson distributed.
(a) What is the probability of no breakdowns in 1 hour?
(b) What is the probability of 2 to 4 breakdowns in a half-hour period?
(c) Is it conceivable that there would be no breakdowns in an 8-hour day?

4.51 Persons arrive at a teller's window at a mean rate of 3 every 6 minutes.
(a) What is the probability that no more than 12 minutes elapse before the next arrival?
(b) What is the probability that more than 2 minutes elapse before a new customer arrives for service?
(c) What is the probability that at least 6 minutes but no more than 10 minutes elapse before the next arrival?
(d) Construct the probability density function for the exponential distribution.

4.52 An auto insurance company has 200,000 policy holders in the 60–65 age group. It has been found that the probability of any individual in that age group being involved in an accident during one calendar year is .00002. What is the probability that 5 persons in that age group might be involved in an accident?

4.53 Random Transit places additional buses into service during the rush hour. As a result, buses arrive at a particular stop every 5 minutes. What is the probability that a person arriving at the stop would have to wait for a bus **(a)** at least 5 minutes? **(b)** 3 to 6 minutes?

4.54 During the peak hours on a Friday evening, shoppers arrive at the checkout counter of Economy Foods, Inc., at the rate of 3 every 8 minutes.

(a) What is the probability that during the next 20 minutes, 6 to 10 persons inclusive would arrive?

(b) What is the probability that 12 minutes or more elapse between arrivals?

4.55 In a normal distribution, determine the following.

(a) $P(1.05 < Z < 1.97)$

(b) $P(Z > -1.67)$

(c) $P(Z < 1.92)$

(d) Find z_0 such that $P(Z < z_0) = .33$

4.56 The probability that a fire alarm is false is .20.

(a) What is the probability that in a series of 15 alarms, there will be no more than 2 false alarms? at least 4 false alarms? between 5 and 7 (inclusive) false alarms?

(b) In a series of 6 alarms, what is the probability that there will be 2 that are false?

(c) In a series of 100 alarms, what is the probability that between 15 and 30 inclusive will be false?

4.57 A manufacturer of bolts has found that her product is normally distributed with a mean length of 2.5 inches and a standard deviation of .05 inch.

(a) What is the probability that a randomly selected bolt will exceed 2.58 inches?

(b) Bolts that are less than 2.40 inches must be scrapped. In a lot of 1200, how many could be expected to be scrapped?

(c) What is the minimum length for a bolt to be in the top 20% of the distribution of all lengths?

4.58 Automobiles arrive at a turnpike toll booth at the average rate of 4 per minute.

(a) What is the probability that 6 cars will arrive in any 1-minute period?

(b) What is the probability of at least 2 arrivals in 1 minute?

(c) What is the probability that no more than 2 cars will arrive in a 30-second period?

4.59 A study at the emergency ward of a hospital indicated that the mean number of admissions between 6:00 and 8:00 P.M. on a Monday is 4.5 and is assumed to be Poisson distributed.

(a) What is the probability that 10 cases would be admitted during that time period?

(b) What is the probability that there would be at least 7 but no more than 13 cases?

(c) What is the standard deviation with respect to the number of admissions during that time period?

(d) What is the probability that there would be less than 24 minutes between admissions?

(e) What is the probability that there would be more than 2 hours between admissions?

4.60 The number of employee-hours required by the Worth Construction Company to assemble and finish its Model S prefabricated home is normally distributed, with a mean number of employee-hours equal to 400 and a standard deviation of 40 employee-hours.

(a) What is the probability that assembling and finishing a home will take between 380 and 425 employee-hours?

(b) What is the minimum number of employee-hours necessary to assemble and finish 97.72% of the homes?

4.61 A certain type of car is purported to average 30 miles to the gallon with normal driving and a standard deviation of 5 miles per gallon. Mileage is presumed to be normally distributed.

(a) What is the probability of a car getting 36 miles or more per gallon?
(b) What is the probability that a car will get between 22.4 and 26 miles per gallon?
(c) What is the maximum mileage that can be registered to be in the middle 50% of all mileages?

4.62 The St. Lawrence Seaway is dotted with several locks that enable ships to move from ocean level to lake level. At the Snell lock, ships arrive at the mean rate of 6 per 2-hour period.

(a) What is the probability of having 8 ships arrive between 6:00 and 8:00 A.M.?
(b) What is the probability of between 4 and 8 ships (inclusive) arriving?
(c) What is the probability that a ship will wait more than 20 minutes to get through the lock?
(d) What is the probability that a ship will wait between 1 and 3 hours?

4.63 The lengths of steel bars produced by the Angular Steel Company are normally distributed with mean $\mu = 28.5$ feet and standard deviation $\sigma = .6$ foot.

(a) What is the probability that a steel bar is less than 27.0 feet long? Explain the meaning of your probability statement.
(b) What is the probability that a steel bar is between 26.9 and 30.8 feet long?
(c) What is the probability that a steel bar is more than 30.0 feet long?
(d) What is the probability that a steel bar is between 27.0 and 28.0 feet long?
(e) The probability is .75 that a steel bar will be more than ______ feet long.
(f) The middle 50% of the steel bars are between ______ and ______ feet long.

4.64 How many questions would a professor have to put on a multiple-choice exam, when there are 4 choices for each question, to be 99.9% sure that a student who makes a random guess on each question misses at least one half of the questions?

4.65 It is stated that 64% of a certain population prefers a particular deodorant. Suppose 450 persons are randomly selected from the population and are interviewed.

(a) What is the expected number of persons who prefer the deodorant?
(b) What interval would encompass 99.74% of the possible responses—that is, the number out of 450 who prefer the product?
(c) With reference to your answer of part (b), what conclusions can you draw if only 250 prefer the deodorant?

4.66 Newspapers and TV bring us daily coverage of the presidential primaries. Information from an informal survey of registered Republican voters is shown in the following table.

	Candidate A	*Candidate B*	*Other*
Men	4	2	1
Women	2	3	1

(a) If 4 persons were chosen from the group to have an impromptu debate, what is the probability that they all would be for Candidate B?
(b) If 3 persons were setting out to campaign, what is the probability that no more than one would be a woman?

(c) What is the probability that a panel of 5 persons will be composed of 2 women favoring Candidate B, 2 men favoring Candidate A, and 1 person favoring another candidate?

(d) A random variable Y represents the number of men favoring Candidate A. If 6 persons are selected from the total group, what values would you assign to Y?

(e) Construct the distribution of the random variable Y.

(f) What are the mean and the standard deviation for the distribution of Y?

4.67 Batch-type computer projects are submitted at the in–out window of the computer center at the mean rate of 20 per hour and follow a Poisson distribution.

(a) What is the probability that more than 6 minutes pass between projects submitted?

(b) What is the probability that 10 projects arrive in a 15-minute period?

(c) What is the probability that more than 10 arrive in a 30-minute period?

(d) What is the probability that between 3 and 15 minutes elapse between the submission of projects?

(e) What is the standard deviation for the distribution of time between project arrivals?

4.68 One of the highest-grossing movies to date is *Gone with the Wind*. It is deemed that 7 out of 10 regular moviegoers (those seeing at least one movie per month) have seen the movie.

(a) In a group of 15 regular moviegoers, what is the probability that exactly 7 would not have seen the movie?

(b) What is the probability that at least 12 but less than 19 have not seen the movie, when 25 regular moviegoers are questioned?

(c) If a sample of 10 regular moviegoers is taken, what is the probability that exactly 6 have seen the movie?

(d) Draw the probability distribution for the random variable X, the number of persons that have seen *Gone with the Wind,* when $n = 5$.

(e) For the 7 who have seen the movie, 3 have seen it more than once. What is the probability that if 5 regular moviegoers were questioned, there would be 1 that had seen the movie more than once, 2 that had seen it once, and 2 that had not seen it?

4.69 Cases of lettuce containing 40 heads are known to have an average of 94% good heads. What is the probability that the next case will contain four or more bad heads?

4.70 The union of the XYZ Corporation is about to select a negotiation team of four to meet with management regarding a new contract. Information about the eligible persons is given in the following table.

	Prior Experience	*No Experience*
Male	5	1
Female	2	3

(a) How many different union negotiating teams can there be?

(b) How many negotiating teams containing only those with prior experience can there be?

(c) What is the probability that the negotiating teams with prior experience contain only men?

(d) What is the probability of the negotiating team having two men and two women?
(e) What is the probability that the team contains no more than two persons with no prior experience?

4.71 An automatic bottling machine does not properly seal 2% of the bottles. What is the probability that in the next 5 cases (24 bottles per case) 3 or more bottles will not be sealed properly? What is the probability that in the next 20 cases, there will be **(a)** more than 10 bottles not sealed properly and **(b)** less than 4?

4.72 A distribution center completes orders through a manual picking process. The average time to complete an order is 8.28 minutes with a standard deviation of 3.12 minutes. The distribution of times is deemed to be normal.
(a) Does it seem likely that it would take more than 15 minutes to pick a complete order? Why or why not?
(b) If the company processed 2000 orders a day, how many would be expected to take between 6 to 12 minutes to complete?
(c) Orders that take less than 3 minutes to pick generally have some items back ordered. What percentage of all orders would be incomplete?
(d) What are the maximum and minimum times to pick an order and be within two standard deviations of the mean?

4.73 It has been observed that an average of 3.2 telephone calls are received by an information operator in a 4-minute period.
(a) What is the probability that 6 calls would arrive in a 4-minute period?
(b) What is the probability that at least 5 but less than 11 calls would arrive in an 8-minute period?
(c) What is the probability that more than 3 minutes would elapse between information calls?
(d) What is the probability that at least 4 minutes but less than 12 minutes would elapse between incoming information calls?
(e) What is the average time between incoming calls?

4.74 The age of the depositor varies somewhat by the location of a bank's branches. At two such locations, the probability of being a depositor over 60 years of age was .3 and .4, respectively. If five customers are observed at each bank, what is the probability that two at each location would be under 60 years of age?

4.75 The purchasing department of Ardmore Company has dealt with several suppliers, as indicated in the following table.

	Supplier Location		
Supplier Status	***Local***	***Out of Town***	***Foreign***
Prior	4	2	2
New	3	2	2

(a) If 4 orders are awarded randomly this month, what is the probability that at least 3 will go to prior suppliers?
(b) If 6 orders are awarded randomly, what is the probability that there would be 2 local, 2 out of town, and 2 foreign suppliers?
(c) What is the average number of orders awarded to prior suppliers when 5 orders are awarded randomly?

4.76 A gas station attendant has found that 4 out of 10 customers desire to have their oil checked when they get gas.

(a) What is the probability that of the next 15 customers, only 4 will desire to have the oil checked? more than four?

(b) If the attendant services 20 customers in the next hour, what is the probability that at least 6 but no more than 12 would ask to have their oil checked?

(c) It is also known that 2 of those customers who have their oil checked purchase oil. The average purchase per customer is 1.2 quarts. If oil costs 50¢ a quart, how much money can be expected from the sale of oil when 200 customers stop for gas during the day?

4.77 At a grocery store, it has been observed that 40% of the customers have bottles to return for credit.

(a) If 10 persons enter the store, what is the probability that 6 of them return bottles?

(b) If 15 persons enter the store, what is the probability that more than 4 but less than 9 return bottles?

(c) What is the average number of customers returning bottles in one day if 228 customers come to the store? What is the standard deviation?

(d) For every 4 out of 10 customers that return bottles, 2 of them return quart bottles and 2 return the small size. If 6 people enter the store, what is the probability that 2 have quart bottles to return, 3 have small bottles, and one has no bottles?

(e) If 58 persons come to the store on any given morning, what is the probability that 30 have bottles to return? 30 or more?

4.78 Murdock Manufacturing produces a line of ball bearings. Sixty percent of their production centers on a ball bearing that averages .27 inch in diameter with a standard deviation of .02 inch.

(a) The machines producing ball bearings will be stopped and adjusted when the diameter exceeds .311 inch. What is the probability that a machine will have to be adjusted?

(b) Each ball bearing is inspected with a go–no go gauge, and only those that fall between .235 and .307 inches will be passed. In a run of 100,000, how many could be expected to be available for sale?

(c) To what extent must the lower limit—as indicated in (b)—be adjusted if a decision is made to scrap the lowest 1.50%?

4.79 The probability that a husband will remain with his wife through delivery at childbirth is .4. If a hospital has 60 deliveries in one week, what is the probability that for 35 or more of the deliveries, a woman's husband would be present?

4.80 Because doctors tend to maintain a closer watch on mothers during pregnancy, the average weight of babies today is 7 pounds 2 ounces, with a standard deviation of 12 ounces. Birth weights are assumed to be normally distributed.

(a) What is the probability that a baby will weigh over seven pounds?

(b) What is the probability a baby will weigh at least 58 ounces but less than 108 ounces?

(c) Of all the babies, 22% need additional care and nurturing in the hospital because of low weight. What is the maximum a baby could weigh and be in this low-weight group?

4.81 In a lounge, 16 persons are seated together and it is determined that 11 hold driver's licenses. Five persons are from out of state, and it is known that 3 of these people have driver's licenses.

(a) What is the probability that a person has a license and resides out of state?

(b) What is the probability that a person does not have a license or is not from out of state?

(c) If it is known that a person is from out of state, what is the probability that the person does not have a license?

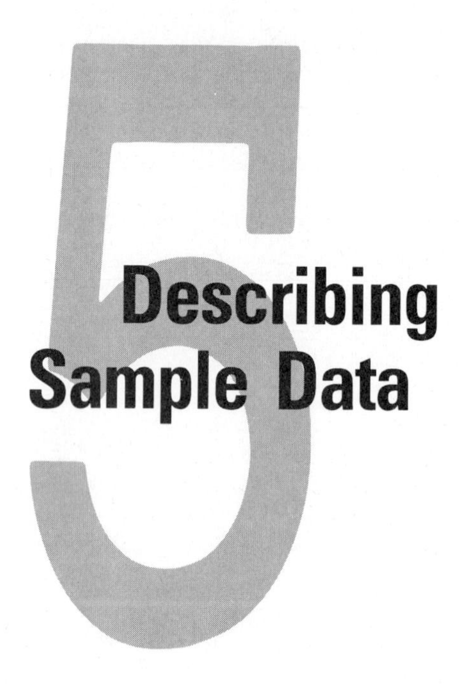

5 Describing Sample Data

5.1 INTRODUCTION

Harry Hotspur, an internal auditor for Carlyle Industries, has selected a sample of 100 accounts from a total of 5000. To date, he has audited all of the accounts in the sample and has recorded the magnitude of the error in each account. Harry realizes that if he were to present the data to John Carlson, his supervisor, in their present form, they will appear as a mishmash of numbers and convey little relevant information. Thus, he must decide how to organize his findings for presentation and what numerical characteristics to calculate in order to help describe the data.

The questions facing Harry are some of the ones to be addressed in this chapter. That is, we consider the situation where a set of data (called a sample) is available, and it is desired both to summarize the results for ease in presentation and to describe the data via certain numerical characteristics.

In previous chapters, we have discussed basic probability, random variables, and probability distributions. In all cases, it was assumed that we were dealing with a total population and that its distribution was completely specified. A probability distribution models the population in the sense that it identifies the frequency with which values or sets of values can be expected to occur. That is, it describes the chance variation within the population. Recall the definition of a population: A *population* is defined to be the total collection or set of items of interest.

The probability distribution for a population is said to be completely specified when the family of the distribution and the numerical values for the parameters of that family are known. In general, even though the population is defined and the family of the distribution is known or able to be approximated, the values for the parameters will be unknown. The major emphasis of statistics is to attempt to determine the best possible values for the parameters and, in some cases, even to attempt to determine the family itself. To determine the exact values of the parameters would require inspection of every item in the population. Most populations of interest are extremely large—some are even infinite. For these populations, it would be far too time consuming and costly to inspect every item. In other instances, the population may be finite and relatively small, but the inspection or testing of the items may require destroying them. For example, consider a manufacturer of missiles for the government. To test the missiles for reliability and accuracy, they must be fired on a test range. Not only is testing costly, but by the time this information is determined, no missiles remain to be sold. Thus, we can see that one is generally restricted to using only a small part of the population when trying to determine values for unknown parameters.

From samples, or small parts of the population, we must attempt to form an idea of the type of probability distribution that governs the random variation in the population. As we will see in this chapter, estimating the shape of the probability distribution entails making graphical representations of the sample data. Estimating the location and dispersion of the population requires calculating certain descriptive characteristics. In later chapters, we make probability statements about the values of these characteristics and then reach decisions based upon the magnitudes of those probabilities.

5.2 TYPES OF SAMPLES

DEFINITION 5.1 Any portion or subset of a population is called a **sample.**

A sample can be selected in a variety of ways. For example, if we wished to ascertain opinion on some political issue, we might poll all of our friends regarding their opinions. But we should not attempt to make any inferences about the population in general from this sample since our friends will quite often have political views similar to our own, which may or may not represent the rest of the voters. Thus, not all types of samples enable one to make inferences. For inferential purposes, statistics requires a particular type of sample, known as a random sample.

DEFINITION 5.2 For a *finite* population, a sample is said to be a **simple random sample** if all samples of the same size have an equal chance of occurring. If the population is of size N and the sample is of size n, then the number of possible distinct samples selected without replacement is C_n^N, and the probable occurrence of any sample is $1/C_n^N$.

DEFINITION 5.3 For an *infinite* population or for a continuous distribution, a sample is said to be a **simple random sample** if each item selected is governed by the same probability function. Note that sampling with replacement from a finite population is equivalent to having an infinite population.

In the majority of our applications, the populations are either infinite or of such magnitude that the sample size relative to the population size is quite small. Definition 5.3 ensures that the sample observations will be independent of each other, and hence, it greatly simplifies the calculation of probabilities, since conditional ramifications will not be a concern.

Definitions 5.2 and 5.3 describe a technique generally known as *simple random sampling*. Other methods of statistical sampling use randomness in varying degrees. Two such techniques are stratified random sampling and cluster sampling.

DEFINITION 5.4 If a population is defined as a group of mutually exclusive classes or strata and a simple random sample is selected from each stratum, then the procedure is called **stratified random sampling.**

DEFINITION 5.5 If a population consists of a large number of mutually exclusive groups or clusters, each comprising relatively few items, and a simple random sample of clusters is selected with each cluster being sampled or totally inspected, then the procedure is called **cluster sampling.**

Sampling is generally used whenever one wishes to estimate a parameter. Simple random sampling admits the possibility of obtaining an extreme or nonrepresentative sample—for example, a set of values from one end of the population. Stratified sampling guarantees that all segments of the population are represented in the sample. Stratified sampling attempts to ensure that the sample obtained will be representative of the population; it also attempts to reduce the variance of an estimator. It should be used when the population has well-defined strata and small variability within strata and wide variability between strata.

Cluster sampling, on the other hand, attempts to reduce the costs of sampling. It should be used if the variability among clusters is small but the costs of sampling between clusters are large. In such cases, cluster sampling may reduce travel, time, and ultimately cost, as contrasted to simple random sampling. Although simple random sampling does not guarantee that the sample will be representative, it enables one to calculate the probability or likelihood of occurrence of an extreme sample. That ability to calculate the degree of extremity of the sample in terms of the probability of its occurrence is the reason that simple random samples are used. In the remainder of the text, we assume that all samples are simple random samples.

The following represent some examples of situations in which stratified and cluster sampling procedures might be used. The personnel director for a large company desires to estimate the total number of employee-hours lost due to accidents among all employees. She knows that laborers, technicians, and administrative/office personnel have different accident rates. To ensure that all three categories or strata are represented, she considers stratified sampling. As a second example of stratified sampling, consider a forester who wants to estimate the total number of farm-acres planted in trees for a state. Knowing that the number of acres of trees varies considerably with the size of the farm, he decides to stratify on farm sizes.

An industry is considering a revision of its retirement program and would like to estimate the proportion of employees who favor the revision. The industry has 76 separate plants scattered throughout the United States. The estimate is needed soon and with a minimum of cost. Thus, it is decided to use cluster sampling with each plant as a cluster. As another example, consider a horticulturist who has 100 cartons of iris bulbs in inventory, with each carton containing 25 bulbs. These bulbs have been in inventory for some time, and she would like to sample a total of 100 bulbs to determine if they are still salable. Since there is a relatively high cost for opening cartons, she decides to use cluster sampling with each carton representing a cluster.

The selection of a random sample can be accomplished in many ways. A random device, such as a roulette wheel or a die, or drawing from a container may be used. For reasonably large finite populations, the more common procedure is to number the items in the population and to use a random-number table to select the items for the sample.

A random-number table represents values selected at random from a standard uniform distribution—that is, a uniform distribution on the interval [0,1]. This distribution has the property that all numbers of the same length (number of digits) are equally likely. Using a random-number table ensures that all members of the population have the same chance of being included in the sample.

To use a random-number table, we simply start reading at some point in the table and move in any direction from that point, using the number of digits required by the population. As each random number is read, the item corresponding to that number becomes a member of the sample. If a number repeats, it is ignored and the next number in line is taken, unless the sampling is to be with replacement.

Appendix Table B.10 is a random-number table. Table 5.1 is a segment of Appendix Table B.10, and the highlighted values represent the numbers selected in Example 5.1.

EXAMPLE 5.1 Suppose a company has 1000 hourly employees and wishes to select a sample of 20 employees to ask them for their opinions on a new pension plan.

TABLE 5.1 A portion of a random-number table

↓				
84920	57170	68330	37732	57290
48442	01758	21745	31119	08581
79069	80967	70163	94036	42167
78513	08625	53525	28019	03857
63714	64842	56131	57792	35123
44744	39391	72415	55232	21415
65297	84173	38901	31790	47555
66659	73449	16043	58331	82802
44719	74443	15337	67382	30599
82389	69488	95381	96182	02133
25934	92164	36703	10690	63481
21815	49139	76886	48347	78130
97143	93845	65103	48210	58179
37390	43778	50391	14937	76870
09275	94921	70852	34208	63638
95802	24480	32672	16598	27824
56393	39094	20449	48776	70064
25680	92291	57582	70899	18986
30560	16394	93477	20148	04431
21116	↑ 33765	48908	89481	32937

Solution: To select the sample, the personnel manager first numbers the employees as follows: 000, 001, 002, 003, . . . , 998, 999. Then going to the section of the random-number table presented in Table 5.1 and using three-digit numbers, the manager starts in the upper left corner and moves in the direction of the arrows to obtain the following sample:

849	484	790	785	637	447	652	666	823	259
218	971	373	092	958	563	256	305	211	337

These numbers are highlighted in the table. Note that the second 447 is a repeat and is ignored, with the next number in line taking its place. The 20 employees corresponding to these numbers are the ones to be surveyed. ■

Precaution should be observed when using a random-number table. We should not always start in the same place nor always move in the same direction from the starting point. To do so would result in the selection of the same sample every time and a loss of randomness. For a continuous population, we randomly select the times or points at which observations are to be taken. For example, suppose that a process control procedure requires that a sample of five consecutive items be selected and checked at a randomly selected time once each day. If the process operates 24 hours per day, the requirement can be met by dividing the day into 48 half-hour segments and then using a random number table to select a number between 00 and 47. Then, the inspector selects the items some time during the corresponding time period for that day.

EXERCISES

5.1 Give four examples of situations in business in which samples would be appropriate and identify the population of interest for each.

5.2 Give an example for which the collection of the sample data would involve destruction of the items.

5.3 Give an example **(a)** in which stratified sampling would be appropriate, and **(b)** in which cluster sampling would be appropriate.

5.3 PRESENTING SAMPLE DATA

Organizing the Data

Once the sample has been selected, what do we wish to do with the resulting data? The purpose of the sample is to attempt to learn something about the distribution of the population from which the sample was selected. Initially, we may desire to present or organize the data. If the sample is small, say 25 or fewer, the data may be put into an array.

DEFINITION 5.6 An **array** is a listing of the sample values in numerical order from smallest to largest. That is, an array would be $x_{(1)}, x_{(2)}, x_{(3)}, \ldots, x_{(n-1)}, x_{(n)}$, where the enclosed number is the position of the observation in the numerical ordering and n is the sample size. Hence, $x_{(1)}$ is the smallest number, and $x_{(n)}$, the largest number in the sample.

EXAMPLE 5.2 Suppose a sample of 6 students had been selected and their grade point averages (GPAs) recorded as $x_1 = 2.94$, $x_2 = 2.26$, $x_3 = 3.17$, $x_4 = 1.92$, $x_5 = 2.58$, $x_6 = 2.67$. The subscripts indicate the order of observation. Form an array.

Solution: An array for this sample is

$$x_{(1)} = 1.92 \qquad x_{(2)} = 2.26 \qquad x_{(3)} = 2.58$$
$$x_{(4)} = 2.67 \qquad x_{(5)} = 2.94 \qquad x_{(6)} = 3.17$$

The subscripts now indicate the position in the numerical ordering, with x_4 becoming $x_{(1)}$, x_2 becoming $x_{(2)}$, x_5 becoming $x_{(3)}$, and so on.

When larger samples have been selected, it is generally too cumbersome to present either the actual observations or their array. In such cases, two possible alternatives are stem-and-leaf plots and frequency tables. A stem-and-leaf plot provides an alternative to the array for larger data sets.

TABLE 5.2 Stem-and-leaf plot for Example 5.3

Stem	Leaves											
0	.33	.49	.88	.92	.92	.95						
1	.06	.07	.09	.15	.20	.24	.30	.30	.36	.50	.79	.88
2	.01	.19	.42	.98								
3	.33	.58	.71	.76	.92	.98						
4	.27	.33	.40	.46	.64	.94	.98					
5	.12	.16	.50	.66								
6	.18	.54	.56	.81	.98							
7	.86	.88										
8	.15	.23	.29	.59								

DEFINITION 5.7 A **stem-and-leaf plot** is an ordered listing of the individual values in a bar-graph design, with the last r digits representing leaves and all preceding digits identifying the stems.

EXAMPLE 5.3 Construct a stem-and-leaf plot for the Penny Ante data in Table 5.3.

Solution: For the stem-and-leaf plot, we let $r = 2$. That is, the last two digits, representing cents, form the leaves, and the dollar digits are the stems, as shown in Table 5.2.

The most common means for the presentation of large data sets is the frequency table.

DEFINITION 5.8 A table identifying mutually exclusive classes and the corresponding number of observations in the sample belonging to each class is called a **frequency table.** The number of observations in the class is known as the **class frequency.**

The classes in a frequency table are generally identified numerically either by class limits or by class boundaries. There is a rather subtle but important distinction between the two concepts.

DEFINITION 5.9 The minimum and maximum stated data values of a class are called **class limits** and may actually have occurred in the sample. The theoretical endpoints, or dividing lines between two consecutive classes, are called **class boundaries** and may or may not coincide with observed values. The distance between the upper and lower boundaries is called the *class width* or *interval.*

The class limits usually are dictated by the accuracy of the measuring instrument. The following example illustrates the difference between the two concepts.

TABLE 5.3 Sales amounts in dollars for 50 randomly selected cash register receipts from Penny Ante

0.92	6.56	5.12	8.59	7.88
1.79	8.15	0.88	4.98	8.23
1.36	1.24	3.33	1.20	0.33
3.58	5.50	7.86	6.81	6.98
0.49	0.92	1.50	1.09	3.76
8.29	4.40	1.88	1.06	5.66
6.54	4.33	4.27	2.42	1.15
2.98	1.07	2.01	0.95	1.30
1.30	3.92	4.64	6.18	4.46
2.19	3.71	5.16	3.98	4.94

EXAMPLE 5.4 Penny Ante, a local novelty store, has randomly selected 50 cash register receipts in an attempt to determine the distribution of the size of a customer's purchase. The 50 observations are given in Table 5.3. Use these data to construct a frequency table.

Solution: In constructing the frequency table for the data in Table 5.3, we use 6 classes with a common class interval of 1.50. In the frequency table, the class is defined by the class limits—that is, possible values to designate the classes. We use a lower limit of 0.01 for the first class, and the table is given in Table 5.4.

For this frequency table, the class limits are observable values, since sales themselves are recorded to the nearest cent. The class boundaries, on the other hand, are nonobservable values, being the theoretical endpoints and occurring halfway between the upper limit of one class and the lower limit of the next class. For our table, the boundaries are 0.005, 1.505, 3.005, 4.505, 6.005, 7.505, 9.005.

As is seen, the frequency table is a neat and concise way of presenting the sample with minimal loss of information. Note that the class boundaries are contiguous, whereas the class limits need not be.

TABLE 5.4 Frequency table for Penny Ante sales

Class	*Class Limits*	*Frequency*
1	0.01–1.50	16
2	1.51–3.00	6
3	3.01–4.50	10
4	4.51–6.00	7
5	6.01–7.50	5
6	7.51–9.00	6

TABLE 5.5 Frequency table for Penny Ante sales, showing class marks, cumulative frequencies, and relative frequencies

(1) Class	*(2)* B_i Class Boundaries	*(3)* f_i Frequency	*(4)* x_i Mark	*(5)* $F(B_i)$ Cumulative Frequency	*(6)* Relative Frequency
1	0.005–1.505	16	0.755	0	.32
2	1.505–3.005	6	2.255	16	.12
3	3.005–4.505	10	3.755	22	.20
4	4.505–6.005	7	5.255	32	.14
5	6.005–7.505	5	6.755	39	.10
6	7.505–9.005	6	8.255	44	.12
		$n = 50$		$F(9.005) = 50 = n$	1.00

NOTE: The entries in column (5) are the sums of the f_i's at the lower boundary.

There are no definite rules governing the number of classes to be used; only guidelines have been suggested by statisticians. One such guideline is to let k, the number of classes, equal approximately the square root of n, the sample size, for samples of 25 or more. Basically, the only rule is that of common sense. If too many classes are used, little is gained over presenting the actual sample data, but if too few are used, too much of the information in the data is lost. This loss of information regarding the individual values may result in two difficulties. First, the grouped data may no longer be representative of the population. Second, numerical characteristics calculated from the table may differ significantly from those calculated with the ungrouped data. If possible, for simplicity, the class intervals should be the same for all classes. Keeping the same intervals is not always possible, but it is imperative that the definition of the classes cover all of the sample values.

Table 5.4 is the basic form of a frequency table. Although the frequencies may be expressed in other formats, there are some general features illustrated by this table. One of these features is the class mark, which is the midpoint of the class or the average of the upper and lower class boundaries. The midpoint is denoted by X_i—that is,

$$X_i = \frac{B_i + B_{i+1}}{2} \tag{5.1}$$

where B_i, B_{i+1} are lower and upper boundaries of the ith class.

Column 4 in Table 5.5 identifies the class marks for the Penny Ante example. The actual usefulness of the class mark becomes apparent later when we use the frequency table for making calculations.

DEFINITION 5.10 The **cumulative frequency** at a point x, denoted $F(x)$, is defined to be the total number of observations less than x—that is,

$$F(x) = \text{number of observations less than } x \tag{5.2}$$

We define $F(x)$ as the less than cumulative frequency, since it is defined at the class boundaries in a frequency table, and the number of values that equal a boundary are indeterminate from the table. Thus, for a frequency table, we have

$$F(B_i) = \sum_{j=1}^{i-1} f_j \tag{5.3}$$

where: B_i = the lower boundary of the ith class

f_j = the frequency of the jth class

The cumulative frequencies for the table in Example 5.4 are shown in column 5 of Table 5.5. One other important form of frequency is the relative frequency.

DEFINITION 5.11 The **relative frequency** of the ith class is defined to be the frequency of the ith class divided by the sample size—that is,

$$i\text{th relative frequency} = \frac{f_i}{n} \tag{5.4}$$

where: $n = \sum_{i=1}^{k} f_i$

k = the number of classes

EXAMPLE 5.5 Using the data from Example 5.4, we want to construct a frequency table showing class marks, cumulative frequencies at the class boundaries, and relative frequencies.

Solution: See columns (4), (5), and (6), respectively, in Table 5.5.

Graphical Representations of the Data

Another method of presenting sample data that is useful in determining an approximate shape for the population distribution is a graphical device known as a histogram. The histograms introduced in Chapter 3 for probability distributions had areas corresponding to probabilities. The changes involved with using sample data are given in the following definition.

DEFINITION 5.12 A **histogram** is a vertical bar graph constructed such that the bases of the rectangles correspond to the class boundaries, and the heights of the rectangles are such that the area inscribed in each is proportional to the corresponding relative frequency.

EXAMPLE 5.6 To illustrate this concept, we construct the histogram for Table 5.5.

Solution: For the histogram, we need the class boundaries B_i and frequencies f_i from the table. The histogram appears in Figure 5.1. Note that the rectangles are contiguous and that the areas inscribed in each rectangle have a proportion of the total area equal to the corresponding relative frequency. That is, the first rectangle has 32% of the total area, the second 12%, the third 20%, and so on. ■

The proportional area phrase in Definition 5.12 implies that if one class is twice as wide as the others, then the height of the rectangle corresponding to it must be decreased accordingly to maintain the proportionality. If one does not adjust the height to reflect the differences in the bases of the rectangles, the histogram becomes misleading.

Two other graphical displays that are helpful for describing sample data are the frequency polygon and the ogive or cumulative frequency graph.

DEFINITION 5.13 A **frequency polygon** is a line graph constructed by plotting the class frequencies versus the class marks and then connecting the points with straight lines. The polygon is completed by including classes of zero frequency at each end of the table. ■

These classes should have widths equal to those of the other classes, with their midpoints being used to close off the polygon. The frequency polygon is illustrated in Figure 5.2, using the Penny Ante sales data.

DEFINITION 5.14 An **ogive, or cumulative frequency graph,** is a nondecreasing line graph constructed by plotting points (x_i, y_i), where x_i is the ith class boundary and y_i is the cumulative frequency at that boundary. The points (x_i, y_i) are then connected by straight lines, assuming that the data are uniformly dispersed within each class. ■

Note that for any $x < B_1$, $F(x) = 0$ and for any $x > B_{k+1}$, $F(x) = n$. We again use the Penny Ante sales data with the ogive given in Figure 5.3.

An important property of the ogive is that it can be used to obtain the sample percentiles and to determine the percentage of the sample that lies below any value x. For example, if we wished to estimate the value having 20% of the sales receipts below it, we project horizontally from 10 (since $\frac{10}{50} = 0.20$) until the ogive is intersected, then drop a vertical line from the intersection to the x-axis, and read off the value. In this case, it appears to be about 1.00. To estimate the percentage falling below a specific value, we simply reverse the procedure. For example, if $x = 5.00$, we erect a vertical line at 5.00, intersecting the ogive, and then project a horizontal line from the intersection back to the $F(x)$-axis. In Figure 5.3, we obtain a value of approximately 35, which corresponds to 70%. Note that we could plot the relative cumulative frequencies rather than the f_i. Doing so would enable us to read the percents and percentiles directly from the graph.

FIGURE 5.1 Histogram for the Penny Ante sales data

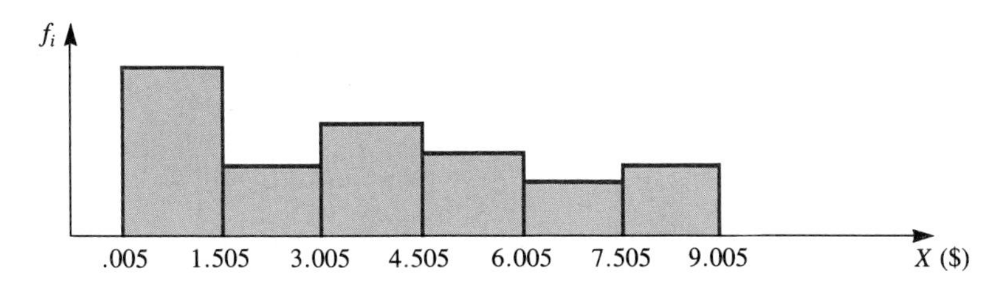

FIGURE 5.2 Frequency polygon for the Penny Ante sales data superimposed over the histogram

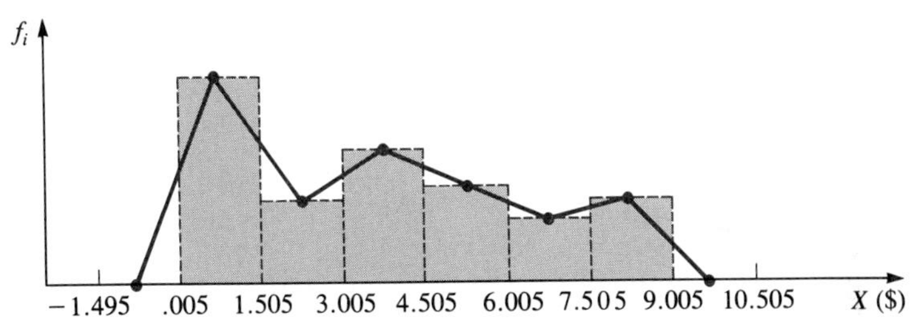

FIGURE 5.3 Ogive for the Penny Ante sales data

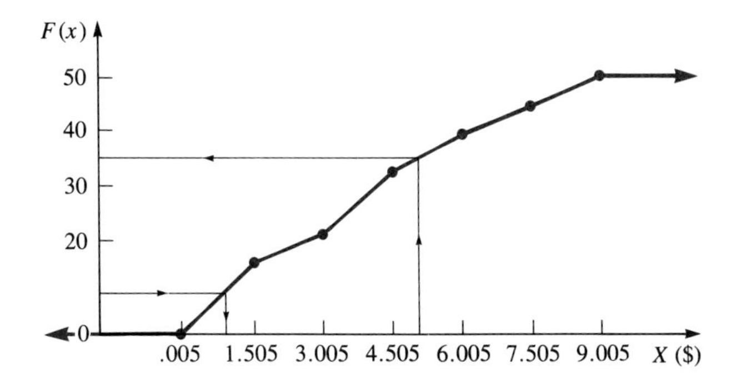

EXERCISES

5.4 Desiring to study the prices of hardbound textbooks at Unicorn University, the student government randomly selected 48 titles from the card file of the bookstore and recorded the listed price for each. The results were as follows.

19.95	22.50	28.63	17.25	16.95	18.35	25.85	23.95
15.49	24.55	27.21	22.95	37.50	21.30	22.95	23.50
22.50	21.95	24.20	23.95	26.10	26.55	27.45	23.95
22.50	21.95	19.80	20.50	17.95	16.25	16.00	22.00
25.05	19.90	18.70	22.50	23.95	24.00	18.95	19.90
31.00	21.50	23.95	16.95	18.50	17.25	33.50	22.95

Construct a frequency table for these data, showing class boundaries, class marks, class frequencies, cumulative frequencies at the boundaries, and relative frequencies.

5.5 Refer to Exercise 5.4 and sketch the histogram, frequency polygon, and ogive corresponding to your table.

5.6 Refer to your graphs in Exercise 5.5.

(a) Estimate the percentage of books having prices below \$25.50.

(b) Estimate the price below which 90% of the textbooks fall.

(c) Estimate the percentage of books having prices between \$20.00 and \$30.00.

5.7 Use the following data, which represent annual salaries in \$1000 of 10 junior executives selected at random from the Custom Car Company, to construct an array and a stem-and-leaf plot.

25.5	23.8	26.7	34.2	31.6	29.7	28.4	26.7	30.1	27.3

5.4 SUMMARIZING SAMPLE DATA

After the data are collected and organized for presentation, it is desirable to extract from the data numerical values that help to identify or describe the sample. In describing sample data, we use measures corresponding to those used in Chapter 3 for populations, as well as some that are more easily determined. An important distinction is that the parameters or values pertaining to the population are constants and hence do not vary within the population. However, sample values or statistics, as we shall call them, are variables and will have different values for different samples even though the samples are from the same population. If the samples are obtained randomly, then the statistics will be random variables.

DEFINITION 5.15 A **statistic** is a number obtained from a sample. ■

To illustrate the difference between parameters and statistics, let us consider persons passing through a turnstile. If it is known that there are equal numbers of males and females, the probability that a randomly selected person is male, π, is .5 and is constant for all selections—that is, it is a parameter. If the number of males and females is unknown, π is still a parameter and constant for all selections, but it has an unknown value. To obtain an estimate of π, we might observe 10 persons and find 4 males. Thus, our estimate of π would be .4, which, based on a sample of 10 observations, is a statistic. To note the variable nature of the statistic, suppose 10 more persons are observed and 7 are males. The parameter π is the same, since it is the same turnstile, but the value of the estimate or statistic based on this sample is .7.

Measures of Location

Many measures can be used to describe sample data; however, we restrict ourselves to the most common ones. We first consider those used to identify the location or center of the data.

DEFINITION 5.16 The **median** of a sample is any number such that no more than 50% of the values are greater than it and no more than 50% are less than it. In practice, it is the middle value in the array and will be denoted $X_{.5}$, where .5 represents the proportion of the data falling below the median.

For consistency, we will use the following equations to calculate the median from ungrouped data. If n is odd,

$$X_{.5} = X_{((n+1)/2)} \tag{5.5}$$

where $((n + 1)/2)$ indicates the position of the value in the array—that is, after the data have been ranked from low to high. If n is even,

$$X_{.5} = \frac{X_{(n/2)} + X_{((n/2)+1)}}{2} \tag{5.6}$$

where $(n/2)$ and $((n/2) + 1)$ also identify positions in the array. Verbally, Equation 5.5 indicates the middle value in the array, and Equation 5.6 indicates the average of the two middle values. For either equation, the data must be arranged in ascending order first. Note that the data can be arranged in descending order without affecting the final result.

EXAMPLE 5.7 Using the data in Example 5.2, we calculate the median GPA for the 6 students in the sample.

Solution: The array was given as 1.92, 2.26, 2.58, 2.67, 2.94, 3.17. Since n is even, Equation 5.6 applies. Thus,

$$x_{.5} = \frac{x_{(3)} + x_{(3+1)}}{2} = \frac{x_{(3)} + x_{(4)}}{2}$$

$$= \frac{2.58 + 2.67}{2} = 2.625$$

Theoretically, any number between 2.58 and 2.67 inclusive will satisfy Definition 5.16, but, for consistency, we use Equations 5.5 and 5.6.

If a seventh student having a GPA of 2.83 is added to the sample, the array becomes 1.92, 2.26, 2.58, 2.67, 2.83, 2.94, 3.17, and $n = 7$. The median is now given by Equation 5.5.

$$x_{.5} = x_{((7+1)/2)} = x_{(4)} = 2.67$$

The procedure for calculating the median must be altered whenever the data have been grouped, since the individual values are not available. First, we define the median class as that class that has the largest cumulative frequency less than or equal to $.5n = n/2$. This class contains the median value, and using information with respect to this class, we perform a linear interpolation according to the following equation. For a frequency table, the equation is

$$X_{.5} = B_{.5} + \left[C_{.5}\left(\frac{.5n - F(B_{.5})}{f_{.5}}\right)\right] \tag{5.7a}$$

where: $B_{.5}$ = the lower boundary of the median class

$C_{.5}$ = the width of the median class

$f_{.5}$ = the frequency of the median class

$F(B_{.5})$ = the cumulative frequency at $B_{.5}$

The median is also known as the *50th percentile*; thus, Equation 5.7a can be generalized to include any percentile X_p by substituting p for .5 in the equation. Most students are familiar with the idea of a percentile, because college entrance exam scores generally indicate the percentile to which the score corresponds. Thus, a score on the mathematics portion that represents the 90th percentile implies that 90% of all students taking that portion of the examination have a lower score.

DEFINITION 5.17 The ***p*th percentile** is defined to be a number such that no more than $p\%$ of the values are less than it and no more than $(1 - p)\%$ of the values are greater than it. (The formula for finding the pth percentile in a frequency table is given in Equation 5.7b.)

For determining the pth percentile from a frequency table, we must first identify the class containing it. We determine the number of observations that must be accumulated by the time X_p is reached, by calculating np. The class in question is then that class having the largest cumulative frequency at its lower boundary that is less than or equal to np. The rationale for this is simple. At that boundary, we are below X_p, but, upon reaching the next boundary, we will have gone past X_p because the cumulative frequency will exceed the required number of observations. Thus, Equation 5.7a is altered to the general form

$$X_p = B_p + \left[C_p\left(\frac{np - F(B_p)}{f_p}\right)\right] \tag{5.7b}$$

where: B_p = the lower boundary of the class containing the pth percentile
C_p = the width of that class
f_p = the frequency of that class
$F(B_p)$ = the cumulative frequency at B_p

For ungrouped data, we linearly interpolate between values in the array. We illustrate Equations 5.7a and 5.7b and Definition 5.17 in the following example.

EXAMPLE 5.8 Using Table 5.5 (page 156), we determine both the median sale and the 80th percentile—that is, the dollar values such that 50% and 80%, respectively, of all sales will be below them.

Solution: To utilize Equation 5.7a, we must first identify the median class. Calculating $.5n$ yields a value of $25 = .5(50)$, and we compare the cumulative frequencies to this value. Doing so, we find that 22 is the largest number less than or equal to 25, and it implies that we need 3 more observations to reach the median. Hence, the median class is the third one. Identifying the values required by Equation 5.7a, we obtain $B_{.5} = 3.005$, $F(B_{.5}) = 22$, $C_{.5} = 1.500$, and $f_{.5} = 10$. Substituting these values in Equation 5.7a gives

$$x_{.5} = 3.005 + \left[1.500\left(\frac{.5(50) - 22}{10}\right)\right]$$

$$= 3.005 + \left[1.500\left(\frac{25 - 22}{10}\right)\right] = 3.455$$

For $x_{.8}$, we first determine the class containing the 80th percentile. This class is that which has the largest cumulative frequency less than or equal to $.8n = .8(50) = 40$.

The fifth class has a cumulative frequency of 39, which satisfies our criterion and implies that we need one more value to reach the 80th percentile. In this case, the numbers required by Equation 5.7b become $B_{.8} = 6.005$, $F(B_{.8}) = 39$, $C_{.8} = 1.500$, and $f_{.8} = 5$. Substituting into Equation 5.7b, we obtain, for the 80th percentile,

$$x_{.8} = 6.005 + \left[1.500\left(\frac{.8(50) - 39}{5}\right)\right]$$

$$= 6.005 + 1.500(.2) = 6.305$$

These values are plotted on the histogram in Figure 5.4.

Another statistic that is occasionally used as a measure of location is the mode.

DEFINITION 5.18 A **mode** of a set of data is the value that occurs most often. If all values occur with the same frequency, then the data are said to have no mode. Note that the mode may not exist or may not be unique for some data.

FIGURE 5.4 Histogram of Penny Ante data showing 50th and 80th percentiles

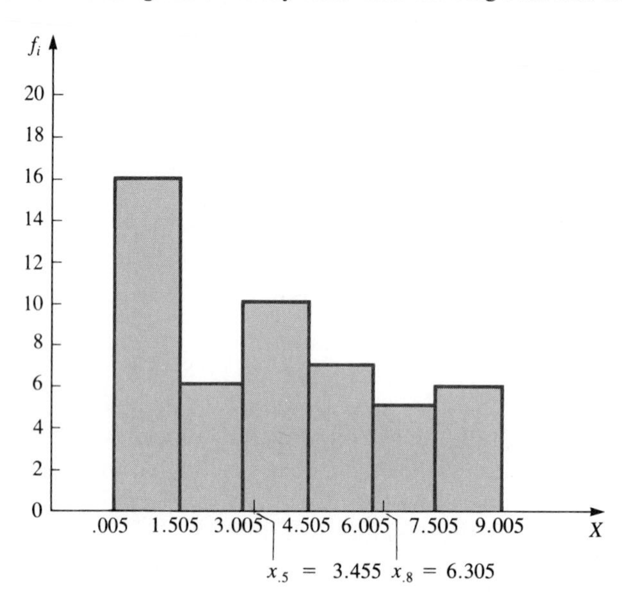

EXAMPLE 5.9 A sample of 10 students was selected from a statistics class, and the number of credit hours for which each is currently enrolled was recorded. The results are

8 13 12 9 9 10 12 15 9 13

Let us determine the mode.

Solution: The mode for this sample is 9 credit hours, which occurs 3 times while no other value occurs more than twice.

EXAMPLE 5.10 Use the Penny Ante sales data in Table 5.3 to illustrate the concept of non-uniqueness.

Solution: These data have two modes since both .92 and 1.30 occur twice, whereas all others occur only once.

EXAMPLE 5.11 Refer to the frequency table for the Penny Ante data in Table 5.5 and consider the mode.

Solution: The crude mode for the grouped data would be .755, which is the mark of the class with the largest frequency (also called the *modal class*).

The last statistic to be considered as a measure of location is also the most commonly used. It corresponds to the expected value or population mean presented in Chapter 3 and identifies the physical balancing point of the data.

DEFINITION 5.19 The arithmetic average of a set of sample data is called the **sample mean** and is denoted by $\overline{X}$ (pronounced X bar). For ungrouped data, it is determined by the equation

$$\overline{X} = \frac{\sum_{i=1}^{n} X_i}{n} \tag{5.8}$$

EXAMPLE 5.12 Calculate the mean GPA for the 6 students of Example 5.2.

Solution: The GPAs were $x_1 = 2.94$, $x_2 = 2.26$, $x_3 = 3.17$, $x_4 = 1.92$, $x_5 = 2.58$, $x_6 = 2.67$. Using Equation 5.8, we get

$$\begin{aligned}\overline{x} &= \frac{2.94 + 2.26 + 3.17 + 1.92 + 2.58 + 2.67}{6} \\ &= \frac{15.54}{6} = 2.59\end{aligned}$$

EXAMPLE 5.13 For the ten students in Example 5.9, calculate the average number of credit hours for which a student is enrolled.

Solution: The sample data are 8, 13, 12, 9, 9, 10, 12, 15, 9, 13. Thus, the sample mean is

$$\begin{aligned}\overline{x} &= \frac{8 + 13 + \cdot\cdot\cdot + 9 + 13}{10} = \frac{110}{10} \\ &= 11 \text{ credit hours}\end{aligned}$$

EXAMPLE 5.14 For the Penny Ante sales data in Table 5.3, determine the amount of an average purchase.

Solution: For these data, $n = 50$. Using Equation 5.8 gives

$$\overline{x} = \frac{187.84}{50} = \$3.76$$

When the data have been grouped into a frequency table, the equation for calculating the sample mean must be altered. The alteration consists of calculating a weighted average of the class marks, using the class frequencies for weights. The equation is

$$\overline{X} = \frac{\sum_{i=1}^{k} X_i f_i}{\sum_{i=1}^{k} f_i} = \frac{\sum_{i=1}^{k} X_i f_i}{n} \tag{5.9}$$

where: k = number of classes in the table

X_i = mark of the ith class

f_i = frequency of the ith class

$\sum_{i=1}^{k} f_i = n$ (the total sample size)

We mentioned earlier that in grouping data some information is lost because the individual identities are no longer available. Equation 5.9 illustrates this point because, in using this equation, we assume that the average within each class equals the class mark. Generally, this is not the case, and the sample mean calculated from a frequency table differs in value from the sample mean calculated using the ungrouped data. Even so, the mark is the best representative of the observations in the class.

EXAMPLE 5.15 From the Penny Ante sales data in Table 5.5, we want to calculate the amount of the average purchase, using Equation 5.9. The frequency table is re-created in Table 5.6, with the appropriate computations.

Solution: Substituting the computations from Table 5.6 into Equation 5.9, we obtain, from column (4),

$$\bar{x} = \frac{183.25}{50} = \$3.665 \quad \text{or} \quad \$3.67$$

TABLE 5.6 Frequency table for Penny Ante sales, showing class marks, frequencies, and their products

Class	B_i *Class Boundaries*	f_i *Frequency*	x_i *Mark*	*(4)* $x_i f_i$ *Product*
1	0.005–1.505	16	0.755	12.080
2	1.505–3.005	6	2.255	13.530
3	3.005–4.505	10	3.755	37.550
4	4.505–6.005	7	5.255	36.785
5	6.005–7.505	5	6.755	33.775
6	7.505–9.005	6	8.255	49.530
		$n = 50$		$183.250 = \sum_{i=1}^{6} x_i f_i$

Note that this value for $\overline{X}$ is slightly smaller than the corresponding value of \$3.76 obtained in Example 5.14. The difference is due to the loss of information from grouping the data.

Each of the three measures of location that we have considered has desirable and undesirable properties. We comment briefly upon these properties.

1. The median and the mode are more easily determined than the mean.
2. The median and the mode are unaffected by extreme values in the sample, whereas the mean may be affected.
3. The mean and particularly the mode are sensitive to grouping, but the median is affected only moderately.
4. For a given sample, the mean has a unique value, whereas the median, by definition, and the mode may not be unique.
5. The mean and the median will have relatively stable values, but the mode may be unstable from one sample to another.
6. The sample mean in most situations has a readily available probability distribution, but the distributions for the other two are sometimes difficult to obtain.

The last property is of prime importance in statistical inference and is the reason that the mean is preferred over all other measures of location in most situations. In some texts, the median and the mode are also considered to be averages, but we will not do so in this text.

EXERCISES

5.8 Refer to Exercise 5.7 and determine **(a)** the average salary, **(b)** the median salary, and **(c)** the modal salary.

5.9 Refer to Exercise 5.4 and calculate, from the ungrouped data, **(a)** the median price of a book, **(b)** the average price of a book, and **(c)** the modal price of a book.

5.10 Refer to Exercise 5.4 and determine **(a)** the median, **(b)** the mean, and **(c)** the 70th percentile for your frequency table. Compare your answers to (a) and (b) with the answers to Exercise 5.9(a) and (b).

5.11 A random sample of 100 individual tax returns gave the following frequency table of deductions claimed under medical expenses.

Deduction Claimed (\$)	***Number of Returns***
$0 \le X < 500$	22
$500 \le X < 1000$	36
$1000 \le X < 2000$	24
$2000 \le X < 3000$	12
$3000 \le X < 5000$	6

(a) Determine the average medical deduction for these returns.
(b) Determine the amount above which 20% of the deductions lie.

Measures of Dispersion

Not only is the central point of the data of interest, but it is imperative that we have some idea of the degree to which the values are spread or dispersed around that point, as was indicated in Chapter 3. Three of the statistics that measure dispersion will be considered: one for its computational ease, the second for its distributional properties, and the third for its practical interpretation.

DEFINITION 5.20 The **sample range,** denoted R, is defined to be the difference between the largest and smallest values, respectively, in the sample data—that is,

$$R = X_{(n)} - X_{(1)} \tag{5.10}$$

where $X_{(n)}$ and $X_{(1)}$ are the largest and smallest values, respectively.

EXAMPLE 5.16 Use the sales data in Table 5.3 to determine the range of the purchases.

Solution: We see that the smallest purchase is \$.33 and the largest \$8.59. Thus, the range, by Equation 5.10, is

$$R = 8.59 - .33 = \$8.26$$

EXAMPLE 5.17 Calculate the range in GPAs for the 6 students of Example 5.2.

Solution: The array of GPAs is 1.92, 2.26, 2.58, 2.67, 2.94, 3.17. Thus, the range is

$$R = 3.17 - 1.92 = 1.25$$

The range is frequently used in situations where computational ease is desired. Disadvantages of the range are its instability as the sample size increases, its susceptibility to extreme values, and the difficulty encountered in trying to obtain the distributions of functions of the range.

In the majority of situations, we use a measure of dispersion comparable to that defined in Chapter 3.

DEFINITION 5.21 For sample data, the **variance,** denoted by s^2, is defined to be the average squared deviation from the sample mean—that is,

$$s_X^2 = \frac{\sum_{i=1}^{n} (X_i - \overline{X})^2}{n - 1} \tag{5.11}$$

We divide by $n - 1$ in Equation 5.11 to obtain an unbiased estimate of the population variance σ_X^2. The concept of unbiasedness is explained in Chapter 7. The rationale for using s_X^2 as the principal measure of dispersion is its distributional properties. Intuitively, if μ_X is unknown, we use $\overline{X}$ as an estimator of it. Since $\overline{X}$ is a variable, there is a resultant loss of precision in estimating σ_X^2. As far as s_X^2 is concerned, the calculation of $\overline{X}$ involves using the sample information once and restricts the amount of information still available for calculating s_X^2. This restriction is necessary because only $n - 1$ *independent* pieces of information are in the numerator of the sample variance. This concept is discussed in more detail in conjunction with the concept of degrees of freedom in Chapter 6.

EXAMPLE 5.18 Calculate the variance of the GPAs for the students of Example 5.12.

Solution: The average GPA for these students was found to be 2.59 with individual values of 2.94, 2.26, 3.17, 1.92, 2.58, 2.67. Substituting these values into Equation 5.11 yields

$$\begin{aligned} s_X^2 &= \frac{(2.94 - 2.59)^2 + (2.26 - 2.59)^2 + \cdots + (2.67 - 2.59)^2}{6 - 1} \\ &= \frac{(.35)^2 + (-.33)^2 + \cdots + (.08)^2}{5} \\ &= \frac{.1225 + .1089 + \cdots + .0064}{5} \\ &= \frac{1.0232}{5} = .2046(\text{GPA})^2 \end{aligned}$$

Note that the units for the sample variance in Example 5.18 are $(\text{GPA})^2$. Like the population variance in Chapter 3, the sample variance is difficult to interpret since it is expressed in units squared. To avoid interpretive difficulties, we obtain a more practical value by taking the square root.

DEFINITION 5.22 The **sample standard deviation** is defined to be the square root of the sample variance—that is,

$$s_X = \sqrt{s_X^2} = \sqrt{\frac{\sum_{i=1}^{n} (X_i - \overline{X})^2}{n - 1}} \tag{5.12}$$

EXAMPLE 5.19 Determine the standard deviation of the GPAs in Example 5.18.

Solution: Using Equation 5.12, we obtain

$$s_X = \sqrt{.2046} = .45$$

Equation 5.11 can be expressed in various forms that simplify the computations considerably. The simplified or computational forms can be derived by utilizing the properties of summation discussed in Chapter 1. One of the computational forms is

$$s_X^2 = \frac{\sum_{i=1}^{n} X_i^2 - \frac{\left(\sum_{i=1}^{n} X_i\right)^2}{n}}{n - 1} \tag{5.13}$$

To illustrate that Equation 5.13 yields the same value as Equation 5.11, we now recompute the variance in Example 5.18, using the computational form.

EXAMPLE 5.20 The six GPAs in Example 5.18 were

2.94 2.26 3.17 1.92 2.58 2.67

We calculate the variance for these data by using Equation 5.13.

Solution: We have already calculated the sum of the values in Example 5.12. It is

$$\sum_{i=1}^{6} x_i = 15.54$$

Additional computations are required to obtain $\sum_{i=1}^{6} x_i^2$.

$$\begin{aligned}\sum_{i=1}^{6} x_i^2 &= (2.94)^2 + (2.26)^2 + \cdot\cdot\cdot + (2.67)^2 \\ &= 8.6436 + 5.1076 + 10.0489 + 3.6864 + 6.6564 + 7.1289 \\ &= 41.2718\end{aligned}$$

Substituting the above values into Equation 5.13 gives

$$\begin{aligned}s_X^2 &= \frac{41.2718 - \dfrac{(15.54)^2}{6}}{6 - 1} = \frac{41.2718 - 40.2486}{5} \\ &= \frac{1.0232}{5} = .2046\end{aligned}$$

Thus, the standard deviation is, as before,

$$s_X = \sqrt{.2046} = .45$$

As in the case of the mean, we must alter Equations 5.11 and 5.13 to calculate the variance when using data summarized in a frequency table. The alteration becomes the weighted average of the squared deviations of the class marks from the mean, with the class frequencies used as weights. Implementing these changes, we obtain

$$s_X^2 = \frac{\sum_{i=1}^{k} (X_i - \overline{X})^2 f_i}{n - 1} \qquad \textbf{(5.14)}$$

and

$$s_X^2 = \frac{\sum_{i=1}^{k} X_i^2 f_i - \frac{\left(\sum_{i=1}^{k} X_i f_i\right)^2}{n}}{n - 1} \tag{5.15}$$

where: X_i = the class mark of the ith class

k = the number of classes in the table

f_i = the frequency of the ith class

We now apply Equations 5.14 and 5.15 to the data of Example 5.15.

EXAMPLE 5.21 For the data in Example 5.15, we calculate the variance, using Equations 5.14 and 5.15. For these data, $\bar{x} = 3.665$.

Solution: Table 5.7a lists the calculations required by Equation 5.14. Substituting the results from the computations in column (5) of Table 5.7a into Equation 5.14, we obtain

$$s_X^2 = \frac{339.3450}{49} = 6.9254(\$)^2$$

Table 5.7b lists the calculations required by Equation 5.15. Substituting the results given in columns (6) and (4) of Table 5.7b yields

$$s_X^2 = \frac{1010.9550 - [(183.25)^2/50]}{50 - 1}$$

$$= \frac{1010.9550 - 671.6112}{49}$$

$$= \frac{339.3438}{49} = 6.9254(\$)^2$$

TABLE 5.7a Computations for the variance of grouped data when Equation 5.14 is

Class	f_i	x_i	$(x_i - \bar{x})^2$	(5) $(x_i - \bar{x})^2 f_i$
1	16	0.755	8.4681	135.4896
2	6	2.255	1.9881	11.9286
3	10	3.755	.0081	0.0810
4	7	5.255	2.5281	17.6967
5	5	6.755	9.5481	47.7405
6	6	8.255	21.0681	126.4086
	$n = 50$			339.3450

TABLE 5.7b Computations for the variance of grouped data when Equation 5.15 is used

Class	f_i	x_i	(4) $x_i f_i$	x_i^2	(6) $x_i^2 f_i$
1	16	0.755	12.080	0.5700	9.1200
2	6	2.255	13.530	5.0850	30.5100
3	10	3.755	37.550	14.1000	141.0000
4	7	5.255	36.785	27.6150	193.3050
5	5	6.755	33.775	45.6300	228.1500
6	6	8.255	49.530	68.1450	408.8700
			183.250		1010.9550

The slight difference in the numerator values of the two equations is due to rounding error. The standard deviation is obtained by taking the square root of either equation.

$$s_X = \sqrt{6.9254} = \$2.63$$

The sample standard deviation conveys the same information about the sample that σ_X does for the population. Specifically, it is the square root of the average squared distance of the observations from $\overline{X}$. Using Tschebycheff's inequality from Chapter 3, we know approximately that at least 75% of the sample values are within a distance of two s_X of $\overline{X}$. The result is approximate, since Tschebycheff's inequality refers to population values.

Had s_X^2 and s_X been calculated using the original or ungrouped data, the values would have been found to be 6.2983 and 2.51, respectively. The differences between these numbers and those above can be attributed to the loss of information from grouping.

EXERCISES

5.12 Refer to Exercise 5.4 and find the range, variance, and standard deviation for the ungrouped data.

5.13 Refer to Exercise 5.4 and calculate the variance and the standard deviation of the data for your frequency table.

5.14 Calculate the variance and the standard deviation of the medical deductions given in Exercise 5.11.

5.15 For the data in Exercise 5.7, calculate the range in the salaries and the variance and the standard deviation of the salaries.

5.5 CODING†

The calculations involved in determining the variance for sample data can become quite laborious. Example 5.21 illustrates this point quite well. Also, because some calculators and computers truncate digits, a significant loss of information sometimes occurs. For example, if the variability in a set of data occurs only in the eighth decimal place, a calculator may give a value of zero for s_X^2. To avoid this loss of information, we may "code" the sample values—that is, make a linear transformation on them.

DEFINITION 5.23 **Coding** consists of subtracting a constant from each value and dividing the results by another constant. That is, we define a new variable U such that

$$U_i = \frac{X_i - a}{b} \tag{5.16}$$

For a frequency table, we set the constant a equal to the mark of the median class and the constant b equal to the common class interval—that is, the difference between consecutive class boundaries. For unequal intervals, the values for a and b will probably be different.

After the class marks have been coded, the mean and variance of the new variable U are calculated from Equations 5.9 and 5.15, respectively, with U_i replacing the corresponding X_i and $\overline{U}$ replacing $\overline{X}$. The resultant values are called $\overline{U}$ and s_U^2 and must be "decoded" to obtain the original mean and variance, $\overline{X}$ and s_X^2. To calculate $\overline{X}$, we simply multiply $\overline{U}$ by b and add a. To obtain s_X^2 requires only multiplication of s_U^2 by b^2. The proofs for these relations are quite simple and instructive. For $\overline{U}$, we have

$$\begin{aligned}
\overline{U} &= \frac{\sum_{i=1}^{k} U_i f_i}{n} = \frac{\sum_{i=1}^{k} \left(\frac{X_i - a}{b}\right) f_i}{n} \\
&= \frac{\frac{1}{b}\sum_{i=1}^{k} (X_i - a) f_i}{n} = \frac{1}{b}\left[\frac{1}{n}\left(\sum_{i=1}^{k} X_i f_i - \sum_{i=1}^{k} a f_i\right)\right] \\
&= \frac{1}{b}\left[\frac{\sum_{i=1}^{k} X_i f_i}{n} - \frac{a}{n}\sum_{i=1}^{k} f_i\right] \\
&= \frac{1}{b}\left[\overline{X} - \frac{a}{n}(n)\right] = \frac{1}{b}(\overline{X} - a)
\end{aligned}$$

† This section is optional.

Solving for $\overline{X}$ gives $\overline{X} = b\overline{U} + a$. For s_U^2, we have

$$
\begin{aligned}
s_U^2 &= \frac{1}{n-1}\sum_{i=1}^{k}(U_i - \overline{U})^2 f_i \\
&= \frac{1}{n-1}\sum_{i=1}^{k}\left[\frac{X_i - a}{b} - \left(\frac{\overline{X} - a}{b}\right)\right]^2 f_i \\
&= \frac{1}{n-1}\sum_{i=1}^{k}\left\{\frac{1}{b}[X_i - a - (\overline{X} - a)]\right\}^2 f_i \\
&= \frac{1}{n-1}\sum_{i=1}^{k}\left\{\frac{1}{b}[X_i - \overline{X}]\right\}^2 f_i \\
&= \frac{1}{n-1}\sum_{i=1}^{k}\frac{1}{b^2}[X_i - \overline{X}]^2 f_i = \frac{1}{b^2}\left\{\frac{1}{n-1}\sum_{i=1}^{k}[X_i - \overline{X}]^2 f_i\right\} \\
&= \frac{1}{b^2}s_X^2
\end{aligned}
$$

Multiplying both sides of the equation by b^2 gives $s_X^2 = b^2 s_U^2$. These relations are summarized in the following equations.

$$\overline{X} = b\overline{U} + a \qquad s_X^2 = b^2 s_U^2 \tag{5.17}$$

EXAMPLE 5.22 Recalculate the mean and variance of the Penny Ante sales data using the recommended coding scheme.

Solution: We first determine a and b, the mark of the median class and the common class interval, respectively. The data, both original and coded, are given in Table 5.8. In Example 5.8, we found the median class to be 3.01–4.50, which implies the constant a is 3.755. The common interval b is 1.500. Hence, the

TABLE 5.8 Coding of Penny Ante sales data

Class Limits	f_i	x_i	$u_i = \dfrac{x_i - 3.755}{1.50}$	$u_i f_i$	$u_i^2 f_i$
0.01–1.50	16	0.755	−2	−32	64
1.51–3.00	6	2.255	−1	− 6	6
3.01–4.50	10	3.755	0	0	0
4.51–6.00	7	5.255	1	7	7
6.01–7.50	5	6.755	2	10	20
7.51–9.00	6	8.255	3	18	54
				− 3	151

coded marks are given by

$$u_i = \frac{x_i - 3.755}{1.500}$$

Thus, $\overline{U}$ is calculated by substituting the u_i's for the x_i's in Equation 5.9. The computations are performed in Table 5.8 with the following result.

$$\overline{u} = \frac{-3}{50} = -.06$$

For s_U^2, we substitute the u_i's for the x_i's in Equation 5.15 to obtain

$$s_U^2 = \frac{151 - \dfrac{(-3)^2}{50}}{50 - 1} = \frac{151 - .18}{49} = \frac{150.82}{49} = 3.07796$$

The values just obtained have no significance in terms of the original data. They must be decoded by using the relationships in Equation 5.17. Substituting the appropriate values gives

$$\overline{x} = b\overline{u} + a = 1.50(-.06) + 3.755 = 3.665 \qquad \text{or} \qquad 3.67$$

and

$$s_X^2 = b^2 s_U^2 = (1.50)^2(3.07796) = 6.9254$$

These values are identical to those obtained previously, but the arithmetic involved was greatly simplified, and the possibility of rounding errors was minimized.

It must be emphasized that in a frequency table only the class marks may be coded. The class frequencies may not be coded or transformed in any way for use in the equations of Sections 5.4 and 5.5. For ungrouped data, a convenient value for a is either the mean or the median, whereas a multiple of 10 is often used for b.

EXERCISES

5.16 What is the recommended coding formula for u_i for your frequency table in Exercise 5.4?

5.17 Use your answer to Exercise 5.16 to actually code the data in your frequency table. Then calculate $\overline{u}$ and s_U and use them to find $\overline{x}$ and s_X.

5.18 Refer to Exercise 5.7 and find the mean and variance, using coding.

5.19 Use a coding formula on the data of Exercise 5.11 to find $\overline{x}$ and s_X.

5.6 OTHER DESCRIPTIVE MEASURES

The measures discussed earlier in this chapter are the ones more commonly used in basic statistical inference. In this section, we present some additional measures that are somewhat more specialized in usage. Among these measures are the geometric mean, the mean deviation, the Pearson coefficient of skewness, the coefficient of kurtosis, and the coefficient of variation.

In many situations, business data are presented as percents or ratios. When data are of this form, the average that is often calculated is the geometric mean.

DEFINITION 5.24 If $X_1, X_2, \ldots, X_n$ represent n observations for a random variable X, then the **geometric mean** is

$$G = \sqrt[n]{X_1 \cdot X_2 \cdot X_3 \cdots X_n} \quad (5.18)$$

that is, the nth root of the product of the values.

Depending on the calculator available, it may be easier to use logarithms to calculate G, since

$$\log_{10} G = \frac{1}{n} \sum_{i=1}^{n} \log_{10} X_i \quad (5.19)$$

The value for G is then determined by taking the antilog.

EXAMPLE 5.23 Suppose that we are interested in calculating an average value for the price-to-earnings (P/E) ratio for companies in a particular industry. A random sample of nine companies had the following P/E ratios.

6.8	12.4	8.2	39.7	14.6	10.5	16.3	58.2	11.1

Solution: First, we calculate the product of the nine ratios, obtaining

$$\prod_{i=1}^{9} x_i = (6.8)(12.4) \cdots (11.1) = 4.43 \times 10^{10}$$

Then,

$$G = \sqrt[9]{4.43 \times 10^{10}} = 15.24$$

From Equation 5.19, the sum of the logarithms (base 10) is

$$\sum_{i=1}^{9} \log_{10} x_i = .8325 + 1.0934 + \cdots + 1.0453 = 10.6465$$

and

$$\log_{10} G = \frac{1}{9}(10.6465) = 1.1829$$

Thus, $G = 15.24$, the same as before.

We note that in Example 5.23, the sample mean is $\bar{x} = 19.76$, which is much larger than G. In general, the geometric mean is less influenced by extremes on the high side than $\bar{X}$ is, but it is influenced more by extremes on the low end. For a set of positive values that are not all equal, the arithmetic mean $\bar{X}$ will always be larger than the geometric mean.

An alternative to the standard deviation that is encountered occasionally in certain areas of application is the mean deviation, also known as mean absolute deviation (MAD). It is often used in inventory control and production management and literally is the average distance of the observations from their mean.

DEFINITION 5.25 The **mean deviation** is the average of the absolute values of the deviations from the sample mean—that is,

$$MD = \frac{1}{n}\sum_{i=1}^{n} |X_i - \bar{X}| \tag{5.20}$$

EXAMPLE 5.24 We use the P/E ratio data of Example 5.23 for our example. The data are

6.8 12.4 8.2 39.7 14.6 10.5 16.3 58.2 11.1

Earlier we calculated $\bar{x}$ to be 19.76.

Solution: Subtracting 19.76 from each of the observations gives the following actual differences.

−12.96 −7.36 −11.56 19.94 −5.16 −9.26 −3.46 38.44 −8.66

The sum of the absolute values of these differences is 116.8, giving a mean deviation of

$$MD = \frac{1}{9}(116.8) = 12.98$$

We note that, for these data, $s = 17.43$. In general, s will be greater than the mean deviation. To interpret the mean deviation, we always consider it relative to the mean. For an approximately normal distribution or for samples from a normal population, we can expect approximately 58% of the values to be within one MD of the mean—that is, within $\overline{X} \pm MD$. We note that small values for the mean deviation indicate that the distribution is compact. That is, the values are concentrated around $\overline{X}$. The mean deviation is used primarily in dealing with small samples in descriptive rather than inferential situations.

In many situations, we are faced with evaluating an assumption about some general property of the distribution from which the sample was selected. For example, most of the procedures in Chapters 7 through 11 assume an underlying normal distribution, while some of the alternative procedures assume only symmetry in the population. Often, we must make this evaluation with a limited amount of information. The next two statistics that we consider can sometimes aid in making these evaluations.

DEFINITION 5.26 The **Pearson coefficient of skewness** is defined as

$$Sk = \frac{\frac{1}{n}\sum_{i=1}^{n}(X_i - \overline{X})^3}{s_L^3} \tag{5.21}$$

where: $s_L = \sqrt{\frac{1}{n}\sum_{i=1}^{n}(X_i - \overline{X})^2}$

The sample coefficient of skewness measures the degree of the lack of symmetry in the sample data. For symmetric distributions, in which the curve to the left of the mean is a mirror image of the curve to the right of it, the coefficient of skewness is 0. If Sk is less than zero, the data are said to be *negatively skewed*. That is, they have a long tail to the left. If Sk is greater than zero, there is a long tail on the right, and the data are said to be *positively skewed*. The standard deviation, s_L, used in Equation 5.21 is equal to $\sqrt{[(n-1)/n]s_X^2}$, where s_X^2 is given by Equation 5.11.

The next statistic measures the degree of peakedness in the data. Its value is interpreted relative to the corresponding value for a normal distribution.

DEFINITION 5.27 The **sample coefficient of kurtosis** is defined as

$$K = \frac{\frac{1}{n}\sum_{i=1}^{n}(X_i - \overline{X})^4}{s_L^4} \tag{5.22}$$

where s_L is the same as in Definition 5.26.

The value for K corresponding to any normal distribution is 3.0, and the normal distributions are called *mesokurtic*. If $K < 3$, the distribution is flatter and more spread out than a normal and is called *platykurtic*. For $K > 3$, the corresponding distribution is more peaked, has a higher proportion of values near the mean than a normal does, and is called *leptokurtic*. Thus, if a sample is from a normally distributed population, we expect a value near 3 for K. The use of the value 3 as a reference point is based on the assumption that the distribution under consideration is not highly skewed in either direction since normal distributions are symmetric. Sampling distributions for Sk and K have been developed in the case of normal populations and can be used to evaluate normality.

EXAMPLE 5.25 A student in an introductory investments class believes that a normal distribution is appropriate for P/E ratios. To prove that this might be presumptuous, we calculate Sk and K for the P/E ratios of Example 5.23.

Solution: The data are 6.8, 12.4, 8.2, 39.7, 14.6, 10.5, 16.3, 58.2 and 11.1. Earlier, we found the mean to be 19.76 and s_X to be 17.43. To find s_L, we use the relation $s_L^2 = [(n - 1)/n]\, s_X^2$, obtaining

$$s_L = \sqrt{\frac{8}{9}(17.43)^2} = 16.43$$

We must calculate both

$$\sum_{i=1}^{9} (x_i - 19.76)^3 \quad \text{and} \quad \sum_{i=1}^{9} (x_i - 19.76)^4$$

to obtain Sk and K. These calculations are presented in Table 5.9.

Then,

$$Sk = \frac{(58985.90)/9}{(16.43)^3} = 1.48$$

and

$$K = \frac{(2404322.34)/9}{(16.43)^4} = 3.67$$

TABLE 5.9 Table of calculations for Example 5.25

x_i	$x_i - 19.76$	$(x_i - 19.76)^3$	$(x_i - 19.76)^4$
6.8	−12.96	−2,176.78	28,211.10
12.4	− 7.36	− 398.69	2,934.35
8.2	−11.56	−1,544.80	17,857.94
39.7	19.94	7,928.22	158,088.62
14.6	− 5.16	− 137.39	708.92
10.5	− 9.26	− 794.02	7,352.65
16.3	− 3.46	− 41.42	143.32
58.2	38.44	56,800.24	2,183,401.10
11.1	− 8.66	− 649.46	5,624.34
		58,985.90	2,404,322.34

Since $Sk = 1.48 > 0$, we see that these data indicate that the population may be skewed to the right. The value for K indicates that the distribution may be leptokurtic.

The final descriptive measure that we consider is useful in trying to compare variability for two different distributions, populations, or samples. For example, if we want to compare the standard deviation for a population of weights to that of a population of heights, we have differing units for the two, which makes them not directly comparable. To eliminate this problem, we convert to a measure of relative variation that has no units associated with it—that is, it represents a pure number.

DEFINITION 5.28 The **sample coefficient of variation** is defined to be the ratio of the sample standard deviation to the sample mean—that is,

$$V = \frac{s_X}{\overline{X}} \qquad \textbf{(5.23)}$$

EXAMPLE 5.26 We would like to compare the variability in sales for the Penny Ante novelty store to the variability in the number of credit hours for which students in a statistics class are enrolled. The results for the Penny Ante data are in Example 5.21, and the data on credit hours are in Example 5.9.

Solution: From Example 5.21, we see that, for Penny Ante, $\bar{x} = \$3.67$ and $s_X = \$2.63$. The sample of credit hours is 8, 13, 12, 9, 9, 10, 12, 15, 9, 13, which gives $\bar{x} = 11$ hours and $s_X = 2.31$ hours.

It is meaningless to attempt to compare $2.63 to 2.31 hours. Thus, we calculate the coefficient of variation for each sample and compare them.

$$\text{Penny Ante:} \qquad V_{\text{PA}} = \frac{\$2.63}{\$3.67} = .72$$

$$\text{Credit hours:} \qquad V_{\text{H}} = \frac{2.31 \text{ hrs}}{11 \text{ hrs}} = .21$$

Hence, we now may say that the relative variation in sales for Penny Ante is higher than the relative variation in credit hours for which the students in the statistics class are enrolled.

EXERCISES

5.20 During a discussion in class on normal distributions, a student asked whether it is reasonable to assume that scores on a standardized final exam for a statistics class are normally distributed. From records, we select a random sample of 14 scores and obtain the following results.

68	75	74	83	91	79	74
78	62	85	75	59	76	75

(a) Calculate the mean deviation for these scores and compare the percentage within ± 1 *MD* of the mean to the value expected for a normal distribution.
(b) Calculate the coefficient of skewness.
(c) Calculate the coefficient of kurtosis.

5.21 Using your values from Exercise 5.20, what is your response to the student's question?

5.22 Calculate the coefficient of variation for the sample of P/E ratios in Example 5.23.

5.23 Calculate the coefficient of variation for the sample of exam scores in Exercise 5.20 and compare the variability in P/E ratios in Example 5.23 to the variability in exam scores.

5.7 SUMMARY

In this chapter, we have considered various methods of selecting samples, namely, simple random sampling, stratified sampling, and cluster sampling. For making inferences about the population by using the methods in this text, simple random sampling is necessary; stratified and cluster sampling require special types of analysis. Table 5.10 summarizes the methods of selecting samples as well as other concepts discussed in this chapter.

Once a sample has been selected, then the primary objectives are to organize the data and to summarize the data. These summaries, called statistics, serve to describe the data and to form the basis for inferences to the population. Coming from random samples, the statistics are random variables.

For organizing smaller sets of data, we looked at an array, which is simply an ordering of the observations by numerical value. For large sets of data, we may use a frequency table to summarize the information. The basic frequency table identifies the classes, by either boundaries or limits, and the class frequencies, which are the number of observations belonging to the class. Other items that may be listed in a frequency table are the relative frequency, the cumulative frequency, and the class midpoint or mark. The construction of the table must be such that the classes are mutually exclusive, all observations are included, and a minimal amount of information is lost. A stem-and-leaf plot is an identification by groups of values with the observations in each group put into an array.

The graphical representations we examined in this chapter are the histogram, frequency polygon, and ogive. A histogram is a vertical bar graph where the bases correspond to class boundaries while the areas are proportional to the class frequencies. The frequency polygon merely connects the midpoints of the tops of the rectangles in the histogram. The histogram and the frequency polygon

TABLE 5.10 Summary table

<table>
<tr><td>Methods of Data Collection</td><td>Simple random sampling: Where all items have the same chance of being selected
Stratified sampling: Where a random sample is selected from each stratum
Cluster sampling: Where a random sample of clusters is selected</td></tr>
<tr><td>Organizing Data</td><td>Tabular
Array: An ordering of data
Frequency table: A grouping of the data into mutually exclusive classes
Stem-and-leaf plot: An ordered grouping of data by significant digits
Graphical
Histogram: A vertical bar graph
Frequency polygon: A line graph in which the midpoints of the tops of the rectangles in the histogram are connected
Ogive: A series of straight lines connecting points $(B_i, F(B_i))$, where the B_i are the class boundaries and $F(B_i)$ are the corresponding cumulative frequencies</td></tr>
<tr><td>Descriptive Measures</td><td>Location
Median (50th percentile)
Ungrouped data: $X_{.5} = X_{((n+1)/2)}$ (n odd)
$X_{.5} = \frac{X_{(n/2)} + X_{((n/2)+1)}}{2}$ (n even)
Grouped data: $X_{.5} = B_{.5} + \left[C_{.5}\left(\frac{.5n - F(B_{.5})}{f_{.5}}\right)\right]$
Mean (average value)
Ungrouped data: $\overline{X} = \frac{\sum_{i=1}^{n} X_i}{n}$
Grouped data: $\overline{X} = \frac{\sum_{i=1}^{k} X_i f_i}{n}$
Dispersion
Range (used primarily for small samples)
$X_{(n)} - X_{(1)}$
Variance (average squared distance from the mean)
Ungrouped data: $s_X^2 = \frac{\sum_{i=1}^{n} (X_i - \overline{X})^2}{n - 1}$</td></tr>
</table>

or

$$s_X^2 = \frac{\sum_{i=1}^{n} X_i^2 - \frac{\left(\sum_{i=1}^{n} X_i\right)^2}{n}}{n-1}$$

Grouped data: $$s_X^2 = \frac{\sum_{i=1}^{k} (X_i - \overline{X})^2 f_i}{n-1}$$

or

$$s_X^2 = \frac{\sum_{i=1}^{k} X_i^2 f_i - \frac{\left(\sum_{i=1}^{k} X_i f_i\right)^2}{n}}{n-1}$$

Standard deviation: $s_X = \sqrt{s_X^2}$

Descriptive Measures

Other Measures

Geometric mean (nth root of the product) of n values

$$G = \sqrt[n]{X_1 \cdot X_2 \cdot X_3 \cdot \cdot \cdot X_n}$$

Mean deviation (average distance of the values from $\overline{X}$)

$$MD = \frac{1}{n} \sum_{i=1}^{n} |X_i - \overline{X}|$$

Coefficient of skewness (measures the lack of symmetry in a set of data)

$$Sk = \frac{\frac{1}{n} \sum_{i=1}^{n} (X_i - \overline{X})^3}{s_L^3}; \quad s_L = \sqrt{\frac{1}{n} \sum_{i=1}^{n} (X_i - \overline{X})^2}$$

Coefficient of kurtosis (measures the degree of peakedness in a set of data)

$$K = \frac{\frac{1}{n} \sum_{i=1}^{n} (X_i - \overline{X})^4}{s_L^4}$$

Coefficient of variation (measures the relative variation in a set of data)

$$V = \frac{s_X}{\overline{X}}$$

should approximate the actual distribution of the population. The ogive is a graph of the cumulative frequencies plotted at class boundaries and connected by straight lines.

For describing sample data, we considered measures of location and measures of dispersion. For location, we have the 50th percentile or median, which is the middle of the data in terms of frequency, and the mean, which is the physical center of the data. The mean is the average value and is the most commonly used measure for making inferences. Another measure of central tendency that is reported occasionally is the mode—that is, the value occurring most often.

In measuring the dispersion or spread in sample data, we use the range, variance, and standard deviation. The range is the difference between the largest and smallest values. In most situations, the variance is used to describe the dispersion in the data. It is defined as the average squared distance from the mean and has the drawback of being in units squared. To overcome this, we take the square root to obtain the standard deviation, which is easier to interpret since its units are the same as those of the original measurements.

In the last section, we looked at several descriptive measures that are more specialized in nature. The mean deviation is an alternative to the standard deviation. It is the average of the absolute values of the deviations from the mean. For data that are in the form of ratios or percents, the geometric mean, the nth root of the product of the n values, is generally preferred over the arithmetic (sample) mean. Whenever the shape of the distribution is of interest, we use the coefficients of skewness and kurtosis. The coefficient of skewness measures the degree of lack of symmetry in the data, and the coefficient of kurtosis measures the degree of peakedness (or pointedness) in the data. The final measure, the coefficient of variation, is the ratio of the standard deviation, s_X, to the sample mean. It is used to compare variability in sets of data that are measured in different units.

SUPPLEMENTARY EXERCISES

5.24 In a traffic survey, the number of automobiles passing a specific point was determined for a certain time period. Data for a sample of 50 time periods are presented in the following frequency table.

No. of Autos	*No. of Periods* $= f_i$
0– 4	7
5– 9	20
10–14	13
15–19	6
20–24	4

(a) Construct the histogram for this table.
(b) Calculate the median number of cars passing per time period.
(c) Calculate the mean number of cars passing per time period.
(d) Calculate the standard deviation.
(e) Construct the ogive and use it to estimate the 75th percentile.

5.25 A random sample of six exams from a class gave the following scores: 69, 51, 73, 69, 87, 69. Determine **(a)** the modal score, **(b)** the median score, **(c)** the average score, and **(d)** the variance of the scores.

5.26 There is a rumor that tuition is going to increase next year at a certain university. It is desired to select a random sample of 200 students to obtain an idea of student opinion regarding the possible increase. Explain how the sampling might be accomplished.

5.27 The following frequency table represents the weekly incomes of 50 students randomly selected from a large evening class. (Note that the limits and boundaries are the same. Having them coincide is often done for simplicity.)

Income (\$)	***No. of Students***
$0 \leq X < 50$	7
$50 \leq X < 100$	10
$100 \leq X < 150$	11
$150 \leq X < 200$	15
$200 \leq X < 250$	4
$250 \leq X < 350$	3

(a) Sketch the histogram for these data.
(b) Above what income do 60% of the students earn weekly?
(c) What is the average weekly income?
(d) What is the standard deviation of weekly incomes?

5.28 What are the purposes of cluster sampling? of stratified sampling? What are the differences between the two?

5.29 Prove that an alternative computational form for the variance of ungrouped data is given by

$$s_X^2 = \frac{\sum_{i=1}^{n} X_i^2 - n\overline{X}^2}{n - 1}$$

5.30 Explain why the statistics calculated from a frequency table generally differ in value from the values calculated for the same statistics from the ungrouped data.

5.31 Refer to Appendix Table B.10 and select a random sample of 40 three-digit numbers. As a starting point, use the number in the 7th column and 11th row and then move horizontally to the right from that value.

5.32 Construct a frequency table having 7 classes for your sample in Exercise 5.31.

5.33 The following frequency table contains downtimes in minutes of a certain machine.

Downtime	*Frequency* = f_i	*Mark* = X_i
$0 \le X < 20$	9	10
$20 \le X < 40$	23	30
$40 \le X < 60$	36	50
$60 \le X < 80$	23	70
$80 \le X < 100$	9	90

(a) Construct the histogram, frequency polygon, and cumulative frequency graph (ogive) for this table.

†**(b)** Code the data using the obvious coding.

†**(c)** Find the average downtime, using the coded data.

†**(d)** Find the variance of the downtimes, using the coded data.

(e) Find the median downtime.

5.34 Give examples of situations where one might prefer to use the median rather than the mean as the measure of location.

5.35 The Lux Company makes batteries that are unconditionally guaranteed for two years. A random sample of 100 of these batteries gave the following lives in days (in tabular form).

Number of Days/ Class Boundaries	*No. of Batteries*
$500 \le X < 600$	2
$600 \le X < 700$	10
$700 \le X < 800$	14
$800 \le X < 900$	56
$900 \le X < 1000$	13
$1000 \le X < 1100$	5

(a) Calculate the median life in days.

(b) Calculate the mean life in days.

(c) Calculate the standard deviation of lives.

(d) What is the number of days before which 20% of the batteries failed?

†**5.36** Refer to the table in Exercise 5.35 and code the data. Using the coded data, determine the mean life and the standard deviation of lives. Compare your answers with those obtained in Exercise 5.35.

5.37 A random sample from the records of County General Hospital yielded the following lengths of stay in days

8	14	2	3	10	5	7	4.5	6	13	8	7
6	5	4	4	12	20	1	4	5.5	3	4	8
6	6	9	17	11	5	6	4	12	1	7	5
16	4	7	23	8	6	4	7	5	13	6	12
2	3	5	3	7	6	14	12	10	3	5	2

Calculate the median, mean, and standard deviation of the lengths of stay.

† These questions pertain to the optional section on coding.

5.38 Construct a frequency table for the data in Exercise 5.37, and construct **(a)** the histogram, **(b)** the frequency polygon, and **(c)** the ogive.

5.39 Use the frequency table in Exercise 5.38 to calculate **(a)** the mean length of stay, **(b)** the median length of stay, and **(c)** the standard deviation of lengths of stay.

5.40 If a set of data is skewed to the right—that is, has a long tail on the right—then where will the mean be with respect to the median? Why?

5.41 Give an example of a population for which the geometric mean might be preferred over the sample mean as a measure of location.

5.42 A random sample of classes in a business college showed the following enrollments.

43	28	31	34	41	38	39	36	22	19	45	37
38	36	44	30	29	24	35	33	29	25	17	41
37	32	37	30	21	33	27	20	36	41	48	29
36	27	52	28	33	30	32	39	37	24	13	38
35	29	26	36	40	28	31	31	37	49	28	32

(a) Calculate the average class size for this sample.
(b) Calculate the median class size for this sample.
(c) What is the modal class size?

5.43 For the sample data given in Exercise 5.42, determine **(a)** the range of class sizes and **(b)** the standard deviation of class sizes.

5.44 Construct a frequency table for the data of Exercise 5.42. Use 7 classes and a common class width of 6.

5.45 Refer to Exercise 5.44 and calculate the mean, median, and crude mode for the frequency table.

5.46 Determine the standard deviation for the frequency table of Exercise 5.44.

5.47 Compare your answers for Exercises 5.42 and 5.43 to your answers for Exercises 5.45 and 5.46. If there are any differences, explain why.

5.48 Before the sample in Exercise 5.42 was selected, classes corresponding to large-lecture sections were removed from the population of all classes. Suppose these classes had not been removed from consideration before sampling and that a large-lecture marketing principles class having an enrollment of 316 had been sampled. What would be the effect of including this class on **(a)** the mean, **(b)** the median, **(c)** the range, and **(d)** the standard deviation?

5.49 Which values in Exercise 5.48 are affected the most? Explain why.

5.50 A sample of 25 electronic components gave times to failure (in days) of

44	48	64	51	32	29	48	122	69	73	55	37	18
101	49	74	59	56	62	60	45	52	61	39	57	

(a) Calculate the average time to failure for this sample.
(b) Calculate the median time to failure.
(c) Calculate the standard deviation of times to failure.

5.51 Calculate the sample coefficient of kurtosis for the data of Exercise 5.42.

5.52 Calculate the sample coefficient of kurtosis for the data of Exercise 5.50.

5.53 Refer to Exercises 5.51 and 5.52 and indicate which sample is more likely to have come from a normal population on the basis of the K-values. Is this consistent with what you would have thought without calculating K?

5.54 Calculate Sk for the data of Exercise 5.42.

5.55 Calculate the sample coefficient of skewness for the data of Exercise 5.50.

5.56 What value for Sk would you expect to obtain for a sample from a normal population?

5.57 Compare the values you calculated for Sk in Exercises 5.54 and 5.55 and indicate which sample is more likely to have come from a normal population. Is this consistent with your conclusions in Exercise 5.53?

5.58 Refer to Exercise 5.42 and calculate the mean deviation.

5.59 Calculate the 25th and 75th percentiles for the data of Exercise 5.50.

5.60 Calculate the mean deviation for the data of Exercise 5.50.

5.61 Interpret your answers to Exercises 5.58 and 5.60 and compare your answers to the standard deviations calculated in Exercises 5.43 and 5.50.

5.62 Calculate the 30th and 80th percentiles for the frequency table that you constructed in Exercise 5.44.

5.63 A real estate agent has collected the following data on the selling price of 10 recently sold homes as a percentage of the purchase price.

148.1	132.6	187.7	121.4	98.2
141.6	159.5	169.7	118.9	128.0

Calculate the average of the percentage of selling price to purchase price for these 10 houses, using the geometric mean.

5.64 Calculate the sample mean for the percentages in Exercise 5.63 and compare it to the geometric mean.

5.65 In what situations would it be inappropriate to calculate the geometric mean?

5.66 Data on the distribution of IRA tax returns and contributions by income group for the years 1980 through 1982 are presented in Table 5.11. They are taken from the *FRBNY Quarterly Review* (Summer 1984). Use the data in Table 5.11 to do the following.

(a) Construct a histogram of contributions for each year.
(b) Construct a frequency polygon of contributions for each year.
(c) Construct an ogive of contributions for each year.

5.67 Use the data in Table 5.11 to do the following.

(a) Construct a histogram for the number of tax returns with IRAs for each year.
(b) Construct a frequency polygon for the number of tax returns with IRAs for each year.
(c) Construct an ogive for the number of tax returns with IRAs for each year.

TABLE 5.11 Distribution of IRA tax returns and contributions by income group

Income group (annual income in $)*	*Tax Returns with IRA Contributions (000 of returns)*			*IRA Contributions ($000,000)*		
	1980	***1981***	***1982***	***1980***	***1981***	***1982***
Less than 6,000	16.4	42.9	147.5	13.3	37.2	296.3
Above 6,000 and less than 11,000	99.2	180.2	503.6	78.0	150.3	786.7
Above 11,000 and less than 20,000	310.9	419.7	1,348.7	321.7	489.0	2,388.5
Above 20,000 and less than 30,000	643.8	857.7	2,978.7	799.0	1,122.9	6,068.5
30,000 and more	1,494.1	1,914.8	7,119.5	2,218.9	2,950.8	18,876.1
Total	2,564.4	3,415.1	12,098.0	3,430.9	4,750.2	28,416.0

NOTE: Columns may not add to totals because of rounding.

* Each income group represents 20% of all tax returns filed. The annual income cutoffs of the income groups are approximate.

SOURCE: Internal Revenue Service, *Statistics of Income: Individual Tax Returns (1980–1982)*. The data for 1982 are preliminary.

6 Distributions of Sample Statistics

6.1 INTRODUCTION

Ledbetter Chemicals requires plastic pipe with an average diameter of 5 centimeters (cm) for one of its production operations. A shipment of this pipe has arrived at the receiving department and is waiting for a decision on its disposition by Quality Control. Zeke Underwood, in charge of quality control for incoming materials, has selected a random sample of 50 pieces. His calculations show that the average diameter of the pieces of pipe in the sample is 4.96 cm. To make a decision with respect to the shipment, Zeke must answer the question, "Is this average diameter unreasonably small if the true mean diameter for the shipment is 5.00 cm., or is it an acceptable deviation?"

To answer this question, Zeke must be able to determine the probability of observing a sample average diameter as small as 4.96 centimeters or smaller for a sample of 50 if the true population mean diameter is 5.00 centimeters. To calculate the probability requires the identification of a probability distribution that can be used to describe the chance variation of the sample statistic, $\overline{X}$ in this case.

In this chapter, we start to combine the basic concepts introduced in the preceding chapters. We have considered simple probability and the idea of a random variable and its associated probability distribution, looked at specific types of distributions for describing natural phenomena, and, in the last chapter, introduced the concepts of a sample and the measures to describe it. In some situations, we may have an idea as to the type of distribution that can be used to describe the chance variation within a population. Often we will not. In almost all cases, the specific member of the family of distributions—that is, the values of the parameters—will be unknown. Thus, the primary purpose of sampling is to attempt to learn something about the distribution of the population.

We saw in Chapter 5 how to describe the sample by calculating certain characteristics. Since samples themselves vary, the values of these characteristics also vary from sample to sample. Therefore, if we are attempting to use a sample characteristic to estimate the corresponding population characteristic, we know in advance that the sample value cannot be expected to equal the population value.

Of more use than merely the sample value would be the ability to determine the probability of the sample value's being a specified distance away from the population value. To be able to make such a probability statement requires that we know the type of distribution that can be used to describe, or at least to approximate, the random variation of the sample statistic. The determination of the most common of these distributions is the main topic of this chapter.

6.2 THE CONCEPT OF A SAMPLING DISTRIBUTION

Before continuing, let us consider the following example to illustrate the concept of a sampling distribution.

EXAMPLE 6.1 Assume we wish to buy three 5-inch bolts. To make the purchase, we go to a hardware store, which has only 5 of the bolts in stock. The lengths of the 5 bolts are 4.85, 5.12, 5.08, 4.92, 5.03 inches. We randomly select 3 bolts from the 5 available. What are the mean, variance, and standard deviation of the distribution of average lengths from samples of 3 bolts each?

Solution: The population of interest consists of the 5 available bolts and obviously is finite. Thus, the distribution is discrete. The mean, variance, and standard deviation of the population are

$$\mu_X = \sum_{i=1}^{5} x_i p(x_i)$$

$$= 4.85\left(\frac{1}{5}\right) + \cdot \cdot \cdot + 5.03\left(\frac{1}{5}\right) = 5.00 \text{ in.}$$

$$\sigma_X^2 = \sum_{i=1}^{5} (x_i - \mu_X)^2\, p(x_i)$$

$$= (-.15)^2\left(\frac{1}{5}\right) + \cdot \cdot \cdot + (.03)^2\left(\frac{1}{5}\right) = .0101$$

$$\sigma_X = \sqrt{.0101} = .1006$$

The calculation of $\mu_{\overline{X}}$ and $\sigma^2_{\overline{X}}$, the mean and variance, respectively, of the population of $\overline{X}$'s, requires identifying all possible samples and the corresponding distribution of $\overline{X}$. The $C^5_3 = 10$ possible samples are shown in Table 6.1. Each has a probability of $\frac{1}{10}$ of occurring.

Thus, by direct computation, we obtain

$$\mu_{\overline{X}} = \sum \bar{x}p(\bar{x}) = 5.00 \text{ in.}$$

$$\sigma^2_{\overline{X}} = \sum (\bar{x} - \mu_{\overline{X}})^2 p(\bar{x}) = .0017$$

$$\sigma_{\overline{X}} = \sqrt{.0017} = .0412$$

The values for $\sigma^2_{\overline{X}}$ and $\sigma_{\overline{X}}$ are slightly in error because the $\overline{X}$'s were rounded to three decimals. ■

Note that the values for μ_X, the mean for the population of individual values, and $\mu_{\overline{X}}$, the mean of the population of sample averages, are identical. These means will always be equal because $\overline{X}$ has the property of unbiasedness, which is illustrated in Equation 6.5 and discussed in more detail in Chapter 7. On the other hand, σ^2_X and $\sigma^2_{\overline{X}}$, the respective variances, are not equal. The variance of the population of sample averages will always be smaller than σ^2_X due to the cancellation effect of the averaging process. This concept is illustrated later in Equation 6.7.

The approach taken to determine the distribution of the sample averages and the corresponding values for the mean and variance of the distribution of sample means becomes quite laborious whenever the population of interest is

TABLE 6.1 The samples and the distribution of the resulting sample means for Example 6.1

Sample	$\bar{x}$	$p(\bar{x})$	$\bar{x}\,p(\bar{x})$	$(\bar{x} - \mu_{\overline{X}})^2\,p(\bar{x})$
4.85,5.12,5.08	5.017	.1	.5017	.0000289
4.85,5.08,4.92	4.950	.1	.4950	.0002500
4.85,4.92,5.03	4.933	.1	.4933	.0004489
4.85,5.12,4.92	4.963	.1	.4963	.0001369
4.85,5.12,5.03	5.000	.1	.5000	.0000000
4.85,5.08,5.03	4.987	.1	.4987	.0000169
5.12,5.08,4.92	5.040	.1	.5040	.0001600
5.12,5.08,5.03	5.077	.1	.5077	.0005929
5.12,4.92,5.03	5.023	.1	.5023	.0000529
5.08,4.92,5.03	5.010	.1	.5010	.0000100
			5.0000	.0016974

large. The results of Chapter 3 simplify the determination of the mean and variance, and the next two sections provide a rationale for identifying at least an approximate distribution for sample averages under rather general conditions.

An Example of the Distribution of the Sample Mean

Before proceeding to the main concept of this chapter—one of the most important results in statistical theory—we consider another example. In the example, we select a number of samples from a nonnormal population, calculate the sample means, and construct a frequency table of their values and the corresponding histogram. The purpose is to illustrate the meaning and significance of the Central Limit Theorem, which is introduced in the next section.

EXAMPLE 6.2 We have selected 36 random samples of 4 observations each from a uniform distribution on the interval 0 to 10—that is, from $U(0,10)$. The samples and their corresponding means are presented in Table 6.2.

Solution: A frequency table for the sample means is presented in Table 6.3. As a comparison, we superimpose on the distribution of the population, from which the samples were drawn, the histogram for Table 6.3 and the frequency polygon. These are presented on the same scale in Figure 6.1.

TABLE 6.2 Samples of size 4 from a U (0,10) distribution and their means

Sample	*Observations*	$\bar{x}$	*Sample*	*Observations*	$\bar{x}$
1	3.95,4.63,9.95,0.67	4.80	19	9.47,7.08,9.54,0.13	6.56
2	6.95,2.79,5.05,7.46	5.56	20	2.52,2.00,9.63,9.35	5.88
3	4.93,8.97,7.40,8.47	7.44	21	2.33,0.07,4.82,3.67	2.72
4	8.36,4.78,0.03,8.11	5.32	22	7.14,2.49,6.87,0.08	4.15
5	7.52,9.62,8.29,7.75	8.30	23	5.48,3.45,1.94,5.22	4.02
6	6.14,7.60,2.64,4.65	5.26	24	0.57,1.48,8.05,9.26	4.84
7	7.39,9.39,4.19,8.07	7.26	25	3.45,7.50,1.61,6.08	4.66
8	9.25,3.96,8.69,9.97	7.97	26	7.37,0.77,3.53,5.49	4.29
9	6.02,2.91,5.74,5.73	5.10	27	9.57,2.60,0.68,2.03	3.72
10	5.67,7.04,0.54,8.09	5.34	28	6.62,9.18,0.74,1.37	4.48
11	0.69,0.01,8.85,6.45	4.00	29	1.36,4.85,6.52,8.43	5.29
12	2.52,7.12,4.97,4.30	4.73	30	5.90,0.24,7.44,9.14	5.68
13	1.51,9.16,0.49,4.76	3.98	31	3.11,5.04,6.59,8.55	5.82
14	4.36,6.86,9.86,7.41	7.12	32	1.53,8.27,8.21,2.78	5.20
15	5.22,8.70,0.37,3.43	4.43	33	9.41,8.55,1.98,2.91	5.71
16	4.24,0.63,6.42,1.37	3.17	34	7.86,3.84,9.23,1.69	5.66
17	7.15,8.24,4.23,3.98	5.90	35	1.53,8.41,6.13,6.23	5.58
18	8.81,9.92,3.33,0.81	5.72	36	1.00,4.28,8.67,6.60	5.14

TABLE 6.3 Frequency table of means of samples of size 4 from a $U(0,10)$ distribution

B_i Class Boundaries	f_i Frequency	x_i Mark	Relative Frequency
2.5–3.5	2	3.0	.06
3.5–4.5	8	4.0	.22
4.5–5.5	11	5.0	.30
5.5–6.5	9	6.0	.25
6.5–7.5	4	7.0	.11
7.5–8.5	2	8.0	.06

Note that the empirical distribution of the sample means, as depicted by the histogram and frequency polygon, already could be closely approximated with a normal curve. The approximation is quite accurate even though the original distribution is decidedly nonnormal, but symmetric; the sample size of 4 is quite small, and we have used only 36 repetitions.

Before continuing, we define a term and some notations that are very useful in later discussions. Overall, these are not new notations and a new term, but merely a clarification of the notations and an alternative name for a term with which we are already familiar. We note that, in general, the notations μ_X and σ_X represent the mean and the standard deviation of the distribution of X values. Correspondingly, the notations $\mu_{\overline{X}}$ and $\sigma_{\overline{X}}$ represent the mean and the standard deviation, respectively, for the sampling distribution of $\overline{X}$—that is, of the population of values for $\overline{X}$. The new name that is often used in later chapters for $\sigma_{\overline{X}}$ is *standard error of the mean*. Generally, this is used in reference to an estimator and is simply the standard deviation of that estimator.

FIGURE 6.1 Graphical representation of the $U(0,10)$ distribution, a histogram of 36 means from samples of size 4 drawn from it and the frequency polygon

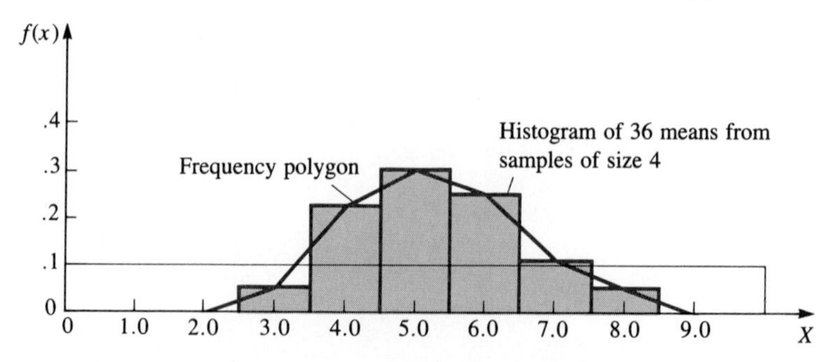

EXERCISES

6.1 Refer to Example 6.2 and Table 6.2.

(a) Find the percentage of the sample means that are within one standard deviation of the population mean—that is, in the range of $\mu_{\overline{X}} \pm \sigma_{\overline{X}}$—within two standard deviations, and within three standard deviations. (Note that $\mu_{\overline{X}} = 5.00$ and $\sigma_{\overline{X}} = 1.4434$.)

(b) What are the corresponding percentages for a normal distribution?

6.2 In Example 6.2, we calculated the sample means and constructed the frequency table and graphs from those values.

(a) Do the same thing for the sample totals—that is, calculate $\Sigma_{i=1}^{4} X_i$ for each sample, construct a frequency table, and construct the corresponding histogram and frequency polygon for your totals.

(b) What do these graphs imply to you?

6.3 THE DISTRIBUTION OF SAMPLE MEANS AND TOTALS

To this point, we have considered the collection, presentation, and description of sample data. Recall that a primary objective is to use the sample in making inferences about the population. Inferences are based on statistics and invariably take the form of probability statements. Thus, to make inferences, it becomes imperative that we are capable of determining: (1) the family to which the distribution of the statistics belongs and (2) the values for the parameters of that family. We shall address these two points in the order of importance.

Linear Combinations of Normal Random Variables

Before considering the general case, we first look at the situation involving normally distributed random variables. We saw in Chapter 4 that a linear transformation of a single normally distributed random variable; namely,

$$Z = \frac{X - \mu_X}{\sigma_X} = \frac{1}{\sigma_X} X - \frac{\mu_X}{\sigma_X}$$

has a normal distribution. In the general format, this equation is written as $Z = aX + b$ with $a = 1/\sigma_X$ and $b = -\mu_X/\sigma_X$. The following theorem extends this result to handle linear combinations of several normally distributed random variables.

THEOREM If $X_1, X_2, \ldots, X_k$ is a set of normally distributed random variables and L is any linear combination of them—that is,

$$L = \sum_{i=1}^{k} a_i X_i \qquad \textbf{(6.1)}$$

where the a_i's are any real numbers, then L will also have a normal distribution.

The proof of this theorem is reasonably complex, but the concept is a simple one. It basically states that if you add or subtract normal random variables or multiply them by constants, the resulting variable also has a distribution that is a member of the normal family.

To utilize this theorem, we must be able to determine the mean and the variance of L. The mean of the distribution is

$$\mu_L = E[L] = E\left[\sum_{i=1}^{k} a_i X_i\right] = \sum_{i=1}^{k} a_i E[X_i]$$

or

$$\mu_L = \sum_{i=1}^{k} a_i \mu_i \tag{6.2}$$

where μ_i is the mean of X_i. The variance can be difficult to determine unless the X_i's are independent random variables. For most situations, this generally will be the case, and the variance of L will be given by

$$\text{Var}(L) = \sigma_L^2 = \sum_{i=1}^{k} a_i^2 \sigma_i^2 \tag{6.3}$$

with standard deviation

$$\sigma_L = \sqrt{\sum_{i=1}^{k} a_i^2 \sigma_i^2} \tag{6.4}$$

Equation 6.2 is obvious and logical, whereas Equation 6.3 generally causes some difficulty for the student. As an example of Equation 6.3, we consider the mean of a random sample from an infinite population. In this case,

$$\overline{X} = \frac{\sum_{i=1}^{n} X_i}{n} = \sum_{i=1}^{n} \frac{1}{n} (X_i)$$

Thus, $\overline{X}$ is a linear combination of independent random variables with $a_i = 1/n$ for all i. Applying Equation 6.3 gives

$$\sigma_{\overline{X}}^2 = \sum_{i=1}^{n} \left(\frac{1}{n}\right)^2 \sigma_{X_i}^2 = \sum_{i=1}^{n} \frac{\sigma_X^2}{n^2}$$

$$= n\left(\frac{\sigma_X^2}{n^2}\right) = \frac{\sigma_X^2}{n}$$

These results can be applied to many business situations. The next example illustrates only one possibility.

EXAMPLE 6.3 The Middleman Company produces a particular item for sale to three customers in different areas of the country. The item is relatively expensive, and

Middleman prefers not to carry any inventory nor do they wish to alienate their customers by failing to fill an order more than 20% of the time. The number of items ordered per month by each customer follows a normal distribution with mean and variance as follows.

Customer A: $\mu_A = 1000$ $\sigma_A^2 = 4000$

Customer B: $\mu_B = 1750$ $\sigma_B^2 = 9000$

Customer C: $\mu_C = 700$ $\sigma_C^2 = 5000$

a. What is the minimum number of items Middleman must schedule for production to meet the 20% requirement on next month's orders?

b. If Middleman decides to produce 3650 items per month, what percentage of the orders can they expect to be unable to fill?

Solution: To answer the questions, we must determine the distribution of the total number of items ordered each month. Let X_A, X_B, and X_C represent the number of items ordered by customers A, B, and C, respectively. Then $L = X_A + X_B + X_C$ will represent the total, which is a form of Equation 6.1 with $k = 3$ and $a_i = 1$ for each i. By the previous theorem, L will have a normal distribution, and by Equation 6.2, the mean is

$$\mu_L = \mu_A + \mu_B + \mu_C = 1000 + 1750 + 700 = 3450$$

Since the customers are located in different areas of the country, it is reasonable to assume that their ordering policies are independent. Thus, Equation 6.3 gives

$$\sigma_L^2 = \sigma_A^2 + \sigma_B^2 + \sigma_C^2 = 4000 + 9000 + 5000 = 18{,}000$$

and

$$\sigma_L = \sqrt{18{,}000} = 134.16$$

The distribution of L would appear as shown in Figure 6.2.

a. We wish to determine the value ℓ such that

$$P(L > \ell) = .20$$

Standardizing, we obtain

$$P(L > \ell) = P\left(\frac{L - \mu_L}{\sigma_L} > \frac{\ell - \mu_L}{\sigma_L}\right) = P\left(Z > \frac{\ell - \mu_L}{\sigma_L}\right) = .20$$

From Appendix Table B.6, $P(Z > .84) = .20$, and thus, $(\ell - \mu_L)/\sigma_L$ must be .84. Equating and substituting values gives

$$\frac{\ell - 3450}{134.16} = .84$$

and solving for ℓ, we have

$$\ell = .84(134.16) + 3450 = 3562.69 \quad \text{or} \quad 3563 \text{ items}$$

FIGURE 6.2 Distribution of total demand (L) per month for an item produced by the Middleman Company and 20% stockout level

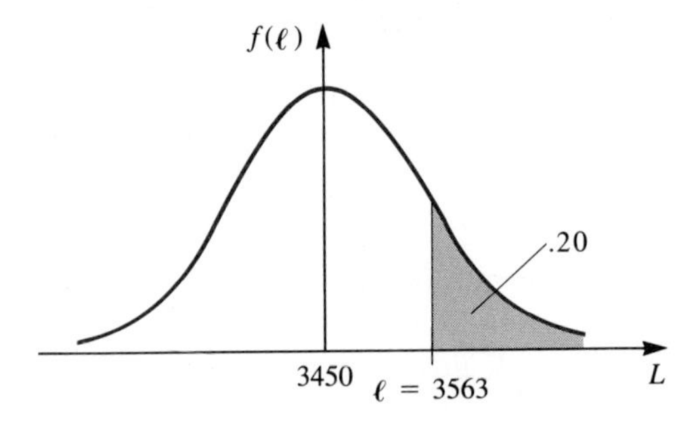

b. The answer uses the distribution of L. Therefore,

$$\begin{aligned} P(L > 3650) &= P\left(\frac{L - 3450}{134.16} > \frac{3650 - 3450}{134.16}\right) \\ &= P\left(Z > \frac{200}{134.16}\right) = P(Z > 1.49) \\ &= .5 - P(0 \le Z \le 1.49) \\ &= .5 - .4319 = .0681 \end{aligned}$$

that is, approximately 7% of the time. See Figure 6.3.

In Example 6.3, we assumed that the quantities ordered had normal distributions. In many situations, the specific type of distribution is not known. The next section provides a tool for dealing with many of those types of situations.

FIGURE 6.3 Distribution of demand per month for the Middleman Company and percent stockout with production of 3650 units

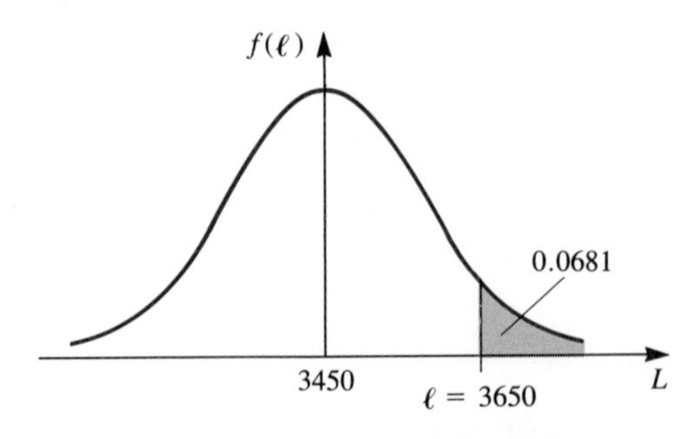

The General Case

Many decisions are based on the sample mean or some function of the sample mean. Thus, it is desirable that we determine the distribution of $\overline{X}$ under the most general conditions possible. The theorem that accomplishes this is the most important single concept in statistical analysis, for it provides the basis upon which most other classical procedures are founded.

CENTRAL LIMIT THEOREM Given random samples from any population with a finite variance, σ^2, as the sample size increases without limit, the distribution of the sample means and of the sample totals approaches a normal distribution.

The beauty of this theorem is that it enables us to use the normal family of distributions and sampling distributions derived from it to make inferences about the population mean *regardless of the form of the original distribution* and contingent only on the size of the sample selected. The actual sample size that will suffice in a particular situation depends on the extent to which the distribution under consideration deviates from a normal. If the original distribution is symmetric, then the normal distribution gives a good approximation even for small sample sizes, as illustrated in Example 6.2. If the population has a highly skewed distribution—that is, has a long tail on one side (Figure 6.4)—then larger sample sizes will be required. In those situations, it generally is recommended that the sample size be at least 30—that is, $n \geq 30$.

Now that we have a distribution for $\overline{X}$, it remains for us to determine values for the parameters of that distribution. In Chapter 4, it was indicated that the parameters associated with the normal family are μ and σ^2, the mean and the

FIGURE 6.4 Triangular distribution with $f(x) = \frac{1}{2} - (x/8)$ and approximate distribution of $\overline{X}$ for samples of size 36 selected from it

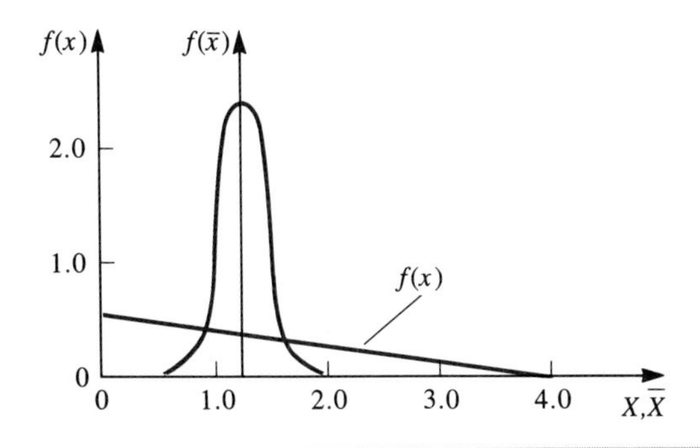

variance, respectively. Thus, it is necessary to determine the expected value of $\overline{X}$ and the variance of the distribution of $\overline{X}$. The mean of a random sample is simply a linear combination of random variables. With this fact in mind, we apply the properties of the expected value operator given in Chapter 3. The results that we use here are merely extensions of the results for two variables to handle situations involving several variables. We have

$$\mu_{\overline{X}} = E[\overline{X}] = E\left[\frac{1}{n}\sum_{i=1}^{n} X_i\right] = \frac{1}{n} E\left[\sum_{i=1}^{n} X_i\right]$$
$$= \frac{1}{n}\sum_{i=1}^{n} E[X_i] = \frac{1}{n}\sum_{i=1}^{n} \mu_X = \frac{1}{n}(n\mu_X)$$

and, hence,

$$\mu_{\overline{X}} = \mu_X \tag{6.5}$$

since $1/n$ is a constant, μ_X is a constant, the expected value of a sum of random variables is the sum of the expected values, and each X_i has mean μ_X.

If the population from which the random sample is selected is infinite, $\overline{X}$ becomes a linear combination of independent random variables. Referring to Equation 6.3, with $a_i = 1/n$, we obtain

$$\text{Var}(\overline{X}) = \sigma_{\overline{X}}^2 = \sum_{i=1}^{n}\left(\frac{1}{n}\right)^2 \sigma_{X_i}^2 = \frac{\sigma_X^2}{n} \tag{6.6}$$

and

$$\sigma_{\overline{X}} = \frac{\sigma_X}{\sqrt{n}} \tag{6.7}$$

If we are sampling from a finite population without replacement, the observations are no longer independent. To correct for this dependence among the observations, we must multiply the right side of Equation 6.6 by what is called the *finite population correction factor* (FPC). Omitting the derivation, which involves combinatorial analysis, the FPC is $(N - n)/(N - 1)$, and for finite populations, Equation 6.6 becomes

$$\sigma_{\overline{X}}^2 = \frac{\sigma_X^2}{n}\left(\frac{N - n}{N - 1}\right) \tag{6.8}$$

where: N = the population size
n = the sample size

EXAMPLE 6.4 A machine produces bolts that are to be 5 inches long. The bolts have an average length of 5.00 inches with a variance of .01. What are the mean, variance, and standard deviation of an $\overline{X}$ calculated from a sample of three bolts selected at random from the process?

Solution: To find $\mu_{\overline{X}}$, we apply Equation 6.5.

$$\mu_{\overline{X}} = \mu_X = 5.00 \text{ inches}$$

The population of interest is the set of all bolts produced by the machine, and the length of a bolt could be any value within some reasonable interval. Thus, the distribution of lengths is continuous, and $\sigma_{\overline{X}}^2$ is given by Equation 6.6 as

$$\sigma_{\overline{X}}^2 = \frac{\sigma_X^2}{n} = \frac{.01}{3} = .0033333$$

and

$$\sigma_{\overline{X}} = \sqrt{.0033333} = .0577 \text{ in.}$$

Before continuing, we illustrate the use of Equations 6.5 and 6.8 by applying them to the situation presented in Example 6.1.

EXAMPLE 6.5 We use the results of this section to determine $\mu_{\overline{X}}$, $\sigma_{\overline{X}}^2$, and $\sigma_{\overline{X}}$ for the data of Example 6.1.

Solution: Using Equations 6.5 and 6.8, we obtain

$$\mu_{\overline{X}} = \mu_X = 5.00$$

$$\sigma_{\overline{X}}^2 = \frac{\sigma_X^2}{n}\left(\frac{N-n}{N-1}\right) = \frac{.0101}{3}\left(\frac{5-3}{5-1}\right) = .00169$$

$$\sigma_{\overline{X}} = \sqrt{.00169} = .0411$$

The current values for $\sigma_{\overline{X}}^2$ and $\sigma_{\overline{X}}$ are the more accurate since less rounding was involved.

Note the difference between the values for $\sigma_{\overline{X}}$ in Examples 6.4 and 6.5. For both situations, $n = 3$, $\mu_X = 5.00$, and σ_X is approximately .1, but the FPC has a significant effect on the final value. Example 6.5 is the exception rather than the rule. The use of the FPC requires that N be known. In most practical situations, N is unknown or much greater than n. Although a generally accepted procedure is to ignore the FPC if $n/N \leq .05$, it is incorporated wherever appropriate throughout the remainder of this text.

We present several examples to illustrate the significance of the Central Limit Theorem. One application, the normal approximation to the binomial distribution, was presented in Chapter 4. The main result we have so far is that

$$\overline{X} \sim N(\mu_X, \sigma_{\overline{X}})$$

Using the standardization formula for a normal distribution, we obtain that

$$\frac{\overline{X} - \mu_{\overline{X}}}{\sigma_{\overline{X}}} = \frac{\overline{X} - \mu_X}{\sigma_{\overline{X}}} \sim N(0,1)$$

Thus, to calculate probabilities for sets of values for a sample mean, we need only calculate

$$Z = \frac{\overline{X} - \mu_X}{\sigma_{\overline{X}}} \tag{6.9}$$

and use Appendix Table B.6 as we did in Chapter 4.

EXAMPLE 6.6 We refer to Example 6.4 and assume the machine is considered to be out of control if the average length of a sample of 3 bolts is less than 4.9 inches or greater than 5.1 inches. What is the probability that the next 3 bolts selected will have a mean inside the control limits?

Solution: It is not likely that the distribution of the lengths will depart significantly from a normal. Thus, even for $n = 3$, applying the Central Limit Theorem should give a very good approximation of the true probability. It has already been determined that, for $n = 3$, $\sigma_{\overline{X}} = .0577$ and $\mu_{\overline{X}} = 5.00$ when the machine is functioning properly. Thus, with the Central Limit Theorem and these results, we have that $\overline{X} \sim N(5.00, .0577)$. Hence,

$$\begin{aligned}
P(4.9 \leq \overline{X} \leq 5.1) &= P\left(\frac{4.9 - \mu_X}{\sigma_{\overline{X}}} \leq \frac{X - \mu_X}{\sigma_{\overline{X}}} \leq \frac{5.1 - \mu_X}{\sigma_{\overline{X}}}\right) \\
&= P\left(\frac{4.9 - 5.0}{.0577} \leq Z \leq \frac{5.1 - 5.0}{.0577}\right) \\
&= P\left(\frac{-.1}{.0577} \leq Z \leq \frac{.1}{.0577}\right) \\
&= P(-1.73 \leq Z \leq 1.73) \\
&= P(-1.73 \leq Z \leq 0) + P(0 \leq Z \leq 1.73)
\end{aligned}$$

Using Appendix Table B.6, we find these values to be

$$P(-1.73 \leq Z \leq 0) = .4582 \quad \text{and} \quad P(0 \leq Z \leq 1.73) = .4582$$

Substitution yields

$$P(4.9 \leq \overline{X} \leq 5.1) = .9164$$

as the probability that a sample of 3 bolts will have an average length in inches between 4.9 and 5.1. This distribution is illustrated in Figure 6.5, along with that of the mean of a sample of 16 bolts, and compared to the distribution of lengths of the individual bolts assuming the distribution of individual bolts is normal. The height of each curve is the indicated value.

EXAMPLE 6.7 For comparison, assume the distribution of bolt lengths is normal and determine the probability that a single bolt will have a length between 4.9 inches and 5.1 inches.

FIGURE 6.5 Comparison of distributions of average bolt lengths for samples sizes of 1, 3, and 16

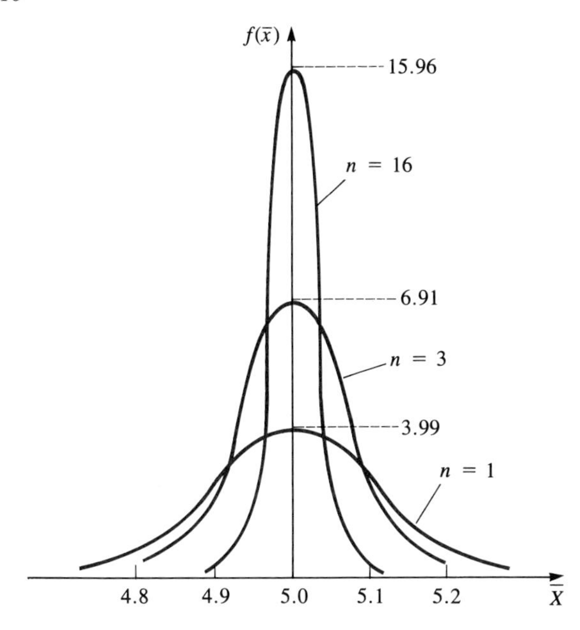

Solution: For a single bolt (see Figure 6.6), $n = 1$, $\mu_X = 5.0$, and $\sigma_X = .1$.

$$P(4.9 \leq X \leq 5.1) = P\left(\frac{4.9 - 5.0}{.1} \leq Z \leq \frac{5.1 - 5.0}{.1}\right)$$
$$= P(-1.00 \leq Z \leq 1.00) = .3413 + .3413 = .6826$$

In this case, we must assume that the population is normally distributed. Note that we have a 68.26% chance of obtaining a single bolt in the range of 4.9 to 5.1 inches while the same interval has a 91.64% chance of occurring when we select 3 bolts and calculate their average length. See Figure 6.7.

EXAMPLE 6.8 Refer to Example 6.6 and determine the probability that the mean for a sample of 16 bolts will be between 4.9 inches and 5.1 inches.

Solution: Under our assumptions, the distribution of the sample mean in this case will now be almost indistinguishable from a normal distribution. For the distribution of these means (see Figure 6.8), the mean and the standard deviation become

$$\mu_{\bar{X}} = \mu_X = 5.00 \quad \text{and} \quad \sigma_{\bar{X}} = \frac{\sigma_X}{\sqrt{n}} = \frac{.1}{\sqrt{16}} = .025$$

FIGURE 6.6 Distribution of the length of a single bolt with probability of a length between 4.90 to 5.10 inches shaded

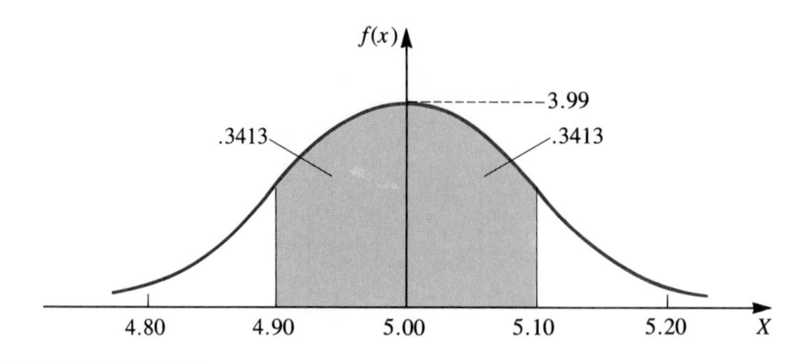

Thus,

$$P(4.9 \le \overline{X} \le 5.1) = P\left(\frac{4.9 - 5.0}{.025} \le Z \le \frac{5.1 - 5.0}{.025}\right)$$
$$= P(-4.00 \le Z \le 4.00) = .50 + .50 = 1.00$$

Note that in these examples, as n increased, the values for the sample mean became more densely packed around the population mean. The condensing of values is illustrated in Figure 6.5 where the distributions of the $\overline{X}$'s for $n =$ 1, 3, and 16 are all shown at once. The compaction was to be expected because

FIGURE 6.7 Distribution of the average length of 3 bolts with probability of an average between of 4.90 to 5.10 inches shaded

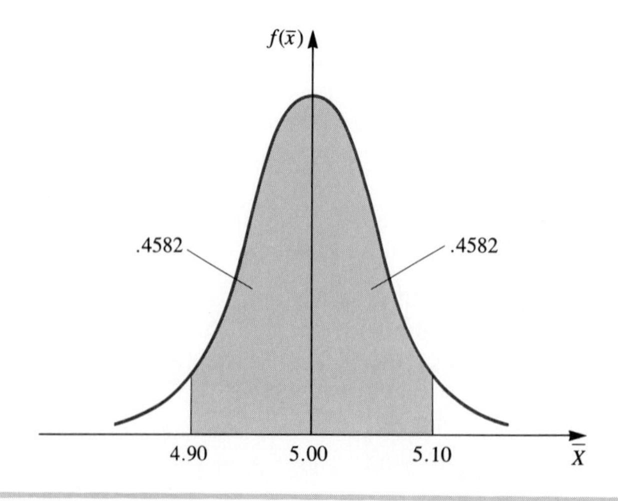

FIGURE 6.8 Distribution of the average length of 16 bolts with probability of an average outside of 4.90 to 5.10 inches being negligible

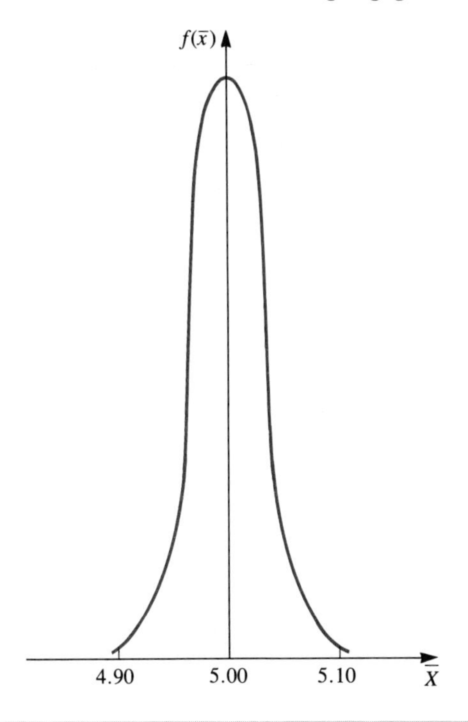

the standard deviation of the sample mean is a decreasing function of the sample size—that is, $\sigma_{\bar{X}} = \sigma_X/\sqrt{n}$ for infinite populations, and as n increases, $\sigma_{\bar{X}}$ decreases. Thus, with more observations, extreme values tend to cancel each other through the averaging process. We also note that if the original population is normal, the Central Limit Theorem is unnecessary, because the distribution of the sample means will be exactly normal regardless of the sample size.

We conclude this section with a few more examples.

EXAMPLE 6.9 Refer to Example 6.2.

a. Determine the probability that a sample of size 4 will have a mean exceeding 6.5.
b. Determine the probability that a sample of size 16 will have a mean in excess of 6.5.

Solution: Since the uniform distribution is symmetric, we see, by inspection, that the mean is 5.00. It can be shown by integration that for the $U(0,10)$ distribution, $\sigma_X^2 = 8.33$.

a. As seen in Figure 6.1, the histogram can be closely approximated with a normal curve, as indicated by the Central Limit Theorem. The mean and the standard deviation of $\overline{X}$ for $n = 4$ are

$$\mu_{\overline{X}} = 5.00 \quad \text{and} \quad \sigma_{\overline{X}} = \frac{\sigma_X}{\sqrt{n}} = \frac{\sqrt{8.33}}{\sqrt{4}} = 1.4434$$

Thus,

$$\begin{aligned} P(\overline{X} > 6.5) &= P\left(Z > \frac{6.5 - 5.0}{1.4434}\right) \\ &= P\left(Z > \frac{1.5}{1.4434}\right) = P(Z > 1.04) \\ &= .5 - P(0 \leq Z \leq 1.04) \\ &= .5 - .3508 = .1492 \end{aligned}$$

b. The only change effected here is the decrease in $\sigma_{\overline{X}}$ resulting from the increase in n. Thus,

$$\sigma_{\overline{X}} = \frac{\sigma_X}{\sqrt{n}} = \frac{\sqrt{8.33}}{\sqrt{16}} = .72$$

and

$$\begin{aligned} P(\overline{X} > 6.5) &= P\left(Z > \frac{6.5 - 5.0}{.72}\right) = P\left(Z > \frac{1.5}{.72}\right) \\ &= P(Z > 2.08) = .5 - P(0 \leq Z \leq 2.08) \\ &= .5 - .4812 = .0188 \end{aligned}$$

These two situations are illustrated in Figure 6.9. ■

FIGURE 6.9 Frequency polygon for $\bar{x}$'s of 36 samples of 4 each from a $U(0,10)$ distribution and the distributions of $\overline{X}$'s for $n = 4$ and 16 from this distribution according to the Central Limit Theorem

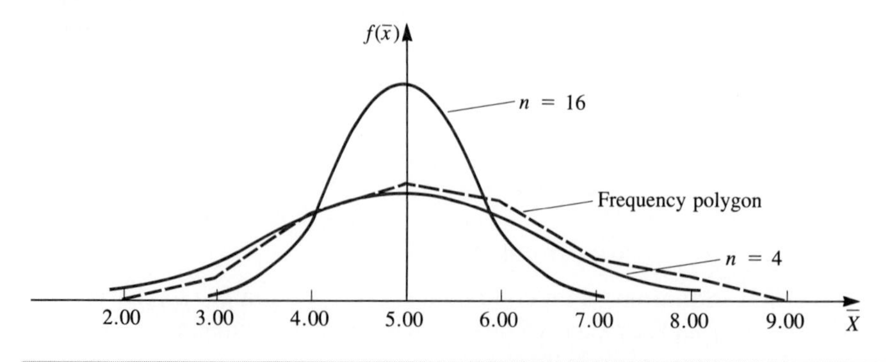

EXAMPLE 6.10 Apex Machinery manufactures large earth-moving equipment. Their sales for a particular machine had been averaging \$450,000 per month with a standard deviation of \$100,000. Six months ago, they were forced to recall some machines because of a safety defect. For the six months since the recall, their sales of this item have averaged \$370,000. What is the probability of Apex's average sales dropping to this point or lower if the recall had no effect on their sales? Sales for this item are nonseasonal and were fairly constant prior to the recall.

Solution: It is necessary to assume that the six months constitute a random sample. If the recall had no effect, then μ_X and σ_X are still \$450,000 and \$100,000, respectively. With $n = 6$, we wish to evaluate $P(\overline{X} \leq 370{,}000)$. We first determine $\mu_{\overline{X}}$ and $\sigma_{\overline{X}}$, and then apply the Central Limit Theorem.

$$\mu_{\overline{X}} = \mu_X = \$450{,}000$$

$$\sigma_{\overline{X}} = \frac{\sigma_X}{\sqrt{n}} = \frac{100{,}000}{\sqrt{6}} = 40{,}824.83$$

$$\begin{aligned} P(\overline{X} \leq 370{,}000) &= P\left(Z \leq \frac{370{,}000 - 450{,}000}{40{,}824.83}\right) \\ &= P\left(Z \leq \frac{-80{,}000}{40{,}824.83}\right) = P(Z \leq -1.96) \\ &= P(Z \leq 0) - P(-1.96 \leq Z \leq 0) = .5 - .4750 = .0250 \end{aligned}$$

Thus, the probability of such a decline in sales due to chance alone is quite small. This probability is shown in Figure 6.10.

In Example 6.10, we see the use of statistics in inferential form with an application to the decision-making process. Apex is faced with assigning a cause for their decline in sales. If all other factors, such as the seasonal effect on sales, levels of advertising and promotion, production capacity, and the like, have

FIGURE 6.10 Distribution of average sales ($n = 6$) for Apex Machinery with the probability of an $\overline{X}$ below \$370,000 shaded

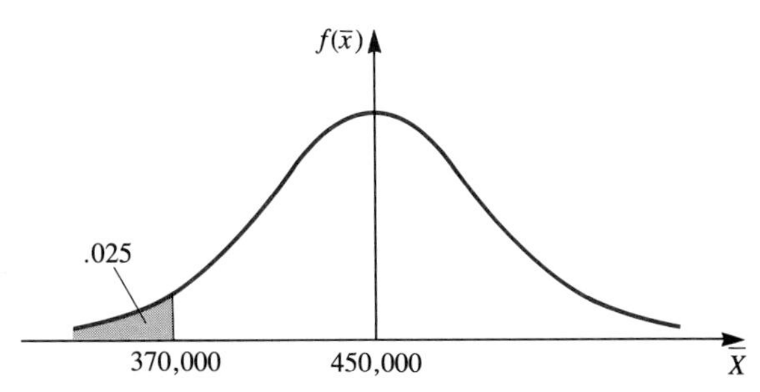

remained constant, then their sales dropped either purely by chance or else due to the effect of the recall. The probability that we calculated assumes that only random or chance variation is at work. Thus, .025 represents the probability of observing average sales of $370,000 or lower due to chance alone—that is, due to routine month-to-month fluctuations in sales. From a decision standpoint, we would have to decide whether or not .025 is unreasonably small. If it is felt that .025 is unreasonably small, we conclude that the sales decline may be attributed to the recall.

The next example illustrates another approach to making inferences. In this case, we will determine a range of values that can be expected to occur with a specified probability when only sampling variation is present. This range of values relates directly to the idea of confidence intervals to be presented in Chapter 7.

EXAMPLE 6.11 A cereal company packages their product in boxes labeled 15 ounces. They actually have an average fill weight of 15.05 ounces and a variance of .0225. If a large number of samples each consisting of 25 randomly selected boxes is taken, between what limits, equidistant from the mean, can we expect 90% of the $\overline{X}$ values to lie?

Solution: Here $\mu_X = 15.05$, $\sigma_X^2 = .0225$, and $n = 25$. We calculate

$$\mu_{\overline{X}} = \mu_X = 15.05 \qquad \sigma_{\overline{X}}^2 = \frac{\sigma_X^2}{n} = \frac{.0225}{25} = .0009$$

Thus,

$$\sigma_{\overline{X}} = \sqrt{.0009} = .03$$

We wish to determine the two values $\bar{x}_1$ and $\bar{x}_2$ symmetric about $\mu_{\overline{X}}$ such that $P(\bar{x}_1 \le \overline{X} \le \bar{x}_2) = .90$. Graphically, the result is shown in Figure 6.11. We know that $\overline{X}$, when standardized, yields the standard normal variable Z. We also know that probability statements are invariant under linear transformations within the parentheses. Hence, the probability statement becomes

$$\begin{aligned} P(\bar{x}_1 \le \overline{X} \le \bar{x}_2) &= P\left(\frac{\bar{x}_1 - \mu_{\overline{X}}}{\sigma_{\overline{X}}} \le Z \le \frac{\bar{x}_2 - \mu_{\overline{X}}}{\sigma_{\overline{X}}}\right) \\ &= P\left(\frac{\bar{x}_1 - 15.05}{.03} \le Z \le \frac{\bar{x}_2 - 15.05}{.03}\right) \\ &= .90 \end{aligned}$$

Since there is a one-to-one correspondence between z-values and probabilities, we can use Appendix Table B.6 to determine the values of Z associated with the transformed limits. By interpolation, these are $z_1 = -1.645$ and $z_2 = +1.645$, where

$$P(z_1 \le Z \le z_2) = P(-1.645 \le Z \le +1.645) = .90$$

FIGURE 6.11 Distribution of average weights for 25 randomly selected cereal boxes with 90% limits shaded

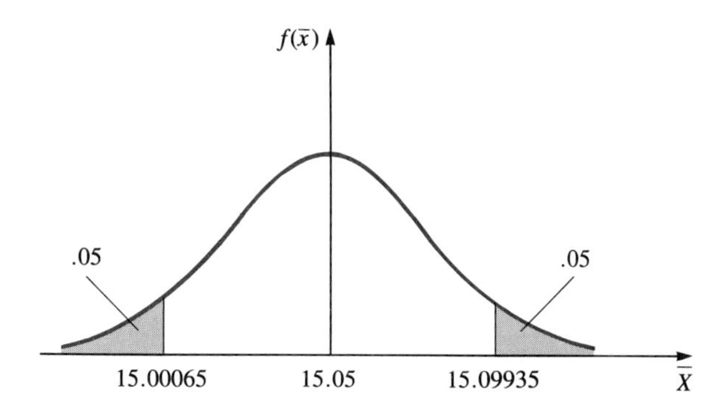

Equating, we obtain

$$\frac{\bar{x}_1 - 15.05}{.03} = -1.645 \qquad \frac{\bar{x}_2 - 15.05}{.03} = +1.645$$

and

$$\begin{aligned}\bar{x}_1 &= 15.05 - 1.645(.03) = 15.05 - .04935 \\ &= 15.00065 \text{ oz}\end{aligned}$$

$$\begin{aligned}\bar{x}_2 &= 15.05 + 1.645(.03) = 15.05 + .04935 \\ &= 15.09935 \text{ oz}\end{aligned}$$

Thus, we conclude that if the fill weights of the boxes average 15.05 ounces and have a variance of .0225 $(\text{oz})^2$, 90% of all samples of size 25 will have average fill weights between 15.00065 ounces and 15.09935 ounces.

So far, we have concerned ourselves only with the Central Limit Theorem as it relates to sample means. The theorem also states that the sums or totals of the sample observations will also have an approximate normal distribution. There are many applications of the Central Limit Theorem with respect to sample totals also, but before proceeding with an example, we obtain two results that enable us to specify the particular member of the normal family. Recall that for $\bar{X}$, we have $\mu_{\bar{X}} = \mu_X$ and $\sigma_{\bar{X}}^2 = \sigma_X^2/n$. If we let T be the sample total, then

$$T = \sum_{i=1}^{n} X_i = n\bar{X}$$

and

$$\mu_T = E[T] = E[n\overline{X}] = nE[\overline{X}]$$

or

$$\mu_T = n\mu_X \tag{6.10}$$

For the variance of T, we have, using Equation 6.3,

$$\sigma_T^2 = n\sigma_X^2 \tag{6.11}$$

EXAMPLE 6.12 The town of Mackenzie, Oregon, uses ashes on its streets in the winter to provide traction. The manager of the town's street department is faced with ordering ashes for the coming season. Looking at his past records, he finds that the town has used an average of 30 tons of ashes a day, and the standard deviation of daily usage has been 11.5 tons. Since the ashes are delivered 5 days after his order, he places each month's order on the 25th of the preceding month. Today is November 25, and he currently has 143 tons of ashes on hand.

a. If the townspeople become irate when the streets are not ashed, how many tons should he order now to have only a 10% chance of incurring the townspeople's wrath in December?
b. What should be his general ordering policy under these conditions?

Solution: To answer these questions, we must make two assumptions that may be questionable. First, we assume that the distribution governing the variation in usage is the same from one day to the next. Second, we require that the usage be independent from day to day. The second assumption satisfies the random sample requirement in the Central Limit Theorem, and the first stipulates that each observation comes from the same population.

a. Making these assumptions and letting T represent the total usage during the planning period, we have, by the Central Limit Theorem, that T has an approximately normal distribution. Applying Equations 6.10 and 6.11 gives

$$\mu_T = 36(30) = 1080 \quad \text{and} \quad \sigma_T^2 = 36(11.5)^2 = 4761$$

We use 36 days since the manager must take into account not only the usage for the 31 days of December but also the usage over the remaining 5 days of November. Therefore, he must cover his expected usage plus a sufficient amount in addition to insure that, if excessive demand occurs, he will run out with a probability of only .10. Thus, we need to determine a value for T such that $P(T > t) \leq .10$. Graphically, the result is shown in Figure 6.12. Since T is approximately a normal random variable, we

FIGURE 6.12 Distribution of monthly usage of ashes, in tons, for Mackenzie, Oregon, with 90% service level shaded

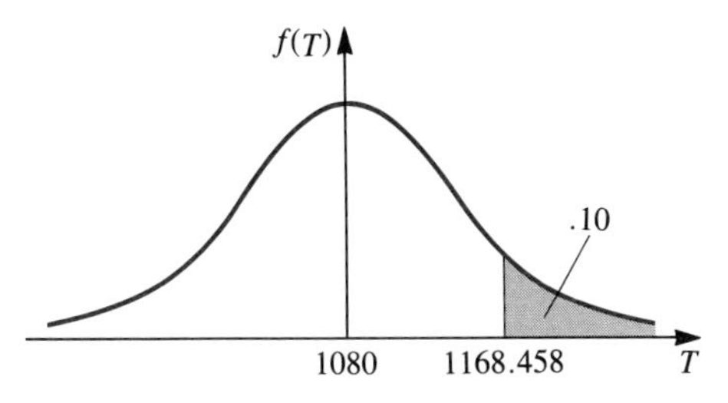

know that

$$P(T > t) = P\left(\frac{T - \mu_T}{\sigma_T} > \frac{t - \mu_T}{\sigma_T}\right)$$
$$= P(Z > z_{.1}) = .10$$

Finding $z_{.1}$ in Appendix Table B.6 and substituting the values for μ_T and σ_T give

$$\frac{t - 1080}{69} = 1.282$$

Solving for t, we find that he needs to order enough to cover a demand of

$$t = 1080 + 1.282(69) = 1168.458 \text{ tons}$$

Since he currently has 143 tons on hand, he must order 1169 − 143 = 1026 tons. It is assumed that whole tons must be ordered, so we rounded up to 1169 to insure that the probability is ≤ .10.

b. The answer to this part follows directly from the answer to part (a) by letting n_i denote the number of days in the next month. Thus, he should order an amount equal to t, where $t = \mu_T + z_{.1}\,\sigma_T - \text{OH}$, where OH is the amount on hand at the time. Thus, the amount to be ordered is

$$30(n_i + 5) + 1.282(11.5)\sqrt{n_i + 5} - \text{OH}$$

Commenting briefly on the assumptions we made in Example 6.12, we note that the independence assumption is most unlikely. But, since we are planning over a 36-day period, it should not affect the results drastically. The assumption of identical distributions is also questionable when extended over the full season. We could lessen the effect of this assumption's being violated by using the mean and the standard deviation of daily demand for the particular month in question. These data should be available from past records.

EXERCISES

6.3 The mean age of day students at Camden University is 20 years and the standard deviation of ages is 1.2 years.

(a) What is the probability that the mean of a random sample of 16 students will be between 19.4 and 20.9 years?

(b) What is the probability that an individual student's age will be in the same range?

6.4 Refer to Example 6.12. Suppose it is now January 26, and the manager must place his order for February. If we assume that it is not a leap year and that the distribution of daily demand is as before, how much should he order if he wishes to be 95% certain of having enough?

6.5 Given a sample size of 144 from a continuous distribution with mean 80 and variance 576, what are the mean, variance, and standard deviation of the distribution of $\overline{X}$?

6.6 One of the products of the Ajax Bolt Company is a lug bolt for large earth-moving equipment. Ajax's main customer for this bolt specifies that it is to be 30 cm long and is to deviate no more than .12 cm from that length 95% of the time. Ajax has just purchased a new machine for cutting the bolts and is attempting to set up limits for their quality control inspectors.

(a) Assume bolt lengths are normally distributed. What standard deviation will give bolts meeting the customer's requirements when $\mu = 30$ cm?

(b) If the inspectors select random samples of 12 bolts each, within what range should 90% of the sample means fall when the machine is operating according to specifications?

6.7 The Nature's Pride Cereal Company fills their 16-ounce boxes with an average weight of 16.1 ounces. The standard deviation of the weights is known to be .18 ounce. The boxes are then packaged into cases containing 24 boxes each.

(a) What is the probability that a case will have a net weight in excess of 388 ounces if box weights are assumed to be normally distributed?

(b) What is the effect on your answer if normality is not assumed?

6.4 DISTRIBUTION OF SAMPLE PROPORTIONS AND TOTALS FOR BINOMIAL POPULATIONS

In the preceding section, we saw that means and totals of sample measurements have distributions that are approximately normal. These results also can be applied to large samples from dichotomous (binomial) populations. We first saw this in Chapter 4 when the normal distribution was used to obtain approximate probabilities for binomially distributed random variables from large samples. The restriction on the sample size is such that both $n\pi$ and $n(1 - \pi)$ must be greater than or equal to 5. We note that the binomial random variable X is actually a sample total that, by the Central Limit Theorem, can be approximated with a normal distribution.

DEFINITION 6.1 A **binomial random variable** is actually a sum of independent Bernoulli random variables each with parameter π, the probability of success. That is,

$$X = \sum_{i=1}^{n} Y_i \tag{6.12}$$

where: $Y_i = 0$ for a failure

$Y_i = 1$ for a success

To identify the specific normal distribution, we recall that for a binomial random variable, $\mu_X = n\pi$ and $\sigma_X^2 = n\pi(1 - \pi)$. This result, in conjunction with the Central Limit Theorem, specifies a binomial total as approximately normal with mean $n\pi$ and variance $n\pi(1 - \pi)$—that is,

$$X \dot{\sim} N(n\pi, \sqrt{n\pi(1 - \pi)}) \tag{6.13}$$

The symbol $\dot{\sim}$ indicates that the distribution is approximate.

For the sample proportion p we have $p = X/n$, where X is the total number of successes and n is the sample size or number of trials. Substituting from Equation 6.12 gives

$$p = \frac{1}{n}\sum_{i=1}^{n} Y_i$$

which is the form of a sample mean. Thus, for large n, the Central Limit Theorem justifies using a normal distribution to approximate the distribution of a sample proportion also. The mean and the variance of a sample proportion obtained through independent trials are

$$\mu_p = E\left[\frac{X}{n}\right] = \frac{1}{n} E[X] = \frac{1}{n}(n\pi) = \pi$$

and

$$\sigma_p^2 = \text{Var}\left(\frac{X}{n}\right) = \frac{1}{n^2}\text{Var}(X) = \frac{1}{n^2}[n\pi(1 - \pi)] = \frac{\pi(1 - \pi)}{n}$$

Therefore, for large n, we obtain that

$$p \dot{\sim} N\left(\pi, \sqrt{\frac{\pi(1 - \pi)}{n}}\right) \tag{6.14}$$

EXAMPLE 6.13 The internal auditing staff of the Culverson Company performs a sample audit of customer accounts each quarter to estimate the number of accounts that are at least 30 days overdue. The last complete audit indicated that 30% of all accounts were in this category. For the current sample audit, the staff plans to randomly select 400 accounts.

a. What is the probability that no more than 110 accounts will be at least 30 days overdue?

b. What is the probability that at least 35% will be 30 days or more overdue?

Solution: To answer these questions, we must assume that the actual proportion is the same as it was in the last complete audit.

a. From Equation 6.13, we see that the total number of accounts is distributed approximately as

$$X \mathrel{\dot{\sim}} N(120, \sqrt{84})$$

Thus,

$$\begin{aligned} P(X \le 110 \mid n = 400, \ \pi = .3) &\doteq P(X \le 110.5 \mid \mu = 120, \ \sigma = 9.1652) \\ &= P\left(Z \le \frac{110.5 - 120}{9.1652}\right) \\ &= P(Z \le -1.04) = .1492 \end{aligned}$$

The symbol $\doteq$ is defined to mean "approximately equal to." Therefore, there is about a 15% chance that no more than 110 accounts will be in this overdue category. The result is shown in Figure 6.13(a). The .5 that was added to 110 is the correction for continuity that was introduced in Chapter 4.

b. In this part, we want to determine

$$P(p \ge .35 \mid n = 400, \ \pi = .3)$$

The correction for continuity here is $.5/n = .5/400 = .00125$. Applying Equation 6.14 gives an approximate answer of

$$\begin{aligned} P(p \ge .35 - .00125 \mid \pi = .3, \ \sigma = .0229) &= P\left(Z \ge \frac{.34875 - .300}{.0229}\right) \\ &= P(Z \ge 2.13) = .0166 \end{aligned}$$

FIGURE 6.13 Normal approximations for the distributions of X and $p = X/n$ for a binomial situation with $n = 400$ and $\pi = .3$

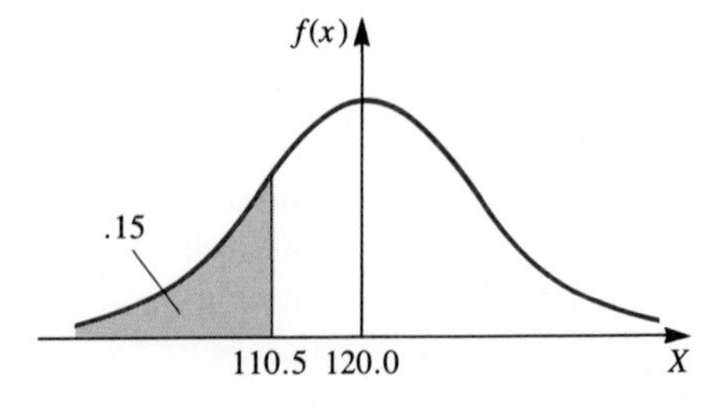

(a) Number of delinquent accounts

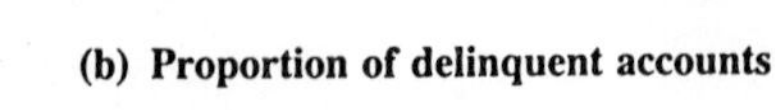

(b) Proportion of delinquent accounts

Thus, there is a 1.66% chance that the proportion of accounts at least 30 days overdue will be at least .35. This result is shown in Figure 6.13(b).

EXERCISES

6.8 In a bottling process, an average of 4% of all bottles is not sealed properly. In a shipment of 2000 bottles, what is the probability that no more than 70 will not be sealed properly?

6.9 Referring to Exercise 6.8, what is the probability that the proportion of bottles that are not sealed properly will be greater than .046?

6.10 It is believed that the Internal Revenue Service (IRS) audits 15% of all persons having incomes less than $40,000 at least once. What is the probability that in a sample of 120 such persons, at least 21 will have been audited at least once?

6.11 If a random sample of 200 is selected from the individuals identified in Exercise 6.10, what is the probability that the proportion of the sample who were audited at least once is at most .12?

6.5 OTHER SAMPLING DISTRIBUTIONS

In this section, we introduce some of the distributions that will play a significant role in later chapters. They are discussed at this time for two reasons. First, they arise only through manipulation of sample data—hence, the term "sampling distributions"—and, second, they are all integrally related to the standard normal distribution.

All distributions discussed previously can be used to describe natural phenomena (often they are called *natural distributions*) as well as to calculate probabilities associated with samples. The three sampling distributions presented in this section are used only to describe certain statistics that are calculated using sample data drawn from normal populations. The procedures based on these distributions are *robust,* which means that slight departures from the normality assumption will not affect the results significantly.

The Chi-Square Distribution

The first of these sampling distributions is used to describe the chance variation of sample variances calculated from random samples from normal populations. Before stating the result, we first define and explain the concept of degrees of freedom.

DEFINITION 6.2 **Degrees of freedom** is defined to be the number of variables minus the number of independent linear restrictions placed upon them.

Intuitively, degrees of freedom is the number of variables to which values may be freely assigned until all values for the remaining variables become determined by the linear restrictions. The following example illustrates this concept.

EXAMPLE 6.14 Assume 5 values have been randomly selected and that it is desired to calculate the variance s_X^2. If the 5 observations are denoted by X_1, X_2, X_3, X_4, X_5, and $\overline{X}$ is assumed to be 10, how many degrees of freedom are available for calculating $\Sigma_{i=1}^{5} (X_i - \overline{X})^2$, the numerator of s_X^2?

Solution: A value of 10 for $\overline{X}$ imposes the linear restriction that

$$\frac{1}{5}\sum_{i=1}^{5} X_i = 10$$

Of our 5 variables, 4 may assume any value, but after they have been assigned, the 5th is automatically determined by the restriction. For example, suppose the following values have been assigned: $x_1 = 3$, $x_2 = -4$, $x_3 = 15$, $x_4 = 10$. At this point, x_5 *must* equal 26 since $\Sigma_{i=1}^{5} x_i = 50$. Thus, there are $n - 1 = 5 - 1 = 4$ degrees of freedom available for calculating the numerator of s_X^2.

We considered the idea of degrees of freedom first because it plays an important part in each of the three sampling distributions. The parameters for all three distributions are degrees of freedom. Thus, to identify which member of a given family is to be used, we need only determine the degrees of freedom.

DEFINITION 6.3 Let Z_i, $i = 1, 2, \ldots, \nu$ (nu), be a set of independent standard normal random variables. That is, each Z_i is $N(0,1)$. Then, the sum of the squared Z_i's will have a **χ^2 (chi-square) distribution** with ν degrees of freedom. That is,

$$\sum_{i=1}^{\nu} Z_i^2 \sim \chi_\nu^2 \qquad \textbf{(6.15)}$$

Earlier, we stated that the Central Limit Theorem is probably the most important single concept in statistics. What may be the second most important result relates to sums of squares and χ^2 distributions. This result is known as Cochran's theorem and provides the basis for all of the inferential procedures in analysis of variance (Chapter 9) and most of those in regression (Chapters 10, 11). We state it here in a simplified form.

COCHRAN'S THEOREM Let Q be a sum of squared random variables such that Q has a χ^2 distribution with ν degrees of freedom. Furthermore, assume that Q can be expressed as the sum of a set of Q_i, $i = 1, 2, \ldots, m$, where each Q_i represents a smaller sum of squares. Then the Q_i's will be independent, with each Q_i being distributed as a χ^2 distribution with ν_i degrees of freedom *if and only if* $\Sigma_{i=1}^{m} \nu_i = \nu$—that is, if and only if the degrees of freedom for the Q_i total the degrees of freedom for Q.

As stated earlier, one of the main uses of the χ^2 distribution is the determination of the distribution of the variance of a random sample taken from a normal population. We illustrate the use of Cochran's theorem by finding the distribution of s_X^2. Let X_i, $i = 1, 2, \ldots, n$, be a random sample from an $N(\mu,\sigma)$ distribution, and let

$$Q = \sum_{i=1}^{n} \left(\frac{X_i - \mu_X}{\sigma_X} \right)^2 = \frac{1}{\sigma_X^2} \sum_{i=1}^{n} (X_i - \mu_X)^2$$

Then, since $(X_i - \mu_X)/\sigma_X$ is an $N(0,1)$ random variable, Q, by Definition 6.3, has a χ^2 distribution with n degrees of freedom. Also,

$$\begin{aligned} \frac{1}{\sigma_X^2} \sum_{i=1}^{n} (X_i - \mu_X)^2 &= \frac{1}{\sigma_X^2} \sum_{i=1}^{n} (X_i - \overline{X} + \overline{X} - \mu_X)^2 \\ &= \frac{1}{\sigma_X^2} \sum_{i=1}^{n} (X_i - \overline{X})^2 + \frac{1}{\sigma_X^2} \sum_{i=1}^{n} (\overline{X} - \mu_X)^2 \\ &= \frac{1}{\sigma_X^2} \sum_{i=1}^{n} (X_i - \overline{X})^2 + \frac{n(\overline{X} - \mu)^2}{\sigma_X^2} \end{aligned}$$

since it is easily shown that

$$\sum_{i=1}^{n} (X_i - \overline{X})(\overline{X} - \mu) = 0$$

The first summation on the right side of the equal sign is simply the numerator for s^2 and has $n - 1$ degrees of freedom. In the second term, $\overline{X}$ is the only random variable and has no restrictions on it; hence, it has one degree of freedom. (This result can also be seen because $[\sqrt{n}\,(\overline{X} - \mu_X)]/\sigma_X$ is an $N(0,1)$ variable, and therefore $[n(\overline{X} - \mu_X)^2]/\sigma_X^2$ is a χ^2 variable with one degree of freedom.) Thus, we have shown that the degrees of freedom on the right side sum to the degrees of freedom for Q—that is, $n = (n - 1) + 1$. By Cochran's theorem, it follows that

$$\frac{1}{\sigma_X^2} \sum_{i=1}^{n} (X_i - \overline{X})^2 \sim \chi_{n-1}^2 \qquad \textbf{(6.16)}$$

Dividing both sides of Equation 6.16 by $n - 1$ gives

$$\frac{1}{\sigma_X^2} \left(\frac{\sum_{i=1}^{n} (X_i - \overline{X})^2}{n - 1} \right) \sim \frac{\chi_{n-1}^2}{n - 1}$$

The term in parentheses is s_X^2, by definition, and we have

$$\frac{s_X^2}{\sigma_X^2} \sim \frac{\chi_{n-1}^2}{n - 1}$$

or

$$\frac{(n - 1)s_X^2}{\sigma_X^2} \sim \chi_{n-1}^2 \qquad \textbf{(6.17)}$$

FIGURE 6.14 Approximate sketches of χ^2 distributions for various degrees of freedom

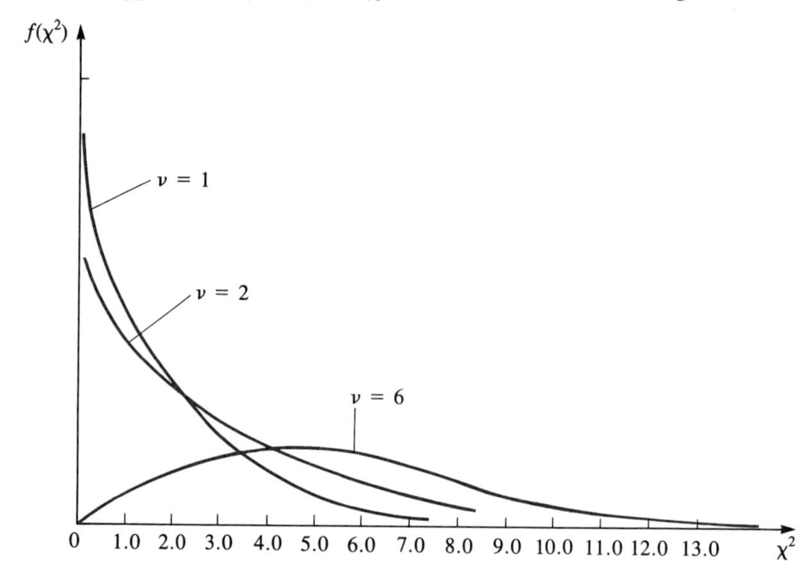

Remember that chi-square is the name of a family of distributions, as are the normal, binomial, and the others discussed in Chapter 4. The chi-square is a one-parameter family, with that parameter being the degrees of freedom. Graphs of χ^2 distributions for various degrees of freedom are presented in Figure 6.14. Unlike the normal, the χ^2 distribution, by definition, must be nonnegative. Values for χ^2 variables at specific probability levels are given in Appendix Table B.8. The table is arranged to show the probability at the top of the column that identifies the area to the right of the χ^2 value. The rows are indexed by the degrees of freedom, and each one gives values for a different member of the family.

EXAMPLE 6.15 Use Appendix Table B.8 to find the following:

a. The value that a χ^2 variable with 5 degrees of freedom will exceed 10% of the time—that is, the number, denoted $\chi^2_{5,.10}$, such that

$$P(\chi^2_5 > \chi^2_{5,.10}) = .10$$

b. The value that a χ^2 variable with 15 degrees of freedom will exceed 95% of the time

c. The probability that a χ^2 variable with 20 degrees of freedom will be less than 9.59

Solution:

a. Graphically, the result would appear as shown in Figure 6.15(a). Turning to Appendix Table B.8, we locate the fifth row and the column headed by .10. At the intersection of the row and column, we find the value for $\chi^2_{5,.10}$ to be 9.24. Therefore,

$$P(\chi^2_5 > 9.24) = .10$$

b. This result is shown in Figure 6.15(b). In Appendix Table B.8, we locate the intersection of the 15th row and the .95 column to find $\chi^2_{15,.95} = 7.26$.

FIGURE 6.15 Graphs of chi-square distributions for Example 6.15

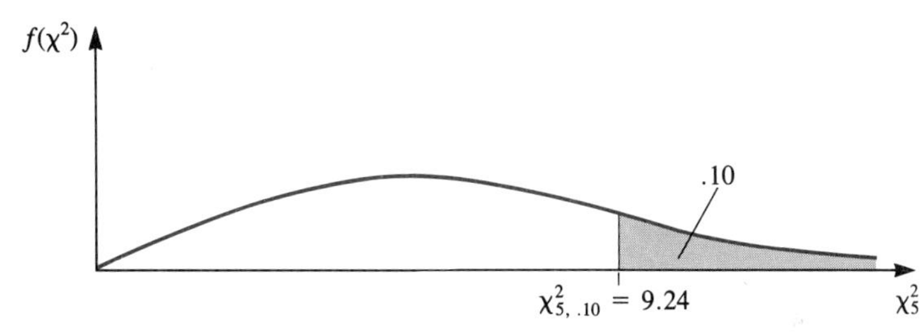

(a) $\nu = 5$

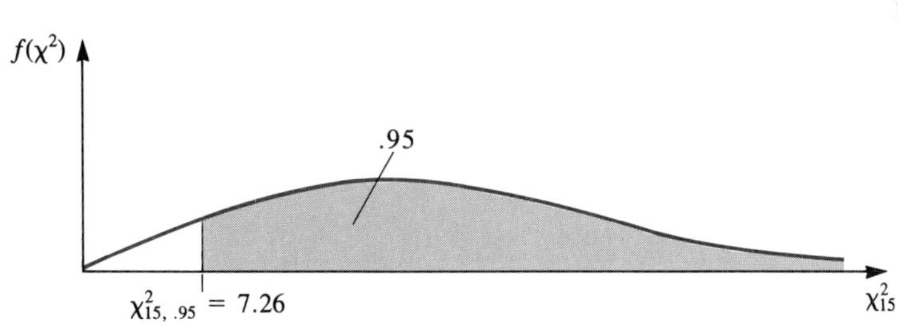

(b) $\nu = 15$

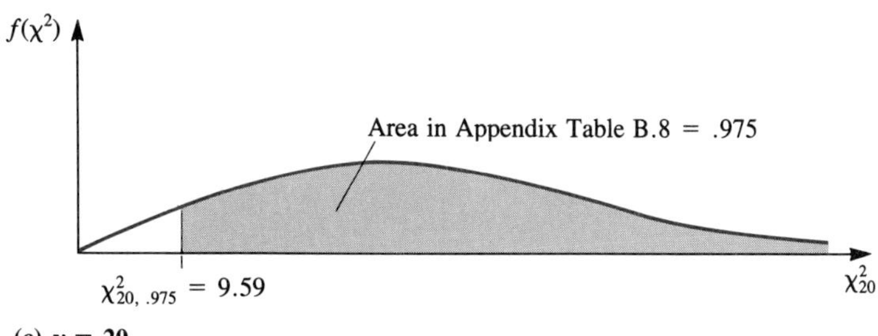

(c) $\nu = 20$

Therefore,

$$P(\chi^2_{15} > 7.26) = .95$$

c. The graph is shown in Figure 6.15(c). Turning to Appendix Table B.8 and scanning the 20th row, we find 9.59 in the column identified by .975. Hence,

$$P(\chi^2_{20} \geq 9.59) = .975$$

and

$$\begin{aligned} P(\chi^2_{20} < 9.59) &= 1 - P(\chi^2_{20} \geq 9.59) \\ &= 1.00 - .975 = .025 \end{aligned}$$

EXAMPLE 6.16 Assume a machine is used to drill tap holes in steel blocks. When working properly, the holes will have a mean diameter of 1.00 inch and a variance of .0001 (inches)^2. When 5 blocks are randomly selected and their tap holes measured, the variance of the diameters is calculated to be .0003319. If the diameters are normally distributed, what is the probability of observing a sample variance as large or larger due to chance alone?

Solution: We wish to determine $P(s_X^2 \geq .0003319)$. It was shown earlier that the variable $[(n-1)s_X^2]/\sigma_X^2$ has a χ^2_{n-1} distribution. Using this result yields

$$\begin{aligned} P(s_X^2 \geq .0003319) &= P\left(\frac{(n-1)s_X^2}{\sigma_X^2} \geq \frac{(n-1)(.0003319)}{\sigma_X^2}\right) \\ &= P\left(\frac{4s_X^2}{.0001} \geq \frac{4(.0003319)}{.0001}\right) \\ &= P(\chi^2_4 \geq 13.28) = .01 \qquad \text{from Appendix Table B.8} \end{aligned}$$

This area is shown in Figure 6.16.

FIGURE 6.16 Distribution of $(n-1)s_X^2/\sigma_X^2$ for variances of samples of size 5

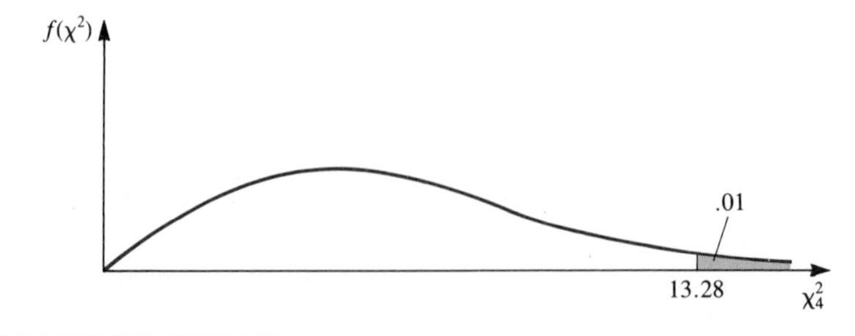

Student's *t*-distribution

The second distribution we consider is a function of the standard normal and the chi-square.

DEFINITION 6.4 A ***t* variable** with ν degrees of freedom is defined to be the ratio of a standard normal variable to the square root of an independent χ_ν^2 variable divided by its degrees of freedom. That is,

$$t_\nu = \frac{Z}{\sqrt{\chi_\nu^2/\nu}} \tag{6.18}$$

Note that the t variable gets its degrees of freedom from the χ^2 variable in the denominator. The distribution of a t variable is quite similar to that of the standard normal variable Z. Recall that the variable $(\overline{X} - \mu_X)/\sigma_{\overline{X}}$ has an $N(0,1)$ distribution and hence is equivalent to a Z variable.

This transformation requires that σ_X be known, which generally will not be the case. When σ_X is unknown, it is replaced by its estimate s_X, which results in a loss of information and precision. With the replacement of σ_X by s_X, the statistic no longer has a standard normal distribution. The new statistic, which is $(\overline{X} - \mu_X)/(s_X/\sqrt{n})$, is equivalent to

$$\frac{\sqrt{n}\,(\overline{X} - \mu_X)}{\sigma_X} \Bigg/ \sqrt{\frac{s_X^2}{\sigma_X^2}}$$

where the numerator is $N(0,1)$ and $s_X^2/\sigma_X^2 \sim \chi_{n-1}^2/(n-1)$. Hence, the statistic has a t-distribution with $n - 1$ degrees of freedom—that is,

$$\frac{\overline{X} - \mu_X}{s_{\overline{X}}} \sim t_{n-1} \tag{6.19}$$

where $s_{\overline{X}} = s_X/\sqrt{n}$. Percentage points of t-distributions are given in Appendix Table B.7 for various degrees of freedom and selected probability levels. As with the χ^2 distribution, degrees of freedom is the parameter of the t family and identifies which member is under consideration. Various members of the t family are sketched in Figure 6.17.

Appendix Table B.7 is identified in the same manner as Appendix Table B.8. That is, each row identifies a different member of the t family, and each column corresponds to a specific probability value. The probability given at the top of a column is the probability that a t variable will exceed the value listed in the body of the table. Being symmetric about zero, the probability of falling below a negative value is the same as the probability of falling above the corresponding positive value.

FIGURE 6.17 Representative sketches for various t-distributions

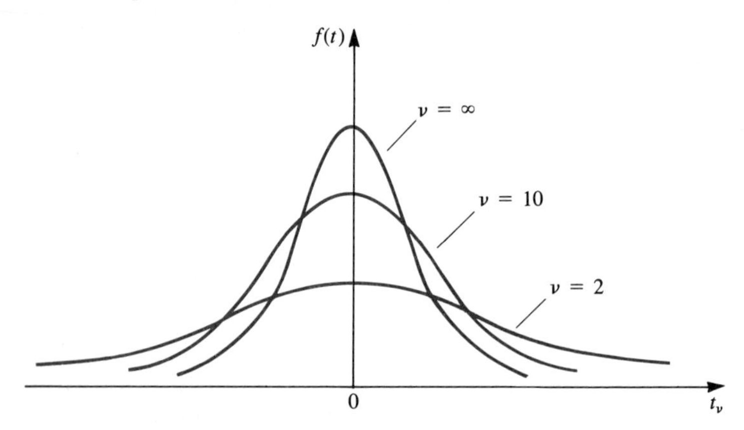

EXAMPLE 6.17 Using Appendix Table B.7, determine the following t values.

a. Find t_0 such that $P(t_7 > t_0) = .05$.
b. Find t_0 such that $P(t_{15} < t_0) = .025$.
c. Find t_0 such that $P(-t_0 < t_{20} < t_0) = .99$.

Solution: Turning to Appendix Table B.7, we determine the unknown values in the following manner. First, determine what the sign will be by sketching a curve and shading in the indicated area. Locate the row in Appendix Table B.7 corresponding to the degrees of freedom and the column corresponding to the appropriate probability value and read the value.

a. The curve is shown in Figure 6.18(a), and we see that for $P(t_7 > t_0) = .05$, t must be positive. Going to the intersection of the 7th row and the column headed by .05, we find

$$t_0 = +1.895 \quad \text{and} \quad P(t_7 > 1.895) = .05$$

b. Figure 6.18(b) shows that for $P(t_{15} \leq t_0) = .025$, t_0 must be negative. Going to the 15th row and the column headed by .025, we have

$$t_0 = -2.131 \text{ and } P(t_{15} < -2.131) = .025$$

c. $P(-t_0 \leq t_{20} \leq t_0)$ is shown in Figure 6.18(c).

FIGURE 6.18 Sketches of t-distributions for Example 6.17

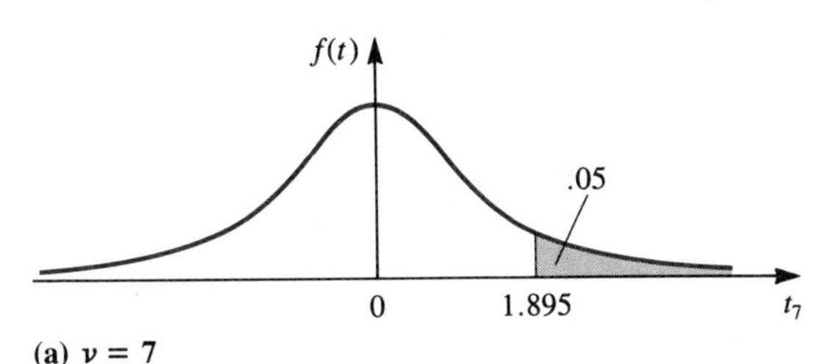

(a) $\nu = 7$

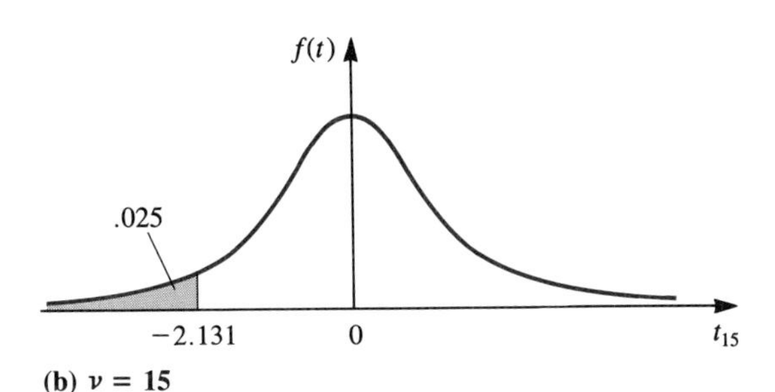

(b) $\nu = 15$

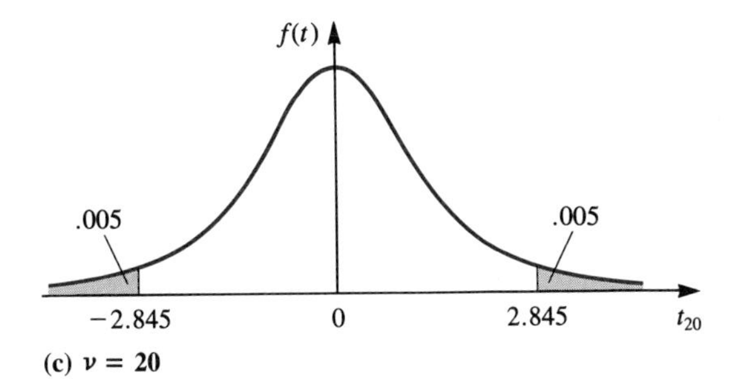

(c) $\nu = 20$

For this problem, we note that the area outside the interval is divided equally between the two tails. Thus, the t_0 value is in the 20th row and the column headed by .005.

$$t_0 = t_{20,.005} = 2.845$$

$$P(-2.845 < t_{20} < 2.845) = .99$$

EXAMPLE 6.18 Refer to Example 6.16. Suppose that the mean diameter of the 5 sample tap holes is 1.0174. The sample variance is .0003319. If the underlying distribution is normal, what is the probability of observing a sample mean this large or larger by chance alone?

FIGURE 6.19 Distribution of $(\overline{X} - \mu_X)/s_{\overline{X}} = t_4$ for Example 6.18

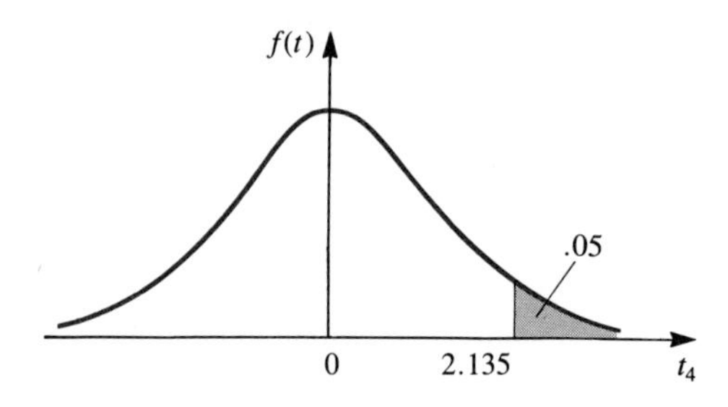

Solution: We need to determine $P(\overline{X} \geq 1.0174)$ and are given that the population is normal. Using the sample standard deviation, we have that

$$\frac{\overline{X} - \mu_X}{s_{\overline{X}}} \sim t_{n-1}$$

In this case,

$$s_{\overline{X}} = \frac{s_X}{\sqrt{n}} = \frac{\sqrt{.0003319}}{\sqrt{5}} = .00815$$

and

$$\begin{aligned} P(\overline{X} \geq 1.0174) &= P\left(t_{5-1} \geq \frac{1.0174 - 1.0000}{.00815}\right) \\ &= P(t_4 \geq 2.135) \\ &\doteq .05 \qquad \text{(from Appendix Table B.7)} \end{aligned}$$

The result is shown graphically in Figure 6.19.

The last distribution we consider arises as a function of two chi-square variables.

The *F*-distribution

DEFINITION 6.5 An ***F* variable** with ν_1 degrees of freedom in the numerator and ν_2 degrees of freedom in the denominator is defined to be the ratio of two independent χ^2 variables divided by their respective degrees of freedom. That is,

$$F_{\nu_1,\nu_2} = \frac{\chi^2_{\nu_1}/\nu_1}{\chi^2_{\nu_2}/\nu_2} \tag{6.20}$$

This family of distributions is of major importance in the comparison of variances for two random samples drawn from normal populations. Referring to Equation 6.17, we see that if the two sample variances are denoted by s_1^2 and s_2^2 and the population variances by σ_1^2 and σ_2^2, respectively, we have

$$\frac{(n_1 - 1)s_1^2}{\sigma_1^2} \sim \chi^2_{n_1-1}$$

and

$$\frac{(n_2 - 1)s_2^2}{\sigma_2^2} \sim \chi^2_{n_2-1}$$

Dividing by respective degrees of freedom gives

$$\frac{s_1^2}{\sigma_1^2} \sim \frac{\chi^2_{n_1-1}}{n_1 - 1}$$

and

$$\frac{s_2^2}{\sigma_2^2} \sim \frac{\chi^2_{n_2-1}}{n_2 - 1}$$

The ratio

$$\frac{s_1^2/\sigma_1^2}{s_2^2/\sigma_2^2} \sim \frac{(\chi^2_{n_1-1})/(n_1 - 1)}{(\chi^2_{n_2-1})/(n_2 - 1)} \sim F_{n_1-1,n_2-1} \tag{6.21}$$

by Definition 6.5.

Like the χ^2 distribution, the F variable, as the ratio of two nonnegative variables, also must be nonnegative. As indicated by the foregoing notation, F is a two-parameter family with the parameters given by the degrees of freedom for the numerator and for the denominator, respectively. Percentage points for F-distributions are given for a probability level of .05 in Appendix Table B.9a and for a probability level of .01 in Appendix Table B.9b. To determine a particular percentage point for an F_{ν_1,ν_2}-distribution, turn to the table corresponding to the appropriate probability. This table will be indexed by two integers. The first integer, the degrees of freedom for the numerator, identifies the column in which the value lies; the second integer, the degrees of freedom for the denominator, identifies the row. The value is then at the intersection of that column and row. Graphically, F-distributions will be similar to those in Figure 6.20. As in Appendix Tables B.7 and B.8, the probability associated with the value is the area to the right. Values for F having corresponding areas of .05 and .01 on the left can be calculated from the relationship

$$F_{\nu_1,\nu_2,1-\alpha} = \frac{1}{F_{\nu_2,\nu_1,\alpha}} \tag{6.22}$$

where $1 - \alpha$ is to the right and will generally be the larger probability. Note that the degrees of freedom are reversed in the two F-values.

FIGURE 6.20 Sketches of F-distributions for various values of ν_1 and ν_2

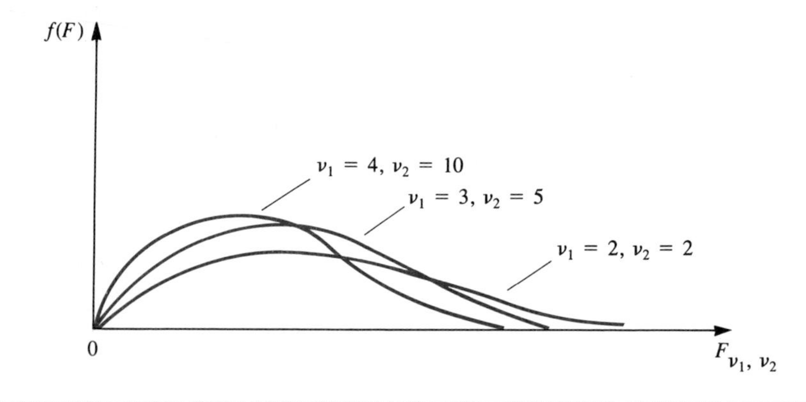

EXAMPLE 6.19 Using Appendix Tables B.9a and B.9b, determine the indicated F values.

a. Find the value F_0 such that an F with $\nu_1 = 3$ and $\nu_2 = 10$ will exceed it 1% of the time—that is, such that $P(F_{3,10} > F_0) = .01$.
b. Find the value that will be exceeded 5% of the time by an F with $\nu_1 = 10$ and $\nu_2 = 8$.
c. Find the value that will be exceeded 95% of the time by an F with $\nu_1 = 5$ and $\nu_2 = 20$.

Solution: Sketches of the distributions are given in Figure 6.21. The values are determined as follows.

a. We want to find F_0 so that $P(F_{3,10} > F_0) = .01$. Since the area is .01, we turn to Appendix Table B.9b and find the intersection of row 10 and column 3. The number is 6.55, and $P(F_{3,10} > 6.55) = .01$.
b. Here the probability statement, $P(F_{10,8} > F_0) = .05$, implies Appendix Table B.9a. The corresponding column and row are 10 and 8, respectively. The value is 3.35, and $P(F_{10,8} > 3.35) = .05$.
c. In this case, we want F_0 such that $P(F_{5,20} < F_0) = .05$. To determine the value, we look up $F_{20,5,.05}$ and use Equation 6.22 to obtain

$$F_{5,20,.95} = \frac{1}{F_{20,5,.05}} = \frac{1}{4.56} = .219$$

Therefore,

$$P(F_{5,20} < .219) = .05$$

FIGURE 6.21 Sketches of F-distributions for Example 6.19

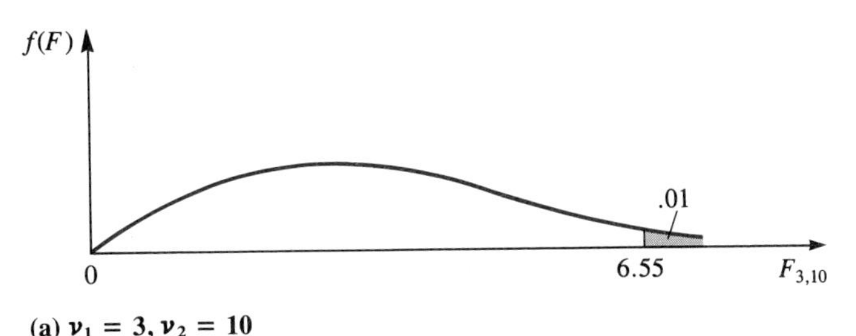

(a) $\nu_1 = 3, \nu_2 = 10$

f(F)

.05

0

3.35

$F_{10,8}$

(b) $\nu_1 = 10, \nu_2 = 8$

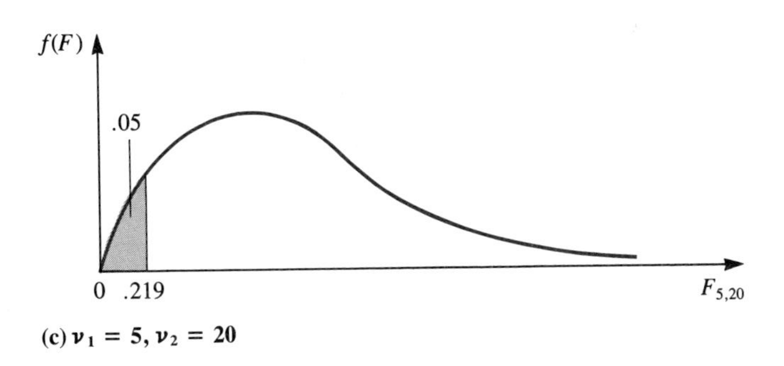

(c) $\nu_1 = 5, \nu_2 = 20$

EXAMPLE 6.20 Suppose the company in Example 6.16 has taken a second sample of 8 blocks, measured the tap holes in them, and obtained a sample variance of .00008. Denote this by s_2^2. The variance of the first sample is .0003319 and is denoted by s_1^2. What is the probability of observing a ratio for s_1^2/s_2^2 as large or larger than that observed by chance alone if a normal population is assumed?

Solution: Since both samples are from the same population, $\sigma_1^2 = \sigma_2^2$. We have that $n_1 = 5$, $s_1^2 = .0003319$, $n_2 = 8$, and $s_2^2 = .00008$. Thus, the observed ratio is $.0003319/.00008 = 4.15$. If the two samples are independent, the ratio of s_1^2/s_2^2 has an F-distribution with 4 degrees of freedom for the numerator and 7 degrees of

FIGURE 6.22 Distribution of $s_1^2/s_2^2 = F_{4,7}$ for Example 6.20

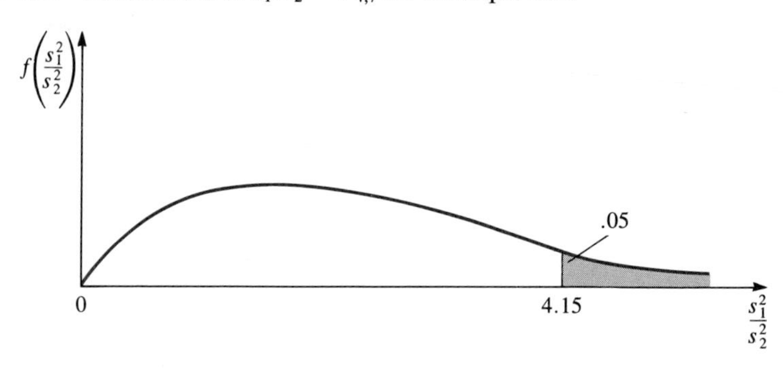

freedom for the denominator. The distribution is shown in Figure 6.22. Thus, from Appendix Table B.9a,

$$P\left(\frac{s_1^2}{s_2^2} \geq \frac{.0003319}{.00008}\right) = P(F_{4,7} \geq 4.15) \doteq .05$$

EXERCISES

6.12 Using Appendix Table B.8, determine the values χ_0^2 such that
- **(a)** $P(\chi_{11}^2 > \chi_0^2) = .05$,
- **(b)** $P(\chi_{25}^2 > \chi_0^2) = .975$,
- **(c)** $P(\chi_8^2 \leq \chi_0^2) = .005$,
- **(d)** $P(\chi_{60}^2 \leq \chi_0^2) = .990$.

6.13 Using Appendix Table B.7, determine the values t_0 such that
- **(a)** $P(t_{15} > t_0) = .05$,
- **(b)** $P(t_{29} < t_0) = .025$,
- **(c)** $P(t_{10} \leq t_0) = .99$,
- **(d)** $P(-t_0 \leq t_{18} \leq +t_0) = .99$.

6.14 Using Appendix Tables B.9a and B.9b, find the values F_0 such that
- **(a)** $P(F_{4,15} > F_0) = .01$,
- **(b)** $P(F_{8,30} > F_0) = .05$,
- **(c)** $P(F_{2,12} \geq F_0) = .95$,
- **(d)** $P(F_{5,6} < F_0) = .01$.

6.6 SUMMARY

In this chapter, we have laid the foundation of many of the applications and inferential procedures in the remainder of the text. So far, we have looked at simple probability concepts, specific probability distributions that can be used to model certain natural phenomena, the idea of sampling to gain information about the population, and, finally, in this chapter, we have determined a means by which we may evaluate probability statements about certain types of sample results. We are now ready to proceed with the main purpose of statistics—to make decisions based upon a set of sample outcomes.

The most important concept of this chapter is the Central Limit Theorem. The theorem states that the distribution of the mean or total of a random sample from almost any population can be approximated with a member of the normal family. This result may not seem like much, but it is a powerful result because many practical situations involve making a decision about the population mean. Since the logical estimate for the population mean is the sample mean, we will be able to use a normal distribution to assess the probability of being in error by a given amount if the sample size is large enough even though the population may be nonnormal.

To be able to use the Central Limit Theorem, we must know the mean and variance of the variable of interest. For sample means, these results were found to be $\mu_{\overline{X}} = \mu_X$ and $\sigma^2_{\overline{X}} = \sigma^2_X/n$ for sampling from infinite populations. If the population is finite, the variance must be multiplied by the finite population correction factor.

In addition, we considered a useful property of the normal family: That linear combinations of normal random variables also have normal distributions. Put more simply, this property says that if normal variables are multiplied by a constant, added to each other or subtracted from one another, the resulting distribution is still normal.

We looked at three other families of distributions in this chapter because they are sampling distributions and arise as functions of normal random variables. The first was the chi-square, which is the distribution of the variance of a random sample from a normal population. An important corollary of the χ^2 distribution is Cochran's theorem, which says that if a sum of squares has a χ^2 distribution and can be broken up into smaller sums of squares whose degrees of freedom total the original degrees of freedom, then the smaller sums of squares will be independent with each having a χ^2 distribution. The second family of sampling distributions is the t, which arises when σ must be replaced by its estimate, s, in the standardization formula. The F-distribution is used to model the ratio of two independent sample variances from normal populations.

The concepts covered in this chapter are summarized for easy reference in Table 6.4.

SUPPLEMENTARY EXERCISES

6.15 Let the numbers 1, 2, 3, 4, 5, 6 constitute a population. If samples of size three are drawn without replacement, there are 20 different possible samples.

(a) List these 20 samples and find the mean and the median for each of them.

(b) Find the mean and the variance of the 20 sample means.

TABLE 6.4 Summary table

	Result	Condition and/or Application
Mean of Sample Means	$\mu_{\bar{X}} = \mu_X$	Inferences about population means
Variance of Sample Means	$\sigma_{\bar{X}}^2 = \dfrac{\sigma_X^2}{n}$	Infinite population
	$\sigma_{\bar{X}}^2 = \dfrac{\sigma_X^2}{n}\left(\dfrac{N-n}{N-1}\right)$	Finite population and nonreplacement sampling
Central Limit Theorem	$\bar{X} \sim$ normal or $\left(\sum_{i=1}^{n} X_i\right) \sim$ normal	$\sigma_X < \infty$; Evaluation of probability statements about sample means or sample totals
	$X \mathrel{\dot\sim} N\left(n\pi, \sqrt{n\pi(1-\pi)}\right)$; $p \mathrel{\dot\sim} N\left(\pi, \sqrt{\dfrac{\pi(1-\pi)}{n}}\right)$ where $p = \dfrac{X}{n}$	$X = \Sigma Y_i$, where Y_i are independent Bernoulli variables with parameter π, and $n\pi \geq 5$, $n(1-\pi) \geq 5$.
Chi-Square Distribution	$\sum_{i=1}^{\nu} Z_i^2 \sim \chi_\nu^2$	Normal population
	and $\dfrac{(n-1)\, s_X^2}{\sigma_X^2} \sim \chi_{n-1}^2$	Distribution of sample variance
Cochran's Theorem	If $Q_\nu = \Sigma_{i=1}^{m} Q_i$, then each $Q_i \sim \chi_{\nu_i}^2$, and the Q_i are independent if and only if $\nu = \Sigma\nu_i$.	$Q_\nu \sim \chi_\nu^2$; Determination of distributions for sums of squares
Student's t	$\dfrac{Z}{\sqrt{\dfrac{\chi_\nu^2}{\nu}}} \sim t_\nu$ or $\dfrac{\bar{X} - \mu_X}{s_{\bar{X}}} \sim t_{n-1}$	Normal population; Accounts for the fact that s is not a constant
F-distribution	$\dfrac{\chi_{\nu_1}^2/\nu_1}{\chi_{\nu_2}^2/\nu_2} \sim F_{\nu_1,\nu_2}$ or $\dfrac{s_1^2/\sigma_1^2}{s_2^2/\sigma_2^2} \sim F_{n_1-1,n_2-1}$	Distribution of the ratio of 2 independent sample variances from normal populations

(c) Find the mean and the variance of the 20 sample medians.

(d) Which measure, mean or median, shows the smaller variation from sample to sample?

6.16 A certain company has 5000 customer accounts. It is believed that the average number of days delinquent per account is 15 days, and the standard deviation is believed to be 10 days. In an effort to substantiate the belief, the manager of accounts receivable has randomly selected 50 accounts and found the average number of days delinquent to be 10.6. What is the probability of observing an $\bar{X}$ this extreme or more so if the belief is true?

6.17 Given a sample of size 16 from a normal distribution with mean 50 and standard deviation 20, what are the mean and the standard deviation of the distribution of the sample means?

6.18 Explain in your own words the meaning of the Central Limit Theorem and its significance.

6.19 The Nature's Pride Cereal Company fills their 16-ounce boxes with an average weight of 16.1 ounces. The standard deviation of weights is known to be .18 ounce. The boxes are then packaged into cases containing 24 boxes each. If a case is randomly selected, what is the probability that the average weight per box will be less than 16.02 ounces?

6.20 Explain why the Central Limit Theorem justifies using a normal distribution to approximate a binomial distribution.

6.21 Sales for Donegal's Department Store have averaged \$12,400 per week and have had a standard deviation of \$2000 per week. In an attempt to increase sales, Donegal's embarked upon a new promotional campaign 12 weeks ago. Since that time, sales have averaged \$13,600 per week.

(a) What is the probability of averaging sales as high as this or higher over the period in question if the promotional campaign is ineffective?

(b) Would you continue the campaign and why? (Assume that sales are normally distributed.)

6.22 When in control, a machine makes parts whose diameters have a standard deviation of .01 inch. A random sample of 16 pieces was taken and found to have a standard deviation of .01479 inch.

(a) What is the probability of observing a standard deviation this large or larger if the machine is in control?

(b) Would you stop the machine? Why?

6.23 Use Appendix Table B.7 to find each of the following unknown values.

(a) $P(t_{10} \leq t_0) = .80.$

(b) $P(t_{22} \geq 1.717) = ?$

(c) $P(-t_0 \leq t_{15} \leq t_0) = .95.$

(d) $P(t_{12} \leq t_0) = .001.$

6.24 The finite population correction factor is ignored when selecting a sample of 50 from a population of 200.

(a) Will the calculated value of $\sigma_{\bar{X}}$ overestimate or underestimate the true value?

(b) By what factor will it do so?

6.25 Refer to Exercise 6.19 and assume that the given standard deviation came from a sample of 24 boxes.

(a) What is the probability that the sample mean will be less than 16.008 ounces?

(b) What is the probability that it will be greater than 16.163 ounces?

6.26 A particular job classification in the rubber industry commands an average wage of \$6.25 per hour. The variance of the wages is .49. A random sample of 28 workers in this job classification of the Safety Rubber Products Company had an average wage of \$5.95.

(a) What is the probability of occurrence of a sample mean this low or lower due to chance alone?

(b) Would you accuse this company of paying subpar wages in this classification? Explain your answer.

****6.27** Using expected values and combinatorials, prove that if a sample of size n is selected from a population of size N, the expected value of $\bar{X}$ is μ_X.

6.28 What happens to the standard error of the mean, $\sigma_{\bar{X}}$, for random samples from an infinite population when the sample size is changed from 16 to 144?

6.29 Conglomerate Corporation has four major divisions. Sales for each division are normally distributed with means and standard deviations as listed in the following table (in millions of dollars). Assume that divisional sales are independent of each other.

	Division			
	A	*B*	*C*	*D*
Avg. Yearly Sales	95	150	300	120
Std. Dev. of Yearly Sales	10.2	20.4	37.3	9.1

For any randomly selected year, what is the probability that Conglomerate's total sales **(a)** will exceed \$700 million, **(b)** will be less than \$600 million, **(c)** will be between \$625 and \$710 million?

***6.30** Show that the square of a t variable having ν degrees of freedom has an F-distribution with 1 degree of freedom for the numerator and ν degrees of freedom for the denominator.

6.31 On a certain standardized exam, it is known that the average score is 550 and the standard deviation of scores is 50.

(a) What is the probability that the average score of 64 randomly selected students who take the exam will exceed 560?

(b) What is the probability that it will be less than 535?

6.32 An elevator has a listed capacity of 2000 pounds, with a safety factor of 500 pounds. Assume that the weights of persons using this elevator are normally distributed with a mean of 165 and a standard deviation of 25 pounds. On a rainy Monday morning, some people arrive late for work and crowd the elevator.

(a) If a total of 12 persons push their way into the elevator, what is the probability that the listed capacity will be exceeded?

(b) If 2 more people, making a total of 14, shove their way into the elevator, what is the probability of exceeding the actual capacity?

(c) Suppose the 14 people already aboard squeeze together to make room for you. Will you enter? Explain your answer.

6.33 Using algebra, prove that the statistic $(\bar{X} - \mu_X)/s_{\bar{X}}$ has a t-distribution with $n - 1$ degrees of freedom.

** Two asterisks indicate a considerably higher level of difficulty.

* One asterisk indicates a moderately higher level of difficulty.

6.34 In attempting to standardize their new safety forces exam, the city of Westville has decided that the passing score will be such that, on the average, 70% of those taking the exam will pass. Using 20 groups of 25 each to determine the passing or "cutting" score, they found that the combined mean was 83 and the standard deviation of the sample means was 1.5. If the population of scores is normally distributed, what should be their "cutting" score?

6.35 A certain portfolio contains stocks of 100 companies listed on the New York Stock Exchange. It is known that these companies have an average earnings per share (EPS) of $2.28 and that the variance of the EPSs is .64. What are the expected value and standard deviation of means of samples of size 15 selected randomly from this portfolio?

6.36 Using Appendix Tables B.8 and B.9, determine the unknown values in the following statements.

- **(a)** $P(\chi_7^2 > \chi_0^2) = .01$.
- **(b)** $P(F_{2,10} > F_0) = .05$.
- **(c)** $P(\chi_{13}^2 < \chi_0^2) = .10$.
- **(d)** $P(F_{4,4} < F_0) = .99$.
- **(e)** $P(F_{4,4} < F_0) = .01$.
- **(f)** $P(\chi_3^2 > \chi_0^2) = .975$.

6.37 The Summot County Auditor's office has found that they frequently run out of paper for their Xerox machine while waiting for an order to arrive, and the office manager has decided that she should order sooner. Her records indicate that daily usage averages 4.8 reams and has a standard deviation of 1.2 reams. It takes 4 days for her order to be delivered after it has been placed.

- **(a)** How many reams of paper should she have on hand when she places an order if she wants only a 5% chance of running out before the order arrives?
- **(b)** What must you assume to answer the question?

6.38 A plastics company has decided to buy a new machine for extruding plastic pipe. It has narrowed the choice of manufacturers of this type of machine down to two, which are equivalent in cost. Of primary interest at this point is the variability in the diameter of the extruded pipe. To further aid in making the choice, they extruded 16 pieces of pipe with each machine and found the sample variances to be .0001 for machine 1 and .00024 for machine 2. If the variability of diameters is the same for both machines, what is the probability of observing sample variances this far apart? (*Hint:* Take the ratio of the larger variance to the smaller variance.)

6.39 Prove that

$$\sum_{i=1}^{n} (X_i - \overline{X})(\overline{X} - \mu_X) = 0$$

***6.40** A population consists of the numbers 2, 5, 7, 10. Random samples of size 3 are to be selected *without* replacement, and the sample means are to be calculated. Construct the probability distribution for the sample means in this situation.

***6.41** Refer to Exercise 6.40 and determine the mean and the standard deviation for the distribution of sample means. Calculate them directly and compare your results to those obtained by using Equations 6.5 and 6.8.

***6.42** Refer to Exercise 6.40 and construct the probability distribution of the sample median for the procedure stated therein.

***6.43** Refer to Exercise 6.42 and calculate the mean and the standard deviation of the distribution of sample medians.

***6.44** Refer to Exercises 6.41 and 6.43 and compare your results.
- **(a)** Does this comparison suggest anything to you about which one you might prefer to use in estimating the original population mean?
- **(b)** If so, which do you prefer and why?

****6.45** Consider the following population consisting of the values 11, 13, 17, 20, 29. If samples of size 3 are to be selected *with* replacement from this population, what is the probability distribution for the sample mean?

***6.46** Refer to Exercise 6.45 and calculate the mean and the standard deviation for the distribution of sample means directly.

***6.47** Refer to Exercise 6.46 and use Equations 6.5 and 6.7. Do the results agree?

***6.48** Refer to Exercise 6.45 and determine the probability distribution for the sample median obtained by that procedure.

***6.49** Calculate the mean and the standard deviation of the distribution of sample medians in Exercise 6.48.

***6.50** Compare the results of Exercises 6.46 and 6.49. If you wanted to estimate the original population mean in this case, would you use the sample mean or the sample median? Give reasons for your answer.

***6.51** The means you calculated in Exercises 6.41 and 6.43 should be the same, and those calculated for Exercises 6.46 and 6.49 should be different. Can you explain why?

***6.52** Redo Exercises 6.40 and 6.41, assuming the samples are selected *with* replacement. Compare your results with those obtained previously, and explain any differences.

***6.53** Redo Exercises 6.45 and 6.46, assuming the samples are to be selected *without* replacement. Compare the results with those obtained previously, and explain any differences.

6.54 Molson Products, Inc., has determined that demand for part #827114 averages 12.60 parts per week and has a variance of 19.32. If Molson randomly selects 9 weeks at which to look at the demand for this part, what is the probability that the average demand for those 9 weeks will be **(a)** less than 10 units and **(b)** between 11 and 14 units?

6.55 Demand for Molson's part #006969 averages 4.53 units per week and has a standard deviation of 1.235 units. Over the past 4 weeks, demand has averaged 6.25 units. If these 4 weeks are considered to be a random sample, what is the probability that demand would average this much or more by chance alone?

6.56 It is rumored that Molson's supplier of part #827114 is going to be hit by a strike that will probably last 10 weeks. Molson wishes to stock up to avoid being affected by this possibility. How many parts should Molson have in inventory at the beginning of the strike so that there is less than a 10% chance of their running out of part #827114 during the strike? (Refer to Exercise 6.54.)

6.57 Molson also obtains part #006969 from the supplier of Exercise 6.56. Determine how many of part #006969 Molson should inventory at the start of the strike to be 99% certain of not being out of stock during the strike. (Refer to Exercise 6.55.)

7 Parameter Estimation

7.1 INTRODUCTION

Margaret Sherwood is the chief internal auditor for the Jenkins Company. Among her assigned duties is the determination of the proportion of ledger accounts containing errors. In the past, Sherwood has painstakingly examined every ledger account to determine the proportion with errors. Recently, each time she has done this task she has mentally rebelled at the lengthy procedure. A few days ago, a colleague mentioned that statistical sampling in auditing was a common phenomenon now and that perhaps Sherwood should try this technique. After due consideration (and after attending a professional development workshop on the subject), Sherwood decides to select from all ledger accounts a random sample of accounts. She plans to check only the sampled accounts for errors. She is aware that the estimate that she obtains will almost certainly not be exactly equal to the true value, but she is comforted by the knowledge that the procedure will also provide an estimate of the size of possible error. She knows that she will be able to expand her original estimate, a single point, into an interval estimate to which a probability level can be associated.

The principal objective of statistical methods is the utilization of sample data to make inferences about the population from which the sample was drawn. Such inferences generally take one of two forms. One type of inference is a test of hypothesis. The fundamental principles of tests of hypotheses are presented in Chapter 8. The second type of inference is the estimation of population parameters, which is examined in this chapter. For example, Margaret Sherwood, after checking the sampled accounts, must use parameter estimation procedures to determine the proportion of ledger accounts containing errors.

7.2 PROPERTIES OF ESTIMATORS

We can obtain estimates of interest to us by means of an estimator.

DEFINITION 7.1 A function of the sample values used to estimate a parameter of the population is an **estimator** for that parameter.

DEFINITION 7.2 An **estimate** is the particular value obtained by evaluating an estimator for a set of sample observations.

We prefer to use estimators that provide estimates that are close to the population parameters. Most such estimators possess specific properties that statisticians consider desirable. We now briefly consider these properties but refrain from any mathematical development that would be more advanced than the general level of this text. In defining these properties, we use the Greek letter theta (θ) to represent a population parameter and $\hat{\theta}$ to represent the estimator of θ.

DEFINITION 7.3 An estimator $\hat{\theta}$ is an **unbiased estimator** of θ if the mean of all possible sample values for the estimator is equal to the true population parameter—that is, if $E[\hat{\theta}] = \theta$.

In Chapter 6, we saw that $E[\overline{X}] = \mu_X$. Therefore, we can say that $\overline{X}$ is an unbiased estimator of μ_X. In Chapter 5, Equation 5.11 for the sample variance s_X^2 has a divisor of $n - 1$ to make s_X^2 an unbiased estimator of σ_X^2.

DEFINITION 7.4 An estimator $\hat{\theta}$ is an **efficient estimator** of θ if it is unbiased and if it has a sampling distribution with smaller variance than any other unbiased estimator.

As an example, in a normal population, the sample mean is more efficient than the sample median since the variance of $\overline{X}$, known to be σ_X^2/n, is smaller than the variance of the median, which can be shown to be approximately $\pi\sigma_X^2/2n$.

DEFINITION 7.5 An estimator $\hat{\theta}$ is a **consistent estimator** of θ if, as the sample size gets larger, the probability that $\hat{\theta}$ approaches θ is closer and closer to 1.

In other words, an estimator is consistent if it provides estimates closer and closer to the population value as the sample size increases. The estimator $\overline{X}$, for example, satisfies the property of consistency.

DEFINITION 7.6 An estimator $\hat{\theta}$ is a **sufficient estimator** of θ if all the information available in the sample regarding θ is utilized in the estimator $\hat{\theta}$.

The sample mean $\overline{X}$ is an example of a sufficient estimator of μ_X.

Statisticians usually consider an estimator as *good* if it possesses the four properties of unbiasedness, efficiency, consistency, and sufficiency. Nearly all estimators used in subsequent portions of this text are good estimators.

While the four properties are all desirable, the one overriding consideration is the availability of a tabulated distribution for the estimator. As we see in the next section, it is generally of interest to make a statement about how far off our estimate is likely to be from the true value. To do so requires that we have a tabulated distribution that at least approximates the sampling distribution of the estimator. Without such a distribution, we are either unable to determine the likelihood of being near the true value, or else we are faced with a very complicated mathematical derivation. The results of Chapter 6 provide us with distributions that can be used to at least approximate the sampling distributions of the majority of the estimators in this text.

7.3 POINT ESTIMATES VERSUS INTERVAL ESTIMATES

In the previous section, we considered certain desirable properties of estimators. Let us now focus upon the use of estimators. As previously stated, our purpose is to use sample data to estimate unknown population parameters.

A rationale for using sample data instead of examining all items of a population to determine the value of a specific parameter was given in Chapter 5. We recall that we usually sample for economic reasons, such as saving time or money. Sometimes we sample because the process involved in obtaining a value destroys the item.

DEFINITION 7.7 A single number computed from sample information and used as an estimate of the true population parameter is known as a **point estimate.**

Accordingly, $\hat{\theta}$, calculated from sample data, is a single point along a number scale and is a point estimate of the population value θ. Such an estimate may be entirely satisfactory for some purposes. However, our knowledge of sampling fluctuations tells us that the point estimate is most likely not equal to the true value and, therefore, may not be very useful without some measure of the possible error in the estimate. Thus, a reasonable question to ask is, "How close do you think the estimate is to the true value?" In attempting to answer such a question, we are led to an interval estimate.

DEFINITION 7.8 The interval estimate constructed from the point estimate by using the appropriate sampling distribution so that $1 - \alpha$ is the likelihood that the interval includes the population parameter is known as a $(1 - \alpha) \times 100\%$ **confidence interval.**

We are interested, then, in accompanying a value for $\hat{\theta}$ with some interval around $\hat{\theta}$ along with a measure of assurance that the interval does contain the parameter θ. Denoting confidence limits by CL, we can frequently, but not always, write the interval estimator of θ in the form

$$CL = \hat{\theta} \pm \varepsilon \qquad \textbf{(7.1)}$$

In Equation 7.1, $\hat{\theta}$ is the point estimator, and ε represents a measure of the possible error. The measure of assurance, attached to the interval and expressed as a probability, is called the confidence coefficient and affects the magnitude of ε.

DEFINITION 7.9 The **confidence coefficient,** $1 - \alpha$, is the probability with which we can assert that an interval includes the parameter we are estimating.

The endpoints of the interval estimate are called the *confidence limits*. To be more specific, Equation 7.1 would become

$$CL = \hat{\theta} \pm z_{\alpha/2}\sigma_{\hat{\theta}}$$

if the Z statistic were the appropriate one to use for a $(1 - \alpha) \times 100\%$ confidence interval around $\hat{\theta}$. The upper limit of the interval is given by $\hat{\theta} + z_{\alpha/2}\,\sigma_{\hat{\theta}}$, the lower limit by $\hat{\theta} - z_{\alpha/2}\,\sigma_{\hat{\theta}}$. The symbol $z_{\alpha/2}$ is a value for Z such that the area under the normal curve to the right is $\alpha/2$. Thus, the area between $-z_{\alpha/2}$ and $z_{\alpha/2}$ is $1 - \alpha$, the confidence coefficient. It is clear that the larger the probability represented by $1 - \alpha$, the larger is the value of Z and the wider is the confidence interval. Refer to Figure 7.1.

It is also possible to obtain one-sided confidence intervals. To do so, we identify whichever limit is of interest and use the corresponding tabular value for α rather than $\alpha/2$.

FIGURE 7.1 Illustration of calculation of confidence limits

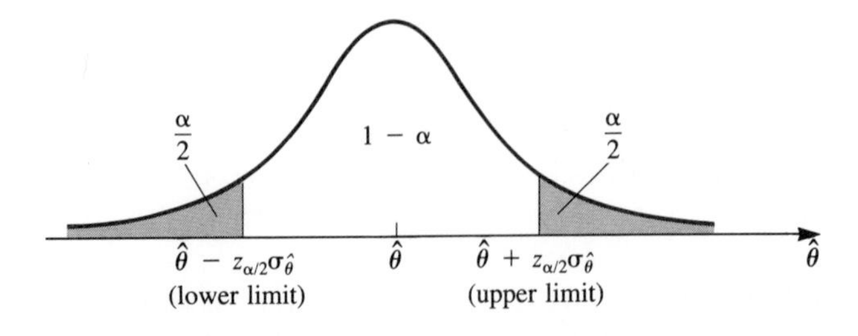

The reason for stating that Equation 7.1 does not always provide the form of the confidence interval should be explained. If the required statistic has a distribution that is not symmetric, such as chi-square, for example, the point estimate is not in the center of the resulting confidence interval. Although the calculation results in an interval, it is not obtained as $\hat{\theta} \pm \varepsilon$.

Expressing estimates in the form of an interval is a common practice. Many government statistics are reported in this format. For example, the Bureau of Labor Statistics may report that the number of unemployed individuals in the United States is 7.8 ± 0.6 million. The point estimate for the number of unemployed persons is 7.8 million. The lower limit of the interval estimate is 7.2 million people, and the upper limit is 8.4 million people.

The justification for the general procedure given for determining confidence intervals lies in statistical theory which is beyond the scope of this text. From a practical viewpoint, the approach is based on the fact that good estimators of the type described in Section 7.2 are approximately normally distributed when n is large.

The procedure outlined in this section is illustrated in detail later in the chapter. That is, certain parameters of interest are considered, the estimator to provide a point estimate is given, and the formula to expand the point estimate into an interval estimate is stated.

7.4 ESTIMATING A POPULATION MEAN

Point Estimate

Probably no problem is more important or occurs more often than estimating the mean value for a population. As an estimator of the population mean μ_X, we use the sample statistic $\overline{X}$, computed according to the equation

$$\hat{\mu}_X = \overline{X} = \frac{\sum_{i=1}^{n} X_i}{n} \tag{7.2}$$

which is the definition of a sample mean as given by Equation 5.8.

The estimator $\overline{X}$ has all the desirable properties of a good estimator referred to in Section 7.2. The resulting estimate is then the point estimate of μ_X. If we must name a single number to describe μ_X or to estimate it, we would use the value of $\overline{X}$ obtained from Equation 7.2.

EXAMPLE 7.1 In a certain continuous industrial process, thousands of ball bearings are produced each hour. We desire to make an hourly check with regard to the mean diameter of the bearings. If the mean diameter is too far from some standard or desired value, appropriate remedial action will be taken. It is not feasible to measure the diameter of every bearing produced, but measurements for a sample of bearings are easily obtained. For a random sample of 4 ball bearings from the last hour's production, the diameters are 1.508, 1.496, 1.504, and 1.500 inches. Determine a point estimate of the mean diameter of all ball bearings.

Solution: Using Equation 7.2, we determine the mean of the sample to be

$$\bar{x} = \frac{\sum_{i=1}^{n} x_i}{n} = \frac{1.508 + 1.496 + 1.504 + 1.500}{4}$$

$$= \frac{6.008}{4} = 1.502 \text{ inches}$$

Thus, 1.502 inches is the point estimate of the population mean μ_X.

Interval Estimate When Population Standard Deviation Is Known

To expand a point estimate into an interval estimate and help convey information about how close we believe the estimate is to the true value, we need to calculate an interval obtained as the point estimate plus and minus some quantity. The appropriate equation to use is

$$CL = \bar{X} \pm z_{\alpha/2}\sigma_{\bar{X}} \tag{7.3}$$

However, before using Equation 7.3, we would like to digress from our main discussion long enough to present the necessary justification for it.

In Chapter 4, one of the principal distributions considered was the normal distribution. Among other things, we learned how to find a z value so that the area under the normal curve between $-z$ and z was equal to any specified amount. For an area of .95, for example, the required z would be 1.96. Consequently, it is appropriate to write

$$P(-1.96 < Z < 1.96) = .95 \tag{7.4}$$

What we are indicating with this equation is that 95% of all z values will fall between -1.96 and 1.96, or that the probability is .95 that a z value will be between -1.96 and 1.96.

From Chapter 6, we also recall the Central Limit Theorem, which essentially states that sample means have distributions that are approximately normally distributed. The mean of the distribution of sample means is $\mu_{\bar{X}} = \mu_X$, and the standard deviation is $\sigma_{\bar{X}}$, which is $\sigma_X/\sqrt{n}$ if an infinite population is assumed. Often, $\sigma_{\bar{X}}$ is referred to as the standard error of the mean.

If μ_X is known, we can determine that $(1 - \alpha) \times 100\%$ of all $\bar{X}$ values are expected to be within $\mu_X \pm z_{\alpha/2}\sigma_{\bar{X}}$. This fact is illustrated in Figure 7.2. We see that $\pm 1.96\sigma_{\bar{X}}$ is the maximum distance from μ_X that we expect $\bar{X}$ to be 95% of the

FIGURE 7.2 Distribution of $\overline{X}$-values with 95% limits for $\overline{X}$

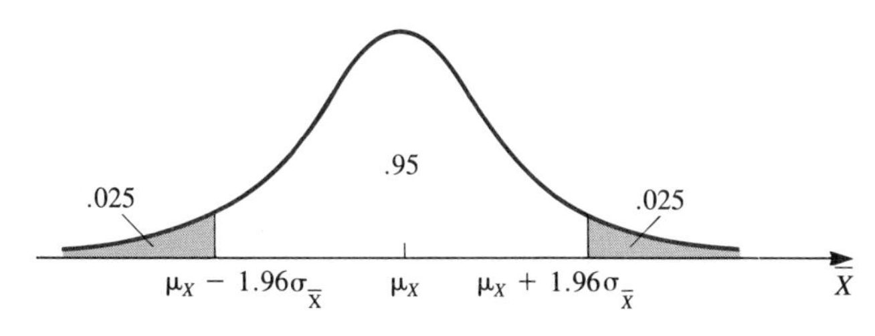

time. For the present situation, μ_X is not known, but regardless of its value, we still expect $\overline{X}$ to fall within $\pm z_{\alpha/2}\sigma_{\overline{X}}$ of it $(1 - \alpha) \times 100\%$ of the time.

In Equation 7.4, we now replace Z by $(\overline{X} - \mu_X)/\sigma_{\overline{X}}$ and obtain

$$P\left(-1.96 < \frac{\overline{X} - \mu_X}{\sigma_{\overline{X}}} < 1.96\right) = .95$$

If each term of the inequality is multiplied by $\sigma_{\overline{X}}$, the expression becomes

$$P(-1.96\sigma_{\overline{X}} < \overline{X} - \mu_X < 1.96\sigma_{\overline{X}}) = .95$$

Then, if $-\overline{X}$ is added to each term, we have

$$P(-\overline{X} - 1.96\sigma_{\overline{X}} < -\mu_X < -\overline{X} + 1.96\sigma_{\overline{X}}) = .95$$

Multiplying each term by -1, we obtain

$$P(\overline{X} + 1.96\sigma_{\overline{X}} > \mu_X > \overline{X} - 1.96\sigma_{\overline{X}}) = .95$$

or, equivalently,

$$P(\overline{X} - 1.96\sigma_{\overline{X}} < \mu_X < \overline{X} + 1.96\sigma_{\overline{X}}) = .95 \qquad \textbf{(7.5)}$$

If the general symbol Z is substituted for 1.96 in Equation 7.5, and $1 - \alpha$ for .95, we then see the similarities between Equations 7.3 and 7.5. When the standard deviation is known, a confidence interval for μ_X can be obtained by adding to and subtracting from the point estimate $\overline{X}$ the quantity $z_{\alpha/2}\sigma_{\overline{X}}$. Thus, the lower limit of the interval is $\overline{X} - z_{\alpha/2}\sigma_{\overline{X}}$ and the upper limit is $\overline{X} + z_{\alpha/2}\sigma_{\overline{X}}$.

The preceding development used a Z statistic based on the Central Limit Theorem. As described in Chapter 6, regardless of the form of the original distribution, there is usually no problem in using Z for making inferences about the population mean as long as the sample size is 30 or more. If the original distribution is symmetric or nearly so, the distribution of the sample means should be approximately normal for sample sizes less than 30. If we know that the original distribution is skewed, we are not as comfortable in using a Z

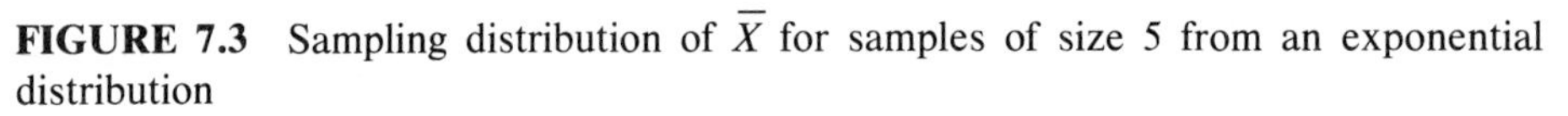

FIGURE 7.3 Sampling distribution of $\overline{X}$ for samples of size 5 from an exponential distribution

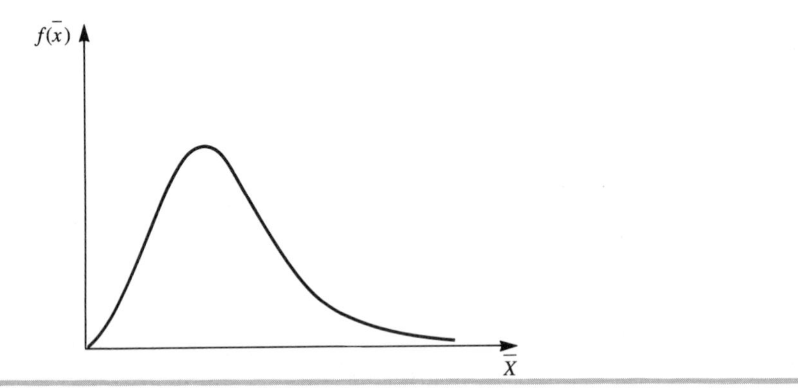

statistic when the sample size is less than 30. In such circumstances, about the best we can say is that the less skewed the distribution and the larger the sample, the better the approximation provided by the normal distribution. However, it is interesting to consider sampling from the highly skewed exponential distribution. Figure 7.3 shows the sampling distribution of $\overline{X}$ for samples of size 5 drawn from such a distribution. Note that even for the relatively small sample size, the distribution of the sample mean looks somewhat like a normal distribution.

As we saw in Chapter 6, it is not necessary to rely on the Central Limit Theorem to provide normality if the original distribution is known to be normal. In such a case, the distribution of sample means is exactly normal regardless of the sample size.

As indicated previously, $\sigma_X/\sqrt{n}$ is used for $\sigma_{\overline{X}}$, the standard error of the mean, if samples are drawn from an infinite population. The same formula would apply if sampling were done with replacement from a finite population. However, in practice, most sampling is without replacement from a finite population, and the finite population correction factor $(N - n)/(N - 1)$, first considered in Chapter 6, is applicable. In such cases, the standard error of the mean becomes

$$\sigma_{\overline{X}} = \frac{\sigma_X}{\sqrt{n}} \sqrt{\frac{N - n}{N - 1}}$$

and it should be used in calculating a confidence interval.

EXAMPLE 7.2 Consider the point estimate obtained in Example 7.1 and determine a 95% confidence interval if the population standard deviation is known to be .004 inches. Assume that the diameters are normally distributed.

Solution: Using Equation 7.3, we have

$$CL = 1.502 \pm (1.96)\frac{.004}{\sqrt{4}} = 1.502 \pm .00392$$

The 95% confidence interval for the unknown population mean as calculated from the sample data has a lower limit (LL) of 1.49808 and an upper limit (UL) of 1.50592.

We speak of the interval we obtained in Example 7.2 as being a 95% confidence interval, but we need to make clear just what is meant by this statement. Clarification of Equation 7.5 is also required. Let us consider drawing many samples of size n from a normal distribution having mean μ_X and standard deviation σ_X. If we calculate a confidence interval for each sample according to Equation 7.3, then it is true that approximately $(1 - \alpha) \times 100\%$ of the intervals calculated include the true value while approximately $\alpha \times 100\%$ of the intervals do not include μ_X. In the long run, the probabilities should be exactly $1 - \alpha$ and α, respectively.

This concept is illustrated in Figure 7.4, where $1 - \alpha = .90$. Note that 9 out of 10 of the lines, representing confidence intervals, constructed around the points, representing $\overline{X}$ values, include the vertical line, representing μ_X. One of the intervals does not include the true value. We also note that all of the intervals

FIGURE 7.4 Illustration of confidence interval concept

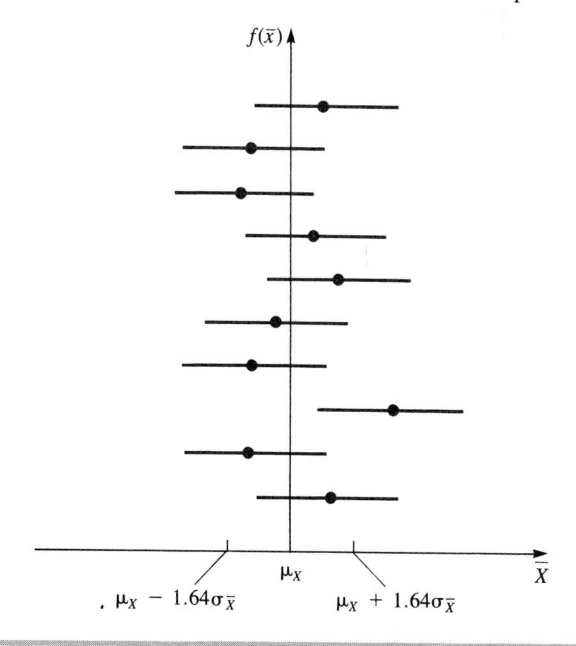

in Figure 7.4 have the same width, which, in general, will be true in repeated sampling when σ_X is known and n and $1 - \alpha$ remain the same.

In most practical problems, we take only one sample, and an interval estimate is of necessity based on the results from the one sample. We therefore attach a probability value of $1 - \alpha$ to the single sample's results as a measure of the degree of belief that the interval contains the true mean. In other words, we have faith in a procedure that leads to correct statements $(1 - \alpha) \times 100\%$ of the time. We therefore say that a particular result at hand has a probability of $1 - \alpha$ attached to it.

As an interpretation or explanation of the meaning of the confidence interval in Example 7.2, we would say that we are 95% certain that the interval from 1.49808 to 1.50592 includes the true value, and hence our degree of confidence regarding our particular result is 95%.

Even though it may seem the same, note that we did not say that we are 95% certain that the true population mean μ_X is in the interval from 1.49808 to 1.50592. An explanation should make the difference clear. The population mean μ_X is a constant. It has no variability. Either it is in or it is not in any particular interval; hence, the only allowable probability values regarding μ_X are 1 and 0. Which value is correct will usually be unknown to us because the value of μ_X is unknown. If we make the statement that we are $(1 - \alpha) \times 100\%$ certain that a particular interval includes the true value, we are placing the emphasis on what actually varies from problem to problem—that is, the interval and not the true value. The previous statement is the correct interpretation of the confidence interval concept.

As was evidenced in the solution to Example 7.2, we see that the form of Equation 7.5 is potentially confusing, since it appears that μ_X is the variable and that the probability that μ_X lies between $\overline{X} - z_{\alpha/2}\sigma_X/\sqrt{n}$ and $\overline{X} + z_{\alpha/2}\sigma_X/\sqrt{n}$ is $1 - \alpha$. As we have seen, this statement makes no sense. We do have $(1 - \alpha) \times 100\%$ confidence that a particular interval does cover the true mean. However, the measure of confidence is $1 - \alpha$ because, before the sample was drawn, $1 - \alpha$ was the probability that the interval we were going to construct would include the true mean.

In the development that led to Equation 7.5, the .95 given in Equation 7.4 was a true probability. In Equation 7.5, it is not a true probability but is a measure of the confidence we have in the truth of the statement. Therefore, to avoid the potential confusion implied by Equation 7.5, we refrain from using such equations in the remainder of the text. In the confidence intervals that we calculate, we use the form $\hat{\theta} \pm \varepsilon$ and/or make reference to the lower limit and the upper limit, *LL* and *UL,* respectively.

Another statement that we can make related to the result of Example 7.2 pertains to the magnitude of the error. The amount that was added to and subtracted from the sample mean was .00392. It will be convenient to refer to this amount as "error." As a statement of interpretation, we could say that we are 95% confident that the size of the error made in estimating μ_X does not exceed .00392.

EXAMPLE 7.3 The Johnson Savings and Loan Association is interested in estimating the mean amount of its outstanding loans. From past experience, the officers know that the standard deviation is \$120. Determine a 92% confidence interval for the population mean if a random sample of size 64 has a sample mean of \$650.

Solution: Although the form of the distribution is not specified, the Central Limit Theorem allows us to use the normal distribution to make the required inference. The population standard deviation σ_X is known to be \$120. To two decimal places, the required value of Z is 1.75. Therefore, the confidence interval requested is

$$CL = 650 \pm (1.75)\frac{(120)}{\sqrt{64}} = 650 \pm (1.75)\frac{(120)}{8} = 650 \pm 26.25$$

The lower limit of the interval is \$623.75 and the upper limit is \$676.25. We are 92% certain that the interval from \$623.75 to \$676.25 includes the mean amount of outstanding loans in the population. We can also say that we are 92% confident that the size of the error made in estimating μ_X does not exceed 26.25.

The statement of the Central Limit Theorem in Chapter 6 applies to sample totals as well as sample means. Formulas for the mean and variance of T, the sample total, were given by Equations 6.10 and 6.11, respectively. The interval estimate for the sample total is given by

$$CL = n\overline{X} \pm z_{\alpha/2}\sqrt{n}\,\sigma_X$$

If it is desired to estimate the total of a finite population, the estimator is $N\overline{X}$, the variance is

$$\frac{N^2\sigma_X^2}{n}\left(\frac{N-n}{N-1}\right)$$

and an interval estimate for the population total is given by

$$CL = N\overline{X} \pm z_{\alpha/2}\frac{N\sigma_X}{\sqrt{n}}\sqrt{\frac{N-n}{N-1}} \tag{7.6}$$

Throughout this section, we have assumed that the standard deviation of a population is known while the population mean is unknown. Theoretically, the population mean must be known to calculate the population standard deviation. However, the assumption we have made, though not realistic for most problems, can be justified for certain types of data. For example, in an industrial production process for which observations have been collected over a long time, it may be known that the standard deviation remains nearly constant while the mean changes over time.

Interval Estimate When Population Standard Deviation Is Unknown

In the first portion of this section, we considered estimating the mean of a population with a point estimate $\overline{X}$. We then expanded the point estimate into an interval estimate, making use of the Z statistic from the normal distribution under the assumption that the population standard deviation σ_X is known.

In most practical problems, we will not be fortunate enough to know σ_X. In such a situation, we are not justified according to theory in using the random variable Z. However, the t-distribution, described in Chapter 6, does provide the appropriate distribution, assuming the underlying distribution is normal. The t statistic is based on s_X, which is an estimate of σ_X and is calculated from sample data.

The calculation of the point estimate $\overline{X}$ will still be made using Equation 7.2. However, in expanding the point estimate into an interval estimate, we replace the z value with a corresponding t value and σ_X with s_X, resulting in the formula

$$CL = \overline{X} \pm t_{n-1,\alpha/2}\, s_{\overline{X}} \tag{7.7}$$

where $s_{\overline{X}} = s_X/\sqrt{n}$ if the population is infinite.

EXAMPLE 7.4 Consider again the data of Example 7.1. Assume the standard deviation σ_X is not known and determine a 95% confidence interval for the population mean. Assume observations are normally distributed.

Solution: We need to estimate σ_X from the sample data. For this purpose, we calculate s_X^2 according to Equation 5.11 and then take the square root. For our problem,

$$\begin{aligned} s_X^2 &= \frac{\sum_{i=1}^{n} (X_i - \overline{X})^2}{n - 1} \\ &= \frac{(1.508 - 1.502)^2 + \cdots + (1.500 - 1.502)^2}{4 - 1} \\ &= \frac{.000080}{3} = .00002667 \end{aligned}$$

and

$$s_X = \sqrt{.00002667} = .00516$$

As discussed in Chapter 6, the value of t needed for the confidence interval will depend on the confidence coefficient, $1 - \alpha$, and the number of degrees of freedom, $n - 1$, for this type of problem. For a 95% confidence interval, the appropriate t value for three degrees of freedom is 3.182. The confidence limits will be

$$CL = 1.502 \pm (3.182)\frac{(.00516)}{\sqrt{4}} = 1.502 \pm .0082$$

Thus, the lower limit of the interval is 1.494, and the upper limit is 1.510. The statement that we make is that we are 95% sure that the interval from 1.494 to 1.510 includes the true population mean.

We note that the interval obtained in Example 7.4 is much wider than the interval obtained in Example 7.2, where σ_X was assumed to be known. Part of the explanation for the wider interval is that s_X was larger than σ_X. But, nearly as much is due to the t value we used instead of the z value. In a sense, the larger interval is part of the penalty we pay for not knowing σ_X.

If it is appropriate to use the finite population correction, the necessary adjustment to the standard error of the mean can easily be made. When using s_X, the *FPC* becomes $(N - n)/N$ instead of $(N - n)/(N - 1)$. Equation 7.7 is then

$$CL = \overline{X} \pm t_{n-1,\alpha/2}\frac{s_X}{\sqrt{n}}\sqrt{\frac{N-n}{N}}$$

Equation 7.7 is frequently used for calculating confidence intervals for μ_X when the population is not normal. The justification for this usage lies in statistical theory, which shows that the t-distribution is approximately correct as long as the population distribution is not highly skewed and the sample size is not less than 30. If the normal distribution cannot be assumed, then the confidence coefficient will be only approximately $1 - \alpha$. In general, the farther the distribution is from normal, the farther the true confidence coefficient may be from $1 - \alpha$. The exact amount of the discrepancy will usually be unknown.

We should also mention that if the sample size is large and the population standard deviation must be estimated from sample data, it is a fairly common practice to use a Z statistic instead of the correct t statistic. For this purpose, samples of size 30 or more are usually considered as large. The justification given for using Z instead of t in such cases is that there is little difference between the two statistics, and it is easier to use Z. In this text, we adhere to what is theoretically correct and let the reader decide when and if to compromise on this point.

EXERCISES

7.1 A company packaging frozen orange juice desires to produce cans having a mean weight of 12 ounces. Determine a point estimate for the population mean if a random sample of 10 cans has the following weights in ounces.

11.84	12.04	12.16	11.97	12.02	11.88	12.00	11.95	11.98	12.08

7.2 A random sample of 8 of a certain class of stocks has the following annual yields in percent.

6.4 10.5 7.6 5.8 12.6 9.3 14.2 11.2

What is a point estimate for the mean annual yield for all stocks of this type?

7.3 In a normal population, the standard deviation is known to be 3.6 units. Determine a 90% confidence interval for the population mean when a random sample of size 400 has a mean of 72.8.

7.4 Rework Exercise 7.3, assuming the sample size is 4 instead of 400. Note the effect on the width of the confidence interval.

7.5 A population of unknown form has a standard deviation of 4.8 units. Determine a 98% confidence interval for the population mean if a random sample of size 36 has a mean of 24.5.

7.6 Consider the information given in Exercise 7.5 and determine the confidence interval requested if the population size is known to be 185. Compare the width of the interval with that obtained in Exercise 7.5.

7.7 Refer to Example 7.3. Determine a 92% confidence interval for totals resulting from samples of size 64.

7.8 A random sample of size 9 of a certain part used in constructing radar units has a mean length of .65 inch and a standard deviation of .018 inch. Determine a 90% confidence interval for the true mean length, assuming the lengths are normally distributed. Write one sentence explaining the meaning of the resulting interval.

7.9 A farmer samples 16 plots to determine the mean yield of wheat in bushels per acre. The observations have a mean of 32.8, a standard deviation of 2.4, and are known to be normally distributed. Determine 95% confidence limits for the true mean yield.

7.10 The Zimcoe Company produces washers to fit bolts of a certain size. A random sample of 64 washers has a mean inside diameter of .402 inch and a standard deviation of .008 inch. Calculate 95% confidence limits for the true mean diameter of all washers.

7.11 A sample of 100 individuals attending the annual meeting of the American Statistical Association showed a mean age of 48.6 years and a standard deviation of 6.4 years. What is a 95% confidence interval for the mean age of all persons attending the meeting?

7.12 Consider Exercise 7.11 and determine the required confidence interval if the total number of individuals attending the meeting was 1000.

7.5 ESTIMATING A BINOMIAL PROPORTION

Frequently, we are interested in estimating a binomial parameter. Typical questions to be answered in binomial situations are the following: What proportion of defective items is being produced? What percentage of the employees are in favor of a new pension plan? What is the proportion of television viewers watching a certain program? These constitute only a few examples where we are interested in estimating π, the binomial proportion. Utilizing the point estimate, we can expand this into an interval estimate for some desired confidence level.

As our estimator of π, we use

$$p = \frac{X}{n} \tag{7.8}$$

where: X = the number of successful occurrences of the binomial variable
n = the total number of trials

The estimator p has the desirable properties we usually seek in an estimator (see Section 7.2).

To develop a confidence interval from the point estimate, it is necessary that we have information on σ_p, the standard deviation of the estimator p, or, as it is often called, the standard error of the proportion. By the Central Limit Theorem, the sampling distribution of a sample proportion is approximately normally distributed for large n. The mean of the distribution is π, and the standard deviation is

$$\sqrt{\frac{\pi(1-\pi)}{n}}$$

The latter quantity must be estimated by

$$\sqrt{\frac{p(1-p)}{n}}$$

since π is not known. Since this value is obtained from the sample data, the t-distribution would be appropriate. However, in most applications the sample size is at least 100, and the standard normal distribution is used. We will adhere to tradition in this instance.

As a rule of thumb, the approximation involved here is satisfactory if $n\pi \geq 5$ and $n(1-\pi) \geq 5$. Of course, to check whether the normal approximation is satisfactory, it is necessary to use p because π is unknown. The resulting formula for the confidence interval, assuming an infinite population, is given by

$$CL = p \pm z_{\alpha/2}\sqrt{\frac{p(1-p)}{n}} \tag{7.9}$$

In sampling from a finite population, the interval estimate for a proportion is

$$CL = p \pm z_{\alpha/2}\sqrt{\frac{p(1-p)}{n}\left(\frac{N-n}{N}\right)}$$

EXAMPLE 7.5 A magazine publisher who wishes to estimate what proportion of subscribers to the magazine have fixed incomes randomly samples 400 subscribers and finds that 64 of them have fixed incomes. Estimate π and calculate a 98% confidence interval for the true proportion.

Solution: We check first to see whether both $np \geq 5$ and $n(1 - p) \geq 5$ are satisfied. Since $np = 64$ and $n(1 - p) = 336$, we are justified in using Equation 7.9 to determine the required confidence interval. Using Equation 7.8, the estimate of π is

$$p = \frac{64}{400} = .16$$

Thus, the best estimate of the proportion having fixed incomes is .16.

To calculate a 98% confidence interval, we substitute in Equation 7.9 and obtain

$$CL = .16 \pm 2.33 \sqrt{\frac{(.16)(.84)}{400}}$$

$$= .16 \pm (2.33)(.0183) = .16 \pm .043$$

The lower confidence limit is .117, and the upper limit is .203.

Since 98% of all the possible intervals constructed in the manner used in Example 7.5 contain the true population proportion, we say that we are 98% sure that the interval from .117 to .203 contains the true proportion of subscribers having fixed incomes. Once again, we note that π is a parameter and not a variable. In this situation, as in all confidence interval problems, it is the interval not the population parameter that varies.

EXERCISES

7.13 In a random sample of 100 television viewers, the Gallup poll found that 36 were watching a particular program. Determine a 99% confidence interval for the proportion of all viewers watching the program.

7.14 The marketing research department of a major oil company has been asked by top management to estimate the proportion of customers who prefer self-service gasoline stations. A random sample of 600 customers is taken, and 240 are in favor of self-service stations. Determine a 92% confidence interval for the true proportion favoring self-service and interpret the resulting interval.

7.6 DETERMINATION OF SAMPLE SIZE

We now digress from the direct computation of confidence intervals to consider a related topic. One of the first questions that usually arises in a statistical investigation is, "How large a sample should I take?" The answer to the question must, of course, be related to the objectives of the research. Two considerations are necessary before a solution can be obtained. Someone must decide how large an error in the estimate can be tolerated and how large a risk there should be of exceeding that error. These decisions, sometimes made by the statistical analyst, really should be made by the investigator, because that person is the one who has the objectives most clearly in mind.

Let us first consider problems where it is desired to estimate a mean value. We refer to Equation 7.3 and designate the quantity $z_{\alpha/2}\,\sigma_{\bar{X}}$, or $z_{\alpha/2}\,\sigma_X/\sqrt{n}$, as ε, the magnitude of the error that can be tolerated. Thus,

$$\varepsilon = \frac{z_{\alpha/2}\,\sigma_X}{\sqrt{n}}$$

Solving this equation for n yields

$$n = \left(\frac{z_{\alpha/2}\,\sigma_X}{\varepsilon}\right)^2 \qquad \textbf{(7.10)}$$

The size of the risk involved will determine the z value needed in Equation 7.10. The other quantity required to obtain a solution for n is obviously the value of σ_X. If σ_X is known, then it must be true that we have *a priori* information about the population being studied. Although this is possible, it is not likely. What is much more reasonable is that we have good information about σ_X, but not necessarily perfect knowledge. This information may come from past studies similar to the current problem. For example, variability in retail sales may remain about the same from year to year even though the mean value may change. Thus, σ_X from last year's study may be quite appropriate to use for this year's study. If no useful past information is available, it may be necessary to spend a small portion of the total resources available to conduct a small pilot study for the primary purpose of estimating σ_X. Another alternative is to have a knowledgeable individual specify maximum and minimum values for the population. Then for an approximately normal population, we know that 99.7% of the values are within $\pm 3\sigma_X$. Thus, a rough estimate of σ_X is $(X_{\max} - X_{\min})/6$.

EXAMPLE 7.6 A large oil company would like to estimate the mean annual automobile mileage for the holders of its credit cards. In doing so, management would like to be 95% sure of being within 300 miles of the true mean mileage. In former studies of a similar nature, the standard deviation was estimated to be 1200 miles. What size sample is required?

Solution: Substituting appropriate values into Equation 7.10, we have

$$n = \left(\frac{(1.96)(1200)}{300}\right)^2 = 61.47$$

or 62 credit card holders.

In sample size problems, we depart from the usual rules of rounding. That is, if the solution of the equation yields an answer that is not an integer initially, such as 61.47 in Example 7.6, we take as our solution the next higher integer regardless of the decimal value. This rule assures that we are doing the job required. If the decimal were less than .5 and we were to round down, we would not be satisfying the conditions of the problem by a small amount. The philosophy we follow is to be conservative—that is, to assure that the confidence coefficient is at least as large as that stated.

We now turn to the binomial problem. Suppose we are interested in estimating the proportion possessing a certain characteristic. If we follow the same general procedure as we did in determining the sample size formula for a mean, we would solve for n in the equation

$$\varepsilon = z_{\alpha/2}\sqrt{\frac{\pi(1-\pi)}{n}}$$

obtaining,

$$n = \left(\frac{z_{\alpha/2}}{\varepsilon}\right)^2 \pi(1-\pi) \tag{7.11}$$

But the equation in this form requires knowledge of π, the very thing we wish to estimate by taking a sample of size n. To resolve this conflict, it is necessary to assume a value for π. If absolutely nothing is known about π, then it is customary to assume that the population is as variable as possible. This occurs when $\pi = .5$, thus making $\pi(1-\pi) = .25$, the largest value possible for $\pi(1-\pi)$. In this situation, Equation 7.11 can be rewritten as

$$n = \left(\frac{z_{\alpha/2}}{2\varepsilon}\right)^2 \tag{7.12}$$

In some problems, a limit on π may be known even if the exact value is not. In such cases, Equation 7.11 can be used to determine n. For example, it may be known that π cannot possibly exceed .2. Therefore, knowing that the product $\pi(1-\pi)$ cannot be larger than .16, we could substitute this value into Equation 7.11. In this case, the solution for n would be given by

$$n = \left(\frac{z_{\alpha/2}}{\varepsilon}\right)^2 (.16)(.84)$$

EXAMPLE 7.7 Suppose the owners of a large company wish to estimate the proportion of employees favoring a proposed pension plan. They would like to be 97% sure of being within .06 of the true proportion. What sample size is required?

Solution: If we assume that the company has no knowledge of the true proportion, then substituting in Equation 7.12, we have

$$n = \left(\frac{2.17}{2(.06)}\right)^2 = 327.01 \quad \text{or} \quad 328 \text{ employees}$$

If there is information that π could not exceed .2, then we would substitute in Equation 7.11 and obtain

$$n = \left(\frac{2.17}{.06}\right)^2 (.2)(.8) = 209.28 \quad \text{or} \quad 210 \text{ employees}$$

Both examples for determining sample size when estimating a mean or a proportion assume that the finite population correction factor can be ignored. This assumption is justified if the population is infinite or if sampling is done with replacement when the population is finite. In most practical problems, the population is finite, and sampling is done without replacement. In such cases, it is appropriate to use the finite population correction factor. However, as we saw in Chapter 6, the value of the correction factor for $\sigma_{\bar{X}}$ given by $\sqrt{(N - n)/(N - 1)}$ is close to 1 whenever N, the population size, is quite large and n, the sample size, is quite small in relation to N. In cases where n is large relative to N, perhaps 5% or more, then Equations 7.10–7.12 should be modified.

If the finite population correction is used, Equation 7.10 becomes

$$n = \frac{N z_{\alpha/2}^2 \sigma_X^2}{(N - 1)\varepsilon^2 + z_{\alpha/2}^2 \sigma_X^2} \tag{7.13}$$

Equation 7.11 becomes

$$n = \frac{N z_{\alpha/2}^2 \pi(1 - \pi)}{(N - 1)\varepsilon^2 + z_{\alpha/2}^2 \pi(1 - \pi)} \tag{7.14}$$

Equation 7.12 becomes

$$n = \frac{N z_{\alpha/2}^2}{4(N - 1)\varepsilon^2 + z_{\alpha/2}^2} \tag{7.15}$$

EXERCISES

7.15 The Gruwald Testing Service has been asked by a client to test the mean breaking strength of wire cable. It is known that the standard deviation of breaking strengths is 12 pounds. The client desires to be 98% sure that the sample mean is within 4 pounds of the true mean. What size sample should the testing service take?

7.16 An automobile association has determined from past studies that the standard deviation of the amount of gasoline used annually by its members is 30 gallons. The association believes that the level of consumption may have changed due to the energy crisis, but it feels that the standard deviation has not changed. How large a sample should the association take to be 99.5% certain of estimating the current consumption level to within 4 gallons of the true value?

7.17 The chief auditor for the Stoanno Company desires to estimate for a large stack of current invoices the proportion containing errors. If he has no idea of the true proportion, what size sample should he take to be 99% sure that the sample estimate will not differ from the true proportion by more than .05?

7.18 The student government of Utopian University wishes to estimate the proportion of students who favor the administration's establishing the position of ombudsman on campus. What size sample should be selected if it is desired to assert with a probability of .97 that the sample proportion will not differ from the true proportion by more than .03?

7.19 What size sample is required to estimate a proportion that is known to be .3 or smaller? Assume we are willing to be 95% sure of being within .05 of the true proportion.

7.20 If a population has 200 individuals and the standard deviation is known to be 10, what size sample is required to be 99% sure of being within 2 units of the population mean?

7.21 In a population of size 500, what sample size is required to be 96% sure of estimating the true proportion with an error no larger than .05?

7.7 INTERVAL ESTIMATES FOR THE VARIANCE OF A NORMAL DISTRIBUTION

We have seen that in most practical problems the parameters of a population are unknown and must be estimated from sample data. Up to this point, the focus has been on estimation of the mean. However, the variance of a population must also be estimated in most cases. In many industrial processes, obtaining information about the variance may be the primary objective. Hence, it is quite reasonable to determine a confidence interval estimate for the variance.

From Chapter 5, we know that the point estimate of σ_X^2, the population variance, is given by s_X^2, where

$$s_X^2 = \frac{\sum_{i=1}^{n} (X_i - \overline{X})^2}{n - 1}$$

To expand the point estimate s_X^2 into an interval estimate, we need an appropriate distribution. This distribution, the χ^2 distribution, was also presented in Chapter 6. An underlying assumption is that the sample of size n has been drawn from a normal distribution. The quantity $(n - 1)\, s_X^2/\sigma_X^2$ then follows a χ^2 distribution with $n - 1$ degrees of freedom.

To construct a confidence interval with a confidence coefficient of $1 - \alpha$, we first write

$$P\left(\chi^2_{n-1,1-(\alpha/2)} < \frac{(n - 1)s_X^2}{\sigma_X^2} < \chi^2_{n-1,\alpha/2}\right) = 1 - \alpha$$

where, for example, $\chi^2_{n-1,\alpha/2}$ is the value for a chi-square variable with $n - 1$ degrees of freedom that has $\alpha/2$ of the area to the right of it. Algebraic manipulation of this expression gives

$$LL = \frac{(n - 1)s_X^2}{\chi^2_{n-1,\alpha/2}} \tag{7.16}$$

as the lower limit of the confidence interval and

$$UL = \frac{(n-1)s_X^2}{\chi^2_{n-1,1-(\alpha/2)}} \tag{7.17}$$

as the upper limit. The resulting interval estimate of σ_X^2 will not be symmetric around s_X^2 since the chi-square distribution is not symmetric.

To transform an interval estimate for the variance into an interval estimate for the standard deviation, one need only take the square root of both limits.

EXAMPLE 7.8 Suppose that a textile manufacturer is interested in the variability of the tensile strength of cotton yarn. Data are assumed to be normally distributed. A random sample of size 11 has a variance of 60. What is a 90% confidence interval for the true population variance?

Solution: Substituting in Equations 7.16 and 7.17, we obtain

$$LL = \frac{(10)(60)}{18.31} = 32.77$$

and

$$UL = \frac{(10)(60)}{3.94} = 152.28$$

In Example 7.8, we are 90% sure that the interval from 32.77 to 152.28 includes the true population variance of the tensile strength of cotton yarn. We have this degree of confidence because we know from theoretical considerations that 90% of intervals calculated in this manner will include the true value. Hence, a probability of .90 expresses the degree of belief in the result based on a single sample.

EXERCISES

7.22 The Techstat Corporation has calculated from a sample of size 21 that the variance s_X^2 for the time required to perform a certain operation is 4.8 minutes squared. Calculate 95% confidence limits for the population variance, assuming times are normally distributed.

7.23 The Parfax Corporation manufactures a certain precision part. A recent random sample of 11 of the parts had a variance of .0006 millimeter squared. Construct a 99% confidence interval for the population variance, assuming observations are approximately normal.

7.24 A random sample of 16 cartons of the detergent Superkleen showed a standard deviation of .24 ounce for the filled weight. Determine a 90% confidence interval for the population standard deviation. Assume weights are normally distributed.

7.25 Refer to Exercise 7.23. Determine 99% confidence limits for the population standard deviation.

7.26 Rework Exercise 7.23 using a confidence coefficient of .95.

7.8 OTHER INTERVAL ESTIMATES

In preceding sections, we have examined appropriate equations to obtain point estimates and interval estimates for various parameters. Specifically, we have considered methods for a mean, a total, a proportion, and a variance. We believe that these are the parameters that are likely to be of the most interest in estimation problems.

However, without going into detailed methods, we should note two other possible estimates of interest: estimating the difference between two population means and estimating the difference between two population proportions. Although problems involving these estimates do sometimes occur in practice, we believe that in such cases the primary inference problem will be in testing hypotheses. Therefore, detailed coverage of these types of problems is reserved for Chapter 8, which deals with tests of hypotheses—especially Section 8.5 and the latter part of Section 8.6.

7.9 SUMMARY

In this chapter, we have considered estimation of population parameters, one of the two forms of statistical inference. Table 7.1 presents a summary of the most important expressions having to do with parameter estimation.

We first examined a point estimate—that is, a single number computed from sample information to estimate a population parameter. We then expanded the appropriate equation for the point estimate of a parameter into an interval estimate. In cases where a Z or t statistic was used, we added to and subtracted from the point estimate an amount that may be designated as "error." The error is dependent upon the variability of the estimator and a confidence coefficient reflecting how sure we want to be of including the true value in our interval. A typical interpretation of a confidence interval would be that "we are $(1 - \alpha) \times 100\%$ sure that the interval from $\hat{\theta} - \varepsilon$ to $\hat{\theta} + \varepsilon$ includes θ." In the preceding statement, ε indicates error. When a chi-square statistic was used, the format of $\hat{\theta} \pm \varepsilon$ could not be used for determining confidence limits since the chi-square distribution is not symmetric.

We also considered four properties of estimators: unbiasedness, efficiency, consistency, and sufficiency. Appropriate equations for obtaining point estimates and interval estimates for means, totals, proportions, and variances were examined. In the case of means, instances where the population standard deviation is known and where it is unknown were both considered. For means, totals, and proportions, required modifications when the population size is

TABLE 7.1 Summary table

	Parameter	Point Estimator	Confidence Limits
Parameter Estimation	μ_X	$\overline{X}$	$CL = \overline{X} \pm z_{\alpha/2}\, \sigma_{\overline{X}}$ if σ_X is known or $CL = \overline{X} \pm t_{n-1,\alpha/2} s_{\overline{X}}$ if σ_X is estimated by s_X
	T	$n\overline{X}$	$CL = n\overline{X} \pm z_{\alpha/2} \sqrt{n}\, \sigma_X$
	Total of a finite population	$N\overline{X}$	$CL = N\overline{X} \pm z_{\alpha/2} \dfrac{N\sigma_X}{\sqrt{n}} \sqrt{\dfrac{N-n}{N-1}}$
	π	p	$CL = p \pm z_{\alpha/2} \sqrt{\dfrac{p(1-p)}{n}}$
	σ_X^2	s_X^2	$LL = \dfrac{(n-1)s_X^2}{\chi^2_{n-1,\alpha/2}}$ and $UL = \dfrac{(n-1)s_X^2}{\chi^2_{n-1,1-(\alpha/2)}}$

	Parameter Estimated	Assumptions	Equation
Sample Size Determination	Mean	Infinite population	$n = \left(\dfrac{z_{\alpha/2}\, \sigma_X}{\varepsilon}\right)^2$
		Finite population	$n = \dfrac{N z_{\alpha/2}^2\, \sigma_X^2}{(N-1)\varepsilon^2 + z_{\alpha/2}^2\, \sigma_X^2}$
	Proportion	Infinite population, general case	$n = \left(\dfrac{z_{\alpha/2}}{\varepsilon}\right)^2 \pi(1-\pi)$
		Finite population, general case	$n = \dfrac{N z_{\alpha/2}^2\, \pi(1-\pi)}{(N-1)\varepsilon^2 + z_{\alpha/2}^2\, \pi\,(1-\pi)}$
		Infinite population, $\pi = .5$	$n = \left(\dfrac{z_{\alpha/2}}{2\varepsilon}\right)^2$
		Finite population, $\pi = .5$	$n = \dfrac{N z_{\alpha/2}^2}{4(N-1)\varepsilon^2 + z_{\alpha/2}^2}$

	Property	Description
Desirable Properties of Estimators	Unbiased	The expected value of $\hat{\theta}$ is equal to θ.
	Efficient	$\hat{\theta}$ is unbiased and has smallest variance.
	Consistent	$\hat{\theta}$ becomes closer to θ as the sample size increases.
	Sufficient	$\hat{\theta}$ uses all information in the sample regarding θ.

known were presented. Equations for determining required sample size to estimate means and proportions were also presented. It should be noted that equations for interval estimates in the chapter rely heavily on the Central Limit Theorem.

SUPPLEMENTARY EXERCISES

7.27 **(a)** What are the four properties of a "good" estimator?
(b) Do one or more of these seem more important than the others?

7.28 Why do we usually take a sample from a population instead of examining all items in the population?

7.29 Why are we usually interested in expanding a point estimate into an interval estimate?

7.30 What statistic would be the appropriate one to use in constructing a confidence interval for a normal population mean in a case where the sample size is 10 and the population standard deviation is known?

7.31 Distinguish between "estimator" and "estimate."

7.32 Why are most confidence intervals for μ_X symmetric around the sample value?

7.33 If you repeatedly calculate $(1 - \alpha) \times 100\%$ confidence intervals, how often would you expect to obtain an interval that would not include the true value?

7.34 Explain why we prefer to say we are $(1 - \alpha) \times 100\%$ sure that "the interval includes the true value" rather than "the true value is in the interval."

7.35 Does it seem paradoxical that a large confidence coefficient results in a wide interval and hence, in a sense, less information concerning the parameter to be estimated?

7.36 Theoretically, what length interval would be required if we wanted to be 100% sure of including the true value?

7.37 Verify Equation 7.13.

7.38 For a random sample of 10 workers, the standard deviation of hourly wages was 85¢. Determine a 95% confidence interval for the population standard deviation of hourly wages. Assume that wages are approximately normally distributed.

7.39 A sample of 16 Unicorn cigarettes was tested for nicotine content. Resulting data, recorded in milligrams and believed to be from a normal distribution, are

17.9	21.2	24.6	26.1	19.7	16.4	22.8	15.9
22.7	23.4	17.4	21.4	20.0	18.3	25.2	22.2

Calculate 95% confidence limits for the population mean, and write one sentence explaining the meaning of your confidence interval.

7.40 How large a sample would be required to estimate the mean time per haircut in Barker's Hair Salon if it is known that the standard deviation is 1.2 minutes and an estimate is desired within .5 minute with 98% confidence?

7.41 The daily output of the production of the Tremark Company has a standard deviation of 24 units. Determine a 99% confidence interval for the true mean output if a random sample of size 80 shows a mean of 134 units.

7.42 In a survey of 2000 families, it was determined that 483 families plan to buy a new automobile from next year's production. Determine a 96% confidence interval for the true proportion of families planning to buy a new automobile.

7.43 Suppose that we wish to estimate the proportion of defective parts turned out by a certain machine. How large a sample would be needed to be able to assert with a probability of .90 that our sample proportion will differ from the true proportion by less than .04 if the management knows that the proportion defective will not be larger than .2?

7.44 It is desired to sample from a population for which the standard deviation is known to be 40. What size sample should be taken to assure with a probability of .80 that the estimate of the population mean would not be in error by more than 2 units?

7.45 The Lightview Company produces light bulbs, using a process having a known standard deviation for length of life of 40 hours. If a random sample of 20 light bulbs has a mean length of life of 950 hours, determine a 94% confidence interval for the true population mean.

7.46 The Johnson Company produces a certain type of thread. In a random sample of 16 pieces of thread, the variance of the breaking strength was found to be 3.5 ounces squared. Determine a 95% confidence interval for the population variance if breaking strengths are normally distributed.

7.47 What size sample is required to estimate the proportion of eligible voters who favor the candidacy of Judge Peters? Assume it is desired to be within .04 of the true proportion with a confidence coefficient of .90.

7.48 A random sample of 15 ball bearings produced by the Steinen Corporation had a variance of .00002 inch squared. Determine a 95% confidence interval for the population variance. Assume that the data are normally distributed.

7.49 A random sample of 600 citizens in a certain city showed that 360 favored the annexation of a smaller adjoining community. Find 95% confidence limits for the proportion of the population favoring annexation.

7.50 A local television station wished to estimate the mean time that children watched the cartoons shown on Saturday mornings. A sample of 100 children was selected at random. The arithmetic mean and the standard deviation of the sample were, respectively, 120 minutes and 20 minutes. Determine a 90% confidence interval for the mean viewing time of all children.

7.51 A campus organization wishes to estimate the proportion of students who favor conversion of the academic year from 3 quarters to 2 semesters. What size sample should it select if it wants to assert with a probability of .85 that the sample proportion will not deviate from the true proportion by more than .03?

7.52 The Burton Tire Company wants to estimate the average amount of money its customers spend on tires per purchase. They take a random sample of 101 sales receipts for the year and find the average to be $75 and the standard deviation to be $25. Estimate the mean of the population with a confidence interval using $\alpha = .05$.

7.53 A real estate appraiser wishes to determine the proportion of homes in Norka that have undergone some form of rehabilitation. She took a sample of 100 homes at random and found that 20 satisfied the definition of rehabilitation. What is an interval estimate using a 90% confidence coefficient?

7.54 The Greatrock Tire Company desires to know what proportion of its radial tires in use are suffering from undetected tread separation. To estimate the proportion, it plans to take a random sample from its customers of record and wants to be 95% sure of being within .03 of the correct value. What size sample should they take?

7.55 Verify Equation 7.15.

7.56 A local bank wants to estimate the average balance in the checking accounts of noncommercial customers near the end of the month. They believe balances are normally distributed. The bank's management wants to be 95% confident that their interval will include the true mean. A sample of 25 produced a mean of $174 and a standard deviation of $52. What is the interval?

7.57 The officers of Rutledge National Bank are wondering what the standard deviation of incomes is for the bank's customers who take out home improvement loans. Incomes are approximately normally distributed. They take a random sample of 25 loan applications from the files and find a sample standard deviation of $1000. If they wish a 99% confidence level, what interval do they obtain as an estimate?

7.58 A bank manager is interested in the average amount of cash that individuals have remaining in their checking accounts at the end of the month. If he wishes to estimate with a confidence coefficient of .94 and if the standard deviation is known to be $25, what size sample should he take to be within $5 of the true average?

7.59 A town contains 1600 registered voters. Just prior to election day, 576 of them were sampled to determine their opinion on issue 2. Construct the 88% confidence interval for the true proportion favoring issue 2 if, in the sample, 288 said they favored issue 2.

***7.60** Refer to Example 7.3. Assume that the association has a total of 2000 outstanding loans.

(a) Estimate the total amount for all of the outstanding loans for the association.

(b) Determine a 92% confidence interval based on the point estimate from (a).

7.61 The Hillyer Appraising Company wishes to make its yearly estimate of the average value of one-half acre lots in Somerset County. The company's client requires them to be within $1000 of the correct price. From past data, the standard deviation is believed to be $5000. Using a confidence coefficient of 88%, what size sample should be taken?

7.62 The standard deviation of lengths of wire produced by a certain industrial process is 2 centimeters. A random sample of 25 lengths has a mean of 72.8 centimeters. What is a 98% confidence interval for the true mean length if the normal distribution is believed to apply?

7.63 Verify Equation 7.14.

7.64 Elkins Associates, Certified Public Accountants, wish to estimate the mean error per transaction for their client. A random sample of 50 transactions produced a mean error of −$8.60 and a standard deviation of $2.40.

(a) Calculate a 98% confidence interval for the mean error of all transactions.

(b) Write a sentence explaining the meaning of your confidence interval.

7.65 The Haines Investment Company wants to estimate the proportion of its customers who own more than one home. They want to be within .06 of the true proportion with a confidence coefficient of .995. What size sample should be drawn?

---

* An asterisk indicates a higher level of difficulty.

7.66 A broker randomly sampled 20 income-producing stocks and found that the mean annual yield was 7.64% and the standard deviation was 1.86%. Calculate a 90% confidence interval for the mean annual yield of the population of stocks from which the sample was drawn. Assume that yields are approximately normally distributed.

7.67 Suppose it is desired to estimate the mean distance traveled by college professors during the summer vacation period. Assume that the population standard deviation is known to be 200 miles, and it is desired to be 92% sure of being within 20 miles of the true mean. What size sample should be taken?

8 Testing of Hypotheses

8.1 INTRODUCTION

Vaughn Fredrick, the vice-president of Vendit, a vending machine company, is in charge of selecting good locations for vending machines that dispense candy and gum. In the past, machines have been placed in good sales locations based on a hunch. However, Fredrick thinks the company should be more scientific in its approach. After reading some trade journals, he learned that other firms have concluded that candy and gum machines are profitable where the potential patrons carry an average of at least $.75 in change. The president of Vendit would like to locate a vending machine in the building of the business college of Bayside University. Fredrick suggests that a survey be taken among the students to see if they do carry at least $.75 in change. To survey all the students is out of the question, so it is decided to take a sample. The results of the sample indicated that students carried an average of $.70 in change, with a standard deviation of $.25. Since the average figure is below the suggested value, it is decided that a vending machine will not be placed in the business college building. Fredrick isn't really satisfied with the decision because he questions whether the results of any random sample can produce an average figure of $.75. Further, if a random sample has a mean of $.70, he isn't sure that it could not have come from a population where the true mean is indeed $.75. There must be a statistical technique that will permit such an analysis.

The use of hypothesis testing would enable Mr. Fredrick to decide whether the true mean amount of change possessed by a student is indeed \$.75. The hypothesis testing procedure uses the sample information that is gathered and analyzes it with respect to the degree of risk that Mr. Fredrick would be willing to accept.

Use of the hypothesis testing procedure permits an analyst to determine if the value assigned to a population parameter is appropriate based on a set of data randomly drawn from a population of interest. In this chapter, attention is given to testing a single population mean and the difference between two population means. We will also test the population proportion and the difference between two population proportions. The final two tests that are covered relate to the population variance and the ratio of two population variances. Following the coverage of the different hypotheses tests, attention is given to an analysis of the types of errors inherent in the hypothesis testing procedure and which errors affect results when the risk is changed or the sample size is altered.

For businesses, numerous opportunities are present in day-to-day operations where knowledge of a parameter can aid in making decisions. For example, if a company maintains a machine that automatically fills cans to contain 12 ounces of product, and a random sample of 50 cans indicates that the average content is 11.92 ounces, is the machine operating properly? Or, suppose a company desires to market a new product and will do so only if more than 30% of the population would use it. Two hundred randomly selected potential customers were questioned about the product, and 36% indicated they would use it. Should the company market the product? Using the hypothesis testing procedure will enable the manager to statistically assess each situation before making a decision.

In estimation, a sample is selected, and a point estimate is calculated. If we desire to indicate the amount of error that is acceptable in making an inference about a population parameter, we construct a confidence interval and determine the upper and lower confidence limits. The measure of the belief that this interval contains the population parameter is specified in advance by the selection of the confidence level $(1 - \alpha)$. In hypothesis testing, we begin with a premise. In the two preceding examples, the premises are that the average actual content is 12 ounces and that no more than 30% of the population would use the new product. When a basic premise is established, an alternative to it must also be established. We then select a value for α. This value indicates the risk that we are willing to take of obtaining a sample that tends to contradict the basic premise, even though that premise is true. That is, α represents the chance of rejecting a true basic premise. Thus, α identifies which sample values are more extreme than we would expect by chance alone if the premise is correct.

Before outlining the procedure we use to perform a hypothesis test, we wish to examine more fully the risks involved in declaring samples to be supportive of the basic premise or not supportive of the basic premise.

8.2 TYPE I AND TYPE II ERRORS

The concepts of risk can be better explained and more fully understood through an illustration. A leading manufacturing firm has just made a contract with a new supplier for one of its basic raw materials. Natural fibers, used in the firm's primary product, are to be purchased with an average breaking strength of 25 pounds. The initial order has been received, and the company desires to know whether it is acceptable or unacceptable.

If we first consider that the shipment contained 100,000 different fibers, it is inconceivable that we could afford the time or expense to test each fiber individually, which would involve breaking each one, thus destroying the fiber and its usefulness. Thus, we must rely on some statistical technique that will allow us to make a decision about the total shipment of fibers after examining only a randomly selected sample of them. In the process of evaluating this sample, we must be aware that some risk is involved. In our case, two distinct possibilities exist: (1) the possibility of accepting the shipment when the average breaking strength is either greater than or less than the desired value of 25 pounds; (2) the possibility of rejecting the shipment when the average breaking strength is the desired value of 25 pounds.

To the individual involved in making business decisions, the risks we have described can be evaluated in dollars and cents. If the firm does not accept the shipment, it may have to find another supplier in a short time, which may be too costly. In addition, the lack of raw material may cause the firm to reduce its scheduled production and thus incur the costs associated with underproduction. If the shipment is accepted by the firm when the average breaking strength is less than 25 pounds, the product may not wear as well and, therefore, not last as long as it should. If such a product were guaranteed, it may incur undue expense because it would need to be replaced due to premature wear. Also, customers may be lost because of the inconvenience caused by the unsatisfactory wear of the product.

On the other end of the scale is acceptance of the fiber when the average breaking strength is greater than 25 pounds. This type of fiber may make the product stiff because of the unnecessary added strength. The product's sales could be substantially reduced because of its overall appearance and texture. Then, the firm will also have the increased expense of the added inventory and, ultimately, a huge loss if the lot of finished goods has to be scrapped.

Aside from the error possibilities that exist, we must also consider that the firm could accept the shipment when the average breaking strength of the fiber is 25 pounds. The firm could also reject the shipment when the average breaking strength of the fiber is not 25 pounds. Obviously, these situations are highly desirable. In the 2-by-2 matrix shown in Table 8.1, each of the four decision possibilities that have been outlined is indicated.

In hypothesis testing, the two correct decision situations will basically cause no problems, but we must establish or determine the degree of risk the firm is willing to take in the case of the two incorrect decision situations. If the firm is

TABLE 8.1 Evaluations for all outcomes of breaking strength of fibers and disposition of shipments

	Average Breaking Strength	
Disposition of Shipment	***25 Pounds***	***Not 25 Pounds***
Accepted	Correct	Incorrect
Not Accepted	Incorrect	Correct

able to determine easily the degree of risk for only one of the incorrect decision situations, which one of the two shown in Table 8.1 is the most likely candidate: not accepting a shipment of fibers where the average breaking strength is 25 pounds or accepting a shipment of fibers where the average breaking strength is not 25 pounds? Clearly, the former case is easier since the desired average is known. This error, for which the firm can easily control the risk—that is, establish a particular percentage of the time it will declare a shipment unacceptable when the average breaking strength is 25 pounds—is called a Type I or α (alpha) error. The other error, whose chance of occurring can be controlled by the firm only for specified breaking strengths other than 25 pounds, is called a Type II or β (beta) error. Table 8.2 shows the 2-by-2 matrix with these two specifically identified decision errors properly labeled.

The establishment of a Type I error by a firm will depend on the financial risk involved. In this example, the risk is evidenced by the willingness of the firm to refuse a shipment whenever sample data, drawn from an acceptable lot, tend to contradict the basic premise. At the same time, it is necessary to guard against the decision to accept a shipment where sample data do not tend to contradict the basic premise, but where the lot is unacceptable.

Our discussion thus far has pertained to a shipment of fibers. However, it is important to understand that the risks evident in Type I and Type II errors

TABLE 8.2 The Type I and Type II errors identified for the possible outcomes corresponding to the disposition of shipments

	Average Breaking Strength	
Disposition of Shipment	***25 Pounds***	***Not 25 Pounds***
Accepted	Correct	Type II error
Not Accepted	Type I error	Correct

are applicable to all hypothesis tests. Before we define these errors in general, we need to understand some other terms that have been alluded to but not specifically identified.

DEFINITION 8.1 A **statistical hypothesis** is a basic premise or assumption about a probability distribution that can be evaluated by using sample data and statistical techniques. ■

For each decision situation, two mutually exclusive hypotheses are stated.

DEFINITION 8.2 A **null hypothesis,** designated as H_0, is a statement about the distribution of a population of interest. The statement is assumed to be true until sample data indicate otherwise. The null hypothesis is never expressed as a strict inequality. ■

The results of evaluation lead us to conclude that the sample data either tend to contradict the null hypothesis or fail to contradict it. When the analysis of sample data contradicts the null hypothesis, we state a conclusion based on the alternative hypothesis, which is actually the one that we would like to conclude in most situations.

DEFINITION 8.3 An **alternative hypothesis,** designated as H_a, is a statement about a population of interest that usually represents all other outcomes not covered by the null hypothesis. Generally, H_a is the hypothesis of interest. ■

We can now view the Type I and Type II errors in a more general setting. Basically, the probability of each error is a conditional probability.

DEFINITION 8.4 A **Type I,** or $\boldsymbol{\alpha}$**, error** is the rejection of the null hypothesis, given that the sample data come from a population where the null hypothesis is correct. The probability of this error occurring is α and is given by

$$\alpha = P(\text{Rejecting } H_0 | H_0 \text{ is true})$$

■

DEFINITION 8.5 A **Type II,** or $\boldsymbol{\beta}$**, error** is the acceptance of the null hypothesis, given that the sample data come from a population where the null hypothesis is incorrect. The probability of this error occurring is β and is given by

$$\beta = P(\text{Accepting } H_0 | H_0 \text{ is not true})$$

■

These errors are identified for the general case in Table 8.3.

The establishment of the Type I error automatically fixes the Type II error for a given sample size, H_0, and sampling statistic. However, we are

TABLE 8.3 Identification of Type I and Type II errors among the possible outcomes concerning H_0

	The State of H_0	
Decision Regarding H_0	***True***	***False***
Accept	Correct	Type II error
Reject	Type I error	Correct

unable to state exactly what this probability is unless we have further knowledge about the population and what H_a should be. In the example we have been following, even if the sample size, H_0, and the sampling statistic are fixed, we cannot determine the Type II error unless we know what the true breaking strength of the shipment measures when it is not 25 pounds. Further discussion of Type I and Type II errors will be covered in the treatment of the operating characteristic and power curves found in Section 8.9.

8.3 HYPOTHESIS TESTING PROCEDURE

Evaluating the truth or falsity of a null hypothesis is based on the analysis of sample data. This analysis is part of a statistical hypothesis testing procedure.

DEFINITION 8.6 A **statistical hypothesis test** is a decision rule that enables us to identify the sample values that lead to rejection of H_0. Through the identification of an appropriate test statistic and its sampling distribution, this decision rule takes the form of a probability statement.

The guidelines and format of a statistical hypothesis test should be stated before a random sample is selected from the population of interest. In the following seven-step method for performing the test, the first four steps are established before selecting the sample, and the last three steps are performed using the selected random sample.

1. State the null hypothesis and the alternative hypothesis.
2. Choose the Type I error level.
3. Identify the test statistic and the appropriate sampling distribution.
4. Establish the decision rule that identifies the set of sample values for which H_0 will be rejected.

Then, collect sample data from the population of interest and perform the last three steps.

5. Evaluate the test statistic using the sample data.
6. Make a statistical decision.
7. Make a managerial decision.

Before performing the test that would be pertinent to the problem outlined in Section 8.2, let us examine each of the steps in greater detail.

Step 1. State the null hypothesis and the alternative hypothesis. In the preceding chapters, we have been concerned about the parameters of a population. These parameters include the mean μ_X and the variance σ_X^2. Confining our discussion to these parameters for the moment, we may want to write the null hypothesis as

$$H_0\text{: } \mu_X = \text{value}$$

or

$$H_0\text{: } \sigma_X^2 \leq \text{value}$$

For each null hypothesis stated, we also have an alternative hypothesis. For the two examples, the alternative hypotheses are

$$H_a\text{: } \mu_X \neq \text{value}$$

and

$$H_a\text{: } \sigma_X^2 > \text{value}$$

The initial question after determining the parameter of interest is, What values are to be associated with the null and alternative hypotheses, respectively? Two factors help to indicate the symbolism to be used. First is the definition of a null hypothesis, which always includes the possibility of no difference. Thus, the null hypothesis will always contain the equals sign and may appear in one of three ways: $=$, $\leq$, or $\geq$. Second, which of these symbols we actually use depends on the circumstances associated with the hypothesis test and on the alternative hypothesis. If there is a desire to conclude that the mean of a population has changed—the implication being that it could have increased or decreased—we have an alternative hypothesis, as indicated by the first H_a above, $\mu_X \neq$ value. Thus, the symbolism for the null hypothesis would be the equal sign, $=$. Desiring to conclude that a population variance has increased indicates that the symbolism for the null hypothesis is equal to or less than, $\leq$, and makes the symbolism of the alternative hypothesis greater than, $>$, as indicated in the second H_a above, $\sigma_X^2 >$ value. Had we desired to conclude that the population variance has decreased, the symbolism for both the null and alternative hypotheses would be reversed—that is, H_0: $\sigma_X^2 \geq \sigma_0^2$ and H_a: $\sigma_X^2 < \sigma_0^2$.

In considering how the symbolism for the null and alternative hypotheses is determined, we note that the hypothesis of interest is usually H_a, but it may be either H_0 or H_a. Thus, the sample data are used either to reject the null hypothesis in favor of the alternative or to accept the null hypothesis as being true. Note that we do not actually accept the null hypothesis; we just fail to reject it. In reality, this is a weaker decision than rejecting H_0 since it actually states that we are unable to do something. That is, even though a difference may have been observed, it is not enough of a difference to allow H_0 to be rejected. On the other hand, rejecting H_0 says that what has been observed is so rare or unusual that it is impossible to believe the null hypothesis is true.

Finally, we observe that the alternative hypothesis identifies whether the hypothesis test is to be one-sided or two-sided. The consequence of being a one-sided or two-sided test is detailed in step 4. However, we need to understand that the alternative hypothesis dictates what the test will be. If the symbolism is $\neq$, the hypothesis test will be two-sided. If the symbolism is either $<$ or $>$, the hypothesis test will be one-sided.

In succeeding chapters, we will find hypotheses that are written statements. For example,

H_0: the data are rectangularly distributed

or

H_0: sales are independent of location

are two hypotheses of this type. For each of these null hypotheses, we again have an alternative that we accept if the sample data tend to contradict the null hypothesis. Further discussion is given to testing hypotheses of this nature as they are encountered.

Step 2. Choose the Type I error level. The selection of the Type I error level—that is, α—establishes the maximum probability that we are willing to reject the null hypothesis when it is really true. Whatever risks are associated with the decision-making process are used in the selection of the Type I error level, which usually ranges between .001 and .10. In practice, the cost of making a Type I error should be considered in the selection of α.

The Type I error level, or α, is also known as the *level of significance*. This latter phrase is commonly used because, as we perform the hypothesis test, we attempt to show that the probable occurrence of this sample does not make it unreasonable, and thus not significantly different from the hypothesized value. If the test does result in the rejection of the null hypothesis, we say that "the test is significant at $\alpha = .05$" or whatever level we have selected.

Step 3. Identify the test statistic and the appropriate sampling distribution. The statistical test to be performed is a probability statement, and, in order for any probability statement to be evaluated, certain items must be known. Thus, to construct any statistical test, the following are needed.

—A statistic whose expected value is known when the null hypothesis is true
—The family of the distribution of the statistic under H_0
—Values for the parameters of the family in the above

To obtain the distribution of a statistic, we must make certain assumptions. Hence, each statistical test has specific assumptions associated with it.

Initially, our discussion deals with procedures whose statistics use one of the families of distributions previously discussed, such as the binomial, normal, and t-distributions. In Section 8.7, we utilize the chi-square (χ^2) distribution; in Section 8.8, the F-distribution. For each test presented, we indicate the assumptions associated with it. For example, if the null hypothesis to be tested is H_0: $\mu_X = \mu_0$, where μ_0 equals a specific value, the obvious test statistic is $\overline{X}$. If the population variance is unknown and other basic assumptions are met, the family of distributions would be t. Further specification of the parameters would dictate the exact t-distribution to be considered. The working statistic in this case is

$$t = \frac{\overline{X} - \mu_X}{s_{\overline{X}}}$$

Step 4. Establish the decision rule to be used. The decision rule in a testing procedure is actually the statistical test since it specifies when to accept and when to reject the null hypothesis. The rule can be phrased in terms of probability, but is more often stated in terms of a corresponding tabular value or values. These values are generally called the critical values for the test statistic and are the cutoffs between accepting and rejecting the null hypothesis. The establishment of the acceptance and rejection regions is based on the first three steps of the testing procedure, which can be readily seen as we pose questions about each step.

1. Are we interested in a one-sided or two-sided test?
2. What is the Type I error level we have selected?
3. What test statistic has been identified and what sampling distribution is appropriate?

With these questions in mind, we state the decision rules for the three basic sets of hypotheses in terms of the probability of occurrence of the sample mean, assuming H_0 to be true.

DECISION RULE For testing $\boldsymbol{H_0}$: $\boldsymbol{\mu_X = \mu_0}$; $\boldsymbol{H_a}$: $\boldsymbol{\mu_X \neq \mu_0}$, we reject H_0 at level α if either

$$P(\overline{X} \geq \bar{x}) < \frac{\alpha}{2} \quad \text{or} \quad P(\overline{X} \leq \bar{x}) < \frac{\alpha}{2}$$

where: $\bar{x}$ = the mean derived from the sample data

In this case, we compare our probability to $\alpha/2$ since the normal distribution is symmetric about μ_X and, if H_0 is true, $\overline{X}$ is as likely to be above μ_0 as below μ_0.

DECISION RULE For testing $\boldsymbol{H_0}$: $\boldsymbol{\mu_X \leq \mu_0}$; $\boldsymbol{H_a}$: $\boldsymbol{\mu_X > \mu_0}$, we reject H_0 at level α if

$$P(\overline{X} \geq \bar{x}) < \alpha$$

since H_a implies that only extremely large values are of interest.

DECISION RULE For testing $\boldsymbol{H_0}$**:** $\boldsymbol{\mu_X \geq \mu_0}$**;** $\boldsymbol{H_a}$**:** $\boldsymbol{\mu_X < \mu_0}$**,** we reject H_0 at level α if

$$P(\overline{X} \leq \bar{x}) < \alpha$$

Each decision rule can be phrased in an equivalent form. These forms are illustrated in Section 8.4.

With the establishment of the guidelines for the test, we are ready to select the sample data. What criteria do we use to determine the size of this sample? Two nonstatistical guidelines are time and expense. For the original problem involving fibers, time may be the most important. If the shipment is deemed unacceptable, we will have to obtain another source of supply. Thus, we must make the determination as quickly as we can, keeping the sample size small. When testing can be accomplished only by destroying the item, we may again want to keep the sample small to minimize expenses.

Statistically, we should be aware that the size of the sample can help us to be more confident of the results for a given value of the statistic. As the sample size increases, we find less and less variability between samples of the same size, and, ultimately, we are provided with a clearer picture of the total population.

Regardless of sample size or guidelines, we should make sure that the sample is randomly selected. Selection of observations on a random basis from large populations ensures independence. This property simplifies the determination of the distribution of the test statistic. Once the sample data are gathered, we are ready to proceed with the remainder of the testing procedure.

Step 5. Evaluate the test statistic using the sample data. In step 3, we identified the test statistic. We also identified the appropriate sampling distribution. In this step, we evaluate the working statistic, using the appropriate sampling distribution. That is, we determine the probable occurrence of the test statistic if H_0 is true. For example, in step 3, suppose we selected $\overline{X}$ as the test statistic and identified the standardized normal distribution as being appropriate. Then,

$$Z = \frac{\overline{X} - \mu_X}{\sigma_{\overline{X}}}$$

The mean of the sample data is substituted for $\overline{X}$, the hypothesized value is substituted for μ_X, and the standard deviation of the mean $\sigma_{\overline{X}}$ is $\sigma_X/\sqrt{n}$, where n is the sample size and σ_X is assumed to be known. Performing the necessary arithmetic operations, we obtain the calculated value for Z.

Step 6. Make a statistical decision. We now desire to know if we are to accept or reject the null hypothesis based on the sample data. Using the probability associated with the z value from step 5, we make an evaluation with respect to the decision rule established in step 4. This decision rule indicates the probability of obtaining a sample value as extreme or more extreme than the one observed if H_0 is true. If the value falls into the acceptance region, the probability is $\geq \alpha$, and we are unable to reject the null hypothesis—that is, we accept H_0. If the value falls into the rejection region, we say there is a significant difference between what we hypothesized or expected under H_0 and what we found from the sample data since the probability $< \alpha$. In this case, we conclude that either a rare event has occurred or else H_0 is not true. Assuming that it is not a rare event, we reject H_0.

Step 7. Make a managerial decision. We are no longer in the field of statistics, but we must recognize that, generally, a hypothesis test was established so that we could make some inference about a population. As a result, we can then make some managerial decision, such as ''accept the shipment'' or ''stop the machine and adjust.'' The reason for using an objective test procedure of this type is to minimize the chance for error that is inherent in a strictly intuitive approach.

In the next three sections, the seven-step procedure is used to illustrate the decision-making process as it relates to the different parameters with which we are concerned.

8.4 TESTING THE SINGLE MEAN μ_X

Standard Deviation Known

This section introduces the first formal statistical test. The parameter of interest is the single population mean μ_X. When testing a single population mean, the null and alternative hypotheses will appear in one of the following forms.

$$H_0: \mu_X = \mu_0 \qquad H_a: \mu_X \neq \mu_0$$
$$H_0: \mu_X \leq \mu_0 \qquad H_a: \mu_X > \mu_0$$
$$H_0: \mu_X \geq \mu_0 \qquad H_a: \mu_X < \mu_0$$

For all three forms, μ_0 is a specified value. As indicated by the alternative hypothesis, the first form represents a two-sided test, and the other two represent one-sided tests. To construct the test, we must make the following assumptions.

1. The sample is selected randomly.
2. The distribution of the population is normal (may be relaxed due to the Central Limit Theorem when the sample is large).
3. The population variance is a known value.
4. The level of measurement in the sample is at least interval.

With these assumptions, we proceed to the three requirements for constructing a test. The obvious statistic for testing μ_X is the sample mean $\overline{X}$. Knowing the value of the sample mean satisfies the first requirement since $\overline{X}$ is an unbiased estimator of μ_X; if H_0 is true,

$$E[\overline{X}] = \mu_0$$

If the second assumption is true, then the statistic $\overline{X}$ will follow a normal distribution. If the second assumption is not true, the Central Limit Theorem says that the distribution will be approximately normal anyway. Therefore, the second requirement is fulfilled. For the final requirement, recall that the parameters of a normal distribution are the mean and the variance. We have already determined that if H_0 is true,

$$\mu_{\overline{X}} = E[\overline{X}] = \mu_0$$

In Chapter 6, we found in general that

$$\sigma_{\overline{X}}^2 = \frac{\sigma_X^2}{n}$$

for an infinite population, and

$$\sigma_{\overline{X}}^2 = \frac{\sigma_X^2}{n}\left(\frac{N - n}{N - 1}\right)$$

for a finite one. We note that when sampling without replacement from a finite population, the Central Limit Theorem still provides a means for obtaining approximate probabilities, though the approximations are not as accurate as for an infinite population. Thus, we have a statistic with a completely specified distribution under these assumptions, and we can calculate the probability of obtaining a sample value as extreme or more extreme than the one observed. That is, we are able to determine if the observed value is an unreasonable one when $\mu_X = \mu_0$.

Since the working statistic uses the standard error, the two equations become

$$\sigma_{\overline{X}} = \frac{\sigma_X}{\sqrt{n}}$$

for the infinite population, and

$$\sigma_{\overline{X}} = \frac{\sigma_X}{\sqrt{n}}\sqrt{\frac{N - n}{N - 1}}$$

for a finite one.

We now repeat the question posed in Section 8.2 and determine, through the hypothesis testing procedure, whether the shipment of fibers is to be deemed acceptable or unacceptable.

EXAMPLE 8.1 A leading manufacturing firm has just made a contract with a new supplier for one of its basic raw materials. Natural fibers, used in the firm's primary product, are to be purchased. The fibers are to have an average breaking strength of 25 pounds and a standard deviation of 4 pounds. The initial order has been received, and the company wants to decide if it is acceptable at $\alpha = .05$.

Solution: Follow the seven-step procedure previously outlined.

1. We state the null and alternative hypotheses.

   $$H_0: \mu_X = 25 \qquad H_a: \mu_X \neq 25$$

   Since we are concerned that the fibers are neither greater than nor less than 25, we have a two-sided alternative.
2. We choose the Type I error level: $\alpha = .05$.
3. Since the parameter is μ_X, the test statistic is $\overline{X}$, and since the standard deviation is known, we can consider the normal distribution to be reasonable. The working statistic is

   $$Z = \frac{\overline{X} - \mu_X}{\sigma_{\overline{X}}}$$

   The normal distribution can be selected because the standard deviation is known to us, and the Central Limit Theorem implies that the distribution of $\overline{X}$ will be approximately normal.

The form of the decision rule that is most often used illustrates the use of what is called the standardized statistic. The rule involves standardizing the sample value and then comparing it to a value from a table. For this problem, we desire to know the limits for $\overline{X}$, measured in the number of standard deviations from μ_X, that fix the chance of making a Type I error at .05. Since we are working with a normal distribution, it should be remembered that Appendix Table B.6 is expressed in terms of the number of standard deviations above the mean. Thus, the probability statements for rejection of H_0 are

$$P(Z < -z_0) < .025 \quad \text{and} \quad P(Z > z_0) < .025$$

Graphically, we want z_0 such that 5.0% of the area under the curve is in the rejection region. But we know from Appendix Table B.6 that

$$P(Z < -1.96) = .025 \quad \text{and} \quad P(Z > 1.96) = .025$$

4. We can state the decision rule: Reject H_0 if $z < -1.96$ or $z > 1.96$. The establishment of the decision rule determines the acceptance and rejection regions. As shown in Figure 8.1, the area falling between $-z_0$ and $+z_0$ is the acceptance region, and the two shaded areas denote the rejection region.

Having established decision rules, the quality control department randomly selects 36 fibers. Each fiber is tested, and the average breaking strength for the sample is found to be 24.28 pounds.

5. We can now evaluate the test statistic.

   $$z = \frac{24.28 - 25}{4/\sqrt{36}} = \frac{-.72}{4/6} = -1.08$$

FIGURE 8.1 Identification of the acceptance and rejection regions for the standardized normal and α = .05

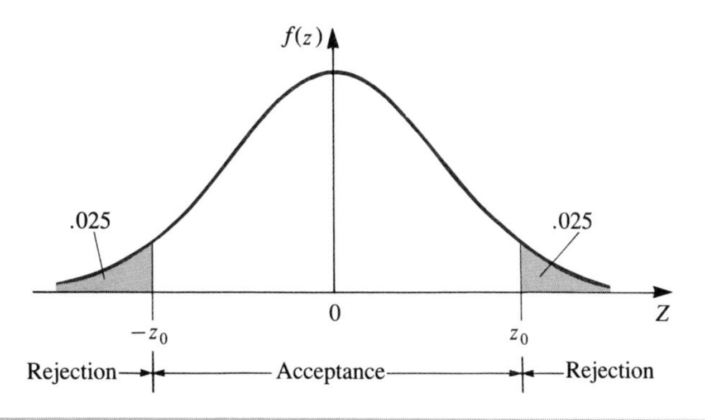

6. In making a statistical decision, we are unable to reject H_0 because $-1.96 < -1.08 < 1.96$, which indicates that 24.28 is not an unreasonable value for $\overline{X}$ based on a sample of 36 when $\mu_X = 25$.
7. We make a managerial decision and accept the shipment.

Since our sample does not statistically contradict the specifications desired, we infer that the total shipment meets these standards and is therefore acceptable.

Another form of the decision rule is to determine the largest and smallest values that are reasonable for $\overline{X}$, given a particular level for α and $\mu_X = \mu_0$. This technique is illustrated in the following example.

EXAMPLE 8.2 Using the data from Example 8.1, establish a decision rule by determining the critical values for $\overline{X}$.

Solution: To establish this decision rule, we determine the maximum and minimum values for $\overline{X}$ from the probability statements

$$P(\overline{X} < \bar{x}_{\text{LC}}) = .025 \quad \text{and} \quad P(\overline{X} > \bar{x}_{\text{UC}}) = .025$$

where $\bar{x}_{\text{LC}}$ represents the lower critical value and $\bar{x}_{\text{UC}}$ represents the upper critical value. As shown in Figure 8.2, the values of $\bar{x}_{\text{LC}}$ and $\bar{x}_{\text{UC}}$ are the limits of the region within which the sample mean must fall to accept H_0. We note that these limits merely represent an interval estimate for $\overline{X}$ rather than an interval for μ_X, as was illustrated in Chapter 7. Thus, we have

$$P(\overline{X} < \bar{x}_{\text{LC}}) = P\left(\frac{\overline{X} - 25}{4/\sqrt{36}} < \frac{\bar{x}_{\text{LC}} - 25}{4/\sqrt{36}}\right) = P\left(Z < \frac{\bar{x}_{\text{LC}} - 25}{4/6}\right) = .025$$

FIGURE 8.2 Identification of the acceptance and rejection regions for critical values of $\overline{X}$

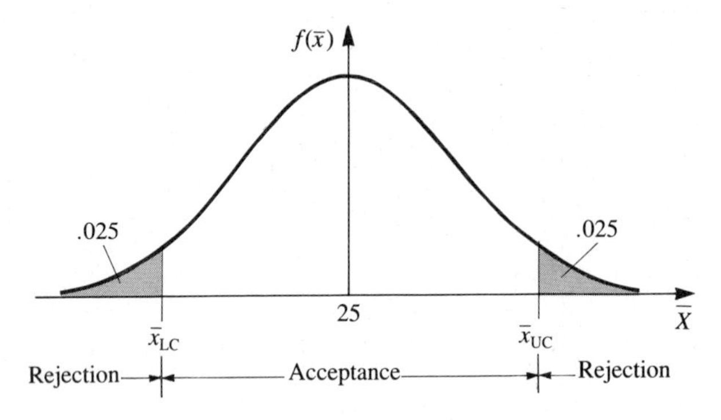

and

$$P(\overline{X} > \bar{x}_{\text{UC}}) = P\left(\frac{\overline{X} - 25}{4/\sqrt{36}} > \frac{\bar{x}_{\text{UC}} - 25}{4/\sqrt{36}}\right) = P\left(Z > \frac{\bar{x}_{\text{UC}} - 25}{4/6}\right) = .025$$

But, $P(Z < -1.96) = .025$ and $P(Z > 1.96) = .025$. Therefore,

$$\frac{\bar{x}_{\text{LC}} - 25}{4/6} = -1.96 \quad \text{and} \quad \frac{\bar{x}_{\text{UC}} - 25}{4/6} = 1.96$$

Solving for $\bar{x}_{\text{LC}}$ and $\bar{x}_{\text{UC}}$, we obtain

$$\bar{x}_{\text{UC}} = 25 + 1.96(.667) = 26.31$$
$$\bar{x}_{\text{LC}} = 25 - 1.96(.667) = 23.69$$

Any value of $\overline{X}$ greater than 23.69 and less than 26.31 is acceptable, and the decision rule can be restated as

$$\text{Reject } H_0 \text{ if } \overline{X} < 23.69 \text{ or if } \overline{X} > 26.31$$

Since the sample value is 24.28, we accept H_0 and reach the same conclusion as before. ■

One final procedure is available to establish a decision rule. It was briefly outlined in Section 8.3. Essentially, it is based on the probable occurrence of obtaining a test statistic if H_0 is true. We select a random sample, determine the value for the test statistic, and find the probability of obtaining such a value if H_0 is true. This probability is then compared to the chosen level of α. The calculated probability is generally referred to as a p value. This approach is very useful because many statistical packages for computers automatically provide a p

value, and the investigator need not have tables available for determining the critical value.

If the decision rule is based on p values, then, basically, there is only one statistical test. That test is simply to reject H_0 if the probability of the observed result is so small that it becomes inconceivable to think that the sample could have come from the hypothesized population. The value for α defines how unlikely the sample must be. Thus, with a specified value for α, the test becomes

$$\text{Reject } H_0 \text{ if } p = P(\text{a sample as extreme as observed or more so}|H_0) < \alpha$$

Note that to be able to calculate this p value, we must be able to completely specify a probability distribution. For this reason, assumptions are required for every statistical procedure. Each probability model or distribution has specific assumptions associated with its development. The assumptions stated with a test procedure are necessary to satisfy the assumptions inherent in the probability model. In most situations, the terms *Z test, t test,* and *F test,* to name a few, refer either to the probability model used to obtain the p value or to the individual(s) who developed the procedure for doing so. As illustrated, there are many forms that a decision rule can take, but the basic test is to reject H_0 if the p value is smaller than α. Example 8.3 illustrates the decision rule based on the determination of the p value.

EXAMPLE 8.3 Using the data from Example 8.1, evaluate the lot by determining the appropriate p value.

Solution: The decision rule would be stated as

$$\text{Reject } H_0 \text{ if } P(\overline{X} < \overline{x}) < .025 \text{ or if } P(\overline{X} > \overline{x}) < .025$$

The value of $\overline{X}$ from the sample is 24.28. We substitute this value into the decision rule and calculate the corresponding probabilities.

$$P(\overline{X} < 24.28) = P\left(\frac{\overline{X} - 25}{4/\sqrt{36}} < \frac{24.28 - 25}{4/\sqrt{36}}\right) = P(Z < -1.08) = .1401$$

and

$$P(\overline{X} > 24.28) = P\left(\frac{\overline{X} - 25}{4/\sqrt{36}} > \frac{24.28 - 25}{4/\sqrt{36}}\right) = P(Z > -1.08) = .8599$$

The p value is twice the smaller of the two probabilities; here, the p value = .2802. In this case, the probability (.1401) is greater than .025 and leads to acceptance of H_0 and the total shipment. A graphical representation of the probabilities is shown in Figure 8.3. It is quite obvious from the graph that the p value exceeds .05 since both probabilities exceed .025.

Any of the three procedures discussed in this section is acceptable, and all are equivalent in the decision-making process. The three hypothesis sets and decision rules based on z-values follow.

FIGURE 8.3 Graphical representations for Example 8.3

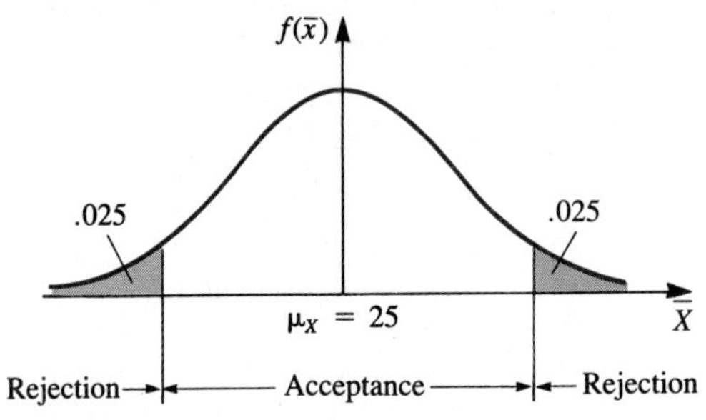

(a) Acceptance and rejection regions

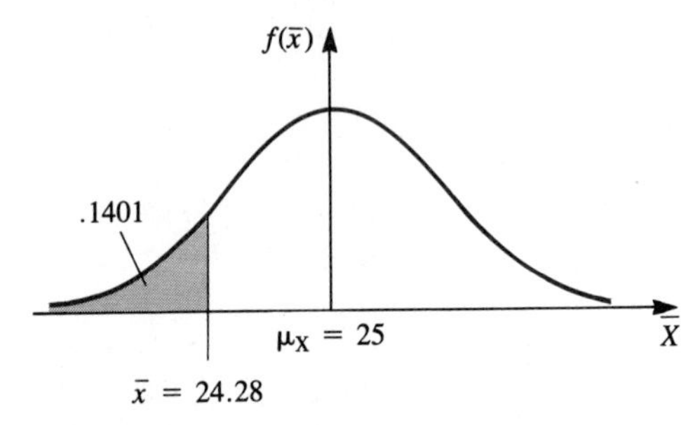

(b) $p(\overline{X} < 24.28)$

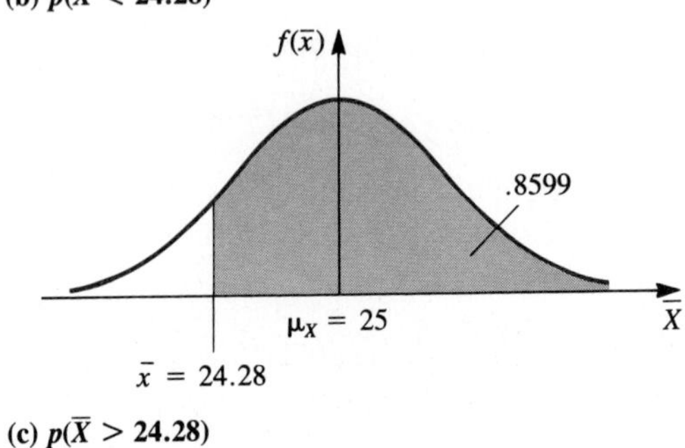

(c) $p(\overline{X} > 24.28)$

DECISION RULES The hypotheses and the appropriate rejection rules for testing the **single population mean μ_X, standard deviation known,** are

$H_0\colon \mu_X = \mu_0 \qquad H_a\colon \mu_X \neq \mu_0$
Reject H_0 if $Z > z_{\alpha/2}$ or $Z < -z_{\alpha/2}$

$H_0\colon \mu_X \leq \mu_0 \qquad H_a\colon \mu_X > \mu_0$
Reject H_0 if $Z > z_{\alpha}$

$H_0\colon \mu_X \geq \mu_0 \qquad H_a\colon \mu_X < \mu_0$
Reject H_0 if $Z < -z_{\alpha}$

Standard Deviation Unknown

We continue testing the single mean, but we introduce a change in the assumptions. The assumptions are as follows.

1. The sample is randomly selected.
2. The distribution of the population is normal (may be relaxed).
3. The population variance is an unknown value.
4. The level of measurement in the sample is at least interval.

The first two requirements for constructing a test are fulfilled in the same manner as before. However, as we observe the parameters of the normal distribution, we note that, if H_0 is true,

$$\mu_{\overline{X}} = E[\overline{X}] = \mu_0$$

but the variance must be estimated from sample data. Thus, using s_X^2 as an unbiased estimator of σ_X^2, we have

$$s_{\overline{X}}^2 = \frac{s_X^2}{n}$$

and the associated standard error is

$$s_{\overline{X}} = \frac{s_X}{\sqrt{n}}$$

Since the variance is unknown, the evaluation of the working statistic $\overline{X}$ will be made by using the family of t-distributions. The following example illustrates the hypothesis testing procedure where the standard deviation is unknown.

EXAMPLE 8.4 The average grade on a particular standardized final exam in statistics previously had been 70. After instituting a new teaching method, instructors believed that this average should have increased. A random sample of 25 students subjected to the new method had an average score of 73 and a standard deviation of 10. For $\alpha = .05$, is the belief supported?

Solution: Again, we employ the seven-step procedure with comments inserted to clarify what is being done.

1. H_0: $\mu_X \le 70$; H_a: $\mu_X > 70$. The phrase ". . . instructors believed that this average should have increased" has caused us to select a one-sided alternative. The only way we would not accept the null hypothesis is when the sample mean is unreasonably high for a μ_X of 70, which is exactly what must occur if we are to conclude an increase.
2. $\alpha = .05$.
3. The statistic is $\overline{X}$, and $t = (\overline{X} - \mu_X)/s_{\overline{X}}$. The t-distribution is selected because it is necessary to estimate the standard deviation from the sample data.

FIGURE 8.4 Indication of acceptance and rejection regions and critical t value for Example 8.4

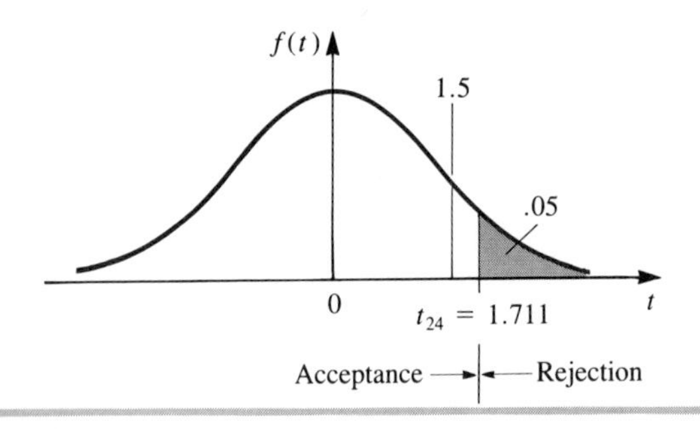

4. Reject H_0 if $t > 1.711$. Using the t-distribution means we need the degrees of freedom—in this case, $25 - 1 = 24$. Thus, the $P(t > t_{24}) = .05$ becomes $P(t > 1.711) = .05$ using Appendix Table B.7. This result is illustrated in Figure 8.4.
5. We have

$$t = \frac{73 - 70}{10/\sqrt{25}} = \frac{3}{10/5} = 1.5$$

6. We are unable to reject H_0 because $1.5 < 1.711$. As we can see from Figure 8.4, the calculated value of 1.5 for t definitely falls in the acceptance region.
7. We conclude that the belief does not appear to be supported since 73 is not an unreasonable value for $\overline{X}$ when $\mu_X = 70$. ■

As we can see, the testing of a single mean can be accomplished by using either the normal or the t-distributions. The choice is based on the assumptions that are satisfied for the particular situation under consideration.

If we reflect on the last example, it illustrates an interesting case. The calculated t value of 1.5 is close to the tabular t value of 1.711. If it were 1.701, what would our decision be? In the strictest statistical sense, we would still accept the null hypothesis because the calculated t of 1.701 would still be a part of the acceptance area. If the tester is still doubtful, more evidence could be gathered that might support the alternative hypothesis. Recall from Section 8.3, in the discussion of sample size, that the larger the sample, the more representative it may be of the population, and, thus, it may provide more substantive results.

tion—that is, assume population variances are unknown *but equal*—it is necessary to estimate the common value from the sample data. The requirements for constructing the test remain the same. In observing the parameters of the normal distribution, we note that

$$E[\overline{X}_1 - \overline{X}_2] = \mu_1 - \mu_2$$

but it is necessary to decide what estimator to use for the common variance, σ_X^2, of the two populations. Note that either s_1^2 or s_2^2 could be used for estimating σ_X^2 since both are unbiased, but using only one discards the information available in the other. The *pooled estimator of the variance* is

$$s_p^2 = \frac{(n_1 - 1)s_1^2 + (n_2 - 1)s_2^2}{n_1 + n_2 - 2} \tag{8.4}$$

and

$$s_{\overline{X}_1 - \overline{X}_2}^2 = \frac{s_p^2}{n_1} + \frac{s_p^2}{n_2} = s_p^2\left(\frac{1}{n_1} + \frac{1}{n_2}\right) \tag{8.5}$$

Because the variances are estimated for each population sampled, it is necessary to reflect this by weighting each sample variance by its degrees of freedom, as shown in Equation 8.4. If we factor s_p^2 out of Equation 8.5 and take the square root, we have the direct solution of the estimated standard deviation for the distribution of the difference of two sample means, as shown in Equation 8.6.

$$s_{\overline{X}_1 - \overline{X}_2} = \sqrt{\frac{(n_1 - 1)s_1^2 + (n_2 - 1)s_2^2}{n_1 + n_2 - 2}}\sqrt{\frac{1}{n_1} + \frac{1}{n_2}} \tag{8.6}$$

Estimation of the standard deviation dictates that to determine the probable occurrence of an observed difference in the sample means we must use the family of t-distributions. Thus, the probability that a value for $\overline{X}_1 - \overline{X}_2$ will deviate from $\mu_1 - \mu_2$ by at least some observed amount is determined by using a t-distribution with $n_1 + n_2 - 2$ degrees of freedom. The degrees of freedom, $n_1 + n_2 - 2$, represent the amount of pooled information about σ_X^2 from the two samples. For a given value of α, we can use Appendix Table B.7 to find t_0 such that

$$P(t > t_0) = \frac{\alpha}{2} \quad \text{and} \quad P(t < -t_0) = \frac{\alpha}{2}$$

Letting t_{calc} denote the t value corresponding to the observed difference, we reject H_0 if either

$$P(t > t_{\text{calc}}) < \frac{\alpha}{2} \quad \text{or} \quad P(t < t_{\text{calc}}) < \frac{\alpha}{2}$$

Evaluating the probability statement is equivalent to rejecting H_0 if either

$$t_{\text{calc}} > t_0 \quad \text{or} \quad t_{\text{calc}} < -t_0$$

where

$$t_0 = t_{(n_1+n_2-2),\alpha/2}$$

and

$$t_{\text{calc}} = \frac{(\bar{x}_1 - \bar{x}_2) - (\mu_1 - \mu_2)}{s_{\bar{X}_1-\bar{X}_2}} \tag{8.7}$$

The use of Equation 8.6 is based on the assumption that $\sigma_1^2 = \sigma_2^2$. In situations where this assumption might be unreasonable, it can be tested by the procedure of Section 8.8 under the assumption of normal distributions. If the null hypothesis of equal variances is not rejected, then we have no reason to doubt the assumption. If H_0 is rejected, Equation 8.6 should not be used. In that case, a workable exact test does not exist, but an approximate procedure can be obtained by using

$$s'_{\bar{X}_1-\bar{X}_2} = \sqrt{\frac{s_1^2}{n_1} + \frac{s_2^2}{n_2}}$$

rather than Equation 8.6. Using this equation in Equation 8.7 yields a statistic having an approximate t-distribution, with degrees of freedom given by the closest integer to

$$\nu = \frac{\left(\frac{s_1^2}{n_1} + \frac{s_2^2}{n_2}\right)^2}{\left[\left(\frac{s_1^2}{n_1}\right)^2\left(\frac{1}{n_1+1}\right)\right] + \left[\left(\frac{s_2^2}{n_2}\right)^2\left(\frac{1}{n_2+1}\right)\right]} - 2$$

If the normality assumption is in doubt, see Chapter 14 for an alternative procedure.

EXAMPLE 8.6 The Leirum Manufacturing Company is considering the purchase of light bulbs in bulk from two suppliers. A sample of 10 bulbs is randomly selected from supplier A and tested. The mean life of the bulbs from supplier A was 1210 hours, and the standard deviation was 52 hours. Twelve bulbs were selected from supplier B, and the test results indicated a mean life of 1270 hours and a standard deviation of 25 hours. Can we conclude at the .05 level of significance that the mean life of the bulbs is the same?

Solution:

1. H_0: $\mu_A - \mu_B = 0$; H_a: $\mu_A - \mu_B \neq 0$.
2. $\alpha = .05$.
3. $t = \dfrac{(\bar{X}_A - \bar{X}_B) - (\mu_A - \mu_B)}{s_{\bar{X}_A-\bar{X}_B}}$.
4. Reject H_0 if

$$t_{\text{calc}} > t_{(10+12-2),.025} = 2.086$$

FIGURE 8.6 Indication of acceptance and rejection regions and critical values for t in Example 8.6

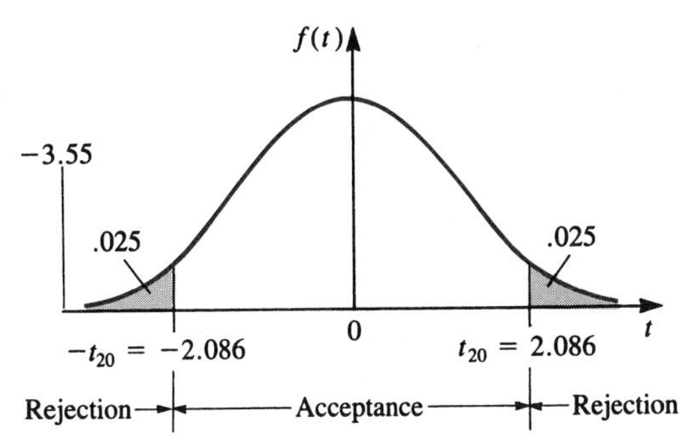

or

$$t_{\text{calc}} < -t_{(10+12-2),.025} = -2.086$$

as shown in Figure 8.6.

5. $$t_{\text{calc}} = \frac{(1210 - 1270) - 0}{16.91} = -3.55$$

where

$$s_{\bar{X}_A - \bar{X}_B} = \sqrt{\left(\frac{9(52)^2 + 11(25)^2}{10 + 12 - 2}\right)\left(\frac{1}{10} + \frac{1}{12}\right)} = 16.91$$

6. Reject H_0 because $-3.55 < -2.086$.
7. On the basis of the sample information, it appears that the mean lives differ. The conclusion means that the actual difference between the sample means is too large to conclude that the difference came from a distribution where the mean is 0. Having made this discovery, the question we ask is: Which bulbs have the longer average life span? To answer this question, we already have stated that the average lives differ, and since there are only two suppliers of interest, the only logical conclusion is that the bulbs from supplier B appear to last longer on the average. Therefore, for the present, the Leirum Company should buy its bulbs from supplier B.

In Example 8.6, a two-sided test was used to show that there was a difference in the mean lengths of life spans for the light bulbs from the two suppliers. The one-sided tests may also be appropriate when testing for a difference between two means and it is necessary to use the pooled estimate for the standard deviation.

DECISION RULES The hypotheses and appropriate rejection rules for testing the **difference between two means, standard deviations unknown but assumed equal,** are

$H_0\colon \mu_1 - \mu_2 = \mu_0 \qquad H_a\colon \mu_1 - \mu_2 \neq \mu_0$
Reject H_0 if $t_{calc} > t_{(n_1+n_2-2),\alpha/2}$ or if $t_{calc} < -t_{(n_1+n_2-2),\alpha/2}$

$H_0\colon \mu_1 - \mu_2 \leq \mu_0 \qquad H_a\colon \mu_1 - \mu_2 > \mu_0$
Reject H_0 if $t_{calc} > t_{(n_1+n_2-2),\alpha}$

$H_0\colon \mu_1 - \mu_2 \geq \mu_0 \qquad H_a\colon \mu_1 - \mu_2 < \mu_0$
Reject H_0 if $t_{calc} < -t_{(n_1+n_2-2),\alpha}$

In the hypotheses, μ_0 often is zero, but it could be another specific value. When testing for a specific difference and for the one-sided tests, caution must be exercised when expressing the order for the two means.

In Chapter 7, the presentation of the estimation of population parameters using a confidence interval did not include an interval estimate for the difference between two population means. Since the distribution for the difference between two population means has been fully specified, we now introduce an equation for the interval estimate for this difference.

$$(\overline{X}_1 - \overline{X}_2) \pm z_{\alpha/2}\sqrt{\frac{\sigma_1^2}{n_1} + \frac{\sigma_2^2}{n_2}} \tag{8.8}$$

Equation 8.8 obviously represents the situation where the standard deviations are known. If the standard deviations are unknown, we substitute a t value for the z and use the pooled estimate for the variance. The pooled estimate for the variance is expressed in Equation 8.4.

$$(\overline{X}_1 - \overline{X}_2) \pm t_{(n_1+n_2-2),\alpha/2}\sqrt{\frac{(n_1-1)s_1^2 + (n_2-1)s_2^2}{n_1+n_2-2}}\sqrt{\frac{1}{n_1} + \frac{1}{n_2}} \tag{8.9}$$

EXAMPLE 8.7 The Leirum Manufacturing Company of Example 8.6 desires to know a 95% confidence interval for the difference between the mean lives for supplier B versus supplier A.

Solution: The data indicate that the standard deviations are unknown, so it is necessary to use the t-distribution with the pooled estimate of the variance, as indicated in Equation 8.9. That is,

$$(1270 - 1210) \pm t_{(12+10-2),.025}\, s_{\overline{X}_B - \overline{X}_A}$$

where: $s_{\overline{X}_B - \overline{X}_A} = 16.91$ (from Example 8.6)

Thus, we have

$60 \pm (2.086)(16.91)$

60 ± 35.27

$LL = 24.73$

$UL = 95.27$

Therefore, we are 95% confident that the interval (24.73, 95.27) contains the true mean difference in average life.

In Examples 8.6 and 8.7, the samples drawn from the populations of interest were randomly and independently selected. However, we might desire to refine the estimate of the error variance by changing the design concept in the random selection of sample information. This topic is discussed next.

Using Paired Samples

Attempting to determine whether a training program for a sales staff has an effect on the average number of sales could be accomplished by randomly selecting individuals from the sales force before training and randomly selecting a different group after training. Using these two samples, we then proceed in the manner described in Section 8.5. However, an alternative to that procedure would be to design the test so that individual salespersons are randomly selected from the sales force. Then, sales information is gathered from each individual before training, and additional sales information is gathered from the same persons after training. In essence, we are randomly selecting pairs of subjects. The before and after experiment is a special case of the paired concept. Other examples would be randomly selecting individuals and having each rate two different products; randomly selecting different surfaces and testing the adherence quality of two different paints on each surface; or randomly selecting pairs of individuals having similar IQs in each pair and randomly assigning one in each pair to one teaching method and the other to a second method, with a common exam being administered at the end.

In using the paired design, we are attempting to match the experimental subjects on some attribute before the experiment is conducted. What we are trying to do is remove the variability of the secondary variable from the estimate of the error variance σ_X^2. Since the experimental units in a pair are identical with respect to the controlled attribute, any differences in pairs can be attributed more confidently to chance variation and the factor of interest, such as the effect of training indicated in the earlier example.

Even though data are gathered in pairs, we are primarily interested in the differences that exist within the pairs of data. The test of interest has null and alternative hypotheses as follows.

$$H_0\colon \mu_1 - \mu_2 = d_0 \qquad H_a\colon \mu_1 - \mu_2 \neq d_0$$

where d_0 is a specific value representing the difference between the population means and is often zero. To construct this test, we make the following necessary assumptions.

1. A random sample of matched pairs is selected.
2. The distribution of the differences is normal, or each population is normal.

3. The variance for the distribution of differences is unknown.
4. The level of measurement is at least interval.

The statistic testing $\mu_1 - \mu_2$ is the sample mean difference $\bar{d}$. This difference is estimated by the equation

$$\bar{d} = \frac{\sum_{i=1}^{n} d_i}{n} \tag{8.10}$$

where d_i represents the difference between the observations in the ith pair, and n is the total number of pairs. Likewise, the estimate of the standard deviation for the distribution of differences is

$$s_d = \sqrt{\frac{\sum_{i=1}^{n} (d_i - \bar{d})^2}{n - 1}} \tag{8.11}$$

or, as an alternative,

$$s_d = \sqrt{\frac{\sum_{i=1}^{n} d_i^2 - \frac{\left(\sum_{i=1}^{n} d_i\right)^2}{n}}{n - 1}} \tag{8.12}$$

Thus, the estimated standard deviation of the sample mean difference will be

$$s_{\bar{d}} = \frac{s_d}{\sqrt{n}} \tag{8.13}$$

Since it is necessary to estimate the standard deviation from sample data, we determine the probable occurrence of $\bar{d}$ as a reasonable value, when H_0 is true, by using the family of t-distributions.

$$t = \frac{\bar{d} - d_0}{s_{\bar{d}}} \tag{8.14}$$

The value of t, as determined by Equation 8.14, will be compared to a value drawn from Appendix Table B.7 using $n - 1$ as the appropriate degrees of freedom. The procedure we have outlined is demonstrated in the next example.

EXAMPLE 8.8 ■ The industrial engineering department of the Arbor Manufacturing Company anticipates that production in a particular assembly line process could be increased by more than 3 units if the materials were placed in a different configuration. The following data represent the average hourly output of 10 randomly selected employees before and after the change.

	Employee									
	1	*2*	*3*	*4*	*5*	*6*	*7*	*8*	*9*	*10*
Before	42.1	40.2	38.7	43.6	45.0	41.6	40.3	36.1	39.2	41.2
After	49.3	48.6	45.2	48.7	44.2	46.7	43.8	40.1	44.4	49.0

Can we conclude at an α level of .05 that there has been an average increase of more than 3 units as a result of the new configuration?

Solution: Since we desire to test that the new configuration will result in an increase, it is necessary to subtract the observations within a pair in the proper direction—that is, the before values will be subtracted from the after values. Thus, the average difference is

$$\bar{d} = \frac{7.2 + 8.4 + \cdots + 7.8}{10} = 5.2$$

and the standard deviation is

$$s_d = \sqrt{\frac{(7.2 - 5.2)^2 + (8.4 - 5.2)^2 + \cdots + (7.8 - 5.2)^2}{10 - 1}}$$
$$= \sqrt{7.0044} = 2.65$$

We now can perform the hypothesis test.

1. H_0: $\mu_A - \mu_B \leq 3$; H_a: $\mu_A - \mu_B > 3$. The hypothesis is stated in this manner to see if the average difference is significantly larger than 3.
2. $\alpha = .05$.
3. $t = \dfrac{\bar{d} - d_0}{s_{\bar{d}}}$.
4. Reject H_0 if $t \geq t_{10-1,.05} = 1.833$.
5. $t = \dfrac{5.2 - 3}{2.65/\sqrt{10}} = \dfrac{2.2}{.84} = 2.62$.
6. Reject H_0 because $2.62 > 1.833$. As we see in Figure 8.7, the calculated value of $t = 2.62$ lies in the rejection region.
7. We can conclude that the new configuration appears to have resulted in an average increase of more than 3 units. Therefore, all positions should be changed accordingly. ■

In this testing procedure, the experimental units are deemed to be related. Because randomly selected pairs of subjects are used, the value for one observation in a pair is related to the value of the other observation. Thus, the difference within a pair is the specific piece of data used in the analysis. The reason is that if no difference exists between treatments—that is, if H_0 is true

FIGURE 8.7 Indication of acceptance and rejection regions and critical value for t in Example 8.8

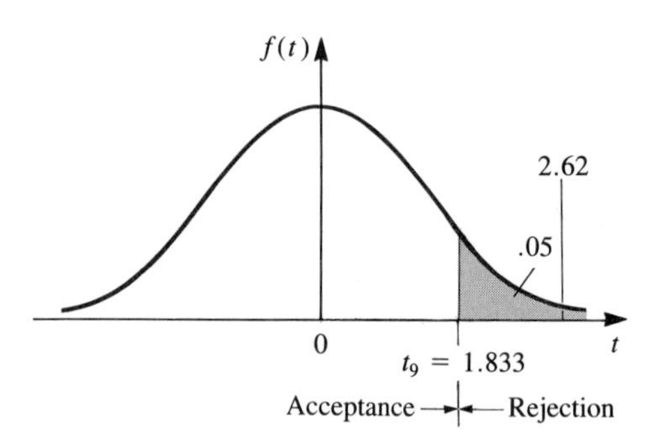

and $d_0 = 0$—then each d_i is expected to be 0, with variations from 0 being attributed to chance or error variation. This test is often referred to as the *paired t test*.

DECISION RULES The hypotheses and appropriate rejection rules for testing the **difference between two means, using paired samples,** are

$H_0\colon \mu_1 - \mu_2 = d_0 \qquad H_a\colon \mu_1 - \mu_2 \neq d_0$
Reject H_0 if $t > t_{n-1,\alpha/2}$ or if $t < -t_{n-1,\alpha/2}$

$H_0\colon \mu_1 - \mu_2 \leq d_0 \qquad H_a\colon \mu_1 - \mu_2 > d_0$
Reject H_0 if $t > t_{n-1,\alpha}$

$H_0\colon \mu_1 - \mu_2 \geq d_0 \qquad H_a\colon \mu_1 - \mu_2 < d_0$
Reject H_0 if $t < -t_{n-1,\alpha}$

For these hypotheses, the value of d_0 is often zero, but it may be any other specified value. The number of pairs of data that are gathered is represented by n.

One final comment regarding the paired t test is a reflection on the dependency between subjects within a pair. If the specific relationship between the sample means is known so that conditional distributions can be determined for use in calculating the probability, an alternative procedure is available, but we judge that it is beyond the scope of this text.

EXERCISES

8.6 The energy crisis has caused many trucking firms to attempt to improve average miles per gallon by controlling driving speed. A random sample of 20 trucks was checked at a driving speed of 65 mph, and the average miles per gallon was 9.87. A random sample of 15 trucks was checked at a speed of 55 mph, and the mean miles per gallon was 13.64. The standard deviations were known to be 2 miles per gallon (65) and 4 miles per gallon (55). At $\alpha = .05$, is the average miles per gallon at 55 mph significantly higher than that at 65 mph?

8.7 The basic operation of Henyon's Bakery runs from Tuesday through Sunday. The volume of business appears to vary from the beginning to the end of the week. Management desires to know if the average sale per customer for the Friday–Sunday customers exceeds by more than \$1.50 the average sale for the Tuesday–Thursday customers. Random samples of 20 were selected from each period, and the average sale was \$1.36 and \$3.27 for Tuesday–Thursday and Friday–Sunday, respectively. From past data, the standard deviations are known to be \$.60 and \$1.10 for the respective time periods. At $\alpha = .10$, what can you conclude?

8.8 With greater career opportunities available for women today, the average age at which they marry would appear to be on the rise. Twenty-one records were randomly selected from the 1975 municipal files and indicated a mean age of 20.2 years and a standard deviation of 4.2 years of age at the time the lady was married. Comparable figures for 31 records randomly selected in 1985 indicated a mean of 22.7 years and a standard deviation of 6.4 years. At $\alpha = .05$, is there evidence of an increase in the mean age to marry during the 10-year period? Discuss the normality assumption.

8.9 A change in the NFL rules regarding impeding receivers went into effect in the 1978 season. It was believed that more scoring would take place as a result. Twenty games were randomly selected from all the games played in 1977 and 20 more from those games played in 1978. The data are as follows.

Year	*Mean Number of Points*	*Standard Deviation*
1977	27.96	7.42
1978	33.14	8.28

At $\alpha = .10$, was there an increase in the mean number of points scored per game?

8.10 In a recent national election, an evaluation of the campaign expenditures (in hundreds of thousands of dollars) by the candidate from 13 randomly selected House seats is given in the following table for the winner and loser.

	Seat												
	1	*2*	*3*	*4*	*5*	*6*	*7*	*8*	*9*	*10*	*11*	*12*	*13*
Loser	6.4	4.9	7.1	8.6	7.6	8.3	7.3	1.7	6.3	11.0	4.5	4.6	3.6
Winner	7.1	8.0	4.9	10.8	10.4	6.3	9.3	3.8	9.3	8.2	10.7	6.0	7.5

At $\alpha = .01$, can we conclude that the mean expenditure by the winners is greater than that of the losers?

8.11 The proponents of a new programming language called NEW insist that programmers can code the same program faster in NEW than in FORTRAN. To attempt to prove their point, they randomly selected 12 programmers from a computer center of a large company and spent one week teaching them NEW. They were then given a flow chart for a problem and asked to code it, first in FORTRAN and then in NEW. The times are given in the following table.

	Programmer											
	1	*2*	*3*	*4*	*5*	*6*	*7*	*8*	*9*	*10*	*11*	*12*
FORTRAN Time	68	47	53	54	61	63	62	49	53	51	50	48
NEW Time	71	43	50	46	62	56	58	51	47	50	50	42

Is the mean time for NEW shorter, as claimed? Use $\alpha = .05$.

8.6 TESTS CONCERNING PROPORTIONS

Single Population Proportion

Chapter 4 introduced the binomial distribution, where the probability of success in a single trial is π. The determination of the probability of x successes in n trials was determined by the binomial equation $p(x) = C_x^n \pi^x (1 - \pi)^{n-x}$. For selected values of n and π, we used Appendix Table B.1 to evaluate the probability. Thus, if we desire to test the null hypothesis and alternative

$$H_0\colon \pi = \pi_0 \qquad H_a\colon \pi \neq \pi_0$$

where π_0 is a specified proportion, we could do this by calculating the probability, under H_0, of obtaining a sample proportion as extreme as the observed value or more so. The sample proportion p is the ratio of successes x to the sample size n. Therefore, we would reject H_0 if

$$P\left(p \geq \frac{x}{n} \,\middle|\, n,\ \pi_0\right) < \frac{\alpha}{2}$$

or

$$P\left(p \leq \frac{x}{n} \,\middle|\, n,\ \pi_0\right) < \frac{\alpha}{2}$$

where x is the observed number of successes. The random variable is $p = X/n$, where X is the binomial random variable. Substituting X/n for p gives the equivalent probability statements

$$P(X \leq x \mid n,\ \pi_0) < \frac{\alpha}{2}$$

or

$$P(X \geq x \mid n,\ \pi_0) < \frac{\alpha}{2}$$

Using Appendix Table B.2, we can make the evaluation for certain combinations of n, π_0. For combinations of n, π_0 not in Appendix Table B.2, we either calculate the probability directly, using

$$P(X \leq x) = \sum_{k=0}^{x} p(x) \quad \text{and} \quad P(X \geq x) = \sum_{k=x}^{n} p(x)$$

where $p(x)$ is given by Equation 4.7, or we use an appropriate approximation. When the assumptions of $n\pi_0$ and $n(1 - \pi_0) \geq 5$ are met and n is large, we can approximate the distribution of proportions with the normal distribution. The test statistic is $p = X/n$, and the standard deviation is

$$\sigma_p = \sqrt{\frac{\pi_0(1 - \pi_0)}{n}} \tag{8.15}$$

if H_0 is true. Therefore, we can determine the probable occurrence of p when the hypothesis is true by using the normal distribution with

$$Z = \frac{\left(p \pm \frac{.5}{n}\right) - \pi_0}{\sigma_p} \tag{8.16}$$

where $\pm\ .5/n$ represents the correction for continuity of Chapter 4. If n is large and either $n\pi_0$ or $n(1 - \pi_0) < 5$, then the Poisson approximation for the binomial distribution may be used.

EXAMPLE 8.9 The Buy-Rite Corporation operates on the assumption that normal turnover of employees should be about 3 or fewer per month per 100 employees. A random sample of 180 employees who were on the payroll last month indicated that 12 are no longer employed by the company. Can we conclude at the .01 level of significance that this turnover rate is abnormal?

Solution: The first step in seeking an answer is to verify the adequacy of the distribution to approximate this proportion problem. The products, $n\pi_0 = 180(.03) = 5.4$ and $n(1 - \pi_0) = 180(.97) = 174.6$, both meet the criteria.

The second concern is the hypothesis to be tested. Based on the statement in the problem, the assumed proportion of turnovers is .03 or less. However, if the proportion is significantly greater than that expected, a problem in hiring practices may be evident. Based on this analysis, we proceed with the test.

1. H_0: $\pi \leq .03$; H_a: $\pi > .03$.
2. $\alpha = .01$.
3. $Z = \dfrac{\left(p \pm \frac{.5}{n}\right) - \pi_0}{\sigma_p}$.
4. Reject H_0 if $Z > z_{.01} = 2.33$.

FIGURE 8.8 Indication of acceptance and rejection regions and critical value for Z in Example 8.9

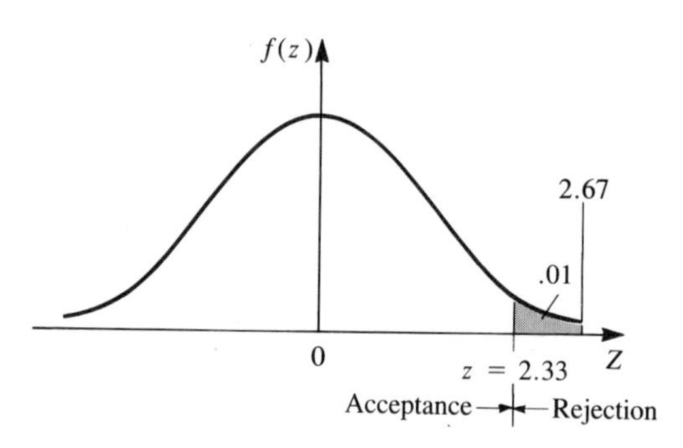

5. $$z = \frac{\left(\frac{12}{180} - \frac{.5}{180}\right) - .03}{\sqrt{\frac{(.03)(.97)}{180}}} = \frac{.0339}{.0127} = 2.67$$
6. Reject H_0 as $2.67 > 2.33$. From Figure 8.8, we see that the value 2.67 falls in the rejection region.
7. Buy-Rite's assumption is not supported by the sample data. The turnover rate appears to have increased, and the company should investigate why this increase has occurred.

The preceding test has been performed on the basis of proportions, but we should also realize that we could obtain the same results by using the number of occurrences with the normal approximation. From Chapter 4, recall that $\mu_X = n\pi$ and $\sigma_X = \sqrt{n\pi(1 - \pi)}$. Thus,

$$Z = \frac{(X \pm .5) - n\pi}{\sqrt{n\pi(1 - \pi)}} \quad \textbf{(8.17)}$$

where X is the number of successes present in the sample of size n and $\pm.5$ is the correction for continuity.

EXAMPLE 8.10 With the information from Example 8.9, perform the hypothesis test, using the normal approximation for the binomial, based on X, the number of successes.

Solution:

1. H_0: $\pi \leq .03$; H_a: $\pi > .03$. The hypotheses remain the same, and for a sample of 180, where the probable occurrence of a turnover is .03, we expect a mean number of turnovers to be $n\pi = 180(.03) = 5.4$. The actual number of turnovers is $x = 12$.
2. $\alpha = .01$.
3. $Z = \dfrac{(X \pm .5) - n\pi_0}{\sqrt{n\pi_0(1 - \pi_0)}}$.
4. Reject H_0 if $Z > z_{.01} = 2.33$. An illustration of the acceptance and rejection regions is shown in Figure 8.8.
5. $z = \dfrac{11.5 - 5.4}{\sqrt{180(.03)(.97)}} = \dfrac{6.1}{2.289} = 2.67$.
6. Reject H_0 because $2.67 > 2.33$.
7. The conclusion is the same as in Example 8.9.

The results of the hypothesis test are the same; in fact, the calculated value of Z is exactly the same. The same value for Z emphasizes that the procedures are different versions of the same sampling distribution and makes either test acceptable when we desire to test the population parameter π.

Regardless of the procedure used to test a population proportion, the hypothesis will take one of the following three forms.

DECISION RULES The hypotheses and appropriate rejection rules for testing the **population proportion** are

H_0: $\pi = \pi_0$ $\quad$ H_a: $\pi \neq \pi_0$
Reject H_0 if $Z > z_{\alpha/2}$ or $Z < -z_{\alpha/2}$

H_0: $\pi \leq \pi_0$ $\quad$ H_a: $\pi > \pi_0$
Reject H_0 if $Z > z_\alpha$

H_0: $\pi \geq \pi_0$ $\quad$ H_a: $\pi < \pi_0$
Reject H_0 if $Z < -z_\alpha$

For each hypothesis set indicated, π_0 is a specified value.

Testing the Difference Between Two Proportions

The basic procedures used to test the population parameter, π, can be extended to testing the difference between proportions for two populations. The hypotheses of interest are

$$H_0: \pi_1 - \pi_2 = 0 \qquad H_a: \pi_1 - \pi_2 \neq 0$$

In the previous section, we saw that if $n\pi$ and $n(1 - \pi) \geq 5$ and n is large, then the distribution of proportions can be approximated with the normal distribution. If we meet the same criteria for two dichotomous populations—that is, $n_1\pi_1$ and $n_1 (1 - \pi_1) \geq 5$, as well as $n_2\pi_2$ and $n_2 (1 - \pi_2) \geq 5$, and both n_1 and n_2 are large—the difference in population proportions can be approximated with a normal distribution. The best estimators of π_1 and π_2 are $p_1 = X_1/n_1$ and $p_2 = X_2/n_2$, respectively. The probable occurrence of an unreasonable difference between the sample proportions can be determined under H_0 as

$$P(Z \geq z_0) < \frac{\alpha}{2} \quad \text{or} \quad P(Z \leq -z_0) < \frac{\alpha}{2}$$

$$\text{where: } z_0 = \frac{(p_1 - p_2) - (\pi_1 - \pi_2)}{\hat{\sigma}_{p_1-p_2}} \tag{8.18}$$

The estimated standard deviation for the distribution of the difference between proportions is

$$\hat{\sigma}_{p_1-p_2} = \sqrt{p(1 - p)\left(\frac{1}{n_1} + \frac{1}{n_2}\right)} \tag{8.19}$$

where n_1 and n_2 represent the respective sample sizes, and

$$p = \frac{X_1 + X_2}{n_1 + n_2} \tag{8.20}$$

For Equation 8.20, the sample values for X_1 and X_2 are the number of successful occurrences in the two samples, respectively. Determining p is based on the fact that if the two sample proportions are hypothesized to be equal, then the linear combination of the sample information will provide the best estimate of the common population proportion. In the event that it is desired to test for an actual difference—that is,

$$H_0\colon \pi_1 - \pi_2 = d_0 \qquad H_a\colon \pi_1 - \pi_2 \neq d_0$$

where d_0 represents the specific difference—then $\hat{\sigma}_{p_1-p_2}$ becomes

$$\hat{\sigma}_{p_1-p_2} = \sqrt{\frac{p_1(1 - p_1)}{n_1} + \frac{p_2(1 - p_2)}{n_2}} \tag{8.21}$$

Equation 8.21 is indicative of the general procedure for determining the variance of the difference of independent random variables as a summation of the individual variances.

EXAMPLE 8.11 The market research department for Michelle, Inc., is attempting to see which type of package is more readily acceptable to its customers. Thus, a survey was conducted of 150 housewives in the Midwest and 120 housewives in the East regarding two different package designs, A and B. In the Midwest, 90 of the housewives preferred A to B; in the East, 60 preferred A to B. At an α level of .02, can we conclude that the package preference differs between the two areas?

Solution: The estimates for π_1 and π_2 are $p_1 = 90/150$ and $p_2 = 60/120$, respectively. Therefore, $150(.6) = 90$ and $150(1 - .6) = 60$ meet the criteria of being ≥ 5. It is also true for the second sample since $120(.5) = 60$ and $120(1 - .5) = 60$. Since, under H_0, we assume the proportions of persons desiring package A and package B are the same in both areas, we need to determine p and $1 - p$. Using Equation 8.20, we have

$$p = \frac{90 + 60}{150 + 120} = \frac{150}{270} = .556$$

and

$$1 - p = 1 - .556 = .444$$

The hypothesis test can now be completed.

1. $H_0\colon \pi_1 - \pi_2 = 0$; $H_a\colon \pi_1 - \pi_2 \neq 0$.
2. $\alpha = .02$.
3. $Z = \dfrac{(p_1 - p_2) - (\pi_1 - \pi_2)}{\sqrt{p(1 - p)\left(\dfrac{1}{n_1} + \dfrac{1}{n_2}\right)}}$.
4. Reject H_0 if $Z > z_{.01} = 2.33$ or if $Z < -z_{.01} = -2.33$, as illustrated in Figure 8.9.
5. $\dfrac{\left(\dfrac{90}{150} - \dfrac{60}{120}\right) - 0}{\sqrt{(.556)(.444)\left(\dfrac{1}{150} + \dfrac{1}{120}\right)}} = \dfrac{.10}{\sqrt{.003703}} = \dfrac{.10}{.0609} = 1.643.$

FIGURE 8.9 Indication of acceptance and rejection regions and critical values for Z for Example 8.11

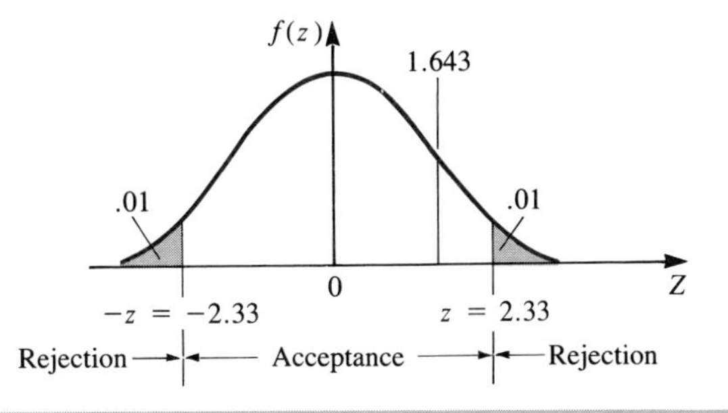

6. Accept because $-2.33 < 1.643 < 2.33$.
7. Based on the sample information, we conclude that the proportions of persons preferring package A to B are not significantly different for the two areas.

The next example illustrates the procedure if the hypothesized difference between the proportions of the two populations is an amount other than zero.

EXAMPLE 8.12 A manufacturer of pollution abatement equipment had randomly sampled 400 of its employees and had found that 250 of them were cigarette smokers. Management was concerned enough to implement a vigorous campaign to reduce this percentage. Each employee received 2 mailings regarding the hazards of smoking. In addition, the plant bulletin boards were blanketed with posters giving the same message. After 3 months, the company questioned 600 employees at random and found that 318 smoked. If the company's goal was to reduce smoking by more than 5 percentage points, was the campaign successful? Let $\alpha = .02$.

Solution: Before proceeding with the solution, we note that it is necessary to assume that the 2 samples are independent of each other. Even though the individuals are randomly selected each time, the samples may not be independent since it is likely that some employees will be in both samples. With random sampling, there should be little effect on the final conclusions, but it is possible that the independence assumption could be violated to the extent that the conclusions might be in error. From the information provided, letting 1 denote before and 2 after, the sample proportions are

$$p_1 = \frac{250}{400} = .625 \qquad 1 - p_1 = 1 - .625 = .375$$

$$p_2 = \frac{318}{600} = .53 \qquad 1 - p_2 = 1 - .53 = .47$$

The standard error is

$$\hat{\sigma}_{p_1 - p_2} = \sqrt{\frac{(.625)(.375)}{400} + \frac{(.53)(.47)}{600}} = \sqrt{.0010011}$$
$$= .03164$$

The hypothesis test can now be performed.

1. $H_0\colon \pi_1 - \pi_2 \le .05$; $H_a\colon \pi_1 - \pi_2 > .05$.
2. $\alpha = .02$.
3. $Z = \dfrac{(p_1 - p_2) - (\pi_1 - \pi_2)}{\sqrt{\dfrac{p_1(1 - p_1)}{n_1} + \dfrac{p_2(1 - p_2)}{n_2}}}$.
4. Reject H_0 if $Z > z_{.02} = 2.054$, as shown in Figure 8.10.

FIGURE 8.10 Illustration of acceptance and rejection regions and critical value for Z for Example 8.12

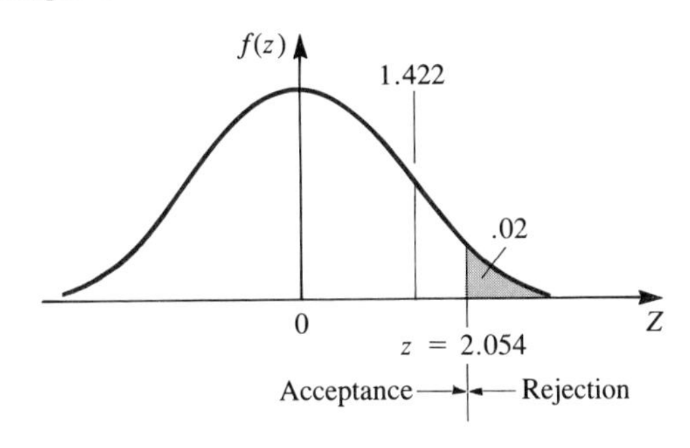

5. $z = \dfrac{(.625 - .53) - .05}{.03164} = \dfrac{.045}{.03164} = 1.422.$
6. Accept H_0 because $1.422 < 2.054$.
7. From the sample information, the firm is unable to conclude that there has been more than a 5% reduction in the number of smokers. Therefore, the campaign does not appear to have been successful.

In Example 8.12, it is important to subtract the sample proportions in the proper order so that the correct difference is compared to the hypothesized difference. The difference indicated in the null hypothesis will establish the direction for the subtraction.

DECISION RULES The hypotheses and appropriate rejection rules for testing the **difference between two population proportions** are

$H_0\colon \pi_1 - \pi_2 = d_0 \qquad H_a\colon \pi_1 - \pi_2 \neq d_0$
Reject H_0 if $Z > z_{\alpha/2}$ or if $Z < -z_{\alpha/2}$

$H_0\colon \pi_1 - \pi_2 \leq d_0 \qquad H_a\colon \pi_1 - \pi_2 > d_0$
Reject H_0 if $Z > z_\alpha$

$H_0\colon \pi_1 - \pi_2 \geq d_0 \qquad H_a\colon \pi_1 - \pi_2 < d_0$
Reject H_0 if $Z < -z_\alpha$

For each of these hypotheses, d_0 can be any specified value, including 0. As indicated earlier, the order of subtraction for π_1 and π_2 is important when the alternative is one-sided. In addition, the estimate for the standard deviation will be based on d_0. When $d_0 = 0$, Equation 8.19 is used, and when $d_0 \neq 0$, Equation 8.21 is used. A confidence interval for the difference between two proportions is given by

$$(p_1 - p_2) \pm z_{\alpha/2}\hat{\sigma}_{p_1-p_2}$$

where $\hat{\sigma}_{p_1-p_2}$ is given by Equation 8.21.

$$where: \; F_{\nu_1,\nu_2,1-(\alpha/2)} = \frac{1}{F_{\nu_2,\nu_1,\alpha/2}}$$

Even though we established a null and alternative hypothesis as a two-sided test, we can convert the two-tailed decision rule into a single statement as follows. We note that s_1^2/s_2^2 is distributed as F_{ν_1,ν_2} and that s_2^2/s_1^2 is distributed as F_{ν_2,ν_1}. But, $s_2^2/s_1^2 = 1/(s_1^2/s_2^2)$ and hence is distributed as $1/F_{\nu_1,\nu_2}$.
Thus,

$$P\left(\frac{s_1^2}{s_2^2} \le F_{\nu_1,\nu_2,1-(\alpha/2)}\right) = P\left(\frac{1}{s_1^2/s_2^2} \ge \frac{1}{F_{\nu_1,\nu_2,1-(\alpha/2)}}\right)$$

$$= P\left(\frac{s_2^2}{s_1^2} \ge \frac{1}{F_{\nu_1,\nu_2,1-(\alpha/2)}}\right)$$

$$= P\left(\frac{s_2^2}{s_1^2} \ge \frac{1}{1/F_{\nu_2,\nu_1,\alpha/2}}\right)$$

$$= P\left(\frac{s_2^2}{s_1^2} \ge \frac{1}{F_{\nu_1,\nu_2,\alpha/2}}\right)$$

That is, we reject H_0 if either

$$\frac{s_1^2}{s_2^2} > F_{\nu_1,\nu_2,\alpha/2}$$

or

$$\frac{s_2^2}{s_1^2} > F_{\nu_2,\nu_1,\alpha/2}$$

Since only one of these ratios can be large, we may restate the decision rule for the two-sided alternative as

$$\text{Reject } H_0 \text{ if } \frac{\text{larger } s^2}{\text{smaller } s^2} = \frac{s_L^2}{s_S^2} \ge F_{\nu_n,\nu_d,\alpha/2}$$

where df_n and df_d are the degrees of freedom for the numerator and denominator variances, respectively.

If we choose to evaluate the null hypothesis and its alternative

$$H_0\colon \sigma_1^2 \le \sigma_2^2 \qquad H_a\colon \sigma_1^2 > \sigma_2^2$$

we reject H_0 only if

$$F = \frac{s_1^2}{s_2^2} > F_{\nu_1,\nu_2,\alpha}$$

However, if we desire to test the null hypothesis and its alternative

$$H_0: \sigma_1^2 \geq \sigma_2^2 \qquad H_a: \sigma_1^2 < \sigma_2^2$$

the F statistic becomes

$$F = \frac{s_2^2}{s_1^2} \tag{8.24}$$

and we reject H_0 if

$$\frac{s_2^2}{s_1^2} > F_{\nu_2,\nu_1,\alpha}$$

EXAMPLE 8.14 Sample standard deviations of expenditures drawn from 25 customers in the grocery department at Kapital and 21 customers at Crowleys were $14.27 and $21.36, respectively. At the .02 level of significance, can we conclude that the variabilities in expenditures of the populations at the 2 stores differ?

Solution: In setting up the null hypothesis and the alternative, the variance for Crowleys will be designated s_1^2, and the variance for Kapital will be s_2^2. Making this designation ensures that the larger sample variance will be in the numerator of the F calculation.

1. $H_0: \sigma_1^2 = \sigma_2^2$; $H_a: \sigma_1^2 \neq \sigma_2^2$.
2. $\alpha = .02$.
3. $F = \dfrac{s_1^2}{s_2^2}$.
4. Reject H_0 if $F > F_{20,24,.01} = 2.74$, as illustrated in Figure 8.12.

FIGURE 8.12 Illustration of acceptance and rejection regions and critical F value for Example 8.14

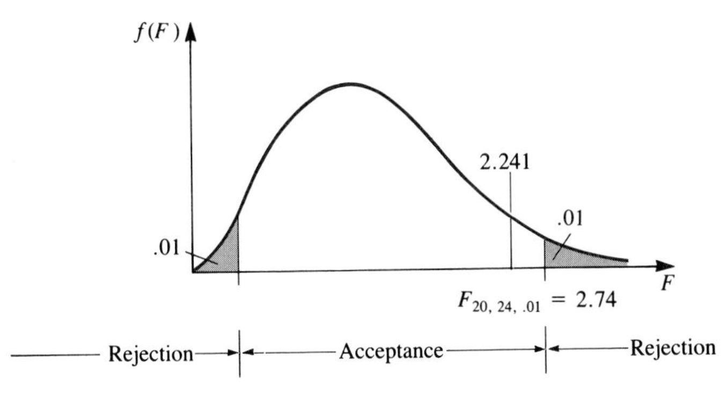

5. $F = \dfrac{(21.36)^2}{(14.27)^2} = 2.241.$
6. Since $2.241 < 2.74$, we accept H_0.
7. The variabilities in expenditures are not significantly different for the populations of buyers at the 2 stores.

A further discussion of the F test is presented in Chapter 9 under the topic of analysis of variance. In this section, we have seen that each F calculation is dependent on the form of the hypothesis.

DECISION RULES The three hypotheses and their appropriate rejection rules for testing **two population variances** are

H_0: $\sigma_1^2 = \sigma_2^2$; H_a: $\sigma_1^2 \neq \sigma_2^2$

Reject H_0 if $s_L^2/s_S^2 \geq F_{\nu_n,\nu_d,\alpha/2}$

H_0: $\sigma_1^2 \leq \sigma_2^2$; H_a: $\sigma_1^2 > \sigma_2^2$

Reject H_0 if $s_1^2/s_2^2 \geq F_{\nu_1,\nu_2,\alpha}$

H_0: $\sigma_1^2 \geq \sigma_2^2$; H_a: $\sigma_1^2 < \sigma_2^2$

Reject H_0 if $s_2^2/s_1^2 \geq F_{\nu_2,\nu_1,\alpha}$

EXERCISES

8.18 A regional survey in 1976 based on a randomly selected sample of 31 persons indicated a mean voting age of 43.7 and a standard deviation of 16.4. In 1985, a similar survey of 41 persons indicated a decrease in the mean age of 1.9 years and a decrease in the standard deviation of 2.1 years. At $\alpha = .02$, can it be concluded that the variability has changed for the voting public?

8.19 In the early 1970s, college graduates were able to take their pick from several job offers available to them at graduation time. A survey of 28 graduates indicated a standard deviation of 4.73 offers. A survey of 25 graduates in the 1980s has seen this figure shrink to 1.86. At the .01 level, has there been a significant decrease?

8.20 Prestige Pastries frequently caters weddings and other festive occasions. There appears to be a greater demand for sheet cakes than for layer cakes, but the variability from week to week is such that management would like to investigate the variability aspect first to see if a difference exists. A random sample of 12 weeks was observed for sheet cakes, and the standard deviation was 9.83 cakes per week. For the layer cakes, a random sample of 9 weeks was observed, and the standard deviation was 13.62 cakes per week. At the .05 level of significance, can the firm conclude there is greater variability in the demand for layer cakes?

8.9 *OC* AND POWER CURVES

Section 8.2 began a discussion of the two types of errors encountered in hypothesis testing, α and β. The establishment of a level for the Type I error automatically fixes the maximum level of the Type II error if the sample size and sampling distribution remain constant. This section now pursues that discussion further to illustrate the determination of specific values for beta.

It is easy to see this with respect to the test of a single mean when the variance is known if we use the form of the decision rule based on the value of $\overline{X}$—that is, on $\bar{x}_c$. For the one-sided alternative, $\mu_X > \mu_0$, the hypotheses are

$$H_0: \mu_X \leq \mu_0 \qquad H_a: \mu_X > \mu_0$$

For a specified value of α, a constant variance σ_X^2, and a sample size n, we saw in Section 8.4 that

$$\bar{x}_c = \mu_0 + z_\alpha \frac{\sigma_X}{\sqrt{n}}$$

Thus,

$$\alpha = P(\overline{X} > \bar{x}_c | \mu_X = \mu_0)$$

and

$$\beta = P(\overline{X} \leq \bar{x}_c | \mu_X = \mu_a)$$

where μ_a is a specified value under H_a. Since $\bar{x}_c$ is fixed by μ_0, α, σ_X, and n, β is then fixed for any value of μ_a. The specific probabilities for α and β are illustrated in Figure 8.13. Note that $\bar{x}_c$ will be the same regardless of the specific value being considered for μ_X under H_a.

FIGURE 8.13 Illustration of α and β for H_0: $\mu_X \leq \mu_0$ and $\bar{x}_c$ fixed by μ_0, σ_X, and n for a value μ_a

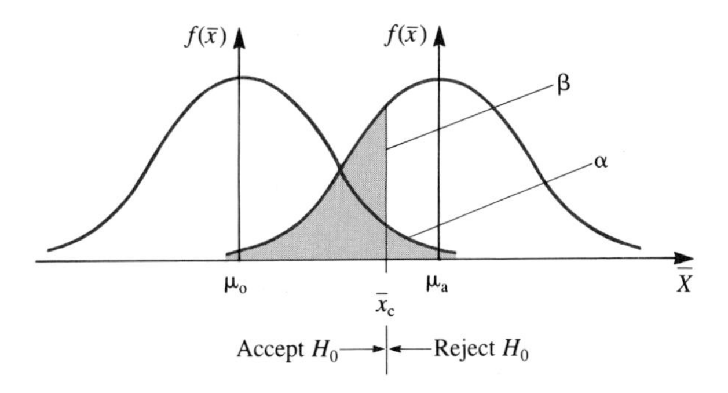

If we return to the problem concerning the fibers originally given in Section 8.2, we can evaluate the Type II error by speculating that the true mean is really something other than the 25 specified under H_0. In Figure 8.14, we have established the acceptance and rejection regions if the theoretical distribution of our fibers has $\mu_X = 25$ and the α level is .05. The critical values separating the

FIGURE 8.14 Identification of β for selected values of μ_X when the hypothesized value is 25, $\sigma_X = 4$, $n = 36$, and $\alpha = .05$

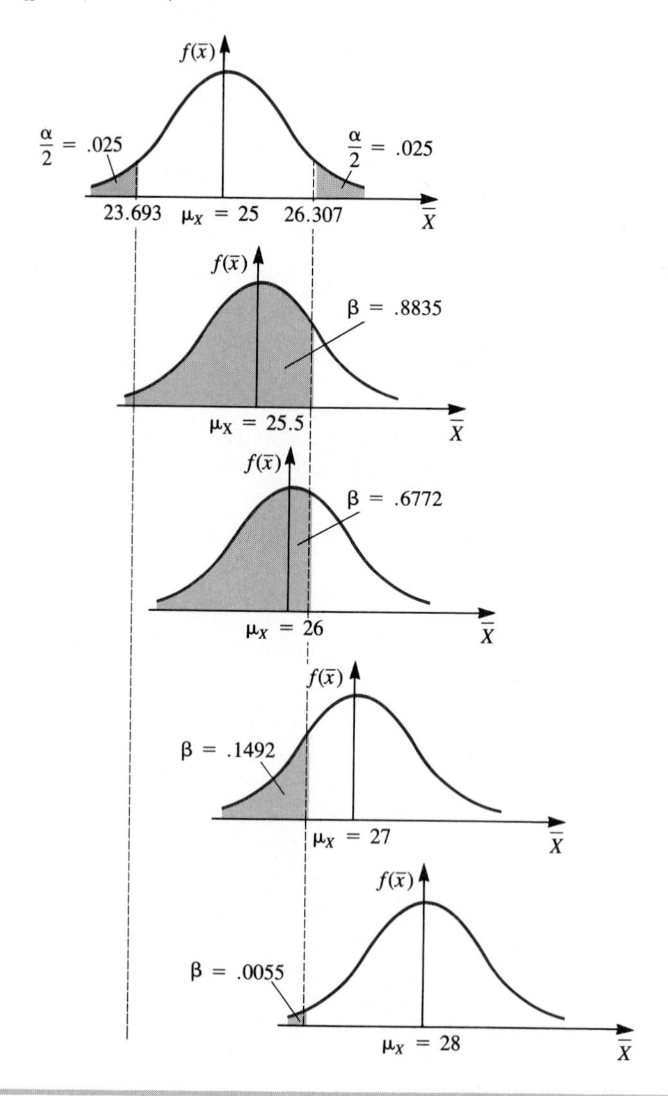

acceptance and rejection regions were previously calculated to be

$$\bar{x}_{\text{UC}} = 25 + 1.96\left(\frac{4}{\sqrt{36}}\right) = 26.307$$

and

$$\bar{x}_{\text{LC}} = 25 - 1.96\left(\frac{4}{\sqrt{36}}\right) = 23.693$$

If we let the true mean assume successive values of 25.5, 26, 27, and 28, the shaded areas represent the probabilities of the corresponding Type II errors. For example, the probability of accepting the hypothesis that $\mu_X = 25$ when it is actually 25.5 is 88.35% and is determined as follows:

$$z = \frac{26.307 - 25.5}{.667} = 1.21 \qquad P(0 \le Z \le 1.21) = .3869$$

$$z = \frac{23.693 - 25.5}{.667} = -2.71 \qquad P(-2.71 \le Z \le 0) = .4966$$

Thus, the β error is $.3869 + .4966 = .8835$. The remaining percentages are calculated in a similar manner.

As can be readily seen, the probability of accepting a false hypothesis decreases rapidly the further the true value is from the hypothesized mean. In Figure 8.15, the levels of β have been plotted versus the different values that have been assumed for the true mean. This plot depicts the *operating characteristic curve* (*OC*), and if μ_X assumes values of 24.5, 24.0, 23.0, and 22.0, we would have a mirror image, making the complete curve symmetric.

FIGURE 8.15 Operating characteristic curve for true values of μ_X when the hypothesized value is $\mu_X = 25$ and $\sigma_{\bar{X}} = .667$

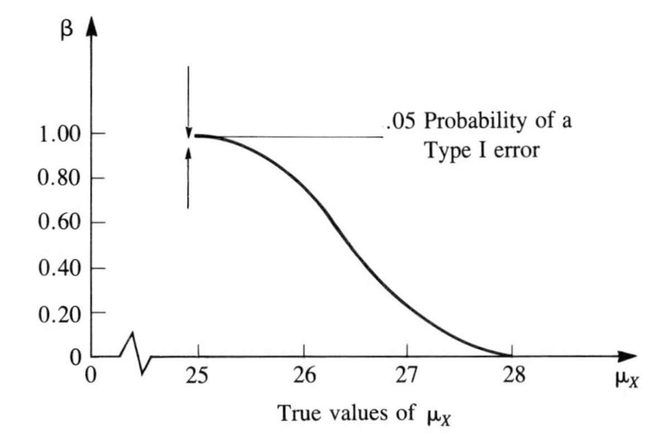

DEFINITION 8.7 The ***OC* curve** is a graph of the probability of committing a Type II error when μ_X takes on values other than our hypothesized value.

The level of the error is measured as the height of the curve at any point we desire to evaluate. The curve may give the impression that the population mean may vary, which is totally untrue. Thus, we should view this curve as the probability of committing an error for alternative possible values of μ_X.

For purposes of comparison, we note that an ideal test procedure would have both α and β equal to zero. That is, it would always lead to acceptance of H_0 if H_0 is true and to rejection of H_0 if H_0 is false. An ideal procedure for testing H_0: $\mu_X = \mu_0$ versus H_a: $\mu_X \neq \mu_0$ would have an *OC* curve as shown in Figure 8.16. Since our procedures use sample information, the ideal is unattainable. The reason for considering the ideal is to provide a standard against which to compare our procedures. The closer the *OC* curve approaches to the ideal curve, the better the procedure is at distinguishing between H_0 and H_a.

The computation of the actual β error levels shown in Figure 8.15 used the standard deviation of the sampling distribution of $\overline{X}$ for $n = 36$. It is obvious that by increasing the sample size, we change the size of the β error. In Figure 8.17, the determination of β errors is shown for sampling distributions with sample sizes of 36, 100, 256.

We can see from Figure 8.17 that if the hypothesis, $\mu_X = 25$, has been accepted when, in fact, $\mu_X = 26$, the β level is substantially decreased as the sample size increases. Since the value of $\sigma_{\overline{X}}$ is changed when n is changed, the sampling distribution becomes more compact as n increases. In Figure 8.18, the operating characteristic curves for two sample sizes are shown for several assumed values of μ_X. The *OC* curves further bear out the reduction of β as n increases with α and σ_X fixed.

Earlier in this chapter, we briefly discussed the statistical decision when a calculated z or t value was close to the tabular value. However, we can evaluate the level of the β error before we actually make the final decision to

FIGURE 8.16 *OC* curve for an ideal test procedure of H_0: $\mu_X = \mu_0$ versus H_a: $\mu_X \neq \mu_0$

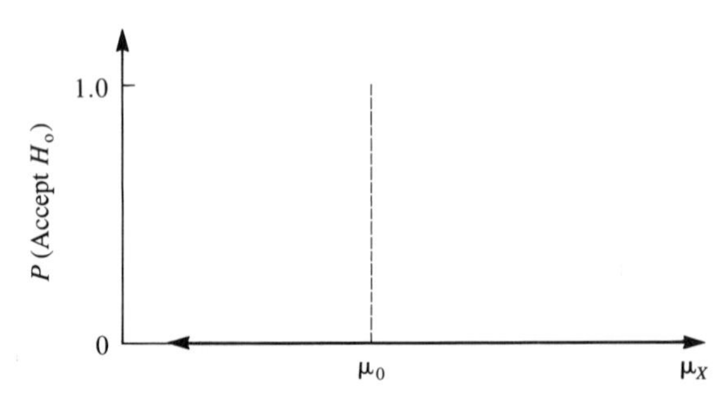

FIGURE 8.17 Identification of β at $\mu_X = 26$ for selected values of n, when the hypothesized value is $\mu_X = 25$, $\sigma_X = 4$, and $\alpha = .05$

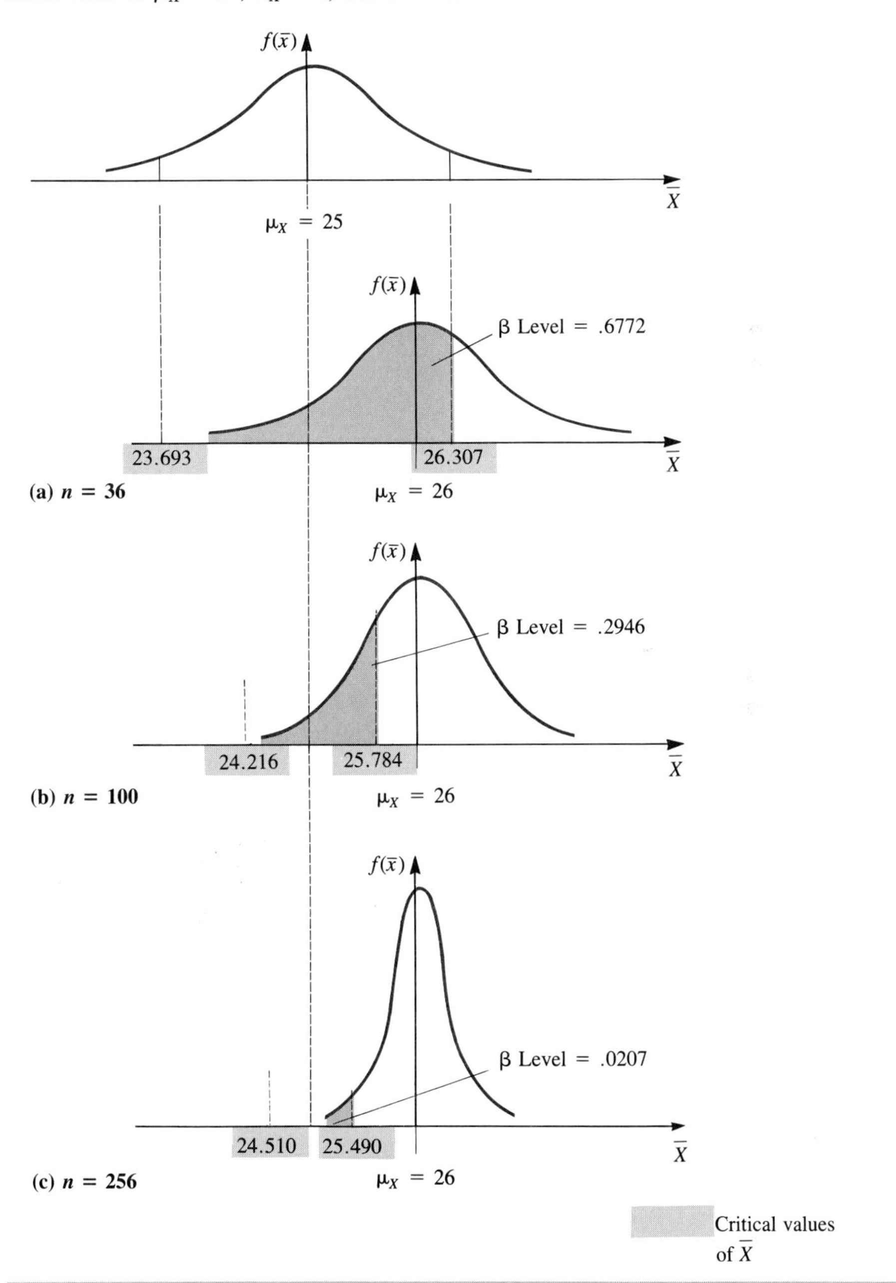

FIGURE 8.18 *OC* curves for $n = 36$ and $n = 100$ when $\mu_X = 25$, $\sigma_X = 4$, and $\alpha = .05$

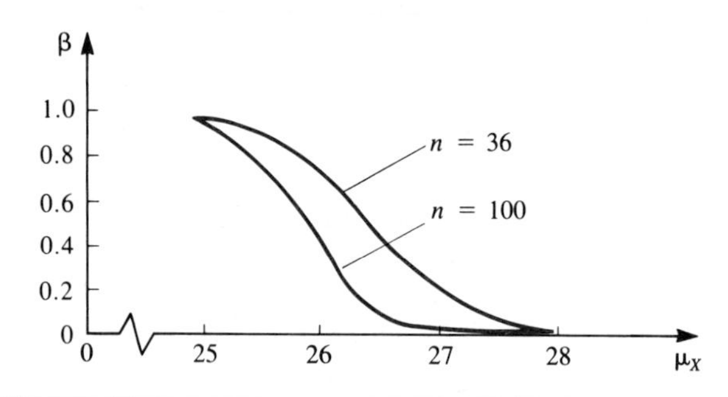

obtain another sample. If the probability of committing a Type II error is small enough to be acceptable for the situation at hand, we could eliminate the expense that would be incurred with selecting another sample and conducting the test again. We might also desire to actually control the level of the β error. Controlling the level of the β error can be done by establishing some β level (for example, .10) and determining the sample size required to satisfy both the α and β levels.

EXAMPLE 8.15 Andrews Electronics manufactures a resistor with a mean of 16 ohms and a standard deviation of 5 ohms. (An "ohm" is a measure of electrical resistance.) The production run is periodically tested to check the quality at $\alpha = .05$. Quality control also wishes to control for a $\beta = .10$, when the true mean may be 18. How large a sample must be taken to satisfy both conditions?

Solution: In seeking the sample size, it is helpful if we draw a picture to depict both the acceptable α and β errors. See Figure 8.19. Since β is to be controlled only for the alternative value of 18, we may concentrate on $\bar{x}_{UC}$. In each case, the value of $\bar{x}_{UC}$ can be calculated as

$$\bar{x}_{UC} = 16 + 1.96\left(\frac{5}{\sqrt{n}}\right)$$

and

$$\bar{x}_{UC} = 18 - 1.28\left(\frac{5}{\sqrt{n}}\right)$$

However, as we see in Figure 8.19, the value for $\bar{x}_{UC}$ is the same for each equation. Thus, the two equations are set equal to each other and solved for n.

$$16 + 1.96\left(\frac{5}{\sqrt{n}}\right) = 18 - 1.28\left(\frac{5}{\sqrt{n}}\right)$$

$$\frac{9.8}{\sqrt{n}} + \frac{6.4}{\sqrt{n}} = 2$$

$$16.2 = 2\sqrt{n}$$

$$\sqrt{n} = 8.1$$

$$n = 65.61$$

Thus, a sample of 66 will be randomly selected from the production run.

In general, the determination of the sample size to simultaneously satisfy specified levels for both α and β is given by

$$n = \left[\frac{(z_{\alpha/2} + z_{\beta})\sigma_X}{\mu_a - \mu_0}\right]^2 \qquad \textbf{(8.25)}$$

where μ_0 is the mean under H_0 and μ_a is the specific value of the mean that is of interest under H_a. The use of this equation assumes that σ_X is the same for both populations and is previously known. For a one-sided alternative, use z_α rather than $z_{\alpha/2}$.

With some modification, Equation 8.25 may be used for determining the maximum sample size required to control both α and β in a one-sample test of a

FIGURE 8.19 Depiction of α and β for Example 8.15

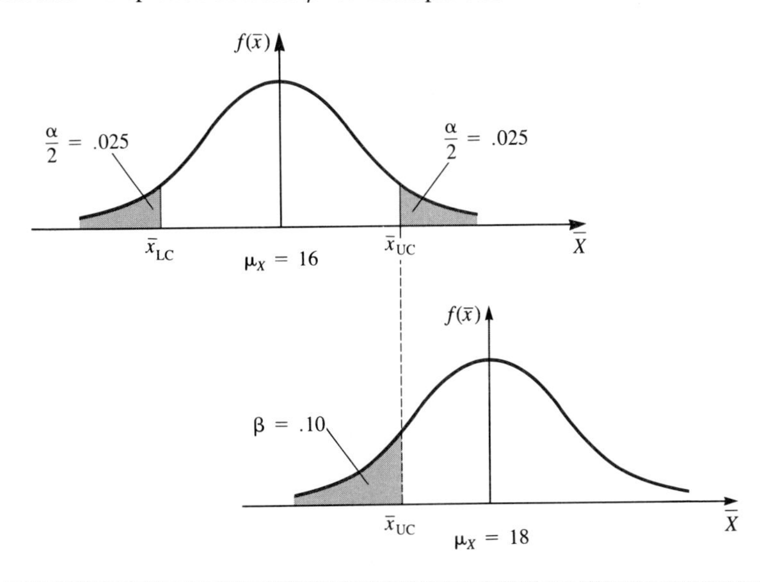

binomial proportion. The modification is necessary because the standard deviation for a binomial random variable is a function of π. Recall from Chapter 7 that the closer π is to .5, the larger σ_Y becomes, and the larger the value used for σ_Y, the larger the resulting sample size. Thus, to calculate σ_Y, we use $\sigma_Y = \sqrt{\pi'(1 - \pi')}$, where π' corresponds to whichever of π_0 or π_a is closer to .5. The random variable Y represents a Bernoulli random variable. Let π_0 be the value of π under H_0, π_a the particular value of interest under H_a, and α and β the maximum probabilities of committing Type I and Type II errors, respectively. Then, for testing a sample proportion, Equation 8.25 becomes

$$n = \left[\frac{(z_{\alpha/2} + z_\beta)\sigma_Y}{\pi_a - \pi_0}\right]^2$$

$$= \left(\frac{z_{\alpha/2} + z_\beta}{\pi_a - \pi_0}\right)^2 (\pi')(1 - \pi') \tag{8.26}$$

where π' equals whichever of π_0 and π_a is closer to .5. For a one-sided alternative, use z_α rather than $z_{\alpha/2}$.

In many instances, it may be desirable to look at the power of the test.

DEFINITION 8.8 The **power of a test** is the ability of the test to detect the alternative and is represented as $1 - \beta$—that is, $1 - \beta = P(\text{Reject } H_0 | H_a \text{ is true})$.

In Figure 8.20, a power curve has been constructed for various values of μ_X under H_a. The actual values of $1 - \beta$ have been derived from the values of β as shown in Figure 8.14. The x-axis variable remains the same as in the OC curve, but the y-axis is now $1 - \beta$. This curve then depicts the probability of

FIGURE 8.20 Power curve for true values of μ_X when the hypothesized value of $\mu_X = 25$, $\sigma_X = 4$, $n = 36$, and $\alpha = .05$

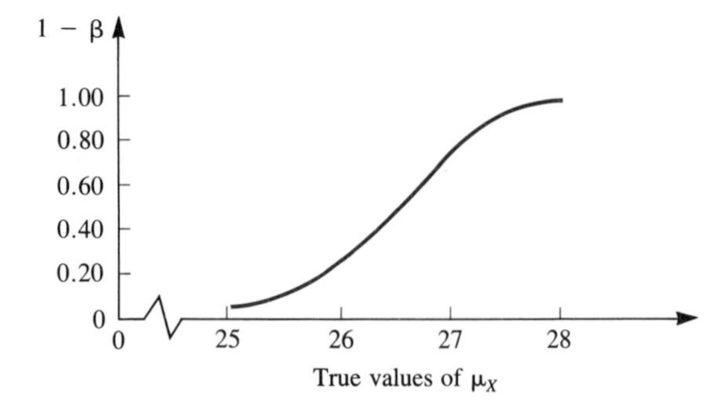

FIGURE 8.21 Identification of β and $1 - \beta$ at $\mu_X = 26$ for selected values of α when the hypothesized value is $\mu_X = 25$, $\sigma_X = 4$, $n = 36$, and $\alpha = .2, .1, .05$, and $.02$

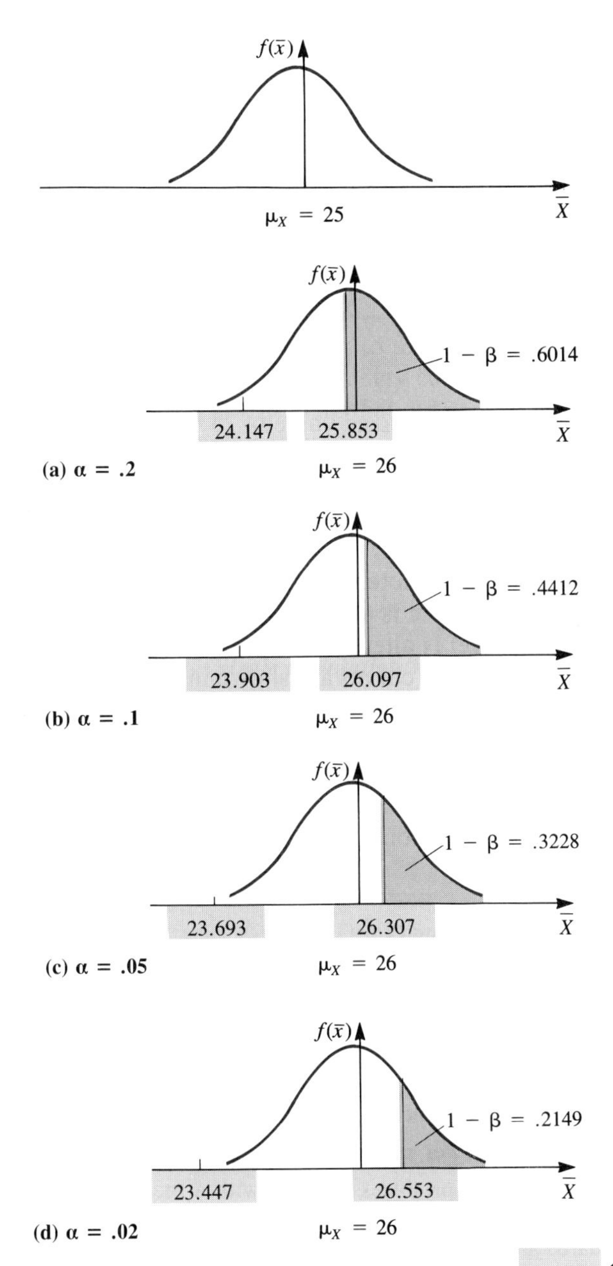

TABLE 8.4 (continued)

(c) Testing a Single Population Proportion and the Difference Between Two Population Proportions Under Prescribed Conditions

Condition	*Statistic (Normal Approximation)*	*Standard Error*
Testing a single proportion using the sample proportion; $n\pi_0$ and $n(1 - \pi_0) \geq 5$	$Z = \dfrac{(p \pm .5/n) - \pi_0}{\sigma_p}$	$\sigma_p = \sqrt{\dfrac{\pi_0(1 - \pi_0)}{n}}$
Testing a single proportion using X, the number of successes; $n\pi_0$ and $n(1 - \pi_0) \geq 5$	$Z = \dfrac{(X \pm .5) - n\pi_0}{\sigma_X}$	$\sigma_X = \sqrt{n\pi_0(1 - \pi_0)}$
Testing for difference in two population proportions; $(\pi_1 - \pi_2) = 0$	$Z = \dfrac{(p_1 - p_2) - (\pi_1 - \pi_2)}{\hat{\sigma}_{p_1-p_2}}$	$\hat{\sigma}_{p_1-p_2} = \sqrt{p(1 - p)\left(\dfrac{1}{n_1} + \dfrac{1}{n_2}\right)}$ $p = \dfrac{X_1 + X_2}{n_1 + n_2}$
Testing for a specific difference in two population proportions; $(\pi_1 - \pi_2) \neq 0$	$Z = \dfrac{(p_1 - p_2) - (\pi_1 - \pi_2)}{\hat{\sigma}_{p_1-p_2}}$	$\hat{\sigma}_{p_1-p_2} = \sqrt{\dfrac{p_1(1 - p_1)}{n_1} + \dfrac{p_2(1 - p_2)}{n_2}}$

(d) Testing a Single Population Variance and the Equality of Two Population Variances

Condition	*Statistic*
Testing a single population variance	$\chi^2 = \dfrac{(n - 1)s_X^2}{\sigma_0^2}$
Testing equality of two population variances	$F = \dfrac{s_1^2}{s_2^2}$ or $F = \dfrac{s_2^2}{s_1^2}$

(e) Determining Sample Size to Control for α and β Errors

Condition	*Equation*
Determining the sample size to control for α and β in a one-sample test of means	$n = \left[\dfrac{(z_{\alpha/2} + z_\beta)\sigma_X}{\mu_a - \mu_0}\right]^2$
Determining the sample size to control for α and β in a one-sample test of proportions	$n = \left[\dfrac{(z_{\alpha/2} + z_\beta)}{\pi_a - \pi_0}\right]^2 (\pi')(1 - \pi')$ π' is the closer of (π_a, π_0) to .5

In Sections 8.7 and 8.8, we considered testing the population parameter called the variance. We again assume the population of interest is normal and that the level of measurement of the sample data is at least interval. The considerations regarding these tests are summarized in Table 8.4d.

In the final section of the chapter, we looked at further coverage of the concept of a Type II error, including analyzing the change in the sample size and also the effect of changes in the α error. We also studied the construction of both the operating characteristic and power curves, and, lastly, the equation for determining the sample size required to control both α and β for a specified value under H_a. These equations are given in Table 8.4e.

SUPPLEMENTARY EXERCISES

8.24 What is meant by statistical inference?

8.25 Why is it necessary to approach hypothesis testing from a systematic point of view?

8.26 **(a)** What two types of errors are encountered when testing hypotheses?
(b) Can we control for both errors at the same time? Explain.

8.27 What is the meaning of the null hypothesis?

8.28 What is the relationship between the sample size and the critical region denoted by the α level?

8.29 **(a)** What is the theoretical consideration regarding the α level?
(b) How is the level determined?

8.30 **(a)** What is the difference between one-tailed and two-tailed hypothesis tests?
(b) How do we determine which test to use?

8.31 What assumptions do we make regarding the population of interest when the t test is employed?

8.32 Distinguish between the independent-samples t test and the paired t test.

8.33 In a test of the difference between 2 proportions, why is the denominator of the test statistic dependent on the hypothesis?

8.34 A tire manufacturer offers tires to its employees at a discount rate through the company store. The company believes that at least 70% of the employees have taken advantage of this offer. A recent check by the employee relations department showed that the 500-car company parking lot contained 320 cars that had at least one tire made by the company.
(a) At an α level of .10, is the company's assumption correct?
(b) Discuss the problems arising from this design.

8.35 The Fox Assembly Company specializes in assembling electronic equipment to the customer's specifications. In addition, 2 lines turn out a regular line of completed assemblies. All new employees were assigned to the 2 regular lines until sufficient knowledge was gained to enable them to move into the specialty areas. In the past, the company had trained a new employee for 24 working hours before assigning the employee a position. The company has now decided to give a new employee a 4-hour orientation and then place him or her with an experienced worker. Six persons have been assigned to positions in this manner. After 2 weeks, their production figures were checked and indicated an average of 21.5 assemblies per worker, with a standard deviation of 4.3 assemblies. These figures were compared with the results of 8 workers trained in the former manner, which were a mean of 24.4 assemblies and a standard deviation of 2.1.

(a) At an α level of .10, can we conclude that the results are the same regardless of training?

(b) What problem exists in this test that may cause you to question the results?

8.36 The variability in age of persons attending college during the 1960s and 1970s was relatively small. Studies indicated the standard deviation was 3.67 years. In the 1980s, individuals are spreading their education over a longer period of time. A recent random sample of 30 students indicated a standard deviation in age of 5.47 years. At an .025 level of significance, can we conclude that there has been an increase in the variability of age for individuals attending college in the 1980s?

8.37 A pharmaceutical company that markets a highly commercial product has not experienced a great amount of success in sales. Ten stores were randomly selected from all of its retail outlets, and the product was rearranged on the shelves. In addition, point-of-purchase displays were added to emphasize the outstanding features of the product. Following are sales figures representing weekly sales before and after the changes were made.

	Store									
	1	*2*	*3*	*4*	*5*	*6*	*7*	*8*	*9*	*10*
Before	27	32	48	36	29	33	37	42	25	31
After	52	44	56	52	41	45	59	64	31	49

At an α level of .025, can we conclude that there has been at least an average increase of 15 units as a result of the changes?

8.38 A manufacturer of a fire-preventing sprinkler system used a solder that was molded over the sprinkler device and that would trigger the system when the melting point was reached. It was essential to have a solder that would melt at a temperature no higher than 130°F. A random sample of 121 triggering devices indicated a mean temperature of 133° and an estimated standard deviation of 26°.

(a) At an α level of .01, are the triggering devices satisfactory?

(b) How do you reflect on the usefulness of the test you have just performed?

8.39 A manufacturer of washing products is concerned about entering the field of cold-water detergents. A survey conducted in 1979 showed that 25 of 250 women interviewed were using such a product. In a recent survey, 128 of 320 women interviewed were using a cold-water detergent. The manufacturer is willing to risk entry into the market if there has been more than a 10% increase in users of cold-water detergents. At an established Type I error level of .05, what action should the manufacturer take?

8.40 The standard deviation of attendance for NFL games (not including playoffs) for 1978 was 4421, rounded to the nearest whole number. In 1984, this figure had risen to 5184 for a sample of 24 games. Is there any difference in variability at the .10 level of significance?

8.41 The personnel department of a large food processor was interested in how the average value of the fringe benefits for each hourly employee compared with employees of the industry. A random sample of 46 employees indicated a yearly rate of $1463 and a standard deviation of \$327. Figures from the industry based on a sampling of 432 persons indicated yearly benefits of \$1527 and a standard deviation of \$289. At an established Type I error level of .05, is the company different from the industry?

8.42 A study within the electronics industry indicated that after 3 years, 1 out of every 3 new college recruits was no longer with the company of original employ. One large firm that was very concerned about the problem had instituted its own program to combat these departures. In addition, a more highly selective screening program had been instituted 3 years ago. The company wants to see how effective the total program has been, and a check of its records shows that 27 persons out of 106 have left the firm for a variety of reasons in the past 3 years.

(a) At an α level of .01, what do you conclude?

(b) Which test is more powerful, one-tailed or two-tailed?

8.43 Coron Products has redistributed its sales territories in an effort to change the variability in sales by their sales force. Prior to the redistribution, a sample of 16 salespersons indicated a dispersion in sales of \$632.27. Since the change in territories, the standard deviation of sales for 9 salespersons was \$512.88. At an α level of .02, can we conclude there has been a change?

8.44 The quality control department of a plastics manufacturer has been rejecting a number of extrusions because measurements have been larger than the standard model. A random sample of 16 selected 2 weeks ago provided the following data: $\bar{x}_1 = .626$ inch, and $s_1 = .005$. A recent sample of 25 provided these data: $\bar{x}_2 = .631$ inch, and $s_2 = .006$. If the mean difference has increased by more than .010 inch, it will be necessary to retool. At an α level of .025, is any action necessary?

8.45 In Exercise 8.44, the test of means utilized the t-distribution because the standard deviations were estimated from sample data. In performing the test, we assumed the variances were equal. At an .10 level of significance, was the assumption justified?

8.46 The Reliable Corporation has a machine that automatically measures the amount of carbon needed in the production of foam glass. For each batch, 500 pounds of ground glass are mixed with at least 20 pounds of finely divided carbon. A check of the machine indicated an average carbon input of 19.368 pounds for 9 batches with a sample standard deviation of .75 pound. Is the machine operating satisfactorily at a .01 level of significance?

8.47 A socially conscious company was concerned about the knowledge of the "rules of the road" of its employees. Twelve employees were randomly selected and given a written test of 100 questions concerning their general knowledge and specific changes that had occurred in the last 5 years. Following are the results.

Person	1	2	3	4	5	6	7	8	9	10	11	12
Test I	82	93	70	68	75	78	84	90	86	73	88	52

The company then asked these 12 employees to attend a class, which was conducted on company premises on company time, to update their knowledge. After four 1-hour sessions were held, another test of comparable difficulty was administered. The results for the same 12 persons are as follows.

Test II	88	92	77	84	82	90	88	90	84	75	82	74

At an α level of 0.05, can we conclude that the updating sessions were helpful in increasing knowledge about driving?

8.48 Hancock Products produces a breakfast cereal that is sold in boxes where the fill weight is labeled 16 ounces. To ensure the proper weight is met, it sets its filling machines at 16.08 ounces. A recent random sample of content weights was gathered. The data are as follows.

16.05	16.11	16.13	16.07
16.10	16.07	16.08	16.10
16.14	16.15	16.09	16.03
16.08	16.12	16.06	16.06
16.11	16.10	16.12	16.11
16.14	16.06	16.11	16.07

(a) Based on these sample data, can the company conclude that the mean fill weight is at least 16.08 ounces? Let $\alpha = .05$.

(b) If the company desires a variability in fill weights of .04 ounce, is this criterion being met? Let $\alpha = .01$.

8.49 A company that sold men's deodorant was concerned whether it should gear its advertising to men who would purchase the product for themselves or to women who would purchase the product for their husbands, boyfriends, or sons. A marketing study indicated that of 120 women interviewed, 72 had made such a purchase at one time or another but not on a regular basis. For the men, 110 out of 190 interviewed had made such a purchase, but they too had not necessarily done it on a regular basis. At a .10 level of significance, what would you propose regarding the advertising campaign?

8.50 Energy Electronics has accepted a shipment of copper wire. The decision by quality control to accept the shipment was based on the acceptance of the null hypothesis $\mu_X = 5$ millimeters, where $\sigma_X = .4$ millimeter, $n = 25$, and $\alpha = .05$. What is the probability of committing a Type II error if the true mean of the shipment is **(a)** 5.1 millimeters, **(b)** 5.4 millimeters? Assume σ_X is unchanged.

8.51 The average age of persons who strongly advocate a certain tax proposal is believed to be no more than 36.7, and the standard deviation is believed to be 5.8 years. What is the probability of rejecting 36.7 if the true mean is 38.0, the level of significance is .10, and a sample of 64 is selected?

8.52 Lockstep Enterprises desires to test a shipment of steel rings that are to have a mean diameter of 1.6 inches or less and a standard deviation of .2 inch. The α level has been established at .04. The company would also like to ensure that, if the true mean of the shipment is 1.65 inches, there is only a 12% chance the shipment would be accepted. What sample size should be randomly selected to satisfy both conditions? (Assume σ_X is unchanged.)

8.53 A beginning typist is expected to average 30 words per minute and have a standard deviation of 5 words per minute after 3 weeks of training. In a beginning class of 140 students, a random sample of 42 students produced an average of 28.2 words per minute. At $\alpha = .10$, does this class appear to be behind schedule?

8.54 The Conway Corporation has perfected a covering for wood to prevent absorption of water when the wood is placed in the ground. For a prescribed area, the expected absorption is 10 grams. In a recent test, 18 pieces of wood were coated and left in the ground for a prescribed period of time. The average absorption was 8.94 grams, and the standard deviation was 1.21 grams. At an α level of .05, are these test results significantly less than the expected absorption level?

8.55 If the critical value approach were taken to test the null hypothesis of Exercise 8.54, what would this value be?

8.56 Often a political candidate attempts to appeal to a particular constituency, such as the ethnic vote or the blue-collar vote. Because of constituency identification, generally 50% of the campaign workers are selected to enhance the particular appeal. One candidate desired to seek the votes of the young marrieds and, particularly, the under 30 age group. A sampling of 278 workers indicated that 152 were under 30. Let $\alpha = .08$.

(a) Do these results differ from the accepted practice?

(b) Is a one-tailed test preferred over a two-tailed test? Why or why not?

8.57 Do Exercise 8.56 by the normal approximation procedure based on the number of successes.

8.58 In a study of local welfare agency records, a random sample of 12 records indicated that the average number of children in families on welfare was 3.27, and the standard deviation was 1.13. Checking records from 10 years ago, a random sample of 10 records showed the average number of children to be 2.61 and the standard deviation to be .87. At $\alpha = .05$, can we conclude there has been an increase in the average number of children in families on welfare?

8.59 The continuing rise in the cost of natural gas has caused many homeowners to better insulate their homes. The standard deviation in monthly gas bills for a standard dwelling place without insulation was $18.82. A recent check of 22 randomly selected homes that have been fully insulated shows a standard deviation in monthly gas bills of $14.92. At an α level of .025, has there been a significant decline in variability?

8.60 The null hypothesis H_0: $\mu_X = 126$ has been accepted at the .10 level of significance, based on a sample of 46 and a known standard deviation of 24. Construct the *OC* curve if the true mean had values of 128, 130, 132, 135, 140, and 143.

8.61 Based on the situation described in Exercise 8.60, construct the power curve if the true mean had values of 109, 112, 117, 120, 122, and 124.

8.62 Hand-dipped 8-inch tapered candles are deemed to burn longer than comparable machine-molded candles. The following data represent samples drawn from the output of the two procedures.

	Sample	*Average Burn (minutes)*	*Standard Deviation*
Hand-dipped	14	132	11
Machine-molded	18	117	18

For a Type I error of .05, what do you conclude?

8.63 To test the hypothesis in Exercise 8.62, the assumption of equal variances was made. Is that assumption supported at $\alpha = .10$?

8.64 Students writing computer programs in GPSS generally require several runs before the program is successful. However, they suspect that the number of runs would diminish in successive projects as their familiarity with the language increases. To test the theory, 8 students are randomly selected from a class, and the number of runs needed to achieve a successful program is recorded for each student for the first project and the sixth project.

	Student							
	1	*2*	*3*	*4*	*5*	*6*	*7*	*8*
Runs (1st Project)	9	6	4	5	8	11	7	6
Runs (6th Project)	4	3	4	6	3	5	5	2

Can we conclude at $\alpha = .10$ that there has been a significant reduction of more than 1.5 in the average number of runs after 6 projects?

8.65 Tapetex, a manufacturer of video tape, has made a change in the ingredients of its tape product called Sonex. It is anticipated that the average tone, measured in rhumes, will be increased by more than 12 rhumes. Variability in the original product was 2.5 rhumes, whereas the new product has a variability of 3 rhumes. A sample of 23 tapes of the old product was tested and 17 of the new, and the average number of rhumes was 91 and 105, respectively. For $\alpha = .07$, has an improvement taken place?

8.66 The Chronicle Bank has found that 10% of its customers have taken out short-term loans of \$200 or more. A rigorous advertising campaign has been conducted in an attempt to increase this percentage. To check on the results of the campaign, the bank randomly selected a sample of 150 customers with checking accounts. If the records indicate that 26 of these customers had taken out loans, what can the bank management conclude? Let $\alpha = .01$.

8.67 Athletic directors generally are under pressure from both alumni and students. In an attempt to ward off conflict, one conducts a survey of both constituencies to determine if the proportion of persons favoring scheduling Roadside University in football is the same. The data are as follows.

	Responses	***Favor Scheduling***
Alumni	282	194
Students	614	392

At an α level of .12, what would you tell the athletic director?

8.68 The automatic process that creates the glass bulb that ultimately becomes an electric bulb has a changing variability with regard to thickness, particularly at the neck. The acceptable level of variability, as measured by σ_X, is not to exceed 6.5 millimeters. In a recent quality control check from a sample of 19, the mean thickness was 10 millimeters, and the standard deviation was 9.13 millimeters. At an α level of .01, does variability exceed the standard?

8.69 The Mohikan Tire Company believes that its tires have an average strength of 500 pounds per square inch. The standard deviation is known to be 50 pounds per square inch. The company is going to take a sample of 100. If the true strength of the tires is 515, what is the probability of a Type II error? The α level is .05.

8.70 Workers that have experienced layoffs in the auto industry were surveyed regarding the reasons for layoff. For company A, 116 persons out of 200 blamed foreign imports for the problem. In company B, 186 out of 280 placed the blame on foreign imports. At an α level of .10, is the proportion of workers in company B more than .05 higher than those in company A on this issue?

8.71 The Bellecose Bank believes that its average first mortgage on loans for this year is more than \$50,000. To test this assumption, the bank takes a random sample of 18 loans, which yields a mean of \$51,648 and a standard deviation of \$3226. At $\alpha = .005$, is the bank's assumption correct? What is the probability of obtaining a sample mean this small if μ_X is actually \$51,000? Assume that the value 3226 can be used for σ_X.

8.72 A certain savings and loan institution wants to advertise that it has the smartest loan officers. It bought copies of a book, *How to Improve Your IQ,* and made all the officers study it. Each officer was tested before and after working with the book. Using the following data, determine if the mean improvement of their IQ scores is more than 2 points? Let $\alpha = .05$.

	Person						
	1	*2*	*3*	*4*	*5*	*6*	*7*
Before	131	142	128	126	119	128	127
After	138	141	135	132	125	132	130

	Person						
	8	*9*	*10*	*11*	*12*	*13*	*14*
Before	131	140	116	121	129	135	132
After	128	138	124	128	132	135	136

8.73 The quality control department of the Burlap Bag Company allows 9 lots (500 bags per lot) to be run on its polyethylene trash bag maker before output is inspected. The average thickness of the polyethylene is to be 6 millimeters, and it must meet that standard with the probability of a Type I error being .05. If the 10th lot does not meet the standard, it is packaged as a lesser-quality item. In addition, random samples are taken from each of the previous 9 lots to determine their disposition. The most recent 10th lot produced a mean and a standard deviation of 5.989 millimeters and 0.1137 millimeter, respectively, from a sample of 120. What action should be taken as a result of this sampling information?

8.74 A hospital has been receiving complaints about cold eggs in the morning from many patients. The distances traveled by the servers create a large variability in the temperature. A new delivery system is now being tested. A random sample of 5 plates is taken from each delivery system, and the temperature is recorded just before serving. Does the new system reduce the variability? Let $\alpha = .01$.

Old	75	60	73	80	77
New	100	95	81	75	92

8.75 For the data in Exercise 8.74, can the hospital claim that the average temperature of eggs under the new delivery system is more than 15° hotter? Pool the sample variances and test at $\alpha = .025$.

8.76 The purchasing agent for a chain of variety stores is contemplating buying tomato plants for the spring planting season. A sample of 100 plants is purchased from Alonzo Nurseries, and 6 are found to be in poor condition. In a random sample of 200 plants purchased from Freizan Brothers, 22 are found to be in poor condition. At an α level of .10, which nursery would you recommend to the purchasing agent?

8.77 The Admiral Tire & Rubber Company believes that more than 70% of its current customers feel that Admiral tires give them "confidence." To test this belief, the company took a random sample of 500 customers, of which 400 said they had "confidence." Does the company appear to be correct in its belief? (α = .05)

8.78 In a particular year, the mean interest rate on automobile loans for a given area was 8.12%, and the standard deviation was 0.4%. Three years later, a survey was conducted to see if the rate had increased by more than 4%. A sample of 121 auto loans was selected at random and yielded a mean rate of 13.28. At α = .07, what do you conclude?

8.79 Argon Metals produces an alloy that has a mean melting point of 1240° and a standard deviation of 76°. The management of the firm judges that the variability is too great. An additive is mixed into the alloy in an attempt to reduce the variability. A random sample of 18 produced a mean melting point of 1238° and a standard deviation of 48°. Was the additive helpful in reducing variability at α = .05?

8.80 From each of 2 different companies Hargrove Enterprises has purchased machines that will produce washers. Because the process is automatic, the company wants to be sure that the quality control standard of .500 inch for the inside diameter and a standard deviation of .008 inch is met. Quality control randomly selects washers from the production of each machine. This information is as follows.

	Machine	
	A	*B*
Sample Size	100	100
Mean Inside Diameter	.502	.498

(a) At α = .05, are the machines producing comparable results?

(b) How would you answer the question of whether or not they are meeting the standard for the average inside diameter?

Assume σ_X = .008 for both.

8.81 Redo Exercise 8.78 based on the probability approach for testing a hypothesis.

8.82 Seagull Savings and Loan wishes to determine if the average value of defaulted mortgages at its Barton branch is greater than at its Carlton branch. Random samples are taken from each branch, and the results are summarized in the following table. If the Barton branch is concluded to have a greater average, the number of mortgages will be reduced. What would you conclude at **(a)** α = .10, **(b)** at α = .005?

	Branch	
	Barton	*Carlton*
Mean Default	24,000	20,000
Standard Deviation	10,000	9,000
Sample Size	25	20

8.83 Is the assumption of equal variances supported in Exercise 8.82?

8.84 Write the null and alternative hypotheses for each of the five situations described.

(a) To determine whether between October of one year and March of the following year, there has been a significant increase in the proportion of persons favoring a particular Democratic candidate.

(b) To determine whether the variability in total points achieved in a quantitative course is significantly less for night students than for day students.

(c) To determine whether the mean wear on pile carpeting is less than the manufacturer's average of .4 inch.

(d) To determine whether the proportion of persons who have reduced their gas consumption is no more than .137.

(e) To determine whether there has been an average decrease of more than 200 pounds in the weight of subcompacts between 1980 and 1986.

9 Analysis of Variance

9.1 INTRODUCTION

When cost reports from the field indicated that material costs, specifically of paint, were increasing, Mary Holich, vice-president of the Rainbow Painting Company, decided to replace several paint sprayers. Mary asked Sally Stanis in the purchasing department to provide data on initial costs and projected lives for several makes of industrial sprayers. After analyzing the data provided by Sally, Mary decided that three brands appeared to be comparable—namely, Ajax, Cook, and Ezon—and superior to all others. Mary concluded that the deciding factor among the three brands was to be the amount of paint used in normal operation. This was to be measured by the thickness of the paint in millimeters. After obtaining the data, the results of each sprayer were to be compared with the other two. Doing the analysis on a 2-by-2 basis seemed time-consuming, and Stanis questioned whether the results would be valid. Testing all of the means at one time and then being able to delineate the sprayers would be most helpful in a situation like this.

The statistical procedure to be discussed in this chapter, analysis of variance (ANOVA), would enable Sally Stanis to determine if the average paint thickness is the same for the three sprayers. The ANOVA procedure analyzes the sample data in such a way that Stanis could conclude at a specific level of confidence that all thicknesses are the same or that at least two of them differ.

The ANOVA procedure, as applied to the one-factor design model indicated in the situation just discussed, is elaborated on in the chapter. The design model is expanded to include two factors, and the ANOVA procedure is viewed in both the multiple observation per cell and single observation per cell settings. Where differences in means are deemed to exist, further analysis is made using the Duncan multiple-range test. The chapter also introduces dot notation and some concepts regarding experimental design. The last two sections introduce computer output for the ANOVA procedure using a specific software package and discuss how to code data for ease of calculations when computer facilities are not available.

Background

In Chapter 8, we investigated a procedure that enabled us to decide if differences existed between two means. However, there are many situations when it may be desirable to decide whether differences exist among three or more means. The procedures established to test for differences between two means become quite cumbersome when extended directly to testing whether differences exist for three or more populations of interest. This readily can be seen from the data shown in Table 9.1. Testing for differences in thickness due to the type of automatic sprayer would require the tests of mean thickness for Ajax versus Cook, Cook versus Ezon, and Ajax versus Ezon. Admittedly, this testing would not be too difficult if the t test for independent samples with each set of data were used. The impact is obvious if, for instance, six automatic sprayers are to be tested. Under such circumstances, it would be necessary to perform 15 individual difference of means tests. The number of difference of means tests required in any situation is the combination of c populations taking two at a time, C_2^c. Note that for this procedure the probability of concluding that at least one mean is different, when in fact none are different, increases significantly.

If α is the level of significance for each test of pairs, then the effective α is equal to or less than $1 - (1 - \alpha)^p$, where p is the number of pairs to be tested.

TABLE 9.1 Thickness of paint in millimeters for three automatic sprayers

Automatic Sprayer		
Ajax	*Cook*	*Ezon*
7.6	7.9	9.4
6.8	7.9	8.4
6.4	7.7	9.6
7.9	8.1	8.8
7.3	7.9	8.6

For 15 pairs, the foregoing procedure is unacceptable. As another example, if we tested five populations, there would be C_2^5 or ten different two-sample tests, and at an α level of .10, the effective α level $\leq 1 - (1 - .10)^{10} = .6513$, which says that the probability may be as high as .65 that we conclude that at least one mean is different when all are identical. After we conclude that differences exist, we may make pairwise comparisons using a procedure that controls the effective α.

As a result of the increase in α with the performance of a battery of t tests, it becomes desirable to utilize instead a procedure known as analysis of variance. This procedure is designed to test for differences among population means by evaluating if the variation among the sample means is larger than is expected by chance alone. To construct this test, we must separate the total variability into various components. This separation is generally known as *partitioning the sum of squares* and is easier to see in a graphical presentation. In Figure 9.1, P_1, P_2, and P_3 represent the populations of interest. The ANOVA procedure tests to see if the populations differ in means or whether there is actually one common population from which all samples have been selected.

FIGURE 9.1 Possibilities for three samples selected

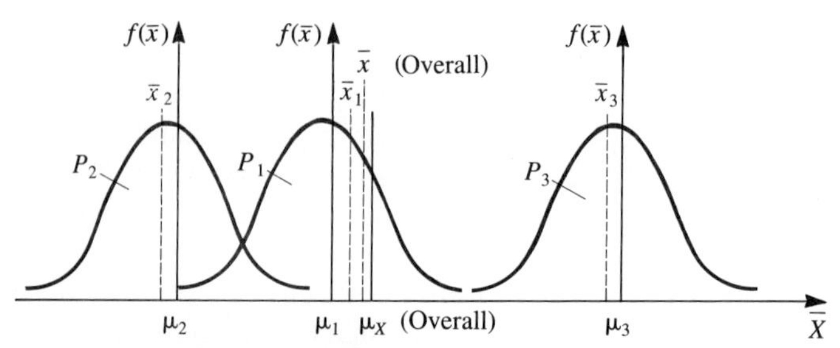

(a) Three populations with different means from each of which a sample has been selected

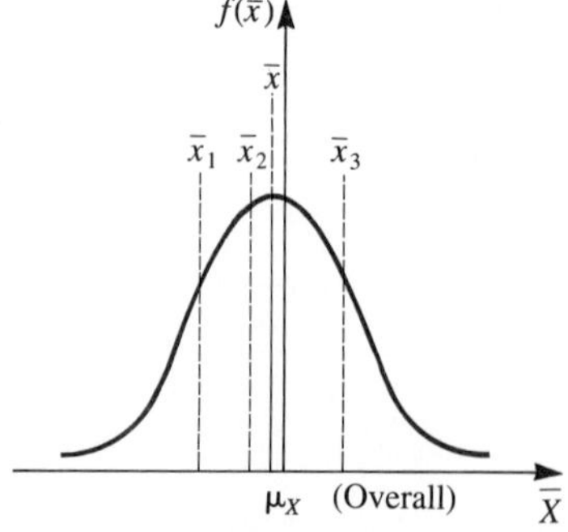

(b) One population from which three samples have been selected

That is, are the μ_j different, or is $\mu_1 = \mu_2 = \mu_3 = \mu_X$? Notationally, we let X_{ij}, $i = 1, \ldots, n_j$, $j = 1, \ldots, J$, represent individual observations. The subscript j identifies the population from which the value comes, and i identifies the observation within the sample. In analyzing the variation of the individual values about the overall mean, μ_X, we separate the deviation of each X_{ij} from μ_X into two parts. These parts represent the deviations of X_{ij} from μ_j and μ_j from μ_X, respectively—that is,

$$X_{ij} - \mu_X = (X_{ij} - \mu_j) + (\mu_j - \mu_X)$$

The first part reflects the random variation of X_{ij} about μ_j, also called random error, and the second part represents the deviation of the mean of population P_j from the overall mean μ_X. This second part is known also as the treatment or differential effect due to population P_j.

Before continuing with the procedures that enable us to estimate the magnitudes of these variabilities, we need to establish some guidelines for the systematic presentation of data that are pertinent to the ANOVA procedure and to discuss the types of experimental designs that can be constructed for investigative purposes. In the next two sections, we look at the guidelines for presenting data and discuss the types of experimental designs.

9.2 DOT NOTATION

Data that are put into a tabular array, such as Table 9.1, can be viewed in a more general sense without regard to the actual numerical value of each item. Often it is desirable to identify each item in the array by its position. Thus, in Table 9.2, each number is shown in its general form; the first subscript denotes the row of the array, and the second denotes the column of the array. As a result, the number 7.3 in Table 9.1 is shown as x_{51} in Table 9.2. The basic notation for any item in an array will be in the form X_{ij}, where i denotes the row and j the column.

Since it may be desirable to find column or row means, it is necessary to establish a notational form that will readily depict this information. A column total can be shown as $\Sigma_{i=1}^{n_j} X_{ij}$, where n_j represents the number of observations in column j. To eliminate the use of the summation sign, we denote a column total by $X_{\cdot j}$, where the dot together with the j indicates that a column total is being referenced. Thus, from Table 9.1, $x_{\cdot 1} = 36.0$, $x_{\cdot 2} = 39.5$, and $x_{\cdot 3} = 44.8$.

TABLE 9.2 General item notation by row and column position

x_{11}	x_{12}	x_{13}
x_{21}	x_{22}	x_{23}
x_{31}	x_{32}	x_{33}
x_{41}	x_{42}	x_{43}
x_{51}	x_{52}	x_{53}

Using this notational form, we can identify a column mean as

$$\bar{X}_{\cdot j} = \frac{X_{\cdot j}}{n_j} \tag{9.1}$$

Again, from the data in Table 9.1, the column means are $\bar{x}_{\cdot 1} = 7.2$, $\bar{x}_{\cdot 2} = 7.9$, and $\bar{x}_{\cdot 3} = 8.96$, respectively.

It is a simple matter to carry these same procedures to the numbers in the particular rows. Thus, the sum of row 1 in Table 9.2 can be denoted as $X_{1\cdot}$, and it equals 24.9. The general form for a row total would be $X_{i\cdot}$, and a row mean becomes

$$\bar{X}_{i\cdot} = \frac{X_{i\cdot}}{n_i} \tag{9.2}$$

where n_i represents the sample size associated with the ith row.

The final extension of this notational form would be a summation over both rows and columns. Thus, the sum of all the data is expressed as $X_{\cdot\cdot}$ and, for the data in Table 9.2, is 120.3. The overall mean, symbolized as $\bar{X}_{\cdot\cdot}$ is

$$\bar{X}_{\cdot\cdot} = \frac{X_{\cdot\cdot}}{n_T} \tag{9.3}$$

where n_T is the total number of items in the array.

9.3 EXPERIMENTAL DESIGN

In the establishment of an experimental design, the investigator is interested primarily in isolating a particular effect or effects in such a manner that inferences can be made with regard to the population in question.

DEFINITION 9.1 A **factor** is defined to be an experimental variable of interest. Each value used in the experiment for the variable or factor is called a **factor level.**

DEFINITION 9.2 If the factor levels used in the experiment represent *all* the levels of interest for that variable, the factor is said to have **fixed effects.** If the levels used in the experiment constitute a random sample of all possible levels for the variable, the factor is called a **random effect.**

In analysis of variance procedures, two basic classifications of effects exist: the fixed effect and the random effect. For example, if we want to know the effect a road surface has on tire wear, the factor being considered is road surface. Levels of that factor may include concrete, brick, dirt, blacktop, and cobblestone. The number of levels is based on the design of the experiment and

TABLE 9.3 One-factor, three-level experimental design

Sales Area		
East	***Central***	***West***
x_{11}	x_{12}	x_{13}
x_{21}	x_{22}	x_{23}
x_{31}	x_{32}	x_{33}
x_{41}	x_{42}	x_{43}

will be at least three in this chapter. Fewer than three would not require ANOVA since procedures from Chapter 8 could be used.

As another example, Table 9.3 gives the data for an experiment designed to test for differences in the mean monthly sales for three different regions. Consideration is given to only a single factor, the sales area, but three levels of the factor are under investigation. If the three areas represent the only areas of interest, then sales area is a fixed effect. The X_{ij} values represent sample data randomly drawn from each area.

The design is easily expanded to a two-factor model if consideration also is given to the season of the year as well as the sales area. This design is illustrated in Table 9.4. For this model, one factor, sales area, has three levels, and the other factor, season of the year, has four levels. Note that season is a fixed effect. Each X_{ij} value is now cross-classified with both the row and column factors. If the number of X_{ij} values is increased for each cross-classification, it enables the experimenter to investigate the effect of combinations of the two factors. In other words, are sales unusually high or low in one sales area for a specific season? The topic of interaction effects is discussed in greater detail in Section 9.6. The number of factors may be increased to three or more. However, the complexity of the design increases with the introduction of each new factor. Some of these designs are introduced in Section 9.8.

TABLE 9.4 Two-factor experimental design, column factor three levels, row factor four levels with one observation per cell

	Sales Area		
Season	***East***	***Central***	***West***
Spring	x_{11}	x_{12}	x_{13}
Summer	x_{21}	x_{22}	x_{23}
Fall	x_{31}	x_{32}	x_{33}
Winter	x_{41}	x_{42}	x_{43}

The method of analysis for the designs thus far illustrated is essentially the same. However, the test statistics and the conclusions drawn from the analysis depend primarily on the type of model under consideration. The term "model" refers to the mathematical representation of the variables or effects contributing to the observation X_{ij}. For example, in a one-factor experiment, the model may be represented as $X_{ij} = \mu_j + \varepsilon_{ij}$, where μ_j is the mean for population P_j and ε_{ij} represents error deviation.

DEFINITION 9.3 If all factors in an experiment have fixed effects, the model is termed a **fixed-effects model.** If all factors are random, the model is a **random-effects model,** and if there are both fixed and random factors present, the model is said to have **mixed effects.**

In very basic terms, we desire to make some comment on the particular levels in a fixed-effects model, whereas in a random-effects model we wish to comment on the population from which the levels of the factor are randomly selected. In Table 9.4, if East, Central, and West are the only sales areas of interest, the model would be classified as a fixed-effects model because all possible seasons are considered also.

The arrangement of the data in Table 9.5 might be viewed as a random-effects model. For example, suppose an experiment is designed to determine if lathe speed has any effect on the life of a cutting tool. The concept of randomness is introduced into this design in the levels of the factor if the three speeds are randomly selected from the population of all possible speeds. Thus, we desire to make an inference about lathe speeds in general. Since the experiment is designed to test if differences exist in mean wear with respect to lathe speed, it is desirable that other factors remain constant for all factor levels. For instance, the lathe, the operator, and the metal type could all affect the wear of the cutting tool. For any factors that cannot be controlled, it is assumed that any differences attributable to them will average out across factors through the random sampling process. If the three indicated speeds are the only ones of interest, the factor

TABLE 9.5 One-factor, three-level experimental design with random effects

Lathe Speeds in RPM		
21	***59***	***113***
x_{11}	x_{12}	x_{13}
x_{21}	x_{22}	x_{23}
x_{31}	x_{32}	x_{33}
x_{41}	x_{42}	x_{43}

becomes a fixed effect. Therefore, although the layout of the experiment identifies the type of experimental design, it does not identify the type of effects present in the model. Throughout the remainder of this chapter, we assume that all factors in a model have fixed effects. For one factor, the difference between fixed- and random-effects models is the interpretation of the results. For two or more factors, the computational procedures are the same up to the construction of the test statistics. In a fixed-effects model, the denominator is the same for all test statistics, whereas in random-effects and mixed-effects models the denominator may change with each test. For a detailed discussion of random- and mixed-effects models see Guenther [1] or Neter, Wasserman, and Kutner [3].

9.4 ONE-WAY ANALYSIS OF VARIANCE

In the introductory section of this chapter, it is stated that in a one-factor model the analysis of variance procedure partitions the total variability into two possible sources. For this procedure, as we indicated earlier, one of the assumptions is that the effects are additive and can be expressed algebraically as

$$X_{ij} = \mu_X + \alpha_j + \varepsilon_{ij} \qquad \text{where } \alpha_j = \mu_j - \mu_X \tag{9.4}$$

To investigate further, we refer to Table 9.6, which shows a total of 20 pieces of data: 5 for each of the four levels of the factor, type of motor vehicle citation. The X_{ij} values in each column represent the age, to the nearest year, of the individual cited. Each of these values represents the accumulation of the separate effects as noted in Equation 9.4. These effects or sources of variability were noted earlier. In the context of the example, the deviation of X_{ij} in any column from the factor level's true mean value μ_j is attributed to the chance error that exists when a sample is randomly selected from a population and is represented

TABLE 9.6 One-factor, four-level design for evaluating the mean age of cited individuals

Type of Motor Vehicle Citation			
Exceeding the Posted Speed	***Red Light or Stop Sign Infraction***	***Crossing Marked Driving Lane***	***Improper or Damaged Equipment***
18	19	24	36
32	41	38	24
24	28	28	56
20	25	40	33
26	37	25	26
μ_j: 24	30	31	35

by ε_{ij}, the error variable. The difference that exists between μ_j and the true overall mean is called a *treatment effect* and is represented by α_j. The use of the word "treatment" has a historical perspective and is an outgrowth of agricultural experiments where some seeds were planted as a control group while other seeds were treated in an effort to affect yield. Thus, differences that exist among column means have become commonly known as treatment effects. If no random deviations exist within each population—that is, all $\varepsilon_{ij} = 0$—and no differences exist between the population means and the overall mean—that is, all $\alpha_j = 0$—then each X_{ij} will have the same value as the overall or common mean, which is designated as μ_X in the model.

Under the assumption of fixed effects, we have that

$$\mu_X = \frac{1}{J}\sum_{j=1}^{J} \mu_j$$

since the populations or treatments of interest are all included in the experiment. Then the differential effect is merely $\mu_j - \mu_X$—that is, $\alpha_j = \mu_j - \mu_X$. Thus, in a fixed-effects, one-factor model, it follows that

$$\sum_{j=1}^{J} \alpha_j = \sum_{j=1}^{J} (\mu_j - \mu_X)$$

$$= \sum_{j=1}^{J} \mu_j - \sum_{j=1}^{J} \mu_X = J\mu_X - J\mu_X = 0$$

With respect to the situation illustrated in Table 9.6, we demonstrate how the actual values might have arisen. We make the following assumptions: First, the average age of all drivers receiving citations for the four violations listed in the table is 30 years. Second, the parameters for the differential effects—that is, the α_j—are in years, $\alpha_1 = -6$, $\alpha_2 = 0$, $\alpha_3 = +1$, $\alpha_4 = +5$. Note that the α_j do sum to zero.

The interpretation of these values is as follows. The value of -6 for α_1 says that recipients of speeding citations are six years younger on the average than the overall age. The value of 0 for α_2 indicates that the mean age of individuals cited for stop signal violations is no different from the average age for all individuals cited in the four categories. Correspondingly, α_3 of $+1$ and α_4 of $+5$ indicate that persons cited for lane violations are one year older, whereas equipment citations are issued to persons five years older on the average. That is, the average age for speeders is 24; for stop signal violators, 30, for lane violators, 31; and for equipment violations, 35. The differences within each column from the corresponding mean are attributed to error or chance variation between individuals.

Also, let $\varepsilon_{11} = -6$, $\varepsilon_{21} = +8$, . . . , $\varepsilon_{44} = -2$, $\varepsilon_{54} = -9$. Then, using the assumption of additive effects gives

$$x_{11} = \mu_X + \alpha_1 + \varepsilon_{11} = 30 + (-6) + (-6) = 18$$
$$x_{21} = \mu_X + \alpha_1 + \varepsilon_{21} = 30 + (-6) + (+8) = 32$$
$$\vdots$$
$$x_{44} = \mu_X + \alpha_4 + \varepsilon_{44} = 30 + (+5) + (-2) = 33$$
$$x_{54} = \mu_X + \alpha_4 + \varepsilon_{54} = 30 + (+5) + (-9) = 26$$

In practice, the values for the parameters μ_X and the α_j as well as the error magnitudes are unknown. Only the final values for the observations will be available. From them, the parameters μ_X and $\mu_j = \mu_X + \alpha_j$, $j = 1, 2, 3, 4$, are estimated, and the results compared. To obtain probability distributions for the statistics used to make these comparisons, we make the following necessary assumptions.

1. Population variances for all levels of a factor are equal ($\sigma_j^2 = \sigma_X^2$ for all j).
2. The error term is normally distributed with zero mean [$\varepsilon_{ij} \sim N(0, \sigma_j^2)$].
3. The error terms are independent (ε_{ij} are independent).
4. All experimental units are randomly selected.
5. All data represent at least an interval level of measurement.

Violations of some of these assumptions do not completely void the use of ANOVA. As a statistical technique, ANOVA is robust, meaning it is insensitive to slight departures from the assumptions. Recent studies have shown that assumption 1 is critical and that even slight departures from it may affect the final results.

In many applications, it is reasonable to believe that the various treatments affect only the average and not the variability between subjects. For example, with human subjects, a teaching method is expected to affect the average response of a class but not the innate human variability. If there are doubts about this assumption, then one of the references should be consulted for a procedure to test it formally. If assumptions 2 and 5 are questionable and the sample sizes are small, then the Kruskal–Wallis procedure of Chapter 15 should be considered. Assumption 4 is under the control of the experimenter and, if satisfied, helps to satisfy 3, but 3 also requires that the errors are not related to the factor levels. If assumption 3 is believed to be violated, then a more advanced text should be consulted.

We illustrate the process of partitioning the sum of squares for sample data with Figure 9.2. For any given X_{ij} value, denoted by the broken line, the total deviation is expressed as $(X_{ij} - \overline{X}_{..})$—that is, the deviation of the observation from the overall mean. This deviation comprises two parts and can be written as

$$(X_{ij} - \overline{X}_{..}) = (X_{ij} - \overline{X}_{.j}) + (\overline{X}_{.j} - \overline{X}_{..})$$

FIGURE 9.2 Graphical representation of variability expressed as deviations, with curves centered at the corresponding estimates of the μ_j

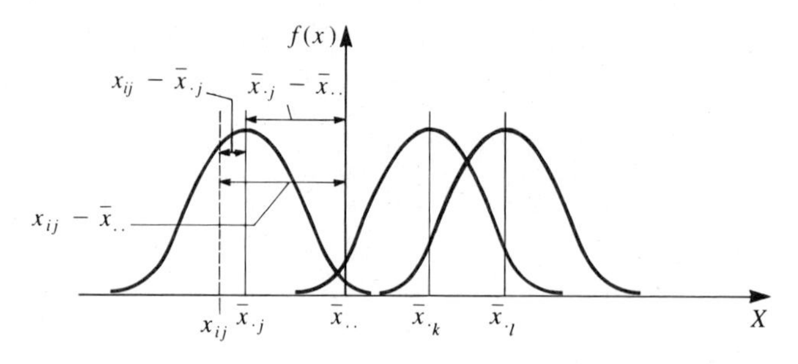

where $(\overline{X}_{.j} - \overline{X}_{..})$ is an estimate of α_j, the treatment or differential effect, and $(X_{ij} - \overline{X}_{.j})$ is an estimate of ε_{ij}, the error term. In the example dealing with driving citations, consider the value for x_{41} of 20. Looking at its deviation from 30, we have

$$(20 - 30) = (20 - 24) + (24 - 30)$$

The sums of squares are obtained by squaring the terms and adding. Doing so gives

$$\sum_{j=1}^{J}\sum_{i=1}^{n_j}(X_{ij} - \overline{X}_{..})^2 = \sum_{j=1}^{J}\sum_{i=1}^{n_j}(X_{ij} - \overline{X}_{.j} + \overline{X}_{.j} - \overline{X}_{..})^2$$

$$= \sum_{j=1}^{J}\sum_{i=1}^{n_j}(X_{ij} - \overline{X}_{.j})^2 + \sum_{j=1}^{J}\sum_{i=1}^{n_j}(\overline{X}_{.j} - \overline{X}_{..})^2 \qquad \textbf{(9.5)}$$

since it can be shown algebraically that

$$2\sum_{j=1}^{J}\sum_{i=1}^{n_j}(X_{ij} - \overline{X}_{.j})(\overline{X}_{.j} - \overline{X}_{..}) = 0$$

In Equation 9.5, J is the total number of levels of the factor being investigated and generally denotes columns. We also can think of J as the number of treatments. The term n_j is the number of observations in the jth column. Each term in Equation 9.5 is called a sum of squares. The leftmost term is called the *sum of squares total* and is denoted by SST—that is,

$$\text{SS}T = \sum_{j=1}^{J}\sum_{i=1}^{n_j}(X_{ij} - \overline{X}_{..})^2 \qquad \textbf{(9.6)}$$

TABLE 9.7 Labeling for an analysis of variance table

Source of Variability	*Degrees of Freedom (df)*	*Sums of Squares (SS)*	*Mean Square (MS)*	*F Ratio*

The *sum of squares among treatment means* is denoted SSA and given by

$$\text{SSA} = \sum_{j=1}^{J} n_j(\overline{X}_{\cdot j} - \overline{X}_{\cdot\cdot})^2 \tag{9.7}$$

The *sum of squares attributed to error* or chance variation is denoted SSE and defined by

$$\text{SS}E = \sum_{j=1}^{J} \sum_{i=1}^{n_j} (X_{ij} - \overline{X}_{\cdot j})^2 \tag{9.8}$$

Each sum of squares is the numerator for the calculation of a variance estimate. The only additional requirement would be the division by the proper degrees of freedom. We are unable to do this in a single step because it would destroy the equality in the expression.

In the ANOVA procedures, the calculation of the sums of squares and the degrees of freedom are summarized in an analysis of variance table. The labeling on such a table is shown in Table 9.7. The mean squares are determined as the sum of squares divided by the degrees of freedom and thus become estimates of the variance if all $\alpha_j = 0$. These calculations represent the statistics that enable us to construct the statistical test.

In Table 9.8, the information appropriate to these headings is shown based on a one-factor model. The determination of the degrees of freedom relies

TABLE 9.8 Symbolic presentation of the analysis of variance table

Source of Variability	*df*	*SS*	*MS*	*F Ratio*
Among treatments	df_A	SSA	$\text{MSA} = \dfrac{\text{SSA}}{\text{df}_A}$	$F = \dfrac{\text{MSA}}{\text{MS}E}$
Error	df_E	SSE	$\text{MS}E = \dfrac{\text{SS}E}{\text{df}_E}$	
Total	df_T	SST		

on the basic premise that a degree of freedom is lost for each parameter estimated. Thus, since only μ_X is estimated, the total degrees of freedom is

$$\text{df}_T = n_T - 1 \tag{9.9}$$

The degrees of freedom among treatment means are

$$\text{df}_A = J - 1 \tag{9.10}$$

since there are J sample means and we must estimate μ_X. The degrees of freedom error is based on the fact that μ_j must be estimated for each population. Hence,

$$\text{df}_E = n_T - J \tag{9.11}$$

where: $n_T = \sum_{j=1}^{J} n_j$

We can see that the degrees of freedom can also be partitioned and are additive.

$$(J - 1) + (n_T - J) = n_T - 1$$

We should not forget that we are testing for differences among individual means, μ_j, of three or more populations—that is, $\mu_1 = \mu_2 = \mu_3 = \cdots = \mu_J$, where $\mu_j = \mu_X + \alpha_j$. Testing for differences in means is accomplished by evaluating whether a significant difference exists between the results of two methods of estimating σ_X^2. The mean squares shown in Table 9.8 are the results to be compared. To compare MSA and MSE, we take the ratio as shown in Equation 9.12.

$$F = \frac{\text{MS}A}{\text{MS}E} \tag{9.12}$$

The value calculated from Equation 9.12 provides a method for testing whether the variability among treatments is larger than expected by chance alone. To begin with, we note that SSA $= \Sigma\Sigma(\overline{X}_{\cdot j} - \overline{X}_{\cdot\cdot})^2$ compares the factor level means to the overall mean. If H_0 is true, these means are expected to be "close" to each other and, in turn, "close" to $\overline{X}_{\cdot\cdot}$, with the deviations representing only chance or error variation. Thus, if H_0 is true,

$$s_{\overline{X}}^2 = \frac{1}{n(J - 1)} (\text{SS}A)$$

is an estimate of σ_X^2/n, or, equivalently, $ns_{\overline{X}}^2 = \text{MS}A$ is an estimate of σ_X^2. On the other hand, if H_0 is not true, then MSA includes both variation due to error and a squared term representing differences between population means. That is, it should be much larger than the true value for σ_X^2. Since MSE is not affected by differing factor level means, it estimates σ_X^2 independently of the status of H_0. Thus, if H_0 is true, MSA and MSE are expected to be close, and if H_a is true, MSA is expected to be much larger than MSE. If the calculated F value is greater than the tabular F value, $F_{J-1,n_T-J,\alpha}$, we can say that this variability would happen less than (100)α% of the time by chance alone. Thus, we conclude that significant differences appear to exist among factor level means.

To determine the distribution of MSA/MSE, we refer to Chapter 6. In Chapter 6, we saw that the ratio of two independent χ^2 variables each divided by its degrees of freedom has an F-distribution. In comparing variances from two random samples, we obtain the independence property through random sampling. In the present situation, the sample variances do not come from separate populations, and we need to utilize Cochran's theorem from Chapter 6. For ease in reference, it is restated in paraphrased form.

COCHRAN'S THEOREM Assume a sum of squares, SST, can be shown to be algebraically equal to a sum of smaller sums of squares, say $\Sigma_j SS_j$. Then, if SST has a χ^2 distribution with ν_T degrees of freedom, the SS_j will be independent, with each having a χ^2 distribution with ν_j degrees of freedom if and only if $\nu_T = \Sigma_j \nu_j$.

Using assumptions 1 through 5, we have, if H_0 is true, that SST is merely the numerator for the variance of a sample from a normal distribution. Thus, as shown in Chapter 6, $\sigma_X^2 SST$ has a χ^2 distribution with $n_T - 1$ degrees of freedom under H_0. Multiplying both sides of Equation 9.5 by σ_X^2 gives $\sigma_X^2 SST = \sigma_X^2 SSE + \sigma_X^2 SSA$. To show that $\sigma_X^2 SSE$ and $\sigma_X^2 SSA$ are independent χ^2 random variables, we need only show that their degrees of freedom add to $n_T - 1$. Since this step was demonstrated earlier, we have, by Cochran's theorem, that

$$MSA = \frac{SSA}{J-1} \sim \frac{\sigma_X^2 \chi_{J-1}^2}{J-1}$$

and

$$MSE = \frac{SSE}{n_T - J} \sim \frac{\sigma_X^2 \chi_{n_T-J}^2}{n_T - J}$$

Thus, under H_0, the statistic

$$\frac{MSA}{MSE} \sim \frac{\sigma_X^2 \chi_{J-1}^2/(J-1)}{\sigma_X^2 \chi_{n_T-J}^2/(n_T - J)} \sim F_{J-1, n_T-J, \alpha}$$

The seven-step procedure for testing a hypothesis, as outlined in Chapter 8, is used in the following examples.

EXAMPLE 9.1 From the study introduced earlier in the chapter, the data are reproduced in Table 9.9. Test for differences in mean paint thickness at $\alpha = .05$.

Solution: For this example, we desire to decide whether the make of the sprayer has any effect on the average thickness of paint. The use of the ANOVA procedure assumes that other major factors affecting the response variable, paint thickness, be held constant so that the only differences being observed are the random effects and treatment effects. Thus, we should note that the viscosity of the paint, the manner in which it was applied, the type of surface to which it was

TABLE 9.9 Thickness of paint in millimeters for three automatic sprayers

Automatic Sprayer		
Ajax	*Cook*	*Ezon*
7.6	7.9	9.4
6.8	7.9	8.4
6.4	7.7	9.6
7.9	8.1	8.8
7.3	7.9	8.6

applied, and the like, were the same for all automatic sprayers. As a result, differences in average paint thickness among sprayers can be attributed to the brand of sprayer and chance variation.

The calculations necessary to complete the ANOVA table are as follows.

$$\text{Total observations} = n_T = 15$$
$$\text{Treatment groups} = J = 3$$
$$\text{Degrees of freedom}_{\text{total}} = n_T - 1 = 15 - 1 = 14$$
$$\text{Degrees of freedom}_{\text{sprayers}} = J - 1 = 3 - 1 = 2$$
$$\text{Degrees of freedom}_{\text{error}} = n_T - J = 15 - 3 = 12$$

To calculate the sums of squares, we need to obtain the column means and the overall mean. From Equation 9.1, the first column mean is

$$\bar{x}_{\cdot 1} = \frac{x_{\cdot 1}}{n_1} = \frac{36.0}{5} = 7.2$$

In a similar manner, $\bar{x}_{\cdot 2} = 7.9$ and $\bar{x}_{\cdot 3} = 8.96$. The overall mean determined from Equation 9.3 is

$$\bar{x}_{\cdot\cdot} = \frac{x_{\cdot\cdot}}{n_T} = \frac{120.3}{15} = 8.02$$

Using these values, we obtain the sums of squares

$$\text{SS}T = \sum_{j=1}^{J} \sum_{i=1}^{n_j} (x_{ij} - \bar{x}_{\cdot\cdot})^2$$
$$= (7.6 - 8.02)^2 + (6.8 - 8.02)^2 + \cdots + (8.6 - 8.02)^2 = 10.464$$

$$SSA = \sum_{j=1}^{J} n_j(\bar{x}_{\cdot j} - \bar{x}_{\cdot\cdot})^2$$

$$= 5(7.2 - 8.02)^2 + 5(7.9 - 8.02)^2 + 5(8.96 - 8.02)^2 = 7.852$$

$$SSE = \sum_{j=1}^{J} \sum_{i=1}^{n_j} (x_{ij} - \bar{x}_{\cdot j})^2$$

$$= (7.6 - 7.2)^2 + (6.8 - 7.2)^2 + \cdots + (8.6 - 8.96)^2 = 2.612$$

The results of these calculations are the entries for the ANOVA table shown in Table 9.10. In addition to the entry of the sums of squares and the degrees of freedom, the mean squares and F ratio have been calculated and are given in the table.

The completed ANOVA table provides the data for testing whether differences in the mean thickness of the paint exist among brands. For a .05 level of significance we have the following:

1. H_0: $\mu_{\text{Ajax}} = \mu_{\text{Cook}} = \mu_{\text{Ezon}}$; H_a: at least 2 means are not equal. An alternative statement of the hypotheses in terms of the differential effects, α_j, is H_0: $\alpha_{\text{Ajax}} = \alpha_{\text{Cook}} = \alpha_{\text{Ezon}} = 0$; H_a: at least 2 α_j differ from zero. Regardless of which statement we use, the acceptance or rejection of H_0 will depend on the outcome of the F test.
2. $\alpha = .05$.
3. $F = \dfrac{\text{MSA}}{\text{MSE}}$.
4. Reject H_0 if $F > F_{2,12,.05} = 3.89$. (Note that even though we have a two-sided alternative, we are concerned with rejection only on the right-hand side due to the definition of the F statistic. The implication of a one-sided test is illustrated in the discussion following this example.)
5. From the ANOVA table the calculated value for F is 18.04.
6. Reject H_0 because $18.04 > 3.89$.
7. At least 2 of the means are not equal, and we conclude that paint thickness differs among these 3 brands of sprayers. Note that the conclusion refers only to these brands since they are the only ones of interest under the fixed-effects assumption. ■

TABLE 9.10 Analysis of variance table for brands of automatic sprayers

Source of Variability	*df*	*SS*	*MS*	*F Ratio*
Among sprayers	2	7.852	3.926	18.04
Error	12	2.612	0.218	
Total	14	10.464		

If the model is correct and the null hypothesis is true, the only reason that the sample means differ is random variation, and MSA is an unbiased estimate of σ_X^2. If the model is correct, the MSE is an unbiased estimate of σ_X^2 independent of H_0 and H_a. Thus, when H_0 is true, MSA/MSE would be expected to be close to 1.00. If H_0 is false, the sample means will differ for two reasons. First, each $\overline{X}_{\cdot j}$ is expected to be near μ_j, and thus variation will be introduced because the μ_j's differ. Second, each $\overline{X}_{\cdot j}$ will deviate from its mean μ_j due to random variation, σ_X^2. The last two statements are illustrated in the equation

$$\overline{X}_{\cdot j} - \mu_X = \underset{\text{(error)}}{(\overline{X}_{\cdot j} - \mu_j)} + \underset{\text{(differences in } \mu_j\text{'s)}}{(\mu_j - \mu_X)}$$

where μ_X is the overall population mean, having $\overline{X}_{\cdot\cdot}$ as its estimator. Thus, if H_0 is false, MSA is an estimator of σ_X^2 plus variation attributed to differences among the μ_j's. In this case, the F ratio estimates $(\sigma_X^2 + D)/\sigma_X^2$, where D is a nonnegative constant. If H_0 is true, $D = 0$; otherwise, $D > 0$, and MSA/MSE should be greater than 1.

In both the alternative hypothesis and in the conclusion, the phrase "at least 2" means not equal is used. Essentially, that phrase is the extent of the statistical inference we can make with regard to the means of the populations. We are not in a position to say that Ajax is better than Cook or that Ezon is better than Cook. If we desire to investigate further, several techniques are available to us. This general subject is investigated further in Section 9.5.

The calculations involved in setting up the analysis of variance table for our example problem were not very difficult. However, as the number of levels to be investigated increases, or as there is an increase in the sample size per level, these calculations become more time consuming. In light of that obstacle, some equations are introduced here that eliminate the squared differences approach. These equations also reduce the chances of rounding errors in the calculations.

The total sum of squares is defined as

$$\sum_{j=1}^{J} \sum_{i=1}^{n_j} (X_{ij} - \overline{X}_{\cdot\cdot})^2$$

If this equation is expanded and the proper terms collected, we can express the total sum of squares as

$$\sum_{j=1}^{J} \sum_{i=1}^{n_j} X_{ij}^2 - \frac{(X_{\cdot\cdot})^2}{n_T} \quad \textbf{(9.13)}$$

The second half of this calculation has become known as the *correction factor* for the overall mean and will be designated as C. Thus, the correction factor is

$$C = \frac{(X_{\cdot\cdot})^2}{n_T} \quad \textbf{(9.14)}$$

and the total sum of squares is

$$SST = \sum_{j=1}^{J} \sum_{i=1}^{n_j} X_{ij}^2 - C \tag{9.15}$$

The sum of squares among treatments can also be expanded and in the collection of the terms is

$$SSA = \sum_{j=1}^{J} \frac{(X_{\cdot j})^2}{n_j} - C \tag{9.16}$$

The sum of squares error, because of the partitioning of the sums of squares, can be determined as the difference between the two calculated values.

$$SSE = SST - SSA \tag{9.17}$$

The SSE also can be determined directly as

$$SSE = \sum_{j=1}^{J} \sum_{i=1}^{n_j} X_{ij}^2 - \sum_{j=1}^{J} \frac{(X_{\cdot j})^2}{n_j} \tag{9.18}$$

If all three sums of squares are calculated directly, it provides a partial check of the calculations. As we proceed through further sections of this chapter, we use the computational forms for calculating sums of squares.

EXAMPLE 9.2 A manufacturer is investigating the purchase of an extrusion-molding machine. Four different vendors have submitted samples based on their company's process. The tensile strengths (psi) for these samples are shown in the following table.

Vendor	*Tensile Strength (psi)*									
A	93	86	90	92	93	92	85	92		
B	76	78	76	79	79	74	72	81	77	81
C	84	80	79	84	87	84	82			
D	83	91	90	91	88	93	89			

At the .01 level of significance, are there any differences in mean tensile strengths among the products of the 4 vendors?

Solution: The calculations for the analysis of variance table are performed using Equations 9.14 through 9.17.

$$C = \frac{(X_{\cdot\cdot})^2}{n_T} = \frac{(2701)^2}{32} = \frac{7{,}295{,}401}{32} = 227{,}981.28$$

$$\begin{aligned}SST &= \sum_{j=1}^{J}\sum_{i=1}^{n_j} X_{ij}^2 - C \\ &= (93^2 + 86^2 + \cdots + 89^2) - 227{,}981.28 \\ &= 229{,}207 - 227{,}981.28 = 1225.72\end{aligned}$$

$$\begin{aligned}SSA_{\text{vendors}} &= \sum_{j=1}^{J} \frac{X_{\cdot j}^2}{n_j} - C \\ &= \left(\frac{723^2}{8} + \frac{773^2}{10} + \frac{580^2}{7} + \frac{625^2}{7}\right) - 227{,}981.28 \\ &= (65{,}341.13 + 59{,}752.90 + 48{,}057.14 + 55{,}803.57) - 227{,}981.28 \\ &= 228{,}954.74 - 227{,}981.28 = 973.46\end{aligned}$$

$$SSE = SST - SSA = 1225.72 - 973.46 = 252.26$$

The direct determination of SSE is

$$\begin{aligned}SSE &= \sum_{j=1}^{J}\sum_{i=1}^{n_j} X_{ij}^2 - \sum_{j=1}^{J} \frac{(X_{\cdot j})^2}{n_j} \\ &= 229{,}207 - 228{,}954.74 = 252.26\end{aligned}$$

which provides a check on the calculations. The ANOVA table can now be completed.

Source of Variability	*df*	*SS*	*MS*	*F ratio*
Among vendors	3	973.46	324.487	36.018
Error	28	252.26	9.009	
Total	31	1,225.72		

Calculation of the values for the mean squares enables us to calculate the ratio to be used in step 5.

1. H_0: $\mu_A = \mu_B = \mu_C = \mu_D$; H_a: at least 2 means are not equal.
2. $\alpha = .01$.
3. $F = \dfrac{MSA}{MSE}$.
4. Reject H_0 if $F_{\text{calc}} > F_{3,28,.01} = 4.57$.
5. From the ANOVA table, F_{calc} is 36.018.
6. We reject H_0 because $36.018 > 4.57$.
7. We conclude that the mean tensile strengths of the moldings appear to differ among the 4 vendors.

In Example 9.2, the rejection of the null hypothesis does not help us to determine which of the vendors' products is the best. From the vendor means,

$$\bar{x}_{\cdot A} = 90.38$$
$$\bar{x}_{\cdot B} = 76.30$$
$$\bar{x}_{\cdot C} = 82.86$$
$$\bar{x}_{\cdot D} = 89.29$$

it should be obvious, since we concluded that at least two differ, that A and B are likely to have different means. Also, it may be that vendors A and D have similar means—that is, drawn from the same population. If the test had been performed for the satisfaction of a purchasing agent, it would be necessary to go beyond the analysis of variance procedure to make this determination. In the next section, we investigate a procedure for evaluating individual differences after H_0 is rejected.

EXERCISES

9.1 A private testing firm has been checking on the amount of illumination that is present in four different types of areas on a college campus. Locations are randomly selected for each type of area, and the light intensity is measured. The data, in lumens per square foot, are shown in the following table.

Campus Areas

Classrooms	*Hallways*	*Laboratories*	*Lounges*
17	3	19	7
16	3	11	7
13	3	19	5
15	4	14	7
15	5	18	6
10	4	18	8
17	7	13	6

At an α level of .05, can the testing firm conclude there are significant differences in the average lumens per square foot due to areas of the campus?

9.2 A company desires to buy several chain saws for use in its organization and has narrowed the choices down to 3 brands. It is decided to test saws of the 3 manufacturers to determine whether one firm provides the best saw with respect to length of service before there is a need to sharpen the chain. The testing procedure was designed to maintain a constant pressure on the saw and to measure the time it takes to cut a standard 4 by 4. When the time to make the cut is increased by 25%, it is deemed necessary to sharpen the chain. Length of service is then defined as the total time necessary to make all the cuts. The data are measured in minutes and shown in the following table.

Manufacturer		
Argo	*Cutrite*	*Quiksaw*
215	278	230
181	256	207
183	281	249
215	299	246
197	294	241
183	287	238
	274	249
	275	
	298	

At a level of significance equal to .01, can the company conclude there are differences in the average lengths of service among the manufacturers? Identify other factors that should be considered in this problem.

9.3 The marketing department of Stannis, Inc., a manufacturer of products for the outdoorsman, desires to compare its sleeping bag, Mountaineer, with those of its competitors. The characteristic to be compared is the amount of kapok filling in ounces. There are presently 3 other companies that compete with Stannis for this market. Sleeping bags of each brand are randomly purchased, and the kapok is removed and weighed. The data in ounces are as follows. Let $\alpha = .05$.

Sleeping Bag			
Mountaineer	*Durorest*	*Hi-Cover*	*Restwell*
47	35	50	40
46	42	50	40
49	35	45	36
50	40	44	36
47	42	49	44
49	40	47	42

What conclusions can be made with respect to the average weights of kapok among the manufacturers?

9.5 THE DUNCAN MULTIPLE-RANGE TEST

The rejection of the null hypothesis in a one-factor analysis of variance does not enable us to make any statistical inferences about the individual means. We can, of course, perform t tests for the difference of means for all the pairs available, providing the α level is adjusted accordingly. One procedure that uses this approach is the Bonferroni. For a discussion of it, see Neter, Wasserman, and Kutner [3]. An alternative to such a procedure is the Duncan multiple-range test.

DEFINITION 9.4 The **Duncan multiple-range test** is concerned with the actual differences between various mean pairs and how each difference compares with a term known as the **least significant range** (LSR). Two factor means are declared to be significantly different if their values differ by more than the LSR.

To utilize the Duncan multiple-range test, one must determine the estimate of the standard error of the mean, $s_{\overline{X}}$. For a single sample, this estimate is the standard deviation divided by $\sqrt{n}$. In this case, the estimate of $\sigma_{\overline{X}}^2$ is the MS*E*. Thus, the estimate of $\sigma_{\overline{X}}$ is

$$s_{\overline{X}} = \sqrt{\frac{\text{MS}E}{n}} \tag{9.19}$$

The term designated MS*E* is the mean square error taken directly from the ANOVA table. The value for n is the sample size on which $\overline{X}$ is based. An assumption for the use of the Duncan multiple-range test is equal sample sizes—that is, $n_1 = n_2 = \cdot\cdot\cdot = n_j = n$. If there are unequal sample sizes, it is necessary to use an alternative procedure, such as the Scheffe multiple-comparison test. See Neter, Wasserman, and Kutner [3].

The LSR is determined as

$$\text{LSR} = r s_{\overline{X}} \tag{9.20}$$

where r is found in Appendix Table B.12. The proper values are obtained from the table based on the level of α, n_2, the degrees of freedom for the estimate of $\sigma_{\overline{X}}^2$, and the difference between the ranks of the two means being compared plus one. That is, $q = (v - w) + 1$, where v is the rank of the larger mean, and w the rank of the smaller mean. In Figure 9.3, a set of means is ranked to illustrate this point, where $\overline{x}_{(i)}$ represents the ith ordered sample mean.

For example, considering the first and last means, the rank difference plus one is $q = (4 - 1) + 1 = 4$, and for the second and last, $q = (4 - 2) + 1 = 3$. The value for n_2 is equal to the degrees of freedom for error found in the ANOVA table—that is, $n_2 = \text{df}_E$. In using the Duncan multiple-range test, we are actually

FIGURE 9.3 Ranked sample means to illustrate the difference between ranks plus one

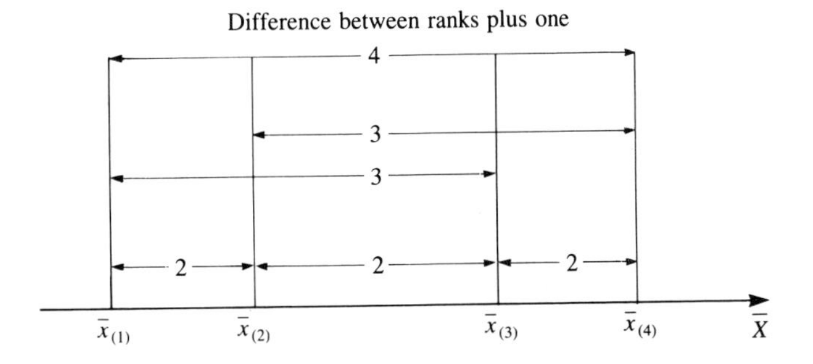

performing a type of t test since the values in Appendix Table B.12 are functions of t values.

This procedure is summarized in the following decision rule. For the hypotheses

$$H_0: \mu_{(v)} - \mu_{(w)} = 0 \qquad H_a: \mu_{(v)} - \mu_{(w)} > 0$$

we reject H_0 if

$$\overline{X}_{(v)} - \overline{X}_{(w)} > r_{q,n_2,\alpha}\, s_{\overline{X}}$$

where $q = v - w + 1$ and $n_2 = \text{df}_E$.

EXAMPLE 9.3 Utilizing the data from Example 9.1, determine which automatic sprayer provides the least thickness of paint. Let $\alpha = .05$.

Solution: The initial step in determining an answer is to order the means of the samples. For the samples, $\bar{x}_{\text{Ajax}} = 7.2$, $\bar{x}_{\text{Cook}} = 7.9$, and $\bar{x}_{\text{Ezon}} = 8.96$. These means are already in ascending order, and, thus, $\bar{x}_{(1)} = 7.2$, $\bar{x}_{(2)} = 7.9$, and $\bar{x}_{(3)} = 8.96$. The standard error of the mean is

$$s_{\overline{X}} = \sqrt{.2187/5} = \sqrt{.0436} = .2088$$

The value for n_2 is obtained from the ANOVA table and equals 12. The number of r values to be obtained from Appendix Table B.12 is equal to the number of factor levels being considered minus 1. Since there are 3 levels in this problem, we need 2 r values, namely, for $q = (3 - 1) + 1 = 3$ and $q = (3 - 2) + 1$ or $(2 - 1) + 1 = 2$. Thus, having n_2, α, and the values for q, we obtain the values of r from Appendix Table B.12.

q	2	3
r	3.08	3.23

The LSR values can now be determined by multiplying each r value by $s_{\overline{X}}$, giving

q	2	3
LSR	.643	.674

We are now ready to compare the differences between means to the corresponding LSR values. We start with the 2 that are the farthest apart in terms of rank differences, namely, $\bar{x}_{(3)} - \bar{x}_{(1)}$, where $q = (3 - 1) + 1 = 3$. We then test all means at the next level of rank differences and continue until all means that are adjacent in rank—that is, $q = 2$—are compared last. Doing so gives

$q = 3$: $\bar{x}_{(3)} - \bar{x}_{(1)} = 8.96 - 7.20 = 1.76$
$1.76 > .674$ (significant)

$$q = 2: \quad \bar{x}_{(3)} - \bar{x}_{(2)} = 8.96 - 7.9 = 1.06$$
$$1.06 > .643 \qquad \text{(significant)}$$

$$\bar{x}_{(2)} - \bar{x}_{(1)} = 7.9 - 7.2 = .7$$
$$.7 > .643 \qquad \text{(significant)}$$

We interpret the results to imply that none of the population means is equivalent. Further, we conclude that the thickness of paint for the Ezon sprayer is more than for either Ajax or Cook and that the thickness of the paint with the Cook sprayer is more than with the Ajax. As a result, we may desire to purchase an Ajax automatic sprayer to minimize the cost of paint to be used, provided the quality of the end result is maintained.

Note that, in general, we could use either the variance from a single sample, s_j^2, or the pooled estimate of the variance, MS*E*, since both are unbiased for $\sigma_{\bar{X}}^2$. The reason for using MS*E* is primarily to build up the degrees of freedom for the variance estimate and thus improve the precision. To understand this reasoning, note that a confidence interval for μ_j involves the *t*-distribution, which gets its degrees of freedom from the variance estimate. If $J = 4$, and $n = 6$, then $\mathrm{df}_E = 20$, and the degrees of freedom for an individual s_j^2 is $n - 1 = 5$. At $\alpha = .05$, for example, we obtain $t_{20,.025} = 2.086$, and $t_{5,.025} = 2.571$. Thus, even though s_j^2 and MS*E* may be quite close in actual value, the interval based on MS*E* will be $2.086/2.571 = .811$ (or 81%), as wide as that based on s_j^2 for a confidence level of 95%.

EXAMPLE 9.4 Students from 4 high schools were given a mathematical proficiency test to see whether the mathematical abilities of students were the same for these schools. Six students were randomly selected from each school, and the scores on a 40-point scale were recorded. The ANOVA test for equality of means rejected the null hypothesis at a .01 level of significance. The mean square error was 7.3085, the degrees of freedom for error was 20, and the sample means were $\bar{x}_{\text{East}} = 21.67$, $\bar{x}_{\text{West}} = 30.67$, $\bar{x}_{\text{North}} = 32.0$ and $\bar{x}_{\text{South}} = 22.5$. What can we conclude at the .01 level of significance?

Solution: Initially, we order the means according to magnitude. Even though we lose the identity of each mean, it is necessary to return to this information for our conclusion. The ordered means are $\bar{x}_{(1)} = 21.67$, $\bar{x}_{(2)} = 22.5$, $\bar{x}_{(3)} = 30.67$, $\bar{x}_{(4)} = 32.0$. The standard error is

$$s_{\bar{X}} = \sqrt{7.3085/6} = \sqrt{1.2180} = 1.104$$

and $n_2 = \mathrm{df}_E = 20$.

From Appendix Table B.12, the *r* values are

q	2	3	4
r	4.02	4.22	4.33

Thus, the LSR values are

q	2	3	4
LSR	4.438	4.659	4.780

Evaluating the differences, we have

$q = 4$: $\bar{x}_{(4)} - \bar{x}_{(1)} = 32.00 - 21.67 = 10.33 > 4.780$ (significant)

$q = 3$: $\bar{x}_{(4)} - \bar{x}_{(2)} = 32.00 - 22.50 = 9.50 > 4.659$ (significant)
$\bar{x}_{(3)} - \bar{x}_{(1)} = 30.67 - 21.67 = 9.00 > 4.659$ (significant)

$q = 2$: $\bar{x}_{(4)} - \bar{x}_{(3)} = 32.00 - 30.67 = 1.33 < 4.438$
$\bar{x}_{(3)} - \bar{x}_{(2)} = 30.67 - 22.50 = 8.17 > 4.438$ (significant)
$\bar{x}_{(2)} - \bar{x}_{(1)} = 22.50 - 21.67 = 0.83 < 4.438$

The significant values are 10.33, 9.50, 9.00, and 8.17, and they indicate that differences exist between means 4 − 1, 4 − 2, 3 − 1, and 3 − 2 but not between 4 − 3 and 2 − 1. Interpreting these results with respect to the high schools, we can say that West and North appear to be comparable schools with respect to student performance on the mathematical proficiency test and that they both appear to be better than East and South, which are comparable to each other. ■

The interpretation of the results may be easier to understand if viewed graphically. In Figure 9.4, the conclusions about the sampling distributions are depicted. Since $\bar{x}_{(1)} = 21.67$ and $\bar{x}_{(2)} = 22.5$ represented East and South schools, respectively, and were found to be not significantly different, by using Duncan's multiple-range test, there is no reason to doubt that these mathematical proficiency scores came from the same population. Thus, the best estimate of this population mean is their combined average. A similar analysis is made for $\bar{x}_{(3)} = 30.67$ and $\bar{x}_{(4)} = 32$, which represent West and North schools, respectively.

FIGURE 9.4 Graphical representation of results of Example 9.4, with curves centered at the averages of means that are not significantly different.

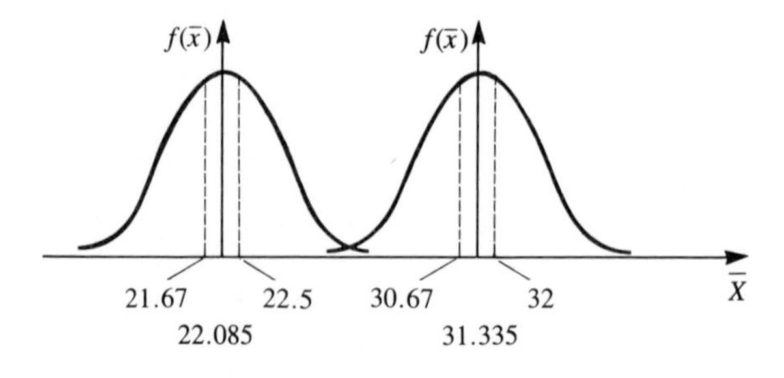

Finally, we can see that the relative position of the West–North mean to the East–South mean indicates a higher mathematical proficiency level.

The Duncan multiple-range test enables us to further evaluate data when the overall test of means is rejected. Its use, however, is restricted to samples of equal size and to making pairwise comparisons a posteriori—that is, after the fact. Other techniques for making a posteriori comparisons are Tukey's, Scheffe's, Newman-Keul's, and the Bonferroni. For a discussion of these procedures and their applicability, see Neter, Wasserman, and Kutner [3].

It is apparent that, even with a one-factor model, many calculations are necessary for a complete analysis. Numerous computer software packages are available that ease the burden of the calculations. Some of these are SPSS, MINITAB, BMD, ADEPT, and SAS. In Section 9.8, the SAS and MINITAB packages are used with some examples to illustrate computer output. Not all of these packages do Duncan's procedure, but each does provide some type of multiple comparison analysis. For a specific package, check the user's manual to see what is provided.

EXERCISES

9.4 Apply Duncan's multiple-range test to the data of Exercise 9.1. Determine the hierarchy of locations with respect to illumination at $\alpha = .05$.

9.5 The ANOVA table is based on Exercise 9.2 and has been determined by using the first 6 items of data for each level of the factor.

Source of Variability	*df*	*SS*	*MS*
Among saw manufacturers	2	22,679.66	11,339.83
Error	15	3,605.67	240.38
Total	17	26,285.33	

The mean lengths of service in minutes are as follows: $\bar{x}_A = 195.67$, $\bar{x}_C = 282.50$, and $\bar{x}_Q = 235.17$. What comment can be made with regard to mean lengths of service before the chain needs to be sharpened at $\alpha = .01$?

9.6 For the data of Exercise 9.3, determine the equalities and differences that may be present with regard to the average weights of the kapok in the sleeping bags at $\alpha = .05$.

9.7 Thirty-three new employees were evenly divided into 3 groups, and each group was introduced to the production operations by a different training procedure. The ANOVA table given would indicate that the 3 methods are dissimilar at the .01 level of significance. If the mean production rates for the 3 procedures are 24.75, 32.83, and 28, respectively, which of the methods is superior?

Source of Variability	*df*	*SS*	*MS*	*F Ratio*
Among methods	2	363.7	181.8	8.90
Error	30	614.0	20.5	
Total	32	977.7		

9.6 TWO-WAY ANALYSIS OF VARIANCE

In the discussion of experimental designs, it was indicated that more than one factor can be investigated at the same time. In this section, we investigate two-factor models with both factors having fixed effects. The two situations to be considered differ in the method of data collection and the number of observations per sample.

Randomized Block Design

In this section, we consider a method of data collection known as the randomized block design. This design is similar to the pairing concept used in Chapter 8. Its purpose is to refine the estimate of the error variance by removing from the SS*E* that variation that can be ascribed to a secondary variable or factor. If this secondary factor has a significant effect on our response and it is left as part of the error term, as a one-factor analysis does, we may mistakenly fail to reject H_0—that is, we increase the chance of a Type II error.

DEFINITION 9.5 A **randomized block design** is an experimental design such that groups of c identical items, called *blocks,* are selected with the c levels of the primary variable, called *treatments,* being assigned randomly to items within each block on a one-to-one basis.

To illustrate the blocking concept, we consider the following situation. The Dekker Corporation is a major producer of seeds for the agricultural industry and has developed a new hybrid corn, K-94. The company has determined that K-94 has superior disease-resistant properties, but it realizes that for successful marketing, it must provide superior yields. Knowing that yield is related to soil type, climate, and fertilization procedures, among other things, Dekker's research team decides to control these procedures and to remove their effects from the error variation. The team decides that it is only interested in comparing K-94 against its four major competitors. In analyzing the primary market areas, Dekker identifies six basic soil type and climate combinations. For each of the six combinations of interest, Dekker randomly selects a 50-acre block of land and verifies the uniformity of the soil conditions within each block. An offer is then made to each owner to provide all seed for the designated block free of charge with the owner using his or her normal fertilizer and cultivation procedures uniformly on the block.

After obtaining the farmers' agreements, the research team divides each block into five 10-acre plots and randomly assigns the five seed types to plots. The farmer is then provided the appropriate seeds when necessary but is not informed as to which is planted where. At the time of harvest, the yield for each 10-acre plot is to be recorded separately at each location. Table 9.11 provides a schematic representation of the experiment. Competitors' seeds are labeled as A, B, C, D.

TABLE 9.11 Schematic representation of a randomized block design for the Dekker Corporation (Treatments are randomly assigned to plots within blocks.)

	Plots				
Block	***1***	***2***	***3***	***4***	***5***
1	A	D	K-94	C	B
2	K-94	C	B	A	D
3	B	A	C	D	K-94
4	D	K-94	A	B	C
5	C	B	D	K-94	A
6	B	K-94	A	C	D

With this type of experimental design, all seed types within a block receive exactly the same soil, climate conditions, fertilization, and cultivation. Thus, within each block, any differences in the final yields among plots can be attributed to chance variation and differences due to seed types. After the experiment is completed, the data are reorganized so that all results for a given treatment are in the same row or column. The reorganization of Table 9.11 is presented in Table 9.12, with the first subscript identifying the growing conditions, and the second identifying the seed type used for that result.

The model for this type of experiment is

$$X_{ij} = \mu_X + \alpha_j + \beta_i + \varepsilon_{ij} \tag{9.21}$$

where $\alpha_j, j = 1, 2, \ldots, J$, represent differential effects due to treatments (seed types), $\beta_i, i = 1, 2, \ldots, I$, represent differential effects due to blocks (growing conditions), and ε_{ij} is a random variable representing chance or error variation.

TABLE 9.12 Reorganization of Table 9.11 for Dekker Corporation experiment to facilitate analysis

	Seed Type				
Block	***A***	***B***	***C***	***D***	***K-94***
1	X_{1A}	X_{1B}	X_{1C}	X_{1D}	X_{1K}
2	X_{2A}	X_{2B}	X_{2C}	X_{2D}	X_{2K}
3	X_{3A}	X_{3B}	X_{3C}	X_{3D}	X_{3K}
4	X_{4A}	X_{4B}	X_{4C}	X_{4D}	X_{4K}
5	X_{5A}	X_{5B}	X_{5C}	X_{5D}	X_{5K}
6	X_{6A}	X_{6B}	X_{6C}	X_{6D}	X_{6K}

The statistical analysis of the data collected using a randomized block design is actually a special case of the more general two-factor model with replication. Thus, discussion of the analysis is delayed until after the presentation of the general case.

Two-Factor Crossed Design with Replication

The design to be considered here is a two-way layout with replication. The term *replication* refers to multiple observations per cell or treatment combination. The concept of a crossed design is defined next.

DEFINITION 9.6 In a two-factor model, the experiment is said to represent a **crossed design** if all levels of one factor occur in combination with all levels of the other factor. This design may be extended to more than two factors.

Occasionally, due to the physical limitations of the experiment, the levels of one factor will be contained completely within levels of another factor. When this situation happens, the factor is said to be *nested,* and the design is referred to as a *nested,* or *hierarchical,* design. Due to the introductory nature of this text, we restrict our discussion to crossed designs that are balanced—that is, involve equal sample sizes. For discussions and analyses of nested designs and/or how to deal with missing observations, see Guenther [1], Meek and Turner [2], or Neter, Wasserman, and Kutner [3].

For the two-factor design with replication, the model is

$$X_{ijk} = \mu_X + \alpha_j + \beta_i + \alpha\beta_{ij} + \varepsilon_{ijk} \quad \textbf{(9.22)}$$

where μ_X, α_j, β_i, and ε_{ijk} are as before. The terms $\alpha\beta_{ij}$, $i = 1, 2, \ldots, I$, $j = 1, 2, \ldots, J$, represent components known as interactions and not products of α_j and β_i.

DEFINITION 9.7 An **interaction** is an atypical or out-of-character response that results when particular levels of the two factors occur in combination.

For example, suppose level 1 gives the highest response for factor A, while level 3 gives the highest response for factor B when each factor is considered individually. Now if level A_1 is run in combination with level B_3, we expect a very high response. If we observe a very low or poor response, it is an atypical result indicating interaction. To be able to measure interaction variation separately from error variation, we must have replication—that is, multiple observations per treatment combination.

Here the term *treatment* refers to the total combination of factor levels received by an experimental unit or subject. Table 9.13 is a representation of a two-factor crossed design with replication. Each cell represents a treatment, and the observations in that cell comprise a sample from a population. For all two-factor designs, K will denote the number of replications or observations per cell.

TABLE 9.13 Two-factor crossed design with replication

	Factor A				
Factor B	A_1	A_2	A_3	A_4	***Total***
	X_{111}	X_{121}	X_{131}	X_{141}	
	X_{112}	X_{122}	X_{132}	X_{142}	
B_1	X_{113}	X_{123}	X_{133}	X_{143}	$X_{1\cdot\cdot}$
	X_{114}	X_{124}	X_{134}	X_{144}	
	$X_{11\cdot}$	$X_{12\cdot}$	$X_{13\cdot}$	$X_{14\cdot}$	Cell totals
	X_{211}	X_{221}	X_{231}	X_{241}	
	X_{212}	X_{222}	X_{232}	X_{242}	$X_{2\cdot\cdot}$
B_2	X_{213}	X_{223}	X_{233}	X_{243}	
	X_{214}	X_{224}	X_{234}	X_{244}	
	$X_{21\cdot}$	$X_{22\cdot}$	$X_{23\cdot}$	$X_{24\cdot}$	Cell totals
Total	$X_{\cdot 1\cdot}$	$X_{\cdot 2\cdot}$	$X_{\cdot 3\cdot}$	$X_{\cdot 4\cdot}$	$X_{\cdots}$

We limit the analysis to cases where both factors represent fixed effects. That is, the levels used in the experiment represent the complete finite population of interest. Thus,

$$\mu_X = \frac{1}{J}\sum_{j=1}^{J} \mu_j = \frac{1}{I}\sum_{i=1}^{I} \mu_i$$

where $\mu_j = \mu_X + \alpha_j$ and $\mu_i = \mu_X + \beta_i$. Thus, it follows that

$$\sum_{j=1}^{J} \alpha_j = \sum_{i=1}^{I} \beta_i = 0$$

when both factors are fixed effects. Also, we have

$$\mu_i = \frac{1}{J}\sum_{j=1}^{J} \mu_{ij} \quad \text{and} \quad \mu_j = \frac{1}{I}\sum_{i=1}^{I} \mu_{ij}$$

where $\mu_{ij} = \mu_X + \alpha_j + \beta_i + \alpha\beta_{ij}$. From these equations, it follows that in a two-factor fixed-effects model,

$$\sum_{i=1}^{I} \alpha\beta_{ij} = \sum_{j=1}^{J} \alpha\beta_{ij} = 0 \qquad \textbf{(9.23)}$$

For a two-factor crossed design, the sums of squares are defined as follows.

$$\text{(Total)} \qquad \text{SS}T = \sum_i \sum_j \sum_k (X_{ijk} - \overline{X}_{...})^2 \tag{9.24}$$

$$\text{(Columns)} \qquad \text{SS}A = \sum_i \sum_j \sum_k (\overline{X}_{\cdot j \cdot} - \overline{X}_{...})^2 \tag{9.25}$$

$$\text{(Rows)} \qquad \text{SS}B = \sum_i \sum_j \sum_k (\overline{X}_{i\cdot\cdot} - \overline{X}_{...})^2 \tag{9.26}$$

$$\text{(Interaction)} \qquad \text{SS}AB = \sum_i \sum_j \sum_k (\overline{X}_{ij\cdot} - \overline{X}_{i\cdot\cdot} - \overline{X}_{\cdot j \cdot} + \overline{X}_{...})^2 \tag{9.27}$$

$$\text{(Error)} \qquad \text{SS}E = \sum_i \sum_j \sum_k (X_{ijk} - \overline{X}_{ij\cdot})^2 \tag{9.28}$$

For triple subscripting, the dot notation is

$$X_{...} = \sum_i \sum_j \sum_k X_{ijk} = \text{overall total} \qquad \text{with } \overline{X}_{...} = \frac{X_{...}}{IJK}$$

$$X_{i\cdot\cdot} = \sum_j \sum_k X_{ijk} = \text{row total} \qquad \text{with } \overline{X}_{i\cdot\cdot} = \frac{X_{i\cdot\cdot}}{JK}$$

$$X_{\cdot j \cdot} = \sum_i \sum_k X_{ijk} = \text{column total} \qquad \text{with } \overline{X}_{\cdot j \cdot} = \frac{X_{\cdot j \cdot}}{IK}$$

$$X_{ij\cdot} = \sum_k X_{ijk} = \text{cell total} \qquad \text{with } \overline{X}_{ij\cdot} = \frac{X_{ij\cdot}}{K}$$

Thus, we see that, as in the one-factor model, SST compares all observations to the grand mean, and SSA compares the column means to the grand mean. SSB compares the row means to the grand mean, and SSAB adjusts cell means for row and column effects and compares the adjusted values to the grand mean. Note that if there are no interaction effects, the term in parentheses in SSAB is expected to differ from 0 only by chance.

To construct test statistics, we need to obtain probability distributions for the mean squares under the various null hypotheses. Obtaining these distributions is done as in the one-factor model by partitioning SST and applying Cochran's theorem. In this case, we have

$$\begin{aligned}
\text{SS}T &= \sum_i \sum_j \sum_k (X_{ijk} - \overline{X}_{...})^2 \\
&= \sum_i \sum_j \sum_k (X_{ijk} - \overline{X}_{i\cdot\cdot} + \overline{X}_{i\cdot\cdot} - \overline{X}_{\cdot j \cdot} + \overline{X}_{\cdot j \cdot} - \overline{X}_{ij\cdot} \\
&\quad + \overline{X}_{ij\cdot} - \overline{X}_{...} + \overline{X}_{...} - \overline{X}_{...})^2 \\
&= \Sigma\Sigma\Sigma (X_{ijk} - \overline{X}_{ij\cdot})^2 + \Sigma\Sigma\Sigma (\overline{X}_{i\cdot\cdot} - \overline{X}_{...})^2 + \Sigma\Sigma\Sigma (\overline{X}_{\cdot j \cdot} - \overline{X}_{...})^2 \\
&\quad + \Sigma\Sigma\Sigma (\overline{X}_{ij\cdot} - \overline{X}_{i\cdot\cdot} - \overline{X}_{\cdot j \cdot} + \overline{X}_{...})^2 \\
&= \text{SS}E + \text{SS}B + \text{SS}A + \text{SS}AB
\end{aligned} \tag{9.29}$$

The preceding equality can be verified by algebra. Before continuing, we consider the assumptions necessary for obtaining the distributions.

1. Variances are homogeneous—that is, $\sigma_{ij}^2 = \sigma_X^2$ for all i, j.
2. $\varepsilon_{ijk} \sim N(0,\sigma_X^2)$.
3. Error terms are independent of all factor levels and each other.
4. The experimental units are randomly selected.
5. All data represent at least an interval level of measurement.

The three null hypotheses are that there are no row effects, no column effects, and no interaction effects. Thus, if the null hypotheses are true, we have one large sample of size IJK from a normal population, and, hence, SST is distributed as $\sigma_X^2\chi_{IJK-1}^2$. To obtain the distributions for the four SSs on the right in Equation 9.29, we first determine the degrees of freedom for each. For the SSE, we have IJK variables and IJ means calculated from them; thus the degrees of freedom are $IJK - IJ = IJ(K - 1)$. For SSA and SSB, the degrees of freedom easily are seen to be $J - 1$ and $I - 1$, respectively. For the interaction term, there are IJ cell means, and we lose one degree of freedom for each row mean and each column mean, but the grand mean has been counted twice since it is part of both the row and column means. To compensate for this double counting, we must add back one degree of freedom—that is,

$$\text{df}_{AB} = IJ - I - J + 1 = (I - 1)(J - 1) \qquad \textbf{(9.30)}$$

Thus, we have

$$\begin{aligned}\text{df}_E + \text{df}_A + \text{df}_B + \text{df}_{AB} &= IJ(K - 1) + (I - 1) \\ &\quad + (J - 1) + (I - 1)(J - 1) \\ &= IJK - 1 = \text{df}_T\end{aligned}$$

Therefore, by Cochran's theorem,

$$SSE \sim \sigma_X^2\chi_{IJ(K-1)}^2$$

and, if the respective null hypotheses are true, each of the other SSs has a $\sigma_X^2\chi^2$-distribution with the appropriate degrees of freedom.

From these results, we can obtain the decision rules for each set of hypotheses. These hypotheses and the corresponding decision rules are presented in Table 9.14, and the ANOVA table is presented in Table 9.15.

We can calculate the sums of squares in Table 9.15 by using Equations 9.24 through 9.28, but it is preferable to use the corresponding computational forms. These forms are generally faster and involve less error due to rounding. They are given next, along with the correction factor for the overall mean.

$$C = \frac{(X_{...})^2}{IJK} \qquad \textbf{(9.31)}$$

TABLE 9.14 Summary of hypotheses and decision rules for two-factor fixed-effects model in a crossed design with replication

Effect of Interest/Hypothesis Set	*Decision Rule*
Column	
H_0: all $\alpha_j = 0$ (no column effect); H_a: some $\alpha_j \neq 0$	Reject H_0 if $\frac{\text{MSA}}{\text{MSE}} > F_{J-1,IJ(K-1),\alpha}$
Row	
H_0: all $\beta_i = 0$ (no row effect); H_a: some $\beta_i \neq 0$	Reject H_0 if $\frac{\text{MS}B}{\text{MS}E} > F_{I-1,IJ(K-1),\alpha}$
Interaction	
H_0: all $\alpha\beta_{ij} = 0$ (no interaction effect); H_a: some $\alpha\beta_{ij} \neq 0$	Reject H_0 if $\frac{\text{MS}AB}{\text{MS}E} > F_{(I-1)(J-1),IJ(K-1),\alpha}$

$$\text{SS}T = \sum_i \sum_j \sum_k X_{ijk}^2 - C \tag{9.32}$$

$$\text{SS}A = \sum_j \frac{(X_{\cdot j \cdot})^2}{IK} - C \tag{9.33}$$

$$\text{SS}B = \sum_i \frac{(X_{i \cdot \cdot})^2}{JK} - C \tag{9.34}$$

$$\text{SS}AB = \sum_i \sum_j \frac{(X_{ij\cdot})^2}{K} - \sum_i \frac{(X_{i\cdot\cdot})^2}{JK} - \sum_j \frac{(X_{\cdot j\cdot})^2}{IK} + C \tag{9.35}$$

TABLE 9.15 ANOVA table for the two-factor, fixed-effects model in a crossed design with replication

Source of Variability	*df*	*SS*	*MS*	*F Ratio*
Column variable (A)	$\text{df}_A = J - 1$	SSA	$\text{MSA} = \frac{\text{SSA}}{\text{df}_A}$	$F = \frac{\text{MSA}}{\text{MSE}}$
Row variable (B)	$\text{df}_B = I - 1$	SSB	$\text{MS}B = \frac{\text{SS}B}{\text{df}_B}$	$F = \frac{\text{MS}B}{\text{MSE}}$
Interaction (AB)	$\text{df}_{AB} = (I-1)(J-1)$	SSAB	$\text{MS}AB = \frac{\text{SS}AB}{\text{df}_{AB}}$	$F = \frac{\text{MS}AB}{\text{MS}E}$
Error	$\text{df}_E = IJ(K-1)$	SSE	$\text{MS}E = \frac{\text{SS}E}{\text{df}_E}$	
Total	$\text{df}_T = IJK - 1$	SST		

$$SSE = \sum_i \sum_j \sum_k X_{ijk}^2 - \sum_i \sum_j \frac{(X_{ij\cdot})^2}{K} \tag{9.36}$$

Note that a computational check is provided by Equation 9.29. Also, since they are sums of squared real numbers, none can be negative.

The format in Tables 9.14 and 9.15 provides all the information necessary to complete the hypothesis tests associated with the two-factor fixed-effects model with replication. We need only make some conclusion with regard to the decision reached for each hypothesis test. That decision, to accept or reject, is based on the comparison of the values in the F ratio column to the values in the F table column. The null hypothesis is rejected when the calculated F ratio is greater than the F table value. The total ANOVA procedure and analysis are illustrated in the next example.

EXAMPLE 9.5 Hannifin Manufacturing Company is establishing a new production line. The Industrial Engineering Department has indicated that 3 different workspace configurations are feasible. In addition, work movements may be accomplished through 4 different methods. In an effort to determine if significant differences exist in either of these 2 factors, an experiment is designed to determine the average number of pieces that can be completed per hour for each of the workspace and work movement combinations. There is also a concern that some of the workspace and work movement combinations may work together atypically to increase or decrease the rate of parts completion. Thirty-six workers are randomly selected from the work force, and 3 are randomly assigned to each combination of workspace configuration and type of work movement. The layout of the design with the results incorporated is given in Table 9.16.

Solution: All the appropriate totals are provided in the statement of the problem. Since multiple observations are present in each cell, a test for interaction is possible, together with the tests for the column effect (type of work movement) and row effect (type of workspace configuration). Using Equations 9.31 through 9.36, we make the following calculations.

$$C = \frac{447.5^2}{36} = 5562.674$$

$$\begin{aligned} SST &= (13.9^2 + 14.5^2 + \cdots + 10.6^2) - 5562.674 \\ &= 5618.33 - 5562.674 = 55.656 \end{aligned}$$

$$\begin{aligned} SSA &= \frac{1}{9}(122.1^2 + 110.2^2 + \cdots + 107.1^2) - 5562.674 \\ &= 5578.719 - 5562.674 = 16.045 \end{aligned}$$

$$\begin{aligned} SSB &= \frac{1}{12}(159.6^2 + 144.7^2 + 143.2^2) - 5562.674 \\ &= 5576.374 - 5562.674 = 13.7 \end{aligned}$$

TABLE 9.16 Hannifin Company data for Example 9.5

Type of Workspace Configuration	*Type of Work Movement* A	B	C	D	$X_{i\cdot\cdot}$
	13.9	12.7	11.9	14.4	
1	14.5	12.5	12.9	13.8	159.6
	14.9	12.2	12.4	13.5	
$X_{1j\cdot}$	43.3	37.4	37.2	41.7	
	14.5	11.7	12.1	10.5	
2	14.6	12.6	11.9	10.9	144.7
	13.4	10.7	10.7	11.1	
$X_{2j\cdot}$	42.5	35.0	34.7	32.5	
	12.3	11.6	12.2	11.4	
3	12.1	12.4	11.4	10.9	143.2
	11.9	13.8	12.6	10.6	
$X_{3j\cdot}$	36.3	37.8	36.2	32.9	
$X_{\cdot j\cdot}$	122.1	110.2	108.1	107.1	447.5

$$\begin{aligned} SSAB &= \frac{1}{3}(43.3^2 + 42.5^2 + \cdots + 32.9^2) - 5578.719 \\ &\quad - 5576.374 + 5562.674 \\ &= 5609.117 - 5578.719 - 5576.374 + 5562.674 \\ &= 16.698 \end{aligned}$$

$$SSE = 5618.33 - 5609.117 = 9.213$$

Using the outcomes from these calculations, the ANOVA and decision tables are given in Table 9.17.

From the solution table, the hypothesis tests indicate that significant differences exist in the average parts per hour processed with respect to types of work movement and with respect to types of workspace configuration. In addition, interaction is present between the 2 factors. That is, certain movement/configuration combinations appear to yield atypical responses; for example, the average for cell (1, D) seems unusually high. Since equal sample sizes exist for both variables, further evaluation can be undertaken by using Duncan's procedure.

Before continuing the analysis using Duncan's multiple-range test, we indicate the value for n with respect to the two-factor crossed design. Recall that n represents the number of observations on which the sample mean is based. In this design, the row means and column means may be based on different sample

TABLE 9.17 ANOVA and decision tables for Example 9.5

Source of Variability	*df*	*SS*	*MS*	*F Ratio*
Between work movements	3	16.045	5.348	$F = \frac{5.348}{.384} = 13.927$
Between configurations	2	13.700	6.850	$F = \frac{6.850}{.384} = 17.839$
Interaction	6	16.698	2.783	$F = \frac{2.783}{.384} = 7.247$
Error	24	9.213	0.384	
Total	35	55.656		

Hypotheses	*F Table*	*Decision*
H_0: all $\alpha_j = 0$; H_a: some $\alpha_j \neq 0$	$F_{3,24,.05} = 3.01$	Reject
H_0: all $\beta_i = 0$; H_a: some $\beta_i \neq 0$	$F_{2,24,.05} = 3.40$	Reject
H_0: all $(\alpha\beta)_{ij} = 0$; H_a: some $(\alpha\beta)_{ij} \neq 0$	$F_{6,24,.05} = 2.51$	Reject

sizes. For row means, each one is based on $n = JK$, whereas for column means, $n = IK$. The determination of n, in turn, gives the corresponding standard errors of the means as

$$\text{(Column means)} \quad s_{\overline{X}} = \sqrt{\frac{\text{MSE}}{IK}}$$

and

$$\text{(Row means)} \quad s_{\overline{X}} = \sqrt{\frac{\text{MSE}}{JK}}$$

Since each cell mean includes any effects due to the particular row or column in which it lies, as well as any interaction effect, the cell means are not directly comparable even if there is no interaction. Thus, Duncan's procedure cannot be used on the cell means themselves. It is possible to correct the cell means for any row or column effects before making comparisons, but the procedures are beyond the scope of this text. For the interested student, the procedures can be found in Guenther [1] or Neter, Wasserman, and Kutner [3].

EXAMPLE 9.6 Using the data given in Example 9.5, test for differences in the mean parts per hour processed between types of work movement and between types of workspace configurations at $\alpha = .05$.

Solution: The mean parts per hour for the 4 different work movements are $\bar{x}_A = 13.57$, $\bar{x}_B = 12.24$, $\bar{x}_C = 12.01$, and $\bar{x}_D = 11.9$. For the workspace configurations, the mean parts per hour are $\bar{x}_1 = 13.3$, $\bar{x}_2 = 12.06$, and $\bar{x}_3 = 11.93$.

The standard error for the column means is

$$s_{\bar{X}} = \sqrt{\frac{.384}{3 \times 3}} = \sqrt{.0427} = .207$$

and for the row means, it is

$$s_{\bar{X}} = \sqrt{\frac{.384}{4 \times 3}} = \sqrt{.032} = .179$$

We begin with the test for differences between the column means. Ordering these means, we have $\bar{x}_{(1)} = 11.9$, $\bar{x}_{(2)} = 12.01$, $\bar{x}_{(3)} = 12.24$, and $\bar{x}_{(4)} = 13.57$. Using $\alpha = .05$, $n_2 = 24$, and q values 2, 3, and 4, we obtain the values for r from Appendix Table B.12.

q	2	3	4
r	2.93	3.08	3.16

Since $n_2 = 24$ is not in the table, the values given represent the linear interpolation of the r values for $n_2 = 20$ and $n_2 = 30$. The multiplication of the r values by $s_{\bar{X}}$ gives the LSR values.

q	2	3	4
LSR	.607	.638	.654

The tests for differences are as follows.

$$\begin{aligned} q = 4:\ & \bar{x}_{(4)} - \bar{x}_{(1)} = 1.67 > .654 && \text{(significant)} \\ q = 3:\ & \bar{x}_{(4)} - \bar{x}_{(2)} = 1.56 > .638 && \text{(significant)} \\ & \bar{x}_{(3)} - \bar{x}_{(1)} = .34 < .638 && \\ q = 2:\ & \bar{x}_{(4)} - \bar{x}_{(3)} = 1.33 > .607 && \text{(significant)} \\ & \bar{x}_{(3)} - \bar{x}_{(2)} = .23 < .607 && \\ & \bar{x}_{(2)} - \bar{x}_{(1)} = .11 < .607 && \end{aligned}$$

In testing for differences between row means, the ordering is $\bar{x}_{(1)} = 11.93$, $\bar{x}_{(2)} = 12.06$, and $\bar{x}_{(3)} = 13.3$, and the r and LSR values are

q	2	3
r	2.93	3.08
LSR	.524	.551

The tests for differences are

$$\begin{aligned} q = 3:\ & \bar{x}_{(3)} - \bar{x}_{(1)} = 1.37 > .551 && \text{(significant)} \\ q = 2:\ & \bar{x}_{(3)} - \bar{x}_{(2)} = 1.24 > .524 && \text{(significant)} \\ & \bar{x}_{(2)} - \bar{x}_{(1)} = .13 < .524 && \end{aligned}$$

The conclusions based on the results of the Duncan multiple-range test indicate that the mean parts per hour is the same for work movement types B, C, and D. A differs significantly from B, C, and D at the .05 level. For the workspace configurations, types 2 and 3 do not differ in the mean parts per hour, but both differ significantly with respect to workspace configuration 1. Therefore, looking at each factor individually, we would choose work movement A and workspace configuration 1. The difficulty with this interpretation is that it does not incorporate the interaction effects. We recommend that the reader consult one of the references before making recommendations in the presence of interaction. ■

Earlier in this section, interaction was stated to exist if a particular level of one variable gives responses that may be considered out of character or pattern when it occurs in combination with a particular level of the other factor. Since interaction did exist between types of work movement and types of workspace configurations, we can expect that mean responses, the average parts per hour, for two or more cells are out of character. In Figure 9.5, the cell means from Example 9.5 are plotted for the types of work movement for each of the workspace configurations. The hypothesis test in Example 9.5 indicated the presence of significant interaction. Generally, we expect the lines associated with each workspace configuration to give parallel segments through the four types of work movement if there is no interaction. We see this is not true for the three lines shown in Figure 9.5, and it is indicative of a possible significant interaction. However, crossed lines of themselves do not necessarily imply that significant interactions are present. Some deviation from parallel will be observed in the sample data due to random variation. For the situation in Example

FIGURE 9.5 Plot of average parts per hour for types of work movement for workspace configurations 1, 2, and 3, with significant interaction present

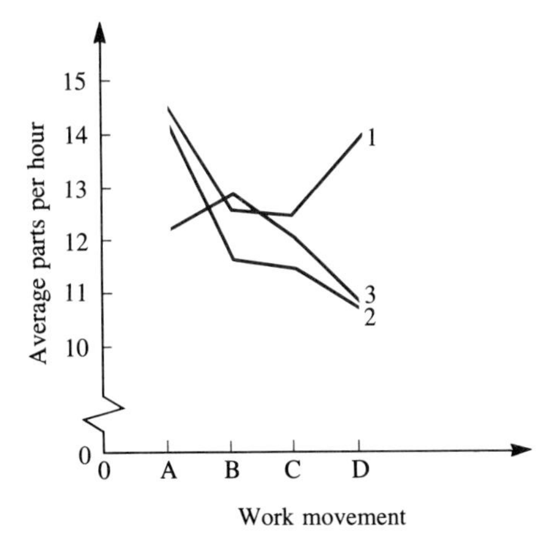

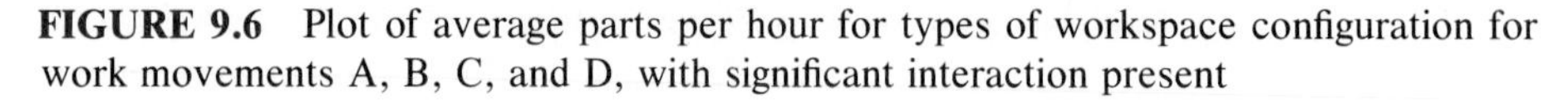

FIGURE 9.6 Plot of average parts per hour for types of workspace configuration for work movements A, B, C, and D, with significant interaction present

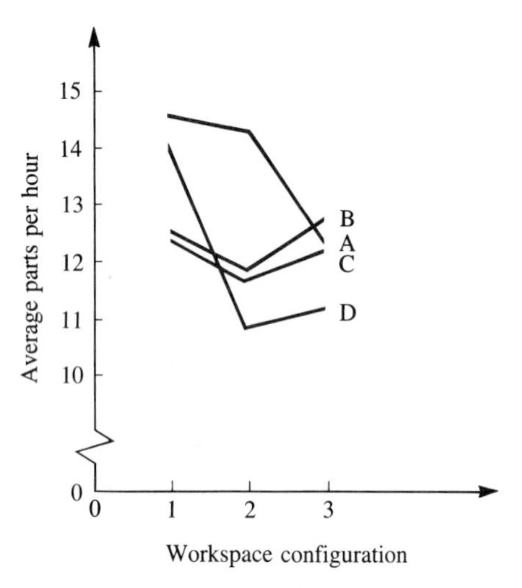

9.5, three values appear to meet this characteristic, D1, A2, and A3. When each of these values is compared to the other two values for the respective work movement, each seems to deviate from the basic pattern. When the relationships are viewed from the opposite direction, as shown in Figure 9.6, the values for D1, A2, and A3 still appear to be out of character with the others for the same workspace configuration. Going beyond the scope of the text, 95% confidence intervals were constructed for the interaction terms after correcting for row and column deviations. It was found that interactions D1, A2, and A3 are significantly different from zero.

As indicated earlier, if there is no interaction present, we expect a plot of the cell means to show approximately parallel line segments. An illustration of this plot is presented in Figure 9.7.

Analysis of Randomized Block Design

The analysis of the data for a randomized block design corresponds to that of a two-factor crossed design with one observation per cell—that is, $K = 1$. Recall that the model in Equation 9.22 is

$$X_{ijk} = \mu_X + \alpha_j + \beta_i + \alpha\beta_{ij} + \varepsilon_{ijk}$$

and, if $K = 1$, ε_{ij1} is simply ε_{ij}. Thus, the terms $\alpha\beta_{ij}$ and ε_{ij} cannot be estimated separately. To see why they cannot, consider the equations for SSE and df_E

FIGURE 9.7 Representative plot of cell means for factor A and for factor B with no interaction present

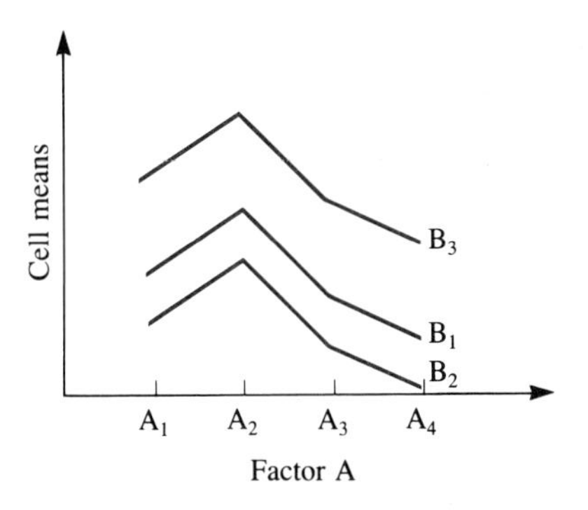

corresponding to the two-factor model. We have, for $K = 1$,

$$\text{df}_E = IJ(K - 1) = 0$$

Also

$$\text{SS}E = \sum_i \sum_j \sum_k (X_{ijk} - \overline{X}_{ij\cdot})^2 = 0$$

since

$$X_{ijk} = \overline{X}_{ij\cdot} \qquad \text{when } K = 1$$

Thus, in a two-factor design with one observation per cell, the two terms, $\alpha\beta_{ij}$ and ε_{ij}, must be combined into one term. For the analysis to proceed under the fixed-effects assumption, we must assume that the interaction term is zero and that what is being observed in the combined term is merely error. That is, to analyze a two-factor design with one observation per cell, we must make the general ANOVA assumptions and also assume that

$$\alpha\beta_{ij} = 0 \qquad \text{for all } i, j \tag{9.37}$$

Then, to calculate a value for SSE, we actually use the equation for SSAB. If the assumption in Equation 9.37 is valid, these deviations reflect only chance or error variation. The definitional equations for the sums of squares in this case are given next.

$$\text{SS}T = \sum_i \sum_j (X_{ij} - \overline{X}_{\cdot\cdot})^2 \tag{9.38}$$

$$\text{SS}A = \sum_i \sum_j (\bar{X}_{\cdot j} - \bar{X}_{\cdot\cdot})^2 \tag{9.39}$$

$$\text{SS}B = \sum_i \sum_j (\bar{X}_{i\cdot} - \bar{X}_{\cdot\cdot})^2 \tag{9.40}$$

$$\text{SS}E = \text{SS}AB = \sum_i \sum_j (X_{ij} - \bar{X}_{i\cdot} - \bar{X}_{\cdot j} + \bar{X}_{\cdot\cdot})^2 \tag{9.41}$$

As usual, the actual calculations are performed with the computational forms.

$$\text{SS}T = \sum_i \sum_j X_{ij}^2 - C \tag{9.42}$$

$$\text{SS}A = \frac{1}{I} \sum_j X_{\cdot j}^2 - C \tag{9.43}$$

$$\text{SS}B = \frac{1}{J} \sum_i X_{i\cdot}^2 - C \tag{9.44}$$

$$\text{SS}E = \text{SS}AB = \sum_i \sum_j X_{ij}^2 - \frac{1}{J} \sum_i X_{i\cdot}^2 - \frac{1}{I} \sum_j X_{\cdot j}^2 + C \tag{9.45}$$

where $C = X_{\cdot\cdot}^2/IJ$.

In the sums of squares, we have used SSA and SSB to denote the column and row factors, respectively. In most situations in this text, columns, A, will correspond to the treatment variable and rows, B, will represent the blocking variable. Be aware, though, that it may be more convenient for presentation purposes to reverse these variables. Regardless of orientation, the treatment variable is always the primary variable of interest, and the blocking variable is of secondary interest. As usual, the computations are summarized in an ANOVA table. A symbolic table with the treatment variable corresponding to columns is shown in Table 9.18. The analysis for a randomized block design is illustrated in the following example.

EXAMPLE 9.7 Eckert Stenographic Service, Inc., has been experiencing problems in the accuracy of work produced by its typing staff. In an effort to pinpoint the source of the problem, the company has decided to investigate whether differences exist in accuracy levels with respect to the make of typewriter. It is suspected that the level of accuracy may also vary with the educational background of the typist. It is desired to remove the variation due to this effect from the error. An experiment is designed to determine the average number of errors per page for 5 pages of straight typing. Three persons are randomly selected from each group with similar educational background and experience and are ran-

TABLE 9.18 Symbolic presentation for a two-way analysis of variance with one observation per cell

Source of Variability	*df*	*SS*	*MS*	*F Ratio*
Among treatments	$J - 1$	$SSTR$	$\frac{SSTR}{J-1}$	$\frac{MSTR}{MSE}$
Among blocks	$I - 1$	$SSBL$	$\frac{SSBL}{I-1}$	$\frac{MSBL}{MSE}$
Error	$(I-1)(J-1)$	SSE	$\frac{SSE}{(I-1)(J-1)}$	
Total	$IJ - 1$	SST		

domly assigned to the 3 makes of typewriters. The data, organized for analysis, are given in the following table.

Prior Formal Education	*Make of Typewriter*		
	Ludlow	*Magnum*	*Kingsway*
High School	1.3	1.4	1.2
Vocational School	1.1	1.0	1.2
Two-Year Secretarial School	.6	.8	.7

At an α level of .05, what can the company conclude?

Solution: In this example, the experiment, as constructed, represents a randomized block design. We see that blocking was done with respect to the prior educational level. Thus, we can remove its effect from the error term when testing for differences due to the make of typewriter. Since this effect is now isolated, we can also test to see if significant differences exist with respect to prior educational level to see if blocking was necessary.

Before testing the hypotheses, it is convenient to construct the ANOVA table. The calculations are determined by using Equations 9.42 through 9.44 and the relation $SSE = SST - (SSB + SSA)$. From the sample information, we obtain each of the following summations.

$$x_{1\cdot} = 3.9 \qquad x_{\cdot 1} = 3.0$$
$$x_{2\cdot} = 3.3 \qquad x_{\cdot 2} = 3.2$$
$$x_{3\cdot} = 2.1 \qquad x_{\cdot 3} = 3.1 \qquad x_{\cdot\cdot} = 9.3$$

$$C = \frac{x_{\cdot\cdot}^2}{IJ} = \frac{9.3^2}{3 \times 3} = \frac{86.49}{9} = 9.61$$

$$SST = \sum_{j=1}^{J} \sum_{i=1}^{I} x_{ij}^2 - C = (1.3^2 + 1.1^2 + \cdots + .7^2) - 9.61$$

$$= 10.23 - 9.61 = .62$$

$$SSA = \sum_{j=1}^{J} \frac{x_{\cdot j}^2}{I} - C = \frac{3.0^2}{3} + \frac{3.2^2}{3} + \frac{3.1^2}{3} - 9.61$$

$$= 3.0 + 3.4133 + 3.2033 - 9.61 = 9.6166 - 9.61 = .0066$$

$$SSB = \sum_{i=1}^{J} \frac{x_{i\cdot}^2}{J} - C = \frac{3.9^2}{3} + \frac{3.3^2}{3} + \frac{2.1^2}{3} - 9.61$$

$$= 5.07 + 3.63 + 1.47 - 9.61 = 10.17 - 9.61 = .56$$

$$SSE = SST - (SSB + SSA) = .62 - (.56 + .0066) = .0534$$

The ANOVA table can now be completed.

Source of Variability	***df***	***SS***	***MS***	***F Ratio***
Among typewriters	2	.0066	.0033	$\frac{.0033}{.01335} = .247$
Among educations	2	.5600	.2800	$\frac{.28}{.01335} = 20.974$
Error	4	.0534	.01335	
Total	8	.6200		

The hypotheses, tests, and decisions for this example are summarized in the following table.

Variable	***Hypotheses***	***Calculated F Ratio***	$F_{.05}$	***Decision***
Typewriter	$H_0: \alpha_1 = \alpha_2 = \alpha_3 = 0$ H_a: some $\alpha_j \neq 0$	0.247	<6.94	Accept
Education	$H_0: \beta_1 = \beta_2 = \beta_3 = 0$ H_a: some $\beta_i \neq 0$	20.974	>6.94	Reject

Thus, we conclude that there appear to be significant differences in the mean numbers of errors due to the type of prior education received by the individual, but no differences appear to exist in the mean numbers of errors with respect to the brand of typewriter used. ■

The conclusions just reached rely on the appropriateness of the assumptions underlying the procedure. In this instance, the distribution of typographical errors is more likely to be Poisson than normal. Since the response is an average, the normality assumption probably is not critical, but for a Poisson distribution, the variance equals the mean, and, hence, the assumption of equal variances is likely to be violated. Thus, these conclusions may be questionable in light of the violation of equal variances.

In addition to the basic ANOVA assumptions, we assumed no interaction between educational level and brand of typewriter. If this assumption is invalid, then the mean square error is inflated since it also includes differences due to interaction. This inflation of the MS*E* increases the chance of a Type II error in the fixed-effects model. There is a procedure to test the assumption of no interaction in this situation. It was developed by Tukey and can be found in Neter, Wasserman, and Kutner [3].

If a null hypothesis is rejected in the overall analysis, differences between specific means for that factor may be investigated by Duncan's procedure. The procedure is identical to that of the last section with $K = 1$. For the column means, we reject H_0: $\mu_{(v)} = \mu_{(w)}$ in favor of H_a: $\mu_{(v)} > \mu_{(w)}$ if

$$\overline{X}_{(v)} - \overline{X}_{(w)} > \text{LSR} = r_{q,n_2,\alpha}\, s_{\overline{X}} \tag{9.46}$$

where $n_2 = (I - 1)(J - 1)$ and $s_{\overline{X}} = \sqrt{\text{MS}E/I}$. For the row means, H_0: $\mu_{(v)} = \mu_{(w)}$ is rejected in favor of H_a: $\mu_{(v)} > \mu_{(w)}$ if

$$\overline{X}_{(v)} - \overline{X}_{(w)} > \text{LSR} = r_{q,n_2,\alpha}\, s_{\overline{X}} \tag{9.47}$$

where $n_2 = (I - 1)(J - 1)$ and $s_{\overline{X}} = \sqrt{\text{MS}E/J}$. As usual, for $v > w$, $q = (v - w) + 1$ and $r_{q,n_2,\alpha}$ is found in Appendix Table B.12.

EXAMPLE 9.8 Classify or group the means of Example 9.7 using Duncan's procedure at $\alpha = .05$ when necessary.

Solution: We consider makes of typewriters first. The term *classify* here indicates that we want to identify which means appear to be from identical populations. Since H_0 was accepted for typewriters, the conclusion is that all are the same as far as the typographical error rates are concerned. That is, $\mu_L = \mu_M = \mu_K$.

Since H_0 was rejected for educational levels, further analysis is required. For rows, we have $n_2 = 4$, $s_{\overline{X}} = \sqrt{.01335/3} = .0667$, $\alpha = .05$, and values for q of 2 and 3. From Appendix Table B.12, we obtain the corresponding values for r, giving

q	2	3
r	3.93	4.01
LSR	.262	.267

For each pair of means, the hypotheses are

$$H_0\colon \mu_{(v)} = \mu_{(w)} \qquad H_a\colon \mu_{(v)} > \mu_{(w)}$$

and H_0 is rejected if $\overline{X}_{(v)} - \overline{X}_{(w)} > \text{LSR}$.

FIGURE 9.8 Graphical representation of average errors for Example 9.8, with curves centered at the corresponding estimates of the population means

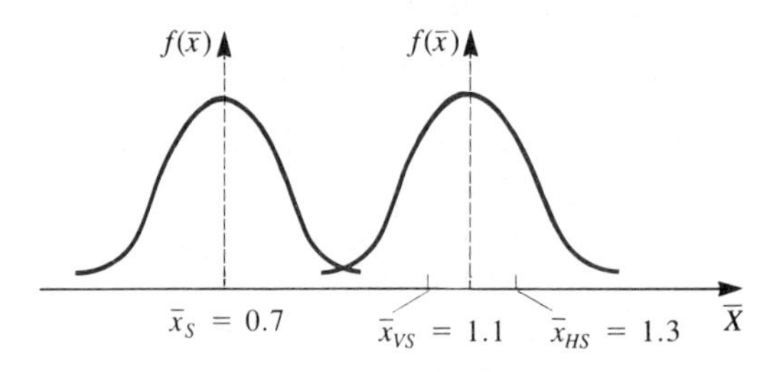

We have $\bar{x}_{(1)} = 0.7$, $\bar{x}_{(2)} = 1.1$, $\bar{x}_{(3)} = 1.3$ and

$$q = 3: \bar{x}_{(3)} - \bar{x}_{(1)} = 1.3 - 0.7 = .6 > .267 \quad \text{(significant)}$$
$$q = 2: \bar{x}_{(3)} - \bar{x}_{(2)} = 1.3 - 1.1 = .2 < .262$$
$$\bar{x}_{(2)} - \bar{x}_{(1)} = 1.1 - 0.7 = .4 > .262 \quad \text{(significant)}$$

where the significant differences are .6 and .4. Thus, we conclude that the error rates for both high-school and vocational-school trained secretaries appear to be higher than the rate for those trained in a 2-year secretarial program, but are similar to each other. The relationship of the means is depicted in Figure 9.8.

Perhaps someone might question whether a particular typist may have had a better performance on one brand of typewriter than another because it was the brand with which the individual was most familiar. If this is true, the assumption of no interaction is violated, and the MSE represents a combination of error and interaction variation rather than error only; thus, we may have accepted the null hypothesis for typewriters incorrectly due to an enlarged MSE and, therefore, have committed a Type II error. If there is any question about the assumption of no interaction, replication should be used so that any interaction effects can be separated from the error term.

EXERCISES

9.8 A company desires to know whether the tensile strength, in pounds per square inch, of its product remains uniform throughout the length of the shift. Three time periods are of interest and are chosen to include 2 hours at the beginning, middle, and end of the shift. From each of the 6 production lines making this product, 1 item is randomly selected within the prescribed time period and tested. The data in pounds per square inch are shown in the following table.

Time Period	*Line*					
	1	*2*	*3*	*4*	*5*	*6*
Beginning	19.7	19.0	18.6	16.3	14.3	12.0
Middle	12.5	17.2	17.3	17.5	24.6	25.1
End	24.4	21.3	20.7	24.5	20.6	20.3

At $\alpha = .05$, is the average tensile strength uniform with respect to lines and the time period of the shift tested? Identify the type of experimental design and any special assumptions necessary for the analysis. Evaluate the assumptions.

9.9 An experiment is designed to evaluate the mean yields in bushels per acre for 3 different kinds of wheat and 6 different kinds of soil. Since it is believed that some wheats do better in certain soils, the experiment was repeated 4 times. What are the sources of variability and the degrees of freedom? Identify the experimental design and the type of effect for each factor.

9.10 Because of the high costs of hospital confinement and the need to free facilities, the average hospital stay for women giving birth has diminished. A recent study was undertaken to determine whether the average confinement was the same for 4 area hospitals. In addition, it was deemed that differences can exist due to the type of birth. Multiple samples were also randomly selected to determine if there was any interaction between the hospital and the type of birth. The following data represent the number of days, to the nearest 10th, based on the time between check-in and checkout.

Type of Birth	*Hospital*			
	Edgewood	*Lincoln*	*Charity*	*General*
	9.5	9.7	7.1	8.7
Cesarean	7.9	9.4	8.4	9.0
	9.1	8.4	7.9	8.9
	2.7	2.6	2.5	2.1
Natural	3.2	2.7	2.0	3.2
	2.0	2.4	2.3	2.5
	3.4	3.4	4.0	3.4
Medically Assisted	3.5	3.4	2.9	3.6
	3.5	3.0	3.0	3.8

(a) At an α level of .01, what can you conclude?
(b) If differences exist within a factor, determine where they lie.
(c) Name at least two other factors that should remain constant in this study.
(d) Evaluate the underlying assumptions.

9.11 A sailmaker desires to know whether sail configuration coupled with sail area can create a significant difference in speed. Since it is logical to assume that speed varies among classes of boats, the sailmaker decides to use all 4 classes on which his sails are used. For the experiment, he randomly selects 5 boats from each class and then randomly assigns sail designs to boats within a class. All boats are then tested on the same water course under the same wind conditions. The average speeds in knots for each design/boat combination are given in the following table.

	Class of Boat			
Sail Design	***Starfish***	***Viking***	***Meteor***	***Sunburst***
1	4.0	5.6	6.1	7.0
2	5.1	5.7	6.4	6.3
3	6.4	7.1	7.5	8.4
4	5.2	5.0	6.2	7.0
5	4.9	4.9	6.2	6.9

(a) At $\alpha = .01$, determine whether differences exist in average speed due to the sail design.
(b) Analyze any differences.
(c) Does class of boat make a difference in average speed at .05?
(d) Identify the design as well as any special assumptions.

9.12 The personnel department of Wheelrite, Inc. has undertaken a study to determine whether average total absences, to the nearest half day, differ significantly with respect to the shift or the length of service, in years, of the employee. One individual is randomly selected from each cross-classification, and the total length of absence for 1 calendar year is recorded.

(a) What can be concluded at $\alpha = .05$?
(b) Evaluate the assumptions underlying your analysis.

	Shift		
Length of Service	***1st***	***2nd***	***3rd***
0–8	14.5	12.0	15.5
9–17	10.0	13.5	12.5
18–25	12.5	12.5	10.0
Over 25	6.5	9.0	10.5

9.13 The following table shows the sum of several replications, $x_{ij\cdot}$, by 4 operators to turn a part on 4 different lathes. The data are in minutes.

	Operator			
Lathe	***Jones***	***Smith***	***Wiley***	***Downs***
1	24.9	30.1	19.4	29.7
2	24.9	29.8	19.1	30.9
3	23.2	28.7	20.2	29.2
4	22.1	29.7	19.1	29.5

$\Sigma\,\Sigma\,\Sigma\, x_{ijk}^2 = 2203.29 \qquad n = 80$

(a) How many replications of the experiment were performed?
(b) Complete the appropriate ANOVA table.
(c) Perform the appropriate hypothesis tests at $\alpha = .01$.
(d) Using Duncan's multiple-range test, determine where differences exist within the factors at $\alpha = .01$.

9.14 An advertising agency is trying to ascertain whether there is a significant difference in the average total time necessary to identify 3 trademarks for each of 5 different product categories and 3 different age groups. The agency wanted to test for interaction between product categories and age, so the experiment was replicated. In the event an individual could not determine the company from the trademark, the maximum time allowable, 10 seconds, was recorded.

	Age Categories		
Product Category	***12–20***	***21–40***	***Over 40***
Cereals	10	15	15
	11	12	12
Airlines	20	17	20
	21	16	19
Automobiles	13	11	15
	17	12	16
Canned Goods	24	23	24
	27	20	26
Drugs	26	27	28
	29	28	28

(a) At an α level of .05, what can the agency conclude?
(b) Evaluate the assumptions of the analysis.

9.7 USING THE COMPUTER FOR ANALYSIS OF VARIANCE

Earlier, it was said that many software packages exist for solving analysis of variance problems. One of these packages is the Statistical Analysis System (SAS). The ANOVA procedure in SAS can handle one-factor designs, blocked designs, and designs whose cell frequencies are proportional to each other. In the latter case, it performs an internal transformation to avoid violating assumptions. After the ANOVA procedure is complete, several techniques are available for comparing the means. Two, Duncan's multiple-range test presented earlier and Scheffe's multiple-comparison procedure, are both illustrated later in this section. Since the Scheffe procedure is not formally presented in this text, the discussion of it is limited to the interpretation of the printout. It is included because the specific example has unequal sample sizes, and Duncan's procedure is inappropriate. The remainder of this section is an analysis of the output from the SAS software package for each of the designs illustrated earlier in the chapter with some MINITAB output provided for comparison.

In Table 9.19, the first observation to be made is the identification of the levels of the variable (1). The ANOVA table is shown (2) and is numerically the same as the table calculated earlier. In this printout, no comparison of the F calculated value to an F tabular value is made. However, significance is shown by the probability statement (3). The statement indicates that the probability an F value would be 18.04 or larger is only .0002. Since $.0002 < \alpha$, we can conclude that the means are not equal, and further analysis is appropriate. The results of the Duncan multiple-range test (4) indicate that the mean thicknesses are different for all three sprayers since each has a different grouping code.

The program output in Table 9.20 represents the results of the one-factor design in Example 9.2, where the sample sizes for the levels are unequal. This point is made in the note at the top of the output. Because the ANOVA program calls for a balanced design, it is necessary to reflect the imbalance at the data input stage. Again, the ANOVA table has the same numerical data as the earlier calculations. The probability statement would indicate that significant differences exist between the mean pounds per square inch for the four vendors since $.0001 < .01 = \alpha$.

Because the sample sizes differ for the four vendors, the Duncan multiple-range test is not appropriate for analyzing the differences in means. However, as can be seen from the output, an analysis was made using the Scheffe multiple-comparison procedure. Since significant differences in means are denoted by ***, we can conclude that the mean pounds per square inch was the same for vendors A and D but differed significantly for all other pairwise comparisons. Note that the starred differences correspond only to pairs where the confidence interval for the difference has both limits on the same side of zero. If the interval includes zero, then the means being compared are not significantly different and are deemed to vary only due to chance.

In Table 9.21, we observe the output corresponding to the randomized block design of Example 9.7. Table 9.21a is a printout from SAS and 9.21b is a printout from MINITAB. In determining if significant differences exist in the mean number of errors with respect to the level of schooling or to the make of typewriter, we need refer only to the shaded section of the printout. From the probability statements, we can see that there are no differences in the mean number of errors with respect to makes of typewriters. However, the other probability statement leads to the conclusion that differences do exist with respect to the educational level achieved. Further testing regarding the means was also undertaken in the program, and these results are shown in Table 9.22. Since the sample sizes were the same for the treatment levels and the block levels, the Duncan multiple-range test was used in the analysis. The first set of data in Table 9.22 compares the mean number of errors for each educational level achieved. From this output, we see that the mean numbers of errors for high-school and vocational-school subjects do not differ, but both differ from the two-year secretarial subjects. In the second set of data, we see that the mean numbers of errors do not differ for the three makes of typewriters. Evidence in the printout of no differences is to be expected since the null hypothesis of no differences was accepted initially.

TABLE 9.19 Sample computer printout for Example 9.1 using SAS Key: ① Variable identification; ② ANOVA table; ③ Probability statement; ④ Results of Duncan multiple-range test

```
                                        SAS
                          ANALYSIS OF VARIANCE PROCEDURE
                              CLASS LEVEL INFORMATION
                         CLASS     LEVELS    VALUES
                         NAME        3       AJAX COOK EZON   ①
                     NUMBER OF OBSERVATIONS IN DATA SET = 15
                                        SAS
                          ANALYSIS OF VARIANCE PROCEDURE                      ②

DEPENDENT VARIABLE: THICKNES
SOURCE                   DF         SUM OF SQUARES         MEAN SQUARE         F VALUE
MODEL                     2             7.85200000          3.92600000           18.04
ERROR                    12             2.61200000          0.21766667
CORRECTED TOTAL          14            10.46400000

SOURCE                   DF              ANOVA SS     F VALUE     PR > F   ③
NAME                      2            7.85200000       18.04     0.0002

                                       SAS
                          ANALYSIS OF VARIANCE PROCEDURE
            DUNCAN'S MULTIPLE RANGE TEST FOR VARIABLE: THICKNES
            NOTE: THIS TEST CONTROLS THE TYPE I COMPARISONWISE ERROR RATE,
                  NOT THE EXPERIMENTWISE ERROR RATE
                            ALPHA=0.05  DF=12  MSE=0.217667
                         NUMBER OF MEANS          2         3
                         CRITICAL RANGE    0.641676  0.672178
            MEANS WITH THE SAME LETTER ARE NOT SIGNIFICANTLY DIFFERENT.

                   DUNCAN   GROUPING              MEAN      N  NAME    ④
                                   A            8.9600      5  EZON
                                   B            7.9000      5  COOK
                                   C            7.2000      5  AJAX
```

TABLE 9.20 Sample computer printout for Example 9.2 using SAS

```
                                                   SAS
                                     ANALYSIS OF VARIANCE PROCEDURE
                                        CLASS LEVEL INFORMATION
                                      CLASS      LEVELS      VALUES
                                      NAME          4        A B C D

                               NUMBER OF OBSERVATIONS IN DATA SET = 40

NOTE: ALL DEPENDENT VARIABLES ARE CONSISTENT WITH RESPECT TO THE PRESENCE OR ABSENCE OF
      MISSING VALUES. HOWEVER, ONLY 32 OBSERVATIONS CAN BE USED IN THIS ANALYSIS.
                                                   SAS
                                     ANALYSIS OF VARIANCE PROCEDURE

DEPENDENT VARIABLE: STRENGTH
SOURCE                   DF        SUM OF SQUARES         MEAN SQUARE         F VALUE
MODEL                     3          973.45803571        324.48601190           36.02
ERROR                    28          252.26071429          9.00931122
CORRECTED TOTAL          31         1225.71875000

SOURCE                   DF              ANOVA SS     F VALUE     PR > F
NAME                      3          973.45803571       36.02     0.0001
```

Continues

TABLE 9.20 Continued

```
                                   SAS
                       ANALYSIS OF VARIANCE PROCEDURE
SCHEFFE'S TEST FOR VARIABLE: STRENGTH
NOTE: THIS TEST CONTROLS THE TYPE I EXPERIMENTWISE ERROR RATE BUT GENERALLY HAS A HIGHER
      TYPE II ERROR RATE THAN TUKEY'S FOR ALL PAIRWISE COMPARISONS.

      ALPHA=0.05  CONFIDENCE=0.95  DF=28  MSE=9.00931
      CRITICAL VALUE OF F=2.94669

COMPARISONS SIGNIFICANT AT THE 0.05 LEVEL ARE INDICATED BY '***'
```

NAME COMPARISON	SIMULTANEOUS LOWER CONFIDENCE LIMIT	DIFFERENCE BETWEEN MEANS	SIMULTANEOUS UPPER CONFIDENCE LIMIT	
A - D	-3.529	1.089	5.708	
A - C	2.899	7.518	12.137	***
A - B	8.842	13.075	17.308	***
D - A	-5.708	-1.089	3.529	
D - C	1.658	6.429	11.199	***
D - B	7.588	11.986	16.384	***
C - A	-12.137	-7.518	-2.899	***
C - D	-11.199	-6.429	-1.658	***
C - B	1.159	5.557	9.955	***
B - A	-17.308	-13.075	-8.842	***
B - D	-16.384	-11.986	-7.588	***
B - C	-9.955	-5.557	-1.159	***

TABLE 9.21a SAS output for Example 9.7

```
                                   SAS
                       ANALYSIS OF VARIANCE PROCEDURE
                          CLASS LEVEL INFORMATION
                       CLASS      LEVELS     VALUES
                       BLOCK        3        1 2 3
                       TRTMENT      3        A B C

                NUMBER OF OBSERVATIONS IN DATA SET = 9
                                   SAS
                       ANALYSIS OF VARIANCE PROCEDURE
```

DEPENDENT VARIABLE:

SOURCE	DF	SUM OF SQUARES	MEAN SQUARE	F VALUE
MODEL	4	0.56666667	0.14166667	10.62
ERROR	4	0.05333333	0.01333333	
CORRECTED TOTAL	8	0.62000000		

SOURCE	DF	ANOVA SS	F VALUE	PR > F
BLOCK	2	0.56000000	21.00	0.0076
TRTMENT	2	0.00666667	0.25	0.7901

TABLE 9.21b MINITAB output for Example 9.7

```
MTB > EXEC 'EX9_7 MINITAB A'
MTB > READ DATA INTO C1-C3
     9 ROWS READ
MTB > END OF DATA
MTB > NAME FOR C2 IS 'BLOCK', C3 IS 'TREAT'
MTB > NOBRIEF
MTB > TWOWAY AOV, OBS IS C1, BLOCK IN C2, TREATMENT IN C3

ANALYSIS OF VARIANCE ON C1

SOURCE         DF        SS        MS
BLOCK           2    0.5600    0.2800
TREAT           2    0.0067    0.0033
ERROR           4    0.0533    0.0133
TOTAL           8    0.6200
MTB > STOP
```

TABLE 9.22 Computer output for Duncan's multiple-range test for the data of Example 9.7 using SAS

```
                              SAS
                 ANALYSIS OF VARIANCE PROCEDURE
DUNCAN'S MULTIPLE RANGE TEST FOR VARIABLE: SCHOOL
NOTE: THIS TEST CONTROLS THE TYPE I COMPARISONWISE ERROR RATE,
      NOT THE EXPERIMENTWISE ERROR RATE

                ALPHA=0.05  DF=4  MSE=.0133333

             NUMBER OF MEANS          2          3
             CRITICAL RANGE    0.262243   0.267793

       MEANS WITH THE SAME LETTER ARE NOT SIGNIFICANTLY DIFFERENT.

           DUNCAN   GROUPING            MEAN       N  BLOCK
                           A         1.30000       3  1
                           A
                           A         1.10000       3  2
                           B         0.70000       3  3

                              SAS
                 ANALYSIS OF VARIANCE PROCEDURE
DUNCAN'S MULTIPLE RANGE TEST FOR VARIABLE: TYPE
NOTE: THIS TEST CONTROLS THE TYPE I COMPARISONWISE ERROR RATE,
      NOT THE EXPERIMENTWISE ERROR RATE

                ALPHA=0.05  DF=4  MSE=.0133333

             NUMBER OF MEANS          2          3
             CRITICAL RANGE    0.262243   0.267793

       MEANS WITH THE SAME LETTER ARE NOT SIGNIFICANTLY DIFFERENT.

           DUNCAN   GROUPING            MEAN       N  TRTMENT
                           A         1.06667       3  B
                           A
                           A         1.03333       3  C
                           A
                           A         1.00000       3  A
```

Tables 9.23a and b show SAS and MINITAB outputs for Example 9.5, where A represents the type of workspace configuration and B represents the type of work movement. From the printout, we can see that the F calculated values are large enough to indicate a small probability of their occurrence strictly due to chance. Thus, we can conclude that column effects are present, as well as row effects. In addition, interaction takes place between the type of work movement and the type of workspace configuration.

In Table 9.24, the row and column means are investigated further by using the Duncan multiple-range test. The output indicates that for the row variable (A), workspace configuration 1 differs significantly from workspace configurations 2 and 3, which are not significantly different. For the column variable (B), type of work movement 1 differs significantly from each of the other three, which do not differ significantly with respect to each other.

The computer printouts shown represent outputs from the SAS and MINITAB programs. The usage of these packages is not meant to be an endorsement of them in particular. Numerous other packages are available that do essentially the same things. Among these are Statistical Package for the Social Sciences (SPSS), various biomedical (BMD) packages, and the Advanced Data Enquiry Package-Timeshared (ADEPT). Each has some advantages and disadvantages, depending on the individual's particular needs and preferences.

TABLE 9.23a SAS output for Example 9.5

```
                                                      SAS
                                         GENERAL LINEAR MODELS PROCEDURE
                                             CLASS LEVEL INFORMATION
                                         CLASS      LEVELS     VALUES
                                         A               3     A1 A2 A3
                                         B               4     B1 B2 B3 B4

                                     NUMBER OF OBSERVATIONS IN DATA SET = 36
                                                      SAS
                                         GENERAL LINEAR MODELS PROCEDURE
DEPENDENT VARIABLE: Y
SOURCE                   DF      SUM OF SQUARES      MEAN SQUARE        F VALUE
MODEL                    11         46.44305556       4.22209596          11.00
ERROR                    24          9.21333333       0.38388889
CORRECTED TOTAL          35         55.65638889

SOURCE                   DF         TYPE I SS      F VALUE    PR > F     DF
A                         2       13.70055556        17.84    0.0001      2
B                         3       16.04527778        13.93    0.0001      3
A*B                       6       16.69722222         7.25    0.0002      6
```

TABLE 9.23b MINITAB output for Example 9.5

```
MTB > EXEC 'EX9_5 MINITAB A'
MTB > READ DATA INTO C1-C3
    36 ROWS READ
MTB > END OF DATA
MTB > NAME FOR C2 IS 'A', C3 IS 'B'
MTB > NOBRIEF
MTB > TWOWAYAOV DATA IN C1, A IN C2, B IN C3

ANALYSIS OF VARIANCE ON C1

SOURCE         DF        SS         MS
A               2    13.701      6.850
B               3    16.045      5.348
INTERACTION     6    16.697      2.783
ERROR          24     9.213      0.384
TOTAL          35    55.656
MTB > STOP
```

TABLE 9.24 Computer output for Duncan's multiple-range test for the data of Example 9.5 using SAS

```
                                        SAS
                          GENERAL LINEAR MODELS PROCEDURE
DUNCAN'S MULTIPLE RANGE TEST FOR VARIABLE: Y
NOTE: THIS TEST CONTROLS THE TYPE I COMPARISONWISE ERROR RATE,
      NOT THE EXPERIMENTWISE ERROR RATE

                          ALPHA=0.05   DF=24   MSE=0.383889

                       NUMBER OF MEANS          2          3
                       CRITICAL RANGE    0.521537   0.547938

             MEANS WITH THE SAME LETTER ARE NOT SIGNIFICANTLY DIFFERENT.

                   DUNCAN   GROUPING             MEAN        N   A
                                   A          13.3000       12   A1
                                   B          12.0583       12   A2
                                   B
                                   B          11.9333       12   A3
```

Continues

TABLE 9.24 Continued

```
                         SAS
              GENERAL LINEAR MODELS PROCEDURE
DUNCAN'S MULTIPLE RANGE TEST FOR VARIABLE: Y
NOTE: THIS TEST CONTROLS THE TYPE I COMPARISONWISE ERROR RATE,
      NOT THE EXPERIMENTWISE ERROR RATE

              ALPHA=0.05  DF=24  MSE=0.383889

       NUMBER OF MEANS          2          3          4
       CRITICAL RANGE    0.602219   0.632705   0.653715

   MEANS WITH THE SAME LETTER ARE NOT SIGNIFICANTLY DIFFERENT.

       DUNCAN   GROUPING              MEAN      N  B
                       A           13.5667      9  B1

                       B           12.2444      9  B2
                       B
                       B           12.0111      9  B3
                       B
                       B           11.9000      9  B4
```

9.8 OTHER EXPERIMENTAL DESIGNS

Due to the intended level of this text, the analysis in this chapter is restricted to experimental designs involving at most two factors. As one would expect, increasing the number of factors increases the difficulties of both computations and analysis. Only a few extensions are mentioned here, and the reader is referred to the references for a comprehensive discussion and analysis.

The two-factor crossed design can be extended to any number of factors. For each factor added, the model gets more complicated. For example, for a three-factor crossed design, the model is

$$X_{ijkl} = \mu + \alpha_j + \beta_i + \alpha\beta_{ij} + \gamma_k + \alpha\gamma_{jk} + \beta\gamma_{ik} + \alpha\beta\gamma_{ijk} + \varepsilon_{ijkl} \qquad \textbf{(9.48)}$$

Denoting the factors by A, B, and C, respectively, the corresponding main effects are α_j, β_i, and γ_k. The terms $\alpha\beta_{ij}$, $\alpha\gamma_{jk}$, and $\beta\gamma_{ik}$ represent the effects associated with the AB, AC, and BC interactions, respectively, and $\alpha\beta\gamma_{ijk}$ represents effects for the ABC interaction. As always, ε_{ijkl} is the random variable representing the error term. If there is replication, each of these factors can be estimated separately and tested for significance. Note that if a fourth factor is added to the design, eight more terms are added to the model in Equation 9.48. For reference, crossed designs are also known as *factorial* designs.

As the randomized block is a special case of the two-factor crossed design, a *Latin square* is a special case of the three-factor crossed design. The randomized block design blocks on a single variable, and the Latin square blocks on two variables. The model for a Latin square is

$$X_{ijk} = \mu + \alpha_j + \beta_i + \gamma_k + \varepsilon_{ijk} \qquad \textbf{(9.49)}$$

As might be expected from the model, the analysis of a Latin square experiment is a special case of that for a three-factor crossed design. The analysis assumes that all interactions in Equation 9.48 are zero.

If a four-factor experiment uses blocking on three variables, the design is called a *Greco-Latin* square. If blocking is used on four variables, then the design is called a *Judeo-Greco-Latin* square. It is obvious that as the number of factors increases, the number of observations required to be able to test interaction effects increases exponentially. Thus, for experiments involving many factors, often only certain subsets of all possibilities are run in order to reduce the cost of experimentation. Some designs that accomplish the testing of subsets are the *incomplete block* and the *fractional factorial* designs. All of the designs mentioned in this section are well beyond the scope of this text. For a thorough treatment of experimental designs in general, see Winer [4].

9.9 CODING OF DATA

It is apparent that the computations involved in the analysis of variance can be quite laborious and cumbersome. With the ready availability of calculators and computers, both PC and larger sizes, cumbersome calculations are not usually a problem. What can be a problem, though, is that practical applications often entail values that are either quite large in magnitude or extremely small in magnitude. In addition, in situations involving very precise measurements, there may be no variability in observations until the fourth or fifth decimal place. In all three of these instances, variability can be introduced into the analysis as well as lost through rounding and truncation errors. One way to avoid these problems is through coding of the data prior to the use of ANOVA.

Coding was introduced and discussed in detail in Chapter 5. By coding, we mean a linear transformation on the observations of the form

$$y_{ij} = ax_{ij} + b$$

where a and b are real numbers. For example, if $a = 1$, then b represents a constant that is either added to or subtracted from each observation. We note that the addition of a constant has no effect on the variance. Thus, the ANOVA table for the coded data, in this case, would be identical to the ANOVA table for the original data. To illustrate, refer to Example 9.1 and subtract 8 from each observation in the data set. Using the transformed data, we calculate

$$\text{SS}T = \sum_i \sum_j y_{ij}^2 - C = 10.47 - .006 = 10.464$$

TABLE 9.25 ANOVA table for transformed data, $y_{ij} = 10x_{ij} - 80$, of Example 9.1

Source of Variability	*df*	*SS*	*MS*	*F Ratio*
Among sprayers	2	785.2	392.60	18.04
Error	12	261.2	21.77	
Total	14	1046.4		

which is the same value obtained with the original data. Try to verify that the other values in the ANOVA table are also unchanged.

If the constant a is a value other than 1, then all sums of squares are changed proportionately. That is, the SSs and MSs are all in the same proportion to the original values. Thus, the ratio of two mean squares—that is, the calculated F—is unaffected by the coding. As an illustration, use the data from Example 9.1, subtract 8 from each value, and multiply the results by 10. In other words, let

$$y_{ij} = 10(x_{ij} - 8) = 10x_{ij} - 80$$

Verify that the resulting ANOVA table is as shown in Table 9.25. In comparing Table 9.25 to Table 9.10, we see that both the sums of squares and mean squares have been multiplied by 100, but that the F ratio is 18.04 in both tables. In general, the entries in the SS and MS columns each change proportionally by a factor equal to a^2.

9.10 SUMMARY

This chapter considered procedures for testing the equality of several means simultaneously. The name analysis of variance (ANOVA) is given to the procedure because the test is performed by comparing the variation among several sample means to the variation within samples. If the samples are all from the same population, then both variations are due to chance or error and, as such, should be approximately equal. Thus, for the model $X_{ij} = \mu + \alpha_j + \varepsilon_{ij}$, if $H_0\colon \mu_j = \mu$ for all j, or, equivalently, $H_0\colon \alpha_j = 0$ for all j is true, the ratio of the two methods of estimating the variance, σ_X^2, should be approximately unity.

To develop the test statistics, we partitioned the total variation, SST, into the two components SSE and SSA. Then, using the assumptions of normal distributions and constant variances, Cochran's theorem was applied to obtain the result that the SSs were independent with each being distributed as a $\sigma_X^2\chi^2$ random variable. Hence, under the null hypothesis, the ratio of the mean squares or variance estimates has an F-distribution.

If the alternative is true, then the variation among sample means reflects both chance variation and a squared differential effect. That is, the variance estimated from the means should be much larger than the variation due to chance, and H_0 is rejected only for large values of F.

The analysis is limited to models that involve fixed effects; the only levels of interest are those in the experiment. For two-factor models, the discussion of the analysis pertains to crossed or factorial designs only. That is, all levels of one factor occur in combination with all levels of the other factor. The general two-factor model for a crossed design is

$$X_{ijk} = \mu + \alpha_j + \beta_i + \alpha\beta_{ij} + \varepsilon_{ijk}$$

where α_j, β_i represent main effects, $\alpha\beta_{ij}$ is an interaction term to explain atypical responses for particular combinations of factor levels, and ε_{ijk} is a random error term. The overall analysis follows the same pattern as the one-factor model. To estimate the variation due to interaction and error separately requires replication, multiple observations per cell.

If there is no replication, the error and interaction terms cannot be separated. In such a case, the analysis requires the additional assumption that all interactions are zero and then uses the interaction SS for the error. A special case requiring the assumption of interactions of zero is the randomized block design. The randomized block design attempts to remove the effect of a secondary variable from the error variation before testing. This secondary variable is called a blocking variable, and the primary variable of interest is called the treatment variable.

In any of the designs, it is often of interest to do further investigation when H_0 is rejected for a main effect. Various procedures are designed to do this investigation, with Duncan's multiple-range test presented here. Duncan's test requires equal sample sizes for the means being compared and is restricted to pairwise comparisons. It cannot be used on cell means to investigate interactions.

The last two sections of the chapter touched on more sophisticated designs, with merely an indication of their uses, and on coding. For a treatment of the advanced designs and random-effects models, consult the references. If the experimental data involve either numbers of large magnitude or numbers with no variability until the third or fourth decimal, for example, coding is advised before performing the analysis; otherwise, significant rounding or truncation errors may result in the computer computations. The equations, hypotheses, and test statistics for this chapter are presented in the tabular summary in Table 9.26.

SUPPLEMENTARY EXERCISES

9.15 The Bradley Manufacturing Company produces pistons for automobile engines. The company has been experimenting with different alloys. Eight pistons for each of the 4 alloys of interest are tested in randomly selected engines. The tests are performed under identical conditions. The amount of wear is recorded in millimeters and shown in the following table.

TABLE 9.26 Summary table

One-Factor ANOVA: Fixed-Effects Model

Model	$X_{ij} = \mu + \alpha_j + \varepsilon_{ij}$
Hypotheses	H_0: $\alpha_j = 0$ for all j; H_a: some $\alpha_j \neq 0$
Assumptions	*Additive effects* $\varepsilon_{ij} \sim N(0, \sigma_j^2)$ $\sigma_j^2 = \sigma_X^2$ for all j (homogeneity of variance) ε_{ij} independent for all i, j at least interval measurement
Sums of Squares	*Definitional* / *Computational* *Total* $SST = \sum_j \sum_i (\bar{X}_{ij} - \bar{X}_{..})^2 = \sum_j \sum_i X_{ij}^2 - C$ *Factor A* $SSA = \sum_j n_j (\bar{X}_{.j} - \bar{X}_{..})^2 = \sum_j \frac{X_{.j}^2}{n_j} - C$ *Error* $SSE = \sum_j \sum_i (X_{ij} - \bar{X}_{.j})^2 = \sum_j \sum_i X_{ij}^2 - \sum_j \frac{X_{.j}^2}{n_j}$ *where*: $X_{..} = \sum_j \sum_i X_{ij}$, $X_{.j} = \sum_i X_{ij}$, $C = \frac{X_{..}^2}{n_T}$, $n_T = \sum_j n_j$
ANOVA Table	see below
Decision rule	Reject H_0 at α if MSA/MSE > $F_{J-1, n_T-J, \alpha}$ = Table F.

ANOVA Table:

Source of Variation	*df*	*SS*	*MS*	*F ratio*	*Table F*
Among Means	$J - 1$	SSA	$\frac{SSA}{J-1}$	$\frac{MSA}{MSE}$	$F_{J-1, n_T-J, \alpha}$
Error	$n_T - J$	SSE	$\frac{SSE}{n_T - J}$		
Total	$n_T - 1$	SST			

Two-Factor ANOVA: Crossed Design with Fixed Effects

Model	$X_{ijk} = \mu + \alpha_j + \beta_i + \alpha\beta_{ij} + \varepsilon_{ijk}$
Hypotheses	H_0: all $\alpha\beta_{ij} = 0$; H_0: all $\beta_i = 0$; H_0: all $\alpha_j = 0$ H_a: some $\alpha\beta_{ij} \neq 0$; H_a: some $\beta_i \neq 0$; H_a: some $\alpha_j \neq 0$
Assumptions	Obvious extensions from one-factor model

Sums of Squares

	Definitional	*Computational*
Total	$SST = \sum_i \sum_j \sum_k (X_{ijk} - \bar{X}_{\cdots})^2$	$= \sum_i \sum_j \sum_k X_{ijk}^2 - C$
Factor A	$SSA = \sum_i \sum_j \sum_k (\bar{X}_{\cdot j \cdot} - \bar{X}_{\cdots})^2$	$= \sum_j \frac{X_{\cdot j \cdot}^2}{IK} - C$
Factor B	$SSB = \sum_i \sum_j \sum_k (\bar{X}_{i \cdot\cdot} - \bar{X}_{\cdots})^2$	$= \sum_i \frac{X_{i \cdot\cdot}^2}{JK} - C$
Interaction	$SSAB = \sum_i \sum_j \sum_k (\bar{X}_{ij\cdot} - \bar{X}_{i\cdot\cdot} - \bar{X}_{\cdot j \cdot} + \bar{X}_{\cdots})^2$	$= \sum_i \sum_j \frac{X_{ij\cdot}^2}{K} - \sum_i \frac{X_{i\cdot\cdot}^2}{JK} - \sum_j \frac{X_{\cdot j \cdot}^2}{IK} + C$
Error	$SSE = \sum_i \sum_j \sum_k (X_{ijk} - \bar{X}_{ij\cdot})^2 =$	$= \sum_i \sum_j \sum_k X_{ijk}^2 - \sum_i \sum_j \frac{X_{ij\cdot}^2}{K}$

where: $X_{\cdots} = \sum_i \sum_j \sum_k X_{ijk}$, $X_{i\cdot\cdot} = \sum_j \sum_k X_{ijk}$, $X_{\cdot j \cdot} = \sum_i \sum_k X_{ijk}$, $X_{ij\cdot} = \sum_k X_{ijk}$, $C = \frac{X_{\cdots}^2}{IJK}$

ANOVA Table

Source of Variation	*df*	*SS*	*MS*	*F ratio*	*Table F*
Column factor	$J - 1$	SSA	$\frac{SSA}{J-1}$	$\frac{MSA}{MSE}$	$F_{J-1, IJ(K-1), \alpha}$
Row factor	$I - 1$	SSB	$\frac{SSB}{I-1}$	$\frac{MSB}{MSE}$	$F_{I-1, IJ(K-1), \alpha}$
Interaction	$(I - 1)(J - 1)$	SSAB	$\frac{SSAB}{(I-1)(J-1)}$	$\frac{MSAB}{MSE}$	$F_{(I-1)(J-1), IJ(K-1), \alpha}$
Error	$IJ(K - 1)$	SSE	$\frac{SSE}{IJ(K-1)}$		
Total	$IJK - 1$	SST			

Effects Being Tested	AB, B, and A
Decision Rule	Reject H_0 if MSAB/MSE > Table F; MSB/MSE > Table F; MSA/MSE > Table F
Randomized Block	Design to separate secondary effects from error term
Analysis	Two-factor with $K = 1$, which requires an assumption of no interaction effects

A Posteriori Analysis of Factor Means

Model	Duncan's multiple-range procedure (used after H_0 is rejected for the factor)
Assumptions	Equal sample sizes for the $\bar{X}$'s being compared General ANOVA assumptions
Procedure	Rank the means being compared: H_0: $\mu_{(v)} = \mu_{(w)}$ H_a: $\mu_{(v)} > \mu_{(w)}$
Decision Rule	Reject H_0 if $\bar{X}_{(v)} - \bar{X}_{(w)} >$ LSR, where LSR $= r_{q,n_2,\alpha}\, s_{\bar{X}}$ with $q = v - w + 1$, $n_2 = df_E$, $s_{\bar{X}} = \sqrt{MSE/n}$; n is the sample size for $\bar{X}$; and $r_{q,n_2,\alpha}$ is found in Appendix Table B.12.

Type of Alloy			
1	*2*	*3*	*4*
2.3	5.9	2.0	4.3
2.4	5.3	3.4	5.2
2.2	4.9	2.7	5.9
3.1	4.8	3.8	4.8
2.5	4.6	3.2	5.0
3.5	5.4	3.7	5.5
3.4	4.9	4.5	4.3
2.7	5.9	3.6	5.0

At an α level of .01, can Bradley conclude that there is equality of mean wear among the alloys?

9.16 In testing the drying time of a commercial wax, an industrial user applied the wax to 4 different surfaces. The drying times to the nearest 10th of a minute are as follows.

Type of Floor Surface			
Wood	*Vinyl*	*Linoleum*	*Slate*
3.6	4.8	4.2	3.6
2.4	3.2	3.6	3.3
4.2	2.9	3.6	4.8
3.7	4.2	2.9	2.2
2.8	2.5	3.4	2.7
3.4	3.0	2.5	
4.7			

(a) At an .05 level of significance, can it be concluded that mean drying times for the wax differs by floor surface?

(b) Why do you think there is such a range in the values for each of the floor surfaces?

9.17 An independent testing bureau has conducted an experiment to see if tread wear for a particular brand of tire would be the same for a variety of road surfaces. Thirty Grandway tires were randomly assigned to 5 different road surfaces. To assure uniformity of wear, the bureau placed all tires being tested on the right rear wheel of comparably weighted cars, and the cars were driven at an average speed of 55 mph. All efforts were made to drive only during nonrainy days. The data, in millimeters of wear, are shown in the following table.

Type of Road Surface				
Macadam	*Concrete*	*Composite*	*Brick*	*Gravel*
10.4	8.6	10.3	11.5	13.8
10.2	9.4	10.1	11.2	12.7
9.9	8.1	9.9	11.8	13.9
9.6	9.3	9.9	12.6	13.9
10.1	8.9	10.8	11.9	12.6
9.3	8.2	10.6	13.2	13.6

(a) For $\alpha = .05$, what can the testing bureau conclude regarding differences in mean tire wear with respect to road surface?

(b) How would the surfaces be grouped if the means are compared?

9.18 The absorbency quality has long been one of the selling points of paper towels. An experiment to test this quality is conducted with 5 brands of interest. Ten sheets are stapled together, weighed, dipped in a tank of water, and weighed again. The differences in weight were recorded to the nearest gram, and the data are shown in the following table for several trials for each brand.

Type of Paper Towel

Sop-Up	*Logos*	*Elegant*	*Bountiful*	*Cleanse*
22.4	20.8	24.2	26.0	21.9
26.0	20.0	26.1	26.1	24.8
26.3	20.9	25.1	29.2	21.6
27.8	21.2	25.4	28.2	22.4
	21.7	24.9	26.8	21.4
	21.4			23.6

At an α level of .01, can we conclude that differences exist among the 5 brands of paper towels with respect to mean absorbency?

9.19 Sixty students in a basic business course were divided into 6 homogeneous groups according to their scores on the mathematical section of the Scholastic Aptitude Test. The students were then administered a remedial quiz worth 60 points. Significant differences between groups were apparent, $F = 72.096$, when an analysis of variance was performed. The ANOVA table follows.

Source of Variability	*df*	*SS*	*MS*
Between groups	5	4293.28	858.66
Error	54	643.17	11.91
Total	59	4936.45	

At an .05 level of significance, where do significant differences lie if the $\bar{x}_j$ values are 21.8, 24.4, 32.2, 37.8, 38.3, and 46.4, respectively?

9.20 Refer to the conclusions of Exercise 9.15. What comment can be made about the different alloys with regard to differences in mean wear of the pistons?

9.21 Through analysis of variance, a subordinate has found that significant differences exist in the performance of the 4 clerks that handle incoming mail invoices for your firm. The clerks had been observed for comparable time periods, and the total number of invoices disposed of by each clerk was recorded. Each invoice requires from 3 minutes to 8 minutes of work. The ANOVA table follows.

Source of Variability	*df*	*SS*	*MS*
Between clerks	3	50.58	16.860
Error	20	34.30	1.715
Total	23	84.88	

The mean number of invoices to the nearest 10th for the 4 clerks were 13.6, 15.4, 14.8, and 17.6, respectively. The subordinate feels that the poorest clerk ought to be dismissed or transferred.

(a) Using $\alpha = .05$, what can you advise him?
(b) Evaluate the underlying assumptions.

9.22 Argo Farm Products is attempting to see if the schedule of pesticide dusting indicates significant differences in wheat yields. One brand of wheat was selected and randomly planted on 5 different plots of ground. All plots were planted in the fall and left to germinate during the winter. All plots were dusted initially in the spring. After that, the schedule varied and is indicated in the following table, together with the average yield in bushels per half acre.

Schedule of Dusting

No Dusting after Initial	*Dusting after Rain*	*Dusting Every 2 Weeks*	*Dusting Monthly*
P_2 22.3	P_1 37.5	P_5 29.0	P_3 24.1
P_4 21.0	P_5 34.2	P_3 27.6	P_2 25.2
P_5 24.3	P_2 36.2	P_4 28.9	P_1 30.6
P_3 23.1	P_4 30.9	P_1 32.5	P_5 27.6
P_1 26.3	P_3 29.9	P_2 30.4	P_4 24.3

At a .05 level of significance, can Argo Farm Products conclude that the schedule of dusting affects the mean yield? Argo also suspected that the plots of ground may have affected the yields.

(a) Block the data for plots and retest at the .01 level of significance for differences in mean yield for dusting schedules.
(b) Also test types of plots.
(c) Where differences exist, evaluate the extent of these differences.

9.23 The advertising manager of a newspaper wishes to determine whether differences exist in the mean numbers of specials offered by the grocery chains in their weekly advertisements in his paper. He will use this information to devise a plan to interest some chains in increasing their weekly space requests. Over a period of 4 weeks, he has collected the following sample data.

	Chain Store			
Week	***Aster***	***Daisey***	***Giant***	***Master***
1	20	52	36	26
2	26	44	40	20
3	32	56	36	32
4	24	36	40	20

Desiring to guard against a Type I error, the manager has established an α level of .01.

(a) What conclusions can he make?
(b) Evaluate the assumptions.

9.24 Century Sales, Inc. analyzed the company absence records for the past year to see if differences existed between men and women and between the days of the week.

(a) Using the data shown in the table, what conclusions can you reach at a .05 level of significance? Values in the table are the average absences per day per employee.

(b) Evaluate all underlying assumptions. Is ANOVA necessary here?

	Day of the Week				
Sex	***Monday***	***Tuesday***	***Wednesday***	***Thursday***	***Friday***
Men	.036	.036	.012	.040	.060
Women	.064	.044	.028	.064	.100

9.25 Three persons are selected from a speed-reading class to see if the number of words that are read vary significantly by the type style that is used. Material was randomly assigned to readers for a prescribed time to block the natural variability that exists between readers. The data in the table are arranged for determining whether differences exist between styles of type as well as the readers that participated. The data represent the average number of words read per minute, to the nearest whole word. Use an α level of .05 to do the analysis. Evaluate all assumptions that are required.

	Style of Type			
Reader	***Courier***	***Italic***	***Script***	***Elite***
1	576	453	538	502
2	565	461	517	524
3	597	497	590	546

9.26 The marketing research department of a household goods manufacturer has gathered data on the sales of its shampoo together with data for 3 of its competitors. At the same time, sales were recorded by the type of container in which the shampoo was packaged. Sales data were gathered at several different stores to allow for the vagaries of consumers. To determine whether a particular shampoo sold better or worse with respect to the container, the market research department collected the data at two different times during the month the study was conducted. All recorded data were for the regular 6-ounce size container only.

	Type of Shampoo			
Container	***Glisten***	***Sheen***	***Glint***	***Dureal***
Tube	62 58	54 58	77 83	62 67
Plastic	84 79	61 56	74 79	59 56
Glass	28 32	18 24	51 47	31 35

(a) At a .05 level of significance, test for all effects.

(b) Determine where specific differences exist.

9.27 An auto dealer was curious about the effect of weather on the sale of cars as well as on the colors requested. Since a variety of colors is available these days, the dealer divided the colors into 4 categories: the light pastels, the glitter group, the dark pastels, and the family conservative group. The data gathered represent vehicles sold during a 4-hour period with a particular weather condition.

	Type of Weather		
	Sunny	*Overcast*	*Rainy*
Light Pastel	2	3	2
	1	4	3
Glitter	3	2	3
	4	2	2
Dark Pastel	3	2	0
	2	3	2
Family Conservative	4	2	2
	3	4	2

(a) At a .01 level of significance what can you tell the curious auto dealer?
(b) Which of the assumptions enumerated in Section 9.4 is likely to be violated?

9.28 In an attempt to evaluate the performance characteristics of 4 brands of calculators, a random sample of 18 students in a quantitative class was selected. All the calculators perform the same functions but, due to design, the operations differ. Each person performed an assigned task, and the performance time, in minutes, is shown in the table. At an α level of .01, can you conclude that the brands differ in mean performance times?

Calculator			
Brand 1	***Brand 2***	***Brand 3***	***Brand 4***
2.8	2.1	2.0	2.9
2.6	2.0	1.4	2.5
3.4	2.4	1.0	2.7
3.0	2.7	1.2	2.2
	2.2	1.4	

9.29 An experiment was conducted to determine if the average weight of the required textbooks used during a semester differed by the curriculum in which a student was enrolled or by the class level achieved. The data shown are to the nearest 10th of a pound. What can you conclude at the .05 level of significance?

	Curriculum				
Grade Level	***Business***	***Liberal Arts***	***English***	***Science***	***Total***
Sophomore	6.4	6.7	8.8	7.6	29.5
Junior	7.1	7.1	9.2	8.2	31.6
Senior	5.7	6.9	8.3	8.0	28.9
Total	19.2	20.7	26.3	23.8	90.0

9.30 Agro, Inc., a fertilizer manufacturer, has been testing different combinations of nitrates, phosphates, and carbonates in an attempt to increase the mean bushels of wheat produced per acre on common plots of ground. The ANOVA table for a recent experiment is shown.

Source of Variability	*df*	*SS*	*MS*
Between fertilizers	4	137.34	34.335
Error	20	64.50	3.225
Total	24	201.84	

The test for equality of means is significant at the .01 level. Where do the differences lie if $x_{\cdot A} = 200.5$, $x_{\cdot B} = 218.5$, $x_{\cdot C} = 221.5$, $x_{\cdot D} = 192.0$, and $x_{\cdot E} = 198$? Assume equal sample sizes.

9.31 The writing ability of college graduates is of some concern to prospective employers. An experiment is conducted to determine whether the grade level of students has a significant effect on common writing errors. Five students are randomly selected from each class level and are asked to write a 500-word essay. Each essay was then evaluated, and the number of each type of error was recorded.

	Type of Error			
Class Level	*Comma Fault*	*Spelling Error*	*Sentence Fragments*	*Subject–Verb Agreement*
	2	5	0	1
	5	13	2	3
Freshman	2	7	1	1
	4	11	2	2
	3	4	1	1
	2	7	3	1
	2	4	1	2
Sophomore	4	8	3	1
	3	6	2	1
	1	0	1	0
	1	2	1	0
	2	0	0	1
Junior	3	6	2	1
	0	2	1	0
	2	4	2	1
	1	2	1	1
	1	0	1	0
Senior	2	4	1	1
	0	2	0	1
	2	3	1	0
	1	0	1	0
	0	4	2	1
Graduate	1	3	1	0
	2	1	0	1
	1	0	0	1

(a) At the .05 level of significance, set up the analysis of variance table and test for differences in the type of error, in the class level, and for interaction.

(b) Evaluate all assumptions. What effect does the evaluation have on your conclusions?

9.32 **(a)** Based on the hypothesis tests performed in Exercise 9.31, determine where differences exist within factors, using Duncan's multiple-range test.

(b) Based on your evaluation in Exercise 9.31, do you feel this is appropriate?

9.33 The Puffy Popcorn Company has popcorn wagons in 3 area shopping malls. The company has kept statistics and knows that daily popcorn sales, in pounds, are normally distributed and that the standard deviation is the same at all 3 malls. It is interested in knowing if the average daily sales differ at the 3 malls, so it takes a random sample of days from each mall. Using $\alpha = .05$, what do you conclude?

Mall		
Valley	***Devil's Dale***	***Flat Acres***
2.3	3.3	4.4
2.2	3.4	5.2
4.5	6.3	5.4
3.8	3.0	4.9
2.8	4.8	5.1
3.1	3.7	
	4.1	

9.34 The Puffy Popcorn Company decided to experiment with the quantity of salt and melted butter used on its popcorn. Trying 4 different quantities of salt and 3 of butter, the company randomly selected popcorn lovers and permitted them to eat all they wanted of their assigned combination. The company then measured how much they consumed in ounces. Let $\alpha = .01$.

(a) Does it appear that interaction exists between the butter and salt?

(b) Are there significant differences with respect to the quantities of salt? Of butter?

	Salt			
Butter	***1***	***2***	***3***	***4***
1	2.4	4.0	5.1	4.8
	2.0	4.6	5.5	4.2
2	3.0	6.1	6.5	5.1
	3.1	5.8	6.0	4.8
3	3.2	5.4	5.6	4.2
	3.5	5.9	5.4	4.0

9.35 A grocery chain purchases eggs from many sources. As a result, the eggs arrive in different containers. Breakage is a problem, and the chain wants to know if the container is a contributor. It also wants to know if the size of the egg makes a difference in the average number of eggs that are broken. Shipments of equal size are checked, and the results are shown in the following table.

Size	*Type of Carton*		
	Styrofoam	***Pressed Paper***	***Plastic***
Small	14	21	23
Medium	13	24	26
Large	21	32	35
X-Large	26	38	41

At $\alpha = .05$, test for any differences and evaluate all assumptions.

9.36 A firm has 2 sources of raw material and 2 machines of different makes to produce a new product. Three observations were obtained for each variable combination, and the results are given in the table. Use a significance level of .05 and analyze completely. The values represent output per shift in 100 pounds.

Machine	*Raw Material*	
	A	***B***
X	7	7
	6	5
	8	9
Y	4	3
	7	3
	4	3

9.37 Information was gathered at the Monmouth Manufacturing Company concerning the number of good parts produced on each of their 4 production lines using 3 grades of incoming materials. The vice-president of manufacturing has stated that material of grade C used in production line 3 results in a significantly greater average number of good parts than material of grade A in production line 3.

(a) Does the manufacturing vice-president have any basis for making this statement? Let $\alpha = .05$.

(b) What effect does an evaluation of the assumptions have on your conclusions?

Production Line	*Material Grade*		
	A	***B***	***C***
1	72	66	80
	76	72	84
2	71	69	97
	75	73	95
3	81	75	100
	83	75	96
4	70	64	95
	72	66	89

9.38 Based on the results determined in Exercise 9.37, determine where differences exist within factors, using Duncan's multiple-range test. Do you think this is appropriate?

9.39 A large corporation recently collected the following information in a study of computer maintenance.

Time	*Computer Number* 1	2	3
8:00–10:00 A.M.	2	1	3
10:01–12:00 A.M.	4	7	1
12:01–2:00 P.M.	2	8	9
2:01–4:00 P.M.	4	8	3
4:01–6:00 P.M.	4	6	5

The tabular values represent the number of repairs required per month over a 1-year period.

(a) At a 5% significance level, are there significant differences among computers and/or the time in which repairs were required?

(b) Evaluate all assumptions.

9.40 A pharmaceutical manufacturer is testing the merits of 4 different diet pills. The following data are the numbers of pounds lost using each type of pill in a 3-month period.

Pill							*Total*
1	15	17	14	11			57
2	15	10	13	17	14		69
3	14	9	7	10	8	7	55
4	10	14	13	13	12		62

(a) Using a significance level of .01, can we conclude there are no differences in the mean numbers of pounds lost with respect to the type of pill?

(b) Comment on the experimental design used here.

9.41 A consumer testing service has conducted 2 complete series of similar tests to determine the life, in coded terms, of television picture tubes from 3 manufacturers, using 2 different sizes of televisions.

Source of Variability	*df*	*SS*	*MS*
Between manufacturers			76
Between sizes		48	
Interaction			
Error			
Total		250	

$$\sum_i \sum_j \sum_k x_{ijk}^2 = 2950 \qquad \sum_i \sum_j x_{ij\cdot}^2 = 5812$$

(a) Complete the table.

(b) Does interaction exist between manufacturer and size at the .05 level?

9.42 A tire manufacturer subjected 3 sets of 3 tires each to wear tests in a laboratory. The 3 sets were constructed of experimental material A, experimental material B, and conventional material C, respectively. The results in thousands of miles of wear are as follows.

Material		
A	***B***	***C***
34	50	35
34	49	42
40	54	34

At an α level of .01, is the mean wear the same for all tires?

9.43 The accuracy of calipers to measure the thickness of an item may depend on the individual using them and on the manufacturer. The quality control department of McCormic Brothers conducted an experiment to see if differences do indeed exist. Four inspectors were chosen and asked to make a measurement of the diameter of a rod, using 3 different calipers.

	Inspector			
Caliper	***1***	***2***	***3***	***4***
A	1.42	1.43	1.41	1.42
B	1.46	1.48	1.46	1.48
C	1.43	1.45	1.42	1.43

At a .01 level of significance, what can you conclude?

9.44 A covering to prevent the absorption of water by wood placed in the ground has been developed by Conway Corporation. The company judges that the extent of the absorption may depend on the type of wood. An experiment was conducted on 3 different kinds of wood. The figures in the data table represent grams of water absorbed.

Type of Wood		
Soft	***Medium***	***Hard***
9.12	6.43	5.01
8.87	6.87	4.82
8.90	6.51	4.96
9.16	6.38	4.87

At α = .01, what can you conclude?

9.45 A major manufacturer of breakfast cereals marketed an established product in 3 different package colors. The test market results for 4 states are given. At a 5% significance level, are there differences among the test market results with respect to colors, to states, or to significant interaction between states and colors? The values represent cases of cereal sold in 100's.

State	Package Color		
	Red	*Yellow*	*Blue*
	64	72	74
California	66	81	51
	70	64	65
	65	57	47
Ohio	63	43	58
	58	52	67
	59	66	58
New York	68	71	39
	65	59	42
	58	57	53
Texas	41	61	59
	46	53	38

9.46 Based on the hypothesis tests in Exercise 9.45, identify differences within factors, using Duncan's multiple-range test.

9.47 The Wonder Company desires to know if the absentee rate varies over the number of years of service and also whether there is a difference between male and female employees. Employee records were randomly selected, and the days absent were noted. Data gathered on 2 separate occasions are shown in the table.

Sex	Years of Service			
	0–5	*6–15*	*16–30*	*Over 30*
Male	9	8	4	9
	12	10	6	7
Female	15	10	8	14
	14	12	12	10

The data are the average number of absences rounded to the nearest whole number.

(a) At an α level of .05, what can you tell Wonder Company about their absences?

(b) Identify possible problems with this experimental design and evaluate the ANOVA assumptions.

9.48 In a test to determine whether family income has an effect on clothing expenditures for a year, the following analysis of variance table resulted. All families were of the same size.

Source of Variability	*df*	*SS*	*MS*
Between incomes	3	86,796	28,932
Error	12	31,200	2,600
Total	15	117,996	

The null hypothesis of equality of expenditures is rejected at $\alpha = .01$. Assume equal sample sizes.

(a) If the mean expenditures per person for the 4 categories, low to high, were 124, 168, 240, and 318, respectively, what else can be concluded?

(b) What other factors do you see that might contribute to the amount of money spent annually on clothing?

9.49 An experiment is designed to test for any differences that may exist among examinations that cover a common set of material. The experimenter anticipates that the grade on the examination (0–100) will represent the data for the experiment, and she intends to use analysis of variance to analyze that data. What problems do you see in using this procedure, and what factors do you think should be controlled?

9.50 Basically, this chapter has limited itself to designs of no more than 2 factors, although more sophisticated designs are discussed in Section 9.8.

(a) In keeping with this section, try designing an experiment that will enable a researcher to evaluate treadwear when the following factors are considered, with the numbers of levels in parentheses: weight of the vehicle (2), speed of the car (3), air pressure of the tire (3), and outside temperature (2).

(b) What types of effects are involved here?

(c) If replication is assumed, how many hypothesis tests are involved?

***9.51** It is stated in the text that

$$\Sigma\,(X_{ij} - \overline{X}_{\cdot j})(\overline{X}_{\cdot j} - \overline{X}_{\cdot\cdot}) = 0$$

Use algebra to prove this.

9.52 (a) Code the data for Exercise 9.44 by multiplying each entry by 100 and then subtracting 650 from each entry.

(b) Do an ANOVA on the resulting data and compare the table to the original one.

REFERENCES

1. Guenther, W. C. *Analysis of Variance*. Englewood Cliffs, NJ: Prentice-Hall, 1964.
2. Meek, Gary, and Stephen Turner. *Statistical Analysis for Business Decisions*. Boston: Houghton-Mifflin, 1983.
3. Neter, John, William Wasserman, and Michael Kutner. *Applied Linear Statistical Models*, 2nd ed. Homewood, IL: Richard D. Irwin, 1985.
4. Winer, B. J. *Statistical Principles in Experimental Design*. New York: McGraw-Hill, 1962.

* An asterisk indicates a higher level of difficulty.

10 Simple Regression and Correlation Models

10.1 INTRODUCTION

June Washington has just been hired by Northcentral Railroad as an accountant. The first day on the job she is informed that railroads have considerable difficulty in separating their total costs (TC) into fixed-cost (FC) and variable-cost (VC) components. June is given, as her first task, the job of finding a fast, efficient method of estimating these two costs.

First, she reasons that a relationship exists between output and total cost and that this relationship is a straight line. Second, she knows that for zero output, the total cost must equal the fixed cost. Therefore, the total cost equation must be of the form $TC = FC + VC \times Output$. If she fits a straight line to the plotted data of total cost versus output, the intercept can be used as an estimate of the fixed costs and the slope can be used as an estimate of the variable cost. She decides to use a regression procedure called least squares estimation to fit a straight line to the data.

Like June Washington, other analysts often need to know whether a useful relationship exists between two or more variables. For instance, a market researcher is able to predict sales of a product for a specific amount of advertising if it is found that a strong relationship exists between these two variables.

The methods of regression provide us with procedures for estimating a specified relationship between two variables and for evaluating the significance of the resulting equation. In addition, correlation methods are used to evaluate the strength of the linear relationship between random variables. Once it has been determined that a strong relationship exists, the estimating equation can be used to improve our predictions and to aid in understanding the nature of the relationship between the variables.

The chapter begins with the subject of regression, which includes choosing the correct model, fitting equations to the data, testing the equations for significance, and using the equations for prediction. We then consider the subject of correlation for use in finding the strength of the relationship.

10.2 CHOOSING A MODEL

If we suspect that a relationship exists between two variables, we must then attempt to identify the nature of the relationship—that is, the function that relates these two variables. The relationship is expressed in the form of a mathematical model. Choosing a proper model starts with the decision as to which is the dependent variable and which is the independent variable.

DEFINITION 10.1 The **dependent variable** is the variable to be predicted.

DEFINITION 10.2 The **independent variable** is the predicting variable.

If the amount of advertising is used to predict the amount of sales, then the amount of advertising is the independent variable and the amount of sales is the dependent variable. The dependent variable is indicated with a capital Y because it is a random variable, and the independent variable is indicated with a lowercase x because it is not a random variable.

The next step in choosing a proper model is to prepare a scatterplot.

DEFINITION 10.3 A **scatterplot** is a graph composed of all the (x, Y) points in the population or sample.

A point is plotted for each pair of x and Y values. Figure 10.1 illustrates a number of scatterplots.

To try to determine the equation that best represents each of these plotted relationships, we first draw the curve that appears to give the best fit for each set of data, and then we attempt to determine the equation of that curve. For the scatterplots in Figure 10.1, it appears that the following equations can be used to estimate the relationships indicated by the sets of points.

a. A straight line: $Y = a + bx$.
b. A parabola: $Y = a + bx + cx^2$.
c. An exponential: $Y = ab^x$.
d. No relationship appears to exist.

FIGURE 10.1 Typical scatterplots

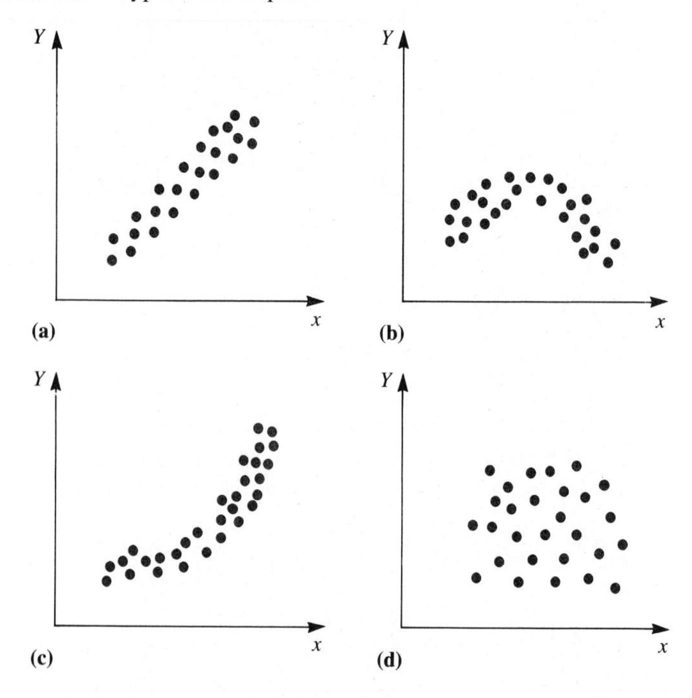

Once we have chosen a form for the curve, the next problem is to estimate the values of the coefficients. For a straight line of the form $Y = a + bx$, we know that a is the value of the Y intercept, or the point at which the line passes through the Y-axis. The b value is the slope of the line and represents the change in Y divided by the change in x ($\Delta Y/\Delta x$). The symbol Δ represents the change in the variable. Determination of the slope by this method is illustrated in the following example.

EXAMPLE 10.1 Determine the equation for the straight line on the scatterplot in Figure 10.2.

Solution:

$$a = Y \text{ intercept} = 1.3$$

$$b = \text{slope} = \frac{\Delta Y}{\Delta x} = \frac{2.5}{4} = .625$$

Therefore, the equation of the line is

$$Y = 1.3 + .625x$$

FIGURE 10.2 Example scatterplot

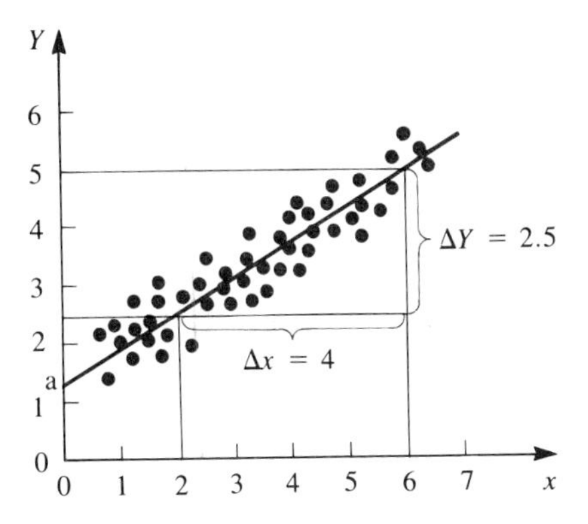

This procedure provides us with a rough estimate of the equation of the relationship. For more precision in this problem and for nonlinear curves, we must resort to more complicated mathematical methods to obtain values for the coefficients.

In many business situations, the relationship that we attempt to model can be approximated by a straight line. At first, this statement would seem suspect, for a large portion of economic and business theory assumes nonlinear relationships. For example, many theories in economics and functional business areas make use of the growth curve, illustrated in Figure 10.3. This curve can be approximated by a cubic equation of the form $Y = a + bx + cx^2 + dx^3$. But, notice that for $a \leq x \leq b$, the curve is approximately linear. In the range a to b, this function could usefully be approximated with a straight line without excessive error and be of use in estimating growth. Since most firms would be operat-

FIGURE 10.3 Typical growth curve

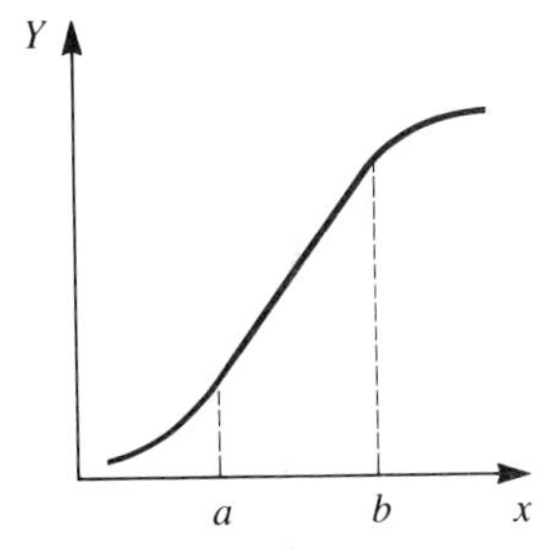

ing in this portion of the curve, the straight line is usually a good choice for a model if the scatterplot appears to be an approximately linear pattern.

The linear model for the population is expressed as

$$Y_i = \alpha + \beta x_i + \varepsilon_i \tag{10.1}$$

where the ε_i represents the vertical distance and direction from an observation to the line for a given value of x. The quantity ε_i is known as the random error term. Figure 10.4 illustrates this model.

For x equal to x_2, the value of Y at the line is found by evaluating the equation of the line at x_2 to obtain $\alpha + \beta x_2$. The actual value of Y at x_2 is y_2. The difference between the two points is ε_2. Rewriting Equation 10.1, for this particular value of x, we have

$$y_2 = \alpha + \beta x_2 + \varepsilon_2$$

Notice that for x_1 and y_1, the error term is negative.

Obviously, if we want to find the value of Y that occurs with a value of x, we must also know how much error exists. If we did not know this, and we treated it as having the value of zero, then the value we would obtain from the equation would not be Y, but rather the value of the point on the line for the given x (such as the points on the line in Figure 10.4). To represent these points on the population line, we use the symbol $\mu_{Y|x}$. The reason for this symbol comes from the concept of expected values. The error term ε is assumed to be a random variable with a mean of zero. Hence, if

$$Y_i = \alpha + \beta x_i + \varepsilon_i$$

then

$$E[Y|x_i] = \alpha + \beta x_i$$

FIGURE 10.4 Illustration of error term

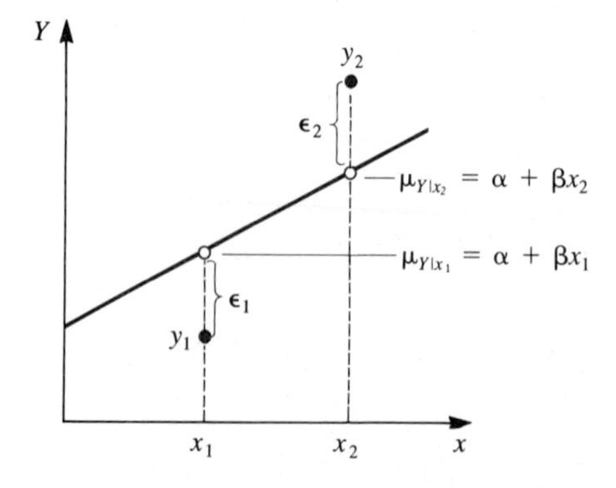

Recall that the expected value of a variable is the mean of that variable. In this case, the expected value is conditional on the value being used for x_i, as illustrated in Figure 10.4; therefore

$$E[Y|x_i] = \mu_{Y|x_i}$$

Finally, we can write

$$\mu_{Y|x_i} = \alpha + \beta x_i \qquad \textbf{(10.2)}$$

which is an alternative way of stating the model of Equation 10.1. Equation 10.2 will be referred to as the equation of the line that best fits the population data or, in short, the population line.

EXERCISES

10.1 The Bigbloom Fertilizer Company produces a fertilizer that increases flower size. The ingredient GRF-7 is the chemical that promotes the bloom size. For experimental reasons, the company tried various amounts of GRF-7 per bottle and observed flower size on treated test subjects. Plot the data and state the type of model that would appear to best describe the relationship.

Amount of GRF-7 (.01 ounce)	*Average Flower Size (millimeters)*
1	10
3	32
2	18
6	61
5	48
4	44

10.2 The population of Cleveland, Ohio, is given in the following table for the years 1820 to 1970 in increments of 10 years. If you were to use a model to predict the present and future population (even though extrapolation is error prone), which equation would appear to be most appropriate? Plot the data first.

Year	*Population*	*Year*	*Population*
1820	606	1900	381,768
1830	1,076	1910	560,663
1840	6,071	1920	796,841
1850	17,034	1930	900,429
1860	43,417	1940	878,336
1870	92,829	1950	914,808
1880	160,146	1960	876,050
1890	261,353	1970	738,956

10.3 Plot the points for Exercise 10.1, draw a straight line through the points that would appear to have the best fit, and estimate the equation of the line.

10.3 CALCULATING THE LEAST SQUARES EQUATION AND STANDARD ERROR OF THE ESTIMATE

To determine the equation $\mu_{Y|x_i} = \alpha + \beta x_i$, which describes the relationship between two variables for a given population, as shown in Figure 10.5a, we need to know the values of x and Y for every element in the population. In most circumstances, this is impossible, so a sample of elements is taken from the population. A straight line is fitted to the sample points, and the parameters α and β are estimated by the statistics a and b, respectively. Thus, the sample equation for the straight line becomes

$$\hat{Y}_i = a + bx_i \quad \textbf{(10.3)}$$

which is an estimator of the population line and is shown in Figure 10.5b.

DEFINITION 10.4 **Linear regression** analysis is the process of quantifying the relationship between a dependent variable and independent variable for the purpose of predicting values for the dependent variable for a specified value of the independent variable. It shows how the independent variable relates to the dependent variable.

Once we have a sample of n observations, we are ready to begin the regression process. Our task is to find the straight line $\hat{Y}_i = a + bx_i$ that best fits the set of points.

DEFINITION 10.5 The best-fitting line is the one that satisfies the **least squares criterion**—that is, the line that minimizes the sum of the squared residuals. The **residual** is the distance of the observed Y_i from the line.

FIGURE 10.5 Equations for population and sample lines

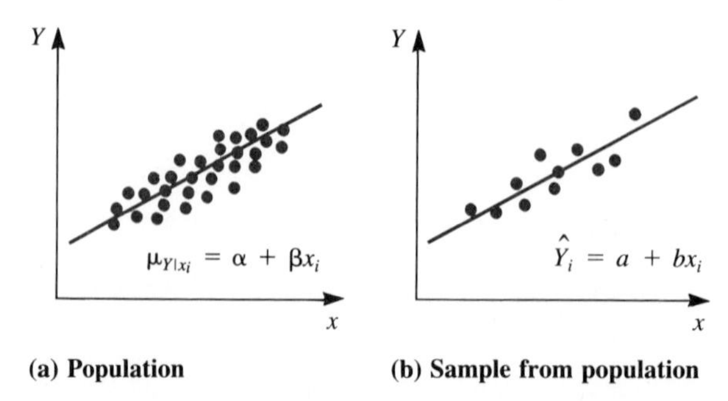

Although other estimation procedures exist, they are generally used for specialized situations. Least squares will be used for regression estimation in this text.

If we were to visually fit a straight line through a set of points and determine its equation, as we did in Example 10.1, we could then estimate the population error term for each point (x_i, Y_i), using the residual e_i, where

$$e_i = Y_i - \hat{Y}_i = Y_i - (a + bx_i) = Y_i - a - bx_i$$

The symbol e_i serves the same function for sample data as ε_i serves for population data, and it is an estimate of ε_i. We then could square the error terms as shown in Equation 10.4.

$$Q = \sum e_i^2 = \sum (Y_i - a - bx_i)^2 \qquad \textbf{(10.4)}$$

Each different line that could be visually fitted to the data would have a different Q. We want the line that gives the smallest possible sum.

$$\min Q = \min \sum e_i^2 = \min \sum (Y_i - a - bx_i)^2$$

Applying partial differential calculus to Equation 10.4 to solve for the values of a and b for the line that minimizes Q, we obtain what are known as the normal equations.[1]

$$an + b\sum x_i = \sum Y_i$$

$$a\sum x_i + b\sum x_i^2 = \sum x_i Y_i \qquad \textbf{(10.5)}$$

Solving for a and b simultaneously gives the equations that we use to calculate the regression line.

$$b = \frac{\sum x_i Y_i - \dfrac{\sum x_i \sum Y_i}{n}}{\sum x_i^2 - \dfrac{(\sum x_i)^2}{n}} \qquad \textbf{(10.6a)}$$

$$a = \frac{\sum Y_i}{n} - b\frac{\sum x_i}{n} \qquad \textbf{(10.6b)}$$

The symbol n represents the number of pairs of observations in the sample.

EXAMPLE 10.2 A certain tire dealer tends to buy as many inches of advertising in the newspaper as the mood strikes him. He is now beginning to think that the

[1] See the appendix at the end of this chapter for the derivation of the normal equations.

FIGURE 10.6 Scatterplot of tire dealer observations

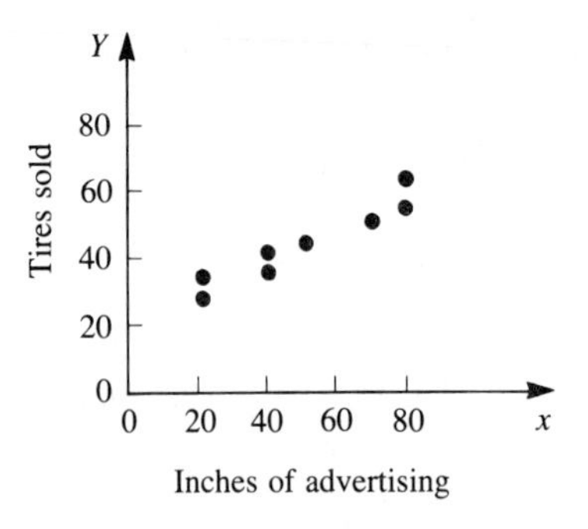

number of tires he has sold in the past is related to the number of inches of advertising he has used. So, he takes a sample of 8 weekly sales reports and compares sales to the number of inches of advertising he used that week. Does it appear that a relationship between these 2 variables might exist?

Inches of Advertising	20	40	50	80	40	20	70	80
Number of Tires Sold	28	38	44	58	42	33	52	65

Solution: We can tentatively decide if a relationship exists by looking at a scatterplot of the data, as shown in Figure 10.6. It is obvious that the tire dealer can control one of the variables, namely, the amount of advertising. Thus, advertising is the independent variable and is plotted on the x-axis. The tire dealer believes that tire sales are related to the amount of advertising. Thus, the variable, tire sales, is the dependent variable and is plotted on the Y-axis. It appears from the scatterplot that a relationship exists and that a straight line may best approximate this relationship.

EXAMPLE 10.3 Using linear regression, find the equation of the straight line that best fits the data points of Example 10.2.

Solution: The original data and the computations to obtain the required summations for use in the equations are given in Table 10.1a. Table 10.1b shows the results using MINITAB. The value for $\sum Y_i^2$ is not needed at this time but will be used in later computations.

TABLE 10.1(a) Calculation of the summations for the data of Example 10.2

Observation i	*Inches of Advertising* x_i	*Number of Tires Sold* y_i	$x_i y_i$	x_i^2	y_i^2
1	20	28	560	400	784
2	40	38	1,520	1,600	1,444
3	50	44	2,200	2,500	1,936
4	80	58	4,640	6,400	3,364
5	40	42	1,680	1,600	1,764
6	20	33	660	400	1,089
7	70	52	3,640	4,900	2,704
8	80	65	5,200	6,400	4,225
	400	360	20,100	24,200	17,310
	Σx_i	ΣY_i	$\Sigma x_i Y_i$	Σx_i^2	ΣY_i^2
	$\bar{x} = 50$	$\bar{y} = 45$			

TABLE 10.1b MINITAB calculations for the data of Example 10.2

```
FILE: OUT10_1   DATA     A    VM/SP CMSL R3.1 6/12/85 PUT8502


1
 MTB > READ DATA INTO C1-C2
       8 ROWS READ
 MTB > END OF DATA
 MTB > NAME FOR C1 IS 'ADVERT', C2 IS 'SALES'
 MTB > BRIEF 2
 MTB > REGRESS C2 ON 1 PREDICTOR C1

 THE REGRESSION EQUATION IS
 SALES = 20.0 + 0.500 ADVERT

                                  ST. DEV.     T-RATIO =
 COLUMN       COEFFICIENT         OF COEF.     COEF/S.D.
                  20.000            2.684          7.45
 ADVERT          0.50000          0.04879         10.25

 S = 3.162

 R-SQUARED = 94.6 PERCENT
 R-SQUARED = 93.7 PERCENT, ADJUSTED FOR D.F.

 ANALYSIS OF VARIANCE

  DUE TO       DF          SS        MS=SS/DF
 REGRESSION     1      1050.0         1050.0
 RESIDUAL       6        60.0           10.0
 TOTAL          7      1110.0

 MTB > STOP
```

The values for a and b are obtained using Equations 10.6a and 10.6b.

$$b = \frac{\sum x_i Y_i - \dfrac{\sum x_i \sum Y_i}{n}}{\sum x_i^2 - \dfrac{(\sum x_i)^2}{n}}$$

$$= \frac{20{,}100 - \dfrac{(400)(360)}{8}}{24{,}200 - \dfrac{(400)^2}{8}}$$

$$= \frac{2100}{4200} = .5$$

$$a = \frac{\sum Y_i}{n} - b\frac{\sum x_i}{n} = \frac{360}{8} - .5\frac{400}{8}$$

$$= 45 - .5(50)$$

$$= 45 - 25 = 20$$

The linear regression equation is

$$\hat{Y}_i = a + bx_i = 20 + .5x_i$$

This equation can easily be drawn on the scatterplot by remembering that it takes only 2 points to determine a straight line. We can use the Y intercept, 20, as one of the points and substitute any value of x into the equation to get another point, say 100.

$$\hat{Y} = 20 + .5(100) = 70$$

We could also use the $\bar{x}$, $\bar{y}$ point (50, 45) since it also falls on the line and is within the range of the data, as shown in Figure 10.7.

FIGURE 10.7 Regression line and scatterplot of data for Example 10.3

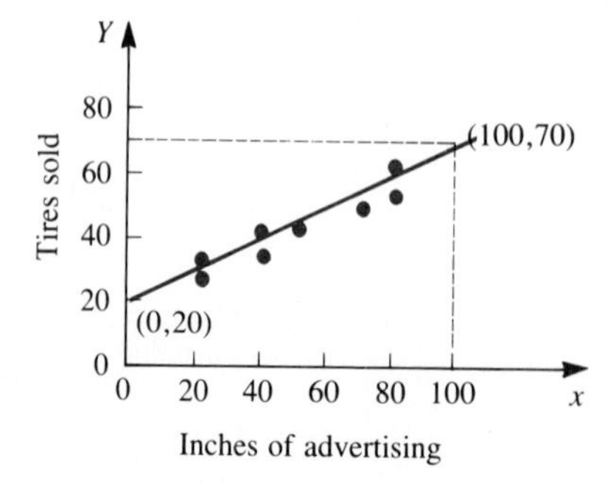

EXAMPLE 10.4 Refer to Example 10.3 and estimate the number of tires the dealer can expect to sell if he does not advertise.

Solution: Failure to advertise would indicate that $x = 0$, in which case, the variable term drops out of the equation, leaving only the constant.

$$\hat{Y} = 20 + .5(0) = 20$$

He would expect to sell 20 tires per week if he did not advertise. But, because $x = 0$ is outside the range of the observed data, he would consider this estimate with caution. Since the data are only for the range 20 to 80, the relationship is assumed to be linear only over that range. He does not know the relationship outside that range, so any extrapolation is risky—that is, subject to an increase in the chance of specific errors.

It is desirable to be able to interpret the meaning of the coefficients a and b. Generally speaking, the value of b indicates the amount $\hat{Y}_i$ changes for each 1 unit change in x_i. The value of a indicates the value of $\hat{Y}_i$ if x_i is set at 0. Specific meanings depend on the context. In the introductory section of the chapter, for example, the value for b represented variable costs and the value for a represented fixed costs.

EXAMPLE 10.5 Refer to the equation developed in Example 10.3 and estimate the number of tires the dealer can expect to sell for each additional inch of advertising.

Solution: If he had been advertising 20 inches, he would expect to be selling

$$\hat{Y} = 20 + .5(20) = 30 \text{ tires}$$

If he now advertised 1 additional inch, he would expect to sell

$$\hat{Y} = 20 + .5(21) = 30.5 \text{ tires}$$

or an increase of .5, which is the value of the slope. Therefore, b is interpreted as the expected increase in tire sales for a 1-inch increase in advertising.

Recall that the purpose of using the least squares method for fitting the line is to minimize the sum of the squared residuals e_i^2—that is, the sum of the squared deviations from the line, $\Sigma(Y_i - \hat{Y}_i)^2$. Using this expression, we can calculate an estimate of the standard deviation, which is also called the standard error of the estimate.

DEFINITION 10.6 In context with respect to regression, the **standard error of the estimate** is the standard deviation of the points about the true regression line $\sigma_{Y|x}$, which is estimated by the square root of the mean square error.

If we do not have access to the entire population to calculate $\sigma_{Y|x}$, we must use sample data to calculate an estimate.

When we use sample data, the sample line $\hat{Y}_i = a + bx_i$ is obtained as an estimator of the population line $\mu_{Y|x_i} = \alpha + \beta x_i$. The residual is then calculated as $e_i = Y_i - \hat{Y}_i$ and used as an estimate of ε_i. To obtain e_i, we must calculate $\hat{Y}_i$, which requires the estimates a and b. Since a degree of freedom is lost for each estimate made and since a and b were found as estimates of α and β, two degrees of freedom are lost in the process. Therefore, the estimator of $\sigma_{Y|x}$ is $s_{Y|x}$ and is calculated as

$$s_{Y|x} = \sqrt{\frac{\Sigma(Y_i - \hat{Y}_i)^2}{n - 2}} \tag{10.7}$$

The numerator is the sum of the squares for the error, Σe_i^2. It is more properly called the sum of the squared residuals. The denominator is the degrees of freedom for the error.

EXAMPLE 10.6 Calculate the standard error of the estimate for the tire problem in Example 10.2.

Solution: First, the residuals must be calculated, squared, and summed. The results are shown in Table 10.2. Recall that the sum of the squared error terms is a minimum for a straight line fitted using the least squares criterion. Therefore, there is no other line that can be fitted to the data for which the sum of the squared error terms would be less than 60. Now we divide by the degrees of freedom for error and take the square root.

$$s_{Y|x} = \sqrt{\frac{\Sigma(Y_i - \hat{Y}_i)^2}{n - 2}}$$

$$= \sqrt{\frac{60}{6}} = \sqrt{10} = 3.162$$

TABLE 10.2 Calculation of the residuals (*e*) for Example 10.6

i	x	y	$a + bx$	$= \hat{y}$	e $y - \hat{y}$	e^2 $(y - \hat{y})^2$	
1	20	28	20 + .5(20)	= 30	−2	4	
2	40	38	20 + .5(40)	= 40	−2	4	
3	50	44	20 + .5(50)	= 45	−1	1	
4	80	58	20 + .5(80)	= 60	−2	4	
5	40	42	20 + .5(40)	= 40	2	4	
6	20	33	20 + .5(20)	= 30	3	9	
7	70	52	20 + .5(70)	= 55	−3	9	
8	80	65	20 + .5(80)	= 60	5	25	
					0	60	Error (residual) Sum of squares

This answer is identified as S on the MINITAB printout in Table 10.1b. Notice that $s_{Y|x}$ increases as the distances of the points from the line become larger.

Equation 10.7 is useful as a definitional form but quite time consuming to use. It is recommended that for calculation purposes, one of the following computational forms of the equation be used.

$$s_{Y|x} = \sqrt{\frac{\sum Y_i^2 - \frac{(\sum Y_i)^2}{n} - b\left(\sum x_i Y_i - \frac{\sum x_i \sum Y_i}{n}\right)}{n-2}} \qquad \textbf{(10.8a)}$$

$$s_{Y|x} = \sqrt{\frac{\sum Y_i^2 - a\sum Y_i - b\sum x_i Y_i}{n-2}} \qquad \textbf{(10.8b)}$$

EXAMPLE 10.7 Calculate the standard error of the estimate for the tire problem in Example 10.2, using Equation 10.8a.

Solution: Using the sums from Table 10.1, we obtain

$$s_{Y|x} = \sqrt{\frac{17{,}310 - \frac{(360)^2}{8} - .5 \times \boxed{2100}}{6}}$$

$$= \sqrt{\frac{17{,}310 - 16{,}200 - 1{,}050}{6}}$$

$$= \sqrt{\frac{60}{6}} = \sqrt{10} = 3.162$$

Notice that the boxed term is the numerator of the equation used for the calculation of b. There is no need to recalculate its value.

EXAMPLE 10.8 Calculate the standard error of the estimate for the tire problem of Example 10.2, using Equation 10.8b.

Solution:

$$s_{Y|x} = \sqrt{\frac{\sum Y_i^2 - a\sum Y_i - b\sum x_i Y_i}{n-2}}$$

$$= \sqrt{\frac{17{,}310 - (20)(360) - (.5)(20{,}100)}{6}}$$

$$= \sqrt{\frac{17{,}310 - 7200 - 10{,}050}{6}}$$

$$= \sqrt{\frac{60}{6}} = \sqrt{10} = 3.162$$

Notice that, except for $\sum Y_i^2$, all of the terms in Equations 10.8a and 10.8b were calculated previously. Thus, the only extensive calculation required at this point is the obtaining of $\sum Y_i^2$.

EXERCISES

10.4 The Wilson Metal Works production supervisor wishes to determine the relationship between the heat applied to the metal used in making files and the hardness resulting from the heat. Assume that a straight line is a reasonable model and determine the equation.

Heat (°C)	100	200	300	400	500
Hardness	2.1	3.2	3.9	4.8	5.1

10.5 Refer to Exercise 10.4. How hard would the production supervisor expect the metal to be if it was heated to 450°C?

10.6 Refer to Exercise 10.4. How much will the hardness increase for each 10° increase in temperature?

10.7 Calculate the standard error of the estimate of the data in Exercise 10.4.

10.4 ASSUMPTIONS FOR INFERENCE

The linear regression model stated in Equation 10.1 is

$$Y_i = \alpha + \beta x_i + \varepsilon_i$$

where x_i is assumed to be a nonrandom variable and ε_i is a random variable with a mean of zero. To obtain distributions for inferential purposes, we need additional assumptions about ε. Before stating them, we comment on the assumptions for x.

1. No measurement error in x: The x values are measured accurately. In practical applications, it is enough that the measurement error is negligible.
2. Nonrandom x: The x values are known constants whose value is either selected and controlled or is readily available. For each x, there is a distribution of possible Y values that occur randomly through the error term ε.

Figure 10.8 illustrates this concept. In many business applications, x is unavoidably random because we may not be able to control those values available for use. Estimation, testing, and prediction results will still hold, however, and only interpretation of the results will be influenced. In this case, the model is actually the correlation model discussed in Section 10.11. The difference between nonrandom and random x is basically the difference between regression models and correlation models. For a more detailed discussion, see Neter, Wasserman, and Kutner [1], pp. 83–84.

FIGURE 10.8 Nonrandom x, random Y

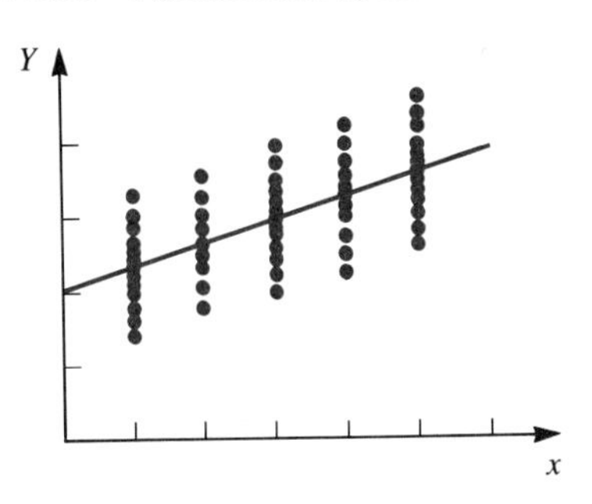

We turn now to the assumptions necessary for making inferences about Y and about the model in general.

1. Normality: The error terms ε_i for each x_i are normally distributed with a mean of 0. This distribution is illustrated in Figure 10.8.

If the distribution of the Y values illustrated in Figure 10.8 represents the population of points, we would expect most of the points to be close to the line, which would be the case in a normal distribution. The distance from each point to the line is ε, and since ε is normally distributed, so is Y because it is a linear function of a normal random variable. Normality directly implies that the Y data must be interval.

2. Homoscedasticity: For each x_i, the variances for the distribution of ε_i and, hence, Y_i are the same—that is, $E(\varepsilon_i^2) = \sigma^2$ for all x_i.
3. Independence: The error (ε) associated with one observation is independent of the errors associated with any other observations and of the x values.

In summary, we have that the random variables denoted by ε_i are each independently and identically distributed as $N(0, \sigma^2)$. With respect to each Y_{x_i}, we state that

$$Y_{x_i} \sim N(\mu_{Y|x_i}, \sigma^2)$$

for each x_i.

The development of the hypothesis tests to be used later is based on these assumptions. Where certain assumptions are violated, transformations are available that will correct for some types of violations. More advanced books should be consulted in these cases. See, for instance, Snedecor and Cochran [2], pp. 290–292.

EXERCISES

10.8 The statistician for the Norka Bulls (NB) professional basketball team is sure that a basketball player's height is related to the number of points scored in a game. From NB's opponents, he randomly selects 10 players, and using the points they scored against NB in their last meeting, as shown below, he calculates the equation

$$\text{Points} = -194.6 + 2.57 \text{ (height in inches)}$$

Height	77	84	78	82	77	80	79	80	79	80
Points	4	22	5	16	5	10	7	11	6	11

He now wants to estimate how many points the 7-foot 3-inch center from NB's next opponent will score. What assumptions does the statistician appear to be violating? How could he have adjusted his data collection to avoid these problems?

10.5 TESTING THE MODEL FOR ADEQUACY

One of the assumptions we make when we fit a straight line to a set of data is that a straight line, rather than a curved line, is the appropriate model. We may be wrong. The following is a method for testing if the wrong model was hypothesized. In particular, the hypotheses of interest are

$$H_0\colon \mu_{Y|x} = \alpha + \beta x \qquad H_a\colon \mu_{Y|x} \neq \alpha + \beta x \tag{10.9}$$

In words, the null hypothesis states that the relationship between x and Y is linear, and the alternative hypothesis states that it is not. Note that under the null hypothesis, β may equal zero, which indicates that the line corresponding to $\mu_{Y|x}$ is parallel to the x-axis and, hence, that Y is independent of the value for x.

Recall from the coverage of analysis of variance in Chapter 9 that the total sum of squares can be partitioned (separated) into two terms: the sum of squares for the treatment and the sum of squares for the error. That is,

$$\begin{array}{ccccc} \Sigma\Sigma(X_{ij} - \overline{X}_{..})^2 & = & \Sigma\Sigma(X_{ij} - \overline{X}_{.j})^2 & + & \Sigma\Sigma(\overline{X}_{.j} - \overline{X}_{..})^2 \\ \text{SS(Total)} & = & \text{SS(Error)} & + & \text{SS(Treatment)} \end{array}$$

In regression problems, the total sum of squares can be broken down similarly. First of all, let us consider the scatterplot in Figure 10.9a. If we do not think that a relationship exists, our best estimate for Y is $\overline{Y}$, no matter what the value of x is. The $\overline{Y}$ line is shown in Figure 10.9a. The deviations of the points from the line are computed by $Y_i - \overline{Y}$, which we will call the total deviation.

If there is some pattern to the scatter of points, a regression line has a better fit than $\overline{Y}$ and also smaller deviations. For a given x, the value of $\hat{Y}$ is calculated from the equation of the line, and the deviation from the line is calculated by $Y_i - \hat{Y}_i$. Figure 10.9b illustrates the regression line and its deviations.

FIGURE 10.9 Partitioning of the deviation $Y_i - \overline{Y}$

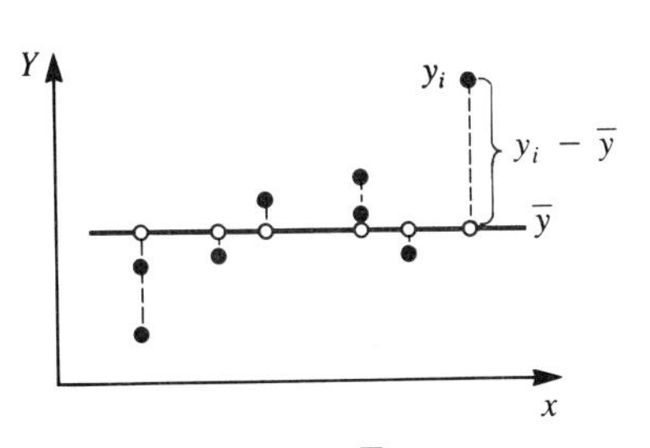

(a) Deviations around $\overline{Y}$

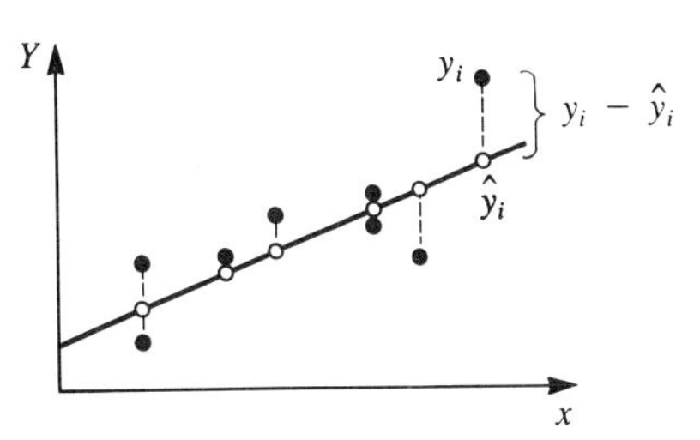

(b) Deviations around the regression line

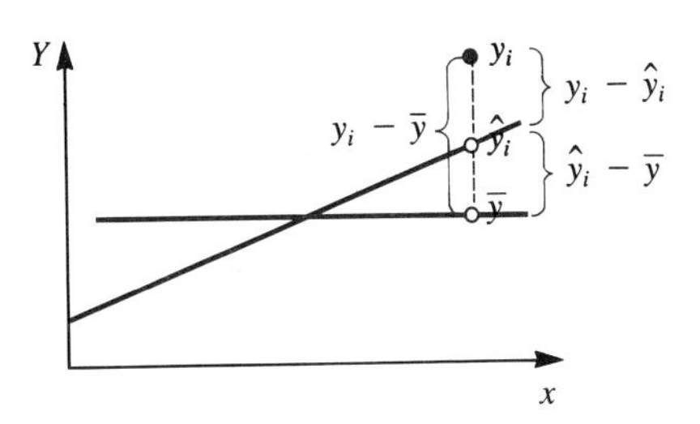

(c) Partitioning a deviation

Finally, looking at Figure 10.9c, we can observe, using the last point to the right in the figure, how the total deviation is partitioned into the deviation around the regression line and the deviation of the regression line from $\overline{Y}$. This partition is expressed as

$Y_i - \overline{Y}$	$=$	$Y_i - \hat{Y}_i$	$+$	$\hat{Y}_i - \overline{Y}$
Total deviation		Deviation around the regression line		Deviation of regression line around mean line

If each kind of deviation is squared and summed, it turns out that the sums of squares are also partitioned.

$$\underset{SST}{\Sigma(Y_i - \overline{Y})^2} = \underset{SSE}{\Sigma(Y_i - \hat{Y}_i)^2} + \underset{SSR}{\Sigma(\hat{Y}_i - \overline{Y})^2}$$

These sums are, respectively, the sums of squares total, error, and regression. The sum of squares regression measures the improvement of fit from using the regression line rather than $\overline{Y}$.

As a further step, notice in Figure 10.9a that for two of the values of x, there are two values of Y. Since replications such as these are possible in practical applications, we go to double subscripting of the Y values, where the j represents the unique values of x and i represents the replications. Therefore, the first two points to the left in Figure 10.9a are both for the same value of x and would be labeled Y_{11} and Y_{21}. Double subscripts require double summations. Therefore, the partitioning equation becomes

$$\begin{aligned} \Sigma\Sigma(Y_{ij} - \bar{Y})^2 &= \Sigma\Sigma(Y_{ij} - \hat{Y}_j)^2 + \Sigma\Sigma(\hat{Y}_j - \bar{Y})^2 \\ \text{SS(Total)} &= \text{SS(Error)} + \text{SS(Regression)} \end{aligned} \tag{10.10}$$

Double subscripting is used here to indicate that there may be more than one observation on Y for some values of x.

If at least two observations are available at some x values, the sum of the squares due to error can be further partitioned into two terms, called the sum of squares for pure error and the sum for squares of lack of fit. These two concepts are defined next.

DEFINITION 10.7 **Pure error** is the error due to chance variation independent of the specified regression model.

DEFINITION 10.8 The **lack of fit** component is that part of the observation that can be attributed to the misspecification of the model and is not chance error.

The additional partitioning is shown in Equation 10.11.

$$\begin{aligned} \Sigma\Sigma(Y_{ij} - \hat{Y}_j)^2 &= \Sigma\Sigma(Y_{ij} - \bar{Y}_j)^2 + \Sigma\Sigma(\bar{Y}_j - \hat{Y}_j)^2 \\ \text{SS(Error)} &= \text{SS(Pure error)} + \text{SS(Lack of fit)} \end{aligned} \tag{10.11}$$

If the model is correct, then both terms, SS*PE* and SS*LF*, are occurring simply due to the error variation. On the other hand, if the incorrect model is specified, SS*PE* is still arising due to chance since its calculation is not based on the model, but SS*LF*, using $\hat{Y}_j$ from the model, contains both error and the lack of fit component.

Therefore, the points may be scattered around the line (error) because the relationship between the variables is not perfect (pure error) and possibly because we are using the wrong model to fit the data (lack of fit). Equation 10.11 implies that both can exist at the same time. Figure 10.10 illustrates both situations. In Figure 10.10(a), the straight line is the wrong model, giving a significant lack of fit component as well as the pure error component. When the correct model is fit to the data points, as shown in Figure 10.10(b), the variation measured by both SS*PE* and SS*LF* is due to chance or error. If the SS*LF* is no larger than we would expect by chance alone, we conclude that the straight line, or whatever model we use, is appropriate. As in Chapter 9, the ratio of the MSs is distributed as an F-distribution if H_0 is true.

FIGURE 10.10 Lack of fit and pure error

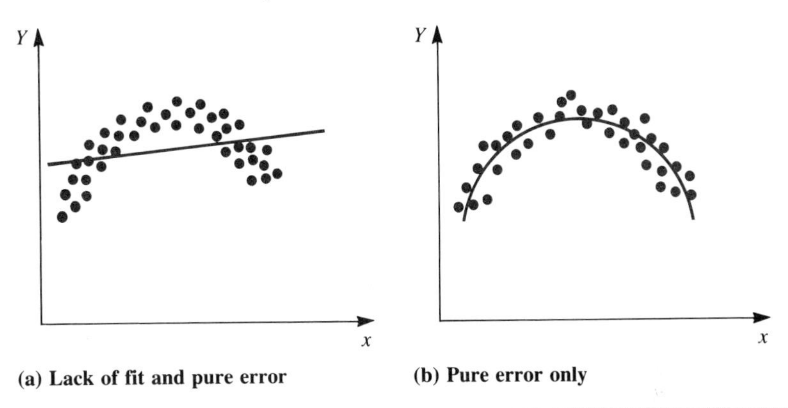

(a) Lack of fit and pure error **(b) Pure error only**

In developing the test, we note that, under the assumptions, if no relationship exists, then SST will have a $\sigma^2\chi^2$ distribution with $n - 1$ degrees of freedom. For the terms on the right in Equation 10.10, the degrees of freedom are easily determined to be $n - 2$ and $2 - 1 = 1$, respectively. Since these degrees of freedom add to $n - 1$, Cochran's theorem implies that SSE and SSR are independently distributed as $\sigma^2\chi^2$ with $n - 2$ degrees and 1 degree of freedom, respectively, if x and Y are actually independent.

In Equation 10.10, if the linear model is correct, then SSE is distributed as $\sigma^2\chi^2$ with $n - 2$ degrees of freedom. For SSPE, we have n variates with c means, giving $n - c$ degrees of freedom, whereas SSLF involves c variables, the $\overline{Y}_j$, with two restrictions—that is, a and b—giving $c - 2$ degrees of freedom. Thus, if the model is correct, SSPE and SSLF are independent and distributed as $\sigma^2\chi^2$ with $n - c$ and $c - 2$ degrees of freedom by Cochran's theorem also. Thus, in deciding if the model is correct, we are deciding whether SSLF is simply occurring due to chance variation or whether something more is being measured, resulting in an inflated value.

To test the model for adequacy, we construct an F test, using the F statistic in Equation 10.12. We recommend summarizing the computations in an ANOVA table first.

$$F = \frac{\text{Mean Square(Lack of Fit)}}{\text{Mean Square(Pure Error)}} = \frac{\text{MS}LF}{\text{MS}PE} \tag{10.12}$$

The sum of squares for pure error calculation is given in Equation 10.13a.

$$\text{SS}PE = \sum_{j=1}^{c} \sum_{i=1}^{n_j} (Y_{ij} - \overline{Y}_j)^2 \tag{10.13a}$$

where: n_j = number of observations for the jth value of x
c = number of uniquely different x values
$\text{df}_{PE} = n - c$, where $n = \sum n_j$

The computational equation for SSPE can be used in place of Equation 10.13a. It is the same as the equation for SSE in a one-factor ANOVA in Chapter 9 and is

$$\begin{aligned} \text{SS}PE &= \sum_{j=1}^{c}\sum_{i=1}^{n_j} Y_{ij}^2 - \sum_{j=1}^{c}\frac{\left(\sum_{i=1}^{n_j} Y_{ij}\right)^2}{n_j} \\ &= \sum_{j=1}^{c}\sum_{i=1}^{n_j} Y_{ij}^2 - \sum_{j=1}^{c} n_j\overline{Y}_j^2 \end{aligned} \tag{10.13b}$$

The sum of squares for lack of fit calculations is

$$\text{SS}LF = \sum_{j=1}^{c}\sum_{i=1}^{n_j} (\overline{Y}_j - \hat{Y}_j)^2 \tag{10.14a}$$

and degrees of freedom are $\text{df}_{LF} = c - 2$. However, by the partitioning shown in Equation 10.11, SSLF is more easily calculated from Equation 10.14b.

$$\text{SS}LF = \text{SS}E - \text{SS}PE \tag{10.14b}$$

We earlier calculated the sum of squares for error in Equation 10.8a as

$$\text{SS}E = \sum_{i=1}^{n} Y_i^2 - \frac{\left(\sum_{i=1}^{n} Y_i\right)^2}{n} - b\left(\sum_{i=1}^{n} x_iY_i - \frac{\sum_{i=1}^{n} x_i \sum_{i=1}^{n} Y_i}{n}\right)$$

with $n - 2$ degrees of freedom.

To be able to use this test, we must replicate some values of x. For example, in the tire problem used earlier, we have two different values of Y for $x = 20$, two for $x = 40$, and two for $x = 80$. Without these replications, there is no way to measure the pure error variation, and we have no degrees of freedom for pure error since $n = c$ and $Y_{ij} = \overline{Y}_j$ when $n_j = 1$.

The procedure for testing the adequacy of the model is as follows.

1. State the hypotheses and the α to be used for testing.
2. Determine the decision rule and critical value.
3. Calculate MSLF and MSPE.
4. Calculate the F ratio.
5. Compare the calculated F to the tabular value.
6. Make the decision and state the conclusion.

In general, the first two steps for the simple linear model are

H_0: linear model is correct

H_a: linear model is incorrect

or, stated as given in Equation 10.9, we can obtain the following decision rule.

DECISION RULE For H_0: $\mu_{Y|x} = \alpha + \beta_x$; H_a: $\mu_{Y|x} \neq \alpha + \beta_x$, reject H_0 if

$$\frac{\text{MS}LF}{\text{MS}PE} > F_{c-2,n-c,\alpha}$$

It is never safe to simply look at the scatterplot of a sample to determine whether the proper model for the population is being used. We need to test the model for adequacy. The procedure is illustrated in Example 10.9.

EXAMPLE 10.9 Visually, it appears that a straight line is indicated in the tire problem given in Example 10.2. Test the data to see if the straight line model seems appropriate at $\alpha = .05$.

Solution: SSE was calculated in Example 10.6 and found to be

$$\text{SS}E = 60$$

$$\text{df}_E = n - 2 = 8 - 2 = 6$$

Next, we calculate SSPE, using Equation 10.13a.

$$\text{SS}PE = \sum_{j=1}^{c} \sum_{i=1}^{n_j} (Y_{ij} - \overline{Y}_j)^2$$

The computations are shown in Table 10.3. The first step is the calculation of the average value of Y for each x shown in Table 10.3(a). Then, the mean of each column is subtracted from each item in the column, and the result is squared, as shown in Table 10.3(b). Finally, in Table 10.3(c), all the values are added to give SSPE. The result is

$$\text{SS}PE = 45$$

with

$$\text{df}_{PE} = n - c = 8 - 5 = 3$$

The calculation of the lack of fit term results in

$$\text{SS}LF = \text{SS}E - \text{SS}PE = 60 - 45 = 15$$

$$\text{df}_{LF} = c - 2 = 5 - 2 = 3$$

TABLE 10.3 Computation of pure error in Example 10.9

(a)

j		*1*	*2*	*3*	*4*	*5*	$c = 5$
x		*20*	*40*	*50*	*70*	*80*	
	i						
y_{ij}	*1*	28	38	44	52	58	$n = 8$
	2	33	42			65	
$\bar{y}_j$		30.5	40	44	52	61.5	

(b)

j		*1*	*2*	*3*	*4*	*5*
x		*20*	*40*	*50*	*70*	*80*
	i					
$(y_{ij} - \bar{y}_j)^2$	*1*	$(28 - 30.5)^2$	$(38 - 40)^2$	$(44 - 44)^2$	$(52 - 52)^2$	$(58 - 61.5)^2$
	2	$(33 - 30.5)^2$	$(42 - 40)^2$			$(65 - 61.5)^2$

(c)

j		*1*		*2*		*3*		*4*		*5*	
x		*20*		*40*		*50*		*70*		*80*	
	i										
$(y_{ij} - \bar{y}_j)^2$	*1*	6.25		4		0		0		12.25	
	2	6.25		4						12.25	
		12.5	+	8	+	0	+	0	+	24.5	= 45

The hypotheses and ANOVA table are given next.

H_0: linear model is correct ($\mu_{Y|x} = \alpha + \beta x$)
H_a: linear model is not correct ($\mu_{Y|x} \neq \alpha + \beta x$)

Reject H_0 if

$$\frac{\text{MS}LF}{\text{MS}PE} > F_{c-2,n-c,\alpha}$$

Source	*SS*	*df*	*MS*
Lack of fit	15	3	5
Pure error	45	3	15
Error	60	6	

FIGURE 10.11 Sampling distribution for the F statistic in Example 10.9

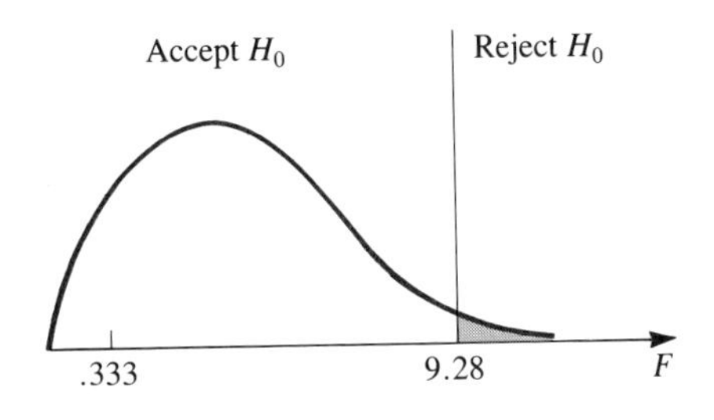

$$F = \frac{\text{MS}LF}{\text{MS}PE} = \frac{5}{15} = .333$$

$$F_{3,3,.05} = 9.28$$

As shown in Figure 10.11, the calculated F falls within the acceptance region. Therefore, we accept the null hypothesis and conclude that there is no reason to doubt that the linear model is appropriate for the tire sales versus inches of advertising data.

In the previous example, we decided that the correct model had been fit to the data. In the following example, we look at a situation where the model seems to be incorrect.

EXAMPLE 10.10 Plastico, Inc., makes plastic pellets that are sold to companies that make plastic products by injection-molding techniques. Over the past few years, the company has been expanding its output. Recently, management has become interested in determining the relationship between output and cost per pound of the product. Since all other factors have been relatively constant over this period, they think that a straight line should adequately represent the data. Since they precisely control output each week to an exact multiple of 1000 pounds and hold the same output level for several weeks, they were able to obtain 2 or 3 points for each output. Test their model for adequacy at $\alpha = .05$.

Output (1000)	1	1	1	2	2	3	3	4	4	4
Cost per pound	10.6	13.0	9.4	9.3	7.1	7.0	8.2	7.4	6.0	7.0

Output (1000)	5	5	6	6	7	7	8	8	9	9
Cost per pound	5.7	7.1	5.3	4.9	5.2	5.8	6.5	6.9	8.4	8.2

Miles Driven	2000	2500	5000	2500	5000	6000	4000	5000
Gallons Used	221	248	493	257	503	610	392	491

(*Hint:* convert miles to thousands of miles. For example, 2500 is 2.5.)

10.10 The Aerotruck Transfer Company wants to determine the relationship between the average speed of its trucks and the miles per gallon (mpg). A trip of 300 miles on a test track is made by various trucks at various speeds where the speed is held constant. Fit a linear model to the given data. At $\alpha = .05$, does the linear model appear to be correct?

	1	*2*	*3*	*4*	*5*	*6*	*7*	*8*
Speed	45	50	60	55	70	65	55	55
mpg	13	15	15	20	13	14	16	18

	9	*10*	*11*	*12*	*13*	*14*	*15*
Speed	65	50	45	65	40	60	70
mpg	11	16	14	16	10	16	12

10.6 INFERENCES ABOUT THE REGRESSION COEFFICIENT

Once it has been decided that the straight line is an appropriate model to use on a set of data, the model can then be tested to determine whether a significant or useful relationship exists between x and Y. Consider the population in Figure 10.14 from which a few points are plotted.

In the population there appears to be no relationship between x and Y, as illustrated by the lack of any pattern. The mechanics of fitting a regression line

FIGURE 10.14 Population and sample regression lines

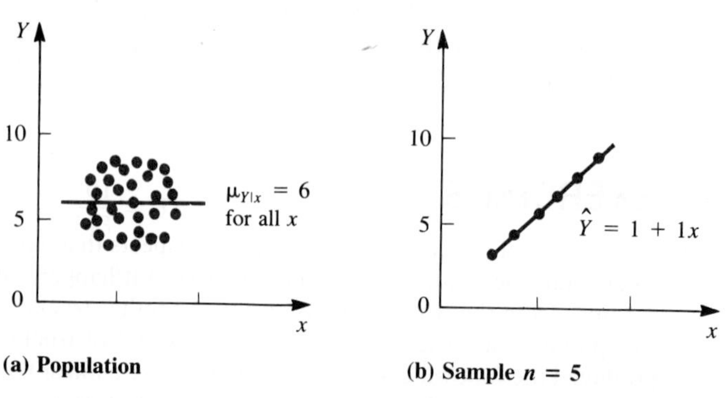

are such that when no relationship exists, the slope β is equal to zero, and the intercept α is equal to μ_Y for all x. That is, the means for all the Y populations are the same. In other words, x contains no useful information about Y. For the illustration in Figure 10.14,

$$\mu_{Y|x} = \alpha + \beta x_i = 6 + 0x_i = 6$$

That is, the line on which the means of populations of Y values lie is parallel to the x-axis. Hence, the slope is zero. Note, though, that it is still a line and does represent a linear model.

Suppose we take a random sample of size 5 from this population. Since any five points can be selected, it is possible that we obtained the particular five points shown, which give the indication that a perfect relationship may exist between x and Y and give the equation

$$\hat{Y}_i = a + bx_i = 1 + 1x_i$$

It is possible, then, that from a population in which no relationship exists, a sample can be taken that indicates that a perfect or near-perfect relationship may exist. But, as the sample size increases, indication of a near-perfect relationship becomes less likely. Therefore, a good argument exists for not taking small samples.

After the parameters in a model have been estimated, the model should always be evaluated for significance—that is, tested to decide whether it is really useful or if b merely differs from zero by chance. In this section, two different but equivalent tests are demonstrated that are used to determine if a relationship exists; these are the analysis of variance F test and the t test for the slope. Both test the null hypothesis that no relationship exists and that, therefore, $\beta = 0$ in the population equation against the alternative that $\beta \neq 0$, and hence, that incorporating the value of x into the estimate improves predictability. In other words, does x really explain a significant part of the variation in Y?

Analysis of Variance

Earlier, it was indicated in Equation 10.10 that the total sum of squares can be partitioned into the sum of squares due to regression and the sum of squares due to error. The conceptual form is given in Equation 10.15, where we are now using a single summation because we are no longer emphasizing replications.

$$\Sigma(Y_i - \overline{Y})^2 = \Sigma(Y_i - \hat{Y}_i)^2 + \Sigma(\hat{Y}_i - \overline{Y})^2 \tag{10.15}$$

The computational form of the equation is after solving for SS(Error)

$$\text{SS(Error)} = \underbrace{\Sigma Y_i^2 - \frac{(\Sigma Y_i)^2}{n}}_{\text{SS(Total)}} - \underbrace{b\left(\Sigma x_i Y_i - \frac{\Sigma x_i \Sigma Y_i}{n}\right)}_{\text{SS(Regression)}}$$

Note that the sum of squares for regression is made up of the deviations between the line $\hat{Y}_i$ and the average $\overline{Y}$. The degrees of freedom associated with the three sums of squares are

Regression	$df_R = 1$, the number of independent variables in the model
Error	$df_E = n - 2$, two degrees of freedom are lost in estimating α and β
Total	$df_T = n - 1$, one degree of freedom lost for $\overline{Y}$

The mean squares are determined as usual by dividing the SS by the corresponding degrees of freedom.

DECISION RULE To test $H_0: \beta = 0$; $H_a: \beta \neq 0$, reject H_0 if

$$\frac{\text{MS}R}{\text{MS}E} > F_{1,n-2,\alpha}$$

EXAMPLE 10.11 For the tire problem of Example 10.2, determine if the equation of the relationship is significant at $\alpha = .05$.

Solution: The hypotheses of interest are

$H_0: \beta = 0$ (no significant relationship exists)
$H_a: \beta \neq 0$ (significant relationship exists)

Repeating the calculations from Example 10.7, we have

$$\text{SS}E = \overbrace{\sum Y_i^2 - \frac{(\sum Y_i)^2}{n}}^{\text{SS}T} - \overbrace{b\left(\sum x_i Y_i - \frac{\sum x_i \sum Y_i}{n}\right)}^{\text{SS}R}$$

$$= 17{,}310 - \frac{(360)^2}{8} - .5(2100)$$

$$= 17{,}310 - 16{,}200 - 1050$$

$$\underbrace{60}_{\text{SS}E} = \underbrace{1110}_{\text{SS}T} - \underbrace{1050}_{\text{SS}R}$$

The degrees of freedom are

Regression	$df_R = 1$
Error	$df_E = n - 2 = 8 - 2 = 6$
Total	$df_T = n - 1 = 8 - 1 = 7$

FIGURE 10.15 Sampling distribution for Example 10.11

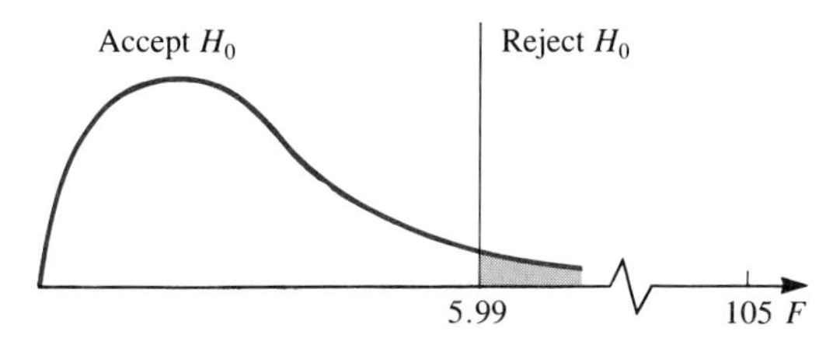

The analysis of variance table can now be completed, and the F test evaluated. These results also can be found on the MINITAB printout in Table 10.1b.

Source	*SS*	*df*	*MS*
Regression	1050	1	1050
Error	60	6	10
Total	1110	7	

$$F = \frac{\text{MS(Regression)}}{\text{MS(Error)}} = \frac{\text{MS}R}{\text{MS}E} = \frac{1050}{10} = 105$$

$$F_{1,6,.05} = 5.99$$

Since $105 > 5.99$, we reject H_0. This result is shown in Figure 10.15. It appears that a linear relationship exists between advertising and sales over the range of 20 to 80 inches of advertising. ■

t Test and Confidence Interval for β

For illustration, assume a population is approximately represented by the scatterplot in Figure 10.16a. If random samples of size 5 are taken from this population, it can be seen that many different sample regression lines are possible, a few of which are illustrated in Figure 10.16b.

These sample regression lines have different slopes, and any one of them is possible when we take a sample. The weaker the extent of the relationship in the population, the more likely it is to obtain a wide range of sample slopes for the population of all possible sample slopes. The standard deviation of all the possible slopes from the population is denoted σ_b, the standard error of the slope, and is given by

$$\sigma_b = \frac{\sigma_{Y|x}}{\sqrt{\sum(x_i - \overline{x})^2}} = \frac{\sigma_{Y|x}}{\sqrt{\sum x_i^2 - \dfrac{(\sum x_i)^2}{n}}} \tag{10.16}$$

FIGURE 10.16 Possible sample regression lines from a population

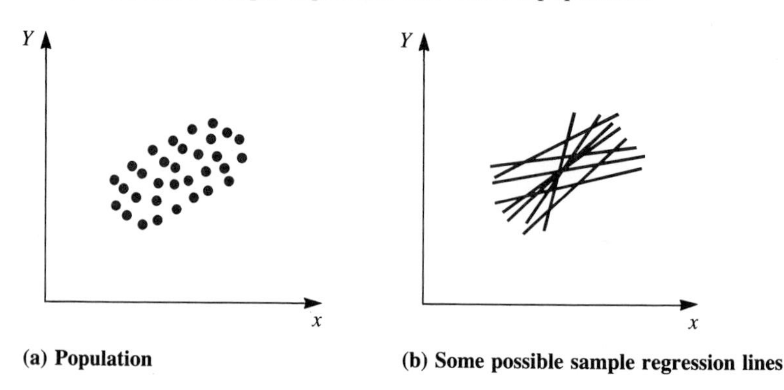

To illustrate the derivation, we note that the equation for b also can be written as

$$b = \frac{\sum (x_i - \bar{x})(Y_i - \bar{Y})}{\sum (x_i - \bar{x})^2} = \frac{\sum (x_i - \bar{x}) Y_i}{\sum (x_i - \bar{x})^2}$$

Now, let $c_i = (x_i - \bar{x})/\sum (x_i - \bar{x})^2$. Then, $b = \sum c_i Y_i$, where the c_i are constants. Thus, b is a linear combination of independent random variables. From Chapter 3, we know that, for such a case,

$$\text{Var}(b) = \text{Var}\,(\sum c_i Y_i) = \sum c_i^2 \,\text{Var}(Y_i)$$

But, $\text{Var}(Y_i) = \sigma_{Y|x}^2$ and $c_i^2 = (x_i - \bar{x})^2/[\sum (x_i - \bar{x})^2]^2$. Hence, the variance of b is given by

$$\sigma_b^2 = \sum \frac{(x_i - x)^2 \sigma_{Y|x}^2}{[\sum (x_i - \bar{x})^2]^2} = \frac{\sigma_{Y|x}^2}{[\sum (x_i - \bar{x})^2]^2} \sum (x_i - \bar{x})^2$$

$$= \frac{\sigma_{Y|x}^2}{\sum (x_i - \bar{x})^2}$$

In addition to the variance of b, under the assumptions, the Y's are normally distributed. Thus, b is a linear combination of normal random variables and has a normal distribution itself.

Since $\sigma_{Y|x}$ is not usually known, but its estimate $s_{Y|x}$ is, the estimated standard error of b is

$$s_b = \frac{s_{Y|x}}{\sqrt{\sum x_i^2 - \frac{(\sum x_i)^2}{n}}} \qquad (10.17)$$

Note that the terms within the radical are exactly the same as in the denominator of the calculation of b. A confidence interval can be established for β using the equation

$$b \pm t_{n-2,\alpha/2} s_b \qquad (10.18)$$

where the error degrees of freedom is the parameter for the t-distribution. Therefore, we compute the upper limit and lower limit as

$$LL = b - ts_b \qquad (10.19a)$$

$$UL = b + ts_b \qquad (10.19b)$$

To test the hypothesis $H_0: \beta = \beta_0$, we may use the t-distribution with the statistic

$$t = \frac{b - \beta_0}{s_b} \qquad (10.20)$$

where β_0 is any real number. If the null hypothesis is that $\beta_0 = 0$, the equation reduces to

$$t = \frac{b - 0}{s_b} = \frac{b}{s_b} \qquad (10.21)$$

and provides an alternative to the analysis of variance F statistic when the alternative hypothesis is $H_a: \beta \neq 0$. The calculated t is compared to $t_{n-2,\alpha/2}$ from Appendix Table B.7.

DECISION RULE To test $H_0: \beta = 0$; $H_a: \beta \neq 0$, reject H_0 if

$$\left| \frac{b - 0}{s_b} \right| > t_{n-2,\alpha/2}$$

EXAMPLE 10.12 Using a t test on the slope, determine if the relationship in Example 10.3 is significant at $\alpha = .05$. If it is significant, construct a 95% confidence interval for the true value of β.

Solution: The hypotheses are

$$H_0: \beta = 0 \qquad H_a: \beta \neq 0$$

From Example 10.7, we know $s_{Y|x} = 3.162$,

$$s_b = \frac{s_{Y|x}}{\sqrt{\sum x_i^2 - \frac{(\sum x_i)^2}{n}}}$$

$$= \frac{3.162}{\sqrt{24{,}200 - \frac{160{,}000}{8}}} = .0488$$

$$t = \frac{b}{s_b} = \frac{.5}{.0488} = 10.25$$

$$t_{6,.025} = 2.447$$

as shown in Figure 10.17. Since $10.25 > 2.447$, we reject the null hypothesis and conclude that a significant positive relationship appears to exist between advertising and sales. The relationship is interpreted to mean that a change in inches of advertising space is associated with a change in the number of tires sold, with both changing in the same direction. The confidence interval is

$$b \pm t_{n-2,\alpha/2} s_b = .5 \pm 2.447(.0488) = .5 \pm .12$$

$$LL = .38 \qquad UL = .62$$

The interval may be interpreted as meaning that we are 95% certain that the interval from .38 to .62 includes the true value of β.

Notice that the F statistic in the analysis of variance test is the square of the t statistic just calculated.

Calculated	***Tabulated***
$F = t^2$	$F_{1,6,.05} = t^2_{6,.025}$
$105 = (10.25)^2$	$5.99 = (2.447)^2$
$105 = 105$	$5.99 = 5.99$

FIGURE 10.17 Sampling distribution for Example 10.12

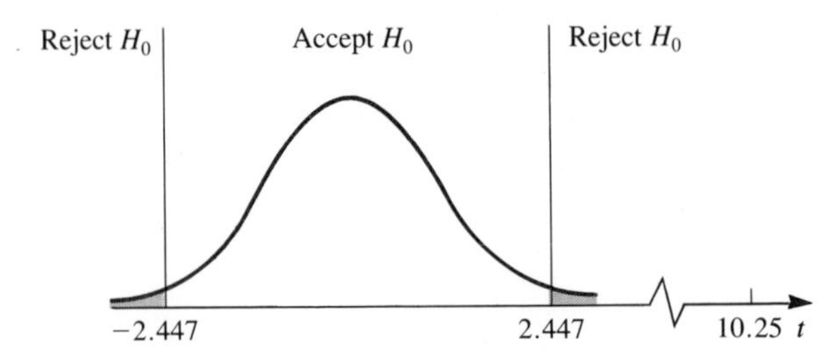

This relationship is true only if the F statistic has one degree of freedom in the numerator and the denominator has the same number of degrees of freedom as the t, which is the case in these procedures. In addition, for the tabular values, note that t is based upon $\alpha/2$ and F is based upon α. Intuitively, α is used for F because a value in either tail of t, when squared, goes into the upper tail of F.

EXERCISES

10.11 Radial tires can fail if they are driven at high speeds while underinflated. A tire firm decides to develop a linear regression model that can be used to estimate the relationship between tire pressure and miles driven at 100 mph before tire failure. The first tire failing on a test car is the one used for the following sample. Data from 8 cars are given.

Pressure	18	12	15	8	11	13	9	16
Distance	12.0	7.2	10.1	2.3	6.2	7.0	4.1	11.7

(a) Calculate the equation.
(b) Test the null hypothesis that $\beta = 0$, using analysis of variance at $\alpha = .05$.
(c) Test the same null hypothesis using the t test.
(d) Show that the 2 tests are equivalent by comparing the F and t.
(e) Estimate the true β for the population regression line by finding a confidence interval for β with $1 - \alpha = .95$.

10.12 A large aquaculture business raises salmon in large tanks. It has been experimenting with a special food additive in hopes of making the fish grow faster. Employees collect a random sample of fish all of the same size and feed each fish different amounts. They then measured the amount of additive each fish ate over a period of 1 year and how many inches each grew. Does there appear to be a useful linear relationship between additive consumption and growth at $\alpha = .05$?

Average Daily Consumption	4.0	5.5	2.0	2.5	3.4	4.1	8.0	5.2	6.1	3.8
Growth	2.3	4.1	3.2	2.2	2.5	2.4	3.2	4.1	3.3	2.8

(a) Calculate the equation.
(b) Test $\beta = 0$, using analysis of variance at $\alpha = .05$.
(c) Test $\beta = 0$, using the t test at $\alpha = .05$.
(d) An alternative test is to calculate a confidence interval for β and then to note if 0 is within the interval. If it is, the null hypothesis that $\beta = 0$ is not rejected. Try this method.

10.7 THE EQUATION AS AN ESTIMATOR

One of the primary uses of the regression equation is for predicting Y at a given value of x. Two cases are possible. In the first, the value of x is within the range of the data for which the equation is developed. The value of x is outside the data range in the second case. These two cases are illustrated in the following example.

FIGURE 10.18 Predicting Y for x within the range of data ($x = 60$) for Example 10.13

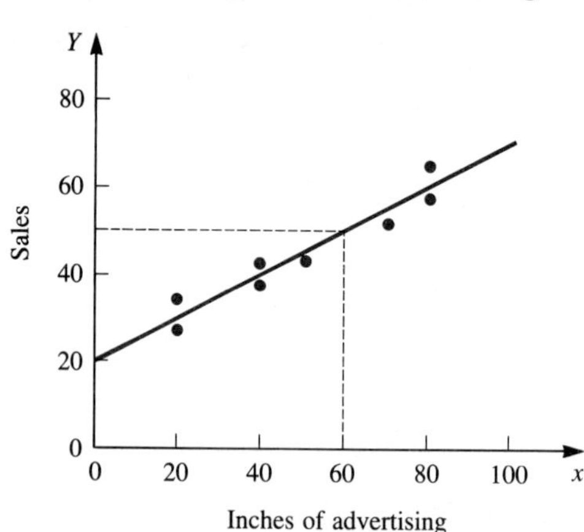

EXAMPLE 10.13 For the tire problem of Example 10.2, predict expected tire sales if

a. The store uses 60 inches of advertising (the first case).
b. The store uses 100 inches of advertising (the second case).

Solution: The equation for $\hat{Y}$ is

$$\hat{Y}_i = 20 + .5x_i$$

a. $\hat{Y} = 20 + .5(60) = 20 + 30 = 50$ tires
b. $\hat{Y} = 20 + .5(100) = 20 + 50 = 70$ tires

These two cases are shown in Figures 10.18 and 10.19, respectively.

This type of prediction is a point estimate. When the point estimate is based on an x beyond the range of the data, the prediction is often called an extrapolation or forecast. Two difficulties exist with point estimates. First, the assumption is made that the relationship remains linear throughout all values of x. Linearity may not be the case, particularly for the extrapolation. Second, no measure of possible error is given. Hence, an interval estimate is a more reasonable approach, for there are a number of sources of error that must be taken into account when making a forecast.

Estimating the Average or Expected Value of Y

Remember that a sample line is just one of many possible sample lines that could be obtained. Therefore, the point estimate is just one of many different point

FIGURE 10.19 Predicting Y for x outside the range of data ($x = 100$) for Example 10.13

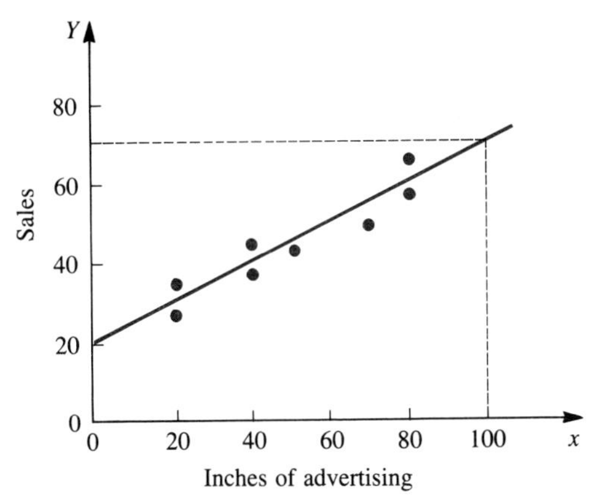

estimates that might occur. Figure 10.20 shows two possible sample lines from a population of values. The wide difference between the two slopes indicates a large standard error of the slope (σ_b), which is a direct result of a large standard error of the estimate ($\sigma_{Y|x}$).

We can see from Figure 10.20 that a wide variation may exist in the point estimate of Y for a given value of x. It is obvious that the range of possibilities at x_2 is wider than at x_1 since it is farther from x_0, the point where the lines cross, than is x_1. Therefore, we can see intuitively that the standard deviation of all the possible point estimates at x_2 is greater than at x_1 since any error in b is magnified when multiplied by x. For example, if $x = 10$ and b is in error by .1, then bx is in error by 1. The standard deviation of $\hat{Y}$ at x_p, the predicting value, is calculated from Equation 10.22. The derivation of Equation 10.22 is fairly simple and is left as a starred exercise.

FIGURE 10.20 Forecasting with two different sample lines from the same population

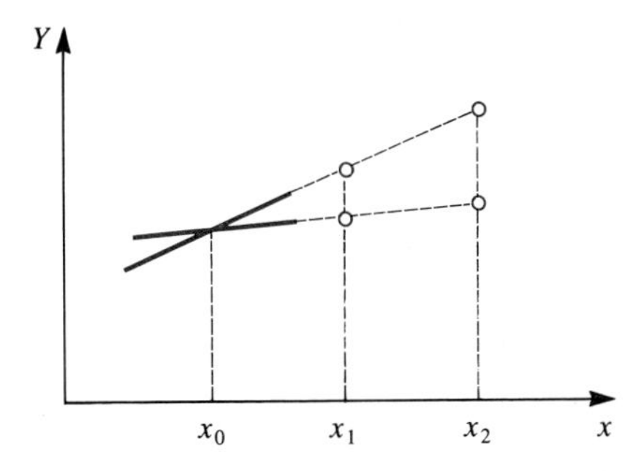

FIGURE 10.21 Confidence interval bands on $\mu_{Y|x}$ over the values of x

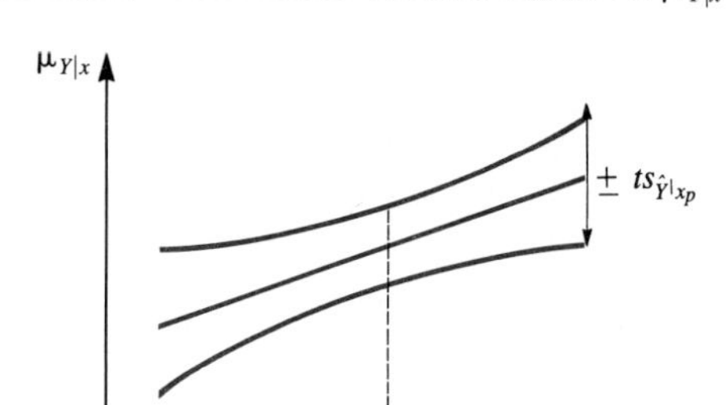

$$\sigma_{\hat{Y}|x_p} = \sigma_{Y|x}\sqrt{\frac{1}{n} + \frac{(x_p - \bar{x})^2}{\sum x_i^2 - \frac{(\sum x_i)^2}{n}}} \tag{10.22}$$

Notice that as x_p gets farther from $\bar{x}$, the standard deviation gets larger. When $\sigma_{Y|x}$ is to be estimated by $s_{Y|x}$, σ is replaced by s.

We desire to estimate $\mu_{Y|x_p}$ for the population—that is, the mean of the distribution of Y values for a given value of x. We make our estimate by using a confidence interval for $\mu_{Y|x_p}$, calculated using Equation 10.23.

$$\hat{Y}_{x_p} \pm t_{n-2,\alpha/2}\, s_{\hat{Y}|x_p} \tag{10.23}$$

The confidence interval based on Equation 10.23 is commonly known as an interval estimate for the average value (expected value) of Y given x_p.

Careful inspection of Equation 10.23 leads us to conclude that, as the distance between x_p and $\bar{x}$ increases, the confidence interval becomes wider. Figure 10.21 illustrates this principle.

EXAMPLE 10.14 For the tire problem of Example 10.7, calculate a 95% confidence interval for the average value of sales for 60 inches of advertising.

Solution: The value for $\hat{Y}$ at $x_p = 60$ is $\hat{Y}_{60} = 20 + .5(60) = 50$. From Appendix Table B.7, $t_{6,.025} = 2.447$, and

$$s_{\hat{Y}|xp} = s_{Y|x}\sqrt{\frac{1}{n} + \frac{(x_p - \bar{x})^2}{\sum x_i^2 - \frac{(\sum x_i)^2}{n}}}$$

$$= 3.162\sqrt{\frac{1}{8} + \frac{(60 - 50)^2}{24{,}200 - (400)^2/8}}$$

$$= 3.162\sqrt{\frac{1}{8} + \frac{100}{4200}}$$

$$= 3.162\sqrt{.1488} = 1.22$$

Thus, the interval is given by

$$\hat{Y}_{x_p} \pm t_{n-2,\alpha/2}\, s_{\hat{Y}|x_p} = 50 \pm 2.447(1.22) = 50 \pm 2.99$$

$$LL = 47.01 \qquad UL = 52.99$$

If we knew the equation of the population line, we would know $\mu_{Y|x_p}$ for 60 inches of advertising. Since we do not, we estimate $\mu_{Y|60}$, using the sample value $\hat{Y}_{60}$, and place a confidence interval around it. This interval is interpreted to mean that we are 95% confident that this interval contains the true value of $\mu_{Y|60}$.

It is possible to test for a specific value for the parameter $\mu_{Y|x}$. The standard procedures for testing a hypothesis can be used with the following three possible sets of hypotheses, using the statistic

$$t = \frac{\hat{Y}_{x_p} - \mu_0}{s_{\hat{Y}|x_p}}$$

DECISION RULES To test H_0: $\mu_{Y|x_p} = \mu_0$; H_a: $\mu_{Y|x_p} \neq \mu_0$, reject H_0 if

$$\left|\frac{\hat{Y}_{x_p} - \mu_0}{s_{\hat{Y}|x_p}}\right| > t_{n-2,\alpha/2}$$

To test H_0: $\mu_{Y|x_p} \geq \mu_0$; H_a: $\mu_{Y|x_p} < \mu_0$, reject H_0 if

$$\frac{\hat{Y}_{x_p} - \mu_0}{s_{\hat{Y}|x_p}} < -t_{n-2,\alpha}$$

To test H_0: $\mu_{Y|x_p} \leq \mu_0$; H_a: $\mu_{Y|x_p} > \mu_0$, reject H_0 if

$$\frac{\hat{Y}_{x_p} - \mu_0}{s_{\hat{Y}|x_p}} > t_{n-2,\alpha}$$

EXAMPLE 10.15 Refer to Example 10.2. The advertising representative for the newspaper claims that if the tire store sets its advertising at 60 inches, it will average sales of more than 48 tires per week. Is this claim believable, using $\alpha = .05$?

Solution: The hypotheses of interest are

$$H_0: \mu_{Y|60} \leq 48 \qquad H_a: \mu_{Y|60} > 48$$

FIGURE 10.22 Sampling distribution for Example 10.15

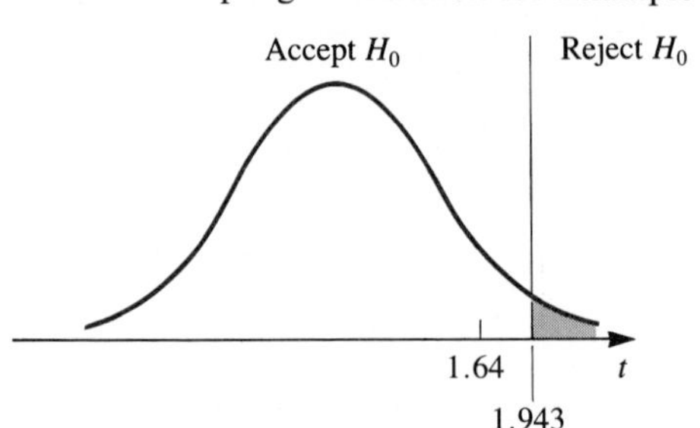

and we reject H_0 if $t > t_{n-2,\alpha}$. We know from previous calculations that $\hat{Y}$ for $X =$ 60 is 50 and $s_{\hat{Y}} = 1.22$.

$$t_{n-2,\alpha} = t_{6,.05} = 1.943$$

$$t = \frac{50 - 48}{1.22} = \frac{2}{1.22} = 1.64$$

as illustrated in Figure 10.22. Since $1.64 < 1.943$, we accept the null hypothesis. The advertising representative's claim is not supported. That is, 50 is not an unreasonably high value if $\mu_{Y|60}$ is really 48.

Predicting an Individual Value for *Y*

If we wish to predict an individual value for Y, the best "guess" is $\mu_{Y|x}$ since Y_x is normal. This mean is unknown, but an estimate of it is available, namely, $\hat{Y}_x$. Therefore, the best estimate for an individual value is also $\hat{Y}_x$. Recall that a point estimate is of better use if we can place bounds on its error. To place bounds, when estimating an individual Y_x value, we must incorporate both the variance of the estimator and the variance of Y_x itself since Y_x is also a random variable and not a parameter, as is $\mu_{Y|x}$.

In Figure 10.23, this concept is illustrated intuitively. The limits on the confidence interval for the average value incorporate the fact that $\mu_{Y|x}$ may be as small as μ_L or as large as μ_U. If $\mu_{Y|x}$ is actually μ_L, the lower bound for Y_x would be $t_{n-2,\alpha/2} s_{Y|x}$ units below it, and, conversely, if $\mu_{Y|x}$ is at μ_U, the bound for Y_x would be the same number of units above it. The composite confidence interval is an interval estimate for an individual value of Y at x_p. It is also popularly known in the literature as the forecast interval.

The equation for the standard deviation of $Y_{x_p} - \hat{Y}_{x_p}$ is

$$\sigma_{Y|x_p - \hat{Y}|x_p} = \sigma_{Y|x} \sqrt{1 + \frac{1}{n} + \frac{(x_p - \bar{x})^2}{\sum x_i^2 - \frac{(\sum x_i)^2}{n}}} \tag{10.24}$$

FIGURE 10.23 Intuitive representation of the interval for an individual value of Y_x

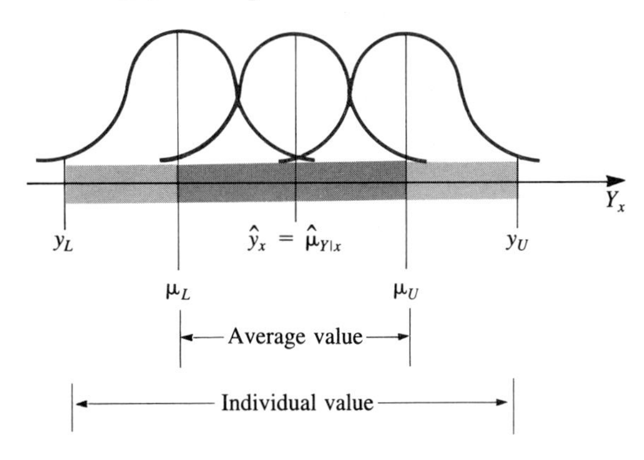

Note that there is an additional 1 under the radical that differentiates it from Equation 10.22. Again, $s_{Y|x}$ estimates $\sigma_{Y|x}$. The confidence interval for an individual value of Y_{x_p} is then calculated as

$$\hat{Y}_{x_p} \pm t_{n-2,\alpha/2} s_{Y|x_p - \hat{Y}|x_p} \qquad \textbf{(10.25)}$$

The confidence interval can be used to place bounds on the value of Y that results for a given x_p. As indicated in Figure 10.24, this confidence interval also increases with the distance of x_p from $\bar{x}$—that is, the estimator has less precision the farther we move away from the mean of the x values. The band is narrowest at $x_p = \bar{x}$.

FIGURE 10.24 Confidence interval bands on Y_x over the values of x

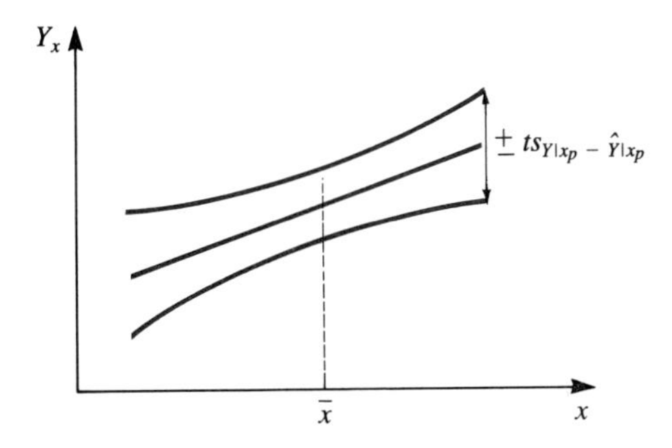

EXAMPLE 10.16 Refer to Example 10.3. If advertising is increased to 60 inches, construct a 95% prediction for the actual sales that result.

Solution: Substituting $x_p = 60$ into the equation for $\hat{Y}$ gives $\hat{Y}_{60} = 20 + .5(60) = 50$.

$$t_{6,.025} = 2.447$$

$$s_{Y|x_p - \hat{Y}|x_p} = s_{Y|x}\sqrt{1 + \frac{1}{n} + \frac{(x_p - \bar{x})^2}{\sum x_i^2 - \frac{(\sum x_i)^2}{n}}}$$

$$= 3.162\sqrt{1 + \frac{1}{8} + \frac{(60 - 50)^2}{24{,}200 - \frac{(400)^2}{8}}}$$

$$= 3.162\sqrt{1 + .125 + .0238}$$

$$= 3.162\sqrt{1.1488}$$

$$s_{Y|x_p - \hat{Y}|x_p} = 3.389$$

The confidence interval is

$$\hat{Y}_{60} \pm t_{n-2,\alpha/2} s_{Y|x_p - \hat{Y}|x_p} = 50 \pm 2.447(3.389) = 50 \pm 8.29$$

$$LL = 41.71$$

$$UL = 58.29$$

Thus we are 95% confident that this interval contains the actual sales (Y) that will result when advertising is 60 inches (x_p).

FIGURE 10.25 Representative confidence interval bands for both average and individual values of Y at a given x

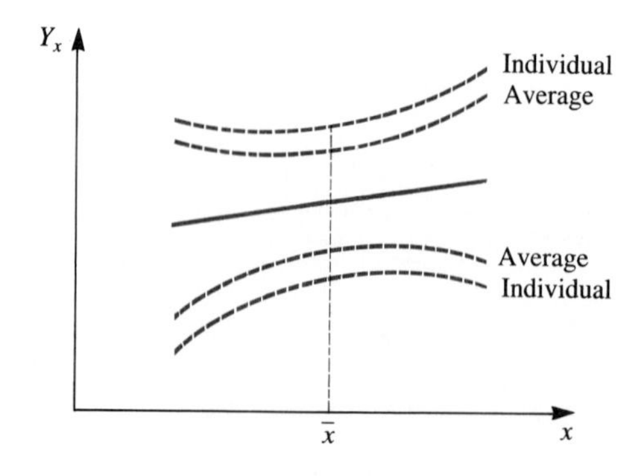

FIGURE 10.26 Relationships unknown beyond range of data

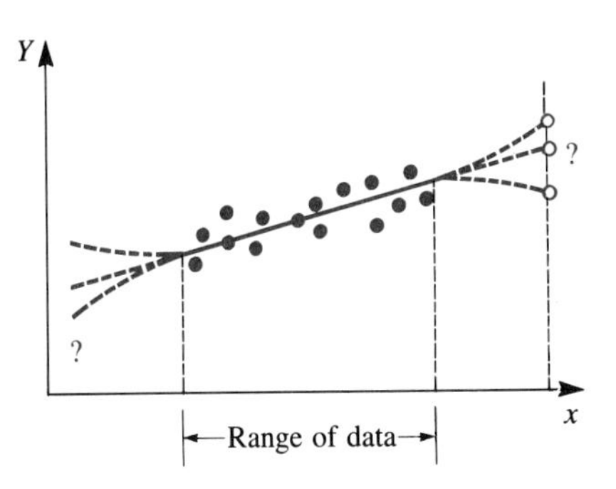

It is instructive to compare the confidence interval bands for the average value with those for the individual value. The comparisons are shown in Figure 10.25. The bands for the individual value lie outside those for the average value. For all values of x, the confidence interval on the individual value is wider.

The preceding predictions assume that all of the assumptions listed in Section 10.4 have been satisfied, especially the assumption that a straight line is the proper relationship. The straight line may have been a good approximation within the range of the data used in the problem. However, there is no assurance that when we predict beyond the range of these data, the linear relationship will still hold. Figure 10.26 illustrates this point.

In this section, we have determined that the farther we project the line beyond the range of the data, the wider the confidence interval and the less likely that the relationship continues to be linear. Therefore, using a regression equation for predicting very far beyond the range of the data may be purely crystal ball gazing.

EXERCISES

10.13 Using the information developed in Exercise 10.11 and a confidence interval, estimate the average number of miles before tire failure if the inflation is 10 pounds per square inch ($\alpha = .05$).

10.14 The tire company of Exercise 10.11 claims that at a speed of 100 mph and a tire pressure of 22 pounds, its tires will average more than 15 miles without failure. Evaluate this claim at $\alpha = .05$.

10.15 Again using the information from Exercise 10.11, predict the actual distance a tire will go before failure if the pressure is 200 pounds per square inch. Place a 95% interval around the prediction.

10.16 What is your reaction to the results in Exercise 10.15?

10.8 ANALYSIS OF RESIDUALS

The residual term as presented in Sections 10.2 and 10.3 is the deviation that exists at any particular value of x_i between the actual value Y_i and the predicted value $\hat{Y}_i = a + bx_i$. The residual is expressed as

$$e_i = Y_i - \hat{Y}_i$$

The residual is the observed error—that is, it is the result of sampling errors and possible misspecification of the model. The actual, but nonobservable, error due to sampling is

$$\varepsilon_i = Y_i - E[Y_i] \tag{10.26}$$

If the model

$$Y_i = \alpha + \beta x_i + \varepsilon_i$$

is correct, then e_i is an estimate of ε_i.

The regression assumptions presented in Section 10.4 are assumptions about the error terms. If these assumptions are satisfied, the tests of significance already presented are applicable. For certain violations of the assumptions, there exist transformations on the data that correct the problem. Otherwise, it may be necessary to turn to other techniques.

In review, the assumptions associated with the error terms ε_i are that they are (a) independent and (b) normally distributed with a mean of zero and a variance of σ^2, where σ^2 is constant for all values of x_i. Also, we are assuming that the linear model is the proper relationship.

If the number of sample points is adequate, it is possible, through inspec-

FIGURE 10.27 A sample residual plot of e_i versus x_i

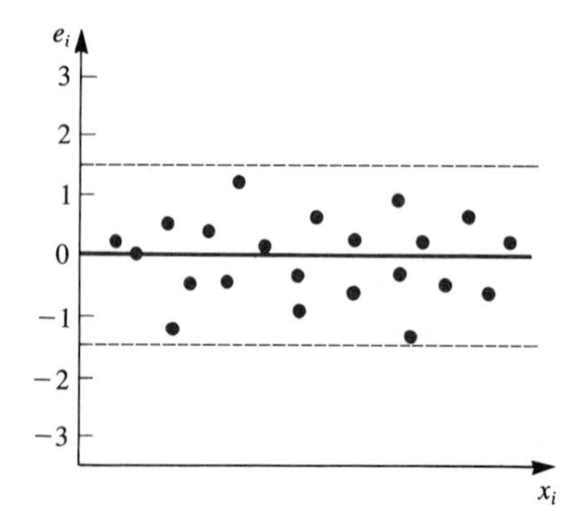

FIGURE 10.28 Some residual plots showing violation of assumptions

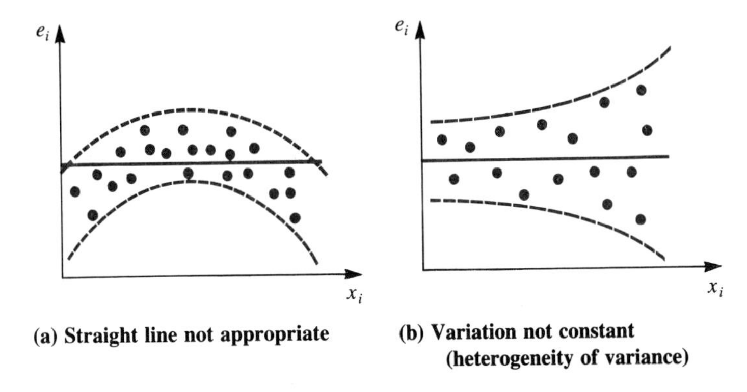

(a) Straight line not appropriate **(b) Variation not constant (heterogeneity of variance)**

tion of various plots of the residual values, to decide if these assumptions might be violated in the data. We are not performing hypothesis tests, but are simply using judgment to see if the graphical patterns correspond to those expected when the assumptions are not violated. A graph of the residual terms as a function of the independent variable x_i should resemble Figure 10.27.

The mean is automatically zero as a result of the least squares technique. When a dashed line is placed above and below the points to enclose them, the two lines are parallel, indicating a constant variance. Since both bands are straight lines running horizontally, it appears that the straight-line model is appropriate. Where these assumptions are violated, plots similar to those in Figure 10.28 are expected to occur.

A graph of residuals plotted versus time as the independent variable may indicate if the residuals are not independent. If, as in Figure 10.29, the sequence

FIGURE 10.29 A residual plot of e_i versus time

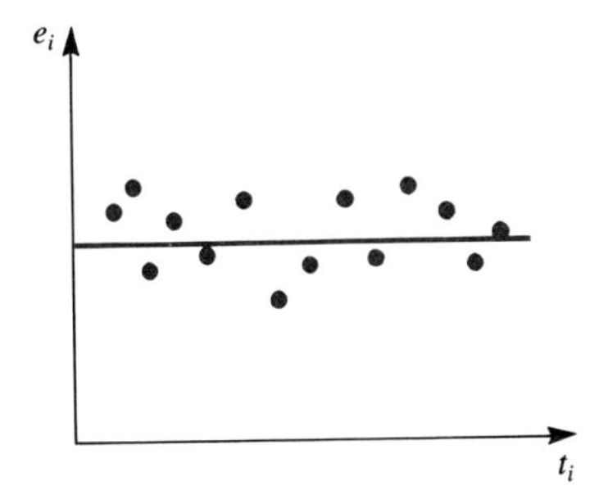

FIGURE 10.30 Residual plot showing error terms that are not independent

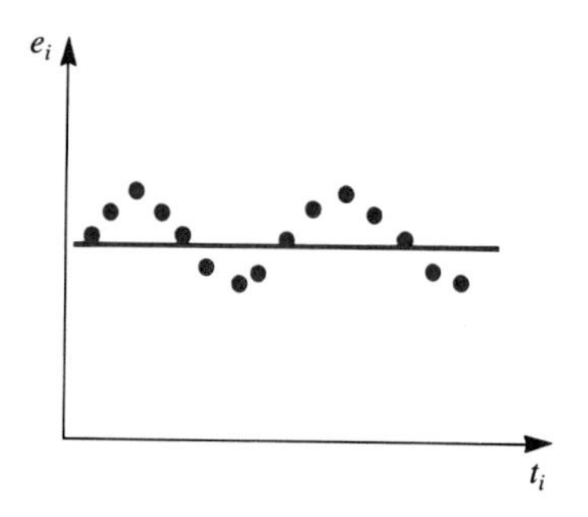

of points is randomly distributed about the line, the terms are considered to be independent of each other over time. A typical violation of this assumption is shown in Figure 10.30. Here, it is obvious that a nonrandom pattern exists. There appears to be a cyclical pattern.

The distribution-free runs test of randomness in Chapter 14 could be applied here. The order in which the data occurred is not always known, making time-related plots impossible in some situations.

The assumption of normality may be investigated by plotting a frequency distribution of the residuals, as shown in Figure 10.31. If, for each x_i in the population, the variances are constant and the distributions are normal, then plotting all the residuals in one frequency distribution should give a distribution that appears approximately normal.

Figure 10.32 shows a plot where the assumption of normality may be violated. The distribution appears to be somewhat uniform. Many other variations are possible.

FIGURE 10.31 Frequency distribution of residuals looking approximately normal

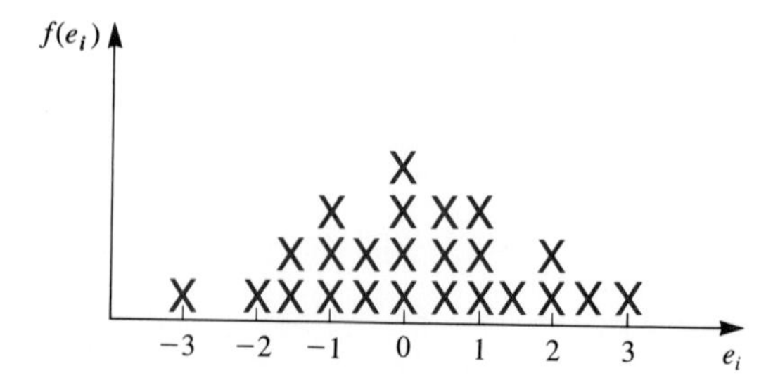

FIGURE 10.32 Nonnormal distribution of residuals

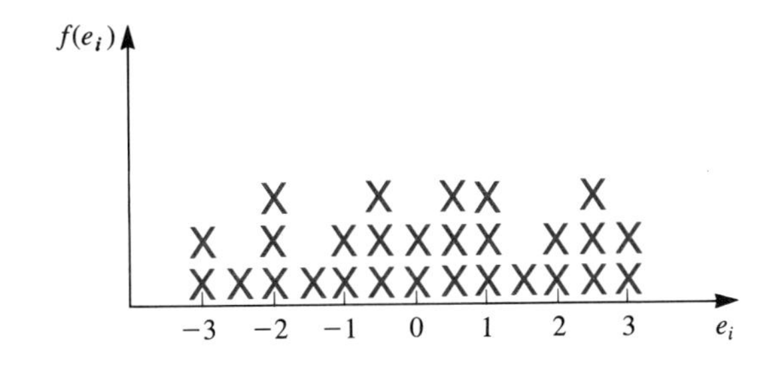

EXAMPLE 10.17 For the tire problem in Example 10.2, plot the residuals and comment on the assumptions. The residuals, calculated in Section 10.3, are shown in Table 10.6.

Solution: A plot of the residuals versus the independent variable, inches of advertising, is shown in Figure 10.33. The limited number of observations prevents drawing any strong conclusions. There is no evidence to suggest that we were wrong in using a straight line, but a hint of the lack of homogeneity of variance seems to be present. Due to the small sample size, we cannot say that this assumption has been violated.

The evidence shown in Figure 10.34 does not tend to suggest a normal distribution. However, these points could, by chance, have come from a normal distribution. Again, with such a small sample, there is insufficient evidence to conclude that the assumption is violated.

TABLE 10.6 Residuals for Example 10.17, using data extracted from Table 10.2

Observation i	*Inches of Advertising* X_i	*Residual* e_i
1	20	−2
2	40	−2
3	50	−1
4	80	−2
5	40	2
6	20	3
7	70	−3
8	80	5

FIGURE 10.33 A residual plot of e_i versus inches of advertising

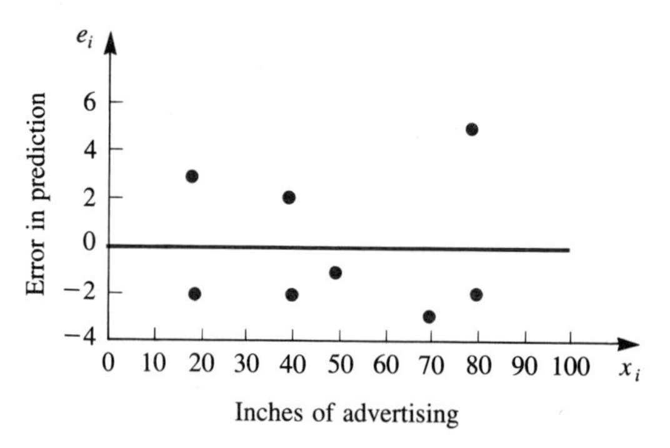

Earlier discussion indicated the plots of the residuals versus time may be used to check the assumption of independence. Because we do not know the time intervals with respect to the observations, we are unable to prepare this plot.

EXAMPLE 10.18 Refer to the Plastico, Inc., problem of Example 10.10 and the computations in Table 10.4a, which shows the residuals. Construct the various types of residual plots and comment on the assumptions. Assume the data were collected in the order presented and in equal increments of time.

Solution: The 3 plots are shown in Figure 10.35. The plot in Figure 10.35a causes us to think that the wrong model has been fit to the data and that the variance is not homogeneous because it appears to decrease as x increases. Lack of fit tends to obscure the analysis of the next two graphs. Had a curved line been

FIGURE 10.34 Frequency distribution of residuals

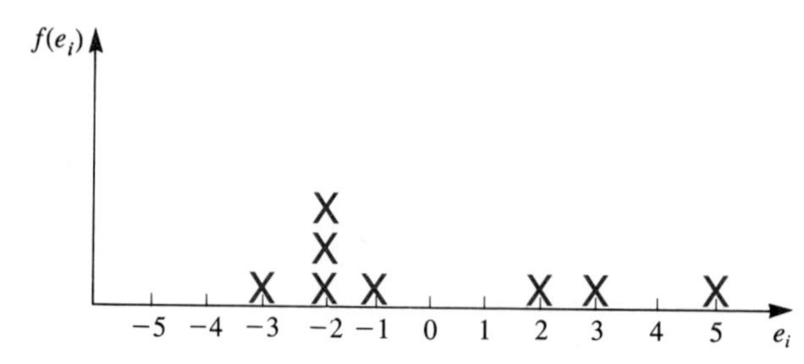

FIGURE 10.35 Residual plots for Example 10.18

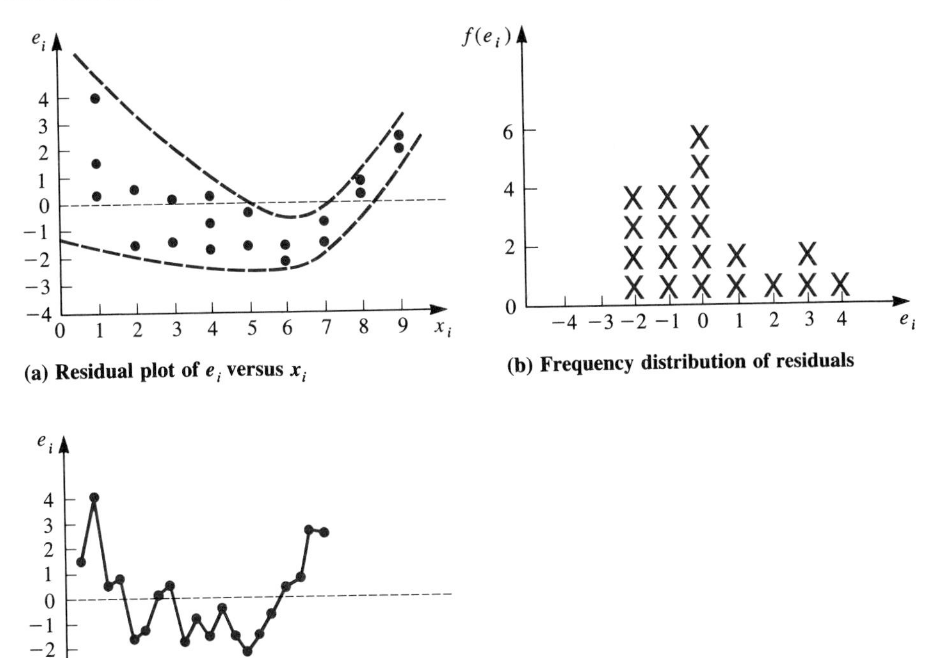

(a) Residual plot of e_i versus x_i

(b) Frequency distribution of residuals

(c) Residual plot over time

fit to the data, the plot in Figure 10.35(b) may have been different and might have suggested that the data are normally distributed about the line. As it is, the data appear to be somewhat skewed. Notice that each point was rounded to the nearest whole number for plotting. In the plot of Figure 10.35(c), we see some pattern in the movement of the points and therefore conclude that they may not be random and that the error terms may not be independent.

EXERCISES

10.17 Look at each of the residual plots illustrated in Figure 10.36, and comment on which assumptions may be violated.

10.18 Plot the 3 types of residual plots and analyze the results, using the data from Exercise 10.11.

FIGURE 10.36 Residual plots for Exercise 10.17

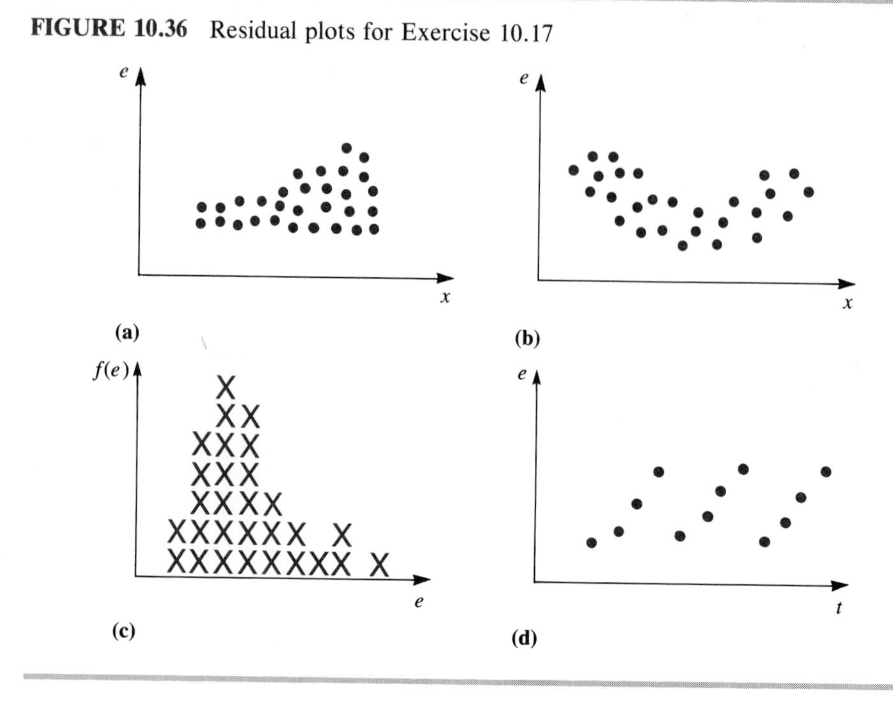

10.9 INTRODUCTION TO LINEAR CORRELATION

Up to this point, the discussion has been limited to the subject of regression. We have learned to evaluate if a relationship exists between two variables, first estimating the proposed equation of the relationship and then testing to decide if the relationship suggested by the data appears to be more than just a chance occurrence.

FIGURE 10.37 Indication of a possible weak relationship

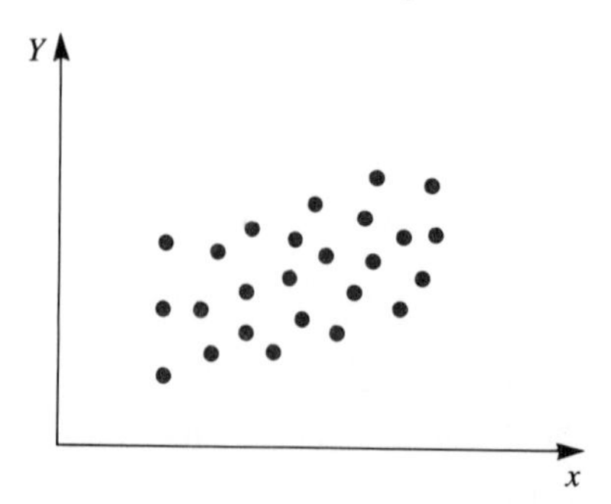

Now we ask the question, How strong is the linear relationship? The analysis required to answer this question is known as correlation. The main difference here is that both X and Y are considered to be random variables with a bivariate normal distribution.

A significant linear relationship can exist between two variables without being a strong relationship. Observe in Figure 10.37 that there appears to be a weak relationship that may not predict Y well, but it may be better than using $\overline{Y}$ as a predictor of Y, which is a common practice when no relationship exists (see Figure 10.38b).

The Coefficient of Determination

One way of measuring the extent of the linear relationship is to calculate the coefficient of determination.

DEFINITION 10.9 The **coefficient of determination** is the proportionate reduction of the total variation in Y associated with the variation of the independent variable X.

The coefficient of determination is symbolized by ρ^2 for the population value and by r^2 for its corresponding estimator. We have

$$0 \leq \rho^2 \leq 1 \quad \text{and} \quad 0 \leq r^2 \leq 1$$

A value of 1 for ρ^2 indicates that a perfect linear relationship exists between X and Y, and that, given a value of X, the value of Y could be predicted perfectly. The scatterplot and fitted line for this type of relationship are shown in Figure 10.38(a). Once again, in this figure, just a few points are used to represent the infinite number of points in the population.

FIGURE 10.38 Extreme degrees of relationship

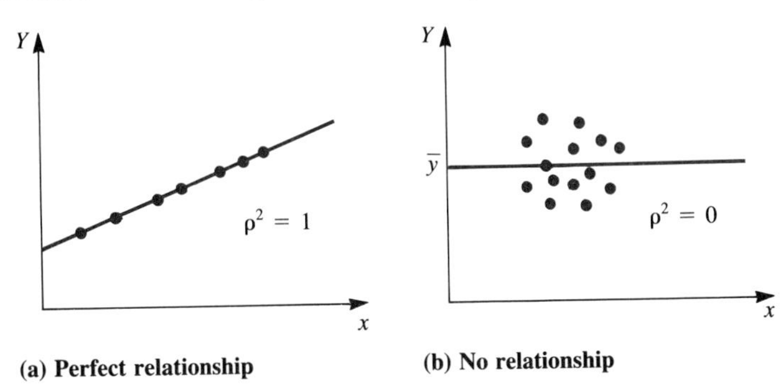

(a) Perfect relationship

(b) No relationship

If no linear relationship exists, as in Figure 10.38(b), the value of ρ^2 is zero, and the slope of the theoretical line is zero.

If a random sample is selected from a population whose ρ^2 is zero, it is rare that the sample r^2 will also be zero, but, as explained earlier, it is possible to appear to have a relationship using sample values where there is none in the population. The value of r^2 may be greater than zero when ρ^2 is zero, although it will usually be small.

Normally a value of r^2 equal to 1 is not expected, for some sampling error always exists. Problems of economic origin typically show r^2 in the .50 to .90 range, whereas r^2 values of .05 to .30 are typical in studies involving human behavior. Predicting human behavior is quite difficult.

If we have already prepared an analysis of variance table, we have available the numbers needed to calculate r^2. Remember that the sum of squares is partitioned as

$$SSR + SSE = SST$$

The regression plus the error sums of squares equals the total sum of squares. Dividing each term by SST, we can find each as a percent of the total.

$$\frac{SSR}{SST} + \frac{SSE}{SST} = 1$$

The formula for r^2 is

$$r^2 = \frac{SSR}{SST} \tag{10.27}$$

which is also

$$r^2 = \frac{\Sigma(\hat{Y}_i - \overline{Y})^2}{\Sigma(Y_i - \overline{Y})^2}$$

That is, r^2 is a measure of the percent of the total variation of Y accounted for by changes in levels of the independent variable.

An alternative form of the equation is

$$r^2 = 1 - \frac{SSE}{SST} \tag{10.28}$$

which can be restated as

$$r^2 = 1 - \frac{\Sigma(Y_i - \hat{Y}_i)^2}{\Sigma(Y_i - \overline{Y})^2}$$

EXAMPLE 10.19

For the tire problem in Example 10.2, determine the percent of variation in tire sales explained by the number of inches of advertising.

Solution: We calculate the coefficient of determination, using Equation 10.27. Recall that the analysis of variance table is

Source	*SS*	*df*	*MS*
Regression	1050	1	1050
Error	60	6	10
Total	1110	7	

$$r^2 = \frac{SSR}{SST} = \frac{1050}{1110} = .946$$

We also could have calculated r^2 by using Equation 10.28.

$$r^2 = 1 - \frac{SSE}{SST} = 1 - \frac{60}{1110}$$

$$= 1 - .054 = .946$$

Note that this value is the first R^2 value given in Table 10.1b. The remaining 5.4% is attributed to variation in tire sales due to factors other than advertising.

Taking another look at the alternative form of the equation for the coefficient of determination, Equation 10.28, we can observe why a perfect relationship gives an r^2 of unity. If the line fits perfectly through all the points, as in Figure 10.38a, then each deviation $Y_i - \hat{Y}_i$ is zero, and, thus, $SSE = \Sigma(Y_i - \hat{Y}_i)^2$ is also zero. Hence, the SSE/SST term in the equation is zero, and the r^2 is 1. Where no relationship exists, SSE equals SST—that is, the total variation is all error variation and the r^2 is zero.

EXERCISE

10.19 Calculate the coefficient of determination for the tire problem of Exercise 10.11.

The Coefficient of Correlation

A measure related to the coefficient of determination is the coefficient of linear correlation.

DEFINITION 10.10 The **coefficient of correlation** is a unitless measure of the strength of the linear relationship between two variables.

It is calculated as the square root of the coefficient of determination multiplied by the sign of the slope of the regression equation. The population coefficient of correlation is represented by the symbol ρ.

$$\rho = (\text{sign } \beta)\sqrt{\rho^2} \qquad \textbf{(10.29)}$$

FIGURE 10.39 Perfect inverse relationship

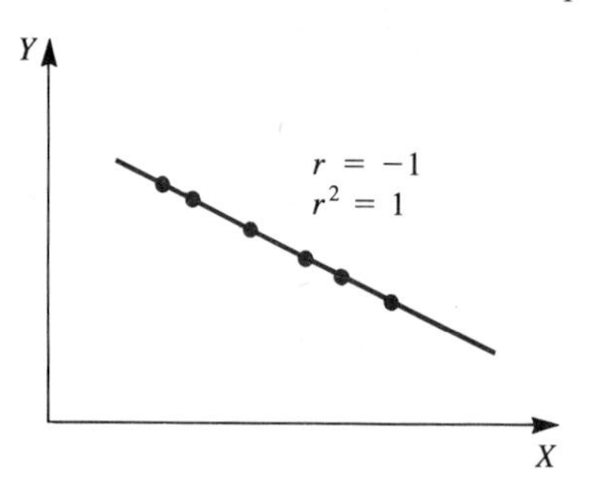

Likewise, for sample values

$$r = (\text{sign } b)\sqrt{r^2} \tag{10.30}$$

The values for ρ and r range from -1 to 1.

$$-1 \leq \rho \leq 1 \qquad -1 \leq r \leq 1$$

The r value measures the strength of a linear regression. The r^2, however, is not restricted to linear relationships. Except for giving the direction of the relationship, the coefficient of correlation has no practical interpretation, unlike the coefficient of determination, which is a percentage measure. Notice that r values may go to -1 rather than zero. The negative sign corresponds to a negative slope in the regression equation. Figure 10.39 shows a perfect relationship with a negative slope, which is indicted by $r = -1$. The corresponding r^2 would be

$$r^2 = (-1)^2 = 1$$

Positive r values are either equal to or larger than their r^2 counterparts.

$$\begin{aligned} r^2 &= 0 & r &= 0 \\ r^2 &= .01 & r &= \pm.10 \\ r^2 &= .25 & r &= \pm.50 \\ r^2 &= .81 & r &= \pm.90 \\ r^2 &= 1 & r &= \pm1 \end{aligned}$$

The larger value for r perhaps accounts for the greater use of r by those whose research activities are primarily in the area of human behavior, where r^2 values tend to be quite small. Economists and business researchers tend to use r^2.

EXAMPLE 10.20 Determine the strength of the relationship between tire sales and number of inches of advertising, using the results of Example 10.19.

Solution:

$$r = (\text{sign } b)\sqrt{r^2} = (+)\sqrt{.946} = .973$$

EXAMPLE 10.21 The equation for predicting the output of chairs per day of a furniture worker versus the machine speed in revolutions per minute is

$$\hat{Y} = 84 - .5x$$

within the range $X = 5$ to 10 revolutions per minute and $r^2 = .49$. Calculate r.

Solution:

$$r = (\text{sign } b)\sqrt{r^2} = (-)\sqrt{.49} = -.70$$

Notice that r is negative because the slope of the line is negative.

Often, we desire to know the correlation between two variables without wanting the equation of the relationship. Rather than having to estimate the line and then to construct the analysis of variance table before being able to calculate r^2 and r, we can calculate r directly by Pearson's formula. This formula is generally used when all that is wanted to be known is the value for r and not the regression equation.

$$r = \frac{\sum X_i Y_i - \dfrac{\sum X_i \sum Y_i}{n}}{\sqrt{\left(\sum X_i^2 - \dfrac{(\sum X_i)^2}{n}\right)\left(\sum Y_i^2 - \dfrac{(\sum Y_i)^2}{n}\right)}} \qquad \textbf{(10.31)}$$

EXAMPLE 10.22 Calculate the coefficient of correlation for the tire problem of Example 10.2 by Pearson's formula.

Solution: The initial computations are found in Table 10.7.

$$r = \frac{20{,}100 - \dfrac{(400)(360)}{8}}{\sqrt{\left(24{,}200 - \dfrac{(400)}{8}^2\right)\left(17{,}310 - \dfrac{(360)}{8}^2\right)}}$$

$$= \frac{20{,}100 - 18{,}000}{\sqrt{(24{,}200) - 20{,}000)(17{,}310 - 16{,}200)}}$$

$$= \frac{2100}{\sqrt{(4200)(1110)}} = \frac{2100}{\sqrt{4{,}662{,}000}}$$

$$= \frac{2100}{2159.1665} = .973$$

which is the same value obtained from Equations 10.27 and 10.30.

TABLE 10.7 Calculation of the summations for Example 10.22

Observation i	*Inches of Advertising* x_i	*Number of Tires Sold* y_i	x_iy_i	x_i^2	y_i^2
1	20	28	560	400	784
2	40	38	1,520	1,600	1,444
3	50	44	2,200	2,500	1,936
4	80	58	4,640	6,400	3,364
5	40	42	1,680	1,600	1,764
6	20	33	660	400	1,089
7	70	52	3,640	4,900	2,704
8	80	65	5,200	6,400	4,225
	400	360	20,100	24,200	17,310
	Σx_i	Σy_i	Σx_iy_i	Σx_i^2	Σy_i^2

If only a correlation coefficient is of interest and neither regression equations nor confidence intervals for predictions are required, then Equation 10.31 is generally used. To determine if the value for r is indicative of a relationship or merely a random deviation from zero requires a hypothesis test.

The assumptions for testing a regression equation were stated earlier. One was that we have random Y but nonrandom x data, although it was pointed out that the F and t tests were valid for random X with only a need to revise the interpretation of the results.

For the correlation coefficient, it is assumed that both X and Y are random variables and that they vary jointly in a distribution known as a bivariate normal distribution, illustrated in Figure 10.40. This distribution can be cut with any plane perpendicular to the X, Y plane, and the exposed cross-section will exhibit a normal distribution. Therefore, for each X, there is a normally distributed subpopulation of Y values, and for each Y, there is a normally distributed subpopulation of X's.

If we are not certain that the assumption of a bivariate normal distribution is satisfied, the t test can still be used, provided that Y is normal. See Snedecor and Cochran [2], p. 195. In addition, it makes no difference which variable is treated as X and which as Y since the correlation coefficient for either pairing is the same.

The null hypothesis for the test of interest is that $\rho = 0$. The t statistic is computed as

$$t = \frac{r - 0}{s_r} \quad \text{where } s_r = \sqrt{\frac{1 - r^2}{n - 2}}$$

The calculated value is compared to the tabular value of t for $\alpha/2$ and $n - 2$ degrees of freedom.

FIGURE 10.40 Bivariate normal distribution showing normal distributions in both the X and Y dimensions

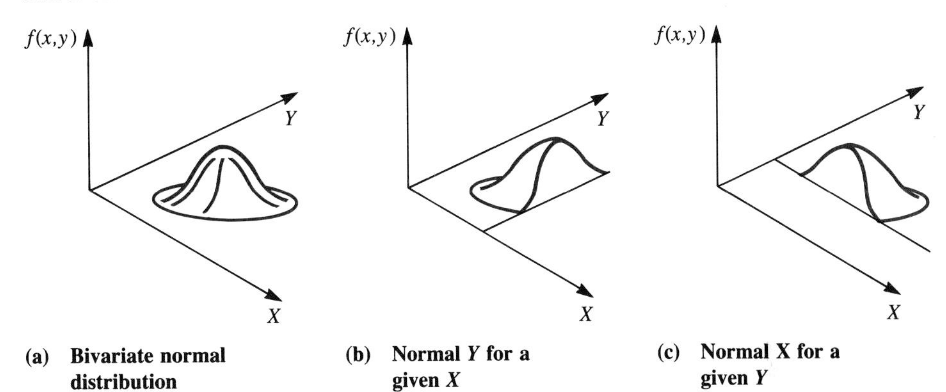

(a) Bivariate normal distribution

(b) Normal Y for a given X

(c) Normal X for a given Y

DECISION RULE To test $H_0: \rho = 0$; $H_a: \rho \neq 0$, reject H_0 if

$$\left| \frac{r\sqrt{n-2}}{\sqrt{1-r^2}} \right| > t_{n-2,\alpha/2} \qquad \textbf{(10.32)}$$

EXAMPLE 10.23 Test the correlation coefficient of Example 10.20 for significance at $\alpha = .05$.

Solution: The hypotheses and decision rule are

$$H_0: \rho = 0 \qquad H_a: \rho \neq 0$$

and we reject H_0 if $t \geq t_{n-2,\alpha/2}$.

$$t = \frac{.973 - 0}{\sqrt{(1 - .946)/(8 - 2)}} = \frac{.973}{\sqrt{.009}}$$

$$= \frac{.973}{.095} = 10.724$$

$$t_{6,.025} = 2.447$$

$10.24 > 2.477$; therefore, we reject H_0, as illustrated in Figure 10.41. The sample correlation is significantly different from 0, so we conclude that correlation exists between tire sales and number of inches of advertising. Other than rounding, this is the same value obtained for t in Example 10.12.

FIGURE 10.41 Sampling distribution for Example 10.23

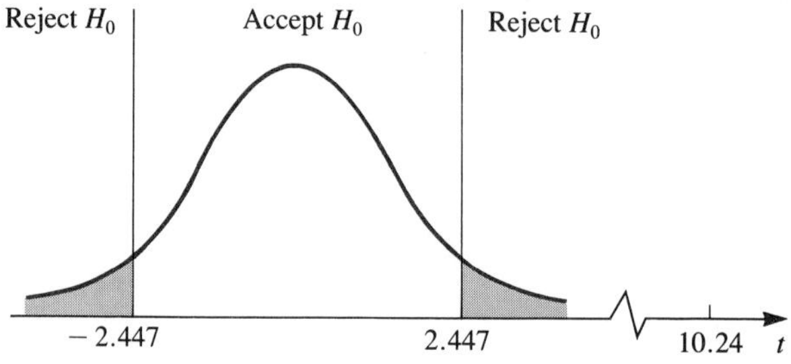

If it is desired to test for values of ρ other than zero, it is necessary to use Fisher's z transformation. The reader may refer to other texts for this test.

EXERCISES

10.20 Using the results of Exercise 10.19, calculate the coefficient of correlation.

10.21 The relationship between price and the quantity sold of a certain cosmetic is $r^2 = .64$. The regression equation is $\hat{Y} = 20.95 - .02x$. Calculate the coefficient of correlation.

10.22 Remember the aquaculture business in Exercise 10.12? Use those data and Pearson's formula to calculate the coefficient of correlation.

10.23 Test the correlation coefficient in Exercise 10.22 to determine if it comes from a population in which no relationship exists. Use $\alpha = .05$.

10.10 SUMMARY

Linear regression is concerned with estimating an equation that is assumed to represent the relationship between an independent and a dependent variable. In this chapter, a straight line is assumed to be the correct model, but this assumption can be tested in certain situations. The linear model is estimated so as to minimize the standard error of the estimate. The standard error of the estimate represents a measure of the spread of the points around the line.

If a sample of observations is used to estimate the population line, a number of assumptions must be satisfied for inferential purposes. Analysis of the residuals provides some hints about how well the assumptions are satisfied. Hypothesis tests for the slope of the line indicate if a relationship is likely to exist in the population. A significant equation may be useful in making point estimates and, what is better, interval estimates for both average and particular values.

Correlation analysis allows us to evaluate the strength of the linear relationship and what percentage of the total variation in Y is explained by the variation in X. The equations for regression and correlation analysis, along with usage, are given in Table 10.8, the summary table.

TABLE 10.8 Summary table

Least Squares Line	$\hat{Y}_i = a + bx_i \qquad b = \dfrac{\sum x_i Y_i - \dfrac{\sum x_i \sum Y_i}{n}}{\sum x_i^2 - \dfrac{(\sum x_i)^2}{n}} \qquad a = \dfrac{\sum Y_i}{n} - b\dfrac{\sum x_i}{n}$
Sums of Squares and Degrees of Freedom	$SST = \sum Y_i^2 - \dfrac{(\sum Y_i)^2}{n}$ $df_T = n - 1$ $SSR = b\left(\sum x_i Y_i - \dfrac{\sum x_i \sum Y_i}{n}\right)$ $df_R = 1 =$ number of independent variables $SSE = SST - SSR = \sum Y_i^2 - a\sum Y_i - b\sum x_i Y_i$ $df_E = n - 2$ $SSPE = \sum_{j=1}^{c} \sum_{i=1}^{n_j} (Y_{ij} - \bar{Y}_j)^2$ $df_{PE} = n - c,$ *where*: $n = \sum n_j$ $SSLF = SSE - SSPE$ $df_{LF} = df_E - df_{PE} = c - 2$
Hypotheses and Test Statistics	Test for linearity: $H_0\colon \mu_{Y\|x} = \alpha + \beta x \qquad H_a\colon \mu_{Y\|x} \neq \alpha + \beta x$ Reject H_0 if $F = \dfrac{MSLF}{MSPE} > F_{c-2,n-c,\alpha}$ (table value) Test for usefulness: $H_0\colon \beta = 0 \qquad H_a\colon \beta \neq 0$ Reject H_0 if $F = \dfrac{MSR}{MSE} > F_{1,n-2,\alpha}$ (table value) Test for a specified nonzero value: $H_0\colon \beta = \beta_0 \qquad H_a\colon \beta \neq \beta_0$ Reject H_0 if $t = \left\| \dfrac{b - \beta_0}{s_b} \right\| > t_{n-2,\alpha/2}$ (table value) *where*: $s_b = \dfrac{s_{Y\|x}}{\sqrt{\sum x_i^2 - \dfrac{(\sum x_i)^2}{n}}}$ $s_{Y\|x} = \sqrt{\dfrac{\sum Y_i^2 - a\sum Y_i - b\sum x_i Y_i}{n - 2}} = \sqrt{\dfrac{SSE}{df_E}}$

TABLE 10.8 (continued)

Interval Estimates	For β $b \pm t_{n-2,\alpha/2} \dfrac{s_{Y\|x}}{\sqrt{\sum x_i^2 - \dfrac{(\sum x_i)^2}{n}}}$ For a point estimate at $x = x_p$ $\hat{Y}_{x_p} = a + bx_p$ For an interval estimate for $\mu_{Y\|x_p}$ $\hat{Y} \pm t_{n-2,\alpha/2}\, s_{Y\|x} \sqrt{\dfrac{1}{n} + \dfrac{(x_p - \bar{x})^2}{\sum x_i^2 - \dfrac{(\sum x_i)^2}{n}}}$ For an interval estimate for Y_{x_p} $\hat{Y}_{x_p} \pm t_{n-2,\alpha/2}\, s_{Y\|x} \sqrt{1 + \dfrac{1}{n} + \dfrac{(x_p - \bar{x})^2}{\sum x_i^2 - \dfrac{(\sum x_i)^2}{n}}}$
Measures of the Strength of a Linear Relationship	Pearson's coefficient of correlation: $r = \dfrac{\sum X_i Y_i - \dfrac{\sum X_i \sum Y_i}{n}}{\sqrt{\left(\sum X_i^2 - \dfrac{(\sum X_i)^2}{n}\right)\left(\sum Y_i^2 - \dfrac{(\sum Y_i)^2}{n}\right)}} = (\text{sign } b)\sqrt{r^2}$ Coefficient of determination: $r^2 = \dfrac{SSR}{SST} = 1 - \dfrac{SSE}{SST}$
Test for a Relationship	$H_0: \rho = 0 \qquad H_a: \rho \neq 0$ Reject H_0 if $t = \left\| \dfrac{r - 0}{s_r} \right\| > t_{n-2,\alpha/2}$ (tabular value) *where*: $s_r = \sqrt{\dfrac{1 - r^2}{n - 2}}$
Assumptions and Evaluations	Assumptions for inferences: Nonrandom x (x values exactly measured) Normality of error terms (also implies that Y values are interval) Zero mean for random variable ε—that is, $E(\varepsilon_i) = 0$ Homogeneity (uniform variance in Y for each x along regression line) Linear model Independence of error terms

TABLE 10.8 (continued)

Assumptions and Evaluations	Evaluation of assumptions: Plot e_i versus x_i to observe whether model is correct and variance is homogeneous. Plot e_i versus t_i to observe whether e_i are independent. Plot e_i versus frequency to observe whether errors are normally distributed.

SUPPLEMENTARY EXERCISES

10.24 The Carchester Carburetor Company uses an assembly line process for building its carburetors. Management has been experimenting with assembly line speed (feet/minute) to attempt to maximize the number of defect-free units assembled per hour. A portion of the collected data is given here. Which of the 4 models in Section 10.2 will probably best fit these data?

Line Speed	2.0	2.8	4.2	6.2	6.9	2.0	3.7	4.8	6.2	7.2
Units Completed	19	31	41	28	26	30	34	36	33	21

10.25 A cosmetic company desires to determine if price influences the sales of a particular perfume. Taking random samples from its records on this product, the company obtains the following data.

Sales ($000,000)	2	3	5	7	8
Price ($/ounce)	13	12	15	16	14

(a) What is the equation of the relationship?
(b) Would lowering the price increase sales? Explain.

10.26 **(a)** Using the information in Exercise 10.25, predict the average sales of cosmetics at a price of $17.00 per ounce. Use a 95% confidence interval.
(b) What is the danger inherent in this type of extrapolation?

10.27 What would be the individual amount of sales for the conditions expressed in Exercise 10.26? Use a 95% confidence interval.

10.28 The Simmons Employment Agency prepared a special test that is given to select management trainees for a certain company. The test has been used for years. Now the company wants Simmons to evaluate the scores received on the tests in relation to the number of years it took for the testees to be promoted to an executive position. The agency used simple linear regression on the following sample data. Is this model appropriate? Use $\alpha = .05$.

Test Score	51	60	84	60	37	90	90	67	78	84
Years to Promotion	2.50	7.00	4.33	2.50	7.33	4.00	5.50	2.25	6.75	3.00

10.29 Schofield Keypunching Associates feels that the number of errors made in keypunching decreases as the number of years of experience increases. A sample of the work of 8 employees is examined and compared to the number of years experience. Visually fit a straight line to the data and determine the equation of the line.

Average Number of Errors	3.5	3.3	2.9	3.2	2.9	3.0	3.3	3.9
Years Experience	1.5	2	4	5	5	4.5	3	1

10.30 The Wilson Automotive Agency desires to estimate the linear relationship between the number of salespeople working on Saturdays and the number of automobiles sold that day. Data collected for a random sample of 6 Saturdays are given.

Number of Salespeople	1	3	5	4	7	10
Number of Autos Sold	1	2	3	2	5	5

(a) What is the estimating equation for the relationship?
(b) Which assumption for inference has obviously been violated?

10.31 **(a)** Test the equation in Exercise 10.30 for significance using analysis of variance. Use $\alpha = .05$.
(b) What effect does the violation of the assumption have on your conclusions?

10.32 The Roberts Company wants to know how closely related the movement of its stock price is to the movement of the Dow-Jones Index. Estimate the strength of the linear relationship, using the data below. Test for significance with an α of .01.

Dow-Jones	891	853	821	842	861	810
Stock Prices	51	50	43	43	45	41

10.33 The Cuyahoga Summit Railroad is having difficulty determining its fixed costs. A recent business school graduate employed in its accounting department suggests the use of linear regression for estimation of the fixed costs. He obtained the following sample of data giving total costs and output.

Total Cost ($000,000)	50	36	54	41	48	63	62	79	65	81
Output (million ton-miles)	12	15	24	30	38	43	50	54	58	74

(a) What estimate of fixed costs is obtained?
(b) What is the value of the standard error of the slope? Interpret both the slope and its standard error in context.

10.34 Test the equation in Exercise 10.33 for significance, using a t test for the slope at $\alpha = .05$.

10.35 Determine a 95% confidence interval for the true value of the slope (β) for the data in Exercise 10.33.

10.36 Analyze the residuals of Exercise 10.33, first graphing the residuals and then commenting on whether or not the assumptions appear to be satisfied.

10.37 For the data in Exercise 10.33, calculate the percent of variation in total cost explained by the ton-miles of output.

10.38 If the railroad in Exercise 10.33 can increase its output to 100, what total cost can the railroad expect to average? Place a 95% confidence interval around the estimate.

10.39 Test the hypothesis that the railroad in Exercise 10.33 will experience a total cost that averages $75 million dollars when output is 100 million ton-miles. Use $\alpha = .05$.

10.40 If the railroad in Exercise 10.33 experiences a period in which its output is 60 million ton-miles, what total cost is predicted for that period? Place a 95% confidence interval around the prediction.

10.41 Shanley's Auto Diagnostic Center sends its new employees to school to learn to diagnose auto engine problems. Believing that the time required to diagnose problems is a function of the number of days of training received, the owner does a simple linear regression on the sample of employee data given. Is this model adequate? Use $\alpha = .01$.

Days Training	2	2	3	3	4	5	5	5
Minutes of Diagnostic Time	26	22	17	20	19	13	14	12

10.42 Using $\alpha = .01$, test the equation in Exercise 10.41 for significance **(a)** using analysis of variance and **(b)** using the t test for the slope.

10.43 Demonstrate that both tests in Exercise 10.42 are equivalent and state the conditions under which this equivalency is true.

10.44 Zoologic Thermometers, Inc. sells Amazon chirp bugs, found only in the upper Amazon River region of Brazil. In its advertising, the company has stated that temperature can be determined by measuring the number of seconds between chirps of the bug. The FTC is now asking for proof. An experiment was conducted by varying the temperature and measuring the time between chirps. The data obtained are

Temperature	15	20	25	30	35	43	46	51	59	62
Time	21	23	28	33	33	36	34	39	43	41
Temperature	66	68	71	74	75	76	77	79	81	84
Time	34	44	40	33	47	42	37	30	25	47

(a) Calculate the straight-line equation.
(b) Test the equation for significance, using $\alpha = .01$.
(c) Plot the residuals. Which assumptions do not appear to be satisfied?

10.45 Calculate the coefficient of determination and the coefficient of correlation for the data of Exercise 10.44.

10.46 A drug manufacturer wants to determine whether a correlation exists between the dosage of its new drug Sanguinol and a patient's body temperature. It collects the following data on a patient in a carefully controlled environment.

Dose (milligrams)	2	4	3	4	2
Temperature (°F)	98.6	101.3	100.6	101.5	99.0
Dose (milligrams)	1	3	3	4	2
Temperature (°F)	98.6	100.8	100.5	102.0	100.0

(a) Calculate the linear regression equation.
(b) Test the equation for significance, using $\alpha = .05$.
(c) Predict the body temperature for a specific 5-milligram dose, using a 95% confidence interval.
(d) Can this result be generalized beyond this patient? Why?

10.47 In the previous problem, determine if a straight line is the appropriate model, using $\alpha = .05$.

10.48 The city of Upperville wants to predict its tax revenues for next year. The data for the last 10 years are

Year	1978	1979	1980	1981	1982	1983	1984	1985	1986	1987
Revenue ($000,000)	1.400	1.502	1.599	1.685	1.812	1.960	2.123	2.400	2.810	2.900

(a) Determine the linear regression equation.
(b) Predict tax revenues for 1988.
(c) What is the coefficient of determination?
(d) Test the equation for significance at $\alpha = .05$.
(e) Do you think any assumptions for the test are violated?

10.49 The telephone company wants to determine how many phones it will install next year in Ohio City. It has data on the total phones in use for each of the past 10 years and data for a few variables that it thinks may be good indicators of phone sales. Which of the following variables appears to be the best predictor if a linear model is used?

Year	*Total Phones*	*Population*	*Housing Permits*	*Number of Businesses*
1977	25,621	105,000	12,400	236
1978	42,851	154,000	12,300	249
1979	53,128	201,000	9,650	321
1980	62,153	232,000	7,250	379
1981	75,821	305,000	20,100	481
1982	85,113	328,000	4,750	521
1983	98,513	402,000	18,400	592
1984	104,776	410,000	2,400	650
1985	111,213	493,000	10,300	740
1986	121,487	548,000	5,210	810

10.50 **(a)** Refer to Exercise 10.49. How many phones could the phone company expect to install in 1990?
(b) Comment on the danger of extrapolation.

10.51 Refer to Exercise 10.49 and, using a regression equation, determine the change in the total phones on the average for each one-person change in the population.

10.52 Refer to Exercise 10.49 and evaluate if the number of housing permits is related to the size of the population.

10.53 **(a)** For the data in Exercise 10.49, predict the number of housing permits that would be issued in 1988 if the population increases to 650,000.
(b) Evaluate the accuracy of this result.

10.54 For the relation of population to total phones, plot the residuals and comment on the underlying regression assumptions. Note that the data are in time sequence.

10.55 Analysts at First Federal Savings and Loan have been wondering if the change in the amount of money saved per month, deposits, is related to the monthly change in the rate of inflation. They randomly select 10 months from the past 5 years of data in order to achieve some degree of independence between the data points.

Change in Inflation Rate	*Change in Deposits*
.006	243,000
.012	−164,000
.003	200,000
−.001	426,000
.010	−174,000
.003	208,000
−.001	581,000
.006	210,000

(a) Calculate the regression equation.
(b) Test the equation for significance at $\alpha = .01$.
(c) Estimate the average change in deposits if inflation increases to . 015 per month. $\alpha = .01$.
(d) What theory, if any, might you develop from this analysis?
(e) Interpret the value of the intercept in your model.

10.56 The president of Japan HiFi Products believes that stereo receivers are purchased not on impulse but after months of consideration. Therefore, changes in advertising expenditures are not expected to be reflected in sales until at least 1 month later. Such a relationship is called a lagged relationship. The following data were collected.

	Advertising ($1000)	*Sales ($100,000)*
Jan.	200	758
Feb.	400	1241
Mar.	300	788
Apr.	150	1621
May	400	913
June	600	624
July	200	1598
Aug.	400	1908
Sept.	300	1005

Calculate 3 equations, 1 using no lag, 1 using a 1-month lag, and 1 using a 2-month lag. Note that the longer the lag, the fewer the number of observations that are available. Decide whether a lagged relationship exists and, if so, the number of months of the lag. (*Hint:* Ignore the 758 and move all other sales up 1 month for a 1-month lag. Then, ignore the 300 for September.)

***10.57** Show that the equation for b can be written as

$$b = \frac{\Sigma(x_i - \bar{x})Y_i}{\Sigma(x_i - \bar{x})^2}$$

* An asterisk indicates a higher level of difficulty.

***10.58** Prove that for $x = x_p$,

$$\sigma^2_{\hat{Y}xp} = \sigma^2 \left(\frac{1}{n} + \frac{(x_p - \bar{x})^2}{\Sigma(x_i - \bar{x})^2}\right)$$

under the regression assumptions. *Hint:* First show that $\hat{Y}_{x_p}$ can be expressed in the form

$$\hat{Y}_{xp} = \bar{Y} + b(x_p - \bar{x})$$

REFERENCES

1. Neter, John, William Wasserman, and Michael Kutner. *Applied Linear Statistical Models,* 2nd ed. Homewood, IL: Richard D. Irwin, 1985.
2. Snedecor, George W., and W. G. Cochran. *Statistical Methods,* 7th ed. Ames, IA: Iowa State University Press, 1980.

APPENDIX

Derivation of the Least Squares Equations

We wish to minimize

$$Q = \Sigma(Y_i - a - bx_i)^2$$

This minimization is accomplished through differentiation, first with respect to a, then with respect to b, and the results set equal to zero as the condition of an optimum.

$$\frac{\partial Q}{\partial a} = -2\Sigma(Y_i - a - bx_i) = 0$$

$$\frac{\partial Q}{\partial b} = -2\Sigma x_i(Y_i - a - bx_i) = 0$$

The -2 terms drop out. Using the rules of the Σ operator from Chapter 1, we get

$$\Sigma Y_i - na - b\Sigma x_i = 0$$

$$\Sigma x_i Y_i - a\Sigma x_i - b\Sigma x_i^2 = 0$$

which, rearranged, give the normal equations

$$an + b\Sigma x_i = \Sigma Y_i$$

$$a\Sigma x_i + b\Sigma x_i^2 = \Sigma x_i Y_i$$

Since this is a quadratic form, second derivatives are not considered, and this is a minimum.

11 Multiple Regression and Correlation Models

11.1 INTRODUCTION

The Victory Valve Company has just created a marketing research group assigned the task of predicting the demand for large industrial valves. In their first meeting, the group makes up a list of factors they believe to be related to sales of the big valves. Included are the general condition of the economy; oil production; building of power plants, chemical plants, and water distribution plants; the amount they advertise; the price they charge; the price their major competitor charges; the number of salespeople; and the growth of the population. But, they are sure that other factors exist, which they have not yet thought of, that are also able to explain the demand for their valves.

They decide to build a mathematical model that can be used for making demand predictions. Their goal is to include the factors that are most related to the dependent variable and that will, together, provide the best model for prediction. The statistical tools they decide to use to build and evaluate this predictive model are multiple regression and correlation.

As with simple linear regression, multiple regression equations can be used for predictive purposes. While a simple linear regression model has only one independent variable, multiple regression models have two or more. More of the variation in the dependent variable may be explained with the additional independent variables in the model.

In this chapter, we examine the development and testing of multiple regression models and review methods for determining the "best" model, constrained by the intended use. We also consider the use of indicator variables for qualitative variables, as well as correlation models.

Throughout, the use of the computer for computations is emphasized. Computation of nonlinear regression equations using the multiple regression computer program is also examined.

11.2 ESTIMATING THE MULTIPLE LINEAR MODEL

It is possible to extend a simple linear regression analysis by adding other independent variables to the model.

DEFINITION 11.1 A regression equation involving several independent variables is called a **multiple linear regression equation** if it is a linear combination of the regression coefficients or parameters—that is, if the equation is of the form

$$Y_i = \alpha + \beta_1 x_{i1} + \beta_2 x_{i2} + \cdots + \beta_k x_{ik} + \varepsilon_i \tag{11.1}$$

where: $\alpha, \beta_1, \beta_2, \ldots, \beta_k$ = parameters

$x_1, x_2, \ldots, x_k$ = nonrandom variables

ε_i = a random error term

Additional variables are included when it is believed that more than one variable is related to the dependent variable. Each additional variable added should reduce the standard error of the estimate; otherwise, the variable does not improve the ability of the model to predict the value of the dependent variable.

It is difficult to visualize the model beyond two dimensions, and it is impossible beyond three. With two independent variables, we are fitting a plane called a **hyperplane** to points in three-dimensional space, as in Figure 11.1.

As with simple linear regression, the constant and coefficients can be estimated by the least squares approach, which is generalized in the following normal equations:

$$\begin{aligned}
na \;&+ b_1\Sigma x_{i1} &&+ b_2\Sigma x_{i2} &&+ \cdots + b_k\Sigma x_{ik} &&= \Sigma Y_i \\
a\Sigma x_{i1} &+ b_1\Sigma x_{i1}^2 &&+ b_2\Sigma x_{i1}x_{i2} &&+ \cdots + b_k\Sigma x_{i1}x_{ik} &&= \Sigma x_{i1}Y_i \\
a\Sigma x_{i2} &+ b_1\Sigma x_{i2}x_{i1} &&+ b_2\Sigma x_{i2}^2 &&+ \cdots + b_k\Sigma x_{i2}x_{ik} &&= \Sigma x_{i2}Y_i \\
& && \vdots && && \vdots \\
a\Sigma x_{ik} &+ b_1\Sigma x_{ik}x_{i1} &&+ b_2\Sigma x_{ik}x_{i2} &&+ \cdots + b_k\Sigma x_{ik}^2 &&= \Sigma x_{ik}Y_i
\end{aligned} \tag{11.2}$$

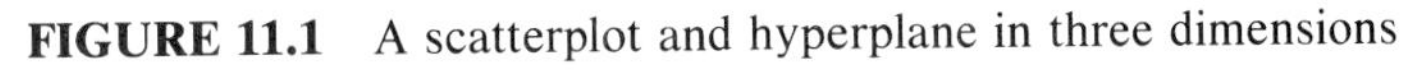

FIGURE 11.1 A scatterplot and hyperplane in three dimensions

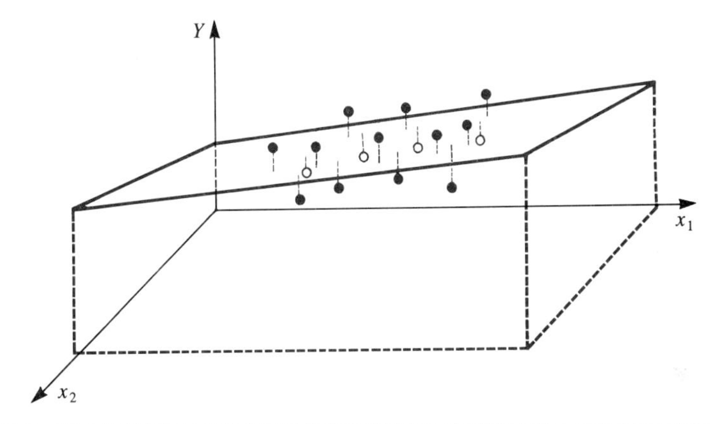

The number of equations increases with the number of variables included in the model. Simultaneous solution of these equations provides the coefficients for the estimating equation

$$\hat{Y} = a + b_1x_1 + b_2x_2 + \cdots + b_kx_k \qquad \textbf{(11.3)}$$

Henceforth, the i subscript is dropped, as in Equation 11.3, unless it is necessary for clarity.

Example 11.1 illustrates the computational procedure for estimating the regression coefficients. The solution of the simultaneous normal equations, while not detailed, could be accomplished by methods of elimination or matrix algebra.[1] The number of observations used in this example is fewer than desirable, but it was limited to reduce the computational effort.

EXAMPLE 11.1 The EPA decides to extend its tests of fuel economy by developing a model that relates driving speed and car weight/power ratio to the miles per gallon obtained. Eight different cars were driven at a constant speed from Chicago to Akron via interstate highway. Estimate the model with the equation

$$\hat{Y} = a + b_1x_1 + b_2x_2$$

The data and some necessary calculations are given in Table 11.1.

---

[1] The coefficients can be found from the following matrix formulation: $\boldsymbol{b} = (\boldsymbol{x}'\boldsymbol{x})^{-1}\boldsymbol{x}'\boldsymbol{Y}$. See Neter, Wasserman, and Kutner [1], p. 238.

TABLE 11.1 Data and calculations for Example 11.1

y *(mpg)*	x_1 *(speed)*	x_2 *(wt/hp)*	x_1^2	x_2^2	x_1x_2	x_1y	x_2y
30	70	20	4,900	400	1,400	2,100	600
10	65	12	4,225	144	780	650	120
15	50	16	2,500	256	800	750	240
25	55	18	3,025	324	990	1,375	450
20	50	14	2,500	196	700	1,000	280
23	45	15	2,025	225	675	1,035	345
18	75	17	5,625	289	1,275	1,350	306
28	60	19	3,600	361	1,140	1,680	532
169	470	131	28,400	2,195	7,760	9,940	2,873

Solution: The normal equations for the two-variable equation are

$$an + b_1\Sigma x_{i1} + b_2\Sigma x_{i2} = \Sigma Y_i$$
$$a\Sigma x_{i1} + b_1\Sigma x_{i1}^2 + b_2\Sigma x_{i1}x_{i2} = \Sigma x_{i1}Y_i$$
$$a\Sigma x_{i2} + b_1\Sigma x_{i2}x_{i1} + b_2\Sigma x_{i2}^2 = \Sigma x_{i2}Y_i$$

Substituting, we obtain

$$8a + 470b_1 + 131b_2 = 169$$
$$470a + 28{,}400b_1 + 7760b_2 = 9940$$
$$131a + 7760b_1 + 2195b_2 = 2873$$

These equations are solved simultaneously, resulting in the model estimate

$$\hat{Y} = -6.924 - .175x_1 + 2.342x_2$$

It is rather difficult, although, as we have seen, not impossible, to calculate the multiple regression equation by hand. However, hand solution is both tedious and potentially inaccurate because of rounding errors. Fortunately, many computer programs are available to perform multiple regression calculations. Therefore, we will approach multiple regression and correlation from the viewpoint of the interpretation of a computer output.

Students should become familiar with the multiple regression program available at their local computer centers or personal computer laboratories. Having a copy of typical output at hand is helpful in reading this section. The computer output examples used in this chapter are from the SAS package and MINITAB. Most computer programs produce similar outputs, although labels and format may vary.

Before going to the interpretation of the computer output, we state the required assumptions for making statistical inferences—that is, for hypothesis tests and confidence intervals. With regard to the terms in Equation 11.1, the x_j's are assumed to be nonrandom variables, and ε_i is a random error term with a zero mean. If the x_j's represent random variables, the procedures and inferences are the same, but the interpretations change slightly. To obtain *F*- and *t*-distributions for the test statistics, we need the following additional assumptions.

1. $\varepsilon_i \sim N(0,\sigma^2)$.
2. Errors are independent of each other and of the values of the x_j's.
3. σ^2 is constant for all levels of the x_j's.
4. The level of measurement of Y is at least interval.

The development of the distributions for the test statistics under these assumptions follows the same line of reasoning used in Chapters 9 and 10. Due to the complexity, the formal presentation is omitted in favor of simply outlining the approach. Specifically, the total sum of squares, $\Sigma(Y - \overline{Y})^2$, is partitioned into terms identified as SS*E* $= \Sigma(Y - \hat{Y})^2$ and SS*Reg* $= \Sigma(\hat{Y} - \overline{Y})^2$, also called SS*Model*, and then Cochran's theorem is applied. Unlike Chapter 10, where the SS*Reg* had only one degree of freedom, the SS*Reg* here has one degree of freedom for each independent variable in the model since $\hat{Y}$ is now given by Equation 11.3. As a result, the SS*Reg* can be partitioned further to permit the investigation of the effects of individual predictors. Keep in mind that the interpretations of the following printouts are based on the preceding assumptions. If any assumptions are violated, then any probabilities associated with the results are incorrect.

Computer programs usually number the variables internally according to the columns of the input data. If on each line of computer input the variables are in the order Y, x_1, x_2, x_3, they will be identified by the computer as variables 1, 2, 3, 4, respectively. Some computer programs report the coefficients on the output using these numbers. Other programs, such as SAS and MINITAB, allow names for the variables.

Because of the way the computer identifies the variables, it is useful to change the subscripting of the equation to agree with the computer system. Thus, the equation becomes $\hat{Y}_1 = a + b_2x_2 + b_3x_3 + b_4x_4$. If the dependent variable had been in column 4, the subscripts would be $\hat{Y}_4 = a + b_1x_1 + b_2x_2 + b_3x_3$.

EXAMPLE 11.2 The tire dealer mentioned in a number of examples in Chapter 10 believes that other variables besides inches of advertising influence tire sales—specifically, the price of the tire on sale for the week listed at the top of the ad and the number of salespersons employed that week. The dealer randomly selects the data for 10 weeks and does a multiple regression analysis. Determine the resulting equation and interpret the coefficients (see Table 11.2).

TABLE 11.2 Data for Example 11.2

Total Sales y_1 *1*	*Inches of Advertising* x_2 *2*	*Price of Sale Tire* x_3 *3*	*Number of Salespersons* x_4 *4*
41	20	40	1
42	40	60	3
59	40	20	4
60	50	80	5
81	50	10	6
80	60	40	6
100	70	20	7
82	70	60	8
101	80	30	9
110	90	40	10

TABLE 11.3a SAS computer output for Example 11.2 Key: ① Analysis of variance table; ② Significance level; ③ Multiple *R*-square; ④ Standard error of the estimate; ⑤ Regression coefficients

```
                                        SAS
DEP VARIABLE: SALES
                              ANALYSIS OF VARIANCE

                             SUM OF          MEAN
        ① SOURCE     DF     SQUARES        SQUARE        F VALUE    ② PROB>F

          MODEL       3  5179.23222    1726.41074         57.814      0.0001
          ERROR       6   179.16778   29.86129594
          C TOTAL     9  5358.40000

        ④ ROOT MSE   5.464549  ③ R-SQUARE        0.9666
          DEP MEAN       75.6    ADJ R-SQ        0.9498
          C.V.       7.228239

                              PARAMETER ESTIMATES

                   PARAMETER        STANDARD       T FOR HO:
 ⑤ VARIABLE  DF     ESTIMATE           ERROR     PARAMETER=0     PROB > |T|

   INTERCEP   1  30.99168251      7.72826889           4.010         0.0070
   ADVERT     1   0.82015209      0.50233951           1.633         0.1537
   PRICE      1  -0.32502376      0.08934962          -3.638         0.0109
   SALESMEN   1   1.84077947      3.85482617           0.478         0.6499
```

Solution: Computer printouts for SAS and MINITAB are shown in Tables 11.3a and b. The coefficients for each variable and the constant term can be found on the computer printouts in the highlighted sections—5 in Table a and 4 in Table b—and under the column labeled PARAMETER ESTIMATE in SAS and COEFFICIENT in MINITAB. Rounding these values to two places, we obtain the equation

$$\hat{Y}_1 = 30.99 + .82x_2 - .33x_3 + 1.84x_4$$

The constant term is difficult to interpret in multiple regression. In general, it is dangerous to put an interpretation on it, other than to say it is a number to make

TABLE 11.3b MINITAB computer output for Example 11.2 Key: ① Analysis of variance table; ② Multiple *R*-square; ③ Standard error of the estimate; ④ Regression coefficients

```
MTB > EXEC 'EX11 2 MINITAB A'
MTB > READ DATA INTO C1-C4
    10 ROWS READ
MTB > END OF DATA
MTB > NAME FOR C1 IS 'SALES', C2 IS 'ADVERT', C3 IS 'PRICE', C4 IS 'SALESMEN'
MTB > BRIEF 2
MTB > REGRESS C1 ON 3 PREDICTOR C2 C3 C4

THE REGRESSION EQUATION IS
SALES = 31.0 + 0.820 ADVERT - 0.325 PRICE + 1.84 SALESMEN

④                                   ST. DEV.    T-RATIO =
 COLUMN        COEFFICIENT          OF COEF.    COEF/S.D.
                  30.992             7.728          4.01
 ADVERT           0.8202            0.5023          1.63
 PRICE          -0.32502           0.08935         -3.64
 SALESMEN          1.841             3.855          0.48

③ S = 5.465

② R-SQUARED = 96.7 PERCENT
  R-SQUARED = 95.0 PERCENT, ADJUSTED FOR D.F.

① ANALYSIS OF VARIANCE

  DUE TO       DF          SS       MS=SS/DF
  REGRESSION    3      5179.2         1726.4
  RESIDUAL      6       179.2           29.9
  TOTAL         9      5358.4

MTB > STOP
```

the model more accurate. A practical interpretation of the regression coefficients is that for each extra inch of advertising, sales increase by an average of .82 tire; for each increase of $1 in the price of the sale tire, sales decrease by an average of .33 tire; and for each additional salesperson, sales increase by an average of 1.84 tires, if the value for the independent variable is within the range of the data and the other two variables are held constant.

After the coefficients are obtained, the model can be used for prediction. Example 11.3 demonstrates this process.

EXAMPLE 11.3 If the store in Example 11.2 uses 50 inches of advertising, puts a tire on sale at $50, and uses 5 salespeople for the week, how many tires on the average would the store be expected to sell that week?

Solution: Substituting $x_2 = 50$, $x_3 = 50$, and $x_4 = 5$ into the equation developed in Example 11.2, we obtain a point estimate of the number of tires sold.

$$\begin{aligned}\hat{Y} &= 30.99 + .82(50) - .33(50) + 1.84(5)\\ &= 30.99 + 41 - 16.5 + 9.20\\ &= 64.69\end{aligned}$$

It is obvious that the equation can be used as an estimator regardless of whether the model is useful—that is, significant—or not. To determine whether the estimator $\hat{Y}$ is an improvement over $\overline{Y}$ requires testing the independent variables for usefulness both as a group and individually. The formal discussion of the test procedures will be presented after the interpretation of the items on the printout has been completed. Keep in mind that all inferences, as well as interpretations of probabilities, are based on the set of assumptions given earlier.

Numerous other items on the computer printout need explanation. In the table of regression coefficients, the label STANDARD ERROR in SAS or ST. DEV. OF COEF. in MINITAB represents the standard deviation of the slope—that is, s_{b_j}. When the coefficient b is divided by s_b, we obtain a t value (see Equation 10.21)

$$t = \frac{b_j}{s_{b_j}} \tag{11.4}$$

where the null hypothesis is that $\beta_j = 0$. The column PROB > |T| indicates the probability of obtaining an absolute value of t greater than this value, thereby giving a two-tailed test. It is included to eliminate the necessity of looking up t values in the t table. If, for example, we are using $\alpha/2 = .025$, a value in this column less than .025 indicates rejection of the null hypothesis.

The standard error of the estimate is the standard deviation representing the spread of the points around the plane. It would normally be represented with the symbol $s_{Y_1|x_2,x_3,x_4}$, but since this notation is cumbersome, we will use the more compact version, $s_{Y|\boldsymbol{x}}$, where $\boldsymbol{x}$ is vector notation representing all independent variables.

$$s_{Y|x} = \sqrt{\frac{SSE}{n - k - 1}} = \sqrt{MSE} \tag{11.5}$$

where: n = number of observations
k = number of independent variables

The degrees of freedom reduces to $n - 2$ in the simple linear regression case since there is only one independent variable. The standard error of the estimate is listed on the computer outputs in Tables 11.3a and 11.3b under the label ROOT MSE, indicated as the highlighted area 4 in SAS, and S in area 3 in MINITAB. It could also be calculated from Equation 11.5 and the mean square error found in the analysis of variance table in highlighted area 1 (Tables 11.3a and 11.3b)

$$s_{Y|x} = \sqrt{\frac{179.17}{6}} = \sqrt{29.86} = 5.46$$

A multiple regression problem must have more observations than it has independent variables. A rule of thumb is that you should have at least five times as many observations other than replicates as you have number of independent variables.

EXAMPLE 11.4 What is the error degrees of freedom for the following problem?

y_1	x_2	x_3	x_4	x_5	x_6
12	31	15	74	2	7
14	33	12	76	5	3
16	35	14	73	4	4

Solution: Here, $n = 3$ and $k = 5$, and error degrees of freedom $= n - k - 1 = 3 - 5 - 1 = -3$. The error degrees of freedom is -3, which, of course, presents an impossible situation since the degrees of freedom must be greater than zero.

The R-SQUARE(D) term on the printout is the **multiple coefficient of determination**, which we symbolize as R^2, rather than the r^2 used for the simple linear case. It is used to measure the percent of the total variation in Y explained by the independent variables included in the model. The value of R^2 always increases or stays the same as additional independent variables are added to the model. It cannot decline. It is calculated as the ratio of the regression sum of squares to the total sum of squares.

$$R^2 = \frac{SSR}{SST} \tag{11.6}$$

The value of R^2 always lies between 0 and 1 inclusive, with 0 indicating no relationship and 1 indicating a perfect relationship within the current data available. Considering again Figure 11.1, we would observe all the points falling on the hyperplane if $R^2 = 1$.

EXERCISES

11.1 Solve the normal equations of Example 11.1 in the text to show that the answer given is correct.

11.2 Now solve the same problem on a computer system to obtain the same answers. It is necessary to enter the original data rather than starting with the normal equations.

11.3 If the EPA is investigating a new car at a speed of 55 mph and a weight-to-horsepower ratio of 15, what will be the expected miles per gallon? Give a point estimate. Use the equation in the text for Example 11.1 or from a computer printout for Exercise 11.2.

11.4 Place a 95% confidence interval around the point estimate of the previous problem. (Do this problem only if the available computer output provides a value for $s_{Y|x_p}$.)

11.5 The following are the observations for a computer regression analysis.

y_1	x_2	x_3	x_4	x_5
10.6	4	3	1	14
13.0	2	7	1	15
14.2	8	9	2	17
15.4	7	9	3	14
16.3	9	8	7	12
14.2	8	7	1	11

(a) Determine the degrees of freedom for **(1)** the error sum of squares, **(2)** the regression sum of squares, and **(3)** the total sum of squares.

(b) Are there sufficient observations? Why?

11.3 TESTING THE EQUATION FOR SIGNIFICANCE

The F test using the analysis of variance is used to determine whether a significant relationship exists between the dependent and independent variables. That is, it tests the overall model for usefulness by testing the complete equation, not the individual variables, for significance. If the equation is significant, individual variables can then be tested for retention in the model by using either an F or a t statistic.

DECISION RULE For H_0: all $\beta_j = 0$ and H_a: some $\beta_j \neq 0$, reject H_0 if **(11.7)**

$$\frac{\text{MS}R}{\text{MS}E} > F_{k,n-k-1,\alpha} \tag{11.8}$$

If the stated model is correct, failure to reject H_0 implies that the model, as a whole, is not significantly better than $\overline{Y}$ as an estimator for $\mu_{Y|\mathbf{x}}$. Hence, no x_j appears to be useful.

EXAMPLE 11.5 Test the equation from Example 11.2 for significance, using $\alpha = .05$.

Solution: H_0: all $\beta_j = 0$ versus H_a: some $\beta_j \neq 0$.

$$F_{k,n-k-1,\alpha} = F_{3,6,.05} = 4.76$$

$$F_{\text{calc}} = 57.81 \text{ from the computer printout}$$

Since $57.81 > 4.76$, we reject the null hypothesis, as illustrated in Figure 11.2. The equation is significant, which implies that at least one β_j is nonzero.

Rejection of H_0 in the overall test implies that at least one independent variable is making a significant reduction in the unexplained or error variation. Thus, we would like to identify these variables. To determine the variables that appear to be making a significant or useful contribution to explaining the variation in Y, we can use either a t statistic or an F statistic. Recall that an F value with one degree of freedom in the numerator is equal to a t value squared. Generally, the choice depends on the statistic calculated by the program used for the analysis. The test is two-tailed since either a positive or a negative relationship can result in a significant reduction in unexplained variation.

DECISION RULES To test the hypotheses

$$H_0: \beta_j = 0 \text{ for each } j \qquad H_a: \beta_j \neq 0 \text{ for each } j \tag{11.9}$$

for a t statistic, reject H_0 if

$$|t_j| > t_{\text{df}_E,\alpha/2} \tag{11.10}$$

for an F statistic, reject H_0 if

$$F_j > F_{1,\text{df}_E,\alpha} \tag{11.11}$$

FIGURE 11.2 Sampling distribution of the F statistic for Example 11.5 if H_0 is true

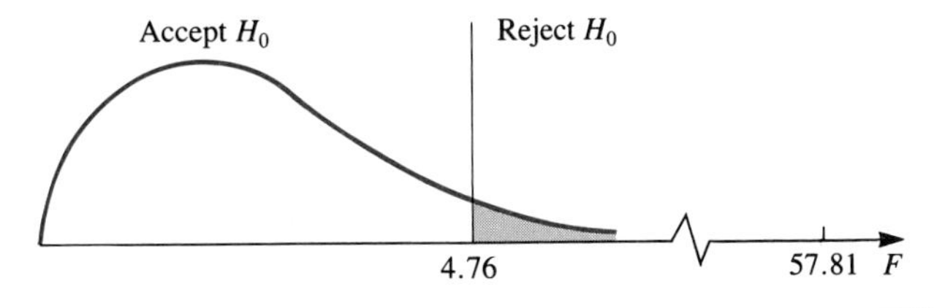

Both decision rules are based on the assumptions given earlier, as well as on an assumption that the independent variables are uncorrelated.

After deciding on a final model, we may want to place a confidence interval around the point estimate. As in Chapter 10, two basic intervals are of interest. The first is a confidence interval for the mean of the Y population at a specific $\boldsymbol{x}_p = (x_{1p}, \ldots, x_{kp})$, and the second is a prediction interval for an observation from the indicated population. The equations for determining the two intervals are similar to the corresponding equations in Chapter 10. In this case, they are

$$\hat{Y}_{x_p} \pm t_{\mathrm{df}_E,\alpha/2} \sqrt{s^2_{\hat{Y}|x_p}} \qquad \text{for } \mu_{Y|x_p} \tag{11.12}$$

and

$$\hat{Y}_{x_p} \pm t_{\mathrm{df}_E,\alpha/2} \sqrt{s^2_{\hat{Y}|x_p} + \mathrm{MSE}} \qquad \text{for } Y_{x_p} \tag{11.13}$$

The term $s^2_{\hat{Y}|x_p}$ is easily calculated for the simple linear model when we are using only one independent variable, but its calculation for a multiple regression model requires a sound knowledge of matrix algebra techniques that are beyond the background assumed for this text. Some, but not all, statistical computer packages will calculate this value. If so, confidence and prediction intervals can be calculated using Equations 11.12 and 11.13.

EXAMPLE 11.6 Since the H_0 was rejected in the test of the overall model, indicating that at least one variable is significant (some $\beta_j \neq 0$), we test each of the individual variables for significance, assuming that the independent variables are uncorrelated with each other.

Solution: Returning to the computer output given in Table 11.3, we find the error degrees of freedom to be 6. Therefore, the tabular value of t is

$$t_{6,.025} = 2.447$$

For each variable, the calculated t values are found on the output and compared to the tabular value. A relationship is said to exist only if the null hypothesis is rejected. Comparing the calculated t to the critical value of the sampling distribution shown in Figure 11.3, we see that only variable 3 appears to be useful for explaining variation in the dependent variable, and variables 2 and 4 should be dropped.

Note that in the example, the phrase "appears to be useful" is in the conclusion. The conclusions are accurate only if the assumption that no correlation exists between the independent variables is true. If it is not, our conclusions are subject to error. Other methods are used to decide if relationships exist. These methods are presented later under the topic of multicollinearity. We can determine whether a significant relationship exists by examining the correlation matrix.

FIGURE 11.3 Sampling distribution for Example 11.6

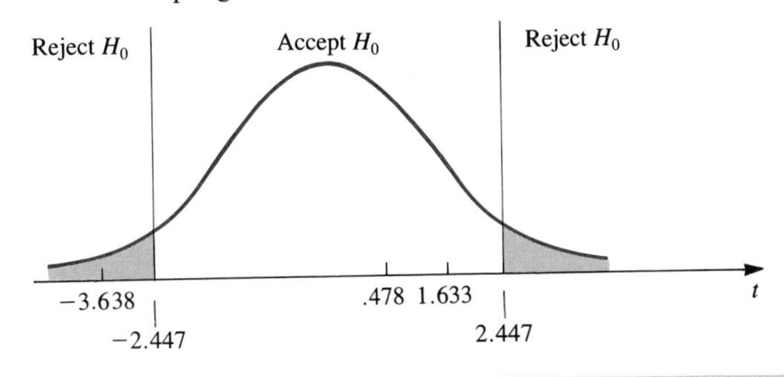

Most computer programs produce a correlation matrix—that is, a collection of the correlation coefficients for every possible pairwise combination of variables. For example, r_{23} is the result of using the Pearson formula, Equation 10.31, letting x be variable 2 and letting Y be variable 3, or vice versa.

EXAMPLE 11.7 Determine the correlation matrix for the data in Example 11.2.

Solution: The computer produced the matrix values shown in Table 11.4. Notice that $r_{23} = r_{32} = -.07$ and that, in general, the numbers above the diagonal are mirror images of the numbers below the diagonal. Typically, one works with either the upper or the lower triangle. We use the lower one.

Multicollinearity

The correlation matrix is used to check for the existence of multicollinearity.

DEFINITION 11.2 **Multicollinearity** is correlation between independent variables.

TABLE 11.4 Correlation matrix for Example 11.7

	Sales 1	*Adv* 2	*Price* 3	*Salespeople* 4
Sales 1	1.00	.94	−.37	.94
Adv 2	.94	1.00	−.07	.98
Price 3	−.37	−.07	1.00	−.13
Salesmen 4	.94	.98	−.13	1.00

It is generally undesirable to develop a model that has extensive multicollinearity. Such a model can lead to the wrong signs on coefficients, t values that lead to acceptance or rejection of variables when the opposite is true, and other quite mysterious behavior in the computations. The problems occur due to rounding errors caused by division by a value very close to zero.

Multicollinearity exists in degrees. Large amounts, such as $r = .8$, may cause the problems just mentioned, whereas small amounts, such as $r = .1$, may show no effects at all. In addition, the pairwise correlation matrix may miss some multicollinearity where one independent variable is linearly related to two or more other independent variables. Hence, we may see the symptoms of multicollinearity without being able to detect it in the correlation matrix.

Remember that a correlation matrix is made up of r values, not r^2 values. Each of the r values can be tested to see if it could have come from a population in which $\rho = 0$. We learned in Chapter 10 that one test for $H_0: \rho = 0$ is based on the t statistic, where the calculated value of t is

$$t = \frac{r - 0}{\sqrt{(1 - r^2)/(n - 2)}} \qquad \textbf{(11.14)}$$

which is compared to the tabulated value for $t_{n-2,\alpha/2}$. If we are given α and n, we can solve for the value of r that corresponds to a tabular value of t. This value, r_{cr}, is referred to as the critical value of r. This equation cannot be used if ρ is not hypothesized to be zero. The decision rule is to reject H_0 if

$$|r| > r_{cr} = +\sqrt{\frac{t^2_{n-2,\alpha/2}}{t^2_{n-2,\alpha/2} + n - 2}} \qquad \textbf{(11.15)}$$

Any value of r in the correlation matrix that falls between $-r_{cr}$ and r_{cr} leads to acceptance of the null hypothesis that no correlation exists between the two variables. For the independent variables, it indicates that no multicollinearity appears to exist.

The larger is the value for n in Equation 11.15, the smaller is the value for r_{cr}. With a large value for n, r_{cr} is quite small and presents a higher chance that the test shows some multicollinearity. Small amounts of multicollinearity are not usually troublesome. Multicollinearity is actually a computational problem rather than a statistical one.

EXAMPLE 11.8 Using the correlation matrix from Example 11.7, determine if significant multicollinearity (correlation between independent variables) exists and state the implications for the t test used in Example 11.6.

Solution: First, we calculate the critical value for r.

$$t_{n-2,\alpha/2} = t_{8,.025} = 2.306$$

$$r_{cr} = \sqrt{\frac{t^2_{8,.025}}{t^2_{8,.025} + n - 2}} = \sqrt{\frac{(2.306)^2}{(2.306)^2 + 10 - 2}} = \pm .6319$$

Multicollinearity appears to exist between variables 2 and 4 ($r_{24} = .98$). It may be that both signs on the coefficients, the coefficient values themselves, and the results of the t test used in Example 11.6 are incorrect because the assumption that the independent variables are uncorrelated does not seem to be satisfied. We may wish to drop one of the two correlated variables from the model.

11.6 Employees at the Cambridge Cork Company are rated each year by their peers and supervisors, and the scores averaged. Scores range from 0 to 10. The personnel department wishes to determine if education and age are correlated with performance ratings. A sample of 10 employees was randomly obtained, and the following data were recorded.

Average Rating	***Years of College***	***Age***
7.8	4	41
7.5	4	43
8.4	5	28
9.1	6	36
7.9	4	24
7.3	4	35
6.4	2	32
4.2	0	29
9.4	7	31
8.3	4	30

(a) Determine the multiple linear regression equation, using all variables.
(b) Calculate a point estimate of the performance of a new employee who has 4 years of college and is 23 years old.
(c) Place a 95% confidence interval around the point estimate if $s_{\hat{Y}|x_p}$ is provided by a computer program.
(d) Test the equation for significance ($\alpha = .05$).
(e) Using the correlation matrix, determine whether the 2 independent variables are really independent.
(f) Use the t test to evaluate each independent variable at $\alpha = .05$. Is the test meaningful? Should either independent variable be dropped from the equation?

11.4 FINDING THE BEST EQUATION

Typically, once the regression equation has been determined to be useful, it is used for one of two purposes: prediction of values for the dependent variable, given values for the independent variables, or examination of the coefficients to observe the sign and amount of change in the dependent variable for a given change in the independent variable.

Best Equation for Prediction

Numerous business activities require predictions for planning and other purposes. Production planners, for example, need an initial sales forecast so that they can set the rates of production on current machines or order additional machines if it appears that sales are about to increase dramatically. The sales manager may use this same prediction to aid in making advertising and pricing decisions. The board of directors may use these predictions to support decisions about building new plants or, perhaps, about closing plants. Therefore, it is necessary that the best possible multiple regression equation be developed for prediction purposes. In developing this equation, we must keep in mind that there is no certainty in predictions, whether done with regression or a crystal ball. We simply try to minimize the chance of error.

Initially, the analyst must attempt to identify those variables that might reasonably be related to the dependent variable and for which data are available or can be collected. We expect that some of these variables are more highly correlated to the dependent variable than others are. Either the simple coefficient of determination or coefficient of correlation, r^2 and r, respectively, may be used to evaluate the strength of this relationship.

As we start with one related variable and add other related variables to the equation, the overall R^2 will increase. Hence, it would appear that the multiple regression equation with the largest R^2 is the best equation for prediction since it explains the largest proportion of the total variation in Y. Also, since R^2 usually increases as more independent variables are included, it would appear that the best equation is always the one with all the available variables included, but it may not be.

As suggested earlier, the best equation could be defined as the one that provides the narrowest confidence intervals around the point estimate, thereby implying that it is the equation with the smallest standard error of the estimate $s_{Y|x}$—that is, $\sqrt{MSE}$.

The obvious approach for finding the equation with the smallest standard error of the estimate is to do a regression for every possible combination of variables, first including all k independent variables, then for all possible combinations of $k - 1$ independent variables, and so on, until we conclude with all regressions using only one independent variable. Example 11.9 illustrates this approach.

EXAMPLE 11.9 Determine the standard errors, R^2, and F for all the possible models of 3, 2, and 1 independent variables and determine which model appears to be best. Use the data from Example 11.2.

Solution: Computer outputs for the model using all 3 variables were given earlier in Table 11.3. Tables 11.5a and b give the outputs for models 2 and 6 using variables 2 and 3 and variable 3 by itself as typical of the outputs. Table 11.6 summarizes these models and the 5 others. Except for model 6, judging by the F values, all the models were significant. Some analysts recommend picking the

TABLE 11.5a SAS and MINITAB computer printouts from the seven possible combinations of variables in Table 11.2: Computer output for model 2 estimated by $\hat{Y} = a + b_2x_2 + b_3x_3$

```
                                              SAS

DEP VARIABLE: SALES
                                      ANALYSIS OF VARIANCE

                              SUM OF          MEAN
          SOURCE     DF      SQUARES         SQUARE        F VALUE        PROB>F

          MODEL       2   5172.42292     2586.21146         97.343        0.0001
          ERROR       7    185.97708    26.56815391
          C TOTAL     9   5358.40000

              ROOT MSE     5.154431       R-SQUARE          0.9653
              DEP MEAN         75.6       ADJ R-SQ          0.9554
              C.V.          6.81803

                                      PARAMETER ESTIMATES

                           PARAMETER         STANDARD      T FOR H0:
    VARIABLE  DF            ESTIMATE            ERROR    PARAMETER=0      PROB > |T|

    INTERCEP   1         28.93553009       6.05341354          4.780          0.0020
    ADVERT     1          1.05644699       0.08161538         12.944          0.0001
    PRICE      1         -0.33882521       0.07974795         -4.249          0.0038
```

```
MTB > EXEC 'EX11_2 MINITAB A'
MTB > READ DATA INTO C1-C4
    10 ROWS READ
MTB > END OF DATA
MTB > NAME FOR C1 IS 'SALES', C2 IS 'ADVERT', C3 IS 'PRICE', C4 IS 'SALESMEN'
MTB > BRIEF 2
MTB > REGRESS C1 ON 2 PREDICTORS C2 C3

THE REGRESSION EQUATION IS
SALES = 28.9 + 1.06 ADVERT - 0.339 PRICE

                                     ST. DEV.     T-RATIO =
COLUMN        COEFFICIENT            OF COEF.     COEF/S.D.
                   28.936               6.053          4.78
ADVERT            1.05645             0.08162         12.94
PRICE            -0.33883             0.07975         -4.25

S = 5.154

R-SQUARED = 96.5 PERCENT
R-SQUARED = 95.5 PERCENT, ADJUSTED FOR D.F.

ANALYSIS OF VARIANCE

 DUE TO       DF             SS        MS=SS/DF
REGRESSION     2         5172.4          2586.2
RESIDUAL       7          186.0            26.6
TOTAL          9         5358.4


MTB > STOP
```

TABLE 11.5b SAS and MINITAB computer printouts from the seven possible combinations of variables in Table 11.2: Computer output for model 6 estimated by $\hat{Y} = a + b_3x_3$

```
                                        SAS
DEP VARIABLE: SALES
                                   ANALYSIS OF VARIANCE

                              SUM OF         MEAN
          SOURCE     DF      SQUARES        SQUARE       F VALUE      PROB>F

          MODEL       1    720.85714     720.85714         1.244      0.2972
          ERROR       8   4637.54286     579.69286
          C TOTAL     9   5358.40000

             ROOT MSE       24.07681     R-SQUARE         0.1345
             DEP MEAN           75.6     ADJ R-SQ         0.0263
             C.V.           31.84763

                                  PARAMETER ESTIMATES

                        PARAMETER         STANDARD       T FOR HO:
     VARIABLE  DF        ESTIMATE            ERROR     PARAMETER=0     PROB > |T|

     INTERCEP   1     92.17142857      16.69744471           5.520         0.0006
     PRICE      1     -0.41428571       0.37151327          -1.115         0.2972
```

```
MTB > EXEC 'EX11_2 MINITAB A'
MTB > READ DATA INTO C1-C4
    10 ROWS READ
MTB > END OF DATA
MTB > NAME FOR C1 IS 'SALES', C2 IS 'ADVERT', C3 IS 'PRICE', C4 IS 'SALESMEN'
MTB > BRIEF 2
MTB > REGRESS C1 ON 1 PREDICTOR  C3

THE REGRESSION EQUATION IS
SALES = 92.2 - 0.414 PRICE

                                  ST. DEV.     T-RATIO =
COLUMN        COEFFICIENT         OF COEF.     COEF/S.D.
                    92.17           16.70          5.52
PRICE             -0.4143          0.3715         -1.12

S = 24.08

R-SQUARED = 13.5 PERCENT
R-SQUARED =  2.6 PERCENT, ADJUSTED FOR D.F.

ANALYSIS OF VARIANCE

 DUE TO       DF            SS       MS=SS/DF
REGRESSION     1         720.9          720.9
RESIDUAL       8        4637.5          579.7
TOTAL          9        5358.4


MTB > STOP
```

TABLE 11.6 Results for seven estimated models

Estimated Model	Standard Error of Estimate	R^2	F	$F_{.05}$
1.* $\hat{Y}_1 = a + b_2x_2 + b_3x_3 + b_4x_4$	5.4645	.9666	57.8	4.76
2.* $\hat{Y}_1 = a + b_2x_2 + b_3x_3$	5.1544	.9653	97.3	4.74
3. $\hat{Y}_1 = a + b_2x_2 + b_4x_4$	9.0578	.8928	29.2	4.74
4. $\hat{Y}_1 = a + b_3x_3 + b_4x_4$	6.0800	.9517	69.0	4.74
5. $\hat{Y}_1 = a + b_2x_2$	9.1212	.8758	56.4	5.32
6.* $\hat{Y}_1 = a + b_3x_3$	24.0768	.1345	1.2	5.32
7. $\hat{Y}_1 = a + b_4x_4$	8.5375	.8912	65.5	5.32

* See Table 11.3 for model 1 computer output and Table 11.5 for model 2 and model 6 computer outputs.

model with the highest value for R^2, of which models 1 and 2 are effectively the same. But, using the method of picking the smallest estimated standard error, model 2 is selected as the best for predictive purposes.

We now have developed a two-variable model to use for making predictions that appears to be better than the overall model, which includes all the variables. A paradox exists, however. Variable 3 alone tested out as not being related to the dependent variable, yet when added to variable 2, it increases the R^2 from .8758 to .9653. The implication is that variable 3 explains an additional .0895 of the variation in the dependent variable, which would mean that it is related to the dependent variable.

One possible reason for this result is illustrated in Figure 11.4. Considering x_3 alone, the sum of squares for error includes the regression sum of squares for variables x_2 and x_4, given that x_3 is in the model. Both terms have large sums of squares, thereby making the error term quite large relative to the sum of squares for the regression. Using the SAS and MINITAB computer printouts of Table 11.5b, we find that variable 3 tests as not being significantly related to the dependent variable.

When x_2 is included in the model with x_3, its influence is removed from the error term and included in the sum of squares for regression. Hence, the sum of squares error is reduced, and x_3 now tests as being significantly related to Y. The relevant SAS and MINITAB computer outputs are shown in Table 11.5a.

As a final comment, we note that other techniques are available for selecting the best model for prediction. Many independent variables are considered for the model, which makes it quite difficult and time consuming to look at every possible combination of variables. Stepwise regression and backward

FIGURE 11.4 Comparison of SSE with X_2 out of and in the model that includes X_3

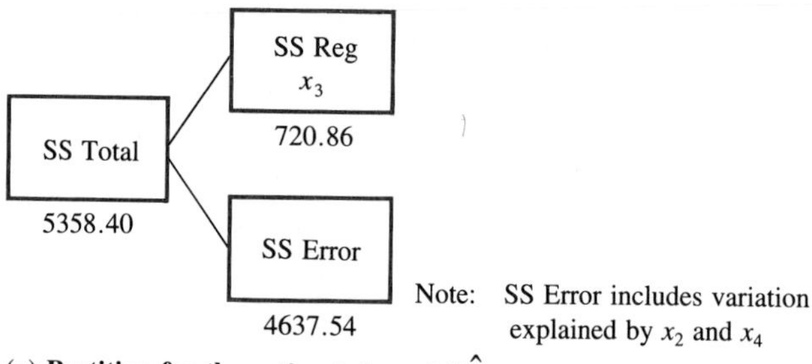

(a) Partition for the estimated model $\hat{Y}_1 = a + b_3x_3$

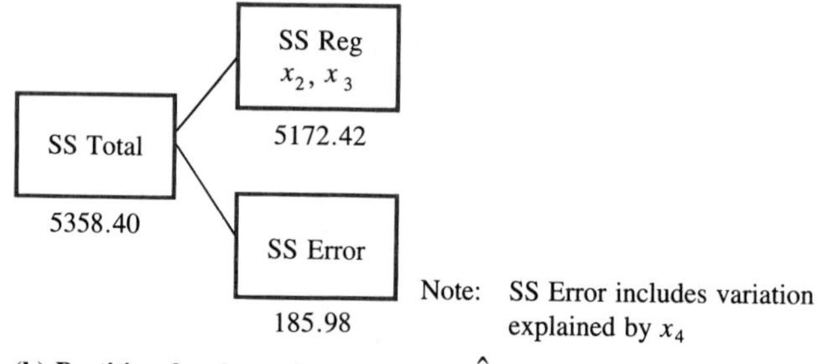

(b) Partition for the estimated model $\hat{Y}_1 = a + b_2x_2 + b_3x_3$

elimination are two such techniques that prove far superior in such a case. Most statistical packages include stepwise analysis. For a detailed discussion of these procedures, see Neter, Wasserman, and Kutner [1].

Best Model for Analysis of Coefficients

Another use of regression models is analyzing the coefficients to aid in decision making. As stated earlier, the coefficient of an independent variable indicates the amount the dependent variable will change for a one-unit change in the independent variable, if other variables do not change. It may be desired, therefore, to determine how many units the independent variable of interest must be changed while holding the others constant to achieve a desired change in the dependent variable. The difficulty in this process is that the coefficient may not correctly reflect this relationship. A significant degree of multicollinearity may induce significant errors in the values of the coefficients. In this instance, the values are incorrect, to the extent in some cases of having the wrong sign. Then they become useless for the intended purpose.

If we want to construct a model to analyze the coefficients, we should develop one with a minimum of multicollinearity. We start with the correlation matrix and determine which independent variables are correlated with each other. Then, for each pair of correlated variables, one member of the pair must be dropped from the model. The first variable to be eliminated should be the one least correlated with the dependent variable—that is, the one with the correlation coefficient having smallest absolute value.

What makes the task more difficult than the previous paragraph indicates is that there are degrees of multicollinearity. In most cases, a little multicollinearity causes little problem, whereas a great deal can result in the errors in the coefficients mentioned earlier. Unless transformations are used to create orthogonal—that is, linearly independent—variables, there will always be some multicollinearity whenever the model contains more than one predictor.

EXAMPLE 11.10 Develop a model that has very little or no multicollinearity, using the data and correlation matrix of Examples 11.2 and 11.7.

Solution: Inspecting the correlation matrix and using the r_{cr} calculated earlier, we learned that the variables 2 and 4 are highly correlated; therefore, 1 of them should be dropped. We refer to the correlation matrix to see which of the two is more highly correlated to variable 1, the dependent variable. We find $r_{12} = .94$ and $r_{14} = .94$. Since there is no difference between them, we could drop either. Either of the resulting models is relatively free of multicollinearity.

$$\hat{Y}_1 = a + b_2x_2 + b_3x_3$$
$$\hat{Y}_1 = a + b_3x_3 + b_4x_4$$

Use of a single independent variable would give a model with no multicollinearity.

A large amount of multicollinearity is the main source of significant rounding errors. Therefore, highly inaccurate answers may be obtained. Some computer programs for regression will not operate properly if a high degree of multicollinearity is present.

EXERCISE

11.7 The manager of the truck fleet for a certain company has been studying the factors that influence the number of hours that a truck motor runs from initial purchase until needing its first overhaul. The manager has gathered data for the following variables for a sample of trucks.

—Hours: time between new and overhaul
—Oil: average miles between oil changes
—Miles: miles truck is driven
—Load: average weight of loads (tons)
—Speed: average speed driven

Hours	*Oil*	*Miles*	*Load*	*Speed*
2095	2005	121,510	15.4	58
1143	9208	46,863	24.8	41
1527	4028	77,877	23.7	51
1483	5814	72,667	24.1	49
1224	8214	51,408	27.2	42
2108	2124	126,480	17.3	60
1883	2654	103,565	14.6	55
1657	3214	87,821	21.4	53
1934	2541	112,172	16.2	58
1396	6234	62,820	25.2	45
1724	2821	94,820	20.3	55
1656	3025	84,456	22.5	51
1821	2418	103,797	14.3	57
1418	6143	62,392	24.5	44

(a) Identify the equation calculated by the computer. Include all variables in the model.
(b) Test the equation for significance at $\alpha = .05$.
(c) Use the computer to determine the correlation matrix. Does there appear to be significant multicollinearity? Between what variables?
(d) Suggest a model that has a low degree of multicollinearity.
(e) Are t tests on the individual variables in the full model meaningful? Why?
(f) Using all possible combinations of the independent variables, obtain regression equations and determine the best equation for predictive purposes. In a tabular display, show, for each model, the independent variables included, the standard error, R^2, variables that a t test indicates should be dropped as being unrelated to the dependent variable, and those variables whose coefficients are negative. The first few entries are

Variables	*Std. Error*	R^2	*t drop*	*Negative*
2 3 4 5	14.0020	.9985	4	2 5
3 4 5	26.9184	.9938	4 5	5
2 4 5	82.1181	.9427	2 4 5	2 4

(g) Which model is best for predictive purposes? Why?
(h) Notice that different models have different variables that are negatively related to the dependent variable. Explain why certain variables are negative in one equation and positive in others.
(i) Notice that, in different models, the t test indicates that different variables are not related to the dependent variable. Why?
(j) What is your conclusion after answering parts (h) and (i)?

11.5 NONLINEAR EQUATIONS

In the previous chapter, we considered fitting a straight line to a set of data, using least squares. We also found that, if the appropriate assumptions were satisfied, we could test to determine if a linear model is appropriate or adequate. If it is not, it may be that a nonlinear equation provides a better fit to the data. In this section, we consider the use of a multiple regression computer program to fit certain types of linear models to a set of data.

DEFINITION 11.3 A **linear model** is a model that is linear in its parameters and not necessarily in its independent variables.

The implication of the definition of a linear model is that nonlinear equations are not necessarily nonlinear models. Linear models are recognized by a series of terms separated by positive and negative signs. Nonlinear models have terms multiplied together. Therefore, a polynomial equation is a linear model even though it may have x^3 or other terms raised to a power. Exponential equations are nonlinear models but may be transformed to a linear form. Equations must be in a linear model form if we are to use the multiple regression computer program to estimate the parameters.

Polynomial models are the easiest to estimate. Consider the parabola and the cubic model. Their graphs are shown in the previous chapter and should be reviewed.

$$Y = \alpha + \beta_1 x + \beta_2 x^2 + \varepsilon \qquad \text{parabola}$$
$$Y = \alpha + \beta_1 x + \beta_2 x^2 + \beta_3 x^3 + \varepsilon \qquad \text{cubic model}$$

Only one independent variable is present in the equation, but it also appears as a squared term in the parabolic equation and cubic equation and as a cubed term in the cubic equation. We treat each of these terms as separate independent variables in using a multiple regression program. The next example illustrates the procedure.

EXAMPLE 11.11 At the Jones Company, employees work alone in completely assembling various models of radios. Management suspects that it takes more than twice as long to assemble radios with twice as many parts because boredom sets in. Production is about to begin on a digital radio with 100 parts. Estimate the average time required to assemble it. Use the following data in preparing the estimate.

Assembly Time	*Number of Parts*
14	50
10	40
31	80
18	62
7	25
13	50
40	90
8	29
16	59
6	21
23	70
11	40

Solution: We consider parabolic and cubic equations, so we take the independent variable, number of parts, and both square and cube the values. The data, ready for entry into a computer, are given in Table 11.7.

The parabolic equation is obtained by entering variable 1 as the dependent variable and variables 2 and 3 as the independent variables. Tables 11.8a and b show the resulting SAS and MINITAB computer printouts.

TABLE 11.7 Data and some computations for Example 11.11

1 Time	*2* Parts	*3* (Parts)^2	*4* (Parts)^3
14	50	2,500	125,000
10	40	1,600	64,000
31	80	6,400	512,000
18	62	3,844	238,328
7	25	625	15,625
13	50	2,500	125,000
40	90	8,100	729,000
8	29	841	24,389
16	59	3,481	205,379
6	21	441	9,261
23	70	4,900	343,000
11	40	1,600	64,000

TABLE 11.8a SAS printout for parabolic model of Example 11.11

```
                                   SAS
DEP VARIABLE: TIME
                            ANALYSIS OF VARIANCE

                           SUM OF          MEAN
          SOURCE     DF    SQUARES         SQUARE        F VALUE      PROB>F

          MODEL       2   1164.41533     582.20767       805.968      0.0001
          ERROR       9   6.50133413    0.72237046
          C TOTAL    11   1170.91667

             ROOT MSE     0.8499238     R-SQUARE        0.9944
             DEP MEAN      16.41667     ADJ R-SQ        0.9932
             C.V.          5.177201

                            PARAMETER ESTIMATES

                         PARAMETER        STANDARD       T FOR H0:
       VARIABLE  DF       ESTIMATE           ERROR     PARAMETER=0     PROB > |T|

       INTERCEP   1     9.43198329      1.55984350           6.047         0.0002
       PARTS      1    -0.25464073      0.06274525          -4.058         0.0028
       PARTS2     1    0.006534396     0.000569116          11.482         0.0001
```

TABLE 11.8b MINITAB printout for parabolic model of Example 11.11

```
MTB > EXEC 'EX11_11 MINITAB A'
MTB > READ DATA INTO C1-C2
    12 ROWS READ
MTB > END OF DATA
MTB > LET C3 = C2**2
MTB > LET C4 = C2**3
MTB > NAME FOR C1 IS 'TIME', C2 IS 'PARTS', C3 IS 'PARTS2', C4 IS 'PARTS3'
MTB > BRIEF 2
MTB > REGRESS C1 ON 2 PREDICTORS C2 C3

THE REGRESSION EQUATION IS
TIME = 9.43 - 0.255 PARTS + 0.00653 PARTS2

                                     ST. DEV.     T-RATIO =
COLUMN        COEFFICIENT            OF COEF.     COEF/S.D.
                   9.432                1.560          6.05
PARTS           -0.25464              0.06275         -4.06
PARTS2         0.0065344            0.0005691         11.48

S = 0.8499

R-SQUARED = 99.4 PERCENT
R-SQUARED = 99.3 PERCENT, ADJUSTED FOR D.F.

ANALYSIS OF VARIANCE

 DUE TO       DF          SS         MS=SS/DF
REGRESSION     2     1164.42           582.21
RESIDUAL       9        6.50             0.72
TOTAL         11     1170.92


MTB > STOP
```

The estimated equation is

$$\hat{Y} = 9.4320 - .2546x_2 + .0065x_3$$

Since $x_2 = x$ and $x_3 = x^2$, we actually have

$$\hat{Y} = 9.4320 - .2546x + .0065x^2$$

Using the F test, we conclude that the equation is significant.

Since the company is planning to produce a radio with 100 parts, management can expect that the employees will require

$$\begin{aligned}\hat{Y} &= 9.4320 - .2546(100) + .0065(100)^2 \\ &= 9.4320 - 25.46 + 65 \\ &= 48.9720\end{aligned}$$

or approximately 49 minutes for the assembly.

We must note that the prediction has been made with an x value that is beyond the range of the data used to estimate the equation. In doing so, we assume that the model is still appropriate. Of course, it may not be, and, therefore, error is added to this prediction.

The cubic equation is determined by using variables 2, 3, and 4 as the independent variables. The equation determined by multiple linear regression is found in the SAS and MINITAB printouts of Tables 11.9a and b, respectively. Hence,

$$\hat{Y} = .120397 + .38499x - .006312x^2 + .000078x^3$$

Again the equation is significant by the F test. If the standard error of this model (.556) is compared to the standard error of the parabola (.85), we conclude that the cubic equation is a better fit for the data and is therefore a better equation for predictive purposes.

Note that before x^3 was introduced into the model, the x^2 term was significant. After the addition of x^3, the x^2 term no longer tests as significant. Multicollinearity between these two variables is the probable cause.

TABLE 11.9a SAS printout for cubic model of Example 11.11

SAS

DEP VARIABLE: TIME

ANALYSIS OF VARIANCE

SOURCE	DF	SUM OF SQUARES	MEAN SQUARE	F VALUE	PROB>F
MODEL	3	1168.44150	389.48050	1258.844	0.0001
ERROR	8	2.47516303	0.30939538		
C TOTAL	11	1170.91667			

ROOT MSE	0.5562332	R-SQUARE	0.9979
DEP MEAN	16.41667	ADJ R-SQ	0.9971
C.V.	3.388223		

PARAMETER ESTIMATES

VARIABLE	DF	PARAMETER ESTIMATE	STANDARD ERROR	T FOR H0: PARAMETER=0	PROB > |T|
INTERCEP	1	0.12039667	2.77580667	0.043	0.9665
PARTS	1	0.38499091	0.18200590	2.115	0.0673
PARTS2	1	-0.006311953	0.003580577	-1.763	0.1159
PARTS3	1	0.000077753	0.000021554	3.607	0.0069

TABLE 11.9b MINITAB printout for cubic model of Example 11.11

```
MTB > EXEC 'EX11_11 MINITAB A'
MTB > READ DATA INTO C1-C2
    12 ROWS READ
MTB > END OF DATA
MTB > LET C3 = C2**2
MTB > LET C4 = C2**3
MTB > NAME FOR C1 IS 'TIME', C2 IS 'PARTS', C3 IS 'PARTS2', C4 IS 'PARTS3'
MTB > BRIEF 2
MTB > REGRESS C1 ON 3 PREDICTORS C2 C3 C4

* NOTE *    PARTS IS HIGHLY CORRELATED WITH OTHER PREDICTOR VARIABLES
* NOTE *   PARTS2 IS HIGHLY CORRELATED WITH OTHER PREDICTOR VARIABLES
* NOTE *   PARTS3 IS HIGHLY CORRELATED WITH OTHER PREDICTOR VARIABLES

THE REGRESSION EQUATION IS
TIME = 0.12 + 0.385 PARTS - 0.00631 PARTS2 +0.000078 PARTS3

                                     ST. DEV.     T-RATIO =
COLUMN       COEFFICIENT             OF COEF.     COEF/S.D.
                   0.120               2.776           0.04
PARTS             0.3850              0.1820           2.12
PARTS2         -0.006312            0.003581          -1.76
PARTS3        0.00007775          0.00002155           3.61

S = 0.5562

R-SQUARED = 99.8 PERCENT
R-SQUARED = 99.7 PERCENT, ADJUSTED FOR D.F.

ANALYSIS OF VARIANCE

 DUE TO      DF             SS        MS=SS/DF
REGRESSION    3        1168.44          389.48
RESIDUAL      8           2.48            0.31
TOTAL        11        1170.92

MTB > STOP
```

The prediction for 100 parts is

$$\begin{aligned}\hat{Y} &= .120397 + .38499(100) - .006312(100)^2 + .000078(100)^3 \\ &= .120397 + 38.499 - 63.12 + 78 \\ &= 53.499\end{aligned}$$

Once again, the warning about predicting beyond the data needs to be repeated. Since the x, x^2, and x^3 variables are all related to each other, polynomial models have an excessive amount of multicollinearity. Therefore, care must be taken in the interpretation of the coefficients. ■

Exponential-type models require a little more effort. Although parabolic and cubic equations are still linear in the parameters, exponentials are not. For example,

$$Y = \alpha\beta^{x+\varepsilon} \quad \text{and} \quad Y = \alpha x^{\beta+\varepsilon}$$

must be transformed before using a multiple linear regression program to estimate the equation because they are nonlinear models.

To convert to linear form, we use the following rules of logarithms.

$$\begin{aligned} &\mathbf{1.}\ \log(ab) = \log a + \log b \\ &\mathbf{2.}\ \log(a/b) = \log a - \log b \\ &\mathbf{3.}\ \log a^b = b \log a \end{aligned} \tag{11.16}$$

Using these rules, we now transform the nonlinear model $Y = \alpha\beta^{x+\varepsilon}$ into a linear model suitable for use with a computer program for multiple linear regression. Natural logarithms may be used (logarithms to the base e), but we will use logarithms to the base 10 in this text.

$$\begin{aligned} Y &= \alpha\beta^{x+\varepsilon} \\ \log Y &= \log(\alpha\beta^{x+\varepsilon}) \\ &= \log \alpha + \log \beta^{x+\varepsilon} \\ &= \log \alpha + (x+\varepsilon)\log \beta \\ &= \log \alpha + x \log \beta + \varepsilon \log \beta \\ &= \log \alpha + (\log \beta)x + (\log \beta)\varepsilon \end{aligned}$$

Letting $Y' = \log Y$, $\alpha' = \log \alpha$, $\beta' = \log \beta$, and $\varepsilon' = (\log \beta)\varepsilon$, we see that the equation is obviously of the linear form

$$Y' = \alpha' + \beta' x + \varepsilon'$$

A similar process is used to convert the equation $Y = \alpha x^{\beta+\varepsilon}$. An exercise for this is included at the end of this section.

Once these equations are converted to log form, the data used for estimating the coefficients must also be converted. For the model $Y = \alpha\beta^{x+\varepsilon}$, the dependent variable in the converted form is $\log Y$ and the independent variable is x. Therefore, the Y values must be converted to $\log Y$, and x will be regressed against $\log Y$ by the computer program giving the estimating equation

$$\log \hat{Y} = \log a + (\log b)x$$

In previous problems, we would at this point test the coefficients for significance. Due to violation of assumptions, however, testing is beyond the scope of this text. Therefore, this discussion on exponential-type models is limited to estimation of the parameters.

EXAMPLE 11.12 Fit the equation $\hat{Y} = ab^x$ to the time and parts data of Example 11.11 as an estimate of the model $Y = \alpha\beta^{x+\varepsilon}$.

Solution: Using logarithms, we must convert the model $Y = \alpha\beta^{x+\varepsilon}$ to the form given earlier.

$$\log Y = \log \alpha + (\log \beta)x + (\log \beta)\varepsilon$$

This equation is estimated from the equation

$$\log \hat{Y} = \log a + (\log b)x$$

and the multiple regression computer program. The dependent variable is log Y, and the independent variable is x. The original data and the log Y values are shown in Table 11.10. The logs were found in Appendix Table B.11. A calculator with a log key could also be used.

Variable 2 was the independent variable, and variable 3, the dependent variable. The printouts obtained from SAS and MINITAB computer programs are shown in Tables 11.11a and b. For the equation

$$\log \hat{Y} = \log a + (\log b)x$$

we have the result

$$\log \hat{Y} = 0.5486 + 0.0116x$$

TABLE 11.10 Data and logarithms to the base 10 for the dependent variable for Example 11.12

1 *Time*	*2* *Parts*	*3* *Log(Time)*
14	50	1.1461
10	40	1.0000
31	80	1.4914
18	62	1.2553
7	25	.8451
13	50	1.1139
40	90	1.6021
8	29	.9031
16	59	1.2041
6	21	.7782
23	70	1.3617
11	40	1.0414

TABLE 11.11a SAS printout for exponential model of Example 11.12

SAS

DEP VARIABLE: LOGTIME

ANALYSIS OF VARIANCE

SOURCE	DF	SUM OF SQUARES	MEAN SQUARE	F VALUE	PROB>F
MODEL	1	0.70391470	0.70391470	2094.846	0.0001
ERROR	10	0.003360221	0.000336022		
C TOTAL	11	0.70727492			

ROOT MSE	0.01833091	R-SQUARE	0.9952
DEP MEAN	1.1452	ADJ R-SQ	0.9948
C.V.	1.600673		

PARAMETER ESTIMATES

VARIABLE	DF	PARAMETER ESTIMATE	STANDARD ERROR	T FOR H0: PARAMETER=0	PROB > |T|
INTERCEP	1	0.54855935	0.01406887	38.991	0.0001
PARTS	1	0.01162287	0.000253944	45.769	0.0001

Continuing with the use of the log tables, we find the antilogs of log a and log b.

$$\log a = 0.5486 \qquad \log b = 0.0116$$
$$a = 3.54 \qquad b = 1.03$$

Hence, the estimated equation is

$$\hat{Y} = 3.54(1.03)^x$$

If we wish to estimate the average time for 100 parts, we have

$$\hat{Y} = 3.54(1.03)^{100}$$

Assuming we do not have an advanced calculator, we must go to the log form of the equation for estimation, which gives

$$\begin{aligned}\log \hat{Y} &= 0.5486 + 0.0116(100)\\ &= 0.5486 + 1.16 = 1.7086\end{aligned}$$

Finding the antilog from the tables, we obtain

$$\hat{Y} = 51.1 \text{ minutes}$$

On the other hand, had we used a scientific calculator, we would have obtained the equation

$$\hat{Y} = 3.5367(1.027)^x$$

TABLE 11.11b MINITAB printout for exponential model of Example 11.12

```
MTB > EXEC 'EX11_12 MINITAB A'
MTB > READ DATA INTO C1-C2
    12 ROWS READ
MTB > END OF DATA
MTB > LET C3 = LOGTEN(C1)
MTB > NAME FOR C1 IS 'TIME', C2 IS 'PARTS', C3 IS 'LOGTIME'
MTB > BRIEF 2
MTB > REGRESS C3 ON 1 PREDICTOR C2

THE REGRESSION EQUATION IS
LOGTIME = 0.549 + 0.0116 PARTS

                                    ST. DEV.     T-RATIO =
COLUMN        COEFFICIENT           OF COEF.     COEF/S.D.
                  0.54856           0.01406         39.00
PARTS           0.0116228         0.0002539         45.78

S = 0.01833

R-SQUARED = 99.5 PERCENT
R-SQUARED = 99.5 PERCENT, ADJUSTED FOR D.F.

ANALYSIS OF VARIANCE

 DUE TO       DF            SS         MS=SS/DF
REGRESSION     1       0.70390          0.70390
RESIDUAL      10       0.00336          0.00034
TOTAL         11       0.70726

MTB > STOP
```

and the estimation of average time as

$$\hat{Y} = 3.5367(1.027)^{100} = 50.77$$

Once again we are adding to our chances of error by predicting outside the range of the data for which the model was estimated. ■

We can see that the use of log tables introduces rounding errors into predictions that we make with equations using logs. Use of a hand calculator that contains log or ln functions is preferable.

Exponential models are not directly comparable to polynomial models through the standard error of the estimate since the exponential model's standard errors are based on the logarithms of the data. Therefore, it is difficult to say whether the cubic or the exponential form is the superior equation for prediction in this case.

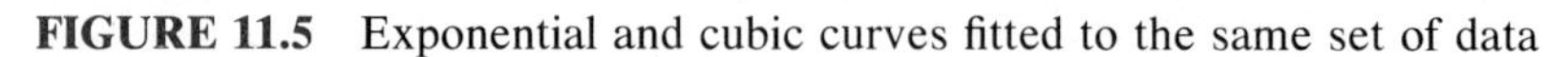

FIGURE 11.5 Exponential and cubic curves fitted to the same set of data

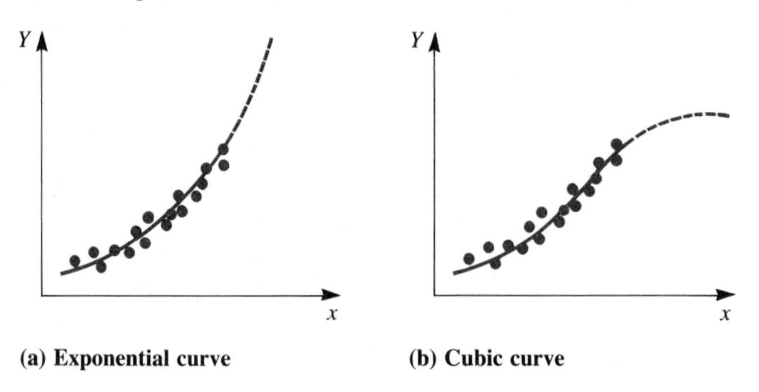

(a) Exponential curve **(b) Cubic curve**

The choice between nonlinear equations should be based on how closely the graph of the equation follows what is expected to happen rather than on a smaller standard error or larger R^2. If it is expected that diminishing returns exist for greater values of the independent variable, a cubic model rather than an exponential model is more likely to be logical, as demonstrated in Figure 11.5. Note that the cubic allows for a downturn, whereas the exponential does not. Therefore, a word of caution is necessary in the use of any equation for predictive purposes: Such use assumes that the model is valid outside the range of the data used in developing the equation, but it may not be a logical assumption.

EXERCISES

11.8 **(a)** Develop the logarithmic form of the equation

$$Y = \alpha x^{\beta + \varepsilon}$$

(b) Estimate the model, using the data in Example 11.11.

(c) Estimate the average assembly time for a radio with 100 parts.

11.9 Population data for Cleveland, given in an earlier chapter, are repeated.

1820	606	1900	381,768
1830	1,076	1910	560,663
1840	6,071	1920	796,841
1850	17,034	1930	900,429
1860	43,417	1940	878,336
1870	92,829	1950	914,808
1880	160,146	1960	876,050
1890	261,353	1970	738,956

(a) Estimate the population of Cleveland for 1980, using the following 5 models.

$$Y = \alpha + \beta x + \varepsilon$$

$$Y = \alpha + \beta_1 x + \beta_2 x^2 + \varepsilon$$

$$Y = \alpha + \beta_1 x + \beta_2 x^2 + \beta_3 x^3 + \varepsilon$$

$$Y = \alpha \beta^{x+\varepsilon}$$

$$Y = \alpha x^{\beta+\varepsilon}$$

(b) Which model is best for predictive purposes? Why?

(c) Find 1980 census figures for Cleveland. How close is the best forecast, identified in part (b), to the actual population figure?

11.6 PARTIAL CORRELATION

The coefficient of multiple determination, R^2, is used to measure the proportion of variation in the dependent variable that is explained by the set of independent variables included in the model. As each new variable is added to the equation, the value for R^2 will increase or stay the same.

DEFINITION 11.4 The **coefficient of partial determination** measures the marginal contribution of a variable to a model when all others are already included. The **partial correlation coefficient** is the square root of the coefficient of partial determination with the sign of the slope of the relationship.

In Chapter 10, when we calculated a simple correlation coefficient, we were not holding other variables that were related to the dependent variable constant. In fact, they were ignored. Their influences, although not measured, affected the value obtained for r. Hence, it is not a measure of the pure relationship between the dependent and independent variables of interest. If all the other variables that are significantly correlated with the dependent variable are included in the model, however, they can effectively be held constant through the use of partial correlation techniques, and a more accurate measure of the influence of the independent variable of interest on the dependent variable can be obtained.

Assume a simple linear relationship between Y and x_3, namely,

$$\hat{Y}_1 = a + b_3 x_3$$

We are about to add x_2 to the model to obtain the equation

$$\hat{Y}_1 = a + b_2 x_2 + b_3 x_3$$

Let $SSE(x_3)$ represent the unexplained or error variation when only x_3 is in the model, and let $SSE(x_2, x_3)$ represent the same when both variables are in the model. The coefficient of partial determination between Y_1 and x_2 with x_3 held constant is

$$r_{12\cdot 3}^2 = \frac{SSE(x_3) - SSE(x_2, x_3)}{SSE(x_3)} \tag{11.17}$$

which represents the relative reduction in the error SS as variable x_2 is added to the model. The partial correlation coefficient is the square root of Equation 11.17 with the proper sign. The notation $r_{12\cdot 3}$ signifies the correlation between the dependent variable and variable x_2 with x_3 held constant. This concept can be expanded to more variables, as is illustrated in Example 11.13.

EXAMPLE 11.13 Using the tire sales data in Example 11.2, calculate the partial correlation coefficient for each independent variable with Y when the other two variables are held constant.

Solution: Using an extension of Equation 11.17 and the data for the SSE from Table 11.12, we obtain

$$r_{12\cdot 34}^2 = \frac{SSE(x_3, x_4) - SSE(x_2, x_3, x_4)}{SSE(x_3, x_4)}$$

$$= \frac{258.7659 - 179.1678}{258.7659}$$

$$= \frac{79.5981}{258.7659} = .3076$$

$$r_{12\cdot 34} = .5546$$

TABLE 11.12 Summary of the sums of squares for models containing various combinations of independent variables for the tire sales problem

	Model	SS*R*	SS*E*
*1.**	2,3,4	5179.2322	179.1678
*2.**	2,3	5172.4229	185.9771
3.	2,4	4784.09	574.31
4.	3,4	5099.6341	258.7659
5.	2	4692.8289	665.5711
*6.**	3	720.8571	4637.5429
7.	4	4775.2824	583.1176

* These data can be found in the computer printouts for model 1 in Table 11.3 and models 2 and 6 in Table 11.5.

$$r_{13\cdot 24}^2 = \frac{SSE(x_2, x_4) - SSE(x_2, x_3, x_4)}{SSE(x_2, x_4)}$$

$$= \frac{574.31 - 179.1678}{574.31}$$

$$= \frac{395.1422}{574.31} = .6880$$

$$r_{13\cdot 24} = -.8295$$

$$r_{14\cdot 23}^2 = \frac{SSE(x_2, x_3) - SSE(x_2, x_3, x_4)}{SSE(x_2, x_3)}$$

$$= \frac{185.9771 - 179.1678}{185.9771}$$

$$= \frac{6.8093}{185.9771} = .0366$$

$$r_{14\cdot 23} = .1913$$

With variables 2 and 3 in the model, notice that variable 4 contributes very little toward the reduction of error or to an increase in the regression sum of squares.

EXERCISE

11.10 Using a computer program for multiple linear regression and the data of Example 11.2, find the strength of the relationship between total sales and inches of advertising while holding price and number of salespersons constant. Compare this figure to the simple correlation between these two variables that can be found in the correlation matrix from the program.

11.7 INDICATOR VARIABLES (DUMMY VARIABLES)

Sometimes we want to include in a regression model variables that are qualitative rather than quantitative. We can do so by including an indicator or dummy variable.

DEFINITION 11.5 An **indicator variable,** or **dummy variable,** represents a qualitative variable in a multiple regression equation and can only assume two states, present or absent. These states are represented by a 1 and 0, respectively.

It is interesting to interpret the meaning of the regression coefficients for indicator variables. Consider a model where x_2 is a quantitative variable and x_3

FIGURE 11.6 Illustration of the meaning of β_3

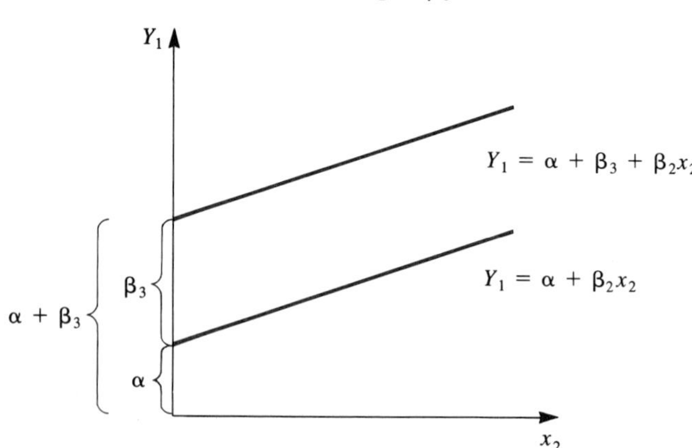

is an indicator variable. Then, consider that x_3 will take on the value of either 0 or 1. The two possible outcomes for Y_1 for a given value of x_2 are

$$Y_1 = \alpha + \beta_2 x_2 + \beta_3(0) = \alpha + \beta_2 x_2$$

and

$$Y_1 = \alpha + \beta_2 x_2 + \beta_3(1) = \alpha + \beta_3 + \beta_2 x_2$$

If $x_3 = 0$, we have one straight line, and if $x_3 = 1$, we have another straight line parallel to the first but with a different intercept. The β_3 coefficient is interpreted to be the difference between the two intercepts, as illustrated in Figure 11.6.

For perfect multicollinearity to be avoided, the number of indicator variables included in the model *must be one less than the number of states or levels* of the qualitative independent variable.

EXAMPLE 11.14 A regression model is to be developed by using the demographic variables of age, income, and geographic location. The geographic location will have as possible outcomes East, South, Midwest, and West. State the model with indicator variables included.

Solution: Since geographical location has 4 possible outcomes, only 3 indicator variables can be used. It does not matter which 3 out of the 4 are chosen. One possibility is

$$Y_1 = \alpha + \beta_2 x_2 + \beta_3 x_3 + \beta_4 x_4 + \beta_5 x_5 + \beta_6 x_6$$

where the quantitative variables are

$$x_2 = \text{age}$$
$$x_3 = \text{income}$$

and the indicator variables are

$$x_4 = \begin{cases} 1 & \text{East} \\ 0 & \text{otherwise} \end{cases}$$

$$x_5 = \begin{cases} 1 & \text{South} \\ 0 & \text{otherwise} \end{cases}$$

$$x_6 = \begin{cases} 1 & \text{Midwest} \\ 0 & \text{otherwise} \end{cases}$$

West is indicated by zeros in all the other indicator variables.

Also, observe that we should not use just 1 indicator variable for geographic location with values such as

$$0 = \text{East}$$
$$1 = \text{South}$$
$$2 = \text{Midwest}$$
$$3 = \text{West}$$

because it implies a quantitative assignment or scaling that may be meaningless. The coefficients β_4, β_5, and β_6 represent the differential effects between East and West, South and West, and Midwest and West, respectively, for any given values of x_2 and x_3. ■

In general, if all independent variables in a model are indicator variables, then we actually have an ANOVA model. When both qualitative and quantitative independent variables are present, the model is called a covariance model. For more detail, see Neter, Wasserman, and Kutner [1].

EXAMPLE 11.15 ■ Two major computer keyboards exist, the common QWERTY organization of keys, which is also used on most typewriters, and the Dvorak organization, which is claimed to produce faster typists. We wish to determine an equation for the relationship between the number of hours of training a new typist has and the speed of typing, considering also the type of keyboard. Determine the estimating equation, using a regression model that includes a dummy variable for the keyboard type, and interpret its coefficient. The data are in Table 11.13.

TABLE 11.13 QWERTY and Dvorak data for Example 11.15

Speed (wpm)	*Hours*	*Keyboard*
10	4	Q
25	14	D
20	17	Q
34	27	D
28	25	Q
22	21	Q
38	32	D
37	37	Q
43	35	D
44	38	D
13	10	Q
18	8	D
16	14	Q
26	18	D
32	32	Q

TABLE 11.14a SAS computer output for the QWERTY/Dvorak keyboard problem in Example 11.15

```
                                   SAS
DEP VARIABLE: SPEED
                          ANALYSIS OF VARIANCE

                       SUM OF         MEAN
SOURCE     DF         SQUARES       SQUARE     F VALUE      PROB>F

MODEL       2      1592.56140    796.28070     664.863      0.0001
ERROR      12     14.37192891   1.19766074
C TOTAL    14      1606.93333

     ROOT MSE        1.094377     R-SQUARE      0.9911
     DEP MEAN        27.06667     ADJ R-SQ      0.9896
     C.V.            4.043264

                          PARAMETER ESTIMATES

                   PARAMETER       STANDARD      T FOR H0:
VARIABLE   DF       ESTIMATE          ERROR    PARAMETER=0     PROB > |T|

INTERCEP    1     5.17733926     0.66473368          7.789         0.0001
HOURS       1     0.85363304     0.02702606         31.586         0.0001
KEYBOARD    1     6.41910612     0.57971202         11.073         0.0001
```

Solution: The model to be estimated is

$$Y_1 = \alpha + \beta_2 x_2 + \beta_3 x_3 + \varepsilon$$

where x_3 is the dummy variable for keyboard type. Since it is assumed from the problem that the Dvorak keyboard will permit faster typing, it was decided to let x_3 be 1 for the Dvorak keyboard and 0 for the QWERTY keyboard. Tables 11.14a and b show the resulting SAS and MINITAB computer printouts. The coefficient for x_3 is 6.419, which is interpreted to mean that for any number of hours of training, the Dvorak keyboard gives approximately a 6.4 words per minute advantage in speed. ■

TABLE 11.14b MINITAB computer output for the QWERTY/Dvorak keyboard problem in Example 11.15

```
MTB > EXEC 'EX11_15 MINITAB A'
MTB > READ DATA INTO C1-C3
    15 ROWS READ
MTB > END OF DATA
MTB > NAME FOR C1 IS 'SPEED', C2 IS 'HOURS', C3 IS 'KEYBOARD'
MTB > BRIEF 2
MTB > REGRESS C1 ON 2 PREDICTORS C2 C3

THE REGRESSION EQUATION IS
SPEED = 5.18 + 0.854 HOURS + 6.42 KEYBOARD

                                  ST. DEV.     T-RATIO =
COLUMN        COEFFICIENT         OF COEF.     COEF/S.D.
                   5.1773          0.6647           7.79
HOURS             0.85363         0.02703          31.59
KEYBOARD           6.4191          0.5797          11.07

S = 1.094

R-SQUARED = 99.1 PERCENT
R-SQUARED = 99.0 PERCENT, ADJUSTED FOR D.F.

ANALYSIS OF VARIANCE

 DUE TO       DF            SS        MS=SS/DF
REGRESSION     2       1592.56          796.28
RESIDUAL      12         14.37            1.20
TOTAL         14       1606.93

MTB > STOP
```

EXERCISE

11.11 A real estate appraiser is developing a regression model to predict selling price in a certain neighborhood. She is planning to use square footage as a variable, but she also wishes to incorporate the type of house (colonial, tudor, or contemporary) and whether the house has a brick front. She obtains the following data.

Area (sq. ft.)	*Type*	*Brick*	*Price ($000)*
2100	Contemp	No	85
2400	Tudor	Yes	110
2396	Colonial	Yes	115
3000	Tudor	Yes	175
2850	Tudor	Yes	160
2500	Contemp	No	135
2000	Colonial	No	79
2200	Colonial	No	89
2600	Contemp	No	98
2400	Colonial	No	105
2900	Tudor	Yes	154

(a) State a model that uses indicator variables.
(b) Estimate the model, using the given data.
(c) Interpret the values of the coefficients and constant term.

11.8 RELATIONSHIP BETWEEN REGRESSION AND ANALYSIS OF VARIANCE

In the study of regression, we have observed that both the dependent and independent variables are quantitative variables. That is, they both assume numerical values. The object of regression analysis is to estimate the relationship between the dependent variable and the independent variables. Analysis of variance models are simply special cases of regression models. The emphasis in ANOVA is on identifying that some relationship exists rather than on actually estimating it. As indicated in the previous section, if all independent variables are indicator variables, the regression model reduces to an ANOVA model.

The independent variables in an analysis of variance model may be qualitative or quantitative. Regression models use the actual values for a quantitative independent variable, but ANOVA models make no use of the actual values. Instead, the ANOVA model uses the values only as indicators. For example, in Table 11.15 the dependent variable, rating of sweetness, is numeric, but the independent variable, type of sweetener, is qualitative. Computationally, analysis of variance is more efficient and does not require postulating a model, justifying its existence as a separate technique, but regression results are usually more informative when using quantitative variables.

TABLE 11.15 Analysis of variance problems

Type of Sweetener			*Amount of Sugar (tsp.)*		
Sugar	*Saccharine*	*None*	*1*	*2*	*3*
2.5	1.3	1.0	2.1	4.3	2.3
2.8	1.7	.9	3.2	3.7	2.2
2.9	1.8	.3	2.8	4.1	2.4
	(a)			(b)	

In the next example, we illustrate, through the use of indicator variables, the regression approach to an ANOVA model.

EXAMPLE 11.16 In Example 9.1, an example of an analysis of variance problem was presented concerning the thickness of paint for different automatic sprayers. It was desired to test the null hypothesis that the make of the sprayer has no effect on the average thickness of paint.

Automatic Sprayer		
Ajax	*Cook*	*Ezon*
7.6	7.9	9.4
6.8	7.9	8.4
6.4	7.7	9.6
7.9	8.1	8.8
7.3	7.9	8.6

Solve this problem by using regression to demonstrate the relationship between ANOVA and regression.

Solution: The regression model to be estimated is

$$Y_1 = \alpha + \beta_2 x_2 + \beta_3 x_3 + \varepsilon$$

where: $x_2 = \begin{cases} 1 & \text{if the sprayer is Ajax} \\ 0 & \text{otherwise} \end{cases}$

$x_3 = \begin{cases} 1 & \text{if the sprayer is Cook} \\ 0 & \text{otherwise} \end{cases}$

and x_2 and x_3 are both zero if the sprayer is Ezon.

The data for entry into a computer become

Y	x_2	x_3
7.6	1	0
6.8	1	0
6.4	1	0
7.9	1	0
7.3	1	0
7.9	0	1
7.9	0	1
7.7	0	1
8.1	0	1
7.9	0	1
9.4	0	0
8.4	0	0
9.6	0	0
8.8	0	0
8.6	0	0

The resulting equation is given along with the standard errors in parentheses.

$$\hat{Y}_1 = 8.96 - \underset{(.2951)}{1.76x_2} - \underset{(.2951)}{1.06x_3}$$

The analysis of variance table obtained from the regression analysis using indicator variables is identical to the table obtained when the analysis of variance was used with these data in Chapter 9.

Source	*Sum of Squares*	*df*	*Mean Square*	*F*
Regression	7.852	2	3.926	18.04
Residual	2.612	12	.218	

EXERCISE

11.12 Rework Exercise 9.1 through the use of regression and indicator variables.

11.9 SUMMARY

Multiple linear regression is a widely applied statistical tool for estimating relationships involving more than one independent variable. Because of the difficulty and inaccuracy of hand solutions, this chapter emphasized a computer approach for estimating the equation and the corresponding statistics. Tests of significance for both the entire model (F) and for individual variables (t) or (F) were discussed.

The subject of multicollinearity was presented with attention to its influence on coefficient values, signs, and t tests. Models can be selected, using the values in the correlation matrix, that have a minimum of multicollinearity and, thus, can be valuable for interpreting the information contained in the coefficients. Choosing the "best" model for prediction purposes was also discussed, with the approach requiring the comparison of standard errors for all possible combinations of independent variables in the regression models.

Certain nonlinear models were estimated by a multiple linear regression computer program after making a linearizing transformation. Indicator, or dummy, variables were introduced to permit the inclusion of qualitative variables in the model. Partial correlation was discussed with respect to evaluating the importance of individual variables in the model by considering the contribution of one variable after the others are already included. Finally, the relationship between regression and analysis of variance was demonstrated. Table 11.16 summarizes the important equations and definitions of Chapter 11.

TABLE 11.16 Summary table

Typical regression model $Y_i = \alpha + \beta_1 x_{i1} + \beta_2 x_{i2} + \cdots + \beta_k x_{ik} + \varepsilon_i$
Confidence intervals (if $s_{\hat{Y}\|x_p}$ available): $\hat{Y}_1 = a + b_2 x_2 + \cdots + b_k x_k$ For $\mu_{Y\|x_p}$: $\hat{Y}_{x_p} \pm t_{\text{df}_E,\alpha/2} s_{\hat{Y}\|x_p}$ For Y_{x_p}: $\hat{Y}_{x_p} \pm t_{\text{df}_E,\alpha/2} s_{Y\|x_p - \hat{Y}\|x_p}$
Standard error of the estimate $s_{Y\|x} = \sqrt{\dfrac{\text{SSE}}{n-k-1}} = \sqrt{\text{MSE}}$ *where*: n = number of observations k = number of independent variables
Multiple coefficient of determination $R^2 = \dfrac{\text{SSR}}{\text{SST}}$
F test for entire equation: H_0: all $\beta_j = 0$; H_a: some $\beta_j \neq 0$ Reject H_0 if $F = \dfrac{\text{MSR}}{\text{MSE}} > F_{k,n-k-1,\alpha}$

TABLE 11.16 (Continued)

t test on slopes (if no significant multicollinearity): H_0: $\beta_j = 0$; H_a: $\beta_j \neq 0$ Reject H_0 if $t = \left\lvert \frac{b_j}{s_{b_j}} \right\rvert > t_{\text{df}_E,\alpha/2}$
Critical value for multicollinearity: H_0: $\rho = 0$; H_a: $\rho \neq 0$ Reject H_0 if $\lvert r \rvert > r_{cr} = +\sqrt{\frac{t^2_{n-2,\alpha/2}}{t^2_{n-2,\alpha/2} + n - 2}}$
Best equation For prediction: combination of variables with smallest standard error For coefficients: model with little or no multicollinearity
Nonlinear equations in x $\hat{Y} = a + bx + cx^2$ parabola $\hat{Y} = a + bx + cx^2 + dx^3$ cubic $\hat{Y} = ab^x$ exponential $\hat{Y} = ax^b$ exponential
Algebra of logarithms $\log(ab) = \log a + \log b$ $\log(a/b) = \log a - \log b$ $\log a^b = b \log a$
Partial correlations: a correlation between two variables in a multiple regression model, holding the remaining variables constant
Indicator variables: qualitative variables that usually assume only two values (1 and 0); use 1 fewer than the number of categories.

SUPPLEMENTARY EXERCISES

11.13 A certain governmental regulatory agency has a plan to determine the miles per gallon (mpg) that new automobiles will achieve. It will use a regression model rather than actually testing the cars. To build the model, the agency takes a sample of new cars and determines the miles per gallon through a test using the size of the engine (CID), weight of the car, and rear axle ratio. The results of the test are as follows.

Mpg	*Size (CID)*	*Weight (1000 lb)*	*Axle Ratio*
26.1	200	3.0	3.25
30.2	150	2.5	2.50
35.3	75	2.2	2.50
20.1	350	3.5	3.25
20.2	395	3.6	3.00
18.2	450	4.1	2.75
19.1	425	4.0	2.50
22.3	400	3.8	3.00
38.4	50	2.2	2.00
35.8	85	2.3	2.00
24.2	250	3.1	2.60
25.1	275	3.4	2.35
21.0	325	3.3	2.80

(a) Estimate the equation for a linear relationship.
(b) Predict the mpg for a new sports car with a 200 CID engine, weighing 2000 pounds, and having an axle ratio of 2.00.

11.14 **(a)** Test the model in Exercise 11.13 for significance, using $\alpha = .05$.
(b) Test each b_j individually to determine if $\beta_j = 0$, using $\alpha = .05$.

11.15 For Exercise 11.13, using various combinations of the independent variables, determine which linear model best predicts miles per gallon for a new car.

11.16 Obtain the correlation matrix for Exercise 11.13 and determine if significant multicollinearity exists between any of the independent variables. If so, what model would you suggest that would have a low degree of multicollinearity?

11.17 What is the strength of the relationship between size (CID) and miles per gallon (mpg) in Exercise 11.13 if weight and axle ratio are held constant?

11.18 Fit a parabolic model between mpg and axle ratio in Exercise 11.13 and determine whether this relationship is significant ($\alpha = .05$).

11.19
```
REGRESSION COEFFICIENTS

VAR      COEFFICIENT(B)          STD ERROR        T

2            .63681               .07307       8.71443
3            .00220               .00037       5.92666
4           -.00738               .01839       -.40131
            -.30608             (CONSTANT)

STANDARD ERROR OF ESTIMATE:   .11692
R SQUARED:                    .96018

ANALYSIS OF VARIANCE

SOURCE        SUM OF SQUARES    DF    MEAN SQUARE       F
REGRESSION       1.97797         3       .65932      48.22902
RESIDUAL          .08202         6       .01367
```

Use the multiple regression computer printout.

(a) Determine the equation.
(b) Determine $s_{Y|x}$.
(c) Determine R^2.
(d) Test the equation for significance at $\alpha = .05$.
(e) Test the individual variables for significance at $\alpha = .05$.

11.20 A real estate agent claims that the price that can be charged for a house in a certain area is related to the age of the house and the number of rooms. The data for a sample of recently sold houses are given. Use the normal equations to obtain the equation for a linear relationship.

Price ($000)	*Age (Years)*	*Rooms*
68	2	8
64	5	7
35	4	5
50	10	7
55	15	10

11.21 In testing the drying time of a commercial wax, an industrial user applied the wax to 4 different surfaces. The drying times to the nearest 10th of a minute are as follows.

Types of Floor Surfaces

Wood	*Vinyl*	*Linoleum*	*Slate*
3.6	4.8	4.2	3.6
2.4	3.2	3.6	3.3
4.2	2.9	3.6	4.8
3.7	4.2	2.9	2.2
2.8	2.5	3.4	2.7
3.4	3.0	2.5	
4.7			

At a .05 level of significance, can it be concluded that floor surfaces make a difference in drying time for the wax? Use regression with indicator variables instead of analysis of variance.

11.22

```
REGRESSION COEFFICIENTS

VAR      COEFFICIENT(B)        STD ERROR            T

 2        -.8374159157        .1367607481      -6.123218301
 3        -.4861022972        .5439340114      -.8936788049
 4        -.2490163726        .1836421782      -1.355986817

STANDARD ERROR OF ESTIMATE:   .6178789014
R SQUARED:                    .9890921618

ANALYSIS OF VARIANCE

SOURCE      SUM OF SQUARES     DF    MEAN SQUARE          F
REGRESSION   69.23645133       3     23.07881711    60.45146277
RESIDUAL      .7635486737      2      .3817743368
```

```
CORRELATION MATRIX

1.0000     -.9888     -.8864      .1309
-.9888     1.0000      .8581     -.2143
-.8664      .8581     1.0000     -.3503
 .1309     -.2143      .3503     1.0000
```

Refer to the computer printout.

(a) Test the equation for significance at $\alpha = .05$.

(b) Test the individual variables for significance at $\alpha = .05$.

(c) Determine which independent variables have a high degree of multicollinearity between them.

(d) Suggest a model that will not have a high degree of multicollinearity.

11.23 The strength of a particular variety of steel wire is nonlinearly related to the amount of carbon added to the alloy. Determine the equation that best fits the data taken from an experiment where various amounts of carbon were used and the resulting steel wires pulled until each broke.

Pounds Required to Break	*Amount of Carbon*
800	1
825	1
875	2
950	3
1050	4
1040	4
1105	5
1150	6
1525	8
2550	10

11.24 An experiment was designed to estimate the relationship between sugar and lemon content of lemonade and how well it is liked. Each respondent scores the lemonade from 0 to 10, awful to great. Lemon and sugar are measured in spoonfuls per 12 ounces of water.

Amount of Lemon	*Amount of Sugar*	*Degree of Liking*
2	1	5.4
2	2	7.1
3	1	6.2
3	2	8.1
4	1	7.3
4	2	8.6

Use the normal equations to obtain the equation for a linear relationship.

11.25 The manager of city buildings wishes to develop a model that she can use to determine the number of gallons of paint required for painting individual buildings. State a linear model for the manager and estimate and evaluate its coefficients based on the following data from previous buildings painted. Is the linear model of any use?

Paint Used	*Building Wall Area (sq. ft.)*	*Condition of Previous Paint*	*Surface Type*
25	10,000	Good	Brick
76	15,200	Poor	Concrete
43.5	18,400	Good	Wood
250	75,000	Poor	Wood
65.5	24,500	Good	Concrete
38.7	16,250	Good	Wood
73.5	18,400	Poor	Brick
242	52,000	Poor	Concrete
202.5	81,000	Good	Brick

11.26 Holding both surface type and condition of previous paint constant in Exercise 11.25, find the strength of the relationship between building wall area and amount of paint used.

11.27 On each of the 4 different research projects, using whatever information is provided, determine which of the 2 equations is better. Show all work where calculations must be performed and/or indicate the reason for the selection. Let $\alpha = .05$.

(a) $\hat{Y} = 2 + \underset{(1)}{3x_2} + \underset{(2)}{5x_3}$ $\quad n = 20$

$\hat{Y} = 4 + \underset{(6)}{8x_4} + \underset{(8)}{7x_5}$

Numbers in ( ) indicate standard errors of β_j's

(b) $\hat{Y} = 6 + 4x_2 + 5x_3$

$\hat{Y} = 10 + 6x_2 + 4x_4$

Correlation Matrix

1	1			
2	.70	1		
3	.80	.75	1	
4	.80	.02	.85	1
	1	2	3	4

(c) $\hat{Y} = 2 + 3x_2 + 5x_3$ $\qquad s_{Y|x} = 12 \quad F = 48$

$\hat{Y} = 3 + 4x_4 + 3x_5$ $\qquad s_{Y|x} = 15 \quad F = 52$

(d) $\hat{Y} = 4 + 5x_2$ $\qquad F = 3.81 \quad n - 2 = 25$

$\hat{Y} = 3 + 2x_3 + 3x_4 + 4x_5$ $\qquad F = 3.25 \quad n - 4 = 23$

11.28 The personnel department of a large firm gives new employees a series of tests that are used to predict success in the company. It is currently giving three tests, the QRI aptitude, the FMTZ personality, and the TPI intelligence. Twelve employees, who have been with the company for years, took the tests to see if test results do correlate with income.

Income	*TPI*	*FMTZ*	*QRI*	*Years Employed*
15,250	126	10	96	2
17,875	132	15	76	3
25,800	141	22	91	5
32,500	138	25	92	10
18,900	110	18	74	5
10,100	120	8	87	1
28,450	125	23	81	11

Income	*TPI*	*FMTZ*	*QRI*	*Years Employed*
16,500	108	14	94	3
24,500	131	21	76	3
16,900	105	16	50	4
26,200	128	22	80	8
21,100	125	20	68	6

(a) Estimate the linear model, using all the variables.
(b) Test the equation for significance at $\alpha = .05$.
(c) Are each of the three test variables significant predictors of income?

11.29 If the personnel department in Exercise 11.28 wishes to use only two tests in future testing sessions, which test should it drop?

11.30 A new employee tested by the personnel department in Exercise 11.28 obtains scores of 135 on the TPI, 20 on the FMTZ, and 90 on the QRI.
(a) Predict the employee's income 10 years from now. Use a 95% confidence interval if $s_{\hat{Y}|x_p}$ is available.
(b) Why is the result you obtained for a prediction obviously incorrect?
(c) If you could assume a constant rate of inflation of 10% per year, could you make a more useful prediction? If so, what would it be?

11.31 The personnel department of Exercise 11.28 desires to find the best model for predicting an employee's income. What model do you suggest?

11.32 The electric company proposes that the monthly electric bill for a residential customer is a factor of the average monthly temperature, the average hours of daylight, the age of the home, the square footage of the home, and the R value of the insulation in the roof. A random sample of bills throughout the year for a sample of customers is selected. The rate for each bill was the same and constant throughout the year. All homes are heated electrically. No house has air conditioning.

Bill	*Ave. Temp.*	*Ave. Hours Daylight*	*Square Footage*	*R of Insulation*	*Home Age*
45.25	54	13.1	2,400	30	3
28.13	64	16.2	1,750	11	40
16.25	56	13.4	1,100	6	50
12.18	60	15.3	1,000	5	42
160.36	25	10.6	2,800	32	2
120.48	48	12.3	1,700	8	48
26.20	62	15.7	1,800	12	10
158.74	11	9.5	2,000	11	42
105.24	38	11.4	1,400	16	35
106.18	47	12.7	1,650	24	13
74.96	19	10.6	1,100	15	31
65.10	31	11.2	1,225	16	30
75.85	10	9.5	1,475	20	15
25.24	60	14.8	1,300	16	28

(a) Estimate the linear model that includes all the variables.
(b) Test the equation for significance ($\alpha = .05$).

11.33 The electric company wants to determine the best revised model for predicting an electric bill. The revised model is to exclude square footage and R of roof insulation because the company cannot easily determine either one for all the homes it serves. Determine the equation for the company. Refer to exercise 11.32.

11.34 If the electric company is able to obtain data on square footage and insulation, can it determine a better model for prediction? If so, what is it? Refer to exercise 11.32.

11.35 The electric company would like to know how much an electric bill decreases for each additional R of insulation put in the roof. What would you suggest? What must be assumed for the interpretation? Refer to exercise 11.32.

11.36 An economist for the trucking industry wants to determine if there are economies of scale in the trucking industry by calculating the long-run cost curve. From a sample of 16 class I motor carriers, he obtains the following total cost and output figures for the year 1971.

Carrier	*Total Cost*	*Output (ton-miles)*
Bender	7,080,604	61,989,765
Branch	43,159,622	324,189,513
Carolina	58,076,940	994,164,746
Cleveland	2,332,702	195,947,225
Consolidated	242,706,391	3,899,341,901
Duff	12,825,259	73,206,480
Eazor	13,493,714	200,634,222
Garrett	49,337,574	707,380,533
Gateway	64,737,415	907,758,819
Hall	2,402,771	2,301,743
Hennis	1,772,026	24,472,571
Imperial	4,344,889	28,491,565
Killion	2,518,984	31,118,662
L and V	1,249,847	1,100,105
Manley	2,783,742	6,878,230
Spector	106,075,838	1,231,421,173

Total cost first must be converted to unit cost by dividing by output. Then, construct a nonlinear regression model and determine the best equation for predicting unit cost as a function of output.

11.37 A quantitative business analysis professor teaching statistics has observed that some students score higher on her exams than others. To attempt to explain this variation, she develops a multiple regression model that considers students' aptitudes, past performance, learning speed, parttime work, and average hours per week studied for all their subjects. Verbal and math aptitude scores come from their college entrance exam. The math grade is from a prerequisite math course.

Final Grade	*Math Aptitude*	*Verbal Aptitude*	*Math Grade*	*Hours Studied*	*Learning Speed*	*Work Parttime*
98	530	620	95	40	Fast	No
76	460	480	86	25	Fast	Yes
54	390	361	64	20	Slow	Yes

Final Grade	*Math Aptitude*	*Verbal Aptitude*	*Math Grade*	*Hours Studied*	*Learning Speed*	*Work Parttime*
79	435	538	72	30	Fast	No
85	491	582	81	38	Slow	No
75	524	515	82	40	Slow	No
63	502	415	75	20	Fast	Yes
92	610	625	84	35	Slow	No
84	595	574	71	42	Slow	No
87	515	585	93	45	Slow	No
73	511	493	88	28	Fast	Yes
68	412	461	72	38	Slow	No
74	423	498	75	25	Slow	Yes
77	495	525	61	30	Fast	No
81	560	547	72	40	Slow	No

Develop the best linear model for predicting grades in this course.

11.38 Working parttime and being a slow learner would be expected to negatively influence a student's grade. Use the data in Exercise 11.37 to answer the following.

(a) On the average, how much lower will students' grades be if they work parttime?

(b) On the average, how much lower will students' grades be if they are slow learners?

11.39 Most students believe that they will have difficulty learning statistics if they have difficulty with mathematics. Do the data in Exercise 11.37 tend to support this view? What is your interpretation?

11.40 A commission of college presidents from private colleges has collected the following data and asked you to develop a linear model to estimate the average number of students a given college should expect to have enrolled. Data were collected from 15 colleges across the country for the following variables.

—Enrollment
—Tuition
—Distance from nearest state university
—Number of junior colleges within 20 miles
—Distance to nearest private college
—Annual tuition at nearest state university
—Population of metropolitan area
—Number of dormitory rooms

Enrollment	*Tuition*	*Dist. State U.*	*Junior College*	*Dist. Pr. Coll.*	*Tuition State U.*	*Metro Pop.*	*Dorm Rooms*
5,023	5250	6	1	20	500	50,000	4000
6,915	4900	5	4	10	700	80,000	3000
10,841	3875	41	3	50	1500	150,000	3000
15,236	2000	50	2	40	2000	200,000	7500
8,417	4000	15	3	15	300	90,000	2000
7,283	4575	18	4	10	1400	94,000	2500
5,146	4950	7	1	12	2500	25,000	2000

Enrollment	*Tuition*	*Dist. State U.*	*Junior College*	*Dist. Pr. Coll.*	*Tuition State U.*	*Metro Pop.*	*Dorm Rooms*
6,814	4800	16	2	5	1400	30,000	2500
11,213	3500	35	3	20	2500	115,000	4250
7,854	4700	15	1	10	1200	20,000	2125
6,843	5000	12	1	8	1500	15,000	2800
10,214	3800	32	4	30	2000	130,000	3850
4,132	6000	4	2	14	800	80,000	2500
5,438	5050	8	2	7	900	50,000	2100
2,124	7400	10	3	3	1000	70,000	1024

Find the best linear equation for predicting the enrollment at private colleges.

11.41 Refer to Exercise 11.40 and develop a good model that has a low degree of multicollinearity.

11.42 Refer to Exercise 11.40. Assume that a president of a private school is able to drop tuition \$1000.

(a) How much would enrollment be expected to increase?
(b) Would this increased enrollment be financially advantageous to the college?
(c) Do you see any weakness in this analysis?

11.43 The Cobb–Douglas production function relates labor input (L) and capital input (C) to production output (P) through the function

$$P = aL^{\alpha}C^{\beta}$$

where α is the elasticity with respect to labor input and β is the elasticity with respect to capital input. The Johnson Company has obtained the following random sample of data for its past operations.

Output	*Labor*	*Capital*
152.1	63.8	386.0
129.1	58.7	363.2
159.9	66.0	408.0
108.8	64.4	354.1
134.7	60.2	365.2
147.8	60.0	373.7
154.3	64.9	396.5
130.9	59.3	359.3
131.4	58.9	359.4

Determine the coefficients and state the estimated function for production output for the Johnson Company.

11.44 Refer to Exercise 11.43 and predict the output for a labor input of 70 and capital input of 450.

11.45 Financial analysts usually consider the price-to-earnings ratio (P/E) of a stock to be related to the compound annual growth rate, the current dividend yield, and the complement of the debt-to-equity ratio, 1 − (D/E). The debt-to-equity factor is approximated in the table with the debt-to-net-worth (D/NW) and 3 different industries are represented.

	P/E	*5-yr. Growth Rate*	*1 − (D/NW)*	*Dividend Yield*	*Industry*
Exxon	7.7	20.5	.815	7.0	Oil
Texaco	8.0	24.5	.615	8.0	Oil
Mobil	6.1	25.5	.612	6.6	Oil
Standard (Ind)	6.8	22.0	.646	5.6	Oil
Merck	14.2	12.5	.855	2.9	Drug
Lilly	11.9	12.5	.966	3.7	Drug
Smithkline	13.9	16.0	.796	2.1	Drug
Schering–Plough	8.6	12.5	.983	3.9	Drug
Goodyear	5.4	10.0	.327	7.7	Rubber
Goodrich	4.6	8.0	.447	6.5	Rubber
Firestone	10.5	9.5	.574	5.5	Rubber
Uniroyal	8.5	7.0	.264	5.2	Rubber

State and estimate a linear model for predicting the price-to-earnings ratio, using all of the variables at $\alpha = .05$.

11.46 Determine the best equation for predicting price-to-earnings ratio, using the data in Exercise 11.45.

11.47 The accounting department in a major corporation is designing a regression model to aid in predicting and understanding the variation in accounts receivable. It is believed that sales, interest rates, and inflation rate should be considered as explanatory variables. A random sample is selected from the files. The data are given in the table. Dollar values are in thousands.

Accounts Receivable	*Sales*	*Interest Rate*	*Rate of Inflation*
25	1024	.07	.05
51	2096	.09	.10
36	1525	.05	.08
56	2514	.11	.13
21	986	.04	.02
27	1042	.05	.04
33	1421	.04	.03
49	1954	.07	.09
46	1825	.06	.08
32	1432	.08	.07
53	2143	.10	.09

Determine the best equation for predicting accounts receivable.

11.48 Refer to Exercise 11.47. If the interest rate increases by 1%, how much is accounts receivable expected to change if the other variables remain constant?

REFERENCE

1. Neter, John, William Wasserman, and Michael Kutner. *Applied Linear Statistical Models*, 2nd ed. Homewood, IL: Richard D. Irwin, 1985.

12 Time Series Analysis and Forecasting

12.1 INTRODUCTION

Levi Haines is a captain on the police force of the city of Hastings. His chief responsibility is to make daily assignments of officers to patrol various sections of the city to aid in traffic direction. His assignments, based almost exclusively on intuition, have not always produced optimum results. Frequently, he has had either too many or too few officers on duty in certain areas. He realizes that there is an abundance of past data on the number of vehicles traveling hourly on city streets. He notes that through time there has been a gradual upward trend in the total amount of traffic, but he also discerns that traffic peaks occur during the morning and afternoon rush hours. He decides to use time series analysis to aid in predicting future traffic volumes. He plans to use this information to provide improved assignments of police officers.

In order to improve his assignments of police officers, Levi must consider an appropriate model that reflects traffic patterns. He must also estimate the various components of the model and then use the estimates to forecast future observations in the series. In this chapter, we are interested in the decomposition of a time series into components that represent the effects of factors that influence the past data. We are also interested in predicting future observations based on information about the components.

12.2 BASIC CONCEPTS AND ASSUMPTIONS

In time series problems, it seems clear that an analysis of past data obtained periodically through time is necessary for predicting future observations. Quantitative measurements on sales, inventories, production, and prices constitute examples of variables that are usually related to time. Good managers analyze past data and, on the basis of that analysis, project into the future as scientifically as possible.

The technique from previous chapters that is the most useful for analyzing time series data is regression analysis. You may recall that in a regression model, the x values are assumed to be known or measured without error. This assumption is satisfied in time series if we let time be the x variable. For time series applications, however, we do not make inferences as we did in prior regression situations, although it is possible to do so. For example, one technique that treats time series as a stochastic process is the Box–Jenkins Model. Since this is an introductory text, the Box–Jenkins technique is not included. Instead, discussion is restricted to least squares estimation of the parameters in the model.

A time series is defined as follows.

DEFINITION 12.1 A **time series** is a sequence of observations of a variable made periodically through time.

The time period itself may be years, months, weeks, days, or any other unit appropriate for the problem at hand. The discussion of time series is restricted to discrete units with equally spaced observations. Cases where data are obtained continuously over time are not considered.

The classical approach to time series analysis assumes that an observation is a composite of trend, seasonal, cyclical, and irregular components. One of the main goals in this chapter is to examine various techniques for estimating these components. First, we briefly examine each component.

DEFINITION 12.2 The **secular trend** is the component of time series data that indicates a steady and gradual long-term growth or decline.

Trend is considered to be the most fundamental or basic of the components. Figure 12.1a depicts a time series having an obvious positive trend component. Many series of interest to a manager are similar in appearance, at least over some restricted portion of time.

DEFINITION 12.3 The **seasonal component** refers to variations in the data associated with the season or month of the year, the day of the week, or some other period of interest within a year.

FIGURE 12.1 The four components of a time series

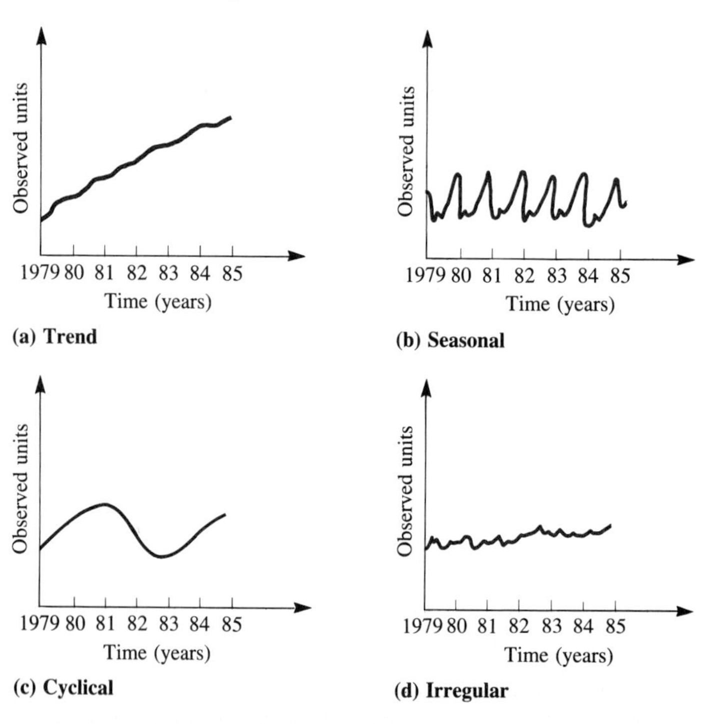

The seasonal variation is assumed to follow a regular pattern. A familiar example of seasonal fluctuation is that department store sales are large during the Christmas season compared to other times of the year. Figure 12.1b depicts a seasonal pattern.

DEFINITION 12.4 The **cyclical component** results from movements or fluctuations of the business cycle or some other source that produces observations that revolve around the trend line in long cycles.

Cyclical movements are longer in time period and more variable in amplitude than seasonal variation and, therefore, do not occur in the same predictable manner as seasonal effects. A portion of a cyclical pattern is indicated in Figure 12.1c.

DEFINITION 12.5 The **irregular** or **random component** measures variation due to chance causes that are usually assumed to be unpredictable and that do not occur in a systematic manner.

Included in the irregular component is variation not accounted for by the other three components. An example indicating irregular variation is given in Figure 12.1d.

For convenience, we designate the four components—trend, seasonal, cyclical, and irregular—by T, S, C, and I, respectively.

One of the classical models assumes that the four components are additive. Algebraically, the additive model is

$$Y = T + S + C + I \tag{12.1}$$

where Y is the observed value of the variable of interest. T is ordinarily expressed in the same units as Y, and S, C, and I are measured in terms of positive or negative deviations from T.

The classical model that most analysts consider more appropriate to use than the additive model is the multiplicative model

$$Y = T \times S \times C \times I \tag{12.2}$$

The values of Y and T are given in the observed units, just as in the additive model, but S, C, and I are expressed in ratios or percents. Actually, it is customary to indicate the value for each of these components as a percent but to express each component as a ratio for use in the equation.

Other models, which are hybrids of the additive and multiplicative models, are sometimes used. We do not consider them here.

The decomposition of time series data into its component parts by means of a classical model has been found to be a very useful procedure for many practical business and economic problems. However, certain limitations and shortcomings of the method should be pointed out. The method of analysis essentially isolates each component one at a time, more or less separately from the other components. Thus, this approach tacitly assumes that the various components are, in a sense, unrelated to each other. However, the validity of this assumption is in question. It seems likely that the forces that affect the cyclical component often contribute to the magnitude of seasonal movements as well. The particular value of the trend component may be the result of many of the same causes that affect long-term cyclical movements. In other words, it is undoubtedly true that the time series components are interrelated.

An approach that estimates components simultaneously rather than individually is appropriate but more complex than can be presented at this level. Econometrics, for example, provides a method of time series analysis that estimates components through a system of simultaneous equations. In general, the econometric approach utilizes a mathematical model to analyze economic situations by means of statistical techniques. The equations are ordinarily based on structural relationships derived from economic theory and technological production information measured over time. Econometric methods have been used more often and with greater success in forecasting levels of the economy than in forecasting for an individual firm. Econometric procedures are beyond the level of this text; thus, the remainder of the presentation assumes the applicability of the classical multiplicative model.

12.3 LINEAR TRENDS

The decomposition of times series data is begun by first isolating the trend component. For now, let us assume that trend can be estimated satisfactorily by a straight line.

In Chapter 10, the equation for a simple linear relationship was $\hat{Y} = a + bx$. In time series analysis, this equation, modified to reflect the trend, becomes

$$\hat{T} = a + bx$$

In addition, a trend equation is not complete without the identification of the origin, the units of x, and the units of Y. Thus, a trend equation can be expressed generally as

$$\hat{T} = a + bx \qquad \textbf{(12.3)}$$

origin = date

x = incremental time units

Y = units associated with the data

If x represents the actual time period of the data, such as 1985, for example, necessary computations become quite cumbersome. However, these complications can be avoided and the calculations simplified by coding the x's. A suggested simple coding scheme is to equate zero with the year or other appropriate point in time, identified as the origin. Thus, the effect of such coding is to predict the trend value as being equal to a when $x = 0$ is substituted in the equation.

Generally, a trend line is used to project into the future, and with the addition of the origin and the units of x and Y, the equation is readily usable. Without this identification, though, it is useless. We note, however, that forecasting the future puts us outside the range of the data upon which the model is based. The dangers of extrapolation were discussed in Chapter 10. The subject is also addressed in Section 12.10.

Several methods are available for determining the trend component of a time series. We consider the simplest procedure first.

DEFINITION 12.6 The **freehand method** of determining a trend line is merely a line-of-sight representation of the graphed data.

The slope of the trend equation is given by

$$b = \frac{\Delta Y}{\Delta x} \tag{12.4}$$

where ΔY represents the change in Y and Δx represents the corresponding change in x. The value of a is the value of Y corresponding to the designation of the origin—that is, when $x = 0$.

EXAMPLE 12.1 Anderson Electronics has experienced gross sales, in thousands of dollars, for the past 6 years as follows.

Year	*Sales ($000)*
1980	1.74
1981	2.11
1982	1.92
1983	2.36
1984	2.20
1985	2.70

Determine the trend line, using the freehand method, and predict sales for 1986.

Solution: Initially the data are plotted. See Figure 12.2. The dashed line placed on the graph is deemed to be representative of the data.

FIGURE 12.2 Sales data for Anderson Electronics and sketch of trend line

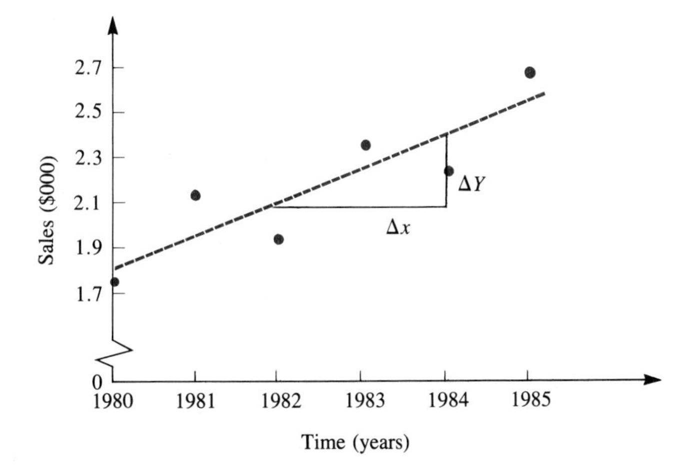

Substituting in Equation 12.4, we obtain the slope.

$$b = \frac{\Delta y}{\Delta x} = \frac{2.4 - 2.1}{84 - 82} = \frac{0.3}{2} = .15$$

If the origin is chosen as 1982, corresponding to $x = 0$, then the value of a is 2.1. The trend equation becomes

$$\hat{T} = 2.1 + .15x_i$$

origin = July 1, 1982

x = 1 year

Y = thousands of dollars

Of course, the meaning of the slope is that sales are increasing, on the average, at \$150 per year. To predict sales for 1986, we must substitute $x = 4$ in the trend equation because a change of 1 in x corresponds to 1 year and 1986 is 4 years later than 1982, the year chosen as the origin. The predicted trend value for 1986 is

$$\hat{T}_{86} = 2.1 + (.15)(4) = 2.1 + .6$$
$$= 2.7 \quad \text{or \$2700 when } Y \text{ units are considered}$$

The freehand procedure may be quite satisfactory for some purposes, but, for most problems, a more objective approach is needed. However, before we look at additional techniques for determining the trend equation, a comment is in order regarding the plotting of data. An observation is considered to be associated with the midpoint of the time period that it represents. In Example 12.1, the observation of 1.74 for 1980 is associated with July 1, 1980, and 2.70 is associated with July 1, 1985. This association explains why the origin was identified as July 1, 1982, in the solution of Example 12.1. As we proceed through this chapter, the date associated with a given data point for a period will be deemed to be at the midpoint of the time period it represents.

We next consider the most elementary objective technique for determining trend.

DEFINITION 12.7 The **method of semiaverages** for determining a trend line is carried out by dividing the data into two halves and then connecting the mean values of the two parts.

The slope is

$$b = \frac{\overline{Y}_2 - \overline{Y}_1}{\Delta x} \tag{12.5}$$

where $\overline{Y}_2$ is the mean of the second half of the data, $\overline{Y}_1$ is the mean of the first half of the data, and Δx is the length of the time period between the two means. The intercept a again represents the Y value corresponding to the choice of the year of origin—that is, when $x = 0$. To depict the trend line, representative of the data, we need only connect the points $\overline{Y}_1$ and $\overline{Y}_2$ plotted on the graph.

EXAMPLE 12.2 The Gremlin Corporation has experienced the following annual sales, in thousands of dollars, for 1980 to 1985.

Year	*Sales ($000)*
1980	1.5
1981	4.4
1982	4.9
1983	8.2
1984	6.7
1985	13.0

What is the trend equation determined by the method of semiaverages?

Solution: Dividing the data into two parts, we obtain the means

$$\bar{y}_1 = \frac{1.5 + 4.4 + 4.9}{3} = \frac{10.8}{3} = 3.6$$

centered at July 1, 1981, and

$$\bar{y}_2 = \frac{8.2 + 6.7 + 13.0}{3} = \frac{27.9}{3} = 9.3$$

centered at July 1, 1984. Thus,

$$b = \frac{9.3 - 3.6}{3} = \frac{5.7}{3} = 1.9$$

where Δx is 3, the length of time between 1981 and 1984. By selecting $\bar{y}_1$ as the value for a, we obtain the trend equation

$$\hat{T} = 3.6 + 1.9x_i$$

origin = July 1, 1981

$x = 1$ year

Y = thousands of dollars

In Figure 12.3, the data, together with the calculated trend line, are shown. We can see that both semiaverages are plotted, and the trend line is determined by connecting these two points.

In time series data, there may be an odd number of observations. In such a case, it is traditional to omit the middle observation when dividing the data into two parts for determining the semiaverages.

FIGURE 12.3 Sales data for the Gremlin Corporation from Example 12.2 and trend line calculated by the method of semiaverages

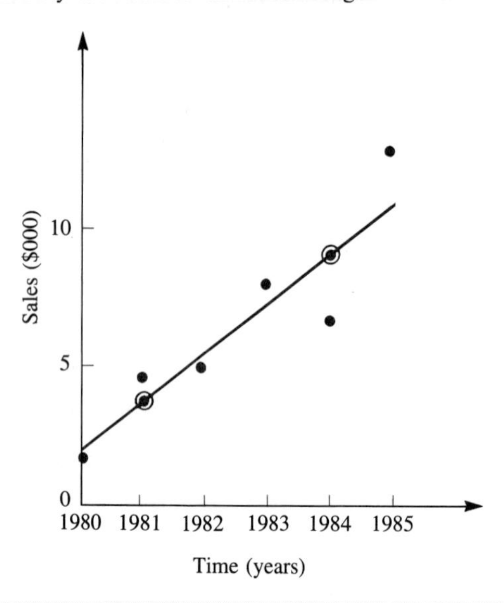

EXAMPLE 12.3 Wonder Products has experienced net sales in thousands of dollars for one of its secondary products as follows.

Year	*Sales ($000)*
1979	46.2
1980	44.3
1981	40.3
1982	39.8
1983	39.4
1984	38.7
1985	36.2

What is the trend equation determined by the method of semiaverages?

Solution: Since there are 7 pieces of data, 39.8, the sales for 1982, will be omitted from the computations. The 2 means are

$$\bar{y}_1 = \frac{46.2 + 44.3 + 40.3}{3} = \frac{130.8}{3} = 43.6$$

centered at July 1, 1980, and

$$\bar{y}_2 = \frac{39.4 + 38.7 + 36.2}{3} = \frac{114.3}{3} = 38.1$$

centered at July 1, 1984. Therefore,

$$b = \frac{38.1 - 43.6}{4} = -\frac{5.5}{4} = -1.375$$

The determination of Δx is 4, the difference between 1980 and 1984.

If $\bar{y}_2$ is selected as a, then the trend equation is

$\hat{T} = 38.1 - 1.375x$

origin = July 1, 1984

x = 1 year

Y = thousands of dollars

The trend equation can be used to predict the trend value for any point in time. For example, the trend value for 1986 would be $\hat{T}_{86} = 38.1 - (1.375)(2) = 38.1 - 2.75 = 35.35$ (thousand dollars). We note that the trend line for Wonder Products is declining, as evidenced by the negative slope. In predicting the trend value for 1986 to be 35.35, we are assuming that the trend will continue to decline.

The method of semiaverages is simple but crude and would likely be used only if an answer were needed quickly and a calculator were not available. By far the most popular and most sophisticated objective procedure used to determine a linear trend equation for time series data is the method of least squares.

DEFINITION 12.8 In determining a trend line, the procedure for finding the line that minimizes the sum of the squares of the vertical deviations of the observed points from the line is the **method of least squares.**

The principles of the method of least squares were presented in detail in Chapter 10 in conjunction with the determination of a linear regression equation. Note that Definition 12.8 is similar to Definition 10.5. The formulas used to determine the slope and intercept were

$$b = \frac{\sum_{i=1}^{n} x_i Y_i - \frac{\sum_{i=1}^{n} x_i \sum_{i=1}^{n} Y_i}{n}}{\sum_{i=1}^{n} x_i^2 - \frac{\left(\sum_{i=1}^{n} x_i\right)^2}{n}} \tag{12.6}$$

and

$$a = \frac{\sum_{i=1}^{n} Y_i}{n} - b\,\frac{\sum_{i=1}^{n} x_i}{n} \tag{12.7}$$

FIGURE 12.4 Coding for a data set with an odd number of yearly observations

78	79	80	81	82	83	84	Time (years)
−3	−2	−1	0	1	2	3	Coded x

Equations 12.6 and 12.7 can also be used for time series data. However, the computations can be simplified by properly coding x. This coding is easily done if observations along the horizontal or x axis are evenly spaced, a requirement that is usually satisfied in time series data.

If the x's are coded such that $\Sigma_{i=1}^{n} x_i = 0$, Equations 12.6 and 12.7 are reduced to

$$b = \frac{\sum_{i=1}^{n} x_i Y_i}{\sum_{i=1}^{n} x_i^2} \tag{12.8}$$

and

$$a = \frac{\sum_{i=1}^{n} Y_i}{n} = \bar{Y} \tag{12.9}$$

To ensure that $\Sigma_{i=1}^{n} x_i = 0$, we must code the center of the data 0. As we see in Figure 12.4, where there is an odd number of observations, such a coding would result in $\Sigma_{i=1}^{n} x_i = 0$. The coding used automatically fixes the origin, July 1, 1981, and the units of $x = 1$ year.

If the number of items in the data set is even, the coding is altered slightly because there is no single middle value in the set of data. This point is illustrated in Figure 12.5. The 0 is placed in the center of the data, and again the placement determines the origin, which is January 1, 1982. The distance from the origin to the first data point fixes the units of $x = 6$ months. This determination is also indicated by the fact that the coding between each year is two 6-month

FIGURE 12.5 Coding for a data set with an even number of yearly observations

79	80	81		82	83	84	Time (years)
−5	−3	−1	0	1	3	5	Coded x

periods. Even though yearly data are illustrated in Figures 12.4 and 12.5, the coding procedure is applicable to any time period that the data represent.

EXAMPLE 12.4 Howard Industries experienced the following annual sales, in millions of dollars, for the period 1980 to 1984 inclusive: 1.0, 3.3, 4.9, 6.3, and 9.5. Estimate the linear trend equation by the method of least squares.

Solution: It is convenient to set up a table to obtain the information needed to solve for a and b.

Year	Y_i	x_i	x_iY_i	x_i^2
1980	1.0	−2	−2.0	4
1981	3.3	−1	−3.3	1
1982	4.9	0	.0	0
1983	6.3	1	6.3	1
1984	9.5	2	19.0	4
Totals	25.0	0	20.0	10

Using these totals and Equations 12.8 and 12.9, we get

$$b = \frac{20}{10} = 2$$

and

$$a = \frac{25}{5} = 5$$

Therefore, the least squares trend line is

$$\hat{T} = 5 + 2x_i$$

origin = July 1, 1982

x = 1 year

Y = millions of dollars

The equation indicates that sales are increasing at $2 million per year, on the average. Obviously, the equation can be used to predict trend. (See Figure 12.6 for a plotting of the data and the trend line.)

EXAMPLE 12.5 The Gidget Company started in business on July 1, 1984. Monthly sales data, in thousands of dollars, for the last half of 1984 were as follows: 0.4, 2.0, 3.3, 5.1, 6.4, 8.6. Determine the linear trend equation.

Solution: It is convenient to make the following table.

Month	Y_i	x_i	x_iY_i	x_i^2
July	0.4	−5	−2.0	25
Aug.	2.0	−3	−6.0	9
Sept.	3.3	−1	−3.3	1
Oct.	5.1	1	5.1	1
Nov.	6.4	3	19.2	9
Dec.	8.6	5	43.0	25
Totals	25.8	0	56.0	70

Therefore,

$$b = \frac{56}{70} = .8$$

and

$$a = \frac{25.8}{6} = 4.3$$

The desired estimate of the trend equation is

$$\hat{T} = 4.3 + 0.8x_i$$

where the origin = October 1, 1984, x units = $\frac{1}{2}$ month, and Y units = thousands of dollars.

Sales are increasing an average of \$800 per $\frac{1}{2}$ month. The value of the trend line for October 1, 1984, is \$4300.

Note that every time we have obtained a trend equation, we have been careful to identify the origin and the units in which the variables are expressed. As stated earlier in the chapter, it is imperative that this information be presented whenever an equation is given. The absence of any part of the identification prevents the potential user of the equation from knowing what x value to substitute to obtain the desired trend for some future time period.

As pointed out earlier, trend equations are usually obtained with the origin in the middle of the data. Frequently, we desire to shift the origin to another point in the series and/or change the units in which x is expressed. A shift in the origin results in a new intercept value, and the slope of the line will be appropriately modified if different units are used. The following examples illustrate some possibilities for modification.

EXAMPLE 12.6 Consider the solution equation from Example 12.4, $\hat{T} = 5 + 2x_i$ with origin = July 1, 1982, x units = 1 year, and Y in millions of dollars. Modify the equation so that the origin = July 1, 1980, while still maintaining x units = 1 year.

Solution: Since we desire to move the origin to a new location, the intercept for the linear trend equation will be changed. Solving the given trend equation at the new location will provide the desired intercept value. Since the units of $x = 1$ year, the movement from July 1, 1982, back to July 1, 1980, means $x = -2$. Therefore,

$$\hat{T}_{7/1/80} = 5 + 2(-2) = 1$$

The new intercept is 1, and the new equation is

$$\hat{T} = 1 + 2x_i$$

with origin = July 1, 1980, x units = 1 year, and Y = millions of dollars.

The equation indicates that sales are increasing at an average of $2 million per year. Since x units were not changed, this slope is, of course, the same as was obtained in Example 12.4. The intercept value of 1 provides the trend value for July 1, 1980. The equation can be used to predict trend for any desired point in time. See Figure 12.6.

It should be obvious that the modified form of the equation is merely an alternative to the original equation. It is certainly not true that one is right and one is wrong. With proper substitution of x values, either equation can be used to predict trend. The following example illustrates this point.

FIGURE 12.6 Data and least squares trend line for Howard Industries (Examples 12.4 and 12.6)

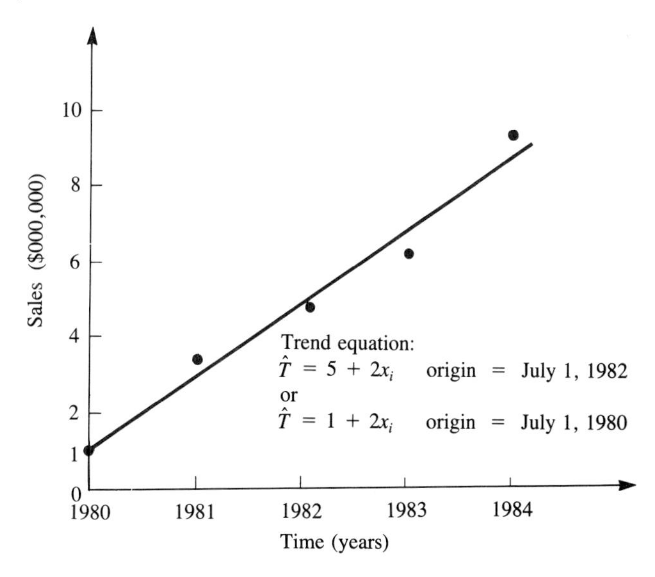

EXAMPLE 12.7 Use $\hat{T} = 5 + 2x_i$, the solution equation of Example 12.4 with origin = July 1, 1982, x units = 1 year, and Y units = millions of dollars, and $\hat{T} = 1 + 2x_i$, the solution equation of Example 12.6 with origin = July 1, 1980, x units = 1 year, and Y units = millions of dollars, and predict the trend value for 1985. Verify that both equations give the same predicted value.

Solution: Substituting first in $\hat{T} = 5 + 2x_i$, we obtain

$$\hat{T} = 5 + (2)(3) = 5 + 6 = 11$$

If we substitute in the modified equation $\hat{T} = 1 + 2x_i$, we obtain

$$\hat{T} = 1 + (2)(5) = 1 + 10 = 11$$

In either case, the predicted trend value is \$11 million. We are assuming that the trend line can be extended beyond the range of the original data.

EXAMPLE 12.8 Consider Example 12.5 and the solution equation $\hat{T} = 4.3 + 0.8x_i$ with origin = October 1, 1984, x units = $\frac{1}{2}$ month, and Y units = thousands of dollars and modify the equation so that the origin = July 15, 1984, and x units = 1 month.

Solution: First, let us calculate the desired intercept value

$$\hat{T} = 4.3 + (0.8)(-5) = 4.3 + (-4) = 0.3$$

In this case, we also desire to change the units in which x is expressed. If the trend line increases .8 units for $\frac{1}{2}$ month, it obviously changes at (2)(.8) = 1.6 units per month. The modified equation thus becomes

$$\hat{T} = 0.3 + 1.6x_i$$

where the origin = July 15, 1984, x units = 1 month, and Y units = thousands of dollars. This equation can be used to predict trend values, as Example 12.9 illustrates. Again it is assumed that the trend continues.

In Example 12.8, the new intercept was found first and then the new slope. It is possible to reverse the sequence. That is, the trend equation would be $\hat{T} = 4.3 + 1.6x_i$ with x units = 1 month, and other identification unchanged. To modify this equation so that the origin = July 15, 1984, we substitute −2.5 for x since July 15 occurs 2.5 months prior to the current origin of October 1, 1984. The resulting equation is $\hat{T} = 0.3 + 1.6x_i$ with origin = July 15, 1984, x units = 1 month, and Y units = thousands of dollars. We see that this equation was obtained in Example 12.8. We conclude that if both the origin and the units of x are to be changed, the order for determining the modified coefficients is optional.

EXAMPLE 12.9 Use $\hat{T} = 4.3 + 0.8x_i$, which is the solution equation of Example 12.5 with origin = October 1, 1984, x units = $\frac{1}{2}$ month, and Y units = thousands of dollars, and $\hat{T} = 0.3 + 1.6x_i$, the solution equation of Example 12.8 with origin = July 15, 1984, x units = 1 month, and Y units = thousands of dollars, and predict the trend value for January 1985.

Solution: Using first the equation for Example 12.5, we have

$$\hat{T} = 4.3 + (.8)(7) = 4.3 + 5.6 = \$9.9 \text{ thousand}$$

From the equation of Example 12.8, we obtain

$$\hat{T} = 0.3 + (1.6)(6) = 0.3 + 9.6 = \$9.9 \text{ thousand}$$

Again we observe that the results are identical, which is, of course, what we expected.

EXAMPLE 12.10 Modify $\hat{T} = 1 + 2x_i$, the equation resulting from Example 12.6 with origin = July 1, 1980, x units = 1 year, and Y units = millions of dollars, so that x units = $\frac{1}{4}$ year and origin = first quarter of 1980.

Solution: Since a change of 2 units corresponds to a period of 1 year, the change per quarter would be $\frac{2}{4} = .5$ unit. Hence, as a first step in the solution, we could write

$$\hat{T} = 1 + .5x_i$$

where the origin = July 1, 1980, x units = quarters, and Y units = millions of dollars.

We desire the equation to have the origin at February 15, 1980, or 1.5 quarters earlier than the present origin. Thus, the intercept value will be

$$1 + (.5)(-1.5) = 1 - .75 = .25$$

The desired equation then becomes

$$\hat{T} = .25 + .5x_i$$

where the origin = February 15, 1980, x units = $\frac{1}{4}$ year, and Y units = millions of dollars.

EXERCISES

12.1 The Morgan Company experienced the following annual sales (in thousands of dollars) beginning with the year 1975.

2 4 10 16 18 20 24 44 52 60

Use the method of semiaverages to determine the trend increment. Also, determine the trend value for 1975 and write the trend equation.

12.2 The following time series provides data on the number of ordinary life insurance policies (in millions) purchased in the United States from 1971 to 1983 (source: *Life Insurance Fact Book,* 1984).

11.28	11.84	12.20	12.76	12.55	13.22	13.68
13.99	14.26	14.75	15.84	15.61	17.74	

(a) Using the method of semiaverages, determine the linear trend equation having the origin = July 1, 1971.

(b) Determine the linear trend equation by the method of least squares, using coding with the origin = July 1, 1977, and x units = 1 year.

(c) Modify the equation from (b) so that the origin = July 1, 1971, and x units = 1 year.

(d) Use equations from (a), (b), and (c) to predict the number of ordinary life insurance policies that will be purchased in 1987.

12.3 Annual sales in thousands of dollars for the Ribold Corporation are as follows for 1979–1984: 4, 7, 12, 18, 20, 23.

(a) Use coding with the origin = January 1, 1982, and x units = 6 months to determine the trend equation by the method of least squares.

(b) Modify the equation from (a) so that the origin = July 1, 1979, and x units = 1 year.

(c) Determine the trend value for 1985.

12.4 The following data represent the total number of government employees at federal, state, and local levels (in millions) for 1977 to 1982 (source: *Statistical Abstract of the United States*): 15.46, 15.63, 15.97, 16.21, 15.97, 15.93.

(a) Use coding with the origin = January 1, 1980, and x units = 6 months and determine by the method of least squares the linear trend equation.

(b) Modify the equation from (a) so that the origin = July 1, 1977, and x units = 1 year.

(c) Predict the number of government employees for 1985.

(d) Does a scatterplot of the data indicate that the choice of a linear trend was reasonable?

12.5 Consider the following annual data for 1978–1984: 3, 5, 8, 8, 10, 14, 15.

(a) Use the method of least squares to determine the linear trend equation having the origin = July 1, 1981, and x units = 1 year.

(b) Modify the equation from (a) so that the origin = July 1, 1978, and x units = 1 year.

(c) Predict the trend value for 1989.

12.4 NONLINEAR TRENDS

The techniques presented in the previous section for determining trend are based on the assumed linearity of this effect. Of course, many problems exist for which such an assumption is unreasonable. This section considers two types of nonlinear trends.

Evidence of nonlinearity ordinarily is provided by scatterplots of the time series data. Nonlinear plots might look, for example, like the graphs of the parabola and the exponential shown in parts (b) and (c) of Figure 10.1. Let us now consider these two types of trend equations.

A parabola is a second-degree polynomial of the form

$$\hat{T} = a + bx + cx^2 \tag{12.10}$$

As an example, see Figure 12.7 for a graph of the parabola $\hat{T} = 3 + 2x + x^2$. In general, the form of the parabola will depend on the signs of the coefficients b and c. The four possibilities are shown in Figure 12.8.

We note that Equation 12.10 is linear in the coefficients. Therefore, the least squares procedures presented in Chapter 11 can be used to find the "best"-fitting parabola. Since there are three coefficients to be determined, there are three normal equations to be solved. The equations follow the model given by Equation 11.2. Specifically, they are

$$an + b\sum_{i=1}^{n} x_i + c\sum_{i=1}^{n} x_i^2 = \sum_{i=1}^{n} Y_i$$

$$a\sum_{i=1}^{n} x_i + b\sum_{i=1}^{n} x_i^2 + c\sum_{i=1}^{n} x_i^3 = \sum_{i=1}^{n} x_iY_i$$

$$a\sum_{i=1}^{n} x_i^2 + b\sum_{i=1}^{n} x_i^3 + c\sum_{i=1}^{n} x_i^4 = \sum_{i=1}^{n} x_i^2Y_i \tag{12.11}$$

FIGURE 12.7 Graph of the parabola $\hat{T} = 3 + 2x + x^2$

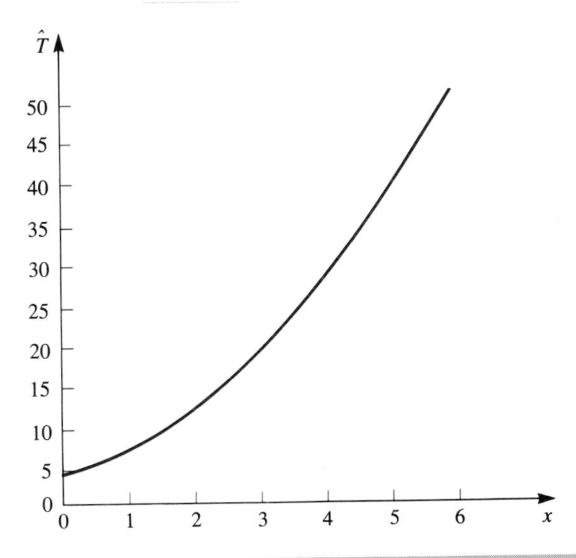

FIGURE 12.8 Representative graphs of parabolas for the various combinations of signs for b and c

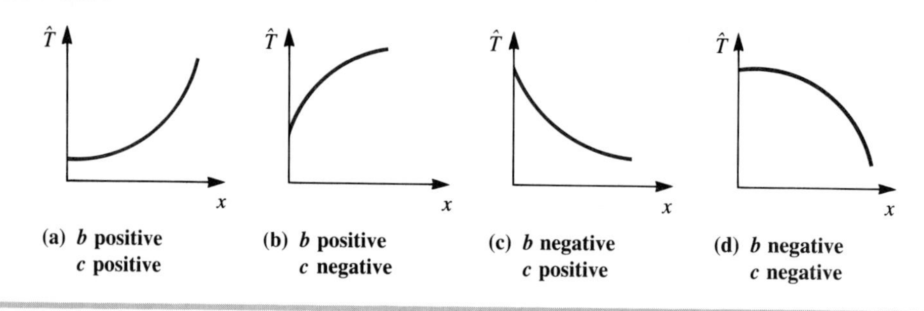

Since it is impossible to observe trend values, we use the actual values, or Y values, in the equations.

If coding of the type described in Section 12.3 is used—that is, the middle of the series is taken as the origin—then all terms involving the sum of odd powers of x go to 0. The resulting solutions to the normal equations are

$$b = \frac{\sum_{i=1}^{n} x_i Y_i}{\sum_{i=1}^{n} x_i^2} \tag{12.12}$$

$$c = \frac{\sum_{i=1}^{n} x_i^2 Y_i - \dfrac{\sum_{i=1}^{n} x_i^2 \sum_{i=1}^{n} Y_i}{n}}{\sum_{i=1}^{n} x_i^4 - \dfrac{\left(\sum_{i=1}^{n} x_i^2\right)^2}{n}} \tag{12.13}$$

and

$$a = \frac{\sum_{i=1}^{n} Y_i - c \sum_{i=1}^{n} x_i^2}{n} \tag{12.14}$$

It is interesting to note that Equation 12.12 for b is the same as Equation 12.8 in the linear model. Since the procedure is so similar to that already illustrated by the linear model, no example is given here.

Polynomials of higher degree than the parabola would have additional terms in the equation. For example, a third-degree polynomial would have the added term dx^3. We might also note that it is possible to force a polynomial to fit the observed data quite closely by adding enough terms. In fact, with n observations, a polynomial of degree $n - 1$ will provide a perfect fit, although the fitting of such a polynomial would ordinarily be very difficult to justify theoretically.

We should also mention that the principle of least squares can be used to fit a polynomial of any degree. The normal equations would contain an additional equation for each added term in the model. The equations for higher-order models are not presented here. The perceptive student will note the pattern formed by the equations and will be able to write those required by a higher-order model.

An exponential trend can be expressed by the equation

$$\hat{T} = ab^x \tag{12.15}$$

As an example, see Figure 12.9 for a graph of the exponential $\hat{T} = (2)(1.2)^x$. If logarithms are taken of both sides of the equation, the resulting equation is linear and can then be solved by the least squares techniques already presented. The equation in logarithmic form is

$$\log \hat{T} = \log a + x \log b \tag{12.16}$$

If the origin is again chosen in the middle of the time series data, the normal equations resulting from the application of the principles of least squares are

$$n \log a = \sum_{i=1}^{n} \log Y_i$$

$$\log b \sum_{i=1}^{n} x_i^2 = \sum_{i=1}^{n} x_i \log Y_i \tag{12.17}$$

FIGURE 12.9 Graph of the exponential $\hat{T} = (2)(1.2)^x$

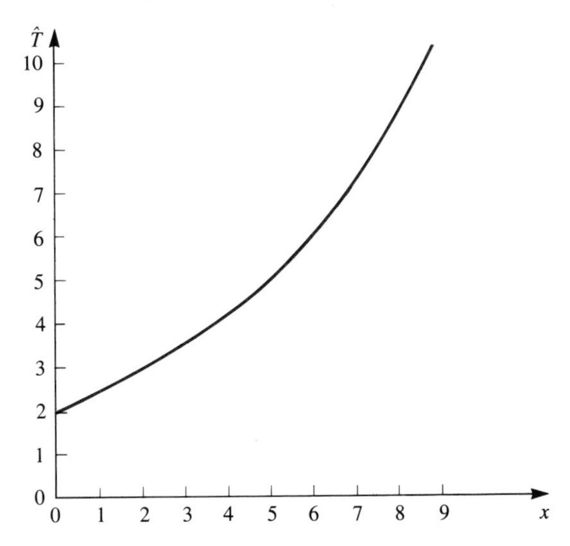

Hence, solutions for a and b are given by the antilogarithms of

$$\log a = \frac{\sum_{i=1}^{n} \log Y_i}{n} \tag{12.18}$$

and

$$\log b = \frac{\sum_{i=1}^{n} (x_i \log Y_i)}{\sum_{i=1}^{n} x_i^2} \tag{12.19}$$

The following example illustrates the required computations.

EXAMPLE 12.11 Craig and Sons have experienced sales of 2.2, 2.3, 2.5, 2.8, and 3.2 in thousands of dollars for 1981 through 1985. Determine the exponential trend equation.

Solution: A tabular format is convenient for the required computations.

Year	Y_i	x_i	$\log Y_i$	$x_i \log Y_i$	x_i^2
1981	2.2	−2	0.3424	−0.6848	4
1982	2.3	−1	0.3617	−0.3617	1
1983	2.5	0	0.3979	0	0
1984	2.8	1	0.4472	0.4472	1
1985	3.2	2	0.5051	1.0102	4
			2.0543	0.4109	10

Substitution in Equations 12.18 and 12.19 results in

$$\log a = \frac{2.0543}{5} = .4109$$

and

$$\log b = \frac{.4109}{10} = .04109$$

In logarithmic form, the desired equation is

$$\log \hat{T} = .4109 + .0411x$$

Using antilogarithms, we obtain $a = 2.5757$ and $b = 1.0992$, and the equation in exponential form is

$$\hat{T} = (2.5757)(1.0992)^x$$

with origin = July 1, 1983, x units = 1 year, and Y units = thousands of dollars. Note that Appendix Table B.11 can be used to obtain logarithms and antilogarithms.

The equation can be used for predictive purposes. For example, if it is desired to predict the trend value for 1986, it is only necessary to substitute 3 for x and solve for $\hat{T}$. The result is 3.42. We see that sales are increasing at an average annual rate of 9.92%.

EXERCISES

12.6 The following time series represents annual data for 1980 through 1984 inclusive: 4, 2, 5, 9, 14.

- **(a)** Fit a parabolic trend to these data by the method of least squares.
- **(b)** Forecast the trend value for 1985.

12.7 Consider the following annual observations for 1980 to 1984 inclusive: 1, 2, 4, 8, 16.

- **(a)** Fit an exponential trend.
- **(b)** What is the average annual growth rate?

12.8 An exponential trend was fitted to a set of data, and the resulting equation in logarithmic form was

$$\log \hat{T} = 0.7782 + 0.0792x_i$$

Assuming that x units = 1 year, find the average annual rate of growth.

12.5 MOVING AVERAGES

The method of moving averages is sometimes used to indicate trend. It differs from the procedures presented in Section 12.3 in that it does not yield a functional form that can be conveniently used for forecasting. However, it would be possible to fit a useful trend curve to the moving average sequence by the method of least squares. This procedure is the recommended one to follow in time series where the seasonal and irregular variations are so large that it is difficult to determine from the observed series if a trend exists.

The method of moving averages is a useful smoothing technique for reducing the influence of seasonal and irregular variations. The method can also be used to reduce cyclical effects if it is possible to determine the period of the cycle. Later, in Section 12.7, moving averages provide a base for calculating seasonal indexes.

DEFINITION 12.9 A **moving average** is a sequence of mean values of a time series over a specified interval of time.

The following steps summarize the procedure for calculating a moving average.

1. Select the appropriate time span and, starting at the beginning of the series, consider the number of observations corresponding to the time span.
2. Sum the observations and record the result opposite the middle observation (or between the two middle observations if there is an even number of values in the time span).
3. Calculate the average value from the sum in the previous step.
4. Replace the first observation in the time span by the first observation not in the time span.
5. Repeat steps 2, 3, and 4 until all observations have been used.

The procedure for finding a moving-average value can also be stated as follows. If the time span has length k, then the moving-average value for the time period j is

$$M_j = \frac{1}{k} \sum_{i=j-[(k-1)/2]}^{j+[(k-1)/2]} Y_i \tag{12.20}$$

where the moving-average value is not possible for time period j, if, for any term in the sum, $j < [(k+1)/2]$ or $j + [(k-1)/2] > n$, where n is the number of the last observation in the series. When using Equation 12.20, note that when k is even, j will take on values halfway between integer values. These points correspond to the points for recording the values. For example, if $k = 4$, then pertinent values for j are 2.5, 3.5, 4.5, and so on.

A recursive formula to obtain M_j is

$$M_j = M_{j-1} + \frac{1}{k}\left[Y_{j+[(k-1)/2]} - Y_{j-1-[(k-1)/2]} \right] \tag{12.21}$$

The sequence of moving averages resulting from the foregoing steps will be smooth compared to the original series. Note that in a series where the time span covers an even number of observations, an additional step of averaging adjacent observations must be performed if it is desired to have the series of moving averages exactly opposite the original values. This step is often called *centering the moving average*. Note also that a disadvantage of the method is that we lose some observations at the beginning and at the end of the series.

EXAMPLE 12.12 Consider the following yearly advertising expenditures, in thousands of dollars, from 1969 to 1984, inclusive, for the Sealcraft Corp. and calculate a 5-year moving average.

3 5 6 4 8 7 9 10 8 11 14 12 9 15 16 20

Solution: It will be convenient to structure the solution in tabular form, as shown in Table 12.1.

TABLE 12.1 Advertising expenditures for the Sealcraft Corp. and moving-average calculation

Year	*Original Value*	*5-Year Moving Total*	*5-Year Moving Average*
1969	3		
1970	5		
1971	6	26	5.2
1972	4	30	6.0
1973	8	34	6.8
1974	7	38	7.6
1975	9	42	8.4
1976	10	45	9.0
1977	8	52	10.4
1978	11	55	11.0
1979	14	54	10.8
1980	12	61	12.2
1981	9	66	13.2
1982	15	72	14.4
1983	16		
1984	20		

Plots of the original data and the 5-year moving average are given in Figure 12.10. It is obvious that the moving-average process has smoothed out irregularities in the original series. Note also in Figure 12.10 that data have been lost at the beginning and at the end of the smoothed series due to the moving-average procedure.

It is possible to use Equations 12.20 and 12.21 to obtain the moving-average values. The procedures to obtain the moving-average value of 6.0 for 1972 are illustrated in Example 12.12.

To use Equation 12.20, we see that $j = 4$ since 1972 is the fourth year in the given time series of Y_i values. The value of k is 5 since we are determining a five-year moving average. The value for $j - [(k - 1)/2]$, the beginning of the sum, is 2, and the value for $j + [(k - 1)/2]$, the end of the sum, is 6. Therefore, substituting in Equation 12.20, we obtain

$$M_4 = \frac{1}{5}(y_2 + y_3 + y_4 + y_5 + y_6)$$

$$= \frac{5 + 6 + 4 + 8 + 7}{5} = \frac{30}{5} = 6.0$$

FIGURE 12.10 Advertising expenditures for Sealcraft Corporation

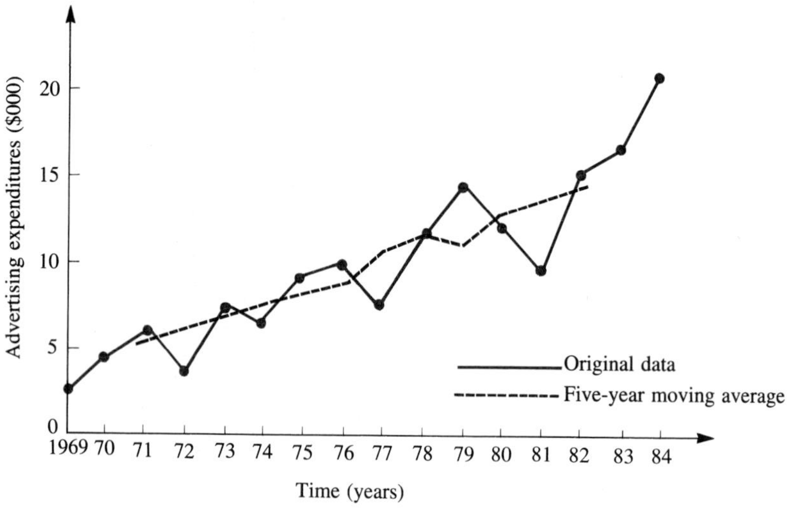

The values of j and k for use in Equation 12.21 are the same—that is, $j = 4$ and $k = 5$. Therefore,

$$M_4 = M_3 + \frac{1}{5}(y_6 - y_1) = 5.2 + \frac{1}{5}(7 - 3)$$

$$= 5.2 + \frac{1}{5}(4) = 5.2 + 0.8 = 6.0$$

EXAMPLE 12.13 Consider the following quarterly sales data of the Zembrach Company for 1982 to 1984, inclusive, and calculate a 4-quarter moving average.

6 14 10 18 14 19 20 28 26 34 35 38

Data are in thousands of dollars.

Solution: Table 12.2 provides the moving averages we seek.

Note that the use of the even period gives us, when averages are centered, a series of values not opposite the original series. To obtain a series having values corresponding to the original series, an additional step, called centering, is necessary. The centered series consists of two-period averages of the uncentered series. For example, the value of 13.00 centered opposite the third quarter

TABLE 12.2 Sales data for the Zembrach Company and moving-average calculation

Year	*Quarter*	*Original Value*	*4-Quarter Moving Total*	*Uncentered 4-Quarter Moving Average*	*Centered 4-Quarter Moving Average*
1982	1	6			
1982	2	14			
			48	12.00	
1982	3	10			13.00
			56	14.00	
1982	4	18			14.62
			61	15.25	
1983	1	14			16.50
			71	17.75	
1983	2	19			19.00
			81	20.25	
1983	3	20			21.75
			93	23.25	
1983	4	28			25.12
			108	27.00	
1984	1	26			28.88
			123	30.75	
1984	2	34			32.00
			133	33.25	
1984	3	35			
1984	4	38			

of 1982 is obtained as the average of 12.00 and 14.00. Also refer to Figure 12.11, which supports the computations given in the table.

Equations 12.20 and 12.21 can be used to obtain the uncentered averages. This fact is illustrated by verifying the value of 14.00 displayed between the third quarter of 1982 and the fourth quarter of 1982 in Example 12.13. Since this value is midway between the third and fourth observations in the original series, $j = 3.5$. In this example, $k = 4$.

The sum of Y_i values in Equation 12.20 goes from $j - [(k-1)/2] = 2$ to $j + [(k-1)/2] = 5$. Therefore,

$$M_{3.5} = \frac{1}{4}(y_2 + y_3 + y_4 + y_5)$$

$$= \frac{1}{4}(14 + 10 + 18 + 14)$$

$$= \frac{1}{4}(56) = 14.00$$

From Equation 12.21 the value of M_j is

$$M_{3.5} = M_{2.5} + \frac{1}{4}(y_5 - y_1) = 12.00 + \frac{1}{4}(14 - 6)$$

$$= 12.00 + \frac{1}{4}(8) = 12.00 + 2.00 = 14.00$$

FIGURE 12.11 Quarterly sales for the Zembrach Company

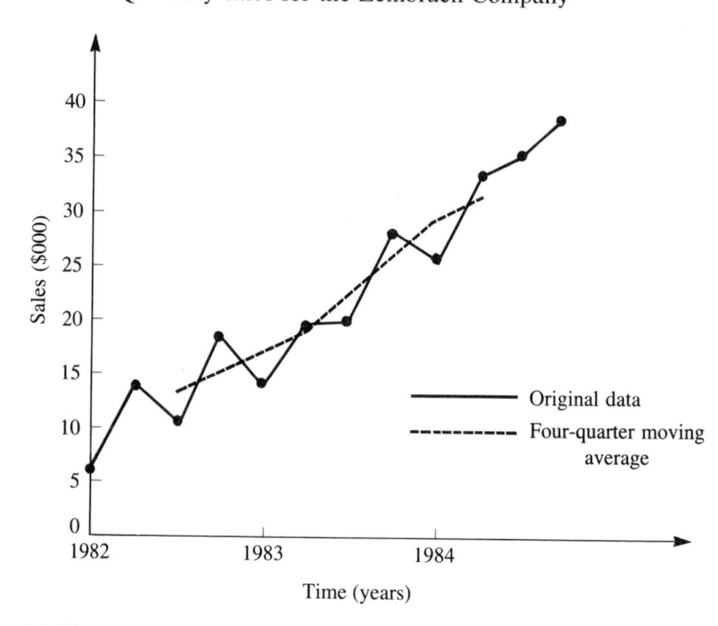

A moving average can be computed for any desired time span. In general, the longer the time period covered, the smoother the resulting values, but more points are lost at each end. The usual length selected is the same length, or approximately the same length, as the period of the component to be smoothed out by the process. For example, in Section 12.7 under the assumption that a seasonal pattern repeats itself every 12 months, we calculate a 12-month moving average to obtain a series that is free of seasonal variations. In fact, our principal use of moving averages will be as a basis for determining seasonal components in a time series.

EXERCISES

12.9 Determine a 5-year moving average for the data of Exercise 12.1.

12.10 Determine a 2-year moving average for the data of Exercise 12.3.

12.6 EXPONENTIAL SMOOTHING

As we observed in the previous section, it is frequently desirable to eliminate erratic movements in time series data. The method of moving averages provides one technique for smoothing time series. Another procedure designed to smooth out fluctuations in time series data and provide a possible forecasting basis is exponential smoothing.

DEFINITION 12.10 A smoothing procedure where recent observations are given larger weights than older observations in such a way that weights given to past data decrease exponentially with time is **exponential smoothing.**

One of the advantages of exponential smoothing over the method of moving averages is that observations are not lost at both ends of the series. Also, computations involved in exponential smoothing are not as tedious as those used in determining moving averages. Another distinct advantage is that there is a functional form for forecasting. In addition, all observations in the series are used in the computations, not just those in the immediate neighborhood, as in determining a moving average. However, unlike the moving-average procedure where observed values are weighted equally, exponential smoothing weights values unequally, with recent observations usually weighted differently than older observations.

The procedure presented here is referred to as first-order or simple exponential smoothing. It yields best results for series that vary randomly over time and have no apparent trend. If a trend is present, the use of simple exponential smoothing results in a series that lags behind the original series. In such a case, second-order or double exponential smoothing would be expected to yield more satisfactory results. As the name implies, two smoothing operations are involved. That is, simple exponential smoothing is done first, and then the resulting smoothed values are smoothed and further adjusted. If a time series is nonlinear over time, third-order or triple exponential smoothing is recommended. Both double and triple exponential smoothing are considered beyond the scope of this text. The interested student should refer to Brown [1].

The general form of the smoothing equation for simple exponential smoothing is

$$S_t = \alpha Y_{t-1} + (1 - \alpha)\, S_{t-1} \tag{12.22}$$

S_t is the smoothed value for time period t and can be looked upon as the predicted value for time period t. The symbol α is referred to as the smoothing constant. Values of α range between 0 and 1, inclusive. Y_{t-1} and S_{t-1} are the observed and smoothed values, respectively, for time period $t - 1$.

If we observe the effect of α in its extremes—that is, 0 and 1—we see what implications its size has on S_t. When $\alpha = 0$, each smoothed value equals the previous smoothed value. In other words, S_t is a constant. However, if $\alpha = 1$, the smoothed value equals the previous actual value. Thus, as α goes from 0 to 1, the weighting emphasis moves from the first observation, as evidenced by S_{t-1}, to the current observation, as evidenced by Y_{t-1}. Therefore, when the actual data to be smoothed have large fluctuations, they can be reduced significantly by selecting a small value for α. The effect of α is demonstrated in Table 12.3. When α is large, we see that the smoothed values more readily experience

TABLE 12.3 The effect of α for a given set of annual data

	α = .1		α = .8	
Year	y_t	S_t	y_t	S_t
1980	16	16	16	16
1981	28	16	28	16
1982	19	17.2	19	25.6
1983	31	17.4	31	20.3
1984	21	18.7	21	28.9
		19.0		22.6

the large fluctuations that are present in the actual data. When α is small, the more distant values have the greater impact. Note that for the sake of simplicity, we assume that $S_1 = Y_1$.

EXAMPLE 12.14 Remove the fluctuations in the following data for 1980 to 1985 by using exponential smoothing with $\alpha = .3$: 2.4, 2.1, 2.2, 1.8, 2.6, 1.9. Forecast the value for 1986.

Solution: The first step is

$$S_{80} = y_{80} = 2.4$$

Other smoothed values are calculated from Equation 12.22. For example, the smoothed value for 1981 is

$$S_{81} = (.3)(2.4) + (.7)(2.4) = .72 + 1.68 = 2.40$$

All of the calculations are summarized in the following table.

Year	y_t	S_t (α = .3)
1980	2.4	2.40
1981	2.1	.72 + 1.68 = 2.40
1982	2.2	.63 + 1.68 = 2.31
1983	1.8	.66 + 1.62 = 2.28
1984	2.6	.54 + 1.59 = 2.13
1985	1.9	.78 + 1.49 = 2.27

The forecast value for 1986 is

$$S_{86} = (.3)(1.9) + (.7)(2.27) = .57 + 1.59 = 2.16$$

The exponential smoothing method can be programmed quite simply for the computer and uses very little computer storage. The only pieces of information that need to be held in computer storage to forecast the current smoothed value are the last actual value, the last smoothed value, and the smoothing constant. For this reason, the method of exponential smoothing has considerable appeal when many time series, such as inventory data, are to be processed.

More discussion on choosing α seems appropriate. Determining α for a particular situation is one of the most difficult parts of exponential smoothing. It may be helpful in focusing on this issue to rewrite Equation 12.22 as

$$S_t = S_{t-1} + \alpha(Y_{t-1} - S_{t-1}) \tag{12.23}$$

In this form, it is clear that the current forecast is equal to the preceding forecast plus an amount equivalent to α multiplied by the last forecast error. Thus, exponential smoothing is known as an "adaptive" procedure because it responds to the error in the previous forecast.

If the time series has large random variability, the forecast error is likely to be large. In this case, it is suggested that a small value of α be used so that the forecast is not unduly affected by the large random variation. If the process is more stable, a larger value of α is preferred. The forecast would then adjust more to forecast errors. Mathematical procedures for determining α are usually based on an analysis of past data and involve choosing α so that the sum of squares of the forecast errors is a minimum.

As stated previously, the method of simple exponential smoothing is most effective for forecasting when applied to series with no trend component. For series containing a trend, it should probably be used for forecasting in the short range only.

EXERCISES

12.11 **(a)** Use exponential smoothing with $\alpha = .5$ to smooth the following series: 12, 10, 13, 16, 14.
(b) Forecast the next value in the series.

12.12 **(a)** Smooth the time series given in Exercise 12.3 using exponential smoothing with $\alpha = .4$.
(b) Forecast the value for 1985. Does simple exponential smoothing appear to be appropriate for this series?

12.7 SEASONAL INDEXES

Most time series data contain regularly recurring or periodic fluctuations in addition to a long-term trend. It is important for planning purposes to have adequate measures of these seasonal patterns, traditionally referred to as seasonal indexes. For example, a manager might be greatly concerned if her company's sales decreased by 5% from June to July. Her concern would be well-founded if July sales are ordinarily 10% greater than those in June. Her concern might turn to joyful optimism if it is traditional for July sales to be 20% less than June sales. Clearly, it is essential that the businessperson have knowledge of seasonal variation.

In general, the seasonal component accounts for variations due to repeating patterns, whether the patterns are repeated on a quarterly, monthly, weekly, daily, or hourly basis. The procedures presented here for estimating seasonal indexes are applicable regardless of the time frame over which the pattern is repeated.

The seasonal component of time series data is usually expressed by a seasonal index number for each recurring time period, such as for each month for monthly data within years or for each day for daily data within weeks. Each seasonal index, traditionally reported in percent, indicates how activity for that period compares to an average period. For example, a value of 109 for May would indicate that, due to seasonal variation, observed values for the month of May are, on the average, 9% higher than the average of all months. Since the seasonal effects are expected to cancel each other over the pattern period, the seasonal indexes, expressed as a percent, should have an average of 100.

One procedure for determining seasonal index numbers is the ratio-to-trend method.

DEFINITION 12.11 The **ratio-to-trend method** is a procedure for estimating seasonal components in time series data by calculating ratios of observed values to trend values determined by least squares procedures.

In this method, a trend line is first determined for the observed series by the method of least squares. Often this determination is done from average period totals rather than from the original period data. The next step is to divide observed values by the corresponding trend values for each period in the series. The resulting values are identified and recorded for each period. At this point, the values are relatively free of trend. What remains is to attempt to minimize cyclical and irregular components, which is done by averaging the values for each period. The resulting series of numbers constitutes a set of unadjusted estimates of the seasonal indexes. The only additional calculation that might be required would be the adjustment to ensure that the seasonal values have an average of 100. With reference to the multiplicative model, $Y = T \times S \times C \times I$, discussed earlier, the estimate of the seasonal component can be expressed as

$$S = \frac{Y}{T \times C \times I} \tag{12.24}$$

EXAMPLE 12.15 The given data show total domestic demand for gasoline in the United States in millions of barrels (source: *Survey of Current Business*) for 1980 to 1982. Use the ratio-to-trend method and determine the seasonal indexes for total domestic gasoline demand. Use the median to minimize cyclical and irregular components.

Month	*1980*	*1981*	*1982*
Jan.	196.8	200.1	185.5
Feb.	192.4	177.1	174.0
Mar.	199.6	196.4	201.2
Apr.	204.9	199.1	207.7
May	209.8	205.9	207.1
June	200.9	212.0	206.1
July	210.2	212.7	211.4
Aug.	207.3	206.9	206.1
Sept.	196.5	200.6	196.9
Oct.	207.8	204.9	198.8
Nov.	187.9	191.9	197.6
Dec.	206.3	207.9	203.6

Solution: We first determine the trend equation from the average monthly figures for each year.

Year	x	y	xy	x^2
1980	−1	201.700	−201.700	1
1981	0	201.292	0	0
1982	1	199.667	199.667	1
Totals	0	602.658	− 2.033	2

$$b = \frac{-2.033}{2} = -1.017$$

$$a = \frac{602.658}{3} = 200.886$$

Therefore, $\hat{T} = 200.886 - 1.017x_i$, where the origin = July 1, 1981, x units = 1 year, and Y units = millions of barrels.

Modifying the equation so that x units = 1 month gives

$$\hat{T} = 200.886 - .085x_i$$

where the origin = July 1, 1981, x units = 1 month, and Y units = millions of barrels.

For convenience, we now change the origin to January 15, 1980, a period of $17\frac{1}{2}$ months earlier than the present origin. Therefore, the required intercept is

$$200.886 - (.085)(-17.5) = 200.886 + 1.483 = 202.369$$

The trend equation is now

$$\hat{T} = 202.369 - .085x_i$$

where the origin = January 15, 1980, x units = 1 month, and Y units = millions of barrels.

Substituting $x_i = 0, 1, 2, \ldots, 35$ in this equation, we obtain the trend values shown in the following table.

Month	*1980*	*1981*	*1982*
Jan.	202.37	201.35	200.34
Feb.	202.28	201.27	200.25
Mar.	202.20	201.18	200.17
Apr.	202.11	201.10	200.08
May	202.03	201.01	200.00
June	201.95	200.93	199.91
July	201.86	200.84	199.83
Aug.	201.78	200.76	199.74
Sept.	201.69	200.67	199.66
Oct.	201.61	200.59	199.57
Nov.	201.52	200.50	199.49
Dec.	201.44	200.42	199.40

The next step is to divide the original data month by month by the corresponding trend values and to express the results as percents. The results are presented in the first 3 columns of Table 12.4.

TABLE 12.4 Percentage ratios of gasoline demand data of Example 12.15 to trend values and calculation of seasonal indexes

Month	*1980*	*1981*	*1982*	*Average (unadjusted seasonal)*	*Adjusted Seasonal Indexes*
Jan.	97.25	99.38	92.59	97.25	97.3
Feb.	95.12	87.99	86.89	87.99	88.0
Mar.	98.71	97.62	100.51	98.71	98.7
Apr.	101.38	99.01	103.81	101.38	101.4
May	103.85	102.43	103.55	103.55	103.6
June	99.48	105.51	103.10	103.10	103.1
July	104.13	105.91	105.79	105.79	105.8
Aug.	102.74	103.06	103.18	103.06	103.1
Sept.	97.43	99.97	98.62	98.62	98.6
Oct.	103.07	102.15	99.61	102.15	102.2
Nov.	93.24	95.71	99.05	95.71	95.7
Dec.	102.41	103.73	102.11	102.41	102.5
				1199.72	1200.0

We now have 3 different seasonal components associated with each month. The next step is to obtain an average value for each month. Either the mean or median can be used. Some analysts believe that the median provides a more effective measure in minimizing irregular variation.

The fourth column in the table is obtained by using the median of the 3 values for each of the months. Since the multiplicative model given by Equation 12.2 assumes that the seasonal indexes have an average of 100% and the ratio-to-trend method does not assure this average, the final step is to adjust the values in the fourth column. This step is often called leveling the index. We note that the sum of the fourth column is slightly less than 1200. Increasing each index by multiplying by the ratio 1200.0/1199.72 results in the final seasonal index. We now have indexes showing how each month of the year compares to an average month with an index of 100.0. We note, for example, that February demand is only 88.0% of the demand in an average month. ■

A second procedure for obtaining seasonal indexes is known as the ratio-to-moving-average method.

DEFINITION 12.12 The **ratio-to-moving-average method** is a procedure for estimating seasonal components in time series data by calculating ratios of observed values to values determined by the moving-average procedure. ■

Many analysts consider this method the most satisfactory for measuring seasonal variation. Hence, it is more widely used for obtaining seasonal indexes than the ratio-to-trend method. Again, the applicability of the multiplicative model is assumed.

The first step in the method is to calculate a centered moving average with length corresponding to the repeating seasonal pattern, thereby eliminating nearly all seasonal variation, probably a large part of irregular variation, and possibly some cyclical variation. The resulting values certainly contain the trend component and possibly some cyclical variation as well. When the original observed values are divided by their respective moving-average values, we essentially obtain estimates of the seasonal indexes with some irregular variation. These values are averaged for each period for the purpose of minimizing irregular variation, with the results being the estimates of the seasonal indexes. The following example using quarterly data illustrates the ratio-to-moving-average method.

EXAMPLE 12.16 ■ The data show sales of the Hancock Company, by quarters, in thousands of dollars, for 1982 to 1985, inclusive. Determine seasonal indexes by the ratio-to-moving-average method. Use the median to obtain the unadjusted seasonal indexes.

Quarter	*1982*	*1983*	*1984*	*1985*
1	20.8	32.0	36.0	45.4
2	32.0	43.2	59.2	72.2
3	30.8	42.2	51.6	63.4
4	29.2	38.2	52.8	57.0

Solution: The first step is to obtain the centered 4-quarter moving average. This computation is displayed in Table 12.5.

As previously explained, the centered moving average is obtained as the mean of 2 successive uncentered moving averages. The last column in the table is obtained by dividing the observed values by the centered moving average. For convenience, the ratios are also displayed in the following table.

Quarter	*1982*	*1983*	*1984*	*1985*	*Median (unadjusted seasonal)*	*Adjusted Seasonal Indexes*
1		90.84	79.87	79.68	79.87	80.4
2		114.36	123.14	122.42	122.42	123.2
3	104.05	107.11	101.03		104.05	104.7
4	90.12	91.17	98.00		91.17	91.7
					397.51	400.0

In the next-to-last column of the table are the median values for each of the quarters. These are then unadjusted seasonal indexes.

The sum of the unadjusted seasonal index column is 397.51. Multiplying each unadjusted index by 400/397.51 results in the final adjusted seasonal indexes given in the last column. This step is necessary for the reasons explained for the ratio-to-trend method.

The seasonal indexes resulting from the ratio-to-moving-average method may be different from the indexes calculated by the ratio-to-trend method. This occurrence does not surprise us because two different methods are involved. We observe that, as a result of the moving-average process, data are lost at the ends of the sequence. In this regard, the ratio-to-trend method has an advantage since no data are lost. Some statisticians think that the ratio-to-trend method should not be used if the data are known to have a strong cyclical factor. This view is based on the belief that the ratio-to-trend method does not remove the cyclical component from the seasonal index.

It is obvious that both methods of calculating seasonal indexes involve a considerable amount of tedious arithmetic. From a practical viewpoint, this is not a serious problem since the computer can be used quite satisfactorily for both methods. In particular, most computer installations have available a standard computer program for the ratio-to-moving-average method. A widely used pro-

TABLE 12.5 Calculation of moving averages for data of Example 12.16 and ratios of sales to moving average values

Year	*Quarter*	*Original Sales*	*4-Quarter Moving Total*	*Uncentered 4-Quarter Moving Average*	*Centered 4-Quarter Moving Average*	*Percentage of 4-Quarter Moving Average*
1982	1	20.8				
	2	32.0				
			112.8	28.20		
	3	30.8			29.600	104.05
			124.0	31.00		
	4	29.2			32.400	90.12
			135.2	33.80		
1983	1	32.0			35.225	90.84
			146.6	36.65		
	2	43.2			37.775	114.36
			155.6	38.90		
	3	42.2			39.400	107.11
			159.6	39.90		
	4	38.2			41.900	91.17
			175.6	43.90		
1984	1	36.0			45.075	79.87
			185.0	46.25		
	2	59.2			48.075	123.14
			199.6	49.90		
	3	51.6			51.075	101.03
			209.0	52.25		
	4	52.8			53.875	98.00
			222.0	55.50		
1985	1	45.4			56.975	79.68
			233.8	58.45		
	2	72.2			58.975	122.42
			238.0	59.50		
	3	63.4				
	4	57.0				

gram developed by the United States Bureau of the Census is based on the ratio-to-moving-average method.

The usefulness of the seasonal index is brought out in later sections. In Section 12.8, we make use of the seasonal index to deseasonalize data. In Section 12.10, we use the seasonal index as a forecasting tool.

EXERCISES

12.13 Estimated quarterly total sales, in billions of dollars, for all retail stores in the United States are given for 1979 to 1982 (source: *Survey of Current Business*).

Quarter	*1979*	*1980*	*1981*	*1982*
1	198.2	218.4	236.3	240.1
2	224.1	233.5	261.7	267.8
3	226.9	241.0	265.0	269.0
4	251.4	269.9	284.6	298.8

Determine seasonal indexes for each quarter by the ratio-to-trend method. Use the median as the measure of central location.

12.14 The Cravens Company has experienced the following sales, in thousands of dollars, for the month of April for 1980 to 1984: 6, 12, 20, 22, 30. If the equation of the trend line is $7 + 1/3x$ with the origin = January 15, 1980, and x units = 1 month, determine the seasonal index for April by the ratio-to-trend method. Use the median as the measure of central location.

12.15 Use the data from Exercise 12.13 and calculate seasonal indexes by the ratio-to-moving-average method.

12.16 Calculate a seasonal index for July only by the ratio-to-moving-average method if the following data are available for the 3-year period 1982 to 1984.

Month	*1982*	*1983*	*1984*
Jan.	6	12	24
Feb.	6	10	16
Mar.	9	11	20
Apr.	10	12	24
May	11	13	28
June	10	14	30
July	9	15	25
Aug.	14	18	32
Sept.	12	19	30
Oct.	11	13	33
Nov.	10	18	29
Dec.	9	19	28

12.8 DESEASONALIZING

DEFINITION 12.13 **Deseasonalizing** is the removal of seasonal variation in time series data by dividing the actual data by the appropriate seasonal index values.

Using the multiplicative model, deseasonalizing is obtained by making the computation $Y/S(100)$. Let us consider an illustration.

If the observed value for September is \$420 and the September seasonal index is 120, then to deseasonalize we divide \$420 by 1.20, obtaining \$350. The deseasonalized value of \$350 indicates what the dollar value would have been if no seasonal effect were present. Note, as was indicated in Section 12.2, the seasonal index is expressed as a percent but must be a pure number for use in a computation. Thus, the value of 1.20 is obtained by dividing 120% by 100%.

The principal use of the deseasonalizing concept is to facilitate making comparisons between periods having different series values and different seasonal indexes. We illustrate this point with an example.

EXAMPLE 12.17 A manager of a company experiencing sales in August and September of \$5500 and \$4715, respectively, might be concerned that the position of the company is declining. If the seasonal indexes for August and September are 110 and 92, respectively, deseasonalize the data and indicate the position of the company.

Solution: Deseasonalizing the August sales value, we have

$$\frac{\$5500}{1.10} = \$5000$$

For the month of September the computation is

$$\frac{\$4715}{.92} = \$5125$$

After deseasonalizing the values for the two months, the company's position looks much better. In fact, the manager would probably look upon it favorably since the deseasonalized value for September is larger than that for August.

It is exactly for the reason indicated in this example that many series published regularly by the United States government are reported in deseasonalized form. In such cases, the term "seasonally adjusted" will usually appear.

EXERCISES

12.17 Consider the data from Exercise 12.14 and the seasonal index for April and determine the deseasonalized value for April 1984.

12.18 Consider the data of Exercise 12.13 and the seasonal index for the fourth quarter and determine the deseasonalized value for the fourth quarter of 1982.

12.9 CYCLICAL AND IRREGULAR COMPONENTS

Both the period and the amplitude of cyclical fluctuations usually vary considerably in magnitude from one cycle to another. For this reason, the cyclical component of a time series is more difficult to measure and forecast than either the trend or the seasonal component. Based on the multiplicative model, it is sometimes possible to isolate a cyclic pattern for the past data, but it is usually not possible to project this into the future.

The seasonal component of a time series can be determined by the ratio-to-moving-average method presented in Section 12.7. If an observed value of Y is then divided by the corresponding seasonal index S, the multiplicative model of Equation 12.2 isolates the trend, cyclical, and irregular components as follows.

$$\frac{Y}{S} = \frac{T \times S \times C \times I}{S} = T \times C \times I$$

Utilizing the method of least squares to estimate T and dividing the values for $T \times C \times I$ by this estimate, we are left with the components $C \times I$. The two previous computations can be combined symbolically as

$$\frac{Y}{T \times S} = \frac{T \times S \times C \times I}{T \times S} = C \times I$$

Usually, the next step in isolating the cyclic component is to employ a weighted moving average, ordinarily of short length, such as three or five periods, to eliminate irregular variation. Weights are usually large for the middle values and small for more remote points. The analyst must choose the length of the moving average and the values for the weights, keeping in mind that the goal is to obtain a smooth cycle but not to eliminate the cyclical effect altogether. The final result of the process will be a series of values expressed as percents, one for each period in the series. These values are sometimes called cyclical relatives and provide a measure of the cyclical variation in the series.

A sinusoidal model involving sines and cosines of certain time periods is sometimes used to estimate the cyclical effect of a time series. Such models usually assume that the cyclical pattern is either constant in period and amplitude or changing at a known rate. Either assumption is not likely to be valid in practical business situations.

The fourth component in the multiplicative model, the irregular variation I, by its very nature cannot be predicted directly. It is possible to obtain a measure of I for each value in the time series by removing the other three components. In the multiplicative model, this computation would be

$$\frac{Y}{T \times S \times C} = \frac{T \times S \times C \times I}{T \times S \times C} = I$$

The procedure for obtaining the cyclical and irregular components is illustrated in Example 12.18. Only a small number of data points are used so that computations are not excessive.

EXAMPLE 12.18 The Cromwell Corporation, using a multiplicative time series model and eliminating trend and seasonal components, has obtained the following series of $C \times I$ components for January through June, inclusive: 110.0, 98.6, 100.8, 104.2, 93.2, 137.0. Determine cyclical relatives using a 1, 4, 6, 4, 1 weighted moving average to smooth out irregular movements. In addition, determine the irregular components for March and April.

Solution: It is convenient to display the data and the required computations in a table as follows.

Month	$C \times I$	*Weighted 5-Month Moving Total*	*Weighted 5-Month Moving Average (C)*	$I = \frac{C \times I}{C}(100)$
Jan.	110.0			
Feb.	98.6			
Mar.	100.8	1619.2	101.2	99.6
Apr.	104.2	1636.8	102.3	101.9
May	93.2			
June	137.0			

The 5-month moving total of 1619.2, for example, is obtained by using the suggested weights on the values for January through May. Specifically,

$$1619.2 = (1)(110.0) + (4)(98.6) + (6)(100.8) + (4)(104.2) + (1)(93.2)$$

Values for C are obtained by dividing the moving totals by the sum of the weights. Thus $101.2 = 1619.2/16$. The method for obtaining the irregular component is indicated in the heading of the I column. For example,

$$99.6 = \frac{100.8}{101.2}(100)$$

As we have noted previously when working with moving averages, it is characteristic that observations are lost at the beginning and at the end of the resulting series. The number of points lost is, of course, a function of the length of the moving average.

EXERCISES

12.19 **(a)** Consider the following information and determine the cyclical relative for February using a 1, 2, 1 weighted moving average to smooth out the irregular movements.

Month	*Sales*	*Trend*	*Seasonal Index*
Jan.	8	8	125
Feb.	11	10	100
Mar.	9	12	75

(b) What is the irregular component for February based on (a)?

12.20 Assume the multiplicative model and determine values for $C \times I$ for the data of Exercise 12.13.

12.10 FORECASTING

Up to this point in the chapter, the emphasis has been on decomposition of a time series into its component parts. We have described in considerable detail methods for identifying the patterns in time series data. Our primary purpose has been to obtain information that is useful in forecasting the future.

Before we consider forecasting procedures, it is appropriate to sound a note of caution. It is obvious that what businesspeople really mean by forecasting is extrapolating beyond the current data into the future. The dangers of extrapolating were pointed out in Chapter 10 relative to regression equations. The same concern exists in the time series context. However, we cannot avoid the problem because forecasting for some future time is precisely what we want to do. The answer is to go ahead and forecast based on information from the past but to be aware that the historical patterns of the past do not hold indefinitely for the future. For example, the businessman and businesswoman need to be alert for the time when the linear trend that has been useful in the past might change in slope or become curvilinear. Also, seasonal indexes may not remain constant. The development of changing dates for holidays to provide three-day weekends could, for many items, affect the seasonal index of that month. Of course, many dangers are associated with forecasting. The intention is not to give an exhaustive list but to give warning that projections very far into the future most likely will prove to be unsatisfactory unless they are based on a series that has been very stable in the past and can be assumed to remain stable in the future.

The multiplicative model provides the basis for forecasting. To predict a value for a particular period, one simply multiplies the various components in the model to obtain the predicted Y. In many practical problems, the only components involved will be the trend and seasonal components. In some cases, a useful measure of the cyclical component may be available.

EXAMPLE 12.19 The Skywell Corporation has developed the following trend equation for total sales.

$$\hat{T} = 8 + 0.25x_i$$

where the origin = January 15, 1979, x units = 1 month, and Y units = thousands of dollars. If the seasonal index for July is 120, what is the forecast of sales for July 1985?

Solution: No information is available on the cyclical component. Therefore, this component and the irregular component will both be given the value of 1 in the multiplicative model.

Since July 1985 occurs 78 months after January 1979, the predicted trend value for July 1985 will be given by

$$\hat{T} = 8 + (0.25)(78) = 8 + 19.5 = 27.5$$

The seasonal index is known to be 120; therefore, the predicted value of total sales for Skywell Corporation for July 1985 is

$$Y = (27.5)(1.20)(1.00)(1.00) = 33 \quad \text{or} \quad \$33{,}000$$

In Section 12.8, it was mentioned that deseasonalizing a series is referred to as seasonally adjusting the series. The term *seasonalizing* sometimes refers to putting the seasonal factor into a forecast—that is, multiplying the trend component by the seasonal component.

EXERCISES

12.21 The trend equation for sales of the Neemax Company is $\hat{T} = 970 + 3x$, where x is in quarters, Y is in thousands of dollars, and the origin = winter quarter of 1983. The sequence of quarters during a year is winter, spring, summer, fall. If the seasonal index for summer is 110, forecast sales for summer 1985.

12.22 If sales are described by a multiplicative time series model, predict the amount of sales when the trend component is 50 units and the indexes for seasonal, cyclical, and irregular components are 90, 120, and 100, respectively.

12.11 SUMMARY

In this chapter, we defined a time series as a sequence of observations made periodically through time. We considered the analysis of time series data from the viewpoint of the multiplicative model with its four components: trend, seasonal, cyclical, and irregular.

We used the method of semiaverages and the method of least squares to estimate long-term linear trends. Least squares principles were also used to estimate parabolic and exponential trends. The methods we examined for smoothing time series data were moving averages and exponential smoothing.

We used both the ratio-to-trend method and the ratio-to-moving-average method to determine seasonal indexes. We also considered the isolation of cyclical and irregular components, as well as the concept of deseasonalizing data. We concluded the chapter with the use of the multiplicative model for forecasting. The main concepts of this chapter are summarized in Table 12.6.

TABLE 12.6 Summary table

Designation of Components	T Trend S Seasonal C Cyclical I Irregular	
Models	Additive: Multiplicative:	$Y = T + S + C + I$ $Y = T \times S \times C \times I$
Estimation of Components	*Trend* Linear Nonlinear Parabolic Exponential	*Estimation Procedure* Semiaverages or least squares Least squares Least squares after logarithmic transformation
Smoothing Procedures	Moving Averages	$M_j = \frac{1}{k} \sum_{i=j-[(k-1)/2]}^{j+[(k-1)/2]} Y_i$
	Exponential smoothing	$S_t = \alpha Y_{t-1} + (1 - \alpha) S_{t-1}$
	Seasonal	$S = \frac{Y}{T \times C \times I}$
	Cyclical	Weighted moving average of values obtained from $\frac{Y}{T \times S}$
	Irregular	Not usually estimated; can be measured by removing other components from Y
Deseasonalizing	$\frac{Y}{S}(100)$	
Forecasting	Substitute components in model $Y = T \times S \times C \times I$	

SUPPLEMENTARY EXERCISES

12.23 The following data show the amount of public school expenditures, in billions of dollars, for 1975 to 1980 (source: *Statistical Abstract of the United States*): 64.8, 70.6, 75.0, 80.8, 86.7, 96.0.

(a) Determine by the method of least squares the linear trend equation for public school expenditures.

(b) Forecast public school expenditures for 1985.

12.24 In a company expecting a long-term upward growth, should there be concern if a month with sales of 18,900 when the seasonal index is 90 is followed by a month with sales of 22,000 when the seasonal index is 110?

12.25 **(a)** Consider the following data and calculate the cyclical relative for October using a 1, 2, 1 weighted moving average to smooth out the irregular movements.

Month	*Actual Value*	*Trend Value*	*Seasonal Index*
Sept.	4	8	125
Oct.	8	10	80
Nov.	12	12	125

(b) What is the irregular component for October based on (a)?

12.26 Using the multiplicative model, predict sales if the trend component is 82 units and the indexes for seasonal, cyclical, and irregular components are 125, 80, and 100, respectively.

12.27 The following time series represents annual data for 1978 to 1984 inclusive: 1, 3, 2, 4, 4, 7, 7.

(a) Use the method of least squares with origin = July 1, 1981, and determine the equation of the linear trend line.
(b) Modify the trend equation from (a) by changing the origin to 1978.
(c) Determine the trend values for 1978 through 1984.
(d) What is the trend value for 1985?
(e) Plot the original values and the trend line.

12.28 Calculate a seasonal index for January only by the ratio-to-moving-average method if the following monthly data are available for the 3-year period 1982 to 1984.

Month	*1982*	*1983*	*1984*
Jan.	8	11	18
Feb.	4	7	10
Mar.	5	8	10
Apr.	6	10	10
May	8	9	9
June	9	10	10
July	10	14	16
Aug.	11	13	15
Sept.	12	13	14
Oct.	11	14	16
Nov.	10	12	18
Dec.	9	10	12

12.29 The following data are available for 1975–1984 inclusive.

13 18 20 16 18 11 16 12 11 10

Use the method of semiaverages and find the annual trend increment.

12.30 Smooth the time series given in Exercise 12.29, using exponential smoothing with $\alpha = .5$. Forecast the value for 1985.

12.31 Calculate a 3-year moving average for the data of Exercise 12.29.

12.32 The following table indicates by month for the years 1979 to 1981 the total egg production on farms in millions of cases of 30 dozen (source: *Survey of Current Business*). Calculate a seasonal index for each month by the ratio-to-trend method, using average monthly figures for each year to determine the trend equation.

Month	*1979*	*1980*	*1981*
Jan.	16.4	16.8	16.7
Feb.	14.7	15.5	15.1
Mar.	16.4	16.5	16.7
Apr.	15.9	15.9	16.0
May	16.3	16.1	16.3
June	15.7	15.5	15.6
July	16.1	15.8	16.1
Aug.	16.1	16.0	16.2
Sept.	15.7	15.9	15.8
Oct.	16.3	16.5	16.4
Nov.	16.1	16.1	16.3
Dec.	16.8	16.8	16.9

12.33 Determine cyclical and irregular components for the data of Exercise 12.32, using a 1, 2, 1 weighted moving average to smooth out irregular movements.

12.34 The Laxton Company has experienced the following sales in millions of dollars for 1979 to 1984: 1, 2, 3, 5, 9, 16.

(a) Fit an exponential trend to the data.

(b) Determine the average annual growth rate of the Laxton Company.

12.35 Smooth the following time series, using exponential smoothing with $\alpha = .3$: 16, 32, 20, 45, 54, 24.

12.36 Determine a seasonal index for January by the ratio-to-trend method if 19, 40, 44, 84, and 85 are observed data for January for 1980 to 1984 inclusive and if the least squares trend line has the equation $\hat{T} = 20 + x_i$, where the origin = January 15, 1980, and x units = 1 month. Use the median as the measure of central location.

12.37 Use the data from Exercise 12.32 and calculate seasonal indexes by the ratio-to-moving-average method.

12.38 Use the seasonal indexes from Exercise 12.37 and deseasonalize the data in Exercise 12.32.

12.39 The trend equation for the Matteson Company is $\hat{T} = 50 + 2x$, where x units = 1 month and origin = January 15, 1980. Assume data are in thousands of dollars. Predict sales for July 1985 if the seasonal index for July is 120.

12.40 The following are annual data (in billions of dollars) for total consumer installment credit in the United States from 1966 to 1983 (source: *Federal Reserve Bulletin*).

76	79	88	97	102	111	127	147	157
164	194	231	274	312	315	336	356	396

(a) Use the method of semiaverages and determine the linear trend equation with the origin = July 1, 1966.

(b) Determine the linear trend equation by the method of least squares, using coding with the origin = January 1, 1975, and x units = 6 months.

(c) Modify the equation resulting from (b) so that origin = July 1, 1966, and x units = 1 year.

(d) Use equations from (a), (b), and (c) to predict total consumer credit for 1985.

12.41 The number of farms (in millions) in the United States from 1949 to 1978 is given in the following series (source: *Statistical Abstract of the United States*).

5.72	5.65	5.43	5.20	4.98	4.80	4.65	4.51	4.37	4.23
4.10	3.96	3.82	3.69	3.57	3.46	3.36	3.26	3.16	3.07
3.00	2.95	2.90	2.86	2.82	2.80	2.77	2.74	2.71	2.67

Fit a parabolic trend to these data by the method of least squares.

12.42 Use the data from Exercise 12.41 and fit an exponential trend.

12.43 Determine a 4-year moving average for the data of Exercise 12.40.

12.44 As a measure of short-term solvency, the Strasburg Manufacturing Company uses a liquidity ratio, called the current ratio, computed by dividing current assets by current liabilities. The company has calculated the current ratio at the end of each quarter beginning in 1981. The ratios are

1.8	2.1	1.9	2.2	2.5	2.3	2.7	2.4
3.2	3.0	2.8	3.2	3.3	2.9	3.6	3.8

(a) Determine the least squares linear trend equation.

(b) Forecast the current ratio for the first quarter of 1985.

12.45 The Price Manufacturing Corporation has the following recent data available on inventory turnover, defined as sales divided by inventories: 10.2, 9.9, 9.8, 9.5, 9.4, 9.6, 9.2, 9.1, 9.3.

(a) Determine the linear trend equation, using the method of least squares.

(b) Forecast the inventory turnover for the next time period.

12.46 Use the data of Exercise 12.45 and fit an exponential trend.

12.47 Calculate a 3-period moving average for the data of Exercise 12.45.

12.48 Smooth the series of Exercise 12.45, using exponential smoothing with $\alpha = .4$. What is the smoothed value for the next time period?

12.49 Use the data of Exercise 12.44 and fit an exponential trend.

12.50 Calculate a 4-quarter moving average for the data of Exercise 12.44.

12.51 Smooth the series of Exercise 12.44, using exponential smoothing with $\alpha = .6$.

12.52 The Lewis Company has experienced the following annual corporate earnings after taxes in billions of dollars beginning with the year 1942.

6.0	6.0	5.8	4.1	5.2	4.8	4.4	5.4	6.8
4.6	7.8	8.0	8.8	9.2	10.4	10.6	8.6	12.2
13.2	17.8	14.6	19.8	16.8	15.6	15.9	16.0	21.7
21.8	20.2	18.1	23.3	22.8	22.7	26.2	28.1	32.4
36.8	40.1	44.2	42.5	44.6	47.8	49.1		

(a) Using the method of semiaverages, determine the linear trend equation with the origin = July 1, 1942.

(b) Determine the least squares linear trend equation, using coding with the origin = July 1, 1963, and x units = 1 year.

(c) Modify the equation resulting from (b) so that origin = July 1, 1942, and x units = 1 year.

(d) Use equations from (a), (b), and (c) to forecast corporate earnings after taxes for 1985.

12.53 Smooth the time series given in Exercise 12.52 using exponential smoothing with $\alpha = .2$. What is the predicted value for 1985? Does this method produce a satisfactory series?

12.54 Calculate a 7-year moving average for the data of Exercise 12.52.

12.55 Fit an exponential trend to the data of Exercise 12.52.

12.56 The Horner Company calculated the trend line for its pasta maker sales to be $\hat{T} = 22 + 6x_i$ with origin = August 31, 1984, and x units = $\frac{1}{2}$ month. Predict sales for March 1985 if the seasonal index for March is 125.

12.57 Deseasonalize the following quarterly data for the Waters Company and comment on the overall progress of the company for the year.

Sales	*Seasonal Index*
182	115
162	100
162	95
158	90

12.58 Find the seasonal indexes for the following quarterly sales of the Gane Manufacturing Corporation.

Quarter	*Sales*	*Quarter*	*Sales*
1	12	1	22
2	15	2	25
3	25	3	35
4	30	4	42
		1	34

12.59 Seasonally adjust the semiannual sales data given for the Tobias Corporation. How is the company doing?

Sales ($000)	*Seasonal Index*
1251	125
1128	75

12.60 Fit a parabolic trend equation to the data given in Exercise 12.40.

12.61 Fit an exponential trend to the data of Exercise 12.40.

REFERENCE

1. Brown, R. G. *Smoothing, Forecasting and Prediction of Discrete Time Series*. Englewood Cliffs, NJ: Prentice-Hall, 1963.

13 Index Numbers: Construction and Applications

13.1 INTRODUCTION

John Case, an engineer for the Collins Manufacturing Corporation, has just received an increase in his annual salary. He is justifiably proud because raises in his company are based on merit. But his joy is mixed with concern. John is also aware that he is living in inflationary times. He knows that many of the purchases he made during the past month were more costly than similar purchases a year ago. He notices in grocery shopping that some food items increase in price every few weeks. He wonders if the amount of his raise will offset rising costs. Perhaps in terms of real wages, he has suffered a loss in spite of the salary increase.

John Case can use the Consumer Price Index (CPI) to determine the real or constant dollar value of his wage increase. In making this determination, he will be using an index number.

In this chapter, the primary emphasis is on the description and calculation of a few of the more important types of price indexes. Some attention is also given to applications of index numbers, such as the determination of the real dollar value of an economic measure.

13.2 BASIC CONCEPTS AND ASSUMPTIONS

DEFINITION 13.1 An **index number** is a measure, expressed in percent, comparing some item of interest for two periods of time.

Index numbers are widely used in business to compare prices, quantities, or values for a commodity or a group of commodities. Index numbers are considered to be indicators of business conditions. They are also helpful in comparing changes among different sectors of the economy or among different geographical areas. The CPI, published regularly by the Bureau of Labor Statistics of the United States Department of Labor, is probably the most familiar example of an index number. The concept of an index number was actually introduced in Chapter 12 by means of the seasonal index, which compares values for some period relative to an average for all periods.

The remainder of this chapter will focus on price indexes. That is, we are interested in a comparison of prices. The concentration on price indexes does not mean to imply that other types of indexes, such as quantity indexes or value indexes, are not important. A well-known and often-used quantity index is the Index of Industrial Production. This index measures changes in volume of production levels for different manufacturing categories. A value index compares values (usually the product of price and quantity). Equations presented here for prices can easily be modified for the other types of index numbers.

The simplest type of price index is called a price relative.

DEFINITION 13.2 A **price relative** is the index number that is the ratio of the price of a single commodity in a given period to the price of that commodity in a designated base period.

DEFINITION 13.3 The **base period** is the designated time period whose prices serve as the standard of comparison for prices in other time periods. The value of an index number in the base period is 100.

The equation to express a price relative is

$$I_{n,0} = \frac{p_n}{p_0}(100) \qquad \textbf{(13.1)}$$

The symbol p_n refers to a price in the year n, or the year of interest, and p_0 relates to a price in the base year. Then, $I_{n,0}$ is the price index comparing a price in the year n to that of the base year, or the year zero.

EXAMPLE 13.1 If the price of a head of cauliflower was $1.25 in 1984 and $1.35 in 1985, what is the price relative that compares the price in 1985 to that of 1984?

Solution: The year designated as the base year is 1984 since the 1985 price is to be compared to that of 1984. The price relative, as indicated by Equation 13.1, is simply the ratio of the price in 1985 to that in 1984, expressed as a percent. The price relative for cauliflower is

$$\frac{1.35}{1.25}(100) = 108$$

An obvious interpretation of the result of Example 13.1 is that the price of cauliflower increased 8% from 1984 to 1985. An index number that is a simple price relative causes no special problems.

Before we consider some of the problems that arise when more than one commodity is of interest, a composite index is defined.

DEFINITION 13.4 An index number constructed to describe changes for a whole class of products is called a **composite index.**

Assume that we are interested in constructing price indexes that would compare retail prices for a certain class of commodities. Since many items are to be included, the desired index number would be a composite index. Let us consider some of the questions that arise in constructing such an index.

1. What is the definition of a retail price?
2. What base period should be designated?
3. Which items should be included?
4. Where should the observations be taken?
5. When should the observations be taken?
6. What units should be used?
7. Should there be weighting of the data?
8. What weights should be used?

All of these questions are important, but we do not pursue them further here. In the following material, we take the point of view that these types of problems have already been solved. This approach enables us to concentrate on the various types of index numbers.

Occasionally, it is necessary to change the base period of a series of indexes. For example, we may wish to compare two index numbers that have different bases. A meaningful comparison is not possible unless the base periods are the same.

DEFINITION 13.5 A change in a series of indexes so that there is a new base period is known as a **shift** in the base.

To shift the base, we divide each index number in the old index by the index for the new base period and express it as a percent.

EXAMPLE 13.2 Price indexes for a certain commodity are 90, 110, and 125 for the years 1975, 1980, and 1985, respectively, with base year 1970. Determine the price indexes for 1975, 1980, and 1985, using 1985 as the base year.

Solution: Each index number in the series with 1970 as the base year is divided by 125, the index number for the new base year. The resulting ratios are then multiplied by 100. The ratios are, respectively, $90/125 = .72$, $110/125 = .88$, and $125/125 = 1$. The desired price indexes with 1985 as the base year are 72, 88, and 100.

EXERCISES

13.1 The prices for a certain item in Johnson's Department Store increased by 20% from 1982 to 1983 and decreased by $33\frac{1}{3}$% from 1983 to 1984. Determine for each year a price relative with 1984 as the base year.

13.2 The given index numbers have the year 1981 as base. Determine a new series with 1984 as the base year.

Year	*Index*
1982	90
1983	108
1984	120
1985	138
1986	144

13.3 SIMPLE AGGREGATIVE INDEX

First, we consider two relevant definitions.

DEFINITION 13.6 An index obtained by summing or aggregating data for one time period and comparing the result with the corresponding aggregate for a base period is called an **aggregative index.**

DEFINITION 13.7 An index is called **simple** if items are not weighted explicitly—that is, all items are considered to be equally important.

The equation to calculate a simple aggregative index is

$$I_{n,0} = \frac{\Sigma p_n}{\Sigma p_0}(100) \tag{13.2}$$

Since we wish to sum prices for a number of items in the numerator and the denominator, to be more precise we should have an additional subscript to indicate that the p's are variable. In addition, the range of summation should be clearly indicated. This obvious relaxation from traditional summation notation, also used in other sections to follow, should cause no problems. Equation 13.2 indicates the following: Sum the prices of all commodities in time period n, divide this total by the sum of the prices of the same commodities in the base period or period 0, and multiply the result by 100. The result is the desired index that compares prices in the year n to those in the base year.

EXAMPLE 13.3 The table lists prices in dollars for several grain products for the years 1982 and 1983 (source: *Survey of Current Business*). Calculate the simple aggregative index comparing prices for grain products in 1983 to prices in 1982.

	Price ($/bushel)	
Commodity	***1982***	***1983***
Barley	2.11	2.35
Corn	2.46	3.15
Oats	2.04	1.89
Rye	3.49	2.51
Wheat	4.25	4.38

Solution: Substituting in Equation 13.2, we find that

$$I_{83,82} = \frac{2.35 + 3.15 + 1.89 + 2.51 + 4.38}{2.11 + 2.46 + 2.04 + 3.49 + 4.25}(100)$$

$$= \frac{14.28}{14.35}(100) = 99.5$$

Thus, the simple aggregative index indicates an overall price decrease of .5% for grain products from 1982 to 1983.

Although the simple aggregative index is the easiest of all composite indexes to compute, it is seldom used in practice, principally because if the prices are expressed in different units, the value of the index will change. This sensitivity to varying units of measure is illustrated in the following example.

EXAMPLE 13.4 A company desires to calculate a simple aggregative index for two commodities to compare prices in 1986 to those in 1979. The price information for these two items is as follows. First, calculate the simple aggregative index using the units indicated. Then, change dollars per pound to dollars per ton and wages per day to wages per week, assuming a 5-day week, and recalculate the simple aggregative index.

Commodity	*Unit*	*Price ($)*	
		1979	*1986*
A	Dollars/pound	.50	.80
B	Dollars/day	29.50	35.20

Solution: Using Equation 13.2, we obtain

$$I_{86,79} = \frac{.80 + 35.20}{.50 + 29.50}(100) = \frac{36}{30}(100) = 120$$

Expressing prices in dollars/ton for commodity A and dollars/week for commodity B, we obtain

Commodity	*Unit*	*Price ($)*	
		1979	*1986*
A	Dollars/ton	1000.00	1600.00
B	Dollars/week	147.50	176.00

Again, using Equation 13.2 gives

$$I_{86,79} = \frac{1600 + 176}{1000 + 147.50}(100) = \frac{1776}{1147.5}(100) = 154.8$$

We note that 154.8 is considerably different from 120. ■

An index that may change values for every change in units is not very satisfactory. Thus, the implication is that the simple aggregative index should be used only when combining commodities that have prices expressed in comparable units. We consider some index numbers that are insensitive to changes in units in the following sections.

EXERCISES

13.3 In the table are wholesale prices for selected farm products for 1976 and 1981 (source: *Statistical Abstract of the United States*).

Product	*Unit*	*Price ($)*	
		1976	*1981*
Wheat	bushel	3.34	4.33
Cotton	pound	.68	.72
Eggs	dozen	.70	.73
Corn	bushel	2.70	3.16
Steers	100 pounds	39.27	63.84

(a) Calculate the simple aggregative composite index that compares 1981 prices to those in 1976.
(b) Recalculate the index requested in (a) after changing the units for steers to pounds.

13.4 Consider the given prices and calculate a simple aggregative index comparing 1986 prices to those in 1973.

Commodity	*Price ($)*	
	1973	*1986*
A	6	9
B	20	18
C	8	10
D	16	12

13.4 ARITHMETIC MEAN OF PRICE RELATIVES

A price index for a single commodity was previously defined as a price relative. It was expressed in Equation 13.1 as $I_{n,0} = (p_n/p_0)(100)$. The use of price relatives when calculating a composite index avoids the problem of changes in units associated with the simple aggregative index. The usual method is to calculate the arithmetic mean of the price relatives.

DEFINITION 13.8 An **arithmetic mean of price relatives** is the simple average of indexes that are price relatives.

The equation for the arithmetic mean of price relatives is

$$I_{n,0} = \sum \frac{p_n}{p_0}\left(\frac{100}{k}\right) \tag{13.3}$$

where k is the number of price relatives.

EXAMPLE 13.5 The following table contains the wholesale prices in dollars per 100 pounds for 3 types of livestock for 1982 and 1983 (source: *Survey of Current Business*). Determine a composite price index for livestock by the arithmetic mean of price relatives method, comparing prices in 1983 to prices in 1982.

Type of Livestock	*Price ($)*	
	1982	*1983*
Beef steers	64.22	62.52
Hogs	55.21	47.73
Lambs	53.03	54.74

Solution: To find the arithmetic mean of price relatives, we substitute in Equation 13.3, which gives the following results.

$$I_{83,82} = \left(\frac{62.52}{64.22} + \frac{47.73}{55.21} + \frac{54.74}{53.03}\right)\left(\frac{100}{3}\right)$$
$$= \frac{97.35 + 86.45 + 103.22}{3}$$
$$= \frac{287.02}{3} = 95.67$$

We conclude that livestock prices decreased 4.33% from 1982 to 1983.

It would be informative to verify that the units problem relating to a simple aggregative index is avoided with the arithmetic mean of price relatives index. The next example demonstrates this point.

EXAMPLE 13.6 Rework Example 13.4, using the arithmetic mean of price relatives rather than a simple aggregative index for each of the 2 cases presented.

Solution: Substituting in Equation 13.3, we obtain for the first case,

$$I_{86,79} = \left(\frac{.80}{.50} + \frac{35.20}{29.50}\right)\left(\frac{100}{2}\right) = \frac{160 + 119.32}{2}$$
$$= \frac{279.32}{2} = 139.66$$

For the second case,

$$I_{86,79} = \left(\frac{1600}{1000} + \frac{176}{147.50}\right)\left(\frac{100}{2}\right) = \frac{160 + 119.32}{2}$$
$$= \frac{279.32}{2} = 139.66$$

We observe from Example 13.6 that the arithmetic mean of price relatives produces the same result regardless of the choice of units. Of course, the only requirement is that the same units must be used for both time periods of interest.

EXERCISES

13.5 Use the data from Exercise 13.3 and calculate the arithmetic mean of price relatives comparing 1981 prices to 1976 prices.

13.6 Consider the data in Exercise 13.4 and calculate the arithmetic mean of price relatives comparing 1986 prices to 1973 prices.

13.5 WEIGHTED AGGREGATIVE INDEXES

Both types of index numbers presented so far, the simple aggregative index given by Equation 13.2 and the arithmetic mean of price relatives given by Equation 13.3, are so-called unweighted indexes. No indication is given as to the relative importance of the various items making up the composite. Therefore, items are assumed to be of equal importance.

In many problems, it is desirable to indicate that some commodities are more important than others in determining the composite index. The obvious way is to assign weights to each item. The weights can be whatever is deemed appropriate for the problem.

If w indicates weights assigned to the various prices, the simple aggregative index becomes, in general form,

$$I_{n,0} = \frac{\Sigma p_n w}{\Sigma p_0 w}(100) \tag{13.4}$$

Such an index is known as a weighted aggregative index.

DEFINITION 13.9 An index number obtained by comparing the weighted aggregate of prices in the year of interest to the corresponding weighted aggregate in the base year is called a **weighted aggregative index.**

The importance of the various commodities in determining the composite weighted aggregative index is frequently given by the quantities produced or consumed. Since quantities are not likely to be the same in both periods, we have a choice as to which year's quantities to use as weights. When quantities of the base year, represented by the symbol q_0, are used as weights, the index is known as a Laspeyres index.

DEFINITION 13.10 A weighted aggregative index that uses quantities of the base year as the weights is known as a **Laspeyres index.**

For the Laspeyres index, the general formula given in Equation 13.4 can be rewritten as

$$L_{n,0} = \frac{\Sigma p_n q_0}{\Sigma p_0 q_0}(100) \tag{13.5}$$

Note that we have used the symbol L to remind us of the name associated with the index.

EXAMPLE 13.7 Below are given 1982 and 1983 prices, in cents per pound, and quantities, in 10,000 metric tons, for certain nonferrous metals (source: *Survey of Current Business*). Calculate a composite index comparing 1983 prices to those in 1982, using a Laspeyres index.

Product	1982 Price	1982 Quantity	1983 Price	1983 Quantity
Aluminum	76	327	78	335
Lead	26	51	22	45
Zinc	38	30	41	26

Solution: To obtain the Laspeyres index, we substitute the data in Equation 13.5 and obtain

$$L_{83,82} = \frac{(78)(327) + (22)(51) + (41)(30)}{(76)(327) + (26)(51) + (38)(30)}(100)$$

$$= \frac{25{,}506 + 1122 + 1230}{24{,}852 + 1326 + 1140}(100)$$

$$= \frac{27{,}858}{27{,}318}(100) = 101.98$$

The price of the three nonferrous metals considered increased by 1.98% from 1982 to 1983 according to the Laspeyres index.

If quantities of the given year n are used as weights, the resulting weighted aggregative index is called a Paasche index.

DEFINITION 13.11 A weighted aggregative index that uses quantities of the year n as the weights is known as a **Paasche index.**

The equation is given by

$$P_{n,0} = \frac{\Sigma p_n q_n}{\Sigma p_0 q_n}(100) \tag{13.6}$$

The symbol P is usually used for this index in recognition of Paasche, the economist who first proposed it.

EXAMPLE 13.8 Using the data from Example 13.7, calculate a Paasche index comparing 1983 prices to 1982 prices.

Solution: From Equation 13.6, we obtain

$$P_{83,82} = \frac{(78)(335) + (22)(45) + (41)(26)}{(76)(335) + (26)(45) + (38)(26)}(100)$$

$$= \frac{26{,}130 + 990 + 1066}{25{,}460 + 1170 + 988}(100)$$

$$= \frac{28{,}186}{27{,}618}(100) = 102.06$$

It is possible that the Laspeyres index and the Paasche index will yield answers that are almost the same, as was the case in Examples 13.7 and 13.8. They would yield identical answers if, for every product, the quantities in the two years were the same. The answers depend on how the quantities compare in the two years and, of course, on the prices themselves. A large difference in the answers given by the Laspeyres and Paasche formulas is illustrated in the fictitious example that follows.

EXAMPLE 13.9 Consider the data and calculate a Laspeyres index and a Paasche index to compare prices in 1986 to those in 1981.

	1981		*1986*	
Commodity	*Price*	*Quantity*	*Price*	*Quantity*
A	10	20	6	480
B	12	16	15	14

Solution: The Laspeyres index is given by

$$\begin{aligned} L_{86,81} &= \frac{(6)(20) + (15)(16)}{(10)(20) + (12)(16)}(100) \\ &= \frac{120 + 240}{200 + 192}(100) \\ &= \frac{360}{392}(100) = 91.84 \end{aligned}$$

The Paasche index is

$$\begin{aligned} P_{86,81} &= \frac{(6)(480) + (15)(14)}{(10)(480) + (12)(14)}(100) \\ &= \frac{2880 + 210}{4800 + 168}(100) \\ &= \frac{3090}{4968}(100) = 62.20 \end{aligned}$$

The difference in the answers is due primarily to the large difference in weights for commodity A.

We note that if index number calculations are to be made annually, information on both prices and quantities must be collected for the Paasche index, whereas only price information is needed for the Laspeyres index. In fact, the denominator of the Laspeyres index remains constant from year to year. In practice, the Laspeyres index is used much more than the Paasche index since the latter requires more data and more computation. It is also easier to interpret changes with the fixed base ($\Sigma p_0 q_0$) provided by the Laspeyres index.

EXERCISES

13.7 The following data are available for calculating a composite index number that compares prices in 1986 to those in 1976.

	1976		*1986*	
Commodity	***Price***	***Quantity***	***Price***	***Quantity***
A	10	4	13	20
B	2	18	1	48
C	8	3	12	16

(a) Determine the Laspeyres index.
(b) Determine the Paasche index.

13.8 Consider the following data for calculating the required weighted aggregative index comparing prices in 1984 to those in 1980.

	1980		*1984*	
Commodity	***Price***	***Quantity***	***Price***	***Quantity***
A	6	12	9	3
B	4	7	2	8

(a) Determine the Laspeyres index.
(b) Determine the Paasche index.

13.6 WEIGHTED MEAN OF PRICE RELATIVES

One of the unweighted indexes we considered was the arithmetic mean of price relatives, given by Equation 13.3. Weights can also be assigned to the price relatives if desired.

DEFINITION 13.12 An index number obtained by comparing weighted price relatives in the year n to the corresponding weighted price relatives in the year 0 is called a **weighted mean of price relatives.**

In general terms, the formula is

$$I_{n,0} = \frac{\sum \frac{p_n}{p_0} w}{\sum w}(100) \tag{13.7}$$

Although weights can be whatever is deemed appropriate, it is customary to use value weights for this type of index. By value, we mean price multiplied by quantity. If value weights for the base year are used, the formula for the weighted mean of price relatives reduces to the Laspeyres index.

$$I_{n,0} = \frac{\sum \frac{p_n}{p_0} p_0 q_0}{\sum p_0 q_0}(100)$$

$$= \frac{\sum p_n q_0}{\sum p_0 q_0}(100) = L_{n,0}$$

If value weights for the given year are used, the index becomes

$$I_{n,0} = \frac{\sum \frac{p_n}{p_0} p_n q_n}{\sum p_n q_n}(100) \tag{13.8}$$

EXAMPLE 13.10 Consider the data in Example 13.7 and calculate a weighted mean of price relatives comparing 1983 prices with those of 1982. Use value weights for 1983.

Solution: Substituting in Equation 13.8, we obtain

$$I_{83,82} = \frac{(78/76)(78)(335) + (22/26)(22)(45) + (41/38)(41)(26)}{(78)(335) + (22)(45) + (41)(26)}(100)$$

$$= \frac{(102.63)(26{,}130) + (84.62)(990) + (107.89)(1066)}{26{,}130 + 990 + 1066}$$

$$= \frac{2{,}880{,}506.4}{28{,}186} = 102.20$$

EXERCISES

13.9 Suppose the following data on prices and quantities for two commodities are available for 1985 and 1986.

	1985		*1986*	
Commodity	***Price***	***Quantity***	***Price***	***Quantity***
A	6	20	3	10
B	5	9	8	15

Comparing 1986 prices to those of 1985, determine **(a)** the simple aggregative index, **(b)** the arithmetic mean of price relatives, **(c)** the Laspeyres index, **(d)** the Paasche index, and **(e)** the weighted mean of price relatives, using value weights for 1985.

13.10 Refer to Exercise 13.7.

(a) Determine the weighted mean of price relatives, using value weights for 1976.

(b) Determine the weighted mean of price relatives, using value weights for 1986.

13.7 CHAIN INDEXES

Index numbers are not always presented with a fixed base period. A comparison of the current period, n, to the previous period, $n - 1$, may be more meaningful. Such a comparison can be accomplished through a link index number or link relative.

DEFINITION 13.13 A **link relative,** $I_{n,n-1}$, is the index number comparing the price in the current period with the price in the previous period.

In effect, each index number has the preceding period as its base. To make comparisons over several periods, we need only multiply the appropriate link relatives, being careful to adjust the decimal point properly. Index numbers constructed in this manner are called chain indexes.

DEFINITION 13.14 Index numbers obtained by multiplying link relatives are called **chain indexes.**

An example of a chain index computation is given by

$$I_{n,n-2} = \frac{I_{n-1,n-2} I_{n,n-1}}{100} \tag{13.9}$$

Values obtained by the chain index procedure should be the same as weighted aggregative indexes when fixed quantities are used as weights. An example of this fact is shown in the following equation, where q_a represents a fixed quantity.

$$I_{n,n-2} = \frac{\Sigma p_n q_a}{\Sigma p_{n-1} q_a} \frac{\Sigma p_{n-1} q_a}{\Sigma p_{n-2} q_a} = \frac{\Sigma p_n q_a}{\Sigma p_{n-2} q_a} \tag{13.10}$$

EXAMPLE 13.11 If the index number comparing 1984 prices to those in 1983 is 104, and the index number comparing 1985 prices to 1984 prices is 105, what is the index number comparing 1985 prices to those in 1983?

Solution: The index numbers 104 and 105 are link relatives. The index number requested can be obtained through multiplication of the link relatives. In computational form, the chain index is given by

$$I_{85,83} = I_{84,83} I_{85,84} = \frac{(104)(105)}{100} = 109.2$$

We conclude that prices were 9.2% higher in 1985 than in 1983.

EXERCISES

13.11 Consider the index numbers in Exercise 13.2.

(a) Determine the link relatives for each year.

(b) Use the link relatives from (a) to determine the index comparing 1986 to 1981.

13.12 The following data are available for the purpose of constructing price indexes.

Commodity	*Price*			
	1984	*1985*	*1986*	*Quantity*
A	4	8	12	5
B	3	2	6	10

(a) Determine the weighted aggregative index comparing 1985 prices to those in 1984 using the given quantities as weights.

(b) Determine the weighted aggregative index comparing 1986 prices to those in 1985 using the given quantities as weights.

(c) Determine the weighted aggregative index comparing 1986 prices to those in 1984 using the given quantities as weights.

(d) Use the answers from (a) and (b) and the link relative method to determine the index comparing 1986 prices to those in 1984. Compare the answer with that from (c).

13.8 APPLICATIONS OF INDEX NUMBERS

The use of index numbers to compare prices in two periods has been illustrated in the foregoing sections. Another principal use of index numbers is to adjust a series of values for the effects of price changes. This procedure is known as *deflating*. The economic measure of interest is divided by the price index to determine the real or constant dollar value of the measure. The CPI is often used as an index for deflating.

The CPI is used so widely that, before illustrating its use in deflating, we look briefly at some of its features. The CPI measures changes in prices of eight general types of goods and services for urban wage earners and clerical workers. It includes over 300 items in the categories of food, housing, apparel, transportation, medical care, personal care, reading and recreation, and other goods and services.

Price data are collected monthly in a sample of stores and cities. Weights for the items in the index are based on consumer expenditure surveys. The CPI is an example of a fixed-weight aggregative index. The actual method of calculation is that of a chain index. That is, index numbers for successive time periods are combined.

Since the index was first started during World War I, there have been major revisions in the method of computation, the specification of the base period (the current base year is 1967), the methods for collecting data, and the items composing the index. For example, many commodities that are important in the expenditures of a family today did not exist when the index was begun. It is almost certain that there will be changes in the future as attempts are made to keep the CPI current in reflecting consumer tastes and preferences.

EXAMPLE 13.12 Suppose an individual's annual salary increased from $12,150 to $17,010 over a given period when the CPI changes from 90 to 125. How does the increase in actual wages compare with the increase in real wages?

Solution: The ratio of new actual wages to old actual wages is

$$\frac{17{,}010}{12{,}150}(100) = 140\%$$

Therefore, the increase in actual wages is 40%. When adjusted for the CPI values, new real wages are

$$\frac{17{,}010}{125}(100) = \$13{,}608$$

Old real wages are

$$\frac{12{,}150}{90}(100) = \$13{,}500$$

The ratio for the two values of real wages is

$$\frac{13{,}608}{13{,}500}(100) = 100.80$$

or an increase of 0.8%. Therefore, considering the purchasing power of the dollar, while actual wages increased 40%, real wages increased only 0.8%.

Index numbers are closely related to the economy in many ways. For example, the CPI provides an automatic wage adjustment for many workers through an escalator clause in collective bargaining agreements. This provision is often called COLA, an acronym for Cost of Living Adjustment. The inclusion of such a clause is to assure workers that their standards of living will not suffer. Many pension benefits and Social Security payments, as well as other government programs, are linked to the CPI. The CPI is undoubtedly the most important economic indicator in the United States.

EXERCISES

13.13 In 1976, the average hourly wage in a certain industry was $10.00. In 1985, it was $12.50. If the price indexes for 1976 and 1985 are 80 and 120, respectively, what is the percent change in real wages from 1976 to 1985?

13.14 Data for a worker's annual income are given below for 3 years, along with a price index for each year. What is the percent change in real income from 1983 to 1985?

Year	*Income ($)*	*Price Index*
1983	13,500	90
1984	15,450	100
1985	16,380	105

13.9 SUMMARY

In this chapter, we examined four basic types of index numbers for comparing prices of items in different periods. These were the simple aggregative index, the arithmetic mean of price relatives, the weighted aggregative index, and the weighted mean of price relatives. In the weighted aggregative index classification, we considered both the Laspeyres index and the Paasche index. We reviewed the equivalence of a Laspeyres index with a weighted mean of price relatives using value weights for the base year, as well as the basic concepts concerning link relatives and chain indexes. We concluded this chapter with the use of index numbers for determining the real dollar value of an economic measure. Table 13.1 provides for a summary of the equations for the index numbers included in this text.

TABLE 13.1 Summary table

Single Commodity Index	Price Relative	$I_{n,0} = \frac{p_n}{p_0}(100)$
Unweighted Composite Indexes	Simple Aggregative	$I_{n,0} = \frac{\Sigma p_n}{\Sigma p_0}(100)$
	Mean of Price Relatives	$I_{n,0} = \Sigma \frac{p_n}{p_0}\left(\frac{100}{k}\right)$
Weighted Composite Indexes	Weighted Aggregative	
	Laspeyres	$L_{n,0} = \frac{\Sigma p_n q_0}{\Sigma p_0 q_0}(100)$
	Paasche	$P_{n,0} = \frac{\Sigma p_n q_n}{\Sigma p_0 q_n}(100)$

TABLE 13.1 (Continued)

	Weighted Mean of Price Relatives	
Weighted Composite Indexes	Base year values	$I_{n,0} = \dfrac{\sum \frac{p_n}{p_0} p_0 q_0}{\sum p_0 q_0}(100) = L_{n,0}$
	Given year values	$I_{n,0} = \dfrac{\sum \frac{p_n}{p_0} p_n q_n}{\sum p_n q_n}$
		Example
Chain Indexes	Link Relative	$I_{n,n-1}$
	Chain Index	$I_{n,n-2} = \dfrac{I_{n-1,n-2} I_{n,n-1}}{100}$
Applications	Deflating	Divide value by index number.

SUPPLEMENTARY EXERCISES

13.15 Consider the data and calculate each weighted aggregative index requested comparing 1984 prices to those of 1983 for **(a)** the Laspeyres index and **(b)** the Paasche index.

	1983		*1984*	
Commodity	***Price***	***Quantity***	***Price***	***Quantity***
A	2	6	3	5
B	4	7	8	10

13.16 Given are price data for a certain commodity and a CPI.

(a) Deflate the prices.

(b) Also, determine the percent change in actual prices from 1970 to 1980 and compare it with the corresponding percent change in real prices.

Year	***Price***	***Price Index***
1970	.15	100
1975	.22	110
1980	.30	120

13.17 Consider the following prices and calculate a simple aggregative index comparing 1984 prices to those in 1981.

Commodity	*Price*	
	1981	*1984*
A	10	7
B	4	5
C	16	12

13.18 Consider the given data for the purpose of calculating an index number comparing 1984 prices to those of 1983.

Item	*1983*		*1984*	
	Price	*Quantity*	*Price*	*Quantity*
A	4	1	5	32
B	10	13	8	20
C	6	11	3	28

Calculate **(a)** the simple aggregative index, **(b)** the arithmetic mean of price relatives, **(c)** the Laspeyres index, **(d)** the Paasche index, **(e)** the weighted arithmetic mean of price relatives using value weights for 1983, and **(f)** the weighted arithmetic mean of price relatives using value weights for 1984.

13.19 In the table are prices in dollars per 1000 board feet for 1976 and 1982 for selected types of wood (source: *Statistical Abstract of the United States*). Calculate a simple aggregative index comparing 1982 prices to 1976 prices.

Wood	*Price*	
	1976	*1982*
Douglas Fir	176.20	118.20
Southern Pine	87.00	127.20
Western Hemlock	79.70	44.50
Oak	43.40	70.80
Maple	36.60	34.30

13.20 Assume the following data are available for calculating index numbers to compare 1986 prices to those in 1968.

Commodity	*1968*		*1986*	
	Price	*Quantity*	*Price*	*Quantity*
A	5	8	6	60
B	20	3	14	10

(a) Determine the Laspeyres index.
(b) Determine the Paasche index.

13.21 Use the data from Exercise 13.19 and calculate the arithmetic mean of price relatives comparing 1982 prices with 1976 prices.

13.22 An index of construction costs for certain years has the following values. Shift the base to 1965 and calculate new index values for each year.

Year	*Index*
1945	56
1950	60
1955	64
1960	72
1965	80
1970	92
1975	100
1980	108

13.23 Consider the data of Exercise 13.17 and calculate the arithmetic mean of price relatives comparing 1984 prices to 1981 prices.

13.24 Average hourly earnings for employees in retail trade are given for 1968 to 1983. Also shown are values of the CPI (source: *Monthly Labor Review*).

Year	*Hourly Earnings*	*CPI*
1968	2.16	104.2
1969	2.30	109.8
1970	2.44	116.3
1971	2.60	121.3
1972	2.75	125.3
1973	2.91	133.1
1974	3.14	147.7
1975	3.36	161.2
1976	3.57	170.5
1977	3.85	181.5
1978	4.20	195.3
1979	4.53	217.7
1980	4.88	247.0
1981	5.25	272.0
1982	5.48	288.6
1983	5.74	297.4

(a) Determine real hourly earnings for each year.

(b) Compare the percent change in actual hourly earnings from 1968 to 1983 with the percent change in real wages for the same period.

(c) Compare the percent change in actual wages from 1982 to 1983 with the percent change in real wages for the same period.

13.25 Refer to Exercise 13.20.

(a) Determine the weighted mean of price relatives, using value weights for 1968.

(b) Determine the weighted mean of price relatives, using value weights for 1986.

13.26 For the following grains, calculate the weighted aggregative index comparing 1984 prices to those in 1975, using quantities of the base year as weights.

Grain	*1975*		*1984*	
	Price	*Quantity*	*Price*	*Quantity*
Rice	50	2	59	4
Wheat	46	3	82	1
Oats	38	5	52	6

13.27 Use the data of Exercise 13.26 and determine a Paasche Index.

13.28 Use the data of Exercise 13.26 and calculate a simple aggregative index.

13.29 Use the data of Exercise 13.26 and calculate the arithmetic mean of price relatives.

13.30 Use the data of Exercise 13.26.

(a) Calculate the weighted mean of price relatives, using value weights for the base year.

(b) Calculate the weighted mean of price relatives, using value weights for 1984.

13.31 The following data are available for purposes of constructing index numbers.

Item	*1972*		*1986*	
	Price	*Quantity*	*Price*	*Quantity*
A	5	12	6	10
B	4	7	3	8
C	1	20	2	18

(a) Calculate a simple aggregative index comparing 1986 prices to those of 1972.

(b) Calculate a weighted arithmetic mean of price relatives comparing 1986 prices to 1972 prices, making use of value weights for 1986.

13.32 The following data are available for the purpose of constructing index numbers.

Item	*1970*		*1980*	
	Price	*Quantity*	*Price*	*Quantity*
A	5	14	7	11
B	8	15	6	16
C	10	21	4	25

(a) Calculate a weighted aggregative index comparing 1980 prices to those in 1970, using quantities of the base year as weights.

(b) Calculate an arithmetic mean of price relatives comparing 1980 prices to those in 1970.

13.33 The sales of Knowlton's Department Store decreased by 20% from 1982 to 1983 and increased by $33\frac{1}{3}$% from 1983 to 1984. Determine an index of sales for each year, using 1982 as base year.

13.34 In a certain industry, the minimum hourly wage rose from 30¢ in 1938 to 75¢ in 1949. Assume the CPI was 50 in 1938 and $83\frac{1}{3}$ in 1949. What was the percent change in real wages from 1938 to 1949?

13.35 A worker's annual income was $11,500 in a year in which the CPI was 92. Two years later, her annual income was $13,024, and the CPI was 112. What is the percent change in her real income?

13.36 The following data are available for the purpose of calculating index numbers.

	1976		*1980*	
Commodity	***Price***	***Quantity***	***Price***	***Quantity***
A	2	8	10	1
B	5	9	4	11
C	3	13	6	7

Calculate a weighted aggregative index comparing 1980 prices to 1976 prices, using **(a)** 1976 quantities as weights and **(b)** 1980 quantities as weights.

13.37 The geometric mean of a set of n numbers, $x_1, x_2, \ldots, x_n$ was defined in Chapter 5 to be $\sqrt[n]{(x_1)(x_2) \cdots (x_n)}$. Consider the data of Exercise 13.36 and calculate a geometric mean of price relatives comparing 1980 prices to 1976 prices.

13.38 Consider the data of Exercise 13.36 and the definition of a geometric mean in Exercise 13.37 and calculate Fisher's ideal index, which is given by the geometric mean of the Laspeyres and Paasche indexes.

13.39 Consider the following data for the purpose of calculating index numbers that compare 1984 prices to those of 1983.

	1983		*1984*	
Item	***Price***	***Quantity***	***Price***	***Quantity***
A	5	12	4	16
B	11	10	9	15
C	6	18	8	14

Calculate **(a)** a simple aggregative index, **(b)** a simple average of price relatives, **(c)** a Laspeyres index, and **(d)** a weighted average of price relatives, using value weights for 1984.

13.40 Consider the following data and determine the index number that compares prices in the year 1984 to prices in the year 1972 for **(a)** simple aggregative index, **(b)** simple average of price relatives, **(c)** Laspeyres index, **(d)** Paasche index, and **(e)** weighted average of price relatives, using value weights for the given year.

	1972		*1984*	
Item	***Price***	***Quantity***	***Price***	***Quantity***
A	4	11	9	12
B	8	7	6	32

13.41 The given data are available for calculating a composite index number that compares prices in 1980 to those in 1970. Calculate the appropriate index number for **(a)** simple aggregative index, **(b)** arithmetic mean of price relatives, **(c)** Laspeyres index, and **(d)** weighted arithmetic mean of price relatives, using value weights for 1980.

Commodity	*1970*		*1980*	
	Price	*Quantity*	*Price*	*Quantity*
A	10	4	13	20
B	2	18	1	48
C	8	3	12	16

13.42 The minimum wage rose from 30¢ in 1938 to \$3.35 in 1981. The consumer price index was 50 in 1938 and 272 in 1981. What is the percent change in real wages from 1938 to 1981?

13.43 Consider the indexes given in Exercise 13.22.

(a) Determine the link relative for each year.

(b) Use the link relatives from (a) to determine the index comparing 1980 to 1945.

13.44 Assume the following data are available for constructing index numbers.

Commodity	*Price*			*Fixed Quantity*
	1982	*1983*	*1984*	
A	4	5	7	2
B	3	6	10	9
C	8	12	11	13

Show that the weighted aggregative index comparing 1984 prices to those in 1982 is the same as the chain index obtained by using link relatives.

14 Distribution-Free Procedures

14.1 INTRODUCTION

Monica Willkom, the internal auditor for the CRP Company, is evaluating inventory levels for one of the manufacturing plants. She has determined that for a specific class of items, the average inventory level should be the equivalent of a six weeks' supply. There are 1842 different items in this class. Hence, it is prohibitive to evaluate the current inventory level of each of these items. Monica decides to select a random sample of 25 of the items and to determine the inventory level for each of the items in the sample. She then calculates the difference between the current inventory level and the average six weeks' demand for each item. Unwilling to assume that these differences can be reasonably approximated with a normal distribution, Monica is in a quandary as to what to do. She needs some procedures of analysis that provide alternatives to those that assume normality.

The procedures of this and the next chapter are designed for situations such as that facing Monica. The current chapter introduces procedures that provide alternatives to most of the techniques of Chapter 8, as well as an alternative to the correlation procedure of Chapter 10. The techniques of Chapter 15 represent alternatives to some of the procedures of Chapter 9.

14.2 BACKGROUND

In the previous chapters, a variety of analytical methods were introduced that have become known as classical statistical procedures. As each method was discussed, the assumptions underlying the application and use, and hence the limitations, of the particular test procedure were indicated. A basic assumption of all the procedures, other than those relating to binomial situations, was either that the population or populations from which we were sampling could be reasonably well approximated with normal distributions or that a large sample was selected. The second assumption inherent in all procedures other than the binomial was that the data or numbers that were collected were "good" measurements. By "good," we mean that any two observations can be compared. We can determine which is numerically larger, and these comparisons are logical and meaningful. That is, an interval level of measurement is assumed.

In numerous practical situations, the normality assumption is suspect, as in Monica's case, and it may be impossible, for any number of reasons, to take a large sample. Second, the data that are observed are often not of an interval level of measurement, and the classical procedures are no longer applicable even for large samples. Thus, it is imperative that procedures with less restrictive assumptions be available. These methods are called *distribution-free* because they make no specific assumptions about the type of family to which the population distribution belongs.

These procedures are also commonly called *nonparametric*. This term is somewhat of a misnomer because it implies that no hypothesis is made about the value of a parameter in the probability function, even though the null hypothesis may concern the value of a parameter. Generally, the only assumption for many of the distribution-free procedures is that the sampled population be continuous; occasionally, the requirements of symmetry and, in the multisample case, of identical shapes are added.

Because distribution-free procedures do not use the actual magnitudes of the observations, they do not test for parameters estimated from them in the same way that classical procedures test for equal means or variances. Generally, distribution-free procedures test for values that can be computed from characteristics of the observations, such as frequency or position in the array. In some cases, the distribution-free procedure is also a test procedure for the classical parameters. For example, if the populations are assumed to be symmetric, then a test procedure for equal medians is also a test procedure for equality of means.

Levels of Measurement

As indicated earlier, the level of measurement inherent in the sample data influences which statistical procedure is applicable. The various levels of measurement were defined and discussed in Chapter 1. A brief review of the three levels that will be of concern to us in this chapter is presented here.

The weakest level of measurement is that associated with categorical data and is called *nominal*. In this case, the sample observations are merely classified into mutually exclusive categories, such as male or female, defective or nondefective. The next higher level of measurement is termed *ordinal*. If the sample observations are able to be ranked with respect to some common attribute, then they are considered to be ordinal. Ordinality implies that a relative position within the set may be assigned to each observation. Most of our previous testing procedures required an *interval* level of measurement. Interval measurement means that for any two observations, we may identify both their relative positions within the sample and the actual distance or difference between them. We note that data that satisfy one level of measurement also satisfy any less-sophisticated level of measurement.

14.3 COMPARISON OF DISTRIBUTION-FREE AND CLASSICAL PROCEDURES

We consider only a few of the points of comparison of distribution-free procedures versus the previous classical tests. Both types of procedures have areas of application in which one is superior to the other. Whether a distribution-free procedure or its classical counterpart should be used depends on the level of measurement, how well the underlying assumptions of the procedure are satisfied by the particular situation under consideration, and the ability of the user to determine the degree of satisfaction. In general, the comparisons are quite favorable for the distribution-free procedures.

In terms of the derivation of corresponding test statistics and their related distributions, the distribution-free procedures are more easily derived since most of them rely only on the simple combinatorial (counting) techniques of Chapter 2. Thus, most distribution-free procedures arise by way of a logical process that can be easily understood by the user. Hence, the user is more capable of understanding the underlying assumptions of the procedure and is less likely to misuse it. At the other extreme, the classical procedures usually require a higher level of mathematics to understand their derivations. Distribution-free techniques generally have an advantage in both computational ease and speed. In most situations, all that is necessary are simple arithmetic operations and ranking when the sample size is less than 30. As the sample size increases, procedures based on ranking procedures become more cumbersome and time consuming than their classical counterparts.

The major argument in opposition to distribution-free procedures is based upon an alleged lack of efficiency in the mathematical sense. This type of efficiency is known as asymptotic relative efficiency (ARE).

DEFINITION 14.1 If T_1 and T_2 are consistent test procedures of a null hypothesis, H_0 against alternative H_a at significance level α, then the **asymptotic relative efficiency** of T_1 to T_2 is the limiting value of the ratio n_2/n_1, where n_1 is the sample size required by test procedure T_1 for the power of T_1 to equal the power of test procedure T_2, based on n_2 observations as n_2 approaches infinity and H_a approaches H_0.

DEFINITION 14.2 A test procedure is said to be **consistent** for a specific alternative H_a if the probability of rejecting the null hypothesis approaches 1 as the sample size approaches infinity and H_a is true.

To help illustrate the concept of relative efficiency, we consider the following situation. Assume that it is desired to test H_0: $\theta = 50$ versus the specific alternative H_a: $\theta = 55$ at $\alpha = .05$. Two test procedures are being considered; call them T_1 and T_2. Furthermore, suppose that to attain a power of $1 - \beta = .90$, with procedure T_2 for the alternative $\theta = 55$, requires a sample size of $n_2 = 40$. To achieve a power of .90 using procedure T_1 requires a sample size of $n_1 = 50$. Thus, in this specific instance, ignoring the asymptotic approach, the relative efficiency of T_1 to T_2 is $n_2/n_1 = 40/50 = .8$. That is, T_1 is 80% as efficient as T_2 for rejecting H_0: $\theta = 50$ with probability of .90 when H_a: $\theta = 55$ is true. If this ratio remains constant and n_2 goes to ∞, then the ARE equals .8 also.

The claim of lower ARE for distribution-free procedures is often based on a comparison of the distribution-free procedure to the classical procedure under conditions satisfying the assumptions of the classical procedure. Under those conditions, distribution-free procedures usually have relative efficiencies slightly less than 1.00 for small samples, with the efficiency decreasing as the sample size increases. The ARE is obtained by letting the sample size become infinite, at which point distribution-free procedures ordinarily have their lowest efficiencies relative to the classical counterpart under classical assumptions. When the procedures are compared under conditions that violate some of the classical assumptions, the distribution-free procedure is often superior to the most efficient classical procedure. Since ARE is a large sample property, it may not have much importance for small- or even moderate-sized samples, which is the recommended arena for distribution-free procedures. When samples are small (say $n \leq 15$), distribution-free procedures are easier, faster, and almost as efficient, even if the classical assumptions are met. At these sample sizes, deviations from the classical assumptions are most critical and generally very difficult to detect. Thus, unless it is known in advance that the classical assumptions are justified, it generally is wiser to use a distribution-free procedure.

In terms of applicability, the distribution-free procedures are obviously superior. Being based on fewer and less elaborate assumptions, they can be applied to all situations where the more restrictive classical assumptions are satisfied, as well as a much larger class of situations where they are not.

14.4 ONE-SAMPLE PROCEDURES

In this section, we consider three procedures that are applicable for analyzing a single sample of observations. The first two are designed to test for a location parameter, with the median being used as the measure of location, and the third is designed to test for randomness of the sample data with respect to the order of observation.

One-Sample Sign Procedure

One of the simplest, safest, and most easily understood tests is the sign procedure. It is based on the binomial distribution, and, as such, it has readily available tables and is intuitively appealing. It is not the most powerful test, but its efficiency is surprisingly high considering the simplistic nature of the information used. This information is merely the algebraic sign of the difference between two numerical values or an identification of the relative position of an observation with respect to a reference point.

The hypothesis to be tested is that the median of the population is equal to some specified value, say $\tilde{X}_0$—that is, one possible hypothesis set is

$$H_0: \tilde{X} = \tilde{X}_0 \qquad H_a: \tilde{X} \neq \tilde{X}_0 \tag{14.1}$$

In Chapter 5, $X_{.5}$ was used to denote the median or 50th percentile. We are using the alternative notation $\tilde{X}$ in this chapter to avoid cumbersome double subscripting in accounting for H_0. Since the median is the 50th percentile, another way of stating the null hypothesis is $P(X > \tilde{X}_0) = P(X < \tilde{X}_0) = .5$ or, by subtracting the median from each value, $P(X - \tilde{X}_0 > 0) = P(X - \tilde{X}_0 < 0) = .5$. The latter statement leads directly to a statement based upon the sign of the difference—that is, P(difference is positive) = P(difference is negative) = .5.

Assumptions of the Sign Procedure. The probability equals .5 when the null hypothesis is true and the following assumptions are satisfied.

1. The observations are from an underlying continuous distribution or $P(X = \tilde{X}_0) = 0$. The basic assumption is still satisfied if the population is discrete, but the median is not an observable value.
2. The observations constitute a random sample.
3. The observations are independent. This assumption is satisfied by randomness if the population is infinite or by sampling with replacement if it is not.
4. At least nominal data are available—that is, each member of the sample can be classified as being either above or below the hypothesized value.

Under these assumptions, we may restate the hypotheses in Equation (14.1) as

$$H_0: \pi_+ = .5 \qquad H_a: \pi_+ \neq .5 \tag{14.2}$$

where $\pi_+ = P(X - \tilde{X}_0 > 0)$. Thus, we intuitively base the test procedure upon the observed number of plus signs. If the null hypothesis is true, then the number of +'s and the number of −'s should be the same, differing only because of random variation. The logical rejection region corresponds to situations in which

FIGURE 14.1 Illustration of possible configurations of observed data points

***********	***********	***********
$\tilde{X} = \tilde{X}_0$	$\tilde{X} < \tilde{X}_0$	$\tilde{X}_0 < \tilde{X}$
(a) True median = $\tilde{X}_0$	**(b) True median < $\tilde{X}_0$**	**(c) True median > $\tilde{X}_0$**

there are either too many or too few +'s. If the true median is less than that specified in H_0, we should observe more −'s than +'s, and if it is greater than $\tilde{X}_0$, we should observe more +'s than −'s. See Figure 14.1. As usual, the measure of too many or too few is the probability of occurrence when H_0 is true.

In Chapter 8, the test procedures were based on standardized statistics, where an extreme value of the statistic corresponded to a very small probability of occurrence. When the calculated value was more extreme than the tabulated value corresponding to α, it implied that the probability of occurrence was less than α, and H_0 was rejected. For the sign procedure, we actually determine the probability of occurrence and compare it to a stated level of significance, which we call the reference α. The procedure is as follows.

1. State the hypotheses and the reference α to be used for testing.
2. Select a random sample from the population under consideration.
3. Subtract the hypothesized value, $\tilde{X}_0$, from every observation, recording only the sign of the difference—that is, let $d_i = x_i - \tilde{X}_0$.
4. Count the number of plus signs, say n_+, and the number of minus signs, say n_-.
5. Depending on whether the hypotheses are one-sided or two-sided, determine which of the following decision rules or test criteria apply.

DECISION RULES If the hypotheses are two-sided—that is, $H_0: \tilde{X} = \tilde{X}_0$; $H_a: \tilde{X} \neq \tilde{X}_0$—then we reject H_0 at the reference α if either

$$P\left(X \leq n_+ | \pi = \frac{1}{2}, n = n_+ + n_-\right) < \frac{\alpha}{2} \tag{14.3}$$

or

$$P\left(X \geq n_+ | \pi = \frac{1}{2}, n = n_+ + n_-\right) < \frac{\alpha}{2} \tag{14.4}$$

If the hypotheses are of the form $H_0: \tilde{X} \leq \tilde{X}_0$; $H_a: \tilde{X} > \tilde{X}_0$, the null hypothesis is rejected at α if

$$P\left(X \geq n_+ | \pi = \frac{1}{2}, n = n_+ + n_-\right) < \alpha \tag{14.5}$$

For the other one-sided hypotheses, H_0: $\tilde{X} \geq \tilde{X}_0$; H_a: $\tilde{X} < \hat{X}_0$, H_0 is rejected if

$$P\left(X \leq n_+ | \pi = \frac{1}{2}, n = n_+ + n_-\right) < \alpha \quad \textbf{(14.6)}$$

To be able to calculate the indicated probabilities, we must know the distribution of the number of positive differences. Under the assumptions, we have a finite number of independent trials from a very large dichotomous population with $\pi_+ = \pi_- = \frac{1}{2}$. We recognize this situation as binomial. Since any observation has a 50% chance of falling above the median, π equals $\frac{1}{2}$. Theoretically, no observations should equal $\tilde{X}_0$, and n should be the sample size. Practically, due to imprecise measurements, values will occasionally equal $\tilde{X}_0$, resulting in a difference of 0. When this situation occurs, we will discard the 0 difference and reduce the sample size correspondingly to $n = n_+ + n_-$. Thus, the probabilities indicated in Equations 14.3 and 14.6 are determined according to the binomial distribution as

$$P\left(X \leq n_+ | \pi = \frac{1}{2}, n = n_+ + n_-\right) = \sum_{x=0}^{n_+} C_x^n \left(\frac{1}{2}\right)^x \left(\frac{1}{2}\right)^{n-x} \quad \textbf{(14.7)}$$

For Equations 14.4 and 14.5, they are

$$P\left(X \geq n_+ | \pi = \frac{1}{2}, n = n_+ + n_-\right) = \sum_{x=n_+}^{n} C_x^n \left(\frac{1}{2}\right)^x \left(\frac{1}{2}\right)^{n-x} \quad \textbf{(14.8)}$$

For appropriate values of n, these probabilities can be found by using Appendix Table B.2.

EXAMPLE 14.1 Consider a situation where we want to compare family income in Fairfield to those reported by the Census Bureau. The median income was reported by the Census Bureau to be $24,580.

Solution: Realizing that the distribution of incomes tends to be highly skewed and that the mean is highly susceptible to extreme values, we decide to employ a distribution-free procedure. For the test, we use a reference α of .10. Knowing that people are highly sensitive about giving their exact incomes, it is decided to simply ask them whether the family income is above or below $24,580; a + is recorded for an income above $24,580, and a − for an income below $24,580. The obvious procedure is a sign test, and the hypotheses to be tested at $\alpha = .10$ are

$$H_0: \tilde{X} = 24{,}580 \qquad H_a: \tilde{X} \neq 24{,}580$$

The results of interviews with 20 randomly selected families were 16 above $24,580 and 4 below. Thus, we have 16 +'s and 4 −'s, and H_0 is to be rejected if either Equation 14.3 or Equation 14.4 is true with $\alpha/2 = .05$. For Equation 14.3, we have

$$P\left(X \leq 16 \mid \frac{1}{2}, 20\right) = \sum_{x=0}^{16} C_x^n \left(\frac{1}{2}\right)^x \left(\frac{1}{2}\right)^{20-x}$$

From Appendix Table B.2, the probability is .9987.

For Equation 14.4, the expression is

$$P\left(X \geq 16 \mid \frac{1}{2}, 20\right) = \sum_{x=16}^{20} C_x^n \left(\frac{1}{2}\right)^x \left(\frac{1}{2}\right)^{20-x}$$

From Appendix Table B.2, the value is .0059.

Since the second probability, .0059, is less than the $\alpha/2$ of .05, we reject H_0 and conclude that incomes in Fairfield differ from those reported nationally. Since they differ, it is obvious from the data that the median can be concluded to be in excess of $24,580. ■

Note that it was unnecessary to actually calculate both of the probabilities in Example 14.1. Since one of them must be larger than the other, it is necessary to calculate only the smaller of the two. In general, for a two-sided test, use Equation 14.3 if $n_+ \leq n/2$ and Equation 14.4 if $n_+ > n/2$, or, equivalently, reject H_0 if $P(X \leq \min(n_+, n_-)|\frac{1}{2}, n) < \alpha/2$, where $\min(n_+, n_-)$ is the smaller of the two numbers.

EXAMPLE 14.2 ■ Suppose it is believed that the town of Sleepy Hollow is economically depressed. To evaluate the belief, experimenters randomly selected 16 families. Of the 16, 3 families indicated incomes above $24,580, 12 indicated incomes below $24,580, and 1 indicated $24,580. Test the belief at a reference α of .05.

Solution: If the belief is true, then incomes for that suburb should be below the national median, and it is desired to test the hypotheses

$$H_0: \tilde{X} \geq 24{,}580 \qquad H_a: \tilde{X} < 24{,}580$$

Since the data available are only nominal, the sign procedure is applicable. The decision rule corresponding to the hypotheses is given by Equation 14.6—that is, we calculate only the probability corresponding to the direction of the alternative. For the data, we have $n_+ = 3$, $n_- = 12$, and 1 zero, which is discarded. The corresponding probability of occurrence is

$$P\left(X \leq 3 | \frac{1}{2}, 15\right) = .0176 < .05$$

Therefore, H_0 is rejected, and we conclude that the median income in Sleepy Hollow is below $24,580. ■

In Examples 14.1 and 14.2, we used a reference α of .10 and .05, respectively. These are only reference values because the distribution of +'s is discrete, and the stated α does not generally correspond to one of the possible values that the variable can assume. The actual or exact α for Example 14.1 is $2(.0059) = .0118$ because the test was two-sided, and the exact α for Example 14.2 is .0176.

Note that if we can assume that the population with which we are concerned has a symmetric distribution, then the mean and median are identical. Thus, the test of the hypothesis $H_0: \tilde{X} = \tilde{X}_0$ is equivalent to testing $H_0: \mu = \mu_0$, and the sign procedure provides an alternative to the one-sample t test discussed in Chapter 8 but with much less restrictive assumptions.

When the sample size increases to the extent that n cannot be found in Appendix Table B.2, the probability is calculated by the normal approximation to the binomial, as presented in Chapter 4.

We note that the sign procedure can be altered to test for any specified percentile. The only change required in the basic procedure is the use of the cumulative probability corresponding to the specified percentile for π rather than the value of $\frac{1}{2}$. For example, if the 75th percentile is of interest, we set $\pi = .75$ rather than .5.

If sampling is without replacement from a finite population of a known size N, then the calculation of the probabilities entails the use of the hypergeometric probability function. The parameters of the appropriate hypergeometric distribution under H_0 are N, $R = .5N$, and n whenever the median is of interest. For any other percentile, say X_π, the parameters, under H_0, are N, $R = \pi N$, and n.

One-Sample Wilcoxon Signed-Rank Procedure

A procedure similar to the sign procedure but with somewhat more restrictive assumptions is the Wilcoxon signed-rank procedure for testing population location. The procedure is based upon ranking the differences of the observations minus the hypothesized median (or mean). The test, based on this procedure, is surprisingly powerful, having an ARE of .955 relative to the one-sample t test, even when the underlying population is normal and, in most cases, an ARE greater than 1.00 when compared to the t test with nonnormal populations. In most situations, where the assumptions of the signed-rank procedure are satisfied, it is either the most powerful procedure or close to it.

Assumptions of Wilcoxon's Signed-Rank Procedure. The assumptions follow.

1. The underlying distribution is continuous and symmetric. Continuity may be relaxed if the population is either infinite or very large.
2. The sample data are of at least an ordinal scale.
3. The observations constitute a random sample and are independent of each other.

Under these assumptions, we can test hypotheses about either the population median or mean since the distribution is assumed to be symmetric. The basic hypotheses for the two-sided case are

$$H_0\colon \tilde{X} = \tilde{X}_0 \qquad H_a\colon \tilde{X} \neq \tilde{X}_0$$

As with the sign procedure, we state a recommended format to be used. The steps to be followed are as listed.

1. State the hypotheses and select the reference α at which they are to be tested.
2. Select a random sample.
3. Subtract the hypothesized value $\tilde{X}_0$ from every observation and record the sign and the difference—that is, let $d_i = x_i - \tilde{X}_0$.
4. Rank the absolute values of the differences from low to high, assigning 1 to the lowest nonzero absolute difference, 2 to the next lowest, and so on, up to n for the highest (discard zero differences).
5. Calculate T_+ and T_-, where

$$\begin{aligned} T_+ &= \text{sum of the ranks of the positive differences and} \\ T_- &= \text{sum of the ranks of the negative differences} \end{aligned} \tag{14.9}$$

The decision rules for the two-sided and various one-sided alternatives are based upon the magnitudes of T_+ and T_-. If $\tilde{X}_0$ is the true median, then the high and low ranks should be evenly divided among the positive and negative differences, differing only due to random variation, and T_+ and T_- should be approximately equal. If $\tilde{X} < \tilde{X}_0$, then there should be more negative differences, and correspondingly more of the high ranks should be assigned to them. Thus, T_- should be greater than T_+ since their sum must equal $n(n + 1)/2$, which is the sum of the first n integers. In this case, n is the number of nonzero differences. Conversely, by the same reasoning, if $\tilde{X} > \tilde{X}_0$, then T_+ should exceed T_-. Appendix Table B.13 gives critical values for the smaller sum corresponding to reference α's of .05, .025, .01, and .005.

6. Depending on the stated alternative, determine which of the following decision rules applies.

DECISION RULES If the alternative is two-sided—that is, $H_a\colon \tilde{X} \neq \tilde{X}_0$—then we reject H_0 at the stated α if

$$T = \min(T_+, T_-) \leq T_{n,\alpha/2} \tag{14.10}$$

For $H_0\colon \tilde{X} \geq \tilde{X}_0$; $H_a\colon \tilde{X} < \tilde{X}_0$, T_+ is expected to be smaller if H_a is true, and we reject H_0 at the stated α if

$$T_+ \leq T_{n,\alpha} \tag{14.11}$$

For $H_0: \tilde{X} \leq \tilde{X}_0$; $H_a: \tilde{X} > \tilde{X}_0$, we expect T_- to be smaller if H_a is true, and we reject H_0 at the stated α if

$$T_- \leq T_{n,\alpha} \tag{14.12}$$

Appendix Table B.13 is indexed by n, the total number of +'s and −'s, and by a reference α. The value of n identifies the row, and α identifies the column in which the critical value $T_{n,\alpha}$ will be found. As before, the stated α is only a reference value since T is a discrete random variable. The true α for a test that rejects H_0 will be somewhat less than the stated α.

EXAMPLE 14.3 Experience indicates that scores on a job aptitude test have been symmetrically distributed with a median of 52. A group of 12 applicants recently took the test with the following results:

61 48 44 57 76 30 65 59 22 72 88 39

Is there any evidence that these applicants have a superior aptitude for the job at $\alpha = .05$?

Solution: Since the scores are symmetrically distributed, a test for the median is also a test for the mean. Since there is doubt about the normality of the distribution due to the preponderance of extreme values, the signed-rank procedure is preferable to a t test. The hypotheses indicated earlier are

$$H_0: \tilde{X} \leq 52 \qquad H_a: \tilde{X} > 52$$

and H_0 is to be rejected at $\alpha = .05$ if

$$T_- \leq T_{n,.05}$$

where n is the number of nonzero differences. The procedure is to subtract 52 from each score, to rank the absolute differences, and to add the ranks corresponding to the negative differences. These steps are summarized in Table 14.1.

Also, T_+ has been calculated to show that $T_+ + T_- = n(n+1)/2$. Here $T_+ + T_- = 46.5 + 31.5 = 78 = 12(13)/2$. From Appendix Table B.13, we have for $n = 12$ and $\alpha = .05$ that $T_{12,.05} = 17$. Since $T_- = 31.5$ is not less than or equal to 17, we accept H_0 and conclude that there does not appear to be sufficient evidence that this group is superior.

We observe that in Example 14.3, there are two absolute differences with a value of 13. Theoretically, ties cannot occur, but in practice they do, either from sampling with replacement from a finite population or from using an imprecise measuring instrument on a continuous population. The general practice for handling ties in ranking procedures is to assign the average of the ranks that would have been assigned to each of the tied values. The two 13's should have received the ranks 6 and 7, but to avoid biasing the statistic with an arbi-

TABLE 14.1 Observations and computations for Example 14.3

Score	*Score* − 52	|*Score* − 52|	*Rank*	*Negative Ranks*	*Positive Ranks*
61	+ 9	9	5		5
48	− 4	4	1	1	
44	− 8	8	4	4	
57	+ 5	5	2		2
76	+24	24	10		10
30	−22	22	9	9	
65	+13	13	6.5		6.5
59	+ 7	7	3		3
22	−30	30	11	11	
72	+20	20	8		8
88	+36	36	12		12
39	−13	13	6.5	6.5	
				$T_- = 31.5$	$T_+ = 46.5$

trary assignment, they were each assigned a rank of 6.5. The next value, 20, then received the rank of 8, and so on.

Note that if there are any doubts about the symmetry of the population, then the sign procedure is to be preferred over the signed-rank procedure. The sign procedure is preferred also if there are serious questions about the precision of the measurements.

For situations in which the sample size is too large for Appendix Table B.13, we may use the normal approximation to the distribution of T. It can be shown, using arithmetic series, that the mean and variance of T are

$$\mu_T = E[T] = \frac{n(n+1)}{4} \quad \text{and} \quad \sigma_T^2 = \frac{n(n+1)(2n+1)}{24}$$

Thus, for large samples, the null hypothesis is rejected if

$$Z = \frac{T - \mu_T}{\sigma_T} \tag{14.13}$$

is less than the appropriate z value from the standard normal table. The alternative hypotheses and appropriate decision rules for the large sample approximation are summarized in the table.

Alternative	*T Value*	***Decision Rule: Reject H_0 if***
$H_a\colon \tilde{X} \neq \tilde{X}_0$	$\min(T_+, T_-)$	$Z < -z_{\alpha/2}$
$H_a\colon \tilde{X} > \tilde{X}_0$	T_-	$Z < -z_\alpha$
$H_a\colon \tilde{X} < \tilde{X}_0$	T_+	$Z < -z_\alpha$

Runs Procedure

An assumption of all previous test procedures has been that a random sample was selected. The runs procedure provides a method for testing the validity of this assumption. The procedure does not have high power, but it is extremely simple and easy to use even for large samples. However, the efficiency may be quite adequate for most situations.

Assumptions of the Runs Procedure. The following are assumptions for the runs procedure.

1. A random sample has been selected.
2. The observations are able to be dichotomized.
3. At least nominal level data are available.

The runs procedure specifically tests whether the order of occurrence of a given set of observations follows a random pattern. For example, suppose an interviewer for a political preference poll is collecting sample data through "people on the street" interviews. The results of the interviews, as indicated by preference for either the Republican or the Democratic candidate, are, in order of occurrence, R R R R R R R R R R D D D D D D D D D D. If it is reported that of the 20 people interviewed, ten prefer the Democratic party and ten prefer the Republican party, the randomness of the sample might not be questioned. On the other hand, if the given sequence is reported, one becomes suspicious very quickly because it seems quite unreasonable that the first ten should all prefer the Republican and the last ten should all prefer the Democrat. The runs procedure enables us to evaluate how unreasonable the sequence is. The format for the procedure is quite simple and is as follows.

1. State the hypotheses to be tested and the level of significance. Ordinarily, the hypotheses will be H_0: random pattern of occurrence versus H_a: nonrandom pattern.
2. List the observations in order of occurrence.
3. Indicate to which of the two classes each observation belongs.
4. Count the number of runs R in the given set of observations, where a run is defined to be an unbroken sequence of elements of the same type.

Nonrandomness in the data can be indicated in two ways. There can be too few runs—that is, elements of the same type tend to be grouped together, resulting in long runs—or too many runs, in which case an element of one type tends to be followed by an element of the opposite type, resulting in very short runs and cycling.

5. Apply the appropriate decision rule.

DECISION RULE For H_0: random occurrence; H_a: nonrandom occurrence, reject H_0 at α if either

$$R \le R_{n_1,n_2,\alpha/2} \quad \text{or} \quad R \ge R_{n_1,n_2,1-\alpha/2} \tag{14.14}$$

Appendix Table B.14 gives upper and lower critical values for R at significance levels of .10, .05, .025. The table is based upon combinatorial formulas and thus is indexed by n_1, the number of elements of one type, and n_2, the number of elements of the other type, where $n_1 \le n_2$. For small values of n_1 and n_2, it may not be possible to reject H_0 for some values of α with any number of runs. When no critical value for R exists, a dash is placed in the corresponding position in the table. It is irrelevant which group of elements is identified as type 1 and which as type 2.

EXAMPLE 14.4 Test the randomness of occurrence of the political survey data given earlier, namely R R R R R R R R R R D D D D D D D D D D. Use $\alpha = .10$.

Solution: The observed sequence is

$$\underbrace{\text{R R R R R R R R R R}}_{1} \qquad \underbrace{\text{D D D D D D D D D D}}_{2}$$

with the indicated hypotheses being

H_0: random pattern $\quad H_a$: nonrandom pattern

For the data, $n_1 = 10$, $n_2 = 10$, and we reject H_0 at $\alpha = .10$ if

$$R \le R_{10,10,.05} \quad \text{or} \quad R \ge R_{10,10,.95}$$

In Appendix Table B.14, we find $R_{10,10,.05} = 6$ and $R_{10,10,.95} = 16$. The observed value of R is 2, and the null hypothesis is rejected since $2 < 6$. We conclude that the pattern of observation appears to be nonrandom. Further checking might reveal that the first 10 persons were interviewed outside the local Republican headquarters, and the second 10 were interviewed outside the Democratic headquarters. This example is to illustrate a point, and it should be noted that a random selection of subjects does not necessitate a random sequence of measurements.

The runs procedure can also be used to test for a trend or a cyclical pattern in the data over time. For alternatives of this type, the median of the sample is determined, and each observation is identified as being above or below the median value, with observations equal to the median being ignored. If an upward trend is present, we expect the initial values to be below the median and later ones to be above the median. If a downward trend is present, these values should be reversed. In either case, we expect a small number of runs. A cyclical pattern is indicated by short alternating sequences of values above and below the

FIGURE 14.2 Illustrations of possible trend and cyclical patterns

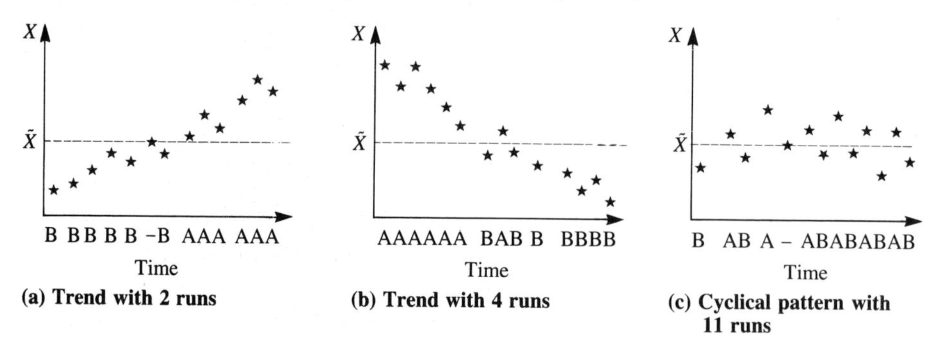

(a) Trend with 2 runs **(b) Trend with 4 runs** **(c) Cyclical pattern with 11 runs**

median. A situation of this type is indicated by a large number of runs. These three situations are illustrated in Figure 14.2a, b, and c, respectively.

As indicated earlier, trend and cyclical alternatives lead to one-sided tests. The hypotheses and decision rules are formalized in the following statements. For trend alternatives, the hypotheses are

H_0: random pattern $\quad H_a$: trend pattern

and we reject H_0 at α if

$$R \leq R_{n_A, n_B, \alpha} \tag{14.15}$$

where n_A is the number of observations above the median and n_B is the number of observations below the median. For the cyclical alternative, we have

H_0: random pattern $\quad H_a$: cyclical pattern

with H_0 being rejected at α if

$$R \geq R_{n_A, n_B, 1-\alpha} \tag{14.16}$$

EXAMPLE 14.5 An investor is considering buying a certain stock and has collected the closing price of the stock for each of the last 13 weeks. The closing prices are

$9\frac{1}{2}$	9	$9\frac{1}{2}$	$8\frac{1}{2}$	9	$8\frac{1}{2}$	8
$7\frac{1}{2}$	$7\frac{1}{4}$	$7\frac{3}{4}$	7	$6\frac{1}{4}$	$6\frac{1}{2}$	

Is there any evidence of trend in the data at $\alpha = .05$?

Solution: The hypotheses to be tested are

H_0: random pattern $\quad H_a$: trend over time

and α has been specified at .05. The median for the data is 8; identifying each observation as above or below yields the sequence A A A A A A – B B B B B B. The dash indicates that the value 8 has been ignored since it is neither above nor below the median. Thus, $n_A = n_B = 6$, and we will reject H_0 at $\alpha = .05$ if

$$R \le R_{6,6,.05} = 3$$

Since the observed number of runs is 2, we reject H_0 and conclude that there appears to be a trend in this stock's price over the period of time in question; in fact, the prices appear to be trending downward.

As n_1 and n_2 approach infinity with their ratio remaining constant, the distribution of R approaches a normal distribution. Thus, for sample sizes too large for n_1, n_2 to be included in Appendix Table B.14, we calculate

$$Z = \frac{R - \mu_R}{\sigma_R} \quad \textbf{(14.17)}$$

where the mean and variance of R are given by

$$\mu_R = \frac{2n_1n_2}{n_1 + n_2} + 1 \qquad \sigma_R^2 = \frac{2n_1n_2(2n_1n_2 - n_1 - n_2)}{(n_1 + n_2)^2(n_1 + n_2 - 1)}$$

The decision rules for the various alternatives when the normal approximation is used are as follows.

Alternative	***Decision Rule: Reject H_0 if***
H_a: Pattern over time	$\lvert Z \rvert > z_{\alpha/2}$
H_a: Trend over time	$Z < -z_\alpha$
H_a: Cycling over time	$Z > z_\alpha$

EXERCISES

14.1 The HiFi Corporation advertises that its electron tubes have a median operating life of at least 1000 hours. A random sample of 25 such tubes showed 15 failing before 1000 hours and 10 after 1000 hours of operation. For $\alpha = .05$, can it be concluded that this sample contradicts the company's advertising claim?

14.2 In 1985, a certain class of stocks on the NYSE were reported to have a median P/E ratio of 12. A random sample of 15 of the stocks in this class recently gave the following P/E ratios.

17 15 17 10 3 21 14 8 16 17 19 36 14 13 9

Would it appear that P/E ratios for this class of stocks have increased since 1985? Use $\alpha = .10$.

14.3 The EPA has stipulated that the median PSI for clean air standards is to be no more than 100. A random sample of 20 days for the city of Smogville showed PSI readings of

111	84	56	122	158	104	92	75	116	187
128	69	90	119	133	107	42	124	169	102

Is this evidence that the city's air is dirtier than the clean air standards recommended at $\alpha = .05$? Assume the distribution of PSIs is not symmetric.

14.4 Test the hypotheses of Exercise 14.3, using Wilcoxon's signed-rank procedure at $\alpha = .05$.

14.5 In Exercise 3.14, it was indicated that the Cleveland Browns average 24 points per game. In a particular 16-game season, their scores were as follows.

23 10 9 17 13 0 24 7 6 37 27 20 16 14 21 6

If the data are random, can it be concluded that the Browns' offense was less potent that season? Use $\alpha = .05$ and assume the distribution of scores to be symmetric.

14.6 Assume that symmetry is reasonable for the population of P/E ratios in Exercise 14.2 and test for a change in the median P/E ratio since 1985, using $\alpha = .10$.

14.7 An $\overline{X}$ chart is used to check the average fill in 4-ounce cough syrup bottles, and the average when the process is in control is known to be 4.02 ounces. The last 12 samples yielded the following means, listed by order of occurrence.

3.96	3.95	3.98	4.01	4.00	3.99
4.04	3.97	4.03	4.05	4.08	4.06

Is there any indication of a pattern developing in the process at a significance level of .10?

14.8 The results of the presidential elections, by party of the winning candidate, are given in order from 1908 through 1984: R D D R R R D D D D D R R D D R R D R R. Can we conclude at $\alpha = .10$ that the winning party occurs randomly?

14.9 Sales, in thousands of units, of a particular item by quarter are given in order of occurrence for the last 4 years.

3 6 4 7 2 8 3 5 3 6 5 8 2 7 1 6

At an α of .05, is there any evidence of a seasonal effect in the sales?

14.5 TWO-SAMPLE PROCEDURES INVOLVING PAIRED DATA

Chapter 8 presented the t test for paired data under the assumptions of normality and interval measurement. This section considers extensions of the sign and signed-rank procedures to the paired data situation as alternatives to that test.

Sign Procedure for Paired Data

The sign procedure of Section 14.4 also can be used with paired samples as a test for identical location parameters. As such, it provides a distribution-free alternative to the paired t test of Chapter 8.

Assumptions for Using the Sign Procedure with Paired Data. For this application, the basic assumptions of the sign procedure are altered as follows.

1. The underlying populations are continuous.
2. The observations constitute a random sample of matched pairs.
3. The pairs are independent of each other.
4. At least nominal data are available—that is, within each pair, it is possible to classify each member of the pair into one of two mutually exclusive categories.

The first assumption implies that, theoretically, ties will not occur. The second assumption states that the items within each pair are identical prior to experimentation and each pair has been randomly selected. The last assumption says that after experimentation is completed, it is at least possible to identify which member of a pair is preferable. For example, if we want to evaluate two shampoos with respect to hair body, then we might use one brand on one side of a person's head and the second brand on the other side. We then compare the two sides and identify which appears to have more body.

Under these assumptions, we are able to test for identical population medians, or means if we can assume symmetry. The procedure consists of subtracting the second observation in each pair from the first—that is, let $d_i = x_{i1} - x_{i2}$—and recording the sign of the difference. The tests that follow correspond to the differences being taken in this direction. If the direction is reversed, the decision rules corresponding to the one-sided alternatives must be altered to compensate for the reversal. The logical derivation then follows the same line of reasoning as presented earlier.

DECISION RULES For hypotheses of the form, H_0: $\tilde{X}_1 = \tilde{X}_2$; H_a: $\tilde{X}_1 \neq \tilde{X}_2$, we reject H_0 at the reference α if either

$$P\left(X \leq n_+ | \pi = \frac{1}{2}, n = n_+ + n_-\right) < \frac{\alpha}{2}$$

or if

$$P\left(X \geq n_+ | \pi = \frac{1}{2}, n = n_+ + n_-\right) < \frac{\alpha}{2}$$

or, equivalently, if

$$P\left(X \leq \min(n_+, n_-) \mid \frac{1}{2}, n\right) < \frac{\alpha}{2}$$

If the hypotheses are H_0: $\tilde{X}_1 \leq \tilde{X}_2$; H_a: $\tilde{X}_1 > \tilde{X}_2$, we reject H_0 if

$$P\left(X \geq n_+ | \pi = \frac{1}{2}, n = n_+ + n_-\right) < \alpha$$

The other possible hypotheses are H_0: $\tilde{X}_1 \geq \tilde{X}_2$; H_a: $\tilde{X}_1 < \tilde{X}_2$, and H_0 is rejected if

$$P\left(X \leq n_+ \middle| \pi = \frac{1}{2}, n = n_+ + n_-\right) < \alpha$$

EXAMPLE 14.6 In an attempt to determine if the color of the package design had an effect upon sales, the advertising department of a particular company placed the product, packaged in 2 different colors, in side-by-side displays in various stores on 16 randomly selected days. Except for the color of the package, the displays were identical. For each day, the number of units sold of each color of package was recorded with the following results.

	Day							
	1	*2*	*3*	*4*	*5*	*6*	*7*	*8*
Color 1	5	8	3	7	4	5	12	4
Color 2	2	6	2	3	9	0	6	4

	Day							
	9	*10*	*11*	*12*	*13*	*14*	*15*	*16*
Color 1	3	8	10	9	7	3	4	8
Color 2	5	7	4	2	4	1	0	3

Can the department conclude at $\alpha = .05$ that sales differ with the color of the package?

Solution: The hypotheses to be tested are

$$H_0: \tilde{X}_1 = \tilde{X}_2 \qquad H_a: \tilde{X}_1 \neq \tilde{X}_2$$

and since there is no indication of normal or even symmetric distributions, a sign test will be used.

Calculating the differences corresponding to the first color minus the second, we observe 13 +'s, 2 −'s, and 1 zero. Discarding the zero and observing that $n_+ > \frac{15}{2}$, our decision rule is to reject H_0 if

$$P\left(X \geq n_+ \middle| \pi = \frac{1}{2}, n = n_+ + n_-\right) < \frac{\alpha}{2}$$

Using Appendix Table B.2, we find

$$P\left(X \geq 13 \middle| \pi = \frac{1}{2}, n = 15\right) = .0037$$

to be less than $\alpha/2$ of .025. Hence, H_0 is rejected, and we conclude that sales appear to differ with package color and that the first color appears to be preferred.

Situations in which this procedure is applicable are identical to those for the paired t. The ideal experiment is of the before-after type. Another way of satisfying the paired assumption is to match the experimental subjects on pertinent variables before treatment. The discussion and words of caution pertaining to the sign procedure in Section 14.4 are applicable here also, as is the normal approximation whenever n is large.

Wilcoxon's Signed-Rank Procedure for Paired Data

If it is logical to add the assumptions of ordinal data and symmetric populations to those of the previous section, then we can use Wilcoxon's signed-rank procedure to test the hypothesis of identical locations—that is, of no treatment effect. It is assumed that the experimental design randomly assigns treatments within a pair. The ARE of the test relative to the paired t test under the assumption of normality is .955. As the sample size decreases, the corresponding efficiency increases toward 1.00.

The format for the procedure is to calculate $d_i = x_{i1} - x_{i2}$ for each pair, where x_{i1} is the first observation in the ith pair and x_{i2} is the second, and then to apply the steps indicated earlier. Once this procedure is accomplished, the following decision rules are applied to the appropriate hypotheses.

DECISION RULES For the two-sided alternative, $H_0\colon \tilde{X}_1 = \tilde{X}_2$; $H_a\colon \tilde{X}_1 \neq \tilde{X}_2$, we reject H_0 at α if

$$T = \min(T_+, T_-) \le T_{n,\alpha/2}$$

where: n = number of nonzero differences

For hypotheses of the form, $H_0\colon \tilde{X}_1 \le \tilde{X}_2$; $H_a\colon \tilde{X}_1 > \tilde{X}_2$, H_0 is rejected at α if

$$T_- \le T_{n,\alpha}$$

The remaining possibility is $H_0\colon \tilde{X}_1 \ge \tilde{X}_2$; $H_a\colon \tilde{X}_1 < \tilde{X}_2$, and we reject H_0 at α if

$$T_+ \le T_{n,\alpha}$$

As with the one-sample Wilcoxon's signed-rank procedure, zero differences are discarded, and tied differences are handled by the average rank procedure.

EXAMPLE 14.7 Let us reconsider the package design problem of Example 14.6. The data that are available are obviously of an ordinal scale, and it might not be unreasonable to assume that sales are symmetrically distributed. With these considerations, we conduct the experiment on 13 more randomly selected days with the following results.

	Day												
	1	*2*	*3*	*4*	*5*	*6*	*7*	*8*	*9*	*10*	*11*	*12*	*13*
Color 1 Sales	7	8	4	3	9	4	8	6	10	5	12	9	1
Color 2 Sales	2	6	1	7	2	4	2	7	2	4	3	8	4

Can it be concluded that sales differ according to color? Use $\alpha = .10$.

Solution: The hypotheses of interest are

$$H_0: \tilde{X}_1 = \tilde{X}_2 \qquad H_a: \tilde{X}_1 \neq \tilde{X}_2$$

and α has been specified to be .10. Scanning the data, we see that there are 12 nonzero differences. Thus, $n = 12$ and H_0 will be rejected if $T = \min(T_+, T_-) \leq T_{12,.05} = 17$. Calculations for the differences and corresponding ranks are as follows.

	Day						
	1	*2*	*3*	*4*	*5*	*6*	*7*
d_i = Difference	5	2	3	−4	7	0	6
$R_{\|d_i\|}$ = Rank	8	4	5.5	7	10		9
Neg. Ranks				7			

	Day					
	8	*9*	*10*	*11*	*12*	*13*
d_i = Difference	−1	8	1	9	1	−3
$R_{\|d_i\|}$ = Rank	2	11	2	12	2	5.5
Neg. Ranks	2					5.5

We find $T_- = 14.5$ and $T_+ = n(n+1)/2 - T_- = 78 - 14.5 = 63.5$. Therefore, $T = \min(T_+, T_-) = 14.5 < 17$, and H_0 is rejected. We conclude, at $\alpha = .10$, that sales appear to differ by color of package and that color 1 is the preferred one. ■

Note that, in this example, the zero difference was discarded. Also, the three 1's each received a rank of 2, the average of the first three ranks, and the two 3's received the rank of 5.5, the average of the fifth and sixth ranks.

Comments relating to the power and applicability of the procedure are identical to those stated in Section 14.4. The large sample approximation of Section 14.4 can also be applied in this situation. The decision rules corresponding to the various alternatives are given in the table for $Z = (T - \mu_T)/\sigma_T$, where μ_T and σ_T are as given in Section 14.4.

Alternative	***T value***	***Decision Rule: Reject H_0 if***
$H_a: \tilde{X}_1 \neq \tilde{X}_2$	$\min(T_+, T_-)$	$Z < -z_{\alpha/2}$
$H_a: \tilde{X}_1 > \tilde{X}_2$	T_-	$Z < -z_{\alpha}$
$H_a: \tilde{X}_1 < \tilde{X}_2$	T_+	$Z < -z_{\alpha}$

EXERCISES

14.10 In a margarine tasting experiment, 16 randomly selected persons were asked to taste each of 2 brands of margarine, Farkay and Crowned, and to identify which was more nearly like butter. The brands were placed in unmarked dishes and were tasted by each individual in a random order. The 16 responses were as follows:

F F C F F F C C Same F C F F C F F

(a) Using $\alpha = .05$, does there appear to be any difference between the 2 brands in their similarity to butter?

(b) What type of experiment was this?

14.11 It is believed that a certain drug retards a person's reaction time. Thus, 13 individuals were randomly selected and their reaction times to a certain stimulus, without the drug and while under its influence, were recorded.

	Person						
	1	*2*	*3*	*4*	*5*	*6*	*7*
Time Without	1.2	2.3	.9	2.4	2.0	2.1	2.2
Time With	2.6	3.1	1.4	2.3	3.5	2.7	2.2

	Person					
	8	*9*	*10*	*11*	*12*	*13*
Time Without	1.6	2.1	2.2	1.5	1.7	2.5
Time With	2.7	1.7	2.6	1.8	3.0	3.2

At an α of .05, is the belief substantiated?

14.12 In an attempt to determine if a difference exists in weathering properties between oil-base and latex exterior paints, the French Girl Paint Company randomly selected 16 houses in various sections of the United States and painted $\frac{1}{2}$ of one side of each house with the exterior latex paint and the other $\frac{1}{2}$ with the exterior oil-base paint. After 2 years, the company's chief chemist inspected each house and rated each half on a 1 to 10 scale (higher score indicating better weathering properties). The chemist did not know in advance which paint was used on which half. His ratings are as follows.

	House							
	1	*2*	*3*	*4*	*5*	*6*	*7*	*8*
Oil-base	8	5	7	4	3	5	8	9
Latex	6	4	5	4	5	2	7	6

	House							
	9	*10*	*11*	*12*	*13*	*14*	*15*	*16*
Oil-base	7	5	6	6	8	4	6	3
Latex	4	8	7	5	5	7	4	5

What conclusion should the French Girl Company reach on the basis of these data if it uses an α of .05 for the test?

14.13 Assume that the distributions of reaction times in Exercise 14.11 are symmetric and test the hypotheses, using Wilcoxon's signed-rank procedure at $\alpha = .05$.

14.14 Assume the distributions in Exercise 14.12 are symmetric and use Wilcoxon's signed-rank procedure to test the hypotheses. Comment on the appropriateness of Wilcoxon's signed-rank procedure in this instance even if the distributions are symmetric.

14.15 The marketing department of Consumer Products, Inc. provides the production department with quarterly sales forecasts in units at the beginning of each quarter. Production then uses these forecasts to set up its production schedule for the coming quarter. The marketing forecast and actual sales (in millions of units) for the last 12 quarters are given

Quarter	*1*	*2*	*3*	*4*	*5*	*6*
Forecast	1.3	2.6	2.8	1.9	1.8	3.0
Actual	1.1	2.0	2.5	2.0	1.3	2.1
Quarter	*7*	*8*	*9*	*10*	*11*	*12*
Forecast	2.9	2.0	2.0	3.1	3.2	1.8
Actual	2.5	2.3	1.3	2.1	2.3	2.6

Using $\alpha = .05$, could the production department accuse the marketing department of being overly optimistic? Do the assumptions of Wilcoxon's signed-rank procedure seem reasonable in this case?

14.6 PROCEDURES INVOLVING TWO INDEPENDENT RANDOM SAMPLES

In the preceding section, we considered procedures for testing for location differences using paired data. This section considers situations where two independent random samples are available. The first procedure is designed to test for location differences, and the second tests for differences in dispersion or variation.

Mann–Whitney Procedure for Identical Distributions

Section 14.5 provided us with procedures for analyzing paired data—that is, the subjects were matched on certain attributes before experimentation. In practice, matching is often impossible to accomplish, and we must resort to selecting two independent random samples. When this situation is true, the Mann–Whitney procedure provides an alternative to the t test given in Chapter 8 for two independent samples. This procedure is a variation of the Wilcoxon rank-sum procedure, and the computational form of its statistic will use the rank sum. In reality, the procedure tests for any differences in the underlying distributions, but it is particularly sensitive to differences in location—that is, in medians or means. It often is referred to as the U test because its statistic is U. As with all statistical procedures, the Mann–Whitney has certain assumptions underlying its development and usage.

Assumptions of the Mann–Whitney Procedure. The assumptions are as follow.

1. Both samples have been randomly selected from the populations of interest. Thus, the samples are independent.
2. The populations are infinite, or else sampling is with replacement from each population.
3. The populations are identical except for possible differences in location.
4. The data are at least ordinal.

In the second assumption, the condition that the population distributions are continuous is stated in some texts. The third assumption may be relaxed, but note that it is possible to reject H_0 with near certainty, even though it is true, for extreme violations of this condition when small samples are used. Most examples illustrating this last point are contrived and would not be expected to occur in practical situations.

Let X_{i1}, $i = 1, 2, \ldots, n_1$, and X_{i2}, $i = 1, 2, \ldots, n_2$, denote the two sets of sample observations, where $n_1 \leq n_2$. The test statistic, U_1, is then defined as

$$U_1 = \sum_{i=1}^{n_2} u_{i1} \tag{14.18}$$

where u_{i1} equals the number of X_1's that are smaller than X_{i2}. Alternatively, we could use $U_2 = \sum_{i=1}^{n_1} u_{i2}$, where u_{i2} is the number of X_2's that are smaller than X_{i1}. The determination of the u_{i1}'s (or u_{i2}'s) is facilitated by first ranking the combined samples from low to high and by identifying each element in the array as to its origin. Then, if H_0 is true and the assumptions are satisfied, all possible configurations of $n_1 X_1$'s and $n_2 X_2$'s are equally likely, with each having probability $1/C_{n_1}^{n_1+n_2}$. Under H_0, the X_1's and X_2's are expected to be interspersed, and U_1 should be approximately equal to U_2, differing only because of random variation. See Figure 14.3a. If the median of the X_1-distribution is less than that of the X_2-distribution, then the X_1 values would be expected to be grouped at the low end of the array, resulting in a disproportionately large value of U_1. See Figure 14.3b. Conversely, if the X_2's are smaller than the X_1's, we expect U_1 to be disproportionately small. See Figure 14.3c. Note that $U_1 + U_2 = n_1 n_2$.

An alternative method for calculating U_1 and U_2 is to replace the observations with their ranks in the overall array and to calculate T_1 and T_2, where

$$T_1 = \text{sum of ranks associated with } X_1 \text{ values}$$

and

$$T_2 = \text{sum of ranks associated with } X_2 \text{ values}$$

FIGURE 14.3 Possible configurations of X_1's and X_2's

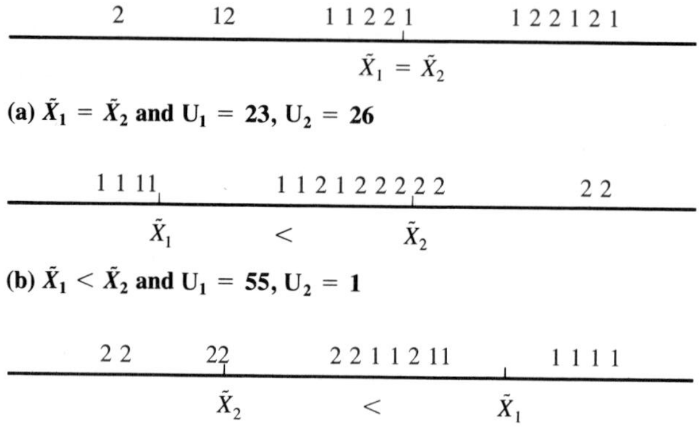

2 12 1 1 2 2 1 1 2 2 1 2 1

$\tilde{X}_1 = \tilde{X}_2$

(a) $\tilde{X}_1 = \tilde{X}_2$ and $U_1 = 23$, $U_2 = 26$

1 1 11 1 1 2 1 2 2 2 2 2 2 2

$\tilde{X}_1 < \tilde{X}_2$

(b) $\tilde{X}_1 < \tilde{X}_2$ and $U_1 = 55$, $U_2 = 1$

2 2 22 2 2 1 1 2 11 1 1 1 1

$\tilde{X}_2 < \tilde{X}_1$

(c) $\tilde{X}_1 > \tilde{X}_2$ and $U_1 = 2$, $U_2 = 54$

Then

$$U_1 = n_1 n_2 + \frac{n_1(n_1 + 1)}{2} - T_1 \tag{14.19}$$

and

$$U_2 = n_1 n_2 + \frac{n_2(n_2 + 1)}{2} - T_2 \tag{14.20}$$

The overall procedure, in summary, is as follows.

1. State the hypotheses to be tested and select a reference α.
2. Select random samples from each of the populations being compared.
3. Rank all the values as if they constitute one large sample of size $n_1 + n_2$.
4. Calculate U_1 and U_2.
5. Apply the appropriate decision rule.

DECISION RULES If the hypotheses in step 1 are $H_0: \tilde{X}_1 = \tilde{X}_2$; $H_a: \tilde{X}_1 \neq \tilde{X}_2$, we reject H_0 at the reference α if

$$U = \min(U_1, U_2) \leq U_{n_S, n_L, \alpha/2} \tag{14.21}$$

For H_0: $\tilde{X}_1 \geq \tilde{X}_2$; H_a: $\tilde{X}_1 < \tilde{X}_2$, reject H_0 if

$$U_2 \leq U_{n_S, n_L, \alpha} \qquad (14.22)$$

If the hypotheses are H_0: $\tilde{X}_1 \leq \tilde{X}_2$; H_a: $\tilde{X}_1 > \tilde{X}_2$, H_0 is rejected if

$$U_1 \leq U_{n_S, n_L, \alpha} \qquad (14.23)$$

Critical values for U are given in Appendix Table B.15. The table is indexed by n_S, the smaller sample size, n_L, the larger sample size, and α, the reference probability of being less than or equal to U. The exact probability will be less than or equal to that stated at the top of the column.

EXAMPLE 14.8 It is desired to determine whether a particular situation comedy show has an appeal factor that differs between men and women. For the evaluation, 7 men and 8 women were randomly selected and asked to view the show in 15 separate rooms. At the conclusion, each person was asked to rate the show on a scale of 0 to 100, with high scores indicating high appeal. The results of the ratings are as follows.

Women's Ratings	24	48	15	26	37	82	33	41
Men's Ratings	57	54	69	77	95	83	88	

Can it be concluded that the show appeals differently to men than to women at $\alpha = .10$?

Solution: The hypotheses indicated in the question are H_0: $\tilde{X}_M = \tilde{X}_W$, H_a: $\tilde{X}_M \neq \tilde{X}_W$, where $\tilde{X}_M$ and $\tilde{X}_W$ denote medians for men and women, respectively, and α is specified to be .10. The corresponding decision rule implies rejection of H_0 if

$$U = \min(U_1, U_2) \leq U_{7,8,.05} = 13$$

Placing the data into an array gives

Value	15	24	26	33	37	41	48	54
Rank	1	2	3	4	5	6	7	8
Sample	W	W	W	W	W	W	W	M

Value	57	69	77	82	83	88	95
Rank	9	10	11	12	13	14	15
Sample	M	M	M	W	M	M	M

Adding, we obtain

$$T_1 = \Sigma R_M = 80 \quad \text{and} \quad T_2 = \Sigma R_W = 40$$

Thus,

$$U_1 = 7(8) + \frac{7(8)}{2} - 80 = 84 - 80 = 4$$

$$U_2 = 7(8) + \frac{8(9)}{2} - 40 = 92 - 40 = 52$$

$$= n_1 n_2 - U_1 = 56 - 4 = 52$$

and

$$U = \min(4,52) = 4$$

Since $4 < 13 = U_{7,8,.05}$, we reject H_0 and conclude that the show's appeal appears to differ according to the sex of the viewer, with men finding it more appealing. Note that it is assumed that interpersonal comparisons can be made in this instance.

If tied values occur, use the average rank procedure that was indicated previously. For sample sizes outside the range of Appendix Table B.15, use the Z statistic, given by $Z = (U - \mu_U)/\sigma_U$, where the mean and variance of U are

$$\mu_U = \frac{n_1 n_2}{2} \quad \text{and} \quad \sigma_U^2 = \frac{n_1 n_2(n_1 + n_2 + 1)}{12}$$

The decision rules based on the normal approximation are presented in the table for the various alternative hypotheses.

Alternative	***U Value***	***Decision Rule: Reject H_0 if***
H_a: $\tilde{X}_1 \neq \tilde{X}_2$	$\min(U_1, U_2)$	$Z < -z_{\alpha/2}$
H_a: $\tilde{X}_1 > \tilde{X}_2$	U_1	$Z < -z_\alpha$
H_a: $\tilde{X}_1 < \tilde{X}_2$	U_2	$Z < -z_\alpha$

The approximation will be good for both n_1 and n_2 greater than or equal to 25 and standard values of α. If n_1 is considerably less than n_2 and α is quite small, the approximation is a poor one.

The ARE of the Mann–Whitney procedure, relative to the comparable t test, under normal conditions is .955 and has been shown to be never less than .864. For rectangular or uniform distributions, this procedure is just as powerful as the t test, giving an ARE of 1.00, and in many situations has an ARE greater than 1.00, possibly approaching infinity.

Note: As mentioned at the beginning of this section, the Mann–Whitney procedure is actually a test of identical distributions. Thus, for it to be purely a test of location differences, the assumptions of equal variances and common shapes must be included with those stated earlier. If these assumptions are not included, the test may become less sensitive to differences in location—for example, if the two populations are skewed in opposite directions over the same range of possible values.

Siegel–Tukey Procedure for Testing Dispersion

In the Mann–Whitney procedure, we assigned ranks to the data in the pooled sample corresponding to each observation's position in the array. Using a different method for ranking, insofar as it remains independent of whether the observation is an X_1 or an X_2, will not affect the distribution of the rank sums T_1 and T_2. Thus, the distribution of U_1 and U_2 is also independent of the method used in assigning ranks. The Siegel–Tukey procedure uses this fact to assign the ranks in such a way that the U statistic becomes sensitive to differences in scale parameters—that is, in variances. As such, it provides an alternative to the F test for equal variances, as presented in Chapter 8.

The assumptions of the procedure are similar to those of the Mann–Whitney procedure. The only change in the assumptions is that the populations have distributions that may differ in variance but are otherwise identical. This assumption implies that the populations have the same median and mean. If this assumption is violated, the results of the test may be questionable. Thus, in using this procedure rather than an F test, we substitute the assumption of identical locations for that of normal populations and interval data.

The hypotheses that we will be testing are the equality of variances and the one-sided alternatives. Except for the method of assigning ranks to the data, the procedure corresponds to that of the Mann–Whitney procedure. First, place the data of the combined samples into an array. Next, rank the data from both ends toward the middle as follows.

1. Assign a rank of 1 to the smallest value and ranks of 2 and 3 to the largest and next largest values, respectively.
2. Assign ranks of 4 and 5 to the second and third smallest values, respectively, and assign ranks of 6 and 7 to the third and fourth largest values, respectively, and continue the pattern until all $n_1 + n_2$ ranks have been assigned, ignoring ties. After all ranks are allocated, check for ties and average the appropriate values where necessary.

After the ranks have been assigned, T_1, T_2, U_1, and U_2 are calculated as before.

If the null hypothesis of equal variances is true and $n_1 = n_2$, we would expect the two sets of observations to be interspersed throughout the array, yielding values for T_1 and T_2, and hence U_1 and U_2, of approximately the same magnitude. If $\sigma_1^2 < \sigma_2^2$, the X_1 values would be expected to be grouped in the middle of the array and to receive the high ranks. Thus, T_1 would be disproportionately large with U_1 correspondingly small. Conversely, for $\sigma_1^2 > \sigma_2^2$, the X_2 values become clustered in the middle, with T_2 becoming large and U_2 becoming small. These situations are illustrated in Figure 14.4.

FIGURE 14.4 Possible configurations resulting when $n_1 = n_2 = 7$

1 221 1 2122 1 2 11 2

(a) $\sigma_1^2 = \sigma_2^2$

2 2 1 2 1 1 1 2 1 2 1 1 2 2

(b) $\sigma_1^2 < \sigma_2^2$

1 1 1 22 1 2222 1 2 1 1

(c) $\sigma_1^2 > \sigma_2^2$

As an example of the ranking procedure, we assign ranks to the configurations in Figure 14.4. Ranking from both ends toward the middle gives

1 4 5 8 9 12 13 14 11 10 7 6 3 2

for all three configurations. The corresponding values for T_1 and T_2 are (a) $T_1 = 50$, $T_2 = 55$; (b) $T_1 = 63$, $T_2 = 42$; (c) $T_1 = 34$, $T_2 = 71$.

From the foregoing considerations, we construct the tests that are appropriate to the hypotheses of interest.

DECISION RULES If we are interested only in detecting a difference, our hypotheses become

$$H_0: \sigma_1^2 = \sigma_2^2 \qquad H_a: \sigma_1^2 \neq \sigma_2^2$$

and we reject H_0 at the stated α if

$$U = \min(U_1, U_2) \leq U_{n_S, n_L, \alpha/2} \tag{14.24}$$

If the indicated hypotheses are

$$H_0: \sigma_1^2 \geq \sigma_2^2 \qquad H_a: \sigma_1^2 < \sigma_2^2$$

we reject H_0 if

$$U_1 \leq U_{n_S, n_L, \alpha} \tag{14.25}$$

For the remaining one-sided case,

$$H_0: \sigma_1^2 \leq \sigma_2^2 \qquad H_a: \sigma_1^2 > \sigma_2^2$$

H_0 is rejected if

$$U_2 \le U_{n_S, n_L, \alpha} \quad \textbf{(14.26)}$$

The following example illustrates the total procedure used in testing one of the hypotheses.

EXAMPLE 14.9 A broker has advised us to consider buying stock in 2 unrelated companies. His report indicates that over the past 20 years, the average price per share is the same for both companies. Stock A returns slightly higher dividends, and since we can only afford to invest in 1, we prefer it, unless it is less stable than stock B. To facilitate making the decision, we randomly select 9 weeks for A and 8 weeks for B from the previous 20 years. The closing prices for A's 9 weeks and B's 8 weeks are as follows.

A	8	$11\frac{1}{2}$	$15\frac{1}{4}$	$10\frac{3}{8}$	$6\frac{3}{4}$
B	12	$10\frac{1}{2}$	$8\frac{1}{4}$	$9\frac{1}{4}$	11

A	$9\frac{7}{8}$	10	$4\frac{1}{2}$	14
B	$8\frac{7}{8}$	$11\frac{1}{4}$	$10\frac{1}{4}$	

For $\alpha = .05$, which stock should be purchased?

Solution: We are interested in the stability of the stocks, which can be measured by the respective variation in closing prices. A small variance indicates a stable stock, and a high variance indicates an unstable one and, since we prefer A unless it is more unstable than B, the hypotheses are

$$H_0\colon \sigma_A^2 \le \sigma_B^2 \qquad H_a\colon \sigma_A^2 > \sigma_B^2$$

With $\alpha = .05$, $n_B = 8$, and $n_A = 9$, the decision rule implies rejection of H_0 if

$$U_B \le U_{8,9,.05}$$

In Appendix Table B.15, we find $U_{8,9,.05} = 18$.

The computations required for calculating U_B follow in the table.

Array	$4\frac{1}{2}$	$6\frac{3}{4}$	8	$8\frac{1}{4}$	$8\frac{7}{8}$
Rank	1	4	5	8	9
R_B				8	9

Array	$9\frac{1}{4}$	$9\frac{7}{8}$	10	$10\frac{1}{4}$	$10\frac{3}{8}$
Rank	12	13	16	17	15
R_B	12			17	

Array	$10\frac{1}{2}$	11	$11\frac{1}{4}$	$11\frac{1}{2}$
Rank	14	11	10	7
R_B	14	11	10	

Array	12	14	$15\frac{1}{4}$
Rank	6	3	2
R_B	6		

Thus, $T_B = 87$ and

$$U_B = n_1 n_2 + \frac{n_1(n_1 + 1)}{2} - T_B$$

$$= 8(9) + \frac{8(9)}{2} - 87$$

$$= 108 - 87 = 21$$

Since $21 > 18$, we accept H_0 and conclude that, on the basis of these samples, it cannot be stated that $\sigma_A^2 > \sigma_B^2$ or that A is less stable. Thus, buy stock A. ■

For sample sizes outside the range of Appendix Table B.15, use the normal approximation with $Z = (U - \mu_U)/\sigma_U$, where

$$\mu_U = \frac{n_1 n_2}{2} \quad \text{and} \quad \sigma_U^2 = \frac{n_1 n_2(n_1 + n_2 + 1)}{12}$$

The decision rules corresponding to the various alternatives when the normal approximation is used are presented in the table.

Alternative	*U*	*Decision Rule: Reject H_0 if*
H_a: $\sigma_1^2 \neq \sigma_2^2$	$\min(U_1, U_2)$	$Z < -z_{\alpha/2}$
H_a: $\sigma_1^2 > \sigma_2^2$	U_2	$Z < -z_\alpha$
H_a: $\sigma_1^2 < \sigma_2^2$	U_1	$Z < -z_\alpha$

EXERCISES

14.16 We wish to test for differences in the frequency of repair for 2 brands of copiers. Six copiers of brand A and 6 of brand B are selected at random, and the number of calls for repair service for each machine is observed. The data are as follows:

Brand A	7	11	9	8	12	6
Brand B	10	5	4	3	7	2

Test for a difference at $\alpha = .05$.

14.17 It has been stated that accounting majors do better in statistics courses than other business students. Therefore, 8 accounting majors and 9 other business students were randomly selected from last quarter's statistics class. Their grades on the common final exam, worth 200 points, are as follows.

Accounting	152	139	176	182	121	86	137	149	
Other	194	178	129	169	94	158	187	145	125

Is the statement supported at $\alpha = .10$?

14.18 Refer to Exercise 14.11 and assume that 26 people were randomly selected and then randomly divided into 2 groups of 13 each. One group was then given the drug and the other was not. Reaction times were determined for the individuals in each group and are as given in Exercise 14.11.

(a) Are the conclusions affected by the change in experimental procedure?

(b) Which procedure do you believe to be appropriate for this type of situation and why?

14.19 Refer to Exercise 14.17. For $\alpha = .05$, does there appear to be any difference in the variability of scores on the exam between accounting majors and others?

14.20 Does there appear to be any difference between the copiers of Exercise 14.16 in terms of the variability of repair frequencies at $\alpha = .10$?

14.21 A quality control inspector has claimed that the night shift at Precision Die Products does not maintain as tight a control on its machine settings as the day shift. To test this claim, the company selected random samples of 9 pieces each from the day shift and from the night shift and measured the diameters of the pieces. The results (in inches) are as follows.

Day	4.990	5.020	4.995	4.998	5.001
Night	4.891	4.898	5.021	5.018	5.019

Day	5.015	4.996	4.991	5.003
Night	4.992	5.024	4.895	5.021

For $\alpha = .05$, does the inspector's claim appear to be supported?

14.7 SPEARMAN'S RANK CORRELATION

We introduced the concept of estimating the strength of the linear relationship between two variables in Chapter 10. The parameter measuring this relationship is the correlation coefficient ρ, which was estimated previously using Pearson's product moment coefficient r, defined as

$$r = \frac{\Sigma X_i Y_i - (\Sigma X_i)(\Sigma Y_i)/n}{\sqrt{[\Sigma X_i^2 - (\Sigma X_i)^2/n][\Sigma Y_i^2 - (\Sigma Y_i)^2/n]}}$$

When we use this equation, we assume that the available data are of an interval level of measurement. Recall that values for ρ are restricted to the interval -1 to $+1$, inclusive, with either -1 or $+1$ indicating a perfect linear relationship and 0 indicating no linear relationship. Since r is calculated from sample data, it is only an estimate of ρ. Using r to test if ρ differs from 0 requires that X and Y have a bivariate normal distribution. If the assumption of joint normality is satisfied, we are able to use the t-distribution for constructing the test.

Spearman derived an alternative procedure, with much less restrictive assumptions, by replacing the actual values in Pearson's formula with their corresponding ranks within each individual set of values. The distribution of the correlation coefficient, denoted r_s, that results from the substitutions can be derived if the following conditions are satisfied.

Assumptions for Testing Spearman's Coefficient. The assumptions are as follows.

1. A random sample of n units is selected.
2. A measurement on each variable, X, Y, is obtained for each unit in the sample.
3. The underlying distribution of each variable is continuous or at least infinite.
4. The data are at least ordinal.

The ARE of the test based upon r_s has a value of .912 when the assumption of normality is satisfied and a value of 1.000 if the distribution is uniform. For other distributions, the ARE varies.

The procedure for calculating r_s is summarized in the following steps. We assume a random sample of n pairs of observations has already been obtained.

1. Rank the n observations corresponding to X among themselves, replacing each value with its rank, R_{X_i}.
2. Rank the $n Y_i$'s among themselves, replacing each with its rank, R_{Y_i}.

Then, r_s is calculated using the equation

$$r_s = \frac{\Sigma R_{X_i} R_{Y_i} - [(\Sigma R_{X_i})(\Sigma R_{Y_i})/n]}{\sqrt{\left(\Sigma R_{X_i}^2 - \dfrac{(\Sigma R_{X_i})^2}{n}\right)\left(\Sigma R_{Y_i}^2 - \dfrac{(\Sigma R_{Y_i})^2}{n}\right)}}$$

which can be simplified to

$$r_s = \frac{\Sigma R_{X_i} R_{Y_i} - [n(n+1)^2/4]}{\sqrt{\left(\Sigma R_{X_i}^2 - \dfrac{n(n+1)^2}{4}\right)\left(\Sigma R_{Y_i}^2 - \dfrac{n(n+1)^2}{4}\right)}} \qquad \textbf{(14.27)}$$

If there are no ties in either the X's or the Y's, the denominator reduces to

$$\Sigma R_{X_i}^2 - \frac{n(n+1)^2}{4} = \frac{n(n^2-1)}{12}$$

since $\Sigma R_{X_i}^2$ and $\Sigma R_{Y_i}^2$ will both equal $[n(n+1)(2n+1)]/6$, which is the sum of the squares of the first n integers.

An alternative method, which simplifies the arithmetic considerably, is available for calculating r_s. If there are no ties in the data, the resulting value will be identical to that obtained by using Equation 14.27. When the number of ties is small and they are handled by the average rank procedure, the two methods yield slightly different values. If there is an excessive number of ties, Equation 14.27 should be used. The alternative procedure is to calculate the differences between corresponding X and Y ranks—that is, $d_i = R_{X_i} - R_{Y_i}$—and then to use

$$r_s = 1 - \left[\frac{6\sum d_i^2}{n(n^2 - 1)}\right] \tag{14.28}$$

Spearman's rank correlation coefficient has a maximum value of 1, which occurs when a perfect positive relationship exists. The maximum can be seen from Equation 14.28 since, in such a situation, $R_{X_i} = R_{Y_i}$ and $d_i = 0$ for all i. Conversely, if there is a perfect negative relation, the smaller X values will correspond to the larger Y values, resulting in $R_{X_i} = n - R_{Y_i} + 1$. That is, if $R_{X_i} = i$, then $R_{Y_i} = n - i + 1$ and $d_i = 2i - (n + 1)$, in which case it can be shown that $\sum d_i^2 = n(n^2 - 1)/3$ with a resultant r_s of -1. If there is no relation between X and Y, there should be no pattern between their respective ranks and the sum of the d_i^2 is expected to be halfway between the extremes—or equal to $n(n^2 - 1)/6$—giving an r_s of zero. Thus, r_s is also restricted to being within the interval from -1 to $+1$ inclusive, and it is expected to be close to the true value of ρ.

Using these considerations, we are able to test for a relationship between X and Y. As has been done throughout this chapter, we condense the seven-step procedure recommended in Chapter 8.

1. State the hypotheses of interest and select the reference α.
2. Select a random sample of n pairs of observations on X and Y.
3. Separately rank the X_i's and Y_i's, assigning a rank of 1 to the smallest X and also to the smallest Y, and so on.
4. Calculate r_s, using Equation 14.28 (14.27 if there is an excessive number of ties).
5. Apply the appropriate decision rule.

DECISION RULES If the hypotheses are $H_0: \rho = 0$; $H_a: \rho \neq 0$, we reject H_0 at the reference α if

$$|r_s| \geq r_{n,\alpha/2} \tag{14.29}$$

If the hypotheses in step 1 are $H_0: \rho \leq 0$; $H_a: \rho > 0$, reject H_0 if

$$r_s \geq r_{n,\alpha} \tag{14.30}$$

For $H_0: \rho \geq 0$; $H_a: \rho < 0$, H_0 is rejected if

$$r_s \leq -r_{n,\alpha} \tag{14.31}$$

Critical values for r_s are given in Appendix Table B.16. The table is indexed by n, the number of pairs of observations, and α, the level of significance. The number, $r_{n,\alpha}$, in the body of the table is the largest value possible such that $P(r \geq r_{n,\alpha}) \leq \alpha$ if no relationship exists. Since the distribution of r_s is symmetric about zero, lower-tail critical values are obtained by merely multiplying the corresponding upper-tail value by -1.

The following example illustrates the computations required for determining the rank correlation coefficient and the test procedure. Equations 14.27 and 14.28 are used for comparative purposes.

EXAMPLE 14.10

It has long been contended that students who study more do better on statistics exams. To test this contention, the statistics department randomly selected 12 students and asked them the number of hours per week that each spent studying statistics and what grade had been attained on the final exam. The results are recorded in the following table.

Student	***1***	***2***	***3***	***4***	***5***	***6***
Hours/Week	6.5	3.0	4.0	5.5	2.5	3.5
Exam Grade	93	62	72	86	50	76

Student	***7***	***8***	***9***	***10***	***11***	***12***
Hours/Week	5.0	2.0	1.5	6.0	4.5	8.0
Exam Grade	83	48	40	95	79	88

a. Calculate r_s, using Equation 14.27.
b. Calculate r_s, using Equation 14.28.
c. Test the contention at $\alpha = .05$.

Solution: The preliminary computations required for both (a) and (b) are given in Table 14.2. The hours per week have been rearranged in ascending order, but the pairings are maintained.

TABLE 14.2 Calculations for Example 14.10

(X) ***Hours/Week***	***(Y)*** ***Exam Grade***	R_X	R_Y	R_X^2	R_Y^2	$R_X R_Y$	$d_i = R_X - R_Y$	d_i^2
1.5	40	1	1	1	1	1	0	0
2.0	48	2	2	4	4	4	0	0
2.5	50	3	3	9	9	9	0	0
3.0	62	4	4	16	16	16	0	0
3.5	76	5	6	25	36	30	−1	1
4.0	72	6	5	36	25	30	+1	1
4.5	79	7	7	49	49	49	0	0
5.0	83	8	8	64	64	64	0	0
5.5	86	9	9	81	81	81	0	0
6.0	95	10	12	100	144	120	−2	4
6.5	93	11	11	121	121	121	0	0
8.0	88	12	10	144	100	120	+2	4
		78	78	650	650	645		10

a. Using Equation 14.27, we have

$$r_s = \frac{\Sigma R_{X_i} R_{Y_i} - [n(n+1)^2/4]}{\sqrt{\left(\Sigma R_{X_i}^2 - \frac{n(n+1)^2}{4}\right)\left(\Sigma R_{Y_i}^2 - \frac{n(n+1)^2}{4}\right)}}$$

$$= \frac{645 - [12(13)^2/4]}{\sqrt{\left(650 - \frac{12(13)^2}{4}\right)\left(650 - \frac{12(13)^2}{4}\right)}}$$

$$= \frac{645 - 507}{\sqrt{(650 - 507)(650 - 507)}} = \frac{138}{143} = .965$$

b. Substituting from the last column on the right into Equation 14.28 gives

$$r_s = 1 - \left[\frac{6(\Sigma d_i^2)}{n(n^2 - 1)}\right] = 1 - \left[\frac{6(10)}{12(143)}\right]$$

$$= 1 - \frac{5}{143} = 1 - .035 = .965$$

Note that since the 2 methods yielded identical values, it is unnecessary to use both equations to calculate r_s. Use Equation 14.27 only if there are numerous ties in either variable.

c. Since the contention is that more study time corresponds to higher grades, the hypotheses are

H_0: $\rho \leq 0$ $\quad$ H_a: $\rho > 0$

and H_0 is rejected at $\alpha = .05$ if

$r_s \geq r_{n,\alpha} = r_{12,.05} = .497$

Therefore, we reject H_0 since $.965 > .497$ and conclude that there appears to be a positive correlation between exam grades and study time.

For values of $n > 30$, the t-distribution may be used to obtain an approximate test. In such situations, the statistic $r_s\sqrt{n-2}/\sqrt{1-r_s^2}$ is distributed approximately as a t with $n - 2$ degrees of freedom. The decision rules for the various alternatives when using the normal approximation are given in the table.

Alternative	*Reject H_0 if*
H_a: $\rho_s \neq 0$	$\lvert t \rvert > t_{n-2,\alpha/2}$
H_a: $\rho_s > 0$	$t > t_{n-2,\alpha}$
H_a: $\rho_s < 0$	$t < -t_{n-2,\alpha}$

EXERCISES

14.22 It has often been claimed that students' performances on the first exam in statistics are indicative of their performances for the rest of the course, so 10 students were randomly selected after the final exam and their scores on both the first exam and the comprehensive final were recorded.

	Student									
	1	*2*	*3*	*4*	*5*	*6*	*7*	*8*	*9*	*10*
First Exam	87	58	94	77	82	43	66	75	72	74
Final Exam	92	73	82	78	80	39	69	79	70	76

Test for the logical relationship between the exam scores at $\alpha = .05$.

14.23 In a local talent contest, the judges were asked to rank the 12 contestants in the voice competition by assigning a 1 to the most talented, 2 to the next most talented, and so on. The rankings for 2 of the judges are as follows.

	Contestant											
	1	*2*	*3*	*4*	*5*	*6*	*7*	*8*	*9*	*10*	*11*	*12*
Judge 1	7	12	4	8	1	3	5	10	2	11	9	6
Judge 2	4	10	2	8	3	5	1	9	6	12	11	7

Can it be concluded that the 2 judges have similar tastes with respect to vocal talent? Use $\alpha = .05$.

14.24 With the increased awareness of the health hazards associated with cigarette smoking, more filter cigarettes are appearing on the market. In evaluating various filters, a researcher collected the following data on the density of the filter and the amount of tars passing through the filter; the more dense the filter, the higher its coded value.

	Cigarette						
	1	*2*	*3*	*4*	*5*	*6*	*7*
Density (Coded)	15	5	7	10	2	14	13
Tar in mg	2.2	12.0	11.6	5.0	16.1	3.3	3.1

	Cigarette					
	8	*9*	*10*	*11*	*12*	*13*
Density (Coded)	6	20	9	11	8	17
Tar in mg	13.1	1.4	5.7	4.8	7.3	2.6

Test for the logical relationship, using an α of .01.

14.8 SUMMARY

This chapter has provided at least one distribution-free alternative for each of the classical one- and two-sample tests of Chapter 8. Also, it has included the runs procedure as a check for randomness and Spearman's rank correlation as an alternative to Pearson's correlation coefficient, which was presented in Chapter 10. By distribution-free is meant a test based on a statistic whose distribution is independent of the specific type of distribution that governs the population. In determining which test to use for a given situation, it is mandatory that we check the assumptions associated with each procedure being considered. In general, the distribution-free procedures are easier to understand, faster and less complicated to use, and more widely applicable than their classical counterparts. The main drawback is that the gain in applicability generally coincides with a loss in power if the classical assumptions are satisfied.

The one-sample sign and Wilcoxon signed-rank procedures provide alternatives to the one sample t test if either normality or an interval level of measurement is questionable. In deciding between the two alternatives, we must give particular attention to the level of measurement that can be associated with the data and to the symmetry assumption. The runs procedure is extremely easy to use and provides a means for checking randomness.

The two-sample procedures for testing for location differences are classified by the type of sampling design used in the collection of the data. For paired data, a randomized block design for two treatments, the sign and the Wilcoxon signed-rank procedures for paired data are options if the assumptions for the paired t test are questionable. The choice is contingent on the level of measurement and symmetry. For two independent random samples, the Mann–Whitney procedure can be used in lieu of the independent samples t test. The t test assumes normality and an interval level of measurement, and the Mann–Whitney procedure requires only infinite populations and rankable data.

When testing equality of variances for nonnormal populations or with data of less than an interval level, we consider the Siegel–Tukey procedure rather than the F test. The Siegel–Tukey procedure uses the same rationale and test statistic as the Mann–Whitney procedure. The difference is in the ranking method. The Mann–Whitney procedure ranks from low to high, but the Siegel–Tukey ranks from both ends toward the middle to obtain a statistic that is sensitive to dispersion differences.

The Spearman rank correlation procedure actually uses Pearson's basic correlation formula, but replaces the observations with their respective ranks. The ranking procedure in this case is to rank the X's within themselves and the Y's within themselves, keeping corresponding X's and Y's identified. As in Chapter 10, this correlation coefficient is restricted to the range -1 to 1, with values near the extremes indicating strong relationships and values near 0 indicating weak or no relationships.

Distribution-free procedures rely primarily on the relative positions of the data points and not on the actual magnitudes. The distribution of the statistic then can be determined through the use of simple combinations and permutations. Therefore, the procedures are easier to understand and simpler to use than their classical counterparts. Comparisons of procedures are generally based on AREs, which necessitates satisfying the assumptions of the more restrictive test. Even so, the distribution-free procedures have surprisingly high AREs whenever the sample sizes are small, in which case violations of assumptions are most critical. Anytime there is doubt as to which of two procedures to use, it is preferable to use the one having the less restrictive assumptions.

Table 14.3 summarizes the procedures with respect to the attribute being tested, type of samples required, the underlying assumptions, and the test statistic of the procedure.

TABLE 14.3 Summary table

		Test Procedures		
		Distribution-Free Procedure	***Assumptions (All assume underlying continuity and random samples.)***	***Classical Test Procedure***
			Location	
Two-sample	***Inde-pendent***	Mann–Whitney	Identical shapes and variances Infinite populations Ordinal data	Independent samples t test
Two-sample	***Paired***	Wilcoxon Signed-Rank	Subjects identical within a pair before treatment Symmetric distributions Ordinal data	Paired t test
Two-sample	***Paired***	Sign	Matched pairs before treatment Nominal data	Paired t test
One-sample		Wilcoxon Signed-Rank	Symmetric distribution Ordinal data	t test
One-sample		Sign	Nominal data	t test
			Dependence	
	Correlation	Spearman's Rank Correlation	Sample of n subjects Measurement for X, Y taken on each subject Ordinal data	Pearson's r and t test
	Random-ness	Runs	Observations can be dichotomized Nominal data	
			Dispersion	
	Two-sample	Siegel–Tukey	Identical location parameters Independent samples Ordinal data	F test

Note that all of the procedures require random samples at some point. If some subjects refuse to be interviewed or drop out of the experiment, then this assumption may be violated. Reducing the sample size accordingly is often done, but the test results may be questionable when that happens. In general, if the assumptions for a procedure are violated, then the probability distribution utilized is incorrect, and the calculated probability is at best a good approximation.

A multitude of distribution-free procedures are available, not only for the types of situations considered here, but also for many other kinds of applications. For a more extensive coverage of distribution-free procedures, see Brad-

TABLE 14.3 (Continued)

Test Statistics

Procedure	***Statistics***
Sign	n_+ = number of positive d_i
One-sample	$d_i = x_i - \tilde{X}_0$
Two-sample	$d_i = x_{i1} - x_{i2}$
Wilcoxon Signed-Rank (rank $\lvert d_i \rvert$)	T_+ = sum of ranks assigned to positive d_i T_- = sum of ranks assigned to negative d_i
One-sample	$d_i = x_i - \tilde{X}_0$
Two-sample	$d_i = x_{i1} - x_{i2}$
Runs	R = number of runs
Mann–Whitney (rank from low to high)	$U_1 = n_1 n_2 + \frac{n_1(n_1 + 1)}{2} - T_1$ $U_2 = n_1 n_2 + \frac{n_2(n_2 + 1)}{2} - T_2$ T_j = sum of ranks assigned to sample j
Siegel–Tukey (rank from ends to middle)	$U_1 = n_1 n_2 + \frac{n_1(n_1 + 1)}{2} - T_1$ $U_2 = n_1 n_2 + \frac{n_2(n_2 + 1)}{2} - T_2$ T_j = sum of ranks assigned to sample j
Spearman's (rank X, Y separately)	$r_s = \frac{\Sigma R_{X_i} R_{Y_i} - [n(n + 1)^2/4]}{\sqrt{\left(\Sigma R_{X_i}^2 - \frac{n(n + 1)^2}{4}\right)\left(\Sigma R_{Y_i}^2 - \frac{n(n + 1)^2}{4}\right)}}$
or (if no ties)	$r_s = 1 - \left[\frac{6 \Sigma d_i^2}{n(n^2 - 1)}\right]$ where $d_i = R_{X_i} - R_{Y_i}$

ley [1] or Siegel [4]. A theoretical approach to this topic is given in Noether [3] and Gibbons [2].

SUPPLEMENTARY EXERCISES

14.25 The average number of pages used per student program at the university computer center has been estimated to be 11.5. A random sample of 20 student programs yielded the following numbers of pages in the printouts:

4	7	16	3	12	26	8	5	3	2
6	9	10	13	15	7	3	22	9	5

For $\alpha = .05$, does the estimate appear to be correct? Check the assumptions of the test you used. Do you feel they are satisfied in this instance?

14.26 To evaluate 2 types of rust inhibitors, researchers randomly selected 25 steel rods. One-half of each rod was coated with Type I, and $\frac{1}{2}$ was coated with Type II. The rods were then subjected to various soil, weather, and climatic conditions for 1 year. At that time, the depth of corrosion was measured on each half of the rod. The results, coded by multiplying by 100, are as follows.

Rod	***1***	***2***	***3***	***4***	***5***	***6***	***7***	***8***	***9***
Type I Depth	13.2	84	51	24	16.7	86	11.6	60	42
Type II Depth	11.6	62	62	12	12.2	91	2.4	111	73

Rod	***10***	***11***	***12***	***13***	***14***	***15***	***16***	***17***	***18***
Type I Depth	54	76.1	57	18.7	71	25.4	14.7	39	33
Type II Depth	50	30.3	22	26.9	60	17.6	1.1	22	37.5

Rod	***19***	***20***	***21***	***22***	***23***	***24***	***25***
Type I Depth	54.1	67.2	46.3	19.8	36.5	27.4	54.9
Type II Depth	27.8	50.6	38.9	20.2	38.3	19.2	31.1

Does there appear to be a difference in the effectiveness of the inhibitors at $\alpha = .01$?

14.27 What is the effect on Spearman's rank correlation coefficient if the ranking method used for the Y variable is reversed from that used for the X variable?

14.28 In a consumer protection survey to determine whether 2 food chains in a certain area sold hamburger of the same quality, researchers bought 1 pound at each chain on each of 10 randomly selected days. Each pound was then fried, and the amount of fat was poured off and weighed. The results, in ounces of fat, are as follows.

Store 1	5.6	3.2	4.7	6.1	4.3	5.2	4.9	5.8	4.1	3.9
Store 2	2.8	4.1	3.0	3.6	3.3	3.8	4.5	3.4	3.7	2.5

Can we conclude, at $\alpha = .10$, that the fat content of the hamburger differs by store?

14.29 Refer to Exercise 14.28. Determine the median fat content for each store. Then subtract $\tilde{X}_1$ from each observation of store 1, and subtract $\tilde{X}_2$ from each observation of store 2.

Now test to decide if one store has more variability in the fat content of its hamburger than the other, using these deviations from the respective medians. Let $\alpha = .10$.

14.30 A survey was conducted to determine if viewers in a certain city believed that during late night shows on TV, the sound level is increased during commercials. In the survey, 100 people were selected at random from the telephone directory and contacted. Of the 100, 58 indicated they believed the sound to be higher during commercials, 10 did not watch the late show, and 32 felt that the sound was not higher. At $\alpha = .02$, can we conclude that more viewers believe that commercials are louder than the regular late night show?

14.31 Refer to Exercise 14.10. Suppose the people had been asked to rate each brand on a 1 to 5 scale (5 indicating very similar to butter), with each person's rating being identified. For example, the first individual might have given Farkay a 4 and Crowned a 2. Would the resulting data change your choice of a test procedure? Why or why not?

14.32 What is the difference between the Mann–Whitney procedure and the Wilcoxon signed-rank procedure for 2 samples?

14.33 The Dow-Jones Industrial Index is supposedly an indicator of market action. To attempt to determine if a relationship exists between the Dow-Jones Index and the stock of Pitcairn Industries, 15 Fridays were selected at random by researchers, and the Dow-Jones and the price of Pitcairn Industries for each Friday were recorded in the table.

Friday	*1*	*2*	*3*	*4*	*5*	*6*	*7*	*8*
Dow	756	872	794	923	869	1,005	798	815
Pitcairn	15.50	13.25	16.00	10.50	18.75	16.25	11.50	10.75

Friday	*9*	*10*	*11*	*12*	*13*	*14*	*15*
Dow	962	956	857	1,126	943	779	980
Pitcairn	14.00	14.25	16.50	9.75	12.25	11.25	8.50

Is there any evidence of a relation between the 2 at $\alpha = .10$?

***14.34** Prove that the mean and variance of Wilcoxon's T statistic are

$$\mu_T = E[T] = \frac{n(n+1)}{4} \quad \text{and} \quad \sigma_T^2 = \frac{n(n+1)(2n+1)}{24}$$

respectively. *Hint:*

$$\sum_{k=1}^{n} k = \frac{n(n+1)}{2} \quad \text{and} \quad \sum_{k=1}^{n} k^2 = \frac{n(n+1)(2n+1)}{6}$$

14.35 A particular brewery indicates on its labels that each bottle of New Sloshingfroth contains 7% alcohol. A random sample of 18 bottles yielded alcoholic contents (rounded to the nearest .1 of a percent) as follows:

6.1	6.5	7.2	6.9	7.0	7.1	6.8	6.6	6.7
7.4	6.4	7.2	7.3	6.3	7.5	7.0	6.0	6.2

If the distribution of alcoholic content can be assumed to be symmetric, can the brewery be accused of mislabeling its product? Use $\alpha = .10$.

---

* An asterisk indicates a higher level of difficulty.

14.36 The closing Dow-Jones averages for the past 12 weeks were

820.1	822.5	830.6	841.3	857.4	860.9
853.2	868.7	881.0	898.8	925.6	943.7

For $\alpha = .05$, can it be concluded that a trend in the market is evident?

14.37 What are the advantages and disadvantages of using distribution-free procedures?

***14.38** Construct a procedure for testing for identical populations, using a runs procedure with 2 independent random samples such that the procedure will be sensitive to differences in location.

14.39 Explain the differences between the sign procedure, Wilcoxon's signed-rank procedure, and the t test for paired data.

14.40 In an attempt to determine if a relation exists between grade point average (GPA) of business graduates and their starting salaries, researchers randomly selected 12 graduates and determined their GPAs and starting salaries. They are presented in the following table, with salaries in $100 per month.

Graduate	*1*	*2*	*3*	*4*	*5*	*6*
GPA	2.56	3.12	2.12	2.86	2.45	2.72
Salary	14.50	15.25	13.75	14.90	14.00	14.25

Graduate	*7*	*8*	*9*	*10*	*11*	*12*
GPA	3.57	2.36	2.50	2.28	3.26	2.94
Salary	16.00	13.75	14.35	13.50	15.50	15.75

Can it be concluded that higher grades command higher starting salaries at $\alpha = .01$?

14.41 It has been stated that there has been a definite increase in GPAs over the past few years. A random sample of 20 students is selected and asked their GPAs to the nearest 10th of a point. In 1965, it was reported that the median GPA was 2.42. The 20 sample GPAs are

2.5	1.5	2.9	2.3	3.1	2.8	2.5	2.6	3.8	2.4
2.5	2.7	3.0	2.2	2.6	2.5	2.8	3.3	1.9	3.5

Using $\alpha = .05$, can we conclude that GPAs have increased since 1965? It is believed that GPAs have a skewed distribution.

14.42 Harry's monthly food expenses, in dollars, for the past 18 months have been, successively,

192	207	198	215	222	241	212	226	252
247	239	269	254	258	235	249	260	268

Assuming that he purchased the same basic commodities in the same quantities each time, are these amounts an indication of rising food prices at $\alpha = .10$?

***14.43** Use the procedure developed in Exercise 14.38 to test the hypotheses in Exercise 14.28.

14.44 Refer to Exercise 14.41 and test the hypotheses of interest using the one-sample t test of Chapter 8. Compare your results to those of Exercise 14.41. What additional assumptions are required here?

***14.45** In an attempt to evaluate 2 marketing plans, researchers selected 24 stores in a certain geographical area. Twelve of these stores were randomly assigned the first plan; the remaining 12 were assigned the second plan. The average weekly sales over the same 1-month period were determined for each of the 24 stores and then ranked from low to high. The ranks assigned to sales from stores using the first plan were 4, 7, 8, 11, 13, 15, 16, 17, 19, 20, 23, 24. Does there appear to be any difference in median sales between the 2 plans at $\alpha = .10$?

***14.46** Does there appear to be any difference in the variabilities of the sales under the 2 plans in Exercise 14.45 at $\alpha = .10$?

***14.47** Refer to Exercises 14.45 and 14.46 and comment on the appropriateness of what you did in Exercise 14.46.

***14.48** For the situation described in Exercise 14.45, the 2-independent-samples t test of Chapter 8 might apply. Explain why it is totally inappropriate for Exercise 14.45 as currently stated and indicate what additional information would be necessary.

14.49 The management at Molson Products, Inc. has found that weekly demand for part #827114 has exceeded 12.1 units 50% of the time in the past. Over the last 10 weeks, demand for this part has been 15, 12, 13, 17, 14, 10, 16, 14, 15, 13. Are these numbers reason for management to think that demand for part #827114 has increased? Use $\alpha = .10$. You may refer to Exercise 3.41 for additional information if you wish.

***14.50** Management at Molson Products also believes that the weekly demand for part #006969 will be less than 5.9 units 75% of the time. For the past 15 weeks, the demand for part #006969 has been 7, 4, 5, 3, 6, 3, 3, 7, 4, 1, 7, 6, 4, 5, 7. Is there any reason to doubt management's belief at $\alpha = .15$?

REFERENCES

1. Bradley, James V. *Distribution-Free Statistical Tests*. Englewood Cliffs, NJ: Prentice-Hall, 1968.
2. Gibbons, Jean D. *Nonparametric Statistical Inference*. New York: McGraw-Hill, 1971.
3. Noether, Gottfried E. *Elements of Nonparametric Statistics:* New York: John Wiley, 1967.
4. Siegel, Sidney. *Nonparametric Statistics for the Behavioral Sciences*. New York: McGraw-Hill, 1956.

15 Chi-Square and *k*-Sample Distribution-Free Procedures

15.1 INTRODUCTION

Tami Moran is a member of the marketing research department of Barstow Products, Inc. Noni Wollis, the vice-president of marketing, asked Tami to determine whether an association exists between types of television commercials and the geographical area of the viewer with respect to product recognition. To do so, Tami prepared four commercials, each using a different style of presentation. All four commercials were aired on network television for one month, with the times of airing being randomized by commercial. At the end of one month, she randomly selected 100 viewers in each of four areas. These viewers were presented with the product and a verbal description of each commercial and asked to identify the commercial with which they most closely associated the product. The responses then were tabulated by commercial according to frequency of occurrence and geographical area. Before making her report to Noni, Tami must analyze a set of data that is obviously nonnormal.

None of the procedures from the earlier chapters apply in this situation, so what does Tami do? As will be seen shortly, her setup is essentially that of a cross-classification that can be analyzed by the chi-square distribution.

The preceding chapter examined the concepts of distribution-free tests for one- and two-sample situations and the various levels of measurement. In this chapter, we consider procedures that, to a great extent, are extensions of Chapter 14. The chi-square distribution is used for three tests requiring at least nominal level data. The Kruskal–Wallis procedure is an extension to k samples of the Mann–Whitney procedure (rank sum) and provides an alternative to the classical one-way analysis of variance procedure based on the F-distribution presented in Chapter 9. Friedman's procedure is the distribution-free analogue of the F test for a randomized block design and reduces to the sign procedure when only two treatments are being compared.

15.2 χ^2 GOODNESS OF FIT PROCEDURE

The procedure presented in this section provides a method for testing if the sample data could have come from some specified distribution. The procedure compares observed frequencies with frequencies that would be expected for the same categories if the null hypothesis is true. Basically, we extend the one-sample sign procedure to more than two categories and apply the Central Limit Theorem. The null hypothesis of interest is that the probability of an observation's being in any particular category is given by the distribution specified in H_0, which, in general, is in the form

$$H_0: F(x) = F_0(x) \qquad H_a: F(x) \neq F_0(x)$$

In words, the null hypothesis states that the sample data are from a population having the distribution identified by F_0. A more workable form of the null hypothesis can be stated by transforming H_0 into an equivalent statement about a multinomial distribution with π_{0i}, the probability of the ith category, determined by F_0. For example, suppose we have hypothesized that the sample came from a normal distribution with a mean of 100 and a standard deviation of 20. If we defined categories by (1) $X < 80$, (2) $80 \leq X < 100$, (3) $100 \leq X < 120$, (4) $X \geq 120$, then using Appendix Table B.6, we could rewrite H_0 in the equivalent form

$$H_0: \pi_1 = .1587 \quad \pi_2 = .3413 \quad \pi_3 = .3413 \quad \pi_4 = .1587$$

and simply test it as a multinomial distribution. Thus, we construct the test under the assumption that the null hypothesis is in the form

$$H_0: \pi_1 = \pi_{01}, \pi_2 = \pi_{02}, \ldots, \pi_k = \pi_{0k} \tag{15.1}$$

where π_{0i} is the probability that an observation will belong to the ith class and k equals the number of classes.

Let O_i denote the observed frequency of the ith class—that is, the number of sample observations belonging to it—and let $E_i = n\pi_{0i}$ (where $n = \Sigma O_i$, the sample size) be the expected frequency under H_0. Then, we wish to compare O_i with E_i for all k categories simultaneously. Since

$$\sum_{i=1}^{k} (O_i - E_i) = \sum_{i=1}^{k} O_i - \sum_{i=1}^{k} E_i = n - n = 0$$

is always true, we cannot use the sum of the deviations as a measure of the total deviation of the sample from H_0. As we did in defining a measure of dispersion, the deviations are squared before summing to obtain

$$\sum_{i=1}^{k} (O_i - E_i)^2 \tag{15.2}$$

This statistic equals zero only when $O_i = E_i$ for all i, indicating perfect agreement with H_0, and increases only with increasing discrepancies between the sample distribution and the distribution specified in the null hypothesis.

To construct the test, we need a statistic with a known or tabulated distribution. The statistic of Equation 15.2 does not meet that condition, but it can be transformed or standardized to obtain a random variable having a χ^2-distribution. For the binomial case, the transformation is easily shown by algebra and is left as a starred exercise for the interested student.

In the general case of k categories, we have that each

$$\frac{O_i - n\pi_{0i}}{\sqrt{n\pi_{0i}}}$$

is approximately a standard normal random variable if H_0 is true. Squaring and summing over all categories give an approximate chi-square random variable with $k - 1$ degrees of freedom. That is,

$$\sum_{i=1}^{k} \frac{(O_i - E_i)^2}{E_i} \sim \chi^2_{k-1} \tag{15.3}$$

One degree of freedom has been lost due to the restriction placed upon the O_i, namely, $\sum_{i=1}^{k} O_i = n$. From our previous considerations, small values for the left-hand side of Equation 15.3 would indicate agreement with H_0, and larger values tend to disagree.

Therefore, we can construct our test procedure under the following assumptions.

Assumptions for χ^2 Goodness of Fit Procedure

1. A random sample is selected.
2. The population is either infinite or the sampling is with replacement.
3. At least nominal (categorical) data are available.

The general form of the hypotheses is

$$H_0\colon F(x) = F_0(x) \qquad H_a\colon F(x) \neq F_0(x)$$

which are rewritten as

$$H_0\colon \pi_1 = \pi_{01}, \pi_2 = \pi_{02}, \ldots, \pi_k = \pi_{0k} \qquad H_a\colon \text{some } \pi_i \neq \pi_{0i} \tag{15.4}$$

where $\Sigma_{i=1}^{k} \pi_{0i} = 1$. With the foregoing assumptions, the test procedure follows the condensed form of the seven-step outline presented in Chapter 8.

1. State the hypotheses and specify α.
2. Define the categories to be used and select n such that $n\pi_{0i} \geq 5$ for all i ($n\pi_{0i} \geq 5$ is the condition for using the normal approximation to a binomial).
3. Select a random sample from the population being considered and count the number of observations O_i belonging to each category.
4. Calculate $\Sigma_{i=1}^{k} (O_i - E_i)^2/E_i$, where $E_i = n\pi_{0i}$.
5. Apply the following decision rule.

DECISION RULE Reject H_0 if

$$\sum_{i=1}^{k} \frac{(O_i - E_i)^2}{E_i} > \chi^2_{k-1,\alpha} \tag{15.5}$$

The following example illustrates this procedure.

EXAMPLE 15.1 Equal Employment Opportunity (EEO) has reported that the work force nationwide is composed of 10% professional, 10% clerical, 30% skilled, 15% service, and 35% semiskilled laborers. A random sample of 100 Akron residents indicated 15 professional, 15 clerical, 40 skilled, 10 service, and 20 semiskilled laborers. At $\alpha = .10$, does the work force in Akron appear to be consistent with the EEO report for the nation?

Solution: The hypotheses indicated by the EEO report are

$$H_0: \pi_1 = .10 \quad \pi_2 = .10 \quad \pi_3 = .30 \quad \pi_4 = .15 \quad \pi_5 = .35$$

and

$$H_a: \text{some } \pi_i \text{ differ from } \pi_{0i}$$

The significance level is specified to be .10. Thus, we will reject H_0 if

$$\sum_{i=1}^{k} \frac{(O_i - E_i)^2}{E_i} > \chi^2_{5-1,.10} = 7.78$$

It will be computationally advantageous to use the tabular format of Table 15.1 for calculating $\Sigma_{i=1}^{k} (O_i - E_i)^2/E_i$. Since $16.43 > 7.78$, we reject H_0 and conclude that Akron's work force appears to deviate from that of the nation as reported by EEO. Although professional, clerical, and skilled categories are somewhat overrepresented in Akron, the major deviation seems to be underrepresentation of semiskilled laborers.

TABLE 15.1 Table of calculations for Example 15.1

Category	O_i	$E_i = n\pi_{0i}$	$O_i - E_i$	$(O_i - E_i)^2$	$\frac{(O_i - E_i)^2}{E_i}$
1	15	10	5	25	2.5
2	15	10	5	25	2.5
3	40	30	10	100	3.33
4	10	15	−5	25	1.67
5	20	35	−15	225	6.43
	100	100	0		$16.43 = \sum_{i=1}^{5} \frac{(O_i - E_i)^2}{E_i}$

As indicated earlier, the χ^2 goodness of fit procedure can be used for any specific family of distributions. Some examples of distributions other than the multinomial are the normal (mentioned earlier), the Poisson, the uniform, and the negative exponential families. When testing a distribution other than a multinomial, we calculate the π_{0i}'s by using the specified distribution. The following example illustrates this situation and demonstrates how the normal probabilities given earlier were obtained from Appendix Table B.6.

EXAMPLE 15.2 It has been reported that salaries of college teachers are normally distributed with a mean of \$34,500 and a standard deviation of \$5000. Figure 15.1

FIGURE 15.1 Distribution of teachers' salaries under H_0 with the corresponding classes identified by number

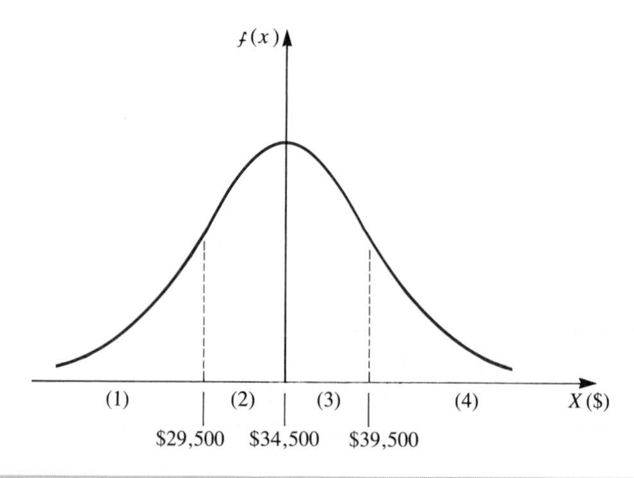

illustrates the distribution graphically, with the categories defined by salary ranges. With these categories, the following frequency table was obtained for a random sample of 100 teachers.

Salary Range	O_i = ***Frequency***
$X < \$29{,}500$	26
$\$29{,}500 \leq X < \$34{,}500$	40
$\$34{,}500 \leq X < \$39{,}500$	30
$X \geq \$39{,}500$	4
	100

On the basis of this sample, can it be concluded that salaries are distributed as specified for $\alpha = .05$?

Solution: The report specifies the distribution to be normal with a mean of $34,500 and a standard deviation of $5000. Thus, the hypotheses are

$$H_0: X \sim N(\$34{,}500; \$5000) \qquad H_a: X \nsim N(\$34{,}500; \$5000)$$

To restate the hypotheses in the multinomial form, we need to calculate the π_{0i}'s.

$$\begin{aligned}\pi_{01} &= P(X < \$29{,}500 | \mu = \$34{,}500,\ \sigma = \$5000)\\ &= P\left(Z < \frac{\$29{,}500 - \$34{,}500}{\$5000}\right)\\ &= P(Z < -1.00) = .5 - .3413 = .1587\end{aligned}$$

$$\begin{aligned}\pi_{02} &= P(\$29{,}500 \leq X < \$34{,}500 | \$34{,}500,\ \$5000)\\ &= P\left(-1 \leq Z < \frac{\$34{,}500 - \$34{,}500}{\$5000}\right)\\ &= P(-1 \leq Z < 0) = .3413\end{aligned}$$

and, similarly, $\pi_{03} = .3413$, $\pi_{04} = .1587$. Thus, the restatement of H_0, H_a is

$$H_0\colon \pi_1 = .1587,\ \pi_2 = .3413,\ \pi_3 = .3413,\ \pi_4 = .1587$$

$$H_a\colon \text{some } \pi_i \neq \pi_{0i}$$

and we will reject H_0 if

$$\sum_{i=1}^{4} \frac{(O_i - E_i)^2}{E_i} > \chi^2_{4-1,.05} = 7.81$$

With the tabular format, the calculations are presented in Table 15.2.

Since 16.86 exceeds 7.81, we reject H_0 and conclude that teachers' salaries do not appear to be distributed as $N(\$34{,}500; \$5000)$. The source of

TABLE 15.2 Calculations for Example 15.2

Category	O_i	E_i	$O_i - E_i$	$(O_i - E_i)^2$	$\frac{(O_i - E_i)^2}{E_i}$
1	26	15.87	10.13	102.62	6.47
2	40	34.13	5.87	34.46	1.01
3	30	34.13	−4.13	17.06	.50
4	4	15.87	−11.87	140.90	8.88
	100	100.00	0.00		16.86

discrepancy lies in the 2 tails. There are many more salaries than expected in the low end and many fewer in the upper end, indicating possible skewness to the right.

Note that in Example 15.2 we did not conclude that a normal distribution was inappropriate but that the specific normal distribution appeared to be incorrect. It is possible to specify a family of distributions or a subset of a family in H_0 rather than a specific member. By subset of a family, we refer to the situation involving multiparameter families where some parameters are specified and some are not. For the normal family, both μ and σ, μ but not σ, σ but not μ, or neither may be specified in H_0. Any unspecified parameters of the hypothesized family must be estimated from the sample data so that the π_{0i}'s may be determined. For example, not specifying μ_X and σ_X for the normal family requires the use of $\overline{X}$ and s_X for μ_X and σ_X, whereas not specifying μ_X in the Poisson family only requires the use of $\overline{X}$ for μ_X. When parameters must be estimated from the sample, there is a resulting loss of precision in the test. This loss is taken into account by a reduction of one additional degree of freedom *for each parameter estimated.* That is, the degrees of freedom for such a test are given by

$$\text{df} = (k - 1) - \text{number of estimated parameters} \qquad \textbf{(15.6)}$$

In Equation 15.6, the value to be used for k is the number of classes used in the frequency table after any necessary adjustments are made to ensure that $n\pi_{0i} \geq 5$ for all i. For example, if an original frequency table had six classes, with the fifth and sixth classes being $120 \leq X < 150$ and $X \geq 150$, n equal to 100 and π_{05}, π_{06} calculated to be .1032 and .0091, respectively, then $n\pi_{06} = .91$. Since $E_6 = n\pi_{06} = .91 < 5$, we would combine the fifth and sixth classes into $X \geq 120$, bringing π_{05} to .1123, with E_5 becoming 11.23. With this combining of classes, k becomes 5 rather than the 6 with which we started.

EXERCISES

15.1 It has been reported that the distribution of grades in higher education today is 20% A's, 30% B's, 35% C's, 10% D's, and 5% F's or W's. In last term's statistics class of 250 students, there were 30 A's, 55 B's, 105 C's, and 35 D's; 25 either failed or withdrew. Could the grade distribution for the statistics class have come from the same distribution as that reported for higher education? Use $\alpha = .005$.

15.2 In an advertising experiment, customers were offered free samples of an item. The item was packaged in 4 different colors; otherwise, everything else was the same and each person could select the color of his or her choice. Of 160 people who took samples, 55 chose red, 42 chose blue, 34 chose yellow, and 29 chose brown. Does the color of the package appear to have any effect on choice? Use $\alpha = .05$.

15.3 In a chrome-plating process, it is believed that the number of defects per unit follows a Poisson distribution. A sample of 100 such units exhibited the following frequency distribution for the number of defects per unit.

Number of Defects Per Unit	*Number of Units*
$0 \leq X \leq 2$	32
$3 \leq X \leq 5$	24
$6 \leq X \leq 8$	20
$9 \leq X \leq 11$	12
$12 \leq X \leq 14$	8
More than 14	4

Is the belief supported at $\alpha = .01$?

15.3 TEST OF SEVERAL BINOMIALS

In this section, we consider an analogue of the ANOVA F test for a randomized design when the populations being considered are dichotomous. That is, we wish to evaluate whether several binomial populations or Bernoulli processes have the same parameter π, the probability of a success on a single trial. One example is the comparison of several suppliers with respect to their proportions of defectives. The assumptions of the procedure are similar to those of the previous section and are restated here.

1. Independent random samples are selected from each of c populations.
2. The populations are very large or sampling is with replacement.
3. The level of measurement is at least nominal so that the data may be dichotomized.

We define the following notation. Let

$n_{\cdot j}$ = size of the sample from the jth population

$n_{i\cdot}$ = total number of items of type i in the combined sample

$$n = \sum_{j=1}^{c} n_{\cdot j} = \sum_{i=1}^{2} n_{i\cdot} = \text{total number of observations}$$

and

O_{ij} = number of observations in the jth sample belonging to category i, $i = 1, 2$

Then, under these assumptions, each sample of size $n_{\cdot j}$ constitutes a random sample from a binomial distribution. Each observation from the jth population has a probability of π_{1j} of belonging to the first category and a probability of $1 - \pi_{1j}$ of belonging to the second category. To test if the c populations are identical, we state the hypotheses of interest as

$$H_0\colon \pi_{11} = \pi_{12} = \cdots = \pi_{1c} \qquad H_a\colon \text{some } \pi_{1j} \text{ differ} \tag{15.7}$$

The rationale for constructing the test is identical to that presented for a goodness of fit procedure. Thus, the statistic

$$\sum_{i=1}^{2}\sum_{j=1}^{c} \frac{(O_{ij} - E_{ij})^2}{E_{ij}}$$

has an approximate χ^2-distribution with $c - 1$ degrees of freedom. In this situation, the E_{ij} are determined by first estimating the common probability of category i, assuming H_0 is true, and then multiplying by the number of observations from population j since the expected value of a binomial variable is $n\pi$. This determination is summarized in the equation

$$E_{ij} = n_{\cdot j}\left(\frac{n_{i\cdot}}{n}\right) = \frac{n_{\cdot j}n_{i\cdot}}{n} \tag{15.8}$$

We reason that if the populations are identical, the best estimate of π_i is given by the number of times category i occurs divided by the total number of trials, as indicated in Chapter 7. The corresponding test procedure for the hypotheses of Equation 15.7 is as follows.

1. State the hypotheses of interest and select α.
2. Select random samples from each of the c populations being considered and count the number of observations of each type in each sample.

For convenience, organize the data in a tabular format, with populations as columns and categories as rows—that is, a $2 \times c$ table.

3. Calculate the E_{ij}'s to check that all are greater than or equal to 5. If any E_{ij}'s are less than 5, larger samples should be selected.

4. Calculate

$$\sum_{i=1}^{2}\sum_{j=1}^{c}\frac{(O_{ij}-E_{ij})^2}{E_{ij}}$$

5. Apply the following decision rule.

DECISION RULE Reject H_0 at α if

$$\sum_{i=1}^{2}\sum_{j=1}^{c}\frac{(O_{ij}-E_{ij})^2}{E_{ij}} > \chi^2_{c-1,\alpha} \tag{15.9}$$

EXAMPLE 15.3 In an effort to determine if a difference exists in the proportions of defective items produced by 3 shifts, the Nite Lite Company randomly selected at various times 120 items during the first shift, 150 during the second shift, and 130 during the third shift. All items were inspected, and the number of defectives was recorded by shift with the following results: 8 on the first, 24 on the second, and 18 on the third. Can it be concluded at $\alpha = .05$ that the proportions differ?

Solution: The null hypothesis is that the proportion of defectives is the same for all shifts—that is,

$$H_0\colon \pi_{D1} = \pi_{D2} = \pi_{D3} = \pi_D \qquad H_a\colon \text{some } \pi_{Dj} \neq \pi_D$$

For clarity the data are reorganized in a 2 × 3 table showing the number of good items, the number of defective items, the shift during which they were produced, and the totals.

	Shift			
	1	***2***	***3***	$n_{i\cdot}$
Defective	8	24	18	50
Good	112	126	112	350
$n_{\cdot j}$	120	150	130	400 = n

Let ij denote the cell or category in the table; the calculations are shown in Table 15.3 [for example, $E_{11} = 120\,(50/400) = 15$ and $E_{22} = 150\,(350/400) = 131.25$]. We have 2 degrees of freedom since, if numbers are specified for any 2 cells in the table, the remaining values are automatically determined by the marginal totals. Thus, the decision rule implies rejection of H_0 if

$$\sum_{i=1}^{2}\sum_{j=1}^{c}\frac{(O_{ij}-E_{ij})^2}{E_{ij}} > \chi^2_{c-1,\alpha} = \chi^2_{2,.05} = 5.99$$

TABLE 15.3 Calculations for Example 15.3

ij	O_{ij}	E_{ij}	$O_{ij} - E_{ij}$	$(O_{ij} - E_{ij})^2$	$\frac{(O_{ij} - E_{ij})^2}{E_{ij}}$
11	8	15.00	−7.00	49.0000	3.27
12	24	18.75	5.25	27.5625	1.47
13	18	16.25	1.75	3.0625	.19
21	112	105.00	7.00	49.0000	.47
22	126	131.25	−5.25	27.5625	.21
23	112	113.75	−1.75	3.0625	.03
	400	400.00	0.00		5.64

For this problem,

$$\sum_i \sum_j \frac{(O_{ij} - E_{ij})^2}{E_{ij}} = 5.64$$

which is less than 5.99. Therefore, we accept H_0 and conclude that the differences in the proportions of defectives between shifts appear to be due to chance variation.

EXERCISES

15.4 In a comparison of the proportion of defectives produced by 4 different machines, researchers obtained the following data. Machine A had 25 defectives in 200 items, B had 10 in 120 items, C had 6 in 140 items, and D had 9 in 160 items. Does the quality of the items produced appear to be the same for all machines? Let $\alpha = .05$.

15.5 A market research department desired to test whether the type of advertising display appeals differently to people in different socioeconomic areas. For its experiment, the market research department selected 1 store in each of 5 different socioeconomic areas in a particular city. Then, it placed the same product on 2 distinctly different displays situated next to each other. To minimize any effect due to location, it alternated the positions of the displays each day. After 6 days, the department tabulated the number of units sold from each display for each store. The data are presented in the following table.

	Store				
Display	***1***	***2***	***3***	***4***	***5***
1	152	112	56	37	63
2	94	108	79	45	137

Do the displays appear to appeal differently to different people? Let $\alpha = .01$.

15.4 TEST FOR SEVERAL MULTINOMIALS (CONTINGENCY TABLES)

It is logical to extend the analysis of the previous section to include situations where it is possible for each item from a given population to assume any one of r categories ($r \geq 2$). We then wish to test to see if the probability of an observation's belonging to any particular category is the same for all populations. That is, we wish to test the hypothesis that c multinomial populations, each having r categories, are identical. In parametric form, the hypotheses of interest are

$$H_0: \begin{cases} \pi_{11} = \pi_{12} = \cdots = \pi_{1c} = \pi_1 \\ \pi_{21} = \pi_{22} = \cdots = \pi_{2c} = \pi_2 \\ \vdots \qquad \cdots \\ \pi_{r1} = \pi_{r2} = \cdots = \pi_{rc} = \pi_r \end{cases} \qquad H_a\text{: at least two equations do not hold} \qquad \textbf{(15.10)}$$

It will be simpler to state these hypotheses in terms of association between categories and populations or equivalently between row and column classifications. The restatement is

$$\begin{aligned} &H_0\text{: no association between row and column variables} \\ &H_a\text{: some association} \end{aligned} \qquad \textbf{(15.11)}$$

It can be seen that the statements in Equations 15.10 and 15.11 are equivalent. If the populations are identical, then differences between cells in any one row would be due only to random sampling and differences in the sample sizes. That is, knowing from which population an item comes does not alter its probability of belonging to any one category. Symbolically,

$$P(X \text{ belongs to category } i | \text{population } j) = P(X \text{ belongs to category } i)$$

which is the definition of independence or no association. The organization of the data is presented in Table 15.4. This representation is also called a contingency table because independence implies that the likelihood of a category is not contingent on the population. Note the similarity to the randomized block design of the two-way analysis of variance presented in Chapter 9. The test presented here is a distribution-free alternative to the two-way ANOVA F test when the data are nominal.

The test of the hypotheses in Equations 15.10 and 15.11 is based on the χ^2-distribution and follows the same intuitive derivation as the previous χ^2 procedures. The underlying assumptions of the procedure are as follows.

1. Independent random samples are selected from each of the c populations.

TABLE 15.4 Representation of data for a contingency table with O_{ij} denoting the observed frequency for the ijth cell

	Population					
Category	*1*	*2*	*3*	...	*c*	*Row Total*
1	O_{11}	O_{12}	O_{13}	...	O_{1c}	$n_{1\cdot}$
2	O_{21}	O_{22}	O_{23}	...	O_{2c}	$n_{2\cdot}$
3	O_{31}	O_{32}	O_{33}	...	O_{3c}	$n_{3\cdot}$
⋮	⋮	⋮	⋮	...	⋮	⋮
r	O_{r1}	O_{r2}	O_{r3}	...	O_{rc}	$n_{r\cdot}$
Column Total	$n_{\cdot 1}$	$n_{\cdot 2}$	$n_{\cdot 3}$	...	$n_{\cdot c}$	$n = \sum_{i=1}^{r} n_{i\cdot} = \sum_{j=1}^{c} n_{\cdot j}$

or **1′.** One large sample of size n is selected and the items classified by population and category.

2. The populations are very large or else sampling is with replacement.

3. The data are at least nominal and can be classified into r mutually exclusive categories.

The difference between assumptions (1) and (1′) is the distinction that is generally made between the test of several multinomials (Equation 15.10), which is also known as a test of homogeneity, and the test for association or dependency in a contingency table (Equation 15.11). The procedure of analysis is the same in either case.

Then, if the foregoing assumptions are satisfied, the statistic

$$\sum_{i=1}^{r} \sum_{j=1}^{c} \frac{(O_{ij} - E_{ij})^2}{E_{ij}}$$

will be approximately distributed as a χ^2 random variable with $(r - 1)(c - 1)$ degrees of freedom. This number is the number of cells to which frequencies may arbitrarily be assigned before all remaining cells become automatically determined by the marginal totals. Under the null hypothesis, the best estimate of π_i is $n_{i\cdot}/n$—that is, the number of items belonging to the ith category divided by the total number of observations. Thus, the expected frequency for the ijth cell is given by

$$E_{ij} = n_{\cdot j}\, p_i = n_{\cdot j}\left(\frac{n_{i\cdot}}{n}\right) \tag{15.12}$$

Following the previous pattern, the test procedure for the hypotheses of either Equation 15.10 or Equation 15.11 is as follows.

1. State the hypotheses of interest (in either form) and select a value for α.
2. Select random samples from each of the c populations being compared and determine O_{ij}, the observed frequency in cell ij.

or 2′. Select one large random sample from the combined populations and determine O_{ij}.

3. Calculate the E_{ij}'s to check that all are greater than or equal to 5. If any E_{ij} are less than 5, either select a larger sample from the appropriate population(s) or combine adjacent categories—that is, rows and/or columns—until the condition is satisfied.
4. Calculate

$$\sum_{i=1}^{r}\sum_{j=1}^{c}\frac{(O_{ij}-E_{ij})^2}{E_{ij}}$$

5. Apply the following decision rule.

DECISION RULE Reject H_0 at α if

$$\sum_{i=1}^{r}\sum_{j=1}^{c}\frac{(O_{ij}-E_{ij})^2}{E_{ij}} > \chi^2_{(r-1)(c-1),\alpha} \tag{15.13}$$

The analysis of an $r \times c$ table is illustrated in the following example.

EXAMPLE 15.4 In a civil rights compliance review of 3 large companies in the same industry and city, the review team randomly selected 100 cards from the personnel files of each company. They then classified each sample in terms of whether the card corresponded to a white male, black male, or female. The resulting data are summarized in the following table.

	Company			
	1	*2*	*3*	*Total*
Females	16	8	27	51
Black Males	36	33	24	93
White Males	48	59	49	156
Total	100	100	100	300

Do the companies appear to differ in the composition of their employees? Use $\alpha = .025$.

Solution: For the form indicated in Equation 15.11, the hypotheses are

H_0: composition of employees is independent of company

H_a: composition is not independent of company

and α is specified to be .025. Since none of the observed values are less than 5 and the $n_{.j}$'s are all equal, neither will any of the expected values be less than 5. (Note that if any O_{ij} had been <5, we would have calculated the E_{ij} at this point.) Thus, our decision rule is to reject H_0 at $\alpha = .025$ if

$$\sum_{i=1}^{3}\sum_{j=1}^{3}\frac{(O_{ij}-E_{ij})^2}{E_{ij}} > \chi^2_{(3-1)(3-1),.025} = 11.14$$

The calculations are presented in Table 15.5. Since 14.64 exceeds 11.14, H_0 is rejected at $\alpha = .025$. We conclude that the percentages of employees in each category appear to differ by company. The major discrepancy occurs with females in company 2 being many fewer than expected, whereas company 3 has many more than expected.

In all three situations involving the chi-square family of distributions, we stipulated that all expected frequencies were to be greater than or equal to 5. The condition is particularly critical for the tests of Sections 15.2 and 15.3. In the present section, this condition is occasionally relaxed to the extent that some of the E_{ij}'s may be less than 5 ("some" is often interpreted to mean less than 10%), but absolutely none may be less than 1. The approach of combining adjacent categories is illustrated in Example 15.5.

TABLE 15.5 Calculations for Example 15.4

ij	O_{ij}	E_{ij}	$O_{ij} - E_{ij}$	$(O_{ij} - E_{ij})^2$	$\frac{(O_{ij} - E_{ij})^2}{E_{ij}}$
11	16	17	−1	1	.06
12	8	17	−9	81	4.76
13	27	17	10	100	5.88
21	36	31	5	25	.81
22	33	31	2	4	.13
23	24	31	−7	49	1.58
31	48	52	−4	16	.31
32	59	52	7	49	.94
33	49	52	−3	9	.17
	300	300	0		14.64

TABLE 15.6 Classification of freshman students by year and intended major

	Year			
Area of Major	*1969*	*1977*	*1985*	*Total*
Business	17	24	34	75
Education	26	20	14	60
Arts	32	26	20	78
Science & Engineering	23	26	26	75
Nursing	2	4	6	12
Total	100	100	100	300

EXAMPLE 15.5 In an attempt to determine if the choice of academic programs has changed over the years, the Registrar's Office at City U randomly selected 100 freshman registrations for each of the years in question and recorded the registrants' intended majors. These data are given in Table 15.6. Is there any indication of a relationship between year of registration and the area chosen? Use $\alpha = .10$.

Solution: The hypotheses in question are

H_0: no association between year and area of major

H_a: some association

and α is specified at .10. Checking the E_{ij}'s first, we find $E_{5j} = 4$ for all 3 years since $E_{5j} = 12(100/300)$. This value is not critically low, but we will regroup for illustrative purposes anyway. Nursing is probably most closely related to the sciences and engineering category. Thus, we will combine those two categories to obtain the revised data in Table 15.7. Our decision rule will be to reject H_0 if

TABLE 15.7 Results of combining science and engineering with nursing in Table 15.6

	Year			
Area of Major	*1969*	*1977*	*1985*	*Total*
Business	17	24	34	75
Education	26	20	14	60
Arts	32	26	20	78
Science, Engineering, Nursing	25	30	32	87
Total	100	100	100	300

TABLE 15.8 Calculations for Example 15.5

Cell	O_{ij}	E_{ij}	$O_{ij} - E_{ij}$	$\frac{(O_{ij} - E_{ij})^2}{E_{ij}}$
1,1	17	25	−8	2.560
1,2	24	25	−1	.040
1,3	34	25	9	3.240
2,1	26	20	6	1.800
2,2	20	20	0	.000
.	.	.	.	.
.	.	.	.	.
.	.	.	.	.
4,3	32	29	3	.310
				13.106

$\chi^2_{\text{calc}} > \chi^2_{6,.10} = 10.64$. Since the sample sizes are the same for all years, each E_{ij} will be given by $n_{i\cdot}/3$. The calculations are shown in Table 15.8.

Since $13.106 > 10.64$, we reject H_0 and conclude that the choice of a major area appears to be related to the year of registration. From 1969 to 1985, business and sciences have increased in intended majors, and arts and education have decreased. ■

EXERCISES

15.6 In an attempt to analyze whether the percent return on stocks is related to type of industry, an investor randomly selected 50 stocks each from the utilities, rubber and chemical products, aerospace, steel, automotive, and fuel industries. He then classified the stocks from each industry by the percent return on investment. This breakdown is shown in the following table.

	Industry					
Percent ROI	*Utilities*	*R and C*	*Aerospace*	*Steel*	*Auto*	*Fuel*
0 to 4	4	26	20	15	18	10
4 to 8	29	19	28	27	19	24
8 or higher	17	5	2	8	13	16

Does it appear that a relationship exists at $\alpha = .01$?

15.7 The Cleveland Transit System (CTS) wishes to analyze customer attitudes toward its service. To do so, it randomly selected people from various neighborhoods and asked questions about the service in that area. One of the questions asked the individuals to indicate whether they believed the buses were consistently on time, 1 to 10 minutes late, or more than 10 minutes late. CTS then categorized the responses by neighborhoods and obtained the following table. Each table entry indicates the number of people belonging to that cell.

Lateness	*East Side*	*West Side*	*South Side*	*Downtown*
On time	12	18	22	25
1–10	28	36	26	14
>10	15	5	9	6

If CTS uses an α of .05, will it conclude that a customer's belief about how well buses maintain their schedules is contingent on the area of residence?

15.5 KRUSKAL–WALLIS PROCEDURE

In Chapter 14, we looked at the Mann–Whitney procedure for unequal location parameters using two independent samples. The Kruskal–Wallis procedure is an extension of the Mann–Whitney procedure to cover situations involving c independent random samples. Thus, it provides a distribution-free alternative to the F test for the one-way design presented in Chapter 9. As is the case with the U test, the Kruskal–Wallis procedure is a test of identical distributions and is particularly sensitive to location differences. For $c = 2$, the test is equivalent to the rank sum form of the Mann–Whitney procedure.

Except for the obvious extensions to cover c populations, the assumptions of the procedure are identical to those listed in Section 14.6. For ease in reference, they are restated here with the indicated changes.

Assumptions of the Kruskal–Wallis Procedure

1. Independent random samples of size n_j are selected from each of the c populations of interest (or are selected before application of the c treatments).
2. The populations are infinite or sampling is with replacement.
3. The underlying distributions are identical with respect to shape and scale.
4. No ties occur between observations in different samples (may be relaxed somewhat).
5. The level of measurement is at least ordinal.

Under these assumptions, the hypotheses to be tested are

$$H_0\colon \tilde{X}_1 = \tilde{X}_2 = \cdot \cdot \cdot = \tilde{X}_c \qquad H_a\colon \text{some } \tilde{X}_j \text{ differ} \tag{15.14}$$

Then if H_0 is true, the distribution of the Kruskal–Wallis statistic H can be determined by considering all possible permutations of the ranks of the r observations in the combined sample. To calculate H, we first rank all observations from 1 to n by increasing order of magnitude and calculate T_j, the sum of the ranks corresponding to the observations in the jth sample, for all $j = 1, 2, \ldots,$

c. Then, H is determined by the equation

$$H = \frac{12}{n(n+1)} \sum_{j=1}^{c} n_j \left(\overline{R}_j - \frac{n+1}{2} \right)^2 \tag{15.15}$$

where: n_j is the number of observations in the jth sample
n is the total number of observations and equals $\sum_{j=1}^{c} n_j$
$\overline{R}_j = T_j/n_j$

The quantity $(n + 1)/2$ in the squared term of Equation 15.15 is the average of the ranks for the combined sample. If H_0 is true, each $\overline{R}_j$ has an expected value equal to $(n + 1)/2$. Thus, since H will equal zero only when all $\overline{R}_j = (n + 1)/2$, a value of zero for H indicates perfect agreement between the null hypothesis and the sample data. Conversely, large values for H will occur only when some of the $\overline{R}_j$'s differ significantly from $(n + 1)/2$, which happens if some samples receive a disproportionate share of either the high ranks or the low ranks. The occurrence of this situation would indicate that the observations corresponding to the aforesaid samples are grouped at the extremes of the array, which implies rejection of the null hypothesis of equal medians. In summary, our rejection region will correspond only to large values of the statistic H.

Equation 15.15 can be shown to be algebraically equivalent to

$$H = -3(n+1) + \left(\frac{12}{n(n+1)} \sum_{j=1}^{c} \frac{T_j^2}{n_j} \right) \tag{15.16}$$

This equation is known as the computational form for H and involves fewer arithmetic operations.

Tables for the exact distribution of the H statistic are limited and exist only for very small sample sizes. For cases with $c = 3$ and all $n_j \leq 5$, critical values corresponding to an α of .1, .05, or .01 can be found in Owen [4]. In other situations, or if Owen's tables are not available, it has been proved that the distribution of H can be approximated with a χ^2-distribution having $c - 1$ degrees of freedom. As the sample sizes increase, the approximation improves and becomes exact in the limiting case.

As usual, we summarize the foregoing procedure and the test in the following steps.

1. State the hypotheses to be tested and select α.
2. Select random samples of size n_j from each of the c populations under consideration or select a random sample of size n from a single population and randomly assign n_j items to receive the jth treatment, $j = 1, 2, \ldots, c$.
3. Construct an array of the combined set of observations and replace each observation with its corresponding position in the array—that is, its rank.
4. Calculate H using Equation 15.16.
5. Apply the following decision rule.

DECISION RULE For hypotheses of the form $H_0: \tilde{X}_1 = \tilde{X}_2 = \cdots = \tilde{X}_c$; H_a: some $\tilde{X}_j$ differ, reject H_0 at α if

$$H > \chi^2_{c-1,\alpha} \tag{15.17}$$

where $\chi^2_{c-1,\alpha}$ is found in Appendix Table B.8.

We illustrate the foregoing procedure with an example.

EXAMPLE 15.6 A professional organization is attempting to determine whether academic salaries differ by geographical area. Six associate professors were randomly selected from each of the following areas of the United States: Northeast, Southeast, Northwest, and Southwest. Each professor was then asked his or her salary for the academic year, with the results being tabulated in thousands of dollars. Some professors refused to disclose their salaries, and the corresponding sample sizes were reduced accordingly. Note that refusals may affect the results and that conclusions reached after reducing the sample sizes may be questionable. We proceed with these words of caution in mind.

Northeast	*Southeast*	*Northwest*	*Southwest*
38.3	37.2	34.1	36.7
37.6	35.3	35.2	35.9
40.1	36.9	36.5	37.4
39.5	34.5	35.7	35.8
39.0	35.1		38.0
	36.4		

For $\alpha = .05$, can it be concluded that salaries of associate professors differ by geographical area?

Solution: The hypotheses indicated by the question imply a test of location. For the median, the hypotheses are

$$H_0: \tilde{X}_{NE} = \tilde{X}_{SE} = \tilde{X}_{NW} = \tilde{X}_{SW} \qquad H_a: \text{some differ}$$

with α specified at .05. Thus, we will reject H_0 if $H > \chi^2_{3,.05} = 7.81$. Placing the data into an array, we have, with areas identified,

34.1	34.5	35.1	35.2	35.3	35.7	35.8
NW	SE	SE	NW	SE	NW	SW
35.9	36.4	36.5	36.7	36.9	37.2	37.4
SW	SE	NW	SW	SE	SE	SW
37.6	38.0	38.3	39.0	39.5	40.1	
NE	SW	NE	NE	NE	NE	

Replacing each observation in the original table with its position in the array, we obtain the table of ranks and the corresponding rank sum for each sample.

	Northeast	*Southeast*	*Northwest*	*Southwest*
	17	13	1	11
	15	5	4	8
	20	12	10	14
	19	2	6	7
	18	3		16
		9		
T_j:	89	44	21	56

Squaring the rank sums gives

$$T_{\text{NE}}^2 = 7921 \quad T_{\text{SE}}^2 = 1936 \quad T_{\text{NW}}^2 = 441$$

$$T_{\text{SW}}^2 = 3136 \qquad \text{with } n = \sum n_j = 5 + 6 + 4 + 5 = 20$$

Then, from Equation 15.16,

$$\begin{aligned} H &= -3(n+1) + \frac{12}{n(n+1)} \sum_{j=1}^{c} \frac{T_j^2}{n_j} \\ &= -3(21) + \frac{12}{20(21)} \left(\frac{7921}{5} + \frac{1936}{6} + \frac{441}{4} + \frac{3136}{5} \right) \\ &= -63 + \frac{12}{420}(1584.20 + 322.67 + 110.25 + 627.20) \\ &= -63 + \frac{12}{420}(2644.32) = -63 + 75.522 \\ &= 12.552 \end{aligned}$$

Since $12.552 > 7.81$, we reject H_0 and conclude that median salaries for associate professors appear to differ by geographical area, although we realize that the refusals may have affected the results. Based on these data, we observe that salaries appear to be highest in the Northeast, followed by the Southwest and Southeast, with the Northwest being lowest, though the latter 2 are probably not significantly different.

Our assumptions were such that, theoretically, ties should not occur. In practice, they do. If there are ties in the data, we will use the average rank procedure of Chapter 14. Ties within the same sample will not affect the value of H or its distribution, but ties between samples will reduce the value of H, altering its distribution, and will make the procedure less sensitive to actual population differences. If numerous between-sample ties occur, it is recommended that the user dichotomize the combined sample in terms of each observation's being above or below the overall median and construct a $2 \times c$ contingency table. If the sample sizes are large enough so that all $E_{ij} \geq 5$, the hypotheses may be tested by the procedure of Section 15.3.

The AREs of the Kruskal–Wallis procedure relative to the F test for a one-way design are identical to those of the Mann–Whitney procedure with respect to the t test for two independent samples. In specific situations, the ARE can be as high as infinity, but it is never lower than .864. If the populations are normally distributed, the ARE is .955, and if the distributions are uniform, it is 1.000.

The Kruskal–Wallis procedure can be adjusted to test H_0 versus one-sided alternatives and has been generalized to handle factorial designs testing both main effects and interactions. For a thorough treatment of the appropriate changes in the procedure, see Bradley [1].

EXERCISES

15.8 In evaluating salaries for a particular job, the hourly rates for that job in cities of varying sizes were determined. The table shows the hourly rate for each city in the random sample as classified by the size of the city.

Size of City (in 1000 persons)				
0–5	*5–10*	*10–50*	*50–250*	*over 250*
5.89	5.94	6.47	6.88	6.72
6.22	6.32	6.26	6.56	7.10
5.96	6.17	6.39	6.65	6.98
6.15	5.92	6.18	6.70	6.84
6.06	6.12	6.44	6.52	7.08

Can we conclude that the size of the city has an effect on hourly rates for this job? Use $\alpha = .05$.

15.9 City U gives its entering MBA candidates a multiple-choice exam on introductory statistics to decide if a student should be exempted from the first course in quantitative analysis. A random sample of 20 students showed the following scores on the exam, with each identified as to the student's undergraduate major.

Business	*Science and Engineering*	*Education*	*Liberal Arts*
72	69	62	29
48	92	31	40
64	77	47	59
83	86	26	36
51	58	50	42

Can it be concluded that the student's preparation in statistics differs by the area of his or her undergraduate major? Use $\alpha = .05$.

15.6 FRIEDMAN'S PROCEDURE

In using the sign procedure for paired data (Chapter 14), we calculated the within-pair differences and counted the resulting number of positive differences. Equivalently, we could have ranked the data within each pair, assigning a 1 to the smaller value and a 2 to the larger value, and used as a statistic the squared difference between the rank sums of the two treatments. Friedman's procedure uses a direct extension of the latter procedure to two or more treatments. It is equivalent to the sign procedure when only two treatments are being considered.

Friedman's procedure consists of ranking the observations within each block (matching set) and adding the ranks associated with each treatment. Then, the test statistic S is calculated by comparing the rank sums for each treatment with the average rank sum. The statistic is

$$S = \sum_{j=1}^{c} \left(T_j - \frac{n(c + 1)}{2} \right)^2 \qquad \textbf{(15.18)}$$

and will have a randomization distribution—that is, determined by considering the results associated with all possible permutations of the c ranks within each matched set, provided the following assumptions are satisfied.

Assumptions for Friedman's Procedure

1. There are available n sets of c objects each, with the elements in each set being similar before treatments.
2. The assignment of treatments to objects within a set is random or, if the same unit receives all treatments, there is no carryover effect. The latter part of the assumption can be approximately satisfied by administering the treatments in a random order.
3. There are no ties within sets (blocks) after treatments (may be relaxed).
4. The level of measurement within each set (block) is at least ordinal.

If the null hypothesis of no treatment effects, $H_0\colon \tilde{X}_1 = \tilde{X}_2 = \cdots = \tilde{X}_c$, is true, then the ranks within each row should occur randomly. Hence, the treatment rank sums, T_j, would be random also and should equal the average rank sum. This average is given by $[n(c + 1)]/2$ since the sum of the ranks, 1 to c, within each row is $[c(c + 1)]/2$, the number of rows equals n, and the total, $[nc(c + 1)]/2$, is divided by the number of treatments c. If $S = 0$, T_j equals $[n(c + 1)]/2$ for all j, and there is perfect agreement with the null hypothesis. Conversely, if, for example, one treatment yields appreciably higher responses, say treatment 1, then it will receive all of the high ranks within each row. The high ranks will be c's, and the corresponding sum would equal nc. The resultant value for S would be larger than should be expected if H_0 is true. Thus, the null hypothesis is rejected for large values of S.

Equation 15.18 is algebraically equivalent to

$$S = \sum_{j=1}^{c} T_j^2 - \frac{c}{4}[n(c+1)]^2 \qquad \textbf{(15.19)}$$

where: c is the number of treatments

n is the number of matched sets

T_j, $j = 1, 2, \ldots, c$, is the sum of the ranks assigned to subjects receiving treatment j

Equation 15.19 is computationally easier than Equation 15.18 and is recommended.

We summarize the overall process for using Friedman's procedure.

1. State the hypotheses of interest and select α.
2. Select a sample of n sets consisting of c matched elements in each.
3. Randomly assign a treatment to each element in each set and record the results, taking care to maintain the matches.
4. Within each set, rank the results in ascending order, replacing each observation with its rank, and calculate S, using Equation 15.19.
5. Apply the following decision rule.

DECISION RULE For the hypotheses $H_0\colon \tilde{X}_1 = \tilde{X}_2 = \cdots = \tilde{X}_c$ $\quad H_a\colon$ some $\tilde{X}_j$ differ, reject H_0 if

$$S \geq S_{c,n,\alpha} \qquad \textbf{(15.20)}$$

where values of $S_{c,n,\alpha}$ are given in Appendix Table B.17 for selected values of c and n at reference α's of .10, .05, .01, and .005.

EXAMPLE 15.7 A canine school is evaluating methods for training dogs. The director of training selected 7 different breeds. Within each breed, litters consisting of 4 males whose parentage was known to be common were chosen, with each male in the litter being assigned to a different training method. All dogs of the same breed worked with the same trainer. At the conclusion of the training period, 3 independent judges were asked to rate the dogs, from low to high, within each breed by response to a complicated series of commands. The 3 ratings for each dog were then averaged, and the results are recorded in the following table.

Breed	Method 1	2	3	4
1	1.33	4.00	1.67	3.00
2	2.33	4.00	2.00	1.67
3	1.67	3.33	2.33	2.67
4	2.67	3.67	1.33	2.33
5	3.00	4.00	2.00	1.00
6	1.00	3.67	2.00	3.33
7	2.00	3.33	1.00	3.67

For $\alpha = .05$, can it be concluded that the training methods differ?

Solution: The hypotheses to be tested at $\alpha = .05$, based on the median as the measure of location, are

$$H_0: \tilde{X}_1 = \tilde{X}_2 = \tilde{X}_3 = \tilde{X}_4 \qquad H_a: \text{some } \tilde{X}_j \text{ differ}$$

The corresponding decision rule implies rejection of H_0 if $S \geq S_{c,n,\alpha} = S_{4,7,.05} = 91$. To calculate the value of S for the data, we first rank the observations within breeds and replace the observed value with its corresponding rank. The ranks under each method are then added to obtain the T_j's.

Breed	Method 1	2	3	4
1	1	4	2	3
2	3	4	2	1
3	1	4	2	3
4	3	4	1	2
5	3	4	2	1
6	1	4	2	3
7	2	3	1	4
T_j	14	27	12	17

Substituting into Equation 15.19, we obtain

$$\begin{aligned} S &= \sum_{j=1}^{c} T_j^2 - \frac{c}{4}[n(c+1)]^2 \\ &= (14^2 + 27^2 + 12^2 + 17^2) - \frac{4}{4}[7(5)]^2 \\ &= (196 + 729 + 144 + 289) - (35)^2 \\ &= 1358 - 1225 = 133 \end{aligned}$$

Since $133 > 91$, we reject H_0 and conclude that the methods appear to differ in effectiveness. Observing that all but one of the top ranks are associated with method 2, we further conclude that it appears to be best, and the other 3 are probably not significantly different.

For values of c and n outside the range of Appendix Table B.17, the statistic $12S/[nc(c+1)]$ is approximately distributed as a chi-square variable with $c-1$ degrees of freedom. In such cases, an approximate test is obtained by rejecting the null hypothesis if

$$\frac{12S}{nc(c+1)} > \chi^2_{c-1,\alpha} \tag{15.21}$$

Friedman's procedure provides an alternative to the F test used in the randomized block design of Chapter 9. When compared to this F test, with the condition of normality satisfied, its ARE varies, depending on the number of treatments being compared. For $c = 2$, the ARE is .637, the same as the sign procedure for paired data, and it increases with the number of treatments to a maximum value of .955. If the underlying distributions are nonnormal, the actual ARE may be either above .955 or below .637 or between the two.

EXERCISES

15.10 In evaluating 3 additives to speed up the curing process of rubber, the research lab of Rubber Products selected a sample of crude rubber from each of 7 different suppliers. Each sample was then split into 3 equal parts, which were then randomly assigned one of the additives and milled under identical conditions. The curing times in minutes are given in the table.

	Supplier						
Additive	***1***	***2***	***3***	***4***	***5***	***6***	***7***
1	24	22	27	31	23	30	28
2	16	18	15	22	26	20	21
3	32	28	30	26	29	25	30

Do the median curing times differ for the additives at $\alpha = .05$?

15.11 Refer to Exercise 15.9 and assume that the scores can be identified also with respect to the student's undergraduate average. We present those scores with this additional information.

	Area			
Average	***Business***	***S&E***	***Education***	***Liberal Arts***
<3.00	72	69	62	29
3.00–3.25	51	58	50	42
3.25–3.50	48	92	31	40
3.50–3.75	64	77	47	59
≥3.75	83	86	26	36

Does it appear that statistics preparation differs by area of major at $\alpha = .05$ with the additional information?

15.7 SUMMARY

In this chapter, we have considered a further class of tests that are identified as distribution free. Except for the goodness of fit procedure, they all fall into the basic category of k-sample procedures. The χ^2 goodness of fit procedure provides us with a means of checking assumptions about the type of population from which we are sampling. The χ^2 procedure for testing several binomials is a special case of the procedure for testing several multinomials, and both may be classified as procedures for testing association between types of categories. The hypotheses are more easily stated and understood in the independence form than in the parametric formulation. In all three cases, the χ^2 procedure compares observed frequencies with those expected when H_0 is true by calculating the squared differences between observed and expected frequencies divided by the expected value and accumulating over all categories. That is, the χ^2 procedures are based on the statistic

$$\sum_i \frac{(O_i - E_i)^2}{E_i}$$

where: O_i = observed frequency of category i

E_i = expected frequency of category i under H_0

The null hypothesis is rejected only if the statistic is too large when compared to the appropriate tabular value.

Both the Kruskal–Wallis and Friedman procedures are extensions of two-sample procedures for testing location differences. The Kruskal–Wallis procedure is an extension of the Mann–Whitney procedure and has the same basic assumptions associated with it. Friedman's procedure relates to randomized block designs and is an extension of the sign procedure for identical locations using paired data. The primary differences between the two are in the levels of measurement and the sampling designs. Kruskal–Wallis requires ordinal measurements throughout and independent random samples, whereas Friedman requires ordinality only within blocks and uses subjects that have been matched before treatment.

The procedures of this chapter, their associated assumptions, classical counterparts, hypotheses, and statistics are presented in Table 15.9.

SUPPLEMENTARY EXERCISES

15.12 In situations where the one-way ANOVA F test can be used, is the Kruskal–Wallis procedure also applicable? Is the converse true?

15.13 What are the differences between the χ^2 procedure for a contingency table and the ANOVA F test for a randomized block design?

TABLE 15.9 Summary table

	Goodness of Fit (One-Sample)	*Association (Either One- or c-Samples)*	*Location* — *Independent Samples*	*Location* — *Matched Sets or Blocks*
Procedure	χ^2 $df = k - 1 - v$ *where:* v = number of parameters to be estimated from the sample	χ^2 $df = (r - 1)(c - 1)$	Kruskal–Wallis (Statistic is H.)	Friedman's (Statistic is S).
Assumptions	Nominal data Infinite populations	Nominal data Infinite populations rc mutually exclusive categories	Ordinal data Equal variances and shapes Infinite populations	Ordinal data within blocks Objects within blocks are identical before treatment Treatments assigned randomly in blocks
Classical Counterpart	None	F test for randomized block (Chapter 9)	F test for 1-way ANOVA (Chapter 9)	F test for randomized block (Chapter 9)
Hypotheses	H_0: $F(X) = F_0(X)$ H_a: $F(X) \neq F_0(X)$	H_0: No association between row and column variables H_a: Some association	H_0: $\tilde{X}_j = \tilde{X}$ all j H_a: some $\tilde{X}_j \neq \tilde{X}$	H_0: $\tilde{X}_j = \tilde{X}$ all j H_a: some $\tilde{X}_j \neq \tilde{X}$
Test Statistic	$\sum_{i=1}^{k} \frac{(O_i - E_i)^2}{E_i}$	$\sum_{i=1}^{r} \sum_{j=1}^{c} \frac{(O_{ij} - E_{ij})^2}{E_{ij}}$	$H = -3(n + 1) + \left[\frac{12}{n(n + 1)} \sum_{j=1}^{c} \frac{T_j^2}{n_j}\right]$	$S = \sum_{j=1}^{c} T_j^2 - \left\{\frac{c}{4} [n(c + 1)]^2\right\}$

15.14 Suppose 100 observations on a particular process have been grouped into a frequency table containing 7 classes. It is desired to test if the data could have come from a distribution belonging to the Poisson family. How many degrees of freedom will the χ^2 statistic have, assuming no regrouping is necessary?

15.15 If you wish to test for association in a 4×7 contingency table, how many degrees of freedom will the test statistic have?

15.16 After a course is over, students often say, "If I'd had Professor So-and-So, I'd have gotten a better grade," implying that grade distributions differ by instructor. In an attempt to evaluate if the proportion of high grades in statistics at Norka U differs by instructor, researchers obtained the following data.

	Instructor			
	Laffin	*Haylor*	*Stunning*	*Cheek*
Grades > C	15	36	20	10
Grades ≤ C	30	44	32	20

Is the students' implication supported at $\alpha = .10$?

15.17 What are the differences between the Kruskal–Wallis and Friedman's procedures?

15.18 In analyzing absenteeism, Rubber Products, Inc. checked its personnel records for the past year and constructed the following table. The table categorizes all employees by type of job and number of days absent during the past year.

	X = Days Absent		
Job Classification	$X \leq 3$	$4 \leq X \leq 7$	$X > 7$
Management	80	17	3
Secretarial	50	30	25
Craftsperson	150	65	35
Service	40	50	30
Laborer	130	138	57

For $\alpha = .01$, can it be concluded that number of days absent is related to job classification?

15.19 The data of Exercise 15.16 were also broken down according to sex and number of days absent. The data showing the number of persons in each of these categories are tabulated.

	X = Days Absent		
Sex	$X \leq 3$	$4 \leq X \leq 7$	$X > 7$
Female	81	57	41
Male	369	243	109

Can we conclude that absenteeism is independent of sex at $\alpha = .025$?

15.20 What advantage does the F test for a randomized block design have over Friedman's procedure?

15.21 A statistics department is attempting to evaluate self-teaching texts. The faculty selected 8 groups of 4 students each. The students in each group were matched in terms of their grades in an introductory mathematics course that was a prerequisite. Each student in a group was assigned randomly to 1 of the 4 texts under consideration. At the end of a specified time period, all students were given a common exam. Group identification, text, and exam scores are given in the table.

Text	Group							
	1	*2*	*3*	*4*	*5*	*6*	*7*	*8*
A	64	91	84	56	60	52	91	75
B	78	82	68	46	70	74	95	77
C	76	85	71	67	75	64	80	82
D	87	79	75	65	78	59	86	69

Do there appear to be any differences in the grades due to texts for $\alpha = .10$?

15.22 One of our assumptions for the regression analysis of Chapter 10 was that the error terms were normally distributed with a mean of zero. The following frequency table of residuals was constructed from a regression analysis that had an MSE of 400.

Class Boundaries	*No. of Residuals*
$X < -30$	10
$-30 \le X < -15$	26
$-15 \le X < 0$	38
$0 \le X < 15$	39
$15 \le X < 30$	24
$X \ge 30$	7

Does it appear that the assumption is satisfied at $\alpha = .05$?

15.23 Before the publicity about the disadvantages of aerosol products, the Daisy-Fresh Company had realized the median weekly sales of its 4 types of deodorant to be about the same in number of units. Recently, the marketing department randomly selected 24 weeks from the past 3 years and then randomly selected 6 weeks from the 24 for each product and recorded its weekly sales in 10,000 units. Can we conclude that the median weekly sales are still equal at $\alpha = .01$?

Type of Deodorant			
Roll-On	*Stick*	*Aerosol*	*Powder*
3.8	3.1	2.0	2.9
3.2	2.7	1.5	1.9
2.5	4.2	2.1	3.3
2.8	3.0	.8	2.2
3.6	2.3	1.3	4.0
2.4	3.4	.4	3.5

15.24 What are the differences between the χ^2 procedure for testing several multinomials and the Kruskal–Wallis procedure?

15.25 In evaluating drill bits from various manufacturers, the Monroe Machine Shop obtained $\frac{1}{2}$ dozen bits from each of 6 manufacturers. The life of each bit was then tabulated in number of holes drilled before it wore out. Five bits were broken by accident, resulting in a reduction of the sample sizes for some manufacturers.

Manufacturer					
A	***B***	***C***	***D***	***E***	***F***
462	498	532	526	540	499
396	486	504	481	508	471
427	510	512	493	490	519
501	429	461	515	513	507
454	473	485		482	500
	469	479			

Are the median lives the same for all manufacturers? Use $\alpha = .025$.

15.26 In a test involving the evaluation of 4 laundry detergents, 5 sheets were soiled in 5 different types of dirt—that is, grease, regular dirt, chocolate, and so on. The sheets were then torn into 4 equal sections, with each section being washed in a different detergent. A housewife was then asked to rank the 4 sections, which had been identified by numbers, as to cleanness, with a 1 corresponding to least clean and 4 to most nearly clean. Using her rankings, given in the table, are there any differences in the ability of the detergents to clean at $\alpha = .10$?

	Dirt				
Detergent	***1***	***2***	***3***	***4***	***5***
1	1	2	3	1	1
2	4	3	4	4	3
3	3	1	2	2	2
4	2	4	1	3	4

15.27 When in control, a production process produces items categorized as good, seconds, reworkable, and scrap in the respective percentages of 80, 10, 7, and 3. A random sample of 200 items from the process had 140 good items, 25 items classified as seconds, 23 reworkable items, and 12 scrap items. At an α of .01, is this sample an indication that the process is out of control?

15.28 With the recent popularity of state lotteries, there is some question about the randomness of occurrence of the winning numbers. In a particular state, numbers are drawn out of a closed basket 1 at a time with replacement, with all numbers supposedly being equally likely on any given draw. The past 100 numbers selected gave five 0's, twelve 1's, nine 2's, six 3's, fifteen 4's, seven 5's, five 6's, thirteen 7's, twelve 8's, and sixteen 9's. Would these results cause you to question the randomness of the lottery when $\alpha = .01$?

15.29 Molson Products, Inc. has a policy of purchasing the home of a transferred executive for a guaranteed price if it cannot be sold for a higher amount within 90 days of the transfer notification date. The guaranteed price is the median of 3 appraisals made independently by company-appointed real estate appraisers. The following are appraisals, in \$1000, for a random sample of 8 houses appraised by the same 3 appraisers.

House	*Appraiser*		
	1	*2*	*3*
1	55.6	55.3	58.4
2	70.1	69.6	72.7
3	68.5	68.1	67.2
4	82.9	80.1	86.9
5	70.0	69.8	72.4
6	75.7	74.7	74.0
7	66.2	66.5	68.3
8	90.5	92.7	97.6

Do the appraisers appear to differ at $\alpha = .10$?

***15.30** An aluminum company constructed the following frequency table of failure times, in days, of a group of 50 electrolytic cells of design A that were installed in one of its smelter pot rooms.

Days to Failure	*No. of Cells*
500– 900	7
900–1,300	11
1,300–1,700	10
1,700–2,100	10
2,100–2,500	5
2,500–2,900	5
2,900–3,400	2

The times to failure for cells of a conventional design are known to have a negative exponential distribution with a mean life of 1300 days. Would you conclude that the same distribution applies for the failure times of cells of design A for $\alpha = .05$?

15.31 A marketing researcher believes that the favorite beverage consumed at meals is related to age. A sample of 100 people produced the following cross-classification.

Beverage	*Age*		
	0–12	*13–30*	*>30*
Milk	15	10	0
Coke	20	20	10
Coffee	5	10	10

Is her belief supported at $\alpha = .10$?

* An asterisk indicates a higher level of difficulty

15.32 The dilation of the pupil of the eye is said to be related to the interest a person has in an object or person. This technique is used in market research to help to select package designs and other objects created to appeal to the visual sense. Three different package designs are tried on 90 subjects and their pupil dilation observed. The frequency of results were tabulated and are shown here. Does the technique appear to have merit? Use $\alpha = .05$.

	Design		
Dilation	***I***	***II***	***III***
None	12	7	5
Some	13	15	8
Much	5	8	17

15.33 The consumer affairs division of the Orange Juice Enjoyers of America performed a taste test of 3 brands of orange juice to see if any differences exist among brands. Eight subjects were randomly chosen, and each subject was asked to taste all 3 brands and to rank them from 1 to 3, with a 1 given to the brand that was liked the most. At $\alpha = .05$, do there appear to be differences among the brands?

Taster	***OJ1***	***OJ2***	***OJ3***
1	1	2	3
2	1	3	2
3	2	1	3
4	1	3	2
5	2	1	3
6	1	2	3
7	1	2	3
8	2	1	3

15.34 Refer to Exercise 15.25. What procedure from an earlier chapter might also be used in that situation? Analyze the data, using the procedure you indicated, compare the results, and indicate which one you believe to be preferable and why.

15.35 The chief auditor of Molson Products has been studying costs of long-distance telephone calls with respect to various departments over the previous 3-month period. In doing so, he selected 8 executives at random from each of the departments of accounting, marketing, and customer relations. The total costs of long-distance calls, in dollars, for each executive over the 3-month period studied were as follows.

Executive	***Accounting***	***Marketing***	***Cust. Relations***
1	83	640	1293
2	41	383	823
3	512	915	149
4	194	219	350
5	79	193	764
6	84	507	243
7	281	621	246
8	301	482	433

Test whether or not the median telephone expenses per executive are the same for the 3 departments at $\alpha = .05$.

15.36 Find a procedure from an earlier chapter that might apply for Exercise 15.35 and use it. Compare your results, indicate which procedure you believe to be preferable, and state why you believe so.

15.37 An accountant is attempting to determine whether a relationship exists between the proportion of delinquent accounts and the sales district in which the account is located. To do so, she selected 100 accounts at random from each of the company's 4 sales districts: Eastern, Midwest, Rocky Mountain/Southwest, and West Coast. The number of delinquent accounts in each of these districts, respectively, was 22, 15, 31, and 28. After analyzing these data at $\alpha = .10$, what should she conclude?

****15.38** Show that for $k = 2$, the binomial case,

$$\sum_{i=1}^{2} \frac{(O_i - E_i)^2}{E_i} = \left[\frac{X - n\pi}{\sqrt{n\pi(1 - \pi)}}\right]^2$$

where $X = O_1$, $n\pi = E_1$, and $\pi = \pi_1$. (This problem requires a moderate amount of algebra.)

15.39 Tami Moran of Barstow Products, Inc. (refer to the Chapter Introduction) has collected her data, which are presented in the table.

	Geographic Area				
Commercial Style	***A***	***B***	***C***	***D***	
1	18	20	33	19	90
2	22	29	16	37	104
3	34	28	22	20	104
4	26	23	29	24	102
	100	100	100	100	400

What should Tami report to her superior, Noni Wollis, concerning an association between type of commercial and geographic area if she uses $\alpha = .01$?

REFERENCES

1. Bradley, James V. *Distribution-Free Statistical Tests*. Englewood Cliffs, NJ: Prentice-Hall, 1968.
2. Gibbons, Jean D. *Nonparametric Statistical Inference*. New York: McGraw-Hill, 1971.
3. Noether, Gottfried E. *Elements of Nonparametric Statistics:* New York: John Wiley, 1967.
4. Owen, D.B. *Handbook of Statistical Tables*. Reading, MA: Addison-Wesley, 1962.
5. Siegel, Sidney. *Nonparametric Statistics for the Behavioral Sciences*. New York: McGraw-Hill, 1956.

** Two asterisks indicate a considerably higher level of difficulty.

16 Introduction to Inference-Free Decision Making

16.1 INTRODUCTION

Every Saturday, Emil Grosbeck of Gros-Pro Concessions, Inc. must make decisions concerning the number of hot dogs he should order for the professional football game to be held on Sunday. In the past, he has just guessed, but now he has hired Joe Zalay of Predictors, Inc., to help him come up with a rule that will provide the optimal ordering decision. Because of the nature of the problem, Joe turns to the techniques of inference-free decision making.

Joe's first step is to acquire from Emil any past records on sales. He also obtains statistics from the weather bureau. Using revenue and cost information supplied by Emil, he computes a payoff table. Then, using the table and the methods of inference-free decision making, he hopes to determine the optimal order for hot dogs for each football game.

In earlier problem situations, the analyst favored a certain action when, through the analysis of sample data, it appeared that a hypothesis about a population parameter could be believed. Otherwise, an alternative is implemented. Two approaches of this decision-making procedure of hypothesis testing have been presented. One approach is classical and largely assumes normally distributed populations; the other is distribution free. If the assumptions underlying the classical procedures are not felt to be satisfied, but the situations being tested are otherwise the same, then distribution-free (nonparametric) techniques are used.

This chapter considers some of the techniques of inference-free decision making. The first technique is the construction of payoff tables and opportunity loss tables. We then take up the decision-making criteria that are used with these tables to decide among alternative actions when sample information is not available. Their application is known as prior analysis. These criteria are followed by the methods of posterior analysis, which use information obtained from sampling or other sources.

16.2 CONDITIONAL PAYOFF AND OPPORTUNITY LOSS TABLES

Another approach to decision making, separate from inferential statistics, is generally known as decision theory. It is popularly referred to as Bayesian statistics because of the use of Bayes' rule in some applications. Since not all applications use this rule, we prefer the term "inference-free."

DEFINITION 16.1 **Inference-free decision making** is a collection of statistical techniques used for decision analysis that do not involve the use of inferential statistical methods.

Inference-free decision making considers certain inputs that are not always a part of the inferential approach, such as subjective probabilities, economic costs, and two or more alternative actions under consideration. Numerous criteria are available for choosing between alternatives.

Most decision-making activities require the choice of one action from a number of possible alternative actions a_i, $i = 1, 2, \ldots, r$.

DEFINITION 16.2 **Alternative actions** are the potential decisions from which the decision maker must choose.

The alternatives usually depend on the occurrence or existence of certain situations or conditions that will occur strictly by chance. These chance occurrences are known as the states of nature and will be designated s_j, $j = 1, 2, \ldots, c$.

DEFINITION 16.3 The **states of nature** are the outcomes that occur by chance and influence the payoffs or losses for any particular alternative action.

Thus, the first step in any inference-free decision-making process is to identify the alternative actions and states of nature that are relevant to the problem. Then, the conditional payoffs can be determined.

DEFINITION 16.4 The **conditional payoff** X_{ij} is the return or gain that will be experienced if a state of nature s_j occurs and action a_i is chosen.

In monetary situations, payoffs will be calculated according to the formula

$$\text{Payoff} = \text{Revenue} - \text{Cost} \qquad \textbf{(16.1)}$$

The term "profit" is not used in place of the word "payoff" unless the cost term includes fixed costs. Often only variable costs are included, and the payoff is the contribution to overhead.

The payoffs for each combination of action and state of nature are computed and displayed in a table called the payoff table. Table 16.1 illustrates the typical payoff table.

EXAMPLE 16.1 A storeowner dealing in electronic surplus has the opportunity to buy surplus radios in lots of 100 for $1300 "sight unseen." Although all the radios are used, he knows that some will look new and can be sold for $25, and the rest will look used and will sell for $5. These lots are classified as good, fair, and poor, based on the percent of used-looking radios in the lot, which is 20%, 50%, and 80%, respectively.

a. State the alternative actions and the states of nature.
b. Determine the payoff function—that is, the formula for calculating the payoff values.
c. Prepare the payoff table.

Solution:

a. The storeowner has 2 alternative actions: buy the lot, and do not buy the lot. The states of nature are good, fair, and poor, the classification based on the percent of used-looking radios in the lot.
b. The payoff function is developed from Equation 16.1.

$$\text{Payoff} = \text{Revenue} - \text{Cost}$$

TABLE 16.1 Typical format for a payoff table

	States of Nature		
Actions	s_1	s_2	s_3
a_1			
a_2			
a_3			

The revenue received depends on the state of nature, given that he buys the lot. If we let θ represent the percent of used-looking radios, then

$$\text{Revenue} = (\text{Number used}) \times (\text{Price used}) + (\text{number new}) \times (\text{Price new})$$

where: number used $= \theta$ (number in lot)

number new $= (1 - \theta)$(number in lot)

Substituting the relevant numbers, we obtain

$$\begin{aligned}\text{Revenue} &= \theta(100)(5) + (1 - \theta)(100)(25)\\ &= 500\,\theta + 2500(1 - \theta) = 2500 - 2000\,\theta\end{aligned}$$

The payoff function, if he buys, is

$$\begin{aligned}\text{Payoff} &= \text{Revenue} - \text{Cost} = 2500 - 2000\,\theta - 1300\\ &= 1200 - 2000\,\theta\end{aligned}$$

If he does not buy, he cannot sell and make a profit. The payoff function becomes

$$\text{Payoff} = 0$$

c. The payoff for each state of nature, given that he buys, is found by substituting the appropriate value for θ in the payoff function.

θ	$1200 - 2000\,\theta = \text{Payoff}$
.2	$1200 - 2000(.2) = 800$
.5	$1200 - 2000(.5) = 200$
.8	$1200 - 2000(.8) = -400$

The resulting payoff table is shown in Table 16.2.

DEFINITION 16.5 **Conditional loss** is a statement of how much less profit or contribution was obtained for a given state of nature and action than would have been obtained if the best action for that state of nature had been chosen.

TABLE 16.2 Payoff table for Example 16.1

	States of Nature		
Actions	*Good* $\theta = .2$	*Fair* $\theta = .5$	*Poor* $\theta = .8$
a_1 *Buy*	800	200	−400
a_2 *Do Not Buy*	0	0	0

In economic terms, this loss is an opportunity loss. To convert a payoff table into an opportunity loss table, we use the following procedure.

1. For each state of nature, select the action that gives the largest payoff. Assign a zero to the corresponding location in the loss table.
2. For each other action in the given state of nature, determine the difference between the largest payoff and the payoff obtained and assign that difference to the corresponding location in the loss table.

EXAMPLE 16.2 Prepare an opportunity loss table from the payoff table in Example 16.1.

Solution: For the state of nature s_1, the best decision is a_1 (buy). Therefore, assign a 0 to location a_1s_1. Since the owner will realize $800 less if he does not buy, assign 800 to location a_2s_1. Following the same procedure for the other 2 states of nature, we obtain the opportunity loss table in Table 16.3.

Notice in Example 16.2 that the losses are not all the type that are represented by red ink. Only the $400 loss is of that type. The other two figures represent profit opportunities that were foregone by not buying the lot.

Many decision problems are of the multiaction type—that is, they have more than two actions. The procedure for developing the payoff and opportunity loss tables is similar.

EXAMPLE 16.3 A small bakery is known for its blueberry muffins. Unsold muffins must be thrown out at the end of the day because they will be too stale to sell the next day. The baker keeps careful records and has noticed that each day she sells either 50, 51, 52, or 53 dozen. She sells the muffins, which have a cost of $1 per dozen to produce, for $1.50 a dozen. What are the payoff and loss tables for the muffins?

Solution: The payoff to be entered into each cell of the table is calculated from Equation 16.1.

$$\text{Payoff} = \text{Revenue} - \text{Cost}$$

TABLE 16.3 Opportunity loss table for Example 16.2

	States of Nature		
Actions	s_1 *Good* $\theta = .2$	s_2 *Fair* $\theta = .5$	s_3 *Poor* $\theta = .8$
a_1 *Buy*	0	0	400
a_2 *Do Not Buy*	800	200	0

TABLE 16.4 Calculations for first column and last row of Example 16.3

a_1s_1	$1.50(50) - 1(50) = 75.00 - 50.00 = 25.00$
a_2s_1	$1.50(50) - 1(51) = 75.00 - 51.00 = 24.00$
a_3s_1	$1.50(50) - 1(52) = 75.00 - 52.00 = 23.00$
a_4s_1	$1.50(50) - 1(53) = 75.00 - 53.00 = 22.00$
a_4s_2	$1.50(51) - 1(53) = 76.50 - 53.00 = 23.50$
a_4s_3	$1.50(52) - 1(53) = 78.00 - 53.00 = 25.00$
a_4s_4	$1.50(53) - 1(53) = 79.50 - 53.00 = 26.50$

The states of nature are the four quantities of demand: 50, 51, 52, and 53 dozen. The alternative actions among which the baker can choose each morning are to bake 50, 51, 52, or 53 dozen muffins. The payoffs are computed by the function

$$\begin{aligned} \text{Payoff} &= \text{Revenue} - \text{Cost} \\ &= \$1.50 \text{ (dozen sold)} - \$1 \text{ (dozen baked)} \end{aligned}$$

The computations for the first column and last row are shown in Table 16.4. The complete payoff table is shown in Table 16.5. Following procedures outlined and demonstrated earlier, we construct the opportunity loss table. See Table 16.6.

An alternative to the payoff table is the decision tree. It has the advantage of showing the logical progression in a decision-making process.

The first set of branches in a decision tree indicates the alternative actions or decisions that can be made. These actions are indicated as the a_i in Figure 16.1. The alternative states of nature for each action follow the junction point, which is referred to as a node. The symbol s_j represents the various states of nature. Following these branches, the payoffs, indicated by X_{ij}, are given. Each node in the tree is either a decision node or a probability node. A square is used to indicate a decision node, and a circle to indicate a probability node.

TABLE 16.5 Payoff table for Example 16.3

	States of Nature (demand)			
Actions (Supply)	s_1 ***50***	s_2 ***51***	s_3 ***52***	s_4 ***53***
a_1 ***50***	25.00	25.00	25.00	25.00
a_2 ***51***	24.00	25.50	25.50	25.50
a_3 ***52***	23.00	24.50	26.00	26.00
a_4 ***53***	22.00	23.50	25.00	26.50

TABLE 16.6 Opportunity loss table for Example 16.3

Actions (supply)	States of Nature (demand)			
	s_1 50	s_2 51	s_3 52	s_4 53
a_1 50	0	0.50	1.00	1.50
a_2 51	1.00	0	0.50	1.00
a_3 52	2.00	1.00	0	0.50
a_4 53	3.00	2.00	1.00	0

FIGURE 16.1 Typical decision tree

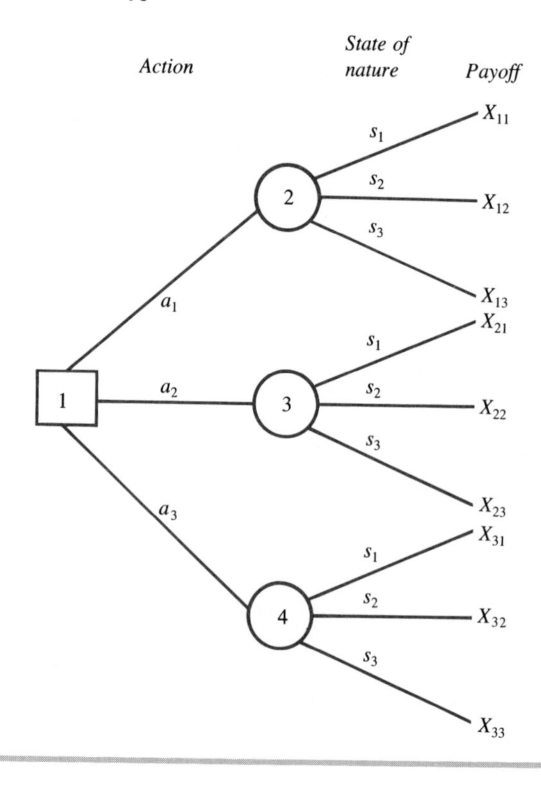

FIGURE 16.2 Decision tree for Example 16.4

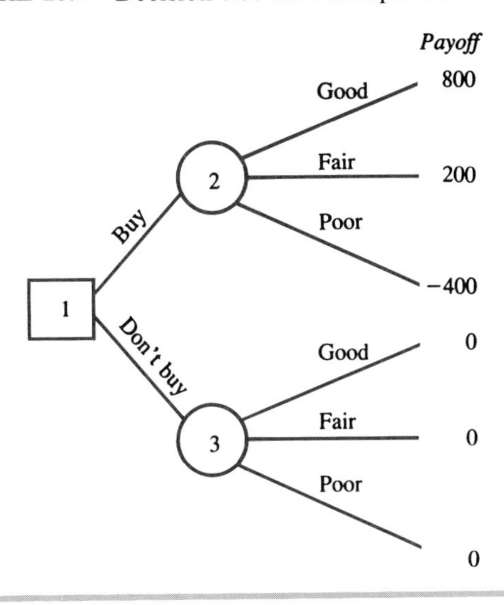

EXAMPLE 16.4 Create a decision tree for Example 16.1. Refer to Table 16.2 for the payoff table.

Solution: The decision tree is shown in Figure 16.2.

EXERCISES

16.1 Arnold Peabody presently owns 40 acres of ground in southern Ohio on which he annually pays a property tax of $200. The land is only suitable for growing wheat, but he isn't sure whether he should till the soil himself or rent it out at $10 an acre plus a yield bonus of 10¢ per bushel. The variability in weather as to excellent, normal, or poor growing season is what concerns him. In an excellent growing season, the yield is 35 bushels per acre, with the normal and poor seasons being 80% and 60% of that total. The wheat can be sold for 60¢ per bushel. If he tills the soil himself, it will be necessary to spend $5 per acre for fertilizer. The probabilities of the 3 seasons are .3, .6, and .1, respectively.

(a) Determine the payoff table.
(b) Determine the loss table.
(c) Create a decision tree for the payoffs.

16.2 A paperboy sells the *Wall Street Journal* at a corner in the business district of a large city. He has a monopoly on sales for this area. He sells each paper for 20¢ and pays 12¢ to the publisher for each paper sold. His records show that he sells 25, 26, 27, 28, or 29 papers each day.

(a) Determine the payoff table.
(b) Determine the loss table.

(c) Create a decision tree for the payoffs.
(d) If he locates a firm that will pay 2¢ for every newspaper he doesn't sell, what will be the payoff table?

16.3 PRIOR ANALYSIS—DECISION CRITERIA

Numerous criteria have been developed for selecting among alternative actions. This section presents some of the more commonly known criteria. Two of the criteria, the maximin and the maximax, are used in conjunction with a payoff table. The minimax, often known as the regret criterion, is applied to an opportunity loss table, and the expected value criterion can be used with either.

Maximin Criterion

The maximin criterion represents a pessimistic viewpoint. It is believed that the state of nature that will occur will be the one that provides the most unfavorable payoff. Therefore, the action is selected that results in the maximum of the minimum payoffs. The maximin procedure is as follows.

1. Determine the minimum payoff for each action.
2. Select the action that provides the largest of these minimum payoffs.

EXAMPLE 16.5 The baker in Example 16.3 must choose 1 of the 4 actions with the payoffs in Table 16.7. If she uses the maximin criterion, how many dozen muffins will she bake?

Solution: The minimum payoffs for each action are

a_1	\$25.00
a_2	\$24.00
a_3	\$23.00
a_4	\$22.00

TABLE 16.7 Payoffs for muffins in Example 16.5

	Demand			
Supply	s_1 50	s_2 51	s_3 52	s_4 53
a_1 50	25.00	25.00	25.00	25.00
a_2 51	24.00	25.50	25.50	25.50
a_3 52	23.00	24.50	26.00	26.00
a_4 53	22.00	23.50	25.00	26.50

of which the largest is for a_1. The baker will therefore choose to bake 50 dozen muffins, since this number will give her the largest payoff for the most pessimistic outcome. ■

Maximax Criterion

The maximax criterion represents an optimistic viewpoint. It is believed that the state of nature that will occur will be the one that provides the most favorable payoff. Therefore, the action is selected that results in the maximum of the maximum payoffs. The maximax procedure is as follows.

1. Determine the maximum payoff for each action.
2. Select the action that provides the largest of these maximum payoffs.

EXAMPLE 16.6 ■ If the baker is an optimist, rather than the pessimist of Example 16.5, how many dozen muffins will she choose to bake, using the maximax criterion?

Solution: The maximum payoffs for each action are found in Table 16.7.

a_1	\$25.00
a_2	\$25.50
a_3	\$26.00
a_4	\$26.50

The largest payoff is for a_4. She will therefore choose to bake 53 dozen muffins and expect to sell all of them. ■

Minimax Criterion

The minimax criterion represents a loss-oriented viewpoint. The objective in its use is to minimize the extent of possible losses. Therefore, the action is selected that minimizes the maximum loss. The minimax procedure is as follows.

1. Determine the maximum loss for each action.
2. Select the action for which the maximum loss is the smallest.

EXAMPLE 16.7 ■ The baker in the preceding examples may concentrate on minimizing her potential losses rather than on maximizing her gains. If she therefore applies the minimax criterion, how many muffins will she choose to bake?

Solution: An opportunity loss table, as shown in Table 16.8, is required. The maximum losses for each action are

a_1	\$1.50
a_2	\$1.00
a_3	\$2.00
a_4	\$3.00

TABLE 16.8 Loss table for muffins in Example 16.7

	Demand			
	s_1	s_2	s_3	s_4
Supply	*50*	*51*	*52*	*53*
a_1 *50*	0	0.50	1.00	1.50
a_2 *51*	1.00	0	0.50	1.00
a_3 *52*	2.00	1.00	0	0.50
a_4 *53*	3.00	2.00	1.00	0

of which the smallest is for a_2. The baker will therefore choose to bake 51 dozen muffins since this number will minimize her potential losses.

Expected Value

Using the expected value concept from Chapter 3, we can write the following equation for the average payoff of the action a_i.

$$E[a_i] = \sum_j X_{ij}\, p(s_j) \tag{16.2}$$

where $p(s_j)$ is the probability of the occurrence of state of nature s_j, and X_{ij} is the payoff. These probabilities will be referred to as prior probabilities.

The expected value criterion specifies that we choose the action that gives us the highest expected (average) payoff, which will henceforth be called expected monetary gain (*EMG*). We will choose the same action if we select the one that gives the lowest expected opportunity loss (*EOL*). Using either the payoff or opportunity loss tables leads to the same conclusion. The expected value procedure is as follows.

1. Calculate the expected value for each action, using the prior probabilities for the states of nature and either the payoffs or opportunity losses.
2. Select the action that has the maximum *EMG* or minimum *EOL*.

EXAMPLE 16.8 The baker has kept records for 2 years on the demand for her muffins and has determined the following prior probabilities for each level of demand.

Demand	50	51	52	53
Probability	.10	.35	.40	.15

If she uses the expected value criterion, how many dozen muffins will she bake each day?

Solution: The *EMG* for each action is

$$EMG_1 = 25.00(.10) + 25.00(.35) + 25.00(.40) + 25.00(.15) = 25.00$$

$$EMG_2 = 24.00(.10) + 25.50(.35) + 25.50(.40) + 25.50(.15) = 25.35$$

$$EMG_3 = 23.00(.10) + 24.50(.35) + 26.00(.40) + 26.00(.15) = 25.18$$

$$EMG_4 = 22.00(.10) + 23.50(.35) + 25.00(.40) + 26.50(.15) = 24.40$$

The maximum *EMG* occurs with a_2, so the baker would choose to bake 51 dozen each morning. In a like manner, the expected opportunity loss is calculated for each action from the loss table.

$$EOL_1 = 0(.10) + .50(.35) + 1.00(.40) + 1.50(.15) = .80$$
$$EOL_2 = 1.00(.10) + 0(.35) + .50(.40) + 1.00(.15) = .45$$
$$EOL_3 = 2.00(.10) + 1.00(.35) + 0(.40) + .50(.15) = .63$$
$$EOL_4 = 3.00(.10) + 2.00(.35) + 1.00(.40) + 0(.15) = 1.40$$

The minimum *EOL* occurs with a_2. This result agrees with the result obtained from the payoff table and demonstrates that either approach will lead to the same decision.

Finding the best action through expected values can also be accomplished through decision trees. The probabilities for each state of nature are indicated on the diagram, as in Figure 16.3, and then expected values are computed for each action, as shown in Figure 16.4.

EXAMPLE 16.9 Repeat Example 16.8 for the baker, this time using decision trees and payoffs to find the action with the largest expected monetary gain.

Solution: First the decision tree is drawn, and the probabilities and payoffs are indicated. The result is Figure 16.5. Then, the expected values for each action are computed, and the results are shown in Figure 16.6. Action a_2 is chosen as having the maximum expected monetary gain. The baker will bake 51 dozen each morning.

Choice of Criteria

How do we determine which of these criteria to use for determining the optimal action? Our choice depends on what information is given. If the probabilities of the states of nature are known, we would use the expected value criterion since it is an objective method. Otherwise, we would use a method that would suit our personal tolerance or philosophy of risk.

FIGURE 16.3 Decision tree with probabilities shown for states of nature

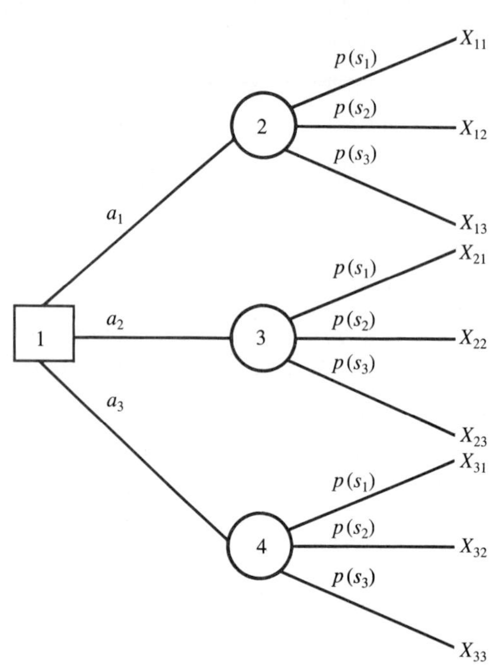

Recognize that in using the expected value criterion, we are treating it with a long-run view. For any individual day, the payoff or loss will be a particular value in the payoff or loss table. But in the long run, where the same decision is made every day, we would be considering the average payoff or average loss.

FIGURE 16.4 Computation of expected values for each action

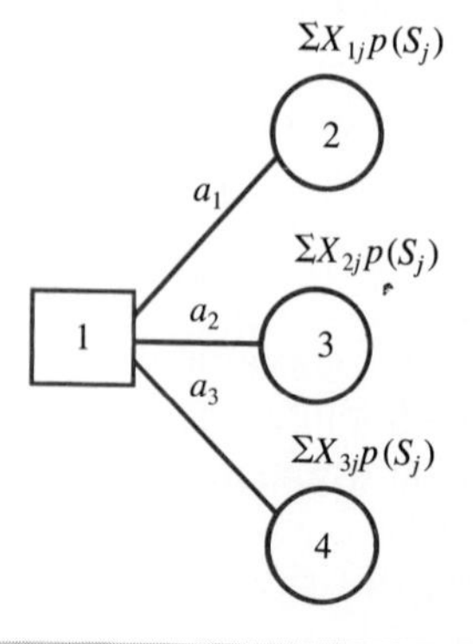

FIGURE 16.5 Decision tree for Example 16.9

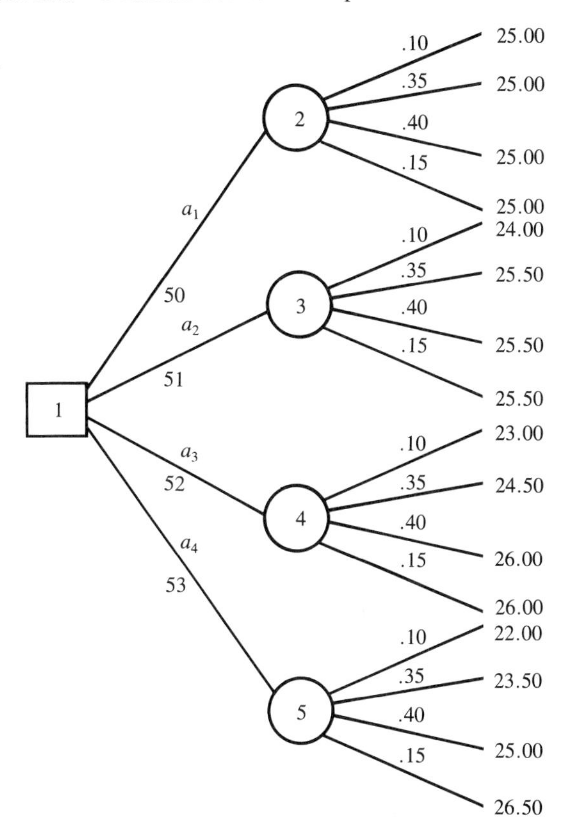

FIGURE 16.6 Expected monetary gain for each action in Example 16.9

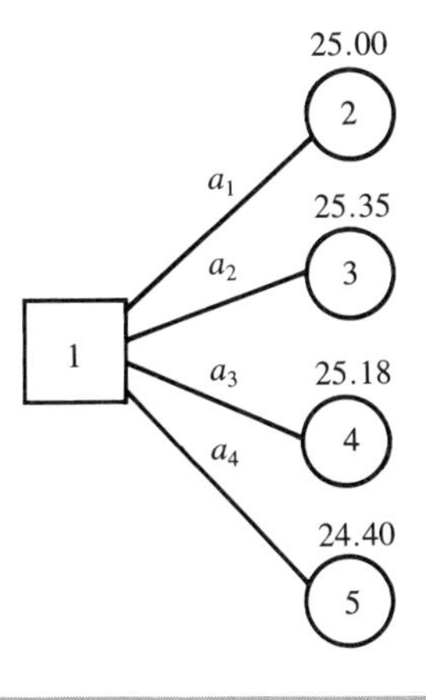

EXERCISES

16.3 Consider the following payoff table where amounts are in dollars and the probabilities for the states of nature are .1, .2, and .7, respectively.

	s_1	s_2	s_3
a_1	4	11	6
a_2	7	5	9
a_3	2	8	10

Which action would you choose according to the **(a)** minimax criterion, **(b)** maximax criterion, **(c)** maximin criterion, **(d)** expected value criterion, and **(e)** expected value, using decision trees?

16.4 For the Arnold Peabody data in Exercise 16.1, which action would you choose, using each of the 4 criteria?

16.5 Using the paperboy data of Exercise 16.2, what action would you choose, using the expected value criterion? Assume the prior probabilities are .15, .20, .30, .25, and .10, respectively.

16.4 EXPECTED VALUE OF PERFECT INFORMATION

Predicting the future is a risky business because it is impossible to be correct consistently. But, pretend for the moment that a decision maker has access to perfect forecasts about the future and, therefore, always knows what state of nature will occur. It would be possible, then, to always select the action that gives the maximum payoff for the predicted state of nature. The expected payoff under perfect information (*EPPI*) can then be calculated by Equation 16.3.

$$EPPI = \sum_j \max(X_{ij})p(s_j) \qquad \textbf{(16.3)}$$

EXAMPLE 16.10 The baker has located a marketing research consultant who can perfectly predict the state of nature to occur each day. What is her expected payoff?

Solution: The maximum payoffs for each state of nature are

s_1	\$25.00
s_2	\$25.50
s_3	\$26.00
s_4	\$26.50

The probability for each state of nature does not change, so the baker's expected payoff under perfect information is

$$EPPI = 25.00(.10) + 25.50(.35) + 26.00(.40) + 26.50(.15)$$
$$= 25.80$$

DEFINITION 16.6 The **expected value of perfect information** (*EVPI*) is a measure of the amount that the expected payoff would increase if the analyst had perfect knowledge as to the state of nature that would occur.

In practice, the *EVPI* represents the maximum that an analyst is willing to pay to know exactly what state of nature will take place. Perfect information is not, however, considered to be an obtainable goal.

The *EVPI* can be calculated as the difference between the *EPPI* and the maximum expected payoff under uncertainty [max(*EMG*)]. The equation is

$$EVPI = EPPI - \max(EMG) \qquad \textbf{(16.4)}$$

If the analysis had been performed with the opportunity loss table, the preceding calculations would not be required for finding *EVPI*. The minimum *EOL* is also the *EVPI*.

$$EVPI = \min(EOL) \qquad \textbf{(16.5)}$$

Note that the *EVPI* represents the maximum that would be paid to the consultant for information concerning the occurrence of the state of nature.

EXAMPLE 16.11 The consultant is requesting $1 for each consultation with the baker concerning muffin demand. What should be the baker's response?

Solution: With perfect information, the baker will always bake the amount demanded and will realize the maximum payoff for each day, giving her, as determined in the previous example, an *EPPI* of $25.80. If, instead, she makes her decision under uncertainty, each day she will bake 51 dozen muffins to achieve a max(*EMG*) of $25.35, as determined in Example 16.8. Therefore,

$$EVPI = 25.80 - 25.35 = .45$$

which is the most the baker would be willing to pay for each consultation. Therefore, the baker would not be willing to pay $1 to the consultant. Notice in Example 16.8 that the minimum *EOL* was .45 for a_2.

$$EVPI = \min(EOL) = .45$$

Hence, the opportunity loss table is preferred over the payoff table because the number of calculations is reduced.

EXERCISES

16.6 Using the Arnold Peabody data of Exercise 16.1, calculate the *EVPI* from both the payoff and loss tables.

16.7 Calculate the *EVPI*, using the paperboy data from Exercise 16.2.

16.5 POSTERIOR ANALYSIS—USE OF SAMPLING OR OTHER INFORMATION

Whenever prior analysis is performed by using expected values, we are working with relative frequency probabilities or, in some cases, subjective probabilities. Where relative frequency probabilities are used, they have been calculated by looking at the number of times each state of nature has occurred in the past. They are based strictly on historical data. Where such data are lacking, the probabilities may be subjective, assigned simply by a hunch or, at best, an educated guess.

Historical probabilities are good, but how much better it would be if we could obtain additional information that could give us a better insight into which state of nature existed in the problem under study. For example, it is not unusual before leaving for school or work that we look at the sky to see if it might rain in the next hour. If we see heavy clouds, we might decide that historically, whenever these conditions have existed, there was a .6 probability that it would rain. But we might also seek additional information. A call to a friend upwind about 20 miles would tell us if it was raining in that location or, if a news broadcast was on television, we could check the radar display for rain. If either of these tasks showed that it was raining upwind, we would update our probability of .6 that it would rain to some higher probability, such as .9. This section presents a method of revising the historical or prior probabilities through sampling and other means.

In some situations, it is possible to take a sample from the population under study. The information obtained can be combined with the prior probabilities to obtain a new set of probabilities that represent a new likelihood of the occurrence of each state of nature. The pooling of the prior probabilities and sample information to obtain what we will call the posterior probabilities is accomplished by Bayes' rule. In Chapter 2, this rule was given by Equation 2.19; it is repeated here as Equation 16.6.

$$P(B_j|A) = \frac{P(B_j)P(A|B_j)}{\Sigma P(B_i)P(A|B_i)} \tag{16.6}$$

The $P(B_i)$ are the prior probabilities and represent the probability of the occurrence of a state of nature. The $P(A|B_i)$ are the conditional probabilities derived from the sampling results and represent the probable occurrence of an event given the state of nature. The $P(B_j|A)$ are the posterior probabilities that represent our updated expectations that each state of nature will occur.

The prior probabilities do not need to be based on historical data. They may be purely subjective estimates resulting from a decision maker's intuition and experience. Bayes' rule, therefore, provides the capability of combining intuition with observation. Once the posterior probabilities are obtained, they are used with either the payoff or opportunity loss tables in the same manner as were the prior probabilities to calculate the expected payoffs or expected oppor-

TABLE 16.9 Payoff table for surplus radios

	Good	*Fair*	*Poor*
a_1 ***Buy***	800	200	−400
a_2 ***Don't Buy***	0	0	0

tunity losses. Let us first of all set up an example and perform a prior analysis. Then, on the following example, we will sample and apply Bayes' rule.

EXAMPLE 16.12 Consider again the problem of the owner of the surplus electronics store. He has the opportunity to buy a lot of 100 radios "sight unseen." His intuition and experience lead him to believe that there is a 10% chance that the lot can be classified as good, 20% as fair, and 70% as poor. The classifications are based on the percent of used-looking radios in the lot—20%, 50%, and 80%, respectively. The payoff table and opportunity loss table are shown in Tables 16.9 and 16.10, respectively. Using prior analysis techniques, determine if the owner should buy the 100 radios, and then calculate the expected value of perfect information.

Solution: Refer to Examples 16.8 and 16.10 for the techniques of solution. So that you can see the problem worked both ways, both the payoff and the loss tables are used in this example. From the payoff table, the expected monetary gains are

$$EMG_1 = 800(.1) + 200(.2) - 400(.7)$$
$$= 80 + 40 - 280 = -160$$

$$EMG_2 = 0(.1) + 0(.2) + 0(.7) = 0$$

The maximum EMG is obtained with action a_2. Therefore, the decision is to not buy the 100 radios. The expected value of perfect information is then obtained thusly.

$$EPPI = 800(.1) + 200(.2) + 0(.7) = 80 + 40 = 120$$

$$EVPI = EPPI - \max(EMG) = 120 - 0 = 120$$

TABLE 16.10 Loss table for surplus radios

	Good	*Fair*	*Poor*
a_1 ***Buy***	0	0	400
a_2 ***Don't Buy***	800	200	0

The storeowner would be willing to pay no more than \$120 to learn to which of the 3 states of nature the lot of radios belongs.

The same solution can be found by using the opportunity loss table.

$$\begin{aligned} EOL_1 &= 0(.1) + 0(.2) + 400(.7) \\ &= 0 + 0 + 280 = 280 \end{aligned}$$

$$\begin{aligned} EOL_2 &= 800(.1) + 200(.2) + 0(.7) \\ &= 80 + 40 + 0 = 120 \end{aligned}$$

The minimum EOL corresponds to a_2; therefore, the decision is to not buy the 100 radios. The minimum EOL is also the expected value of perfect information, so

$$EVPI = \min(EOL) = 120$$

The previous example involved prior analysis only. That is, it used historical probabilities obtained from past events. Let us now consider working with additional information taken from the current situation under study.

To make a decision, we would prefer to know with certainty the current state of nature before picking an action. In the previous example, perfect knowledge would require being able to make a 100% inspection of the merchandise. If a 100% inspection were not allowed, we could consider taking a sample to get some additional information to help us determine what is the current state of nature. This sampling information can be used in conjunction with Bayes' rule to obtain a new, updated set of probabilities to use with the payoff or loss table to make our decision as to the proper action. We can measure how much sampling information contributed to the decision by comparing the $EVPI$ to the expected value of perfect information after sampling ($ELSI$).

DEFINITION 16.7 The **expected loss given sample information** represents the expected value of perfect information after including sampling information in the analysis.

Bayes' rule is used to combine the prior probabilities with the sampling information that will be included in the analysis as conditional probabilities. The posterior, or resulting probabilities, will be used with the payoff or loss table to make the decision.

EXAMPLE 16.13 The seller of the surplus radios will permit the storeowner to take a random sample of 3 radios. If all 3 are used-looking, what decision will the storeowner make now, and what is the expected value of perfect information after having observed the sample?

Solution: Bayes' rule is used for the solution. We will use the special format indicated in Chapter 2. The prior probabilities were given in the previous example.

$$P(B_1) = .1 \qquad P(B_2) = .2 \qquad P(B_3) = .7$$

Since each radio has 2 outcomes, new-looking or used-looking, we are sampling from a binomial distribution where $n = 3$ and $\pi_i = \theta_i$, the percent of used-looking radios in each state of nature. Thus, the conditional probability that 3 used-looking radios can be obtained from a state of nature in which 20% are used-looking is found in Appendix Table B.1.

$$P(X = 3|n = 3, \pi = .2) = P(A|B_1) = .008$$

The other 2 conditional probabilities are found in a like manner.

$$P(X = 3|n = 3, \pi = .5) = P(A|B_2) = .125$$

$$P(X = 3|n = 3, \pi = .8) = P(A|B_3) = .512$$

Applying Bayes' rule with a tabular form, we obtain Table 16.11. Notice that the sum of the joint probabilities represents the marginal probability of obtaining 3 used-looking radios when a sample of 3 is taken.

The posterior probabilities are revised measures of the probabilities of the occurrence of each of the states of nature. Notice that B_3 is now shown to be more likely, and B_1 and B_2 less likely than before the sample. They can be applied

TABLE 16.11 Table of Bayes' rule formulas and computations

	Formulas			
States of Nature	***Prior Probability***	***Conditional Probability***	***Joint Probability***	***Posterior Probability***
B_1	$P(B_1)$	$P(A\|B_1)$	$P(B_1)P(A\|B_1)$	$P(B_1\|A)$
B_2	$P(B_2)$	$P(A\|B_2)$	$P(B_2)P(A\|B_2)$	$P(B_2\|A)$
.	.	.	.	.
.	.	.	.	.
.	.	.	.	.
B_n	$P(B_n)$	$P(A\|B_n)$	$P(B_n)P(A\|B_n)$	$P(B_n\|A)$
	1		$P(A)$	1

	Computations			
States of Nature	***Prior Probability***	***Conditional Probability***	***Joint Probability***	***Posterior Probability***
B_1	.1	.008	.0008	.002
B_2	.2	.125	.0250	.065
B_3	.7	.512	.3584	.933
	1.0		$P(X = 3) = .3842$	1.000

to the opportunity payoff table in the same manner as the prior probabilities were earlier.

$$\begin{aligned} EMG_1 &= 800(.002) + 200(.065) - 400(.933) \\ &= 1.6 + 13 - 373.2 = -358.6 \end{aligned}$$

$$EMG_2 = 0(.002) + 0(.065) - 0(.933) = 0$$

The maximum EMG of 0 leads us to choose action a_2. The owner will not buy the radios. To find the $ELSI$, we calculate as follows.

$$\begin{aligned} EPPI &= 800(.002) + 200(.065) + 0(.933) \\ &= 1.6 + 13 = 14.6 \end{aligned}$$

$$ELSI_3 = EPPI - \max(EMG) = 14.6 - 0 = 14.6$$

The subscript 3 indicates the $ELSI$ for $X = 3$. We will find the $ELSI$ for X values of 0, 1, and 2 a bit later. The same solution can be found from the opportunity loss table.

$$\begin{aligned} EOL_1 &= 0(.002) + 0(.065) + 400(.933) \\ &= 0 + 0 + 373.2 = 373.2 \end{aligned}$$

$$\begin{aligned} EOL_2 &= 800(.002) + 200(.065) + 0(.933) \\ &= 1.6 + 13 + 0 = 14.6 \end{aligned}$$

The minimum EOL corresponds to action a_2. The owner will not buy the radios. The $EVPI$ after sampling is the minimum EOL of sampling information, $ELSI$.

$$ELSI_3 = \min(EOL) = 14.6$$

Although the previous example used the binomial distribution to find the conditional probabilities, other distributions, such as the normal distribution, would be appropriate in other situations. In some applications, the conditional probabilities are not found through sampling but are obtained from other data. Example 16.14 is typical.

EXAMPLE 16.14 A farmer must decide whether to plant corn May 1 or wait until June 1. His payoff is affected by weather conditions between May 1 and June 1 that may be poor or good. The probability of poor weather occurring is .6, of good weather occurring, .4. The opportunity loss table in thousands of dollars is given in Table 16.12. He is planning to hire John Storms, a weather forecaster, to predict the weather for the month of May. In the past, when the weather was poor, he had predicted poor 80% of the time and good 20% of the time. When the weather was good, he had predicted poor 10% of the time and good 90% of the time. If the farmer made his decision without any additional information from John Storms, when will he plant? If Mr. Storms tells him that the weather will be good during May, when will he plant?

TABLE 16.12 Opportunity loss table for corn planting

	Weather	
Plant	***Good***	***Poor***
May 1	0	40
June 1	10	0

Solution: With a prior analysis using expected values, we can answer the first question.

$$EOL_1 = .4(0) + .6(40) = 0 + 24 = 24$$

$$EOL_2 = .4(10) + .6(0) = 4 + 0 = 4$$

The minimum *EOL* corresponds to a_2, so we would decide to wait until June 1 to plant.

If Mr. Storms forecasts for us, we can use the conditional probabilities supplied concerning the accuracy of his forecasts. Since he forecasted that the weather would be good, we need only use the conditional probabilities associated with that outcome. See Table 16.13. Applying these probabilities to the loss table, we find

$$EOL_1 = .75(0) + .25(40) = 0 + 10 = 10$$

$$EOL_2 = .75(10) + .25(0) = 7.5 + 0 = 7.5$$

The minimum *EOL* occurs for a_2; hence, the decision is to wait until June 1 to plant the crop. It appears then that no matter whether Mr. Storms says the weather will be good or bad, we will make the same decision to hold off planting. Therefore, there is no reason to hire him. ■

TABLE 16.13 Bayes' rule computations for weather forecast

	Prior	***Conditional***	***Joint***	***Posterior***
Good	.4	.9	.36	.75
Poor	.6	.2	.12	.25
			.48	

EXERCISES

16.8 A local nursery operator has been buying rose bushes from various suppliers in lots of 1000 for a number of years. These lots are leftovers from the previous year's sales and contain both useless bushes and good bushes. The good ones can be sold at regular price, but the poor ones must be sold at a very low sale price. The operator classifies the lot she receives as poor, fair, and great if they have 20%, 50%, or 80% good bushes, respectively. This year, she is offered the opportunity to take a sample of 3 bushes from the lot before buying. In her sample, 2 of the bushes are bad. Should she buy the lot? The payoff table and prior probabilities are

	Poor	***Fair***	***Great***
Buy	−100	200	500
Don't Buy	0	0	0

Prior Probabilities	
Poor	.7
Fair	.2
Great	.1

16.9 A large canning firm must decide how many loads of fish to buy. The choices are either 10 or 20 loads. The quality of the loads is never known, but for convenience, based on past data, they have been classified into 3 classes according to π, the proportion of fish exceeding 10 inches in length. Prior probabilities for the classes (states of nature) are $p(s_1) = .4$, $p(s_2) = .2$, and $p(s_3) = .4$. The payoff table is as follows. For your information, the firm should buy 20 loads according to the expected value criterion. Also, $EMG_2 = 36$ and $EOL_2 = 4$.

	States of Nature		
	s_1	s_2	s_3
Actions	$\pi = .3$	$\pi = .6$	$\pi = .9$
a_1 ***Buy 10 Loads***	20	30	50
a_2 ***Buy 20 Loads***	10	40	60

(a) Suppose you are allowed to inspect 1 fish at random. If it is longer than 10 inches, how many loads should you buy?

(b) Suppose the fish you inspect is less than 10 inches. How many loads should you buy?

16.10 Mary Szabo builds and manages a large number of apartment houses in the Boston suburbs. She is planning a new building in an area where she currently has none and is trying to decide whether the new building should have 10, 15, or 25 apartments. She figures that there are 3 potential levels of demand: high, medium, and low; she has estimated the probabilities of each to be .40, .40, and .20, respectively. The monthly payoff table is as follows.

Apartment	Demand High	Medium	Low
10	6,000	4,000	4,500
15	8,000	7,000	3,500
25	15,000	6,000	2,000

A research firm is paid to estimate demand. The following data are provided by the firm relative to past experience.

If Event Was	Research Demand Estimate Indicated: High	Medium	Low
High Demand	.80	.15	.05
Medium Demand	.08	.75	.17
Low Demand	.02	.18	.80

If the research indicates low demand, what decision will she make?

16.6 EXPECTED VALUE OF SAMPLE INFORMATION

If a charge is to be made for each sample taken, it is necessary to determine the maximum amount that should be paid. The maximum charge is known as the expected value of sample information ($EVSI$).

DEFINITION 16.8 The **expected value of sample information** represents the difference between the expected value of perfect information before sampling and the average of the expected values of perfect information after sampling.

Equation 16.7 shows the calculation of the $EVSI$.

$$EVSI = EVPI - E[ELSI] \tag{16.7}$$

One calculation for $ELSI$ is the minimum EOL. Therefore, the minimum EOL must be found for each possible outcome of the sample. The marginal probability represents the probability of each of these min EOL occurring. The sum of their products is the average $ELSI$.

EXAMPLE 16.15 Calculate the $ELSI_i$ and marginal probabilities for each of the other 3 sample outcomes not calculated in Example 16.13.

Solution: For each of the following, the conditional probabilities are found in Appendix Table B.1 by using the appropriate X, π, and $n = 3$. See Table 16.14 for the case $X = 0$. The marginal probability is the sum of the joint probabilities.

TABLE 16.14 Bayes' rule computations for $X = 0$

States of Nature	*Prior Probability*	*Conditional Probability*	*Joint Probability*	*Posterior Probability*
B_1	.1	.512	.0512	.626
B_2	.2	.125	.0250	.306
B_3	.7	.008	.0056	.068
	1.0		$P(X = 0) = .0818$	1.000

Applying the posterior probabilities to the opportunity loss table, we obtain

$$\begin{aligned} EOL_1 &= 0(.626) + 0(.306) + 400(.068) \\ &= 0 + 0 + 27.2 = 27.2 \end{aligned}$$

$$\begin{aligned} EOL_2 &= (800)(.626) + (200)(.306) + (0)(.068) \\ &= 500.8 + 61.2 + 0 = 562.0 \end{aligned}$$

The $ELSI_0$ is 27.2. Choose action a_1. Note that, as in previous examples, we could also use the payoff table. Also note that the sum of both the prior and posterior probabilities is 1. See Table 16.15 for the case $X = 1$.

$$\begin{aligned} EOL_1 &= (0)(.213) + (0)(.415) + (400)(.372) \\ &= 0 + 0 + 148.8 = 148.8 \end{aligned}$$

$$\begin{aligned} EOL_2 &= (800)(.213) + (200)(.415) + (0)(.372) \\ &= 170.4 + 83.0 + 0 = 253.4 \end{aligned}$$

The $ELSI_1$ is 148.8. Choose action a_1. See Table 16.16 for the case $X = 2$.

$$\begin{aligned} EOL_1 &= 0(.027) + 0(.212) + 400(.761) \\ &= 0 + 0 + 304.4 = 304.4 \end{aligned}$$

$$\begin{aligned} EOL_2 &= 800(.027) + 200(.212) + 0(.761) \\ &= 21.6 + 42.4 + 0 = 64.0 \end{aligned}$$

The $ELSI_2$ is 64.0. Choose action a_2. The average *ELSI* is the expected value of the *ELSI*. See Table 16.17. Note that the sum of the marginal probabilities in Table 16.17 is 1. Recall from Example 16.12 that the *EVPI* prior to sampling is 120. Therefore, the *EVSI* is

$$EVSI = EVPI - E[ELSI] = 120 - 57.325 = 62.675$$

The storeowner should pay no more than $62.67 to take a sample of 3 radios. ■

Tree diagrams were used in prior analyses. They may be used in Bayesian analysis as well. An additional set of branches is required to indicate the

TABLE 16.15 Bayes' rule computations for $X = 1$

States of Nature	Prior Probability	Conditional Probability	Joint Probability	Posterior Probability
B_1	.1	.384	.0384	.213
B_2	.2	.375	.0750	.415
B_3	.7	.096	.0672	.372
	1.0		$P(X = 1) = .1806$	1.000

TABLE 16.16 Bayes' rule computations for $X = 2$

States of Nature	Prior Probability	Conditional Probability	Joint Probability	Posterior Probability
B_1	.1	.0960	.0096	.027
B_2	.2	.3750	.0750	.212
B_3	.7	.3840	.2688	.761
	1.0		$P(X = 2) = .3534$	1.000

TABLE 16.17 Computations for $E[ELSI]$

X	Action Chosen	ELSI	Marginal Probability	(ELSI) · (Marginal Probability)
0	a_1	27.2	.0818	2.225
1	a_1	148.8	.1806	26.873
2	a_2	64.0	.3534	22.618
3	a_2	14.6	.3842	5.609
			1.000	$E[ELSI] = 57.325$

possible outcome of the sample or other equivalent additional information. Once configured, they can be used to determine the optimal decision at each node and the expected value of sampling information. They can also be used with either payoff or loss table values.

To determine the optimal decisions, we use the following procedure.

1. Working from the right, calculate the expected payoff (or loss) at each probability node and indicate the value above the node.

FIGURE 16.7 Tree diagram for Example 16.16

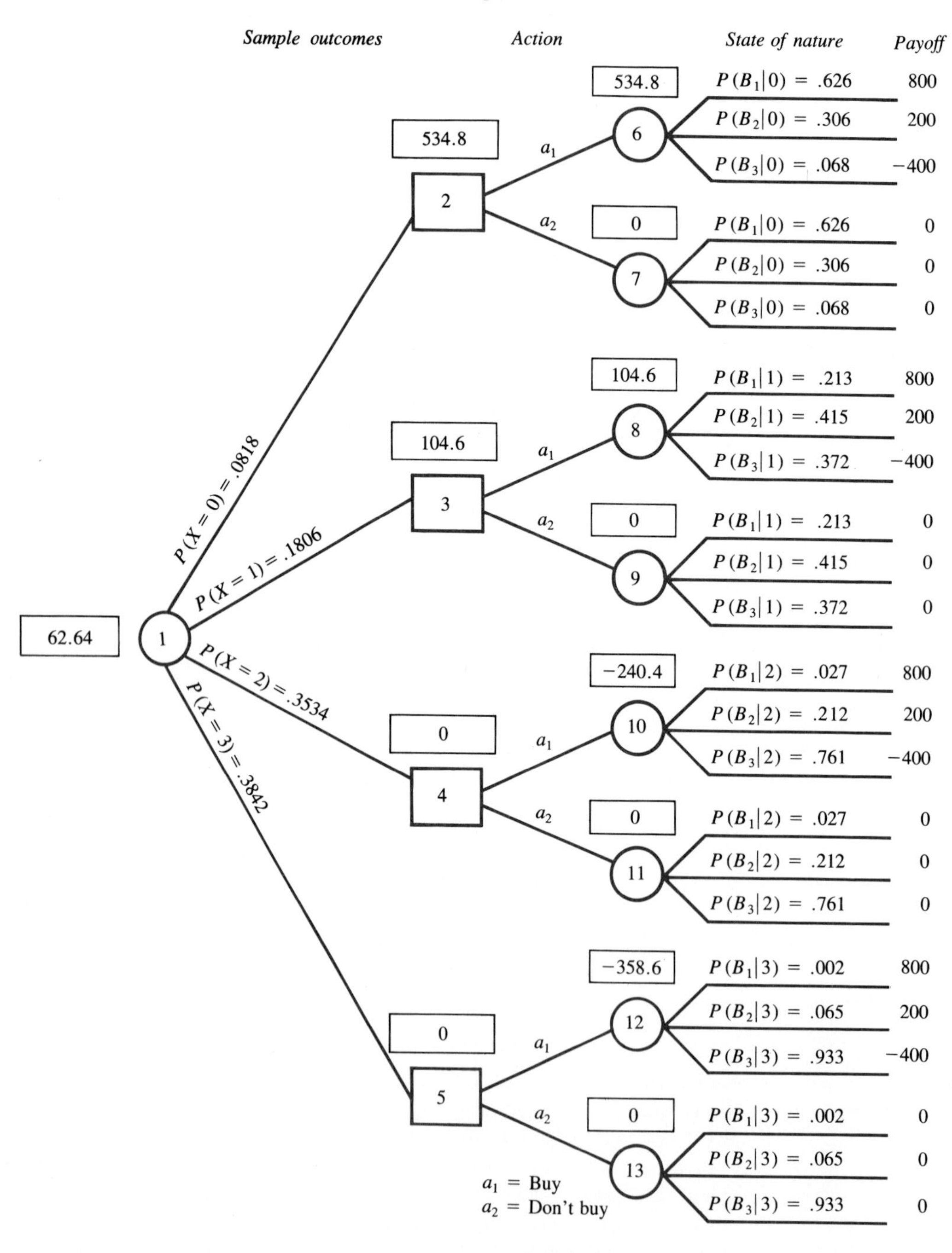

2. Moving left, at each decision node select the optimal decision and indicate its expected value above the node.
3. Finally, moving left to the probability node, calculate the expected value of the alternative decisions. This value represents the *EVSI*.

This procedure is now used in Example 16.16.

EXAMPLE 16.16 Draw a tree diagram for the surplus radio problem of Example 16.13 and indicate the expected values at each node. Use payoff table data and determine the expected value of sampling information.

Solution: The tree diagram is shown in Figure 16.7. Notice that the first node branches into the possible sample outcomes of 0, 1, 2, or 3 used-looking radios. Each of these branches into possible decisions, and each of these branches into the 3 states of nature with the revised, or posterior, probabilities shown. From the figure, the expected values for each decision are calculated. For example, where 0 used-looking radios are sampled, the expected monetary gain is

$$\begin{aligned} EMG_1 &= .626(800) + (.306)(200) + (.068)(-400) \\ &= 500.8 + 61.2 - 27.2 = 534.8 \end{aligned}$$

$$EMG_2 = 0$$

The decision under this sample outcome would be to buy the radios. This calculation can be repeated for the remaining 3 possible outcomes. The last node to the left gives the *EVSI* of 62.64 (slightly different from rounding).

EXERCISES

16.11 Calculate the expected value of sampling information, using the nursery data from Exercise 16.8.

16.12 Calculate the expected value of sampling information, using the canning firm data from Exercise 16.9.

16.13 Calculate the expected value of sampling information, using the apartment decision data of Exercise 16.10.

16.7 UTILITY THEORY

The payoff and loss tables in previous sections used dollars as their basic unit of measurement. With some thought, we would probably agree that not everyone realizes the same value for a given sum of money. For an executive of a small business, $100,000 invested in a new venture may be a significant amount and may require a high probability of success. To the multimillionaire, such a sum would be insignificant and a much lower probability of success would be required. We assume here that the person with the large resources would be willing to take the greater risks. Such an assumption may not be the case and would vary with each executive or decision maker.

If it is accepted that the value of an item should be measured by its utility rather than by its price, then we must turn to the principles of utility theory for a measurement technique.

Utility theory is beyond the scope of this book, but the interested reader is encouraged to read the treatments on the topic by Luce and Raiffa [1] or by Schlaifer [2].

16.8 SUMMARY

In this chapter, we considered decision-making techniques that do not involve statistical inference. Initially, we worked with conditional payoff and opportunity loss tables for given states of nature and possible alternative actions or decisions.

We developed decision criteria for choosing among the alternatives. Included were the maximin, maximax, minimax, and expected value approaches. Emphasis was given to the expected value criterion. These techniques were used with both payoff and loss tables. We examined the expected value of perfect information as a measure of how close the best decision alternative is to optimal decision making.

We also saw that information derived from a sample could be incorporated into the decision process. We reviewed the procedure for determining the value of a sample. Table 16.18 provides a tabular summary of the chapter and presents important formulas.

TABLE 16.18 Summary table

Tables	Conditional payoff tables (Payoff = Revenue − Cost) States of nature are outcomes that are due to chance. Actions are possible decision alternatives. Opportunity loss tables For each state of nature, the best action receives a loss of 0. Other actions are given difference between best payoff and their payoff.
Decision Criteria for Prior Analysis	Maximin—pessimistic Use payoff table. Determine minimum payoff for each action. Select action that gives largest of these minimum payoffs. Maximax—optimistic Use payoff table. Determine maximum payoff for each action. Select action that gives largest of these maximum payoffs.

Decision Criteria for Prior Analysis

Minimax—loss oriented

Use loss table.
Determine maximum loss for each action.
Select action for which maximum loss is the smallest.

Expected value

Use either table.
Calculate expected value for each action.

$$E[a_i] = \sum_j X_{ij} p(s_j) = \begin{cases} EMG & \text{(payoff table)} \\ EOL & \text{(loss table)} \end{cases}$$

$p(s_j)$ = probability for state of nature s_j

Select action that gives (a) largest EMG for payoff table or (b) smallest EOL for loss table.

Expected Value of Perfect Information

Use of payoff table

Calculate $EPPI$.

$$EPPI = \sum_j \max(X_{ij}) p(s_j)$$

Find $\max(EMG)$ using expected value.
Determine $EVPI$,

$$EVPI = EPPI - \max(EMG)$$

Use of loss table

$$EVPI = \min(EOL)$$

Posterior Analysis

Use of sample information

With Bayes' rule,

$$P(B_j|A) = \frac{P(B_j)P(A|B_j)}{\sum P(B_i)P(A|B_i)}$$

In tablular form,

Prior	*Conditional*	*Joint*	*Posterior*
$P(B_1)$	$P(A\|B_1)$	$P(B_1)P(A\|B_1)$	$P(B_1\|A)$
$P(B_2)$	$P(A\|B_2)$	$P(B_2)P(A\|B_2)$	$P(B_2\|A)$
.	.	.	.
.	.	.	.
.	.	.	.
$P(B_n)$	$P(A\|B_n)$	$P(B_n)P(A\|B_n)$	$P(B_n\|A)$
1		$P(A)$	1

Expected Value of Sample Information

$EVSI = EVPI - E[ELSI]$

where: $ELSI$ = expected value of perfect information after sampling

$EVPI$ = expected value of perfect information before sampling

SUPPLEMENTARY EXERCISES

16.14 A builder has purchased a large plot of land and will build either condominiums, small houses, or large houses. If the economy is bad, large houses will make the largest profit. Likewise, with an average economy, the largest profit will be obtained from small houses, and with a good economy, from condominiums.

(a) What are the states of nature?

(b) What are the alternatives?

16.15 Determine the opportunity loss table from the following conditional payoff table.

	s_1	s_2	s_3
a_1	10	8	14
a_2	16	4	12
a_3	6	12	7

16.16 If a well is drilled in eastern Ohio, it may hit oil, gas, or nothing. The probability of obtaining each is .1, .3, and .6, respectively. The conditional payoff table with payoffs in thousands of dollars is

	States of Nature		
Actions	***Oil***	***Gas***	***Nothing***
Drill	100	20	−30
Don't Drill	0	0	0

What decision will be made using the **(a)** maximin criterion, **(b)** maximax criterion, **(c)** minimax criterion, and **(d)** expected value criterion?

16.17 A jeweler must buy diamonds in large lots "as is" when buying in Amsterdam. Lots can be classified as poor, good, and excellent, depending on the percent of perfect diamonds in the lot. A jeweler has the opportunity to buy a lot of 500 diamonds for \$150,000. He can sell perfect stones for \$500 and all others for \$100. He takes a sample of size 2 from the lot and finds 1 is perfect. Probabilities, states of nature, and the opportunity loss table are given in the table (classification based on proportion of perfect stones, θ). Prior probabilities are .3, .5, and .2, respectively.

	Poor ($\theta = .2$)	***Good*** ($\theta = .4$)	***Excellent*** ($\theta = .7$)
Buy	60,000	20,000	0
Don't Buy	0	0	40,000

(a) Would he buy the lot before having sampled?

(b) Should he buy the lot after sampling?

(c) What is the expected value of perfect information?

16.18 For the data in Exercise 16.17, calculate the expected value of sample information.

16.19 The sports information director at a local university must order programs for basketball games in units of 100. He has found that they always sell between 400 and 800 programs at a game with the probability distribution in the following table. Programs cost 50¢ and sell for \$1.

Sales	Probability
400	.1
500	.2
600	.2
700	.3
800	.2

How many will he decide to order for each game if he uses **(a)** maximax, **(b)** maximin, **(c)** minimax, and **(d)** expected value?

16.20 Using the conditional payoff table calculated in Exercise 16.19, calculate the expected value of perfect information. Verify your answer, using the opportunity loss table.

16.21 A real estate developer has the opportunity to buy a plot of land for \$50,000, which, according to zoning regulations, must then be subdivided into 8 lots. Each will be priced at \$10,000. Since the developer must have \$50,000 available 2 months from now for another project, she will not invest if it appears probable that she will take a loss or not sell at least 5 lots within the 2-month time period.

(a) What are the alternatives?
(b) What are the states of nature?
(c) What is the payoff function?
(d) Determine the conditional payoff table.

16.22 Determine the opportunity loss table from the conditional payoff table calculated in Exercise 16.21.

16.23 Two exams have been prepared and are randomly being given to your class and another class. One of the following is true: Both exams are hard, only 1 is hard, or both are easy. You have the option of taking your exam now or postponing it until tomorrow at a cost of $\frac{1}{2}$ a grade point. The payoff table is the following.

	States		
Actions	***Both Hard***	***1 Hard***	***Both Easy***
Take Now	1.0	2.0	3.0
Take Tomorrow	2.0	2.0	2.5

What action will you choose, using the **(a)** maximin criterion, **(b)** maximax criterion, and **(c)** minimax criterion?

16.24 From past experience, you believe the probabilities for the states of nature in Exercise 16.23 are as follows.

State	*Probability*
Both hard	.5
1 hard	.3
Both easy	.2

What decision should you make, using the expected value criterion?

16.25 Calculate the expected value of perfect information, using the conditional payoff table of Exercise 16.23.

16.26 A student visiting another university for a football game discovers it is their homecoming weekend and that many coeds are wearing large mums to honor the event. The next year he considers selling mums outside the stadium the day of the homecoming game at his own school. A little marketing research convinces him that 1% of those attending will buy a mum. At past games, attendance has been 10,000, 15,000, or 20,000 people. He can buy mums for $1.50 each and can sell them for $2.

(a) What are the states of nature?
(b) What are the alternatives?
(c) What is the payoff function?
(d) Determine the conditional payoff table.

16.27 Determine the opportunity loss table from the conditional payoff table calculated in Exercise 16.26.

16.28 Calculate the expected value of perfect information, using the opportunity loss table calculated in Exercise 16.27. Prior probabilities are .2, .4, and .4.

16.29 You are in the pottery business and buy bowls in lots of 100 "as is" for $1800 from a broker. Some bowls have surface irregularities and can only be sold for $10, but perfect bowls can be sold for $30. You classify lots as good, fair, and poor, based on the percent of irregular bowls, which is 20%, 50%, and 70%, respectively. The prior probabilities are .1, .2, and .7, respectively.

(a) Calculate the payoff table.
(b) Will you buy the lot before having sampled?
(c) If you take a sample of size 1 and find a perfect bowl, will you buy the lot?
(d) What is the expected value of perfect information?

16.30 For the data in Exercise 16.29, calculate the expected value of sample information.

16.31 The management of a department store must make a decision regarding the location of a new store. It will be located either in a suburban mall on the edge of the city or downtown. Management is afraid that a gasoline crunch could hurt business and comes up with a payoff table for a 10-year period, based on gasoline availability in the long run.

	Availability of Gas		
	Poor	*Average*	*Excellent*
Mall	100,000	300,000	500,000
Downtown	300,000	250,000	200,000

The probability of gasoline availability of poor, average, and excellent is estimated to be .5, .3, and .2, respectively. What decision should management make?

16.32 The president of the department store of Exercise 16.31 has learned of a weekly letter, called the *Gas Report,* published by a consulting firm. Its consultants are experts in forecasting gasoline availability. The president hires them to forecast the long-run availability, and the consultants report that it will be average. They also supply the following table showing the reliability of their forecasts, which are conditional probabilities.

	Actual Outcome		
Forecast	*Poor*	*Average*	*Excellent*
Poor	.50	.20	.10
Average	.45	.70	.40
Excellent	.05	.10	.50

Use the president's own probability estimates and the appropriate numbers from the table.

- **(a)** Obtain a revised set of probabilities.
- **(b)** Determine the proper decision.

16.33 If the study in Exercise 16.32 had forecasted excellent availability, what decision would the president make?

16.34 Holiday Enterprises has decided to sell Christmas trees as an adjunct to its regular operations. The Scrooge brothers have a load of freshly cut trees they are willing to sell at a cost of $7 per tree. The grapevine has it that the quality of the trees in the past has not been very good. In fact, a load may vary in the percentage of good trees (full branched, well shaped) from as little as 30% to as high as 70%, with most loads having only half the trees classified as good. The occurrence of loads in these 3 categories is 30%, 10%, and 60%, respectively. Holiday figures that if it can sell top-quality trees for $13.00, it could still make money, even though the poorer-quality trees will be sold for $1 under the cost of the tree. Expenses to cover an attendant and other incidental costs will amount to $2 per tree. A truckload contains 40 bundles of trees with 10 trees in each bundle.

- **(a)** Establish the payoff table.
- **(b)** What should Holiday do under **(1)** maximax, **(2)** minimax, and **(3)** expected value?
- **(c)** What is perfect information worth to the company?
- **(d)** The Scrooge brothers said Holiday could randomly sample 2 trees (with replacement). What action should the company take if both are poor quality?

16.35 In Exercise 16.34, if the Scrooge brothers say that for a price, Holiday Enterprises can randomly sample 5 trees, what is the most that it should be willing to pay for the sample?

16.36 A flashbulb manufacturer produces flashbulbs with a defective rate of 1% when the materials are all of proper grade. However, if it uses an inferior batch of materials, which happens about 20% of the time, the defective rate of the bulbs is 20%. A certain local photo store advertises that it will return double what was paid for each bulb that does not flash. Each bulb, for which the dealer pays 15¢, sells for 20¢. She buys them in lots of 10,000 and wonders if she should continue to buy from this manufacturer, the only one from whom she can buy in bulk, since 1 out of 5 lots has a high defect rate.

- **(a)** Set up the payoff table.
- **(b)** Decide, using each of the various criteria, if the dealer should continue to buy from this manufacturer.
- **(c)** Calculate the expected value of perfect information.

16.37 For a payment of $50, the dealer in Exercise 16.36 is allowed to test 10 bulbs from the lot of 10,000. If she then chooses to buy the lot, the 10 bulbs are not replaced by the manufacturer.

- **(a)** Refigure the payoff table for these conditions.
- **(b)** If 2 bulbs from the sample prove to be defective, what will the dealer decide?

16.38 A local bank wishes to introduce automatic tellers at all its branches and, therefore, needs a computer program to perform the functions needed by this system. The bank needs to make a decision whether to write the program in-house or to purchase a package from a software vendor. The vendor claims that only minor modifications will be needed to make the package work on the bank's system. The bank's computer center manager realizes that once into an in-house job, programmers find it easy, difficult, or almost impossible to complete it. In modifying a package, they might find the same situations, so he prepares the following cost table. In modifying a package, the software vendor will help and guarantee success.

	Easy	*Difficult*	*Almost Impossible*
Program	10,000	60,000	100,000
Buy	40,000	50,000	60,000

Whether the bank writes the programs itself or buys a package, experience indicates that the probability that the job will be easy, difficult, or almost impossible is .2, .7, and .1, respectively. What decision should the bank make? Use each of the possible criteria.

REFERENCES

1. Luce, R. D., and H. Raiffa. *Games and Decisions*. New York: John Wiley, 1957.
2. Schlaifer, R. *Analysis of Decisions Under Uncertainty*. New York: McGraw-Hill, 1969.

17 Statistical Process Control

17.1 INTRODUCTION

Mary Johnson, president of Johnson Pressed Metals, stared at the letter she had just received from Universal Motors, her company's biggest customer. Mike Meyerson, from Universal's purchasing department, had written the letter explaining his company's new commitment to quality and the role that vendors would have in making it a reality. He said that Universal would no longer be inspecting incoming parts and materials and that vendors would therefore need to institute procedures that would guarantee that they were shipping only quality parts. The sentence that really caught Mary's attention stated, "If you wish us to continue buying parts from your company, you will have to implement a statistical process control system." Mary had read of this technique in trade journals, but since Johnson Pressed Metals was only a small company, she had not had the money to hire specialists in quality control. The company had, however, tried to turn out the best-quality products it knew how to make and had inspection programs to catch as many defective items as possible before shipment. Now Mary has no choice but to start a statistical process control program if her business is to survive. Since her company does not have the appropriate expertise in-house, she needs to obtain the services of a consultant.

Johnson Pressed Metals is not the only company in this situation. To meet foreign competition, many companies from the smallest to the largest are turning to statistical quality control techniques to help make products and services of the highest quality and are requiring their suppliers to do the same. As Mary Johnson becomes more involved in statistical process control, she will soon be requiring it of her raw material suppliers, just as Universal Motors is requiring it of her company.

This chapter covers the essential elements of a statistical process control program. Emphasis is given to control charts for both variables and attributes. However, attention is also devoted to process capability for variables characteristics and to acceptance sampling procedures.

17.2 BASIC CONCEPTS AND ASSUMPTIONS

Detection Versus Prevention

Until recently, the predominate system of quality control in the United States has been inspection oriented. In the inspection-oriented situation, the product is inspected at various points in the process, and, when the process is completed, the product is given a final inspection. Either every item is inspected (100% inspection) to find defective items or a sample of items is inspected and an inference made about the lot from which the sample was taken. As illustrated in Figure 17.1, defective items may be reworked, if economically or physically feasible, or scrapped. In using this approach, industry came to accept a certain percentage of scrap.

The system of *statistical process control* (SPC) operates under the philosophy of prevention of defects. If the characteristics that cause defective items to be produced can be identified and removed from the process, then only good items are produced and scrap is eliminated. The same machines producing at the same rate then yield a greater number of good items, resulting in greater productivity. The cost per item is thereby decreased. As a result, a more competitive product is produced. With these benefits, it is no wonder that this technique has become so popular in recent years. And, while the following discussion primarily has a manufacturing orientation, SPC is being successfully implemented in service industries as well.

Variation in Production

Every item manufactured has various dimensions that must be held to a certain size. But, because there are many influences on the machinery, there is a variation in size from part to part. These influences may include variations in the materials, the operators' abilities, the environment, and the operation of the machinery.

Two classes of variation are identified for the study of SPC: common and special.

FIGURE 17.1 Scrap in the production process

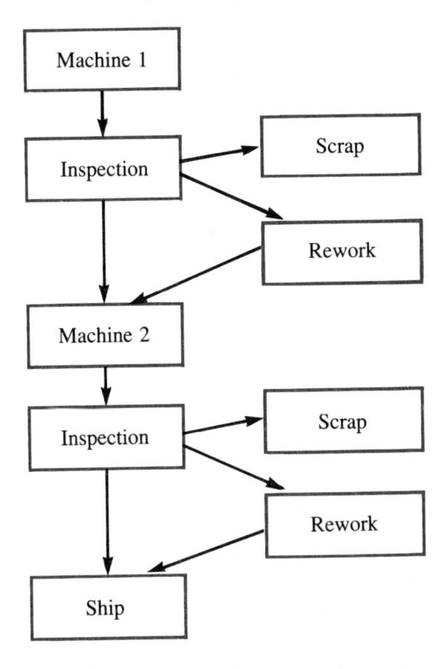

DEFINITION 17.1 **Common cause variation** is the collection of all of the individual causes of variation that are inherent in the process and that cannot be removed without management action.

Although each of the individual causes of variation that make up the total may have widely differing distributions, in combination the distribution is approximately normal. Therefore, whenever we believe that we have only common cause variation in a process, we assume that the process population distribution of the variable being measured is approximately normal. Where only common cause variation exists, it usually is not possible to reduce it without management action, such as the purchase of new machinery that can hold to closer tolerances.

DEFINITION 17.2 **Special cause variation** is the variation due to individual sources of variation that may be statistically identified and removed from the process.

When presented with adequate statistical evidence, operators, inspectors, engineers, and others knowledgeable about the production process often can identify a special cause of variation and can help to eliminate it from the process. Improper adjustment of the machinery, a change in the quality of raw materials, and improper training of a new operator are but a few of the many special causes that may add to the variation in the dimensions or other characteristics of the product.

Quality Control Charts

The object of statistical process control is first to eliminate special causes of variation and then to monitor the process to make sure that the desired process averages and variances are being maintained. Quality control charts are used for this purpose.

Charts can be constructed for both variables and attributes data. Where the process characteristics or variables are measured, we have variables data. Otherwise, if we are counting the number of defective or good items or the number of defects per unit, we have attributes data.

For each type of chart, we use periodic samples, which are also called subgroups, from the process. Subgroups are not necessarily random samples; instead they may be a set of items produced sequentially. The samples may be small or large, depending on the type of chart being used. Analyzing the time series of points representing each sample on the chart makes it possible to determine when a process is not operating properly—that is, when it is out of control.

17.3 CONTROL CHARTS FOR VARIABLES

When variables data are being analyzed, the data are graphed on mean and range charts. For each sample, the mean and the standard deviation are preferred for plotting, but since these techniques often are used by machine operators with a limited knowledge of statistical techniques, the range is used as a measure of variation instead of the standard deviation because it is easy to calculate.

The first step in developing a control chart to be used in plotting the sample means is the collection of a number of samples of a given size. Typically, the sample size is 5 items, but it may be any number from 2 on up. A sample is collected every hour, or other selected interval, until at least 20 samples are accumulated. For each sample, the mean $\overline{X}$ and range R are computed. Then, the average of the individual sample means is computed as an estimate of the actual process mean.

$$\overline{\overline{X}} = \frac{\Sigma \overline{X}_j}{k} \qquad \textbf{(17.1)}$$

where k is the number of subgroups (samples). Next, the average range is calculated.

$$\overline{R} = \frac{\Sigma R_j}{k} \qquad \textbf{(17.2)}$$

Using these two values, we proceed to compute the control chart limits.

By our earlier discussion, we expect the process to be normally distributed or approximately so if the primary causes of variation are common. Then, the distribution of sample averages, even for sample sizes as small as three, are also expected to be approximately normal. From Chapter 6, we know that the mean of the sampling distribution of $\overline{X}$ is the same as the mean of the process distribution and that the standard deviation of the sampling distribution of $\overline{X}$ is equal to the standard deviation of the process divided by the square root of the sample size.

$$\sigma_{\overline{X}} = \frac{\sigma_X}{\sqrt{n}}$$

where σ_X is estimated by s_X as in Chapter 5, giving

$$s_{\overline{X}} = \frac{s_X}{\sqrt{n}} \tag{17.3}$$

Using the symbolism of this chapter, the equation for s_X given in Chapter 5 becomes

$$s_X = \sqrt{\frac{\Sigma\Sigma(X_{ij} - \overline{\overline{X}})^2}{kn - 1}}$$

If we are using the range as a measure of variation, we need to be able to estimate the process standard deviation, using the average range. The average range is always larger than the sample standard deviation, so to obtain an estimate of s_X, we divide $\overline{R}$ by the value d_2 obtained from Appendix Table B.18. The value for d_2 is a function of the size of the sample. The estimate for s_X, denoted $\hat{s}_X$, is

$$\hat{s}_X = \frac{\overline{R}}{d_2} \tag{17.4}$$

One objective of using a mean chart, also called an X-bar chart, is to detect a shift in the process mean. The use of such a chart is essentially a hypothesis test. If a sample value is obtained that is much larger or smaller than the desired process mean, we conclude that the process mean has shifted. Upper and lower control limits are established to help us make this decision. Accordingly, we have the risk of a Type I error—that is, saying that the process mean has shifted when it has not—and of a Type II error—that is, saying that the process mean has not shifted when it has. Usually, the control limits are set at three standard deviations from the desired value of the mean to limit the chance of a Type I error. From tables of the normal curve, we can determine that the probability of saying that the process mean has shifted when it has not is approximately .003. Figure 17.2 illustrates these concepts.

Another way of looking at a quality control chart is as a confidence interval for $\overline{X}$. The upper and lower control limits are located at a distance three

FIGURE 17.2 Illustration of shift in process mean, Type I and Type II errors

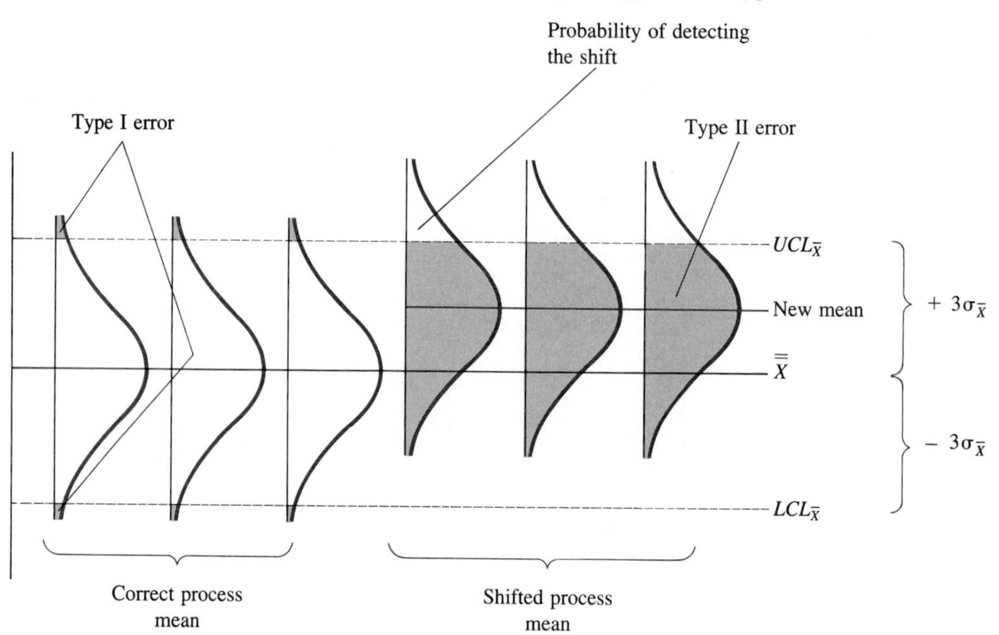

standard deviations from the average sample mean by the equation

$$CL_{\overline{X}} = \overline{\overline{X}} \pm 3\sigma_{\overline{X}}$$

We estimate $\sigma_{\overline{X}}$ with $s_{\overline{X}}$, giving

$$CL_{\overline{X}} = \overline{\overline{X}} \pm 3s_{\overline{X}} \qquad \textbf{(17.5)}$$

Substituting from Equation 17.3 for $s_{\overline{X}}$, we obtain

$$CL_{\overline{X}} = \overline{\overline{X}} \pm 3\left(\frac{s_X}{\sqrt{n}}\right)$$

In computerized applications or where hand calculators are used and difficulty of computation is not a consideration, the standard deviation is computed. However, where computations must be simplified, such as applications where machine operators are doing the calculations, it is preferable to compute the range. Therefore, s_X is estimated by $\overline{R}/d_2$, resulting in

$$CL_{\overline{X}} = \overline{\overline{X}} \pm 3[(\overline{R}/d_2)/\sqrt{n}]$$

Combining the 3, d_2, and $\sqrt{n}$ into a single value, we obtain the value A_2.

$$A_2 = \frac{3}{d_2\sqrt{n}} \tag{17.6}$$

Finally, the equation for calculating the control limits using the range as a measure of variation is

$$CL_{\overline{X}} = \overline{\overline{X}} \pm A_2\overline{R} \tag{17.7}$$

Values for A_2 are found in Appendix Table B.18. The control chart limits are illustrated in Figure 17.3.

A range chart is used in conjunction with an $\overline{X}$ chart to monitor the variability of the process. It is just as undesirable to have the variability increase as it is to experience a change in the process mean.

The range chart is used not only to monitor the process variability but also to help in identifying possible ways to decrease the overall variability. It is the ultimate goal of SPC to reduce the variation around the target value—that is, the desired process mean.

Control limits for monitoring variation are not computed as the average range plus or minus three standard deviations since the sampling distribution for the variance is not normally distributed. Instead, we simply find the values D_3 and D_4 in Appendix Table B.18 and compute the control limits, using Equations

FIGURE 17.3 Limits for mean and range charts

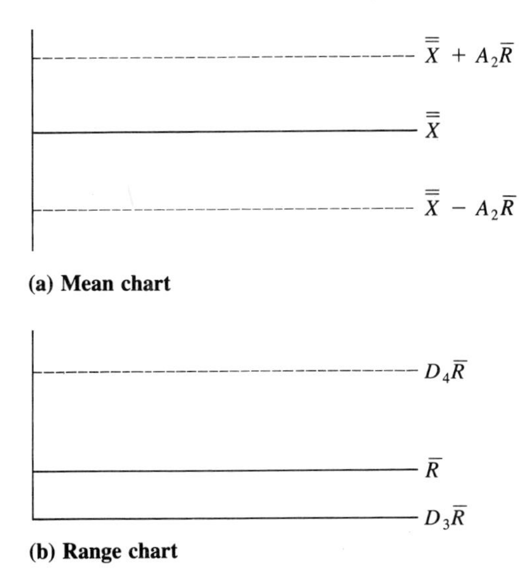

17.8a and 17.8b. These constants are based on the sample size and the relationship between $\overline{R}$ and σ_X for samples from normally distributed populations.

$$LCL_R = D_3\overline{R} \qquad \textbf{(17.8a)}$$
$$UCL_R = D_4\overline{R} \qquad \textbf{(17.8b)}$$

For sample sizes of six or less, the lower control limit is zero.

EXAMPLE 17.1 It is desired to control the quality of a process manufacturing springs used in a certain assembly. Observations to the nearest 10th of an inch for 6 subgroups of 4 observations each are given in the following table. Determine appropriate control limits for both the $\overline{X}$ and R charts and plot the means for each subgroup.

Time					
1	*2*	*3*	*4*	*5*	*6*
2.8	2.8	2.1	2.2	2.0	2.4
2.8	2.7	2.2	2.5	2.7	2.5
2.3	2.8	2.6	2.6	2.1	2.8
2.9	2.5	2.3	2.7	2.0	2.7

Solution: Calculating the values for $\overline{X}$ for each subgroup, we obtain 2.7, 2.7, 2.3, 2.5, 2.2, and 2.6, respectively. The mean of the $\bar{x}$ values, $\bar{\bar{x}}$, is 2.5. R values for the subgroups are .6, .3, .5, .5, .7, and .4, respectively. The mean of the R values, $\overline{R}$, is .5. From Appendix Table B.18, using a subgroup size of 4, we find the values of A_2, D_3, and D_4 to be .729, 0, and 2.282, respectively. Using Equation 17.7, we calculate control limits for the $\overline{X}$ chart as

$$CL = 2.5 \pm (.729)(.5) = 2.5 \pm .36$$
$$UCL_{\overline{X}} = 2.86$$
$$LCL_{\overline{X}} = 2.14$$

For limits on the range, we use Equations 17.8a and 17.8b.

$$UCL_R = (2.282)(.5) = 1.14$$

and

$$LCL_R = (0)(.5) = 0$$

Thus, control limits for the $\overline{X}$ chart are 2.14 and 2.86. For the R chart, the control limits are 0 and 1.14. The control charts are shown in Figure 17.4.

Note in the previous example that we have far fewer than the 25 or so subgroups that are normally used to set up control charts. This reduction in sample size was simply to ease the calculation requirements of the example.

Process control proceeds through a number of steps. The control limits for preliminary control charts are computed with no concern for the current state

FIGURE 17.4 Control charts for Example 17.1

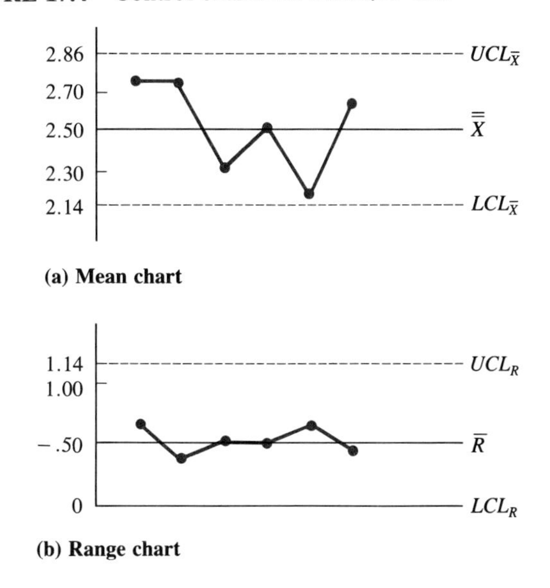

of process control. If some points are observed to be out of control and assignable causes can be found, they are discarded and the limits are recalculated. Otherwise, they are deemed to reflect random variation and should be included. These charts are used for diagnosis to attempt to eliminate special or assignable causes of variation. Control limits are recomputed as the process variation is reduced. When the variation has been reduced to only common variation, the final control limits can be calculated and the charts can be used for ongoing control.

Once the control limits have been established, the next step is to decide if the process mean or variability is out of control—that is, if the process mean has shifted or the variance has changed. Four different conditions will lead to the conclusion that the process is out of control.

1. A point outside the control limits
2. A run of seven or more points, either all above or all below the average line for the chart of interest ($\overline{X}$ or R)
3. A run of seven points all increasing or all decreasing
4. An obviously nonrandom pattern

Figure 17.5 illustrates each of these conditions.

In condition 1, we conclude that a process is out of control if a point is outside the limits since only a .003 chance exists that such a point could come

FIGURE 17.5 Control charts indicating out-of-control conditions

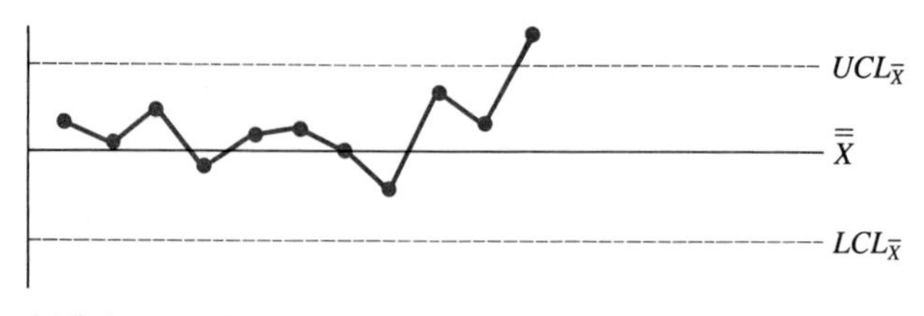

(a) Point out of limits

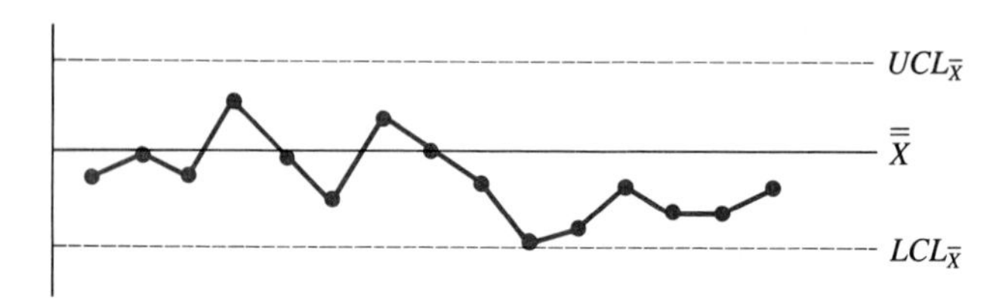

(b) Run of seven points below line

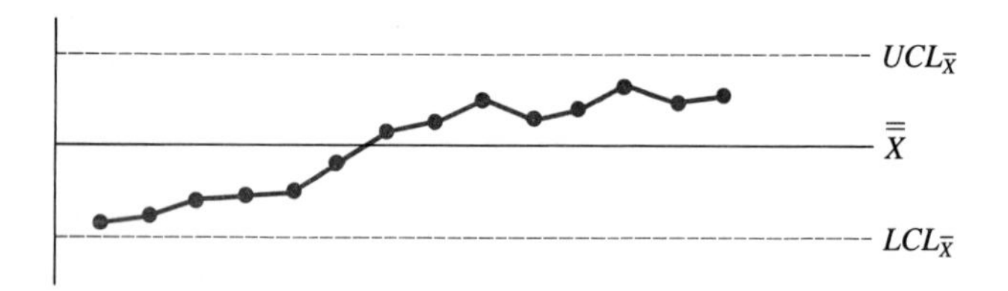

(c) Run of seven points increasing

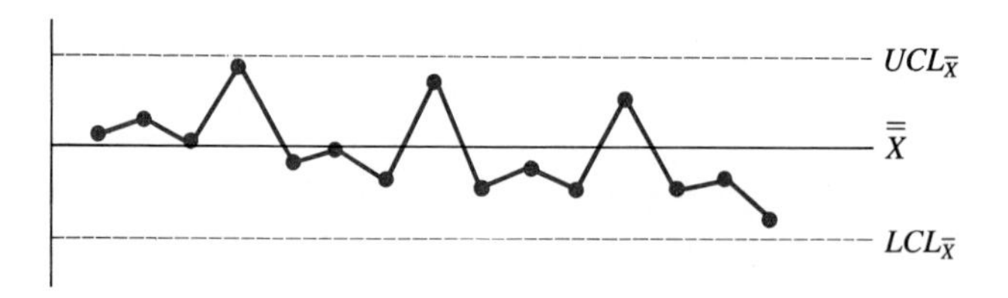

(d) Nonrandom pattern

from a process that is in control. This probability occurs because approximately 99.7% of all the area in a normal curve falls between the three standard deviation limits.

If we observe a run of seven or more points, either all above or below the average line, we conclude that a change has taken place in the process mean or variance. If, as in Figure 17.6, the process mean is on target for seven samples in

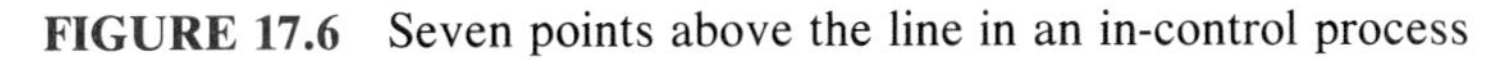

FIGURE 17.6 Seven points above the line in an in-control process

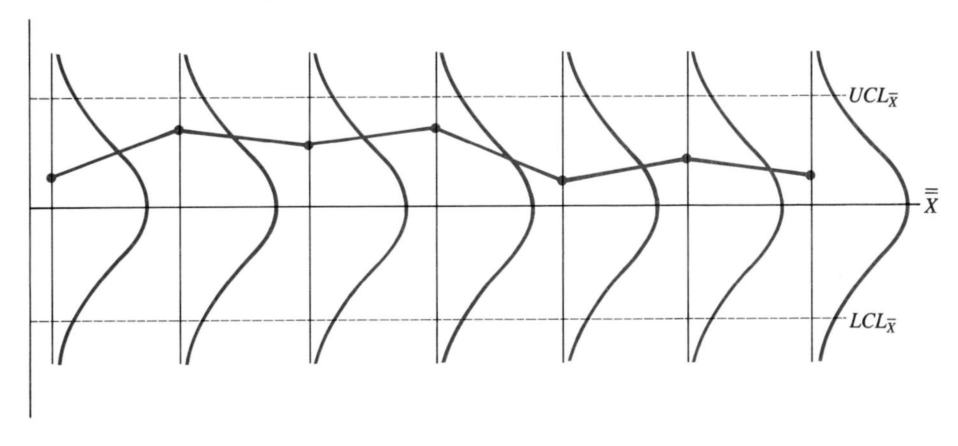

a row, the probability of observing seven sample points above the center line is $(.5)^7 \cong .008$. This probability is so small that we conclude that the process mean has increased. The same arguments hold for seven points below the central line.

We also call the process out of control if we observe a run of seven or more points either all increasing or all decreasing. The calculation of the probability that this run could happen for a series of samples all having the same process mean as in Figure 17.7 is not as simple as the previous calculation. It

FIGURE 17.7 Seven points in a row, each larger than previous point from an in-control process

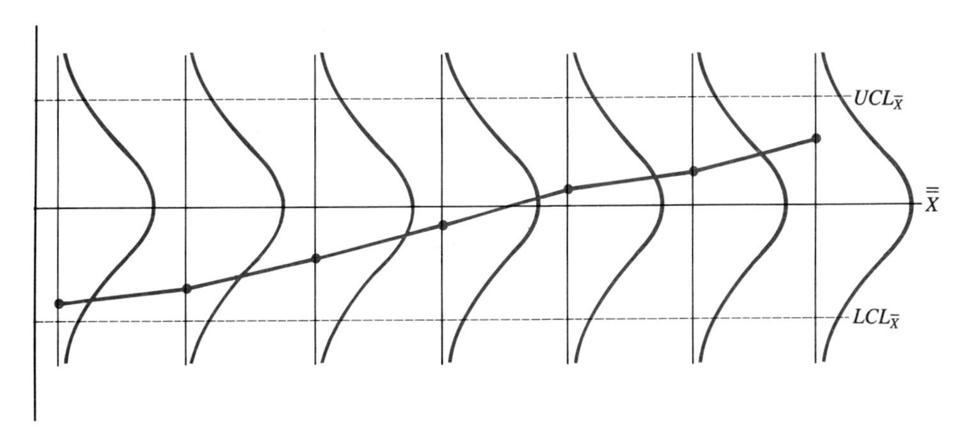

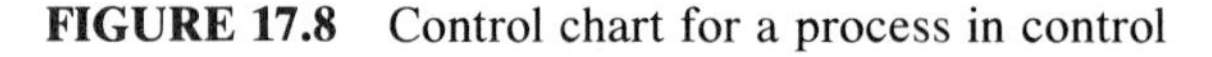

FIGURE 17.8 Control chart for a process in control

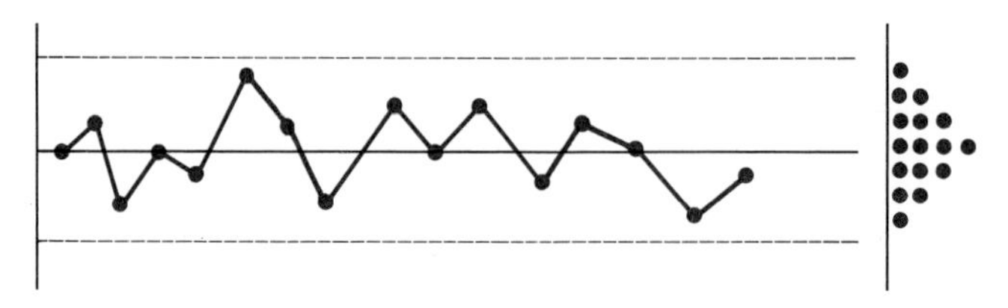

depends on where each point is located as to the probability that the next one will be above it. It is enough to say that this probability is also highly unlikely, and we should conclude that the process mean or variance is changing.

Finally, if the chart has an obviously nonrandom pattern, such as a cycle or a trend, the process is deemed to be out of control.

We have seen the conditions that lead to a conclusion that the process is out of control. How would we expect a chart to appear for a process that is in control? See Figure 17.8. First, we note that the points are scattered randomly about the central line. Extending the points to a histogram at the end of the chart, we note that the shape is approximately normal. Therefore, about 67% of the points are within one standard deviation and 95% are within two standard deviations of the mean.

Example 17.2 For each of the $\overline{X}$ charts shown in Figure 17.9, decide whether the process is in control or not and give the basis for the conclusion.

Solution:

- **a.** This process is out of control because 7 points in a row fall below the center line.
- **b.** This process is out of control because 7 points in a row are each progressively lower than the previous points.
- **c.** This process is out of control because the pattern is nonrandom. It is easy to guess approximately where the next point will be.
- **d.** This process is out of control because a point is outside the control limits.

An interesting use for control charts, besides monitoring the process for shifts in the mean or variation, is the diagnosis of the process by studying the patterns on the charts. Armed with a knowledge about a process and all the variables that affect it, we can use the charts in identifying the special or assignable sources of variation. When these sources are removed, the process variation is reduced and eventually reaches the minimum, which occurs when only common causes of variation remain.

FIGURE 17.9 Control charts for Example 17.2

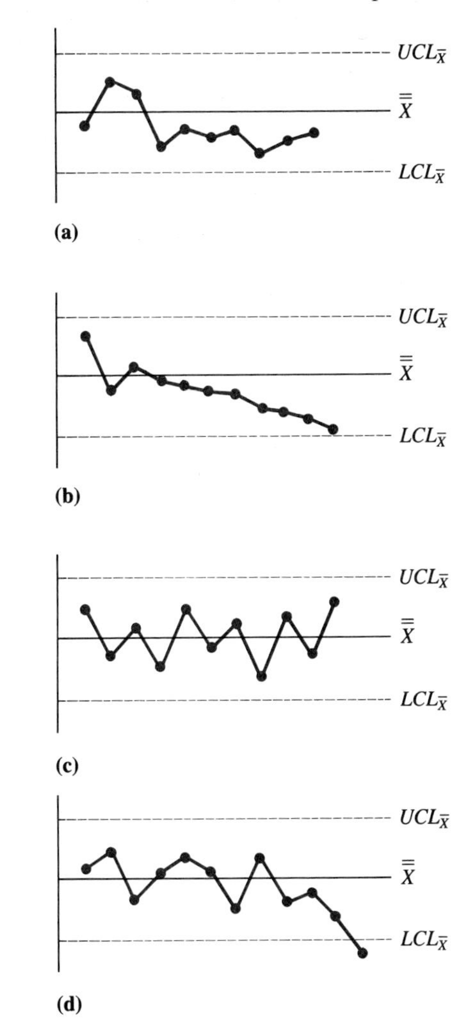

Unfortunately, it is difficult to teach chart reading outside of the context of real processes with which the student is familiar. Texts such as Ott [2] should be consulted.

EXERCISES

17.1 In a particular manufacturing process, it is desired to control the diameter of a disk. Measurements in inches for 5 subgroups of 4 observations each are as follows.

FIGURE 17.10 Control charts for Exercise 17.3

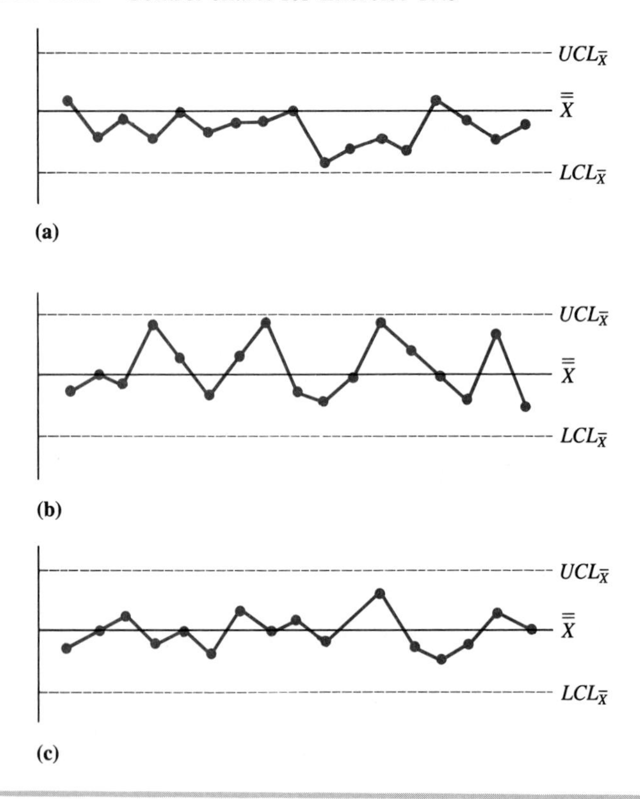

Subgroup				
1	*2*	*3*	*4*	*5*
1.00	0.96	0.99	1.01	0.98
0.99	1.04	1.05	1.04	1.02
1.01	0.98	1.04	1.00	1.01
1.00	1.02	1.04	0.99	0.99

Determine appropriate control limits for this process and comment on the state of control.

17.2 Manufacturers of Flint, a soft drink, desire to establish control for their bottling process. Data in ounces for 4 subgroups of 3 observations each are given as follows.

a. Determine appropriate control limits for this process.

b. Was the process in control? Explain why. Be specific.

Subgroup			
1	*2*	*3*	*4*
8	7	9	10
11	9	9	7
11	8	9	10

17.3 For each of the charts in Figure 17.10, determine if the process is out of control. State why.

17.4 PROCESS CAPABILITY FOR VARIABLES CHARACTERISTICS

When engineers design an item, they must decide on its dimensions. They must specify the dimensions of the item and where this item may fit into another item. Some of the dimensions may be critical. Consider, for example, a rod that must fit snugly through a block, as illustrated in Figure 17.11. If the diameter of the rod is too big, it will not fit. If it is too small, it will fit loosely. The reverse statements are true for the size of the hole in the block. If the diameters of the rod and the hole were precisely manufactured with no variation, a perfect fit would exist every time. However, there is always some variation in a manufacturing process. If the variation is minimal, then most of the time the two parts will fit together.

Although the engineer knows that the diameter of the rod needs to be 1 inch exactly, a variation of a couple of thousandths of an inch is considered

FIGURE 17.11 Two components requiring precision fit

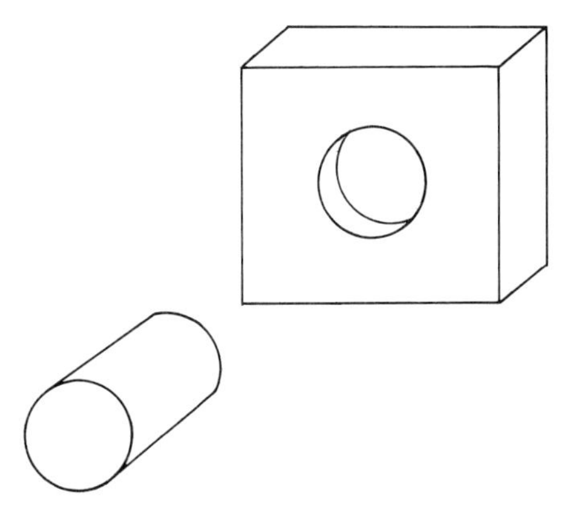

acceptable. Therefore, in designing the part, the engineer will state the largest and smallest sizes permissible. These sizes are known as the specifications. For example, the upper specification limit (USL) for the rod may be 1.0002 inches and the lower specification limit (LSL) may be .9998 inch. The range of these values is called the tolerance. The tolerance for the specified rod would be $1.0002 - .9998 = .0004$ inch. For the rod to be acceptable, it must be manufactured within specifications.

EXAMPLE 17.3 For the rod to fit in the block illustrated in Figure 17.11, the hole must be drilled $1.0001 \pm .0001$ inches. What are the specifications? What is the tolerance?

Solution: The upper specification limit is

$$USL = 1.0001 + .0001 = 1.0002$$

and the lower specification limit is

$$LSL = 1.0001 - .0001 = 1.0000$$

The tolerance is

$$1.0002 - 1.0000 = .0002$$

No single definition of process capability is accepted by all companies, but the least restrictive definition is the following.

DEFINITION 17.3 A process is **capable** if the natural tolerance of the process is less than the design tolerance.

The natural tolerance is defined as six standard deviations and the design tolerance is defined as the upper specification limit minus the lower specification limit. These definitions will be used later in this section.

Process capability studies can only be meaningfully performed on a process that is in control. This state exists only after all special or assignable causes of variation have been removed and only common causes remain. At this point, the process is operating at its best and cannot be improved without capital improvements. A capability study indicates the minimum variation of which the process is capable.

A capability study requires that the variance of the individual items produced by the process be estimated. The required estimate is usually obtained in one of two ways. One method involves taking a large sample of perhaps 100 to 200 items and calculating the sample standard deviation s_X. The other method uses range data from 20 to 40 subgroups from recently compiled range charts, so the total number of individual items is about 100 to 200. We assume here that the subgroup sample size is 5. The average range is calculated, and Equation 17.4 is used to estimate the standard deviation.

EXAMPLE 17.4 The data in Table 17.1 were obtained from $\overline{X}$ and R charts. Estimate the standard deviation of the process.

Solution: Although the number of data points is less than preferred, the calculations can be adequately illustrated. First, the range for each sample is computed.

Time	***8***	***10***	***12***	***2***	***4***	***6***	***8***	***10***
R	.02	.04	.03	.04	.04	.01	.02	.04

The average range is

$$\overline{R} = \frac{.24}{8} = .03$$

An estimate of the process standard deviation now is found from Equation 17.4.

$$\hat{s}_X = \frac{\overline{R}}{d_2} = \frac{.03}{2.33} = .0129$$

where d_2 is found in Appendix Table B.18 for a sample size of 5.

If we treat the data in Table 17.1 as a sample of 40 individual items, we can calculate the standard deviation by the following equation.

$$s_X = \sqrt{\frac{\Sigma(X_i - \overline{X})^2}{n - 1}} = \sqrt{\frac{\Sigma X_i^2 - \frac{(\Sigma X_i)^2}{n}}{n - 1}} = .0136$$

The two results are quite close.

When all assignable or special causes of variation have been removed from the process, all the variation that remains is from the many sources that make up the common variation. Both theory and experimental results lead us to expect that the variation of the measured variable of the individual items is normally distributed with mean μ_X and standard deviation σ_X. We can estimate

TABLE 17.1 Control chart data for Example 17.4

			Time				
8	***10***	***12***	***2***	***4***	***6***	***8***	***10***
1.01	1.03	1.02	1.03	1.04	1.03	1.02	1.01
1.03	1.04	1.05	1.06	1.02	1.04	1.03	1.00
1.02	1.01	1.03	1.03	1.05	1.03	1.02	1.04
1.03	1.05	1.04	1.02	1.01	1.04	1.04	1.04
1.02	1.03	1.05	1.03	1.02	1.04	1.02	1.01

μ_X by using $\overline{X}$ from sample data or $\overline{\overline{X}}$ from the control chart data, and we can estimate σ_X with either $\hat{s}$ or s_X, as demonstrated in Example 17.4.

It is possible to decide if a sample might have come from a process that is normally distributed by summarizing it by using the frequency table or the histogram introduced in Chapter 5. Example 17.5 demonstrates this process.

EXAMPLE 17.5 Is it possible that the data in Example 17.4 were obtained from a process whose output is normally distributed?

Solution: The data are conveniently grouped and tallied, as shown in Table 17.2. There is no reason to doubt that these data come from a normally distributed population.

Using the normal curve tables, we know that approximately 67% of the items produced by a process with normally distributed output have measurements within one standard deviation of the mean, that about 95% are within two standard deviations, and that 99.7% are within three standard deviations.

The least restrictive definition of process capability is one that says that a process is capable if 99.7% of the items produced fall within the upper and lower specification limits—that is, that the process mean is on target and the specification limits are three standard deviations from the mean. This condition is known as 3-sigma capability and is illustrated in Figure 17.12. The design tolerance (USL–LSL) is equal to six standard deviations, the magnitude of the natural tolerance.

Some companies require at least 4-sigma capability. This requirement implies that the specification limits are four standard deviations away from the mean, as shown in Figure 17.13. The tolerance would be equal to eight standard deviations. Again, we note that the process must be centered on the target process mean.

One of a number of possible measures of capability is the process capability percentage (PC%).

$$\text{PC\%} = \frac{6s_X}{\text{tolerance}}(100) \tag{17.9}$$

TABLE 17.2 Frequency table for data in Example 17.4

1.00	X
1.01	XXXXX
1.02	XXXXXXXXX
1.03	XXXXXXXXXXX
1.04	XXXXXXXXX
1.05	XXXX
1.06	X

FIGURE 17.12 Three-sigma capability

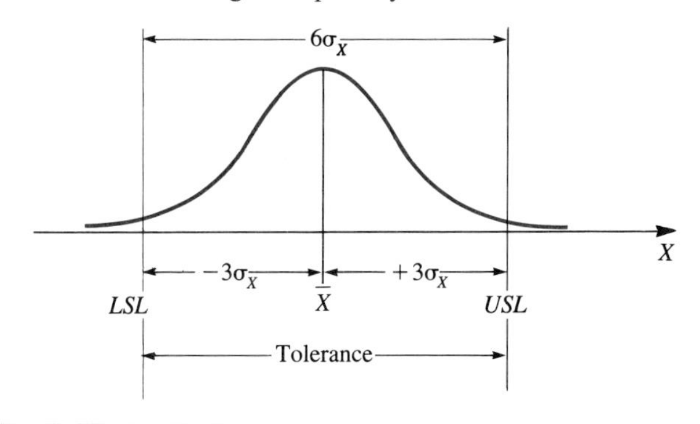

FIGURE 17.13 Four-sigma capability

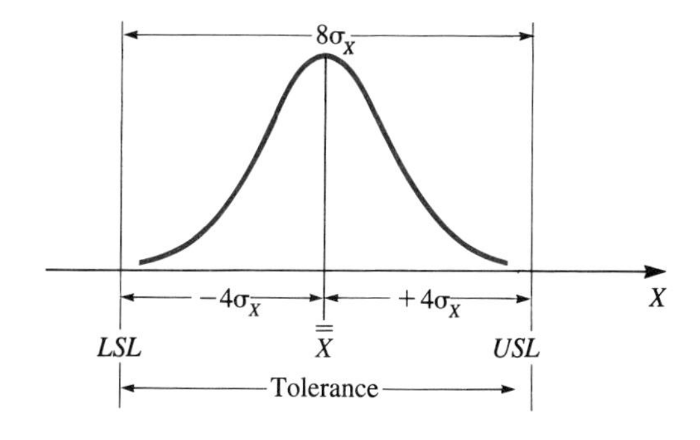

A measure of 100 would indicate 3-sigma capability. A measure of 75 indicates 4-sigma capability.

EXAMPLE 17.6 From the control chart data of Example 17.4, the standard deviation of the process is estimated to be .0129. The target for the process mean is 1.04, and the upper and lower specification limits are 1.08 and 1.00, respectively. For the PC% measure, does the process show 3-sigma capability?

Solution: First, we calculate the tolerance.

$$\text{Tolerance} = USL\text{–}LSL = 1.08 - 1.00 = .08$$

Then, we find PC%.

$$\text{PC\%} = \frac{6(.0129)}{.08}(100) = \frac{.0774}{.08}(100) = 96.75\%$$

FIGURE 17.14 Process mean closer to lower specification limit

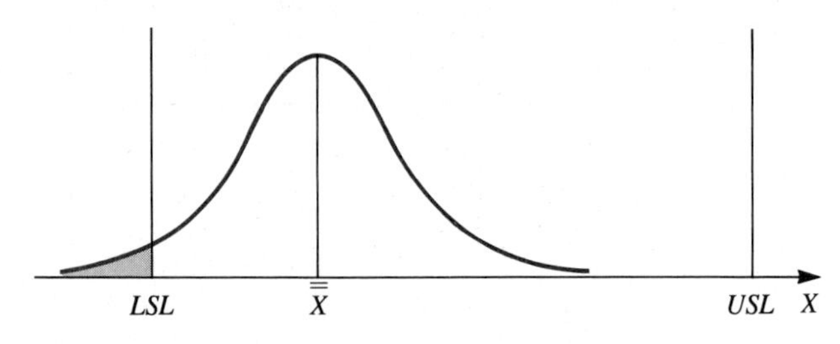

Since the value is between 100 and 75, the process is 3-sigma capable but not 4-sigma capable. The process is called capable only if the customer would accept 3-sigma capability.

Another measure of process capability, C_{pk}, does not assume that the process is on target but adjusts its measure to take into account that the process mean may be closer to one specification limit than the other, as shown in Figure 17.14. The C_{pk} measure is taken in relation to the closest specification limit by using Equation 17.10.

$$C_{pk} = \frac{\min(|z_{USL}|,|z_{LSL}|)}{3} \tag{17.10}$$

where

$$z_{USL} = \frac{USL - \overline{\overline{X}}}{s_X}$$

$$z_{LSL} = \frac{LSL - \overline{\overline{X}}}{s_X}$$

Three-sigma capability gives a value of 1, and 4-sigma capability gives a value of 1.333. Note that the absolute values of the z's are used.

EXAMPLE 17.7 Assume that the target value for a process is 1.04, with specification limits of 1.08 and 1.00. Again, using the data of Example 17.4, the mean $\overline{\overline{X}}$ is 1.0295 and the standard deviation is .0136. Calculate the process capability, using C_{pk}.

Solution: The first step is to find the distance from the process mean to each specification limit in terms of the number of standard deviations.

$$z_{USL} = \frac{1.08 - 1.0295}{.0136} = 3.7132$$

$$z_{LSL} = \frac{1.00 - 1.0295}{.0136} =-2.1691$$

Now, using Equation 17.10, we get

$$C_{pk} = \frac{\min(3.7132,\ 2.1691)}{3} = \frac{2.1691}{3} = .723$$

This process is not capable. With the smaller z, the process has a capability of $2.17\ \sigma_X$.

EXERCISES

17.4 Central Gear Manufacturing Company makes a special gear for which the thickness is a critical dimension and is charted. The data in the table, measured in millimeters, are taken from mean charts that show that the process appears to be in control. Decide if the data might have come from a normal distribution.

Time

8	*10*	*12*	*2*	*4*	*6*
10.4	10.5	10.4	10.7	10.4	10.6
10.6	10.5	10.6	10.3	10.5	10.4
10.4	10.6	10.5	10.2	10.3	10.4
10.3	10.7	10.6	10.5	10.3	10.5

17.5 For the data in Exercise 17.4, estimate the process mean and standard deviation. Estimate the standard deviation 2 ways: once using the total sample of 24 items and once using only the ranges.

17.6 The specification limits for the process in Exercise 17.4 are 10.0 and 10.8.

(a) Calculate both PC% and C_{pk}.

(b) Does the process show at least 3-sigma capability?

(c) Which measure is preferred here, and why?

17.7 A process from which control chart data have been collected used subgroup samples of 5. The specification limits are 2.21 and 2.22. Calculations from the data show $\bar{\bar{X}} = 2.215$ and $\bar{R} = .003$. Find PC% and C_{pk}. At what level is the process capable?

17.5 CONTROL CHARTS FOR ATTRIBUTES

Not all characteristics of an item can be measured. In some cases, the items are categorized as defective or not defective, and, in other cases, the number of defects on an item is counted. In these cases, we are using attribute data rather than variables data.

Where individual items are classified as defective or nondefective, the p chart is appropriate. For each subgroup of sample size n, the number of defective items X is counted. The sample proportion $p = X/n$ is calculated for each subgroup as an estimate of the process proportion π. The sampling distribution for control chart purposes would be based on the binomial distribution, but if sufficiently large samples, say $n \geq 30$, are taken, the binomial can be approximated by the normal distribution if the conditions stated in Chapter 4 are satisfied. Therefore, we limit ourselves to taking large samples for p charts. The estimator for the mean of the sampling distribution of p is $\bar{p}$, which is calculated by Equation 17.11 for n constant for all samples.

$$\bar{p} = \frac{\Sigma p_j}{k} = \frac{\Sigma X_j}{kn} \quad \textbf{(17.11)}$$

where k is the number of subgroups and n is the size of each subgroup sample. The standard deviation of the sampling distribution is estimated by Equation 17.12.

$$s_p = \sqrt{\frac{\bar{p}(1 - \bar{p})}{n}} \quad \textbf{(17.12)}$$

The control limits are calculated by Equation 17.13.

$$CL_p = \bar{p} \pm 3\sqrt{\frac{\bar{p}(1 - \bar{p})}{n}} \quad \textbf{(17.13)}$$

The same four rules that are used with the mean and range charts are used to determine if the process is out of control. In many cases, the lower control limit will be calculated as a negative number. Since it is impossible to have a negative proportion, the lower limit is set equal to zero.

If the point or points indicate an out-of-control situation, a shift in the process proportion defective is indicated. A point below a lower control limit would therefore indicate a decrease in the population proportion defective. This occurrence would ordinarily be viewed favorably. However, a point below the lower limit on a p chart should be investigated to make certain that it is not the result of an error of some type. Also, it should be investigated to attempt to ascertain what was done that was "right." If that factor can be identified, then it should be continued since the result is an improved process.

EXAMPLE 17.8 The following are the results for 25 days of the number of defective units in the daily inspection of 50 electronic devices.

1	4	10	0	4	5	9	14	1	8	2	7	4
5	1	6	1	10	4	3	5	6	0	7	8	

Determine proper control limits for this process. Is the process in control or out of control? Why?

Solution: It is appropriate to use a p chart here since our observations are binomial in nature. Each item is classified as defective or nondefective. $\bar{p}$ is calculated to be

$$\bar{p} = \frac{125}{(25)(50)} = .10$$

and control limits are

$$CL_p = .10 \pm 3\sqrt{\frac{(.10)(.90)}{50}} = .10 \pm .127$$

Hence,

$$UCL_p = .227 \quad \text{and} \quad LCL_p = 0$$

since a negative result is impossible.

The observation to be plotted for the 8th day is 14/50 = .28. The value of .28 is larger than .227, the upper control limit. Therefore, the conclusion is that the process was out of control on that particular day. All other observations, calculated in a similar manner, fall within control limits. The last observation, for example, is 8/50 = .16.

See Figure 17.15 for the plotting of the data in the form of a typical p chart.

The control chart with a point out of control in Figure 17.15 is typical of initial charts when SPC is first begun on a process. An attempt is made to determine why the process was out of control at that point and to eliminate that special cause of variation. Once a process is in control, any out-of-control point requires immediate analysis and correction.

A second type of attribute chart is the c chart, based on the Poisson distribution. It is designed to control the number of defects per unit. A chi-square goodness of fit test can be applied to determine if the data have a Poisson distribution.

For each of k subgroups, we take a single item and count the number of defects on or in the item. The value $\bar{c}$ represents the average number of defects per unit.

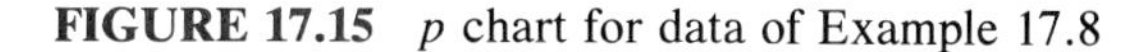

FIGURE 17.15 p chart for data of Example 17.8

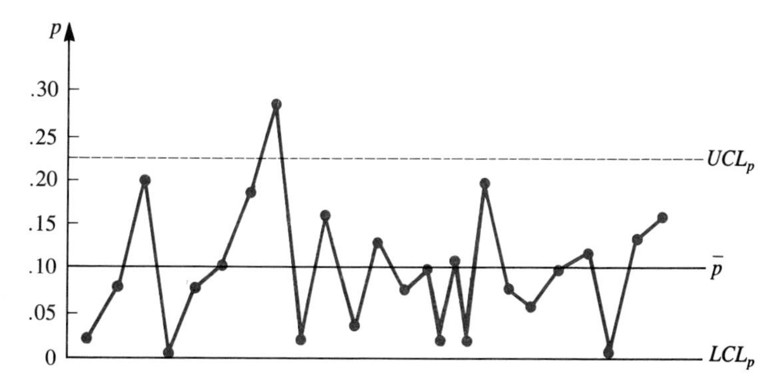

$$\bar{c} = \frac{\Sigma c_j}{k} \tag{17.14}$$

Since the variance is equal to the mean for a Poisson distribution, the standard deviation is given by

$$s_c = \sqrt{\bar{c}} \tag{17.15}$$

The control limits for a c chart are

$$CL_c = \bar{c} \pm 3\sqrt{\bar{c}} \tag{17.16}$$

If the lower control limit is calculated to be negative, it is adjusted to zero since a negative number of defects is impossible.

EXAMPLE 17.9 In an industrial process, metal sheets are manufactured. Each hour, a sheet is randomly selected and examined for surface defects. Suppose that over a period of 20 hours, the following counts of defects per sheet were observed

10	3	4	5	0	2	11	9	8	3
12	5	4	6	13	3	1	13	5	8

Determine appropriate control limits for the process and comment on the state of control.

Solution: A c chart is appropriate since defects per unit is usually considered to have a Poisson distribution. The total number of defects for all 20 sheets is 125. Therefore,

$$\bar{c} = \frac{125}{20} = 6.25$$

and, using Equation 17.16, the control limits are

$$CL_c = 6.25 \pm 3\sqrt{6.25} = 6.25 \pm 7.5$$

Hence,

$$UCL_c = 13.75 \quad \text{and} \quad LCL_c = 0$$

The process is in control since all points fall within the control limits.

Other control charts exist for attributes not covered in this text. Texts on statistical quality control, such as Grant and Leavenworth [1], may be consulted.

EXERCISES

17.8 For quality control of a certain electrical part, random samples of size 50 are taken periodically and the number of defective items determined. The following data show the number of defectives found in the inspection of 25 lots.

5	4	3	3	7	5	1	7	10	3	2	3	7
1	4	2	4	0	6	1	8	1	2	3	4	

Determine appropriate control limits and comment on the state of control.

17.9 The following are the number of defective hair dryer motors found during inspection in the first 10 working days of September in daily samples of 100 motors: 5, 9, 4, 0, 15, 7, 7, 5, 6, 2.

(a) Determine appropriate limits to be used for controlling the quality of the motors.
(b) Was the process in control during this period? Why?

17.10 A manufacturer of woolen goods is trying to control the quality of the product. For this purpose, 25 pieces of 100-yard bolts are checked and the number of defects per piece are counted. The counts are

6	3	1	8	5	2	4	5	9	4	3	3	7
4	3	9	1	6	3	8	5	6	4	5	7	

Was the manufacturing process in control during the production of these bolts?

17.11 A company manufacturing paper is checking its product for defects. A sample sheet is checked each hour, and the number of defects are recorded. Data for 24 consecutive sample sheets are

6	5	8	9	3	2	4	0	3	7	2	5
4	2	1	10	3	18	6	4	5	7	16	6

(a) Determine control limits for this process.
(b) Was the process in control during the collection of these data?

17.6 ACCEPTANCE SAMPLING

Another important tool that has affected modern process control practices is acceptance sampling or, as it is sometimes called, sampling inspection. For many industrial processes, it is an essential part of the sequence of steps required to provide assurance of control.

DEFINITION 17.4 Sampling procedures for inspecting the product in a lot or shipment for the purpose of deciding whether or not to accept the entire batch are known as **acceptance sampling** techniques.

Several stages of a process might utilize acceptance sampling techniques. Inspection of incoming materials to decide whether or not a shipment should be accepted is the most common instance. However, for most processes, it is desirable to assure that the product is acceptable as it leaves one phase of the manufacturing sequence to go on for further processing.

We assume that the product in question is available for inspection in a batch or lot. In some situations, 100% inspection of the lot may be required, but ordinarily a sampling procedure is preferred. There are three primary reasons for preferring sampling over 100% inspection. First, the cost of inspection is less. Second, it is a well-known fact that monotony and fatigue lead to errors of judgment when 100% inspection is used. Third, sampling is a necessity when obtaining the observation results in destroying the item.

The purpose of a sampling plan is to make a decision regarding a lot. That is, it is desired to accept or reject an entire lot based on the results obtained in a sample.

DEFINITION 17.5 A **sampling plan** indicates the number of units to be sampled from a lot and the necessary criteria for making a decision regarding the acceptance or rejection of the lot.

If items in the sample are categorized as acceptable (nondefective) or not acceptable (defective), the variable of interest is obviously discrete, specifically, a hypergeometric variable.

DEFINITION 17.6 Sampling procedures in which the observations are obtained as a result of classifying an item as defective or nondefective are known as **acceptance sampling by attributes.**

If items in the sample are measured on a continuous scale in regard to some characteristic, a decision to accept or reject a lot is usually based on the mean of the sample or on the mean and the standard deviation.

DEFINITION 17.7 Sampling procedures in which the observations are obtained as a result of measurements of a continuous variable are known as **acceptance sampling by variables.**

Both attributes and variables sampling plans will be considered in this section. The first type to be examined is acceptance sampling by attributes.

Acceptance Sampling by Attributes: Single Sampling Plans

As the name implies, in a single sampling plan, the decision to accept or reject a lot is based on the result of one sample. A single sampling plan is traditionally designated by two numbers: the sample size n and the acceptance number c. For example, one single sampling plan is $n = 50$, $c = 1$. For this plan, a random sample of 50 items is selected from the lot, each sampled item is examined to determine if it is defective, and the whole lot is accepted if the count of defective items is one or less and rejected if the count is two or more.

The characteristics of a sampling plan can be studied through its operating characteristic curve, known as the *OC* curve, which was introduced in Chapter 8. In general, we prefer to have a sampling plan accept all good lots and reject all bad lots. If lots are entirely free of defective items—that is, π, the proportion defective, is 0—the count of defectives would be 0 since no defectives would be in the sample of n and the probability of accepting the lot would be 1. This point is plotted as point A on the *OC* curve shown in Figure 17.16. If lots are entirely defective—that is, $\pi = 1$—then the number of defective items in the sample would be 50 and the probability of accepting the lot would be 0. In Figure 17.16, this point is not shown since it is considerably beyond the range of values for π shown in the curve.

Before looking at the *OC* curve in detail, we consider the analogy of the decision-making process for a single sampling plan to the procedure given in Chapter 8 for testing a hypothesis about a proportion. The hypotheses of interest are

$$H_0\colon \pi \leq \pi_0 \qquad \text{and} \qquad H_a\colon \pi > \pi_0$$

In Chapter 8, we reject H_0 only if the sample proportion is quite large, leading to a z value in the extreme right tail of the normal distribution—that is, in the area

FIGURE 17.16 *OC* curve for the single sampling plan of Example 17.10

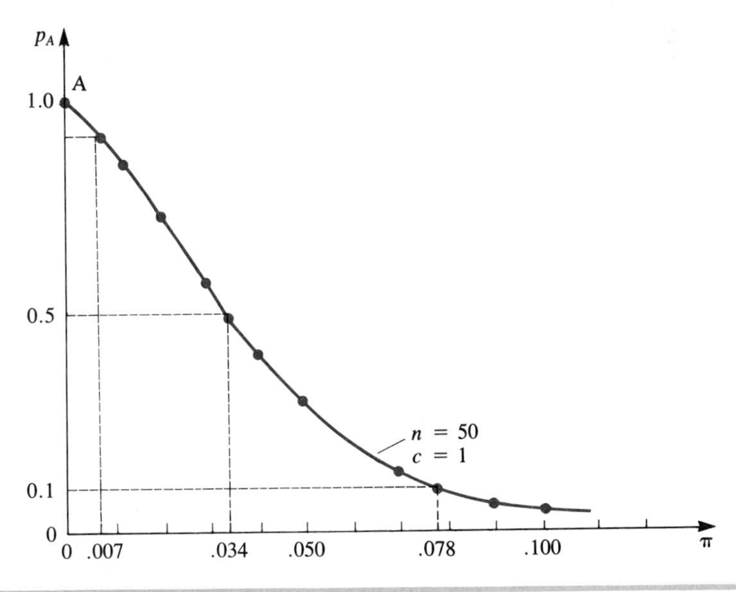

of the curve corresponding to α, the stated significance level or probability of a Type I error. Otherwise, the hypothesis is accepted.

In the acceptance sampling context, π_0 is considered to be the acceptable quality level, and the null hypothesis corresponds to the lot's being good.

DEFINITION 17.8 The **acceptable quality level** (AQL) is the maximum proportion defective that should be accepted with a specified large probability by a sampling plan.

If the proportion defective in the sample is relatively close to π_0, the lot is accepted. If the proportion defective in the sample is much larger than π_0, the lot is rejected. The probability of accepting lots where incoming quality is π can, of course, be calculated for each assumed value of π. The values are equivalent to those designated as corresponding to a Type II error in Chapter 8. The plot of the probabilities is the *OC* curve.

The sampling, assuming finite lots of size N, is really hypergeometric in nature. However, in most practical problems, N is quite large and n is quite small, so the binomial distribution, which assumes that the probability of obtaining a defective is a constant, provides a good approximation.

EXAMPLE 17.10 Determine points on the *OC* curve for the sampling plan $n = 50$, $c = 1$ and plot the curve.

Solution: For the plan $n = 50$, $c = 1$, the probability of acceptance p_A for the binomial distribution is equal to

$$p_A = \binom{50}{0} \pi^0 (1 - \pi)^{50} + \binom{50}{1} \pi^1 (1 - \pi)^{49} \tag{17.17}$$

If $\pi = .01$, for example, the use of Equation 17.17 results in $p_A = .6050 + .3056 = .9106$. For this problem, the required value can also be obtained from Appendix Table B.1 or Appendix Table B.2. Since $n\pi = (50)(.01) = .5$, the approximation provided by the Poisson distribution can be used. Note, however, that there is no need to approximate in this case since the exact binomial probability is easily determined. It is done here to illustrate the closeness of the approximation. Using Appendix Table B.4, we obtain directly

$$p_A = .9098$$

This answer agrees quite closely with that obtained from the binomial distribution.

Table 17.3 presents the results of using the Poisson distribution to calculate p_A, the probability of acceptance of lots with the proportion defective equal to various values of π. The resulting points are also plotted in Figure 17.16. The *OC* curve clearly shows how the probability of accepting lots varies with the proportion defective.

TABLE 17.3 Probability of acceptance of lots of quality π for sampling plan of $n = 50$, $c = 1$ (data used in OC curve in Figure 17.16)

p_A	π
1.0000	0.00
0.9098	0.01
0.7358	0.02
0.5578	0.03
0.4060	0.04
0.2873	0.05
0.1991	0.06
0.1359	0.07
0.0916	0.08
0.0611	0.09
0.0404	0.10
0.0000	1.00

Note that p_A, the probability of acceptance, is equivalent to β, the probability of a Type II error, described in Chapter 8. We recall the definition of Type II error as the acceptance of a null hypothesis when another value, other than the stated one, is true. In the context of sampling inspection, the *OC* curve, really a plot of Type II error probabilities, shows the probability of accepting lots when the quality is at any specified level. For example, it appears from the *OC* curve in Figure 17.16 that lots that have a proportion defective of .034 will have a probability of approximately .5 of being accepted.

It is traditional to identify two specific points on an *OC* curve. These points relate to the probability of rejecting a good lot and the probability of accepting a bad lot.

DEFINITION 17.9 The **producer's risk** is the probability of rejecting a lot having a quality equivalent to the *AQL* value.

DEFINITION 17.10 The **objectionable quality level** (*OQL*) is the proportion defective above which lots should be accepted with a specified small probability by a sampling plan.

DEFINITION 17.11 The **consumer's risk** is the probability of accepting a lot having a quality equivalent to the *OQL* value.

If $1 - \alpha$ is the probability of acceptance, then α is the risk value. For Figure 17.16, the α value is about .05 for rejecting lots with a proportion defective of

FIGURE 17.17 Ideal *OC* curve

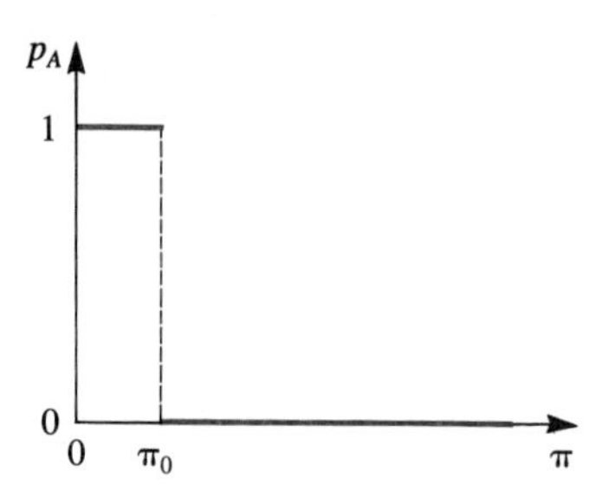

.007. If the proportion defective is .078, the probability is approximately .10 of accepting a lot. In general, consumer's risk is the probability of accepting lots of substandard quality. In the field of quality control and statistics, this specific probability value is, as indicated earlier, designated with the symbol β, corresponding to a Type II error.

If it is desired to accept lots as being good if they have a proportion defective of π_0 or less, then the ideal *OC* curve is horizontal at $p_A = 1$ until $\pi = \pi_0$ is reached. At this point, it drops vertically, and, for values larger than π_0, it is horizontal at $p_A = 0$—that is, it coincides with the horizontal axis. Such a curve is shown in Figure 17.17.

There is no sampling plan with an ideal *OC* curve such as that in Figure 17.17. Such a curve can only be achieved with 100% inspection where no mistakes are made in classifying an item.

It is possible to more nearly approach the ideal *OC* curve by increasing the sample size. The *OC* curve for the plan $n = 200$, $c = 4$ is plotted along with that for $n = 50$, $c = 1$ in Figure 17.18. Notice that the *OC* curve for $n = 200$, $c = 4$ is closer to the ideal *OC* curve.

We have seen that given a sampling plan, it is possible to determine the *OC* curve. We now look briefly at the reverse situation—that is, to determine the sampling plan given the *OC* curve. Specifically, if two points on the *OC* curve are designated, it is possible to determine the corresponding sampling plan.

Let us assume that if the *AQL* proportion defective is π_1, a plan is desired that has $p_A = .95$—that is, the producer's risk, or α, equal to .05. Furthermore, assume that if the *OQL* proportion defective is π_2, the plan should have a probability of acceptance, or consumer's risk, of .10. The determination of the required sampling plan is facilitated by the use of Appendix Table B.19, which shows values for c, the acceptance number, and values of $n\pi$, which are $n\pi_1$ with $p_A = .95$ and $n\pi_2$ associated with $p_A = .10$. The last column of the table gives the ratio of $n\pi_2$ to $n\pi_1$ or, since n is common to both, the ratio of π_2 to π_1. The required c for the sampling plan thus comes from the line in the table having the approximate ratio given by π_2/π_1. Similar tables for other values of the producer's risk and consumer's risk could easily be prepared.

FIGURE 17.18 *OC* curves for two sampling plans with *c* proportional to *n*

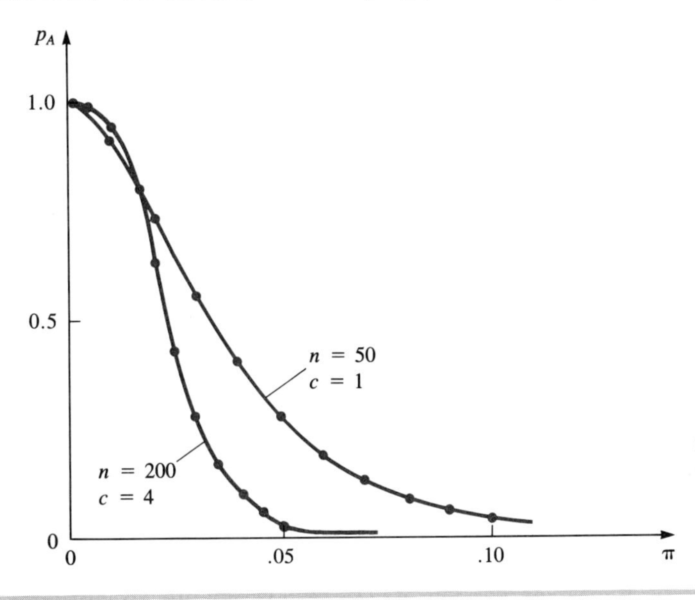

EXAMPLE 17.11 Determine a single sampling plan that rejects lots that have a proportion defective of .015 with probability .05 and accepts lots that have a proportion defective of .06 with probability .10.

Solution: The statement of the problem indicates that $\pi_1 = .015$, $\pi_2 = .06$, $\alpha = .05$, and $\beta = .10$. Therefore, Appendix Table B.19 can be used to obtain the necessary c value. The ratio π_2/π_1 is determined to be

$$\frac{\pi_2}{\pi_1} = \frac{.06}{.015} = 4.00$$

The closest entry in the last column of the table is 4.06, which is associated with $c = 4$. If it is desired to fix the producer's risk at exactly .05, then n is determined by

$$\frac{n\pi_1}{\pi_1} = \frac{1.970}{.015} = 131$$

If the consumer's risk is to be held constant at .10, then n is determined by

$$\frac{n\pi_2}{\pi_2} = \frac{7.99}{.06} = 133$$

A value of 132 for n represents a compromise—that is, the producer's risk is nearly .05 and the consumer's risk is nearly .10. The sampling plan $n = 132$, $c = 4$ essentially satisfies the requirements of the problem.

Let us briefly consider the practical consequences of an inspection decision. The discussion is presented in general terms since specific results often depend on the particular process being inspected.

A Type I decision error is made whenever the sampling plan rejects a good lot. If the inspection decision is to reject the lot and the lot is bad, then no decision error exists. The question of what to do with the lot still remains, however. A rejected lot may be returned to the supplier. If the lot is good, the supplier will most likely be annoyed. This situation can be avoided if it is possible to make 100% inspection of all lots rejected by the sampling plan. If the lot is bad, the long-run effect of returning it to the supplier may be to encourage better control in the production process. A concession in price is sometimes an alternative to returning a lot containing an excessive number of defective items.

If a sampling plan accepts a good lot, then there is no decision error. A Type II error is committed whenever a plan accepts a bad lot. The consequences of this event are largely economic in nature, but the costs are usually difficult to ascertain, especially prior to the sampling. These costs reflect downtime and other delays, as well as possible customer ill will.

Acceptance Sampling by Attributes: Additional Plans

The basic concepts of acceptance sampling by attributes have been illustrated with single sampling plans, where judgment about a lot is based on the results of a single sample of size n. It is reasonable to present such a detailed consideration of single sampling plans since they are the most basic type and are still probably the most widely used. However, certain other types of sampling plans need to be considered briefly.

In a double sampling plan, two different samples are considered. If a lot is very good or very bad, the decision to accept or reject is likely to be made on the first sample. For middle-quality lots, the decision may not be made in the first sample, but it must be made after the second one. An example of a double sampling plan would be $n_1 = 50$, accept if zero defective, reject if three or more defectives. If one or two are defective in the first 50, do not make a decision but sample an additional $n_2 = 50$. If the cumulative number of defectives in the total sample of 100 is three or less, accept the lot. If the cumulative total is four or more, reject the lot. An advantage of a double sampling plan is that the total number of items inspected, on the average, is less than with the corresponding single sampling plan required for a given producer's risk and consumer's risk. In general, the entire OC curve is approximately the same as with the corresponding single sampling plan of size n. The value of n for the corresponding single sample plan is greater than the n_1 of the double sampling plan but less than $n_1 + n_2$.

An extension of the double sampling plan to more than two stages is usually called a multiple sampling plan. At each stage, the decision is made to accept, reject, or continue sampling. A decision to accept or reject must be made when the final stage, usually the seventh, is reached.

Another acceptance sampling by attributes plan that has smaller total inspection than single sampling is sequential sampling. Sequential sampling plans can be represented graphically by linear equations for the acceptance and rejection lines. See Figure 17.19. One of three decisions is made as each unit is inspected. If the plotted point, represented horizontally by the cumulative number inspected and vertically by the cumulative number defective, falls below the acceptance line, the lot is accepted. Similarly, the lot is rejected as soon as a plotted point falls above the rejection line. As long as points fall between the two lines, inspection continues. As might be expected, sequential plans, as well as double and multiple plans, are designed to reduce the cost of sampling.

In circumstances where rejected lots are submitted to 100% inspection, sampling inspection tables by Dodge and Romig can be used to find a plan that minimizes average total inspection for a given average outgoing quality limit (*AOQL*) and a specified process average.

Tables published by the United States Government, officially called Military Standard Sampling Procedures and Tables for Inspection by Attributes, but popularly known as MIL-STD-105D, can be used to easily obtain single, double, or multiple sampling plans. These tables have become the accepted standard in many industries for incoming lots and for final inspection, in part because of their required use in government contracts.

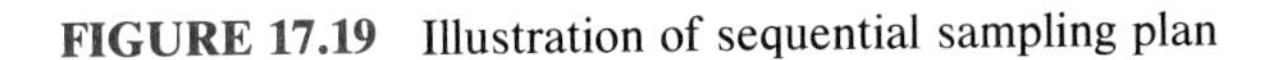

FIGURE 17.19 Illustration of sequential sampling plan

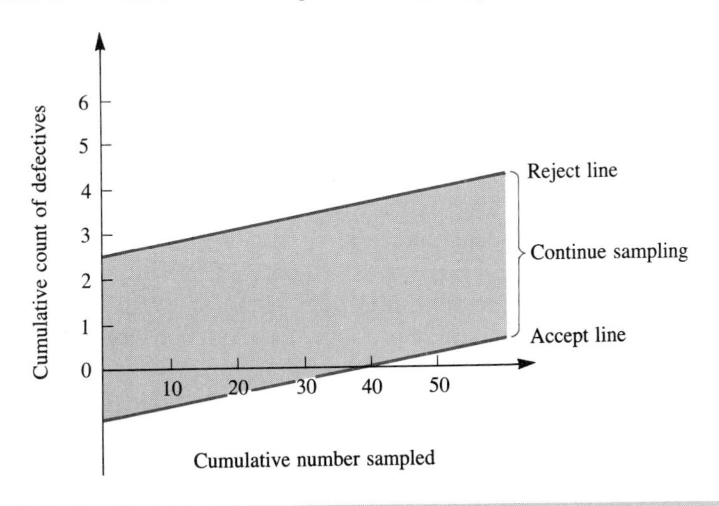

Plans are identified by a code letter determined from a particular combination of lot size and inspection level. The table for this determination is given as Appendix Table B.20. The code letter and *AQL* value, the expression for standard quality, are used for entry into the desired single, double, or multiple sampling table. Each of the three types of samples has a table for three types of inspection provisions: *normal, tightened,* and *reduced*. Only Appendix Table B.21 for single sampling and normal inspection is presented in this text. *AQL* values are read at the top of a column for normal inspection and at the bottom for tightened inspection.

Normal inspection is used when the fraction defective remains within satisfactory limits. If there is evidence of deterioration of the process average, then tightened inspection is invoked. Under a tightened plan, the sample size remains the same but the acceptance number is reduced. The reduced acceptance number has the effect of decreasing the consumer's risk and increasing the producer's risk, thus putting pressure on the producer to improve the product. When the fraction defective is consistently good, reduced inspection is used. Under a reduced sampling plan, the sample size is decreased, resulting in reduced inspection costs. The producer's risk will be decreased slightly, and the consumer's risk will be increased.

There are three general inspection levels. Level I is used when less discrimination is needed, such as when the product to be inspected comes from a supplier with a reputation for a high-quality product. Level II provides standard discrimination and is the normal level used. Level III is used when more discrimination is needed, as when the producer has a reputation for supplying a poor-quality product.

EXAMPLE 17.12 Use MIL-STD-105D tables to determine the appropriate single sampling plan using normal inspection if incoming lots are of size 1000, level II inspection is to be used, and the $AQL = 1.5\%$.

Solution: Appendix Table B.20 is entered for a lot size of 1000 with inspection level II. A code letter of J is indicated. If Appendix Table B.21 is entered with code letter J and $AQL = 1.5$, the suggested sample size is 80 with an acceptance number of 3. A lot is accepted if, in a random sample of 80 items, there are 3 or fewer defective items. The entire lot is rejected if 4 or more defective items are found in the sample.

Generally speaking, the MIL-STD-105D tables have been constructed in such a way that the producer's risk will be close to $\alpha = .05$. The consumer's risk will vary widely, however, depending on which inspection level is used and whether the normal, tightened, or reduced inspection table is used. These points are illustrated in Figure 17.20, which shows approximate *OC* curves, as derived from two points, for various single sampling plans representing combinations of inspection types and inspection levels. The calculations supporting Figure 17.20 are shown in Table 17.4.

FIGURE 17.20 Comparison of *OC* curves for several combinations of inspection level and inspection type (based on single sampling from MIL-STD-105D; supporting data in Table 17.4)

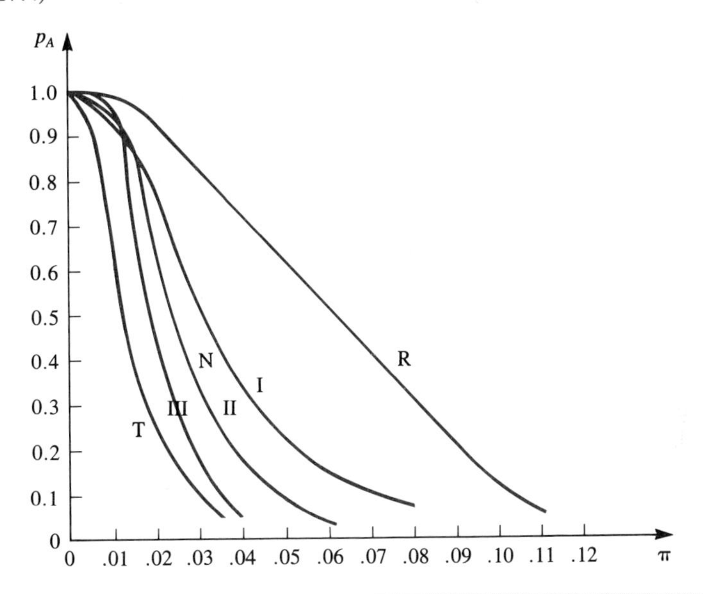

Note, in Figure 17.20, that a change from normal inspection level II to either normal inspection level I or reduced inspection level II yields an *OC* curve that has a larger consumer's risk for a given proportion defective. The combination of reduced inspection level I is not shown, but its *OC* curve would be even farther to the right.

TABLE 17.4 Derivation of two points on *OC* curve for various single sampling plans based on lots of size 5000 and *AQL* = 1.0. N = normal, T = tightened, R = reduced (data plotted in Figure 17.20)

Type	*Inspection Level*	*Code Letter*	*Sample Size*	*Ac*	*Re*	p_A *(for* $\pi_1 = .01$*)*	$n\pi_2$ *(for* $\beta = .10$*)*	π_2
N	II	L	200	5	6	.983	9.30	.046
T	II	L	200	3	4	.857	6.7	.034
R	II	L	80	2*	5	.999	8.0	.10
N	I	J	80	2	3	.953	5.3	.066
N	III	M	315	7	8	.984	11.8	.038

* $Ac = 4$ used to determine p_A.

A change from normal inspection level II to either normal inspection level III or tightened inspection level II produces an *OC* curve that has a smaller consumer's risk for a given proportion defective. The combination of tightened inspection level III would have an *OC* curve even farther to the left than any shown in Figure 17.20.

Acceptance Sampling by Variables

If the characteristic for inspection is measured on a continuous scale, then it is possible to use acceptance sampling by variables. Plans of this type ordinarily have a smaller sample size than their attributes counterparts since each item provides a higher level of information. However, since a measurement is being made on an item instead of merely observing whether it is defective or nondefective, the cost of obtaining the additional information is usually higher.

In the discussion that follows, it will be assumed that the appropriate underlying distribution is of the normal family. Initially, it will also be assumed that the standard deviation of the population, σ_X, is known.

In a single sampling by attributes plan presented previously, the desire to make a decision regarding a lot was put in the context of a hypothesis test. A similar situation exists for acceptance sampling by variables.

Assume the null and alternative hypotheses of interest are

$$H_0\colon \mu \leq \mu_0 \qquad \text{and} \qquad H_\text{a}\colon \mu > \mu_0$$

Also assume that it is desired to determine a sampling plan with a sample of size n and a value of $\overline{X}$, say $\bar{x}_k$, such that the probability of a Type I error for μ_0 is α and the probability of a Type II error for an alternative μ_a is β. This problem is similar to the one posed in Section 8.9, where consideration is given to determining the sample size required to satisfy both the α and β levels in a test of hypothesis. Refer specifically to Example 8.15.

The required sample size, given by Equation 8.25 and modified here to reflect the one-sided alternative given before, is

$$n = \left[\frac{(z_\alpha + z_\beta)\sigma_X}{\mu_\text{a} - \mu_0}\right]^2 \tag{17.18}$$

The value, $\bar{x}_k$, can be obtained from

$$\bar{x}_k = \mu_0 + \frac{z_\alpha \sigma_X}{\sqrt{n}} \tag{17.19}$$

or

$$\bar{x}_k = \mu_\text{a} - \frac{z_\beta \sigma_X}{\sqrt{n}} \tag{17.20}$$

EXAMPLE 17.13 Monroe industries produces a ball bearing with the intent that the mean not exceed 10 millimeters. Assume the standard deviation is known to be 1 millimeter. What size sample should be selected in checking on the manufacturing process if the desired α is .02 and the desired β for an alternative of 11 millimeters is .08? What is the value of the sample mean that identifies the boundary between the acceptance and rejection regions?

Solution: Equation 17.18 can be used to determine n. The necessary values for substitution into the equation are

$$\sigma_X = 1,\ \mu_a = 11,\ \mu_0 = 10,\ z_\alpha = 2.05,\ z_\beta = 1.41$$

$$n = \left[\frac{(2.05 + 1.41)(1)}{11 - 10}\right]^2 = \left[\frac{(3.46)(1)}{1}\right]^2$$
$$= (3.46)^2 = 11.97$$

Thus, a sample of 12 meets the requirements. The value of $\bar{x}_k$ is given by Equation 17.19

$$\bar{x}_k = 10 + \frac{(2.05)(1)}{\sqrt{12}} = 10 + .59 = 10.59$$

or Equation 17.20

$$\bar{x}_k = 11 - \frac{(1.41)(1)}{\sqrt{12}} = 11 - .41 = 10.59$$

Therefore, the required variables sampling plan is to take a random sample of size 12 from the production process, calculate the value for $\bar{X}$, and conclude that the production process is unsatisfactory if the value exceeds 10.59.

The problem of determining an acceptance sampling by variables plan is sometimes presented in a different way. The explanation is given in terms of the information in Example 17.13. Refer to Figure 17.21 for a graphical representation of the development that follows.

Suppose a ball bearing cannot be used if its diameter exceeds 14 millimeters. In other words, we might say that a ball bearing is defective if its diameter exceeds 14 millimeters. Assume we want 1% or fewer of the bearings to exceed 14 millimeters. The value of z that has an area of .01 under the standard normal curve to its right is 2.33. Substituting into the equation

$$Z = \frac{X - \mu_1}{\sigma_X}$$

leads to the value for μ_1. The calculation gives

$$2.33 = \frac{14 - \mu_1}{1}$$

$$\mu_1 = 11.67$$

The quantity μ_1 can be considered the mean of an acceptable process. Clearly, if μ_1 is less than 11.67, the proportion defective larger than 14 will be less than .01.

We also ask, For what proportion defective do we want to reject the lot? Assume this proportion is .05. Then, the corresponding μ_X value, say μ_2, is given by

$$1.645 = \frac{14 - \mu_2}{1}$$

$$\mu_2 = 12.355$$

If the mean is 12.355, then 5% of the bearings will exceed 14 millimeters. Any value farther to the right will have more than a .05 proportion defective. It is obvious that we want to accept if $\mu_X < 11.67$ and reject if $\mu_X > 12.355$. We must now make a decision for the interval between 11.67 and 12.355, which gives us the opportunity to introduce α and β, where α is the risk of rejecting a lot that is good and β is the risk of accepting a lot that is bad. Suppose $\alpha = .02$ and $\beta = .08$.

The two equations to solve for n and the desired value of $\overline{X}$, say $\bar{x}_k$, are

$$z_\alpha = \frac{\bar{x}_k - \mu_1}{\sigma_X/\sqrt{n}}$$

and

$$-z_\beta = \frac{\bar{x}_k - \mu_2}{\sigma_X/\sqrt{n}}$$

If these equations are solved for n, the result is

$$n = \left[\frac{(z_\alpha + z_\beta)\sigma_X}{\mu_2 - \mu_1}\right]^2 \tag{17.21}$$

We note that Equation 17.21 is the same as Equation 17.18 if μ_a is replaced by μ_2 and μ_0 by μ_1.

For the revised ball-bearing problem, the value for n is

$$n = \left[\frac{(2.05 + 1.41)(1)}{12.355 - 11.67}\right]^2 = \left(\frac{3.46}{.685}\right)^2 = (5.05)^2 = 25.5$$

The specified requirements thus lead to a sample size of 26. The value of $\bar{x}_k$ is calculated to be 12.08. The value we are seeking between 11.67 and 12.355 is thus 12.08. If the mean of a sample of 26 is less than 12.08, the lot is to be

FIGURE 17.21 Illustration of the design of acceptance sampling by variables plan

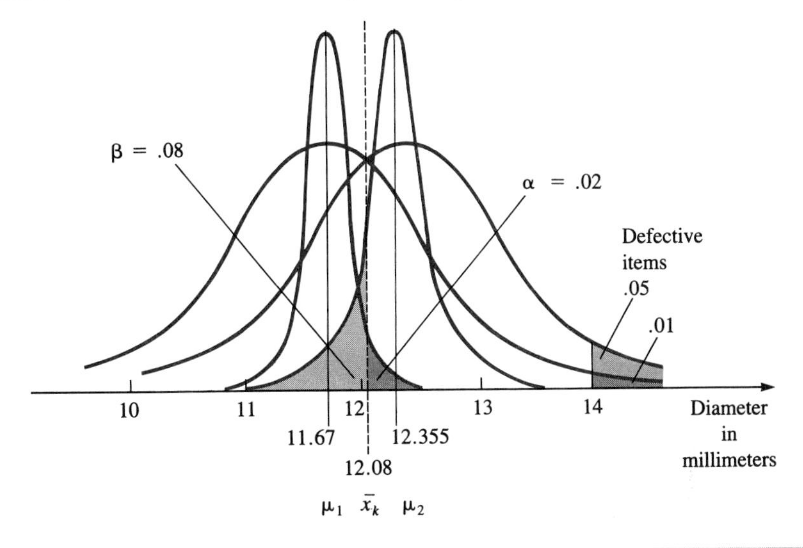

accepted. In so doing, the probability of rejecting a good lot is .02. If the sample mean is larger than 12.08, the lot is to be rejected. In so doing, the probability of accepting a bad lot is .08.

If the z value associated with μ_1 is designated z_1 and the value for μ_2 is denoted as z_2, an alternative but equivalent form of Equation 17.21 is

$$n = \left(\frac{z_\alpha + z_\beta}{z_1 - z_2}\right)^2 \tag{17.22}$$

If it is desired to express the number $\bar{x}_k$ as a z statistic in relation to z_1 and z_2, the following equations are used.

$$z_k = z_1 - \frac{z_\alpha}{\sqrt{n}}$$
$$z_k = z_2 + \frac{z_\beta}{\sqrt{n}} \tag{17.23}$$

In the preceding derivation, the use of Equation 17.23 leads to $z_k =$ 1.924, which is the distance in number of standard deviations between 12.08 and 14. We see that another way of making the decision about a lot is to calculate the number of standard deviations from the sample mean to the limit, in this case, 14. If the calculated number exceeds z_k, 1.924 in this problem, the lot is accepted. Otherwise, the decision is made to reject.

The foregoing presentation assumes the null and alternative hypotheses are $H_0: \mu \le \mu_0$ and $H_a: \mu > \mu_0$. The procedures given here apply equally well for the hypotheses

$$H_0: \mu \ge \mu_0 \quad \text{and} \quad H_a: \mu < \mu_0$$

If there are two-sided specifications, then there may be defectives at both ends of the normal curve. The procedure given is usually modified to include a criterion that the sample standard deviation be within certain bounds. This topic is considered beyond the scope of the text.

The entire presentation on acceptance sampling by variables assumes that σ_X is known. If σ_X is unknown and must be estimated, a larger sample will be required. Equation 17.22 then becomes

$$n = \frac{k^2 + 2}{2}\left(\frac{z_\alpha + z_\beta}{z_1 - z_2}\right)^2 \tag{17.24}$$

where

$$k = \frac{z_\alpha z_2 + z_\beta z_1}{z_1 + z_2} \tag{17.25}$$

Finally, it should be mentioned that many variables sampling plans are contained in MIL-STD-414, the variables sampling counterpart of MIL-STD-105D, which is often used for attributes sampling plans. The procedure for using MIL-STD-414 is similar to that previously described for MIL-STD-105D. First, a table is consulted to obtain a code letter, which is a function of lot size and inspection level. This table is given as Appendix Table B.22. Then, one of the master tables is entered to determine the specific plan. The tables are divided into sections depending on whether data on variability are in the form of the known process standard deviation, the sample standard deviation, or the sample range. There are also divisions relating to whether specification limits are single or double and whether the sampling plan is to be expressed using a z statistic or a fraction defective. The only table presented here is Appendix Table B.23, which assumes a single specification limit, a sample standard deviation, and a z statistic. This table can be used for normal inspection if the *AQL* value is read from the column heading. It can be used for tightened inspection if *AQL* values are read at the bottom of a column. The plan given by any code letter is the sample size and k value. The latter is equivalent to the z_k in the preceding development.

EXAMPLE 17.14

Determine a variables sampling plan if inspection level IV is to be used on lots of 600, where the desired *AQL* is 1.00 and normal inspection is desired.

Solution: Referring to Appendix Table B.22, we find the code letter J. Entering Appendix Table B.23 with code letter J, we find the sample size to be 30 and, for an *AQL* value of 1.00, the k value is 1.86. The sampling plan is thus $n = 30$, $k = 1.86$. A random sample of 30 should be taken, and if the sample mean is more than 1.86 standard deviations from the specified limit, the lot of 600 should be accepted. Otherwise, the lot should be rejected.

EXERCISES

17.12 Determine at least 8 points on the *OC* curve for the attributes sampling plan $n = 100$, $c = 2$. Plot the curve.

17.13 Compare the curve from Exercise 17.12 with the *OC* curves shown in Figure 17.18.

17.14 Determine the attributes sampling plan for which the *OC* curve has $\alpha = .05$ when $\pi_1 = .01$ and $\beta = .10$ when $\pi_2 = .03$.

17.15 Use MIL-STD-105D tables to determine the single sampling plan with normal inspection if incoming lots are of size 2000, level II inspection is to be used, and $AQL = 1.0\%$.

17.16 **(a)** Determine the acceptance sampling by variables plan to satisfy $\sigma_X = 4$, $\mu_a = 20$, $\mu_0 = 18$, $\alpha = .03$, and $\beta = .12$.
(b) What mean value is the boundary between accepting and rejecting the lot?

17.17 Suppose in an industrial process having $\sigma_X = 5$, the upper specification limit is 80. It is desired that the proportion of items exceeding 80 be .02 or smaller for acceptance and .08 or more for rejection. If $\alpha = .01$ and $\beta = .10$, determine the required acceptance sampling by variables plan. Specify n, k, and $\overline{x}_k$. Determine n using both Equation 17.21 and Equation 17.22.

17.7 SUMMARY

In this chapter, we have considered the application of statistical techniques to process control for the purpose of controlling product quality. Although the emphasis was on manufacturing quality, these techniques are adaptable to service quality as well.

Using estimates of the process parameters, we examined control charting techniques for applications where dimensions or other variables for a part could be measured. Periodically, subgroups are taken, and the sample means and ranges are plotted. We reviewed techniques for determining from the charts whether the process means had shifted or the variation had changed.

We also considered measures for determining process capability. These indexes are designed to determine the degree to which individual items are being manufactured within the product specifications. We examined two measures: PC%, which assumes that the process is centered on the required mean, and C_{pk}, which does not make this assumption and is therefore more general.

Control charting techniques for attribute data were also discussed. To plot proportion defectives, the p chart is used. For situations where the data are the numbers of defects per unit, the c chart is implemented.

We concluded this chapter with a consideration of acceptance sampling procedures for deciding whether to accept or reject a lot. Techniques were examined for both sampling by attributes and sampling by variables. In attributes

TABLE 17.5 Summary table

Estimating Process Parameters	Mean: $\overline{\overline{X}} = \frac{\Sigma \overline{X}_j}{k}$ (k = number of subgroups) Standard deviation: $s_X = \sqrt{\frac{\Sigma\Sigma(X_{ij} - \overline{\overline{X}})^2}{kn - 1}}$ or $\hat{s}_X = \frac{\overline{R}}{d_2}$ (for range data) $\overline{R} = \frac{\Sigma R_j}{k}$
Control Charts for Variables	Mean chart control limits: $CL_{\overline{X}} = \overline{\overline{X}} \pm 3s_{\overline{X}}$ (variation measured by the standard deviation) $CL_{\overline{X}} = \overline{\overline{X}} \pm A_2\overline{R}$ (variation measured by the range) $A_2 = \frac{3}{d_2\sqrt{n}}$ Range chart control limits: $LCL_{\overline{R}} = D_3\overline{R}$ $UCL_{\overline{R}} = D_4\overline{R}$
Process Capability Measures	$PC\% = \frac{6s_X}{\text{tolerance}}$ (assumes process centered on required mean) $C_{pk} = \frac{\min(\lvert z_{USL}\rvert, \lvert z_{LSL}\rvert)}{3}$ (general use) $z_{USL} = \frac{USL - \overline{\overline{X}}}{s_X}$ $z_{LSL} = \frac{LSL - \overline{\overline{X}}}{s_X}$
Control Charts for Attributes	p charts for proportion defective: $CL_p = \overline{p} \pm 3\sqrt{\frac{\overline{p}(1 - \overline{p})}{n}}$ $\overline{p} = \frac{\Sigma p_j}{k} = \frac{\Sigma X_j}{kn}$ n = common sample size k = number of subgroups c charts for defects per unit $CL_c = \overline{c} \pm 3\sqrt{\overline{c}}$ $\overline{c} = \frac{\Sigma c_j}{k}$ k = number of subgroups

Acceptance Sampling Concepts	Attributes sampling plans: used with defective/nondefective classification of items Single sampling plan: plan based on a single sample Multiple sampling plan: utilizes one or more samples to reach a decision (n, c): identification of sample size and acceptance number for a single plan p_A: designates probability of accepting a lot *AQL*: acceptable quality level *OQL*: objectionable quality level *OC* curve: plot of p_A versus values for π MIL-STD-105D: government sampling procedures and tables for attributes sampling Variables sampling plans: used when observations are measured on a continuous scale Designation of variables sampling plan: n and $\bar{x}_k$ or n and k MIL-STD-414: Government sampling procedures and tables for variables sampling
Acceptance Sampling Equations	Sample size equations (σ_X known) $n = \left[\frac{(z_\alpha + z_\beta)\sigma_X}{\mu_2 - \mu_1}\right]^2 = \left(\frac{z_\alpha + z_\beta}{z_1 - z_2}\right)^2$ *where:* μ_1 = desired mean μ_2 = objectionable mean z_1 = normal deviate with area π_1 to the right z_2 = normal deviate with area π_2 to the right Equations for $\bar{x}_k$ $\bar{x}_k = \mu_0 + \frac{z_\alpha \sigma}{\sqrt{n}}$ $\bar{x}_k = \mu_a - \frac{z_\beta \sigma}{\sqrt{n}}$ Equations for z_k $z_k = z_1 - \frac{z_\alpha}{\sqrt{n}}$ $z_k = z_2 + \frac{z_\beta}{\sqrt{n}}$ Sample size equation (σ_X unknown) $n = \frac{k^2 + 2}{2}\left(\frac{z_\alpha + z_\beta}{z_1 - z_2}\right)^2$ *where:* $k = \frac{z_\alpha z_2 + z_\beta z_1}{z_1 + z_2}$

sampling, emphasis was given to single sampling plans. Characteristics of a single sampling plan were studied through its *OC* curve. We examined the procedures for determining a single sampling plan with two points on the *OC* curve specified. Several multiple sampling plans designed to reduce the average number of items inspected were considered, as was the use of MIL-STD-105D tables for determining an attributes sampling plan. Finally, we reviewed procedures for determining a variables sampling plan, and for using MIL-STD-414 tables to determine a variables plan. Table 17.5 provides a summary of the major concepts and equations presented in the chapter.

SUPPLEMENTARY EXERCISES

17.18 Use MIL-STD-105D tables to determine a single sampling by attributes plan with normal inspection if incoming lots are of size 6000, level I inspection is to be used, and AQL = 2.5%.

17.19 Determine points on the *OC* curve for the single sampling plan $n = 150$, $c = 3$ and plot the curve.

17.20 Curtice Industries receives parts from the Morgan Company for use in an assembly process. The quality control manager for Curtice Industries desires to institute an attributes sampling plan that accepts 95% of the time lots having a proportion defective of .012 and accepts 10% of the time lots having a proportion defective of .03. What sampling plan should be used?

17.21 The following data are measurements in inches of the diameter of ball bearings. Determine appropriate control limits for the process.

Subgroup			
1	*2*	*3*	*4*
1.01	1.01	1.00	1.00
1.03	1.01	1.01	1.01
1.02	1.05	1.02	1.02
1.02	1.03	1.01	1.06
1.05	1.02	1.03	1.02
1.01	1.02	1.00	1.02

17.22 Determine points on the *OC* curve for the single sampling by attributes plan $n = 60$, $c = 2$ and plot the curve.

17.23 The number of defects is counted on items being produced in a certain industrial process. Each hour an item is inspected. The counts of defects for 20 hours are as follows:

3	3	1	4	13	5	6	8	5	13
4	3	13	2	6	3	5	14	5	4

Determine appropriate control limits for the process.

17.24 One hundred items from the production of a certain part of an electrical device are inspected daily to determine the number defective. The count of defectives for 20 days is as follows.

3	6	9	7	8	8	17	8	3	8
2	11	5	18	7	4	2	10	11	6

Determine appropriate control limits.

17.25 In an industrial process, thin metal sheets are manufactured that are periodically sampled for surface defects. Suppose that in a random sample of 20 sheets, the following counts of defects were observed.

4	1	2	2	0	1	7	4	4	1
8	2	2	2	8	1	0	8	2	4

(a) Determine appropriate control limits for the process.
(b) Comment on the state of control.

17.26 Determine a single sampling plan for which the *OC* curve has $\alpha = .05$ when $\pi_1 = .02$ and $\beta = .10$ when $\pi_2 = .055$.

17.27 Radio tubes are checked to see whether they are defective. Each day, 300 items are sampled randomly, and the total number of defectives is recorded. Observations obtained in this manner for 10 days are 12, 18, 15, 24, 7, 12, 5, 21, 18, and 18. Determine appropriate control limits for the process.

17.28 Determine an attributes single sampling plan for which the *OC* curve has $\alpha = .05$ when $\pi_1 = .018$ and $\beta = .10$ when $\pi_2 = .12$.

17.29 Use MIL-STD-105D tables to determine a single sampling plan with normal inspection if incoming lots are of size 100, level II inspection is to be used, and $AQL = .65$.

17.30 Thompson Manufacturing produces calculator batteries. The management of the company is particularly concerned about the life of the batteries and decides to set a lower specification limit of 500 hours. Assume $\sigma_X = 20$. Also, assume that the desired proportion of items below 500 for acceptance purposes is .015 or less and that the proportion for rejection purposes is .075. If $\alpha = .05$ and $\beta = .10$, what is the required acceptance sampling by variables plan? Specify n, k, and $\bar{x}_k$.

17.31 Determine sufficient points on the *OC* curve for the attributes sampling plan $n = 80$, $c = 3$ to make an accurate plot of the curve. Plot the curve.

17.32 A machine fills boxes of Korn Krunch cereal. Four boxes are selected randomly each time period, and the contents are weighed. The weights of the sample boxes are recorded in the following table in coded form for 6 time periods.

Time Period

1	*2*	*3*	*4*	*5*	*6*
5	3	8	6	3	2
9	8	4	9	6	3
6	5	4	4	3	4
4	4	4	9	4	3

Determine appropriate control limits for the process and comment on the state of control.

17.33 **(a)** Determine the acceptance sampling by variables plan to satisfy $\sigma_X = 2.4$, $\mu_a = 17.5$, $\mu_0 = 16$, $\alpha = .02$, and $\beta = .05$.
(b) What mean value is the boundary between the acceptance and rejection of lots?

17.34 Assume that 1 glass bottle is selected hourly at random from a manufacturing process. Data for 25 hours on the number of air bubbles observed are

15	7	5	5	3	4	1	5	4	4	2	4	4
3	2	4	15	14	9	2	7	16	2	4	3	

(a) Determine appropriate control limits for the process.
(b) Is the process in control? Why?

17.35 Use MIL-STD-105D tables to determine the single sampling by attributes plan with normal inspection if incoming lots are of size 5000, level III inspection is to be used, and $AQL = .25\%$.

17.36 The Hobart Foundary makes castings for the automotive industry. The diameter of a certain hole should be 2 inches plus or minus .02 inch. A sample of 20 castings yielded the following data.

2.015	1.991	2.006	1.993
2.011	1.997	2.002	1.995
1.975	2.006	1.994	2.005
2.003	2.107	2.012	1.997
1.998	1.984	1.988	2.008

Does it appear that the process is capable?

17.37 The Johnson Engine Company is making a piston with a process mean of 2.00 and a standard deviation of .06. A sample of 4 pistons is taken every hour, and the mean and range are plotted on the appropriate charts. If the process mean shifts 1 standard deviation larger, what is the probability of detecting the shift in the next sample taken?

17.38 We have the control chart means and ranges for 8 subgroups. Each subgroup sample is 5 items. Find the C_{pk} and PC% indexes if the upper and lower specification limits are 1.52 and 1.51, respectively.

	Time							
	8	***9***	***10***	***11***	***12***	***1***	***2***	***3***
$\bar{x}$	1.514	1.519	1.515	1.517	1.514	1.516	1.515	1.518
R	.010	.008	.004	.006	.008	.006	.004	.008

REFERENCES

1. Grant, Eugene L. and Richard S. Leavenworth. *Statistical Quality Control,* 5th ed. New York: McGraw-Hill, 1980.
2. Ott, Ellis. *Process Quality Control.* New York: McGraw-Hill, 1975.

APPENDIX A
Some Basic Calculus Procedures

DIFFERENTIAL CALCULUS

Basic Concepts

In differential calculus, we are primarily concerned with the slope or rate of change of a function or curve at any point on the curve. The slope is expressed by means of a derivative—that is, a function that is "derived" from the original function. The value of the derivative at the point of interest is equivalent geometrically to the slope of the tangent line at the point.

The most frequent use of the derivative in statistics is to locate the maximum or minimum of a function, at which points, the slope of the tangent line is zero. If y is a function of x [$y = f(x)$], a maximum or minimum is obtained by finding the derivative of y with respect to x, setting the derivative equation equal to zero, and solving for x. It is also true that the first derivative is zero at some points of inflection where the function changes from concave upward to concave downward, or vice versa. However, this occurrence is of little interest to us in this text.

It is possible to extend the concept of the derivative to functions of several variables by means of the partial derivative. If y is a function of x and z [$y = f(x,z)$], then the partial derivative of y with respect to x, $\partial y/\partial x$, is the rate of change of y with respect to x with z held constant. See the Appendix to Chapter 10 for an application of the partial derivative.

Formulas

Rules have been developed for obtaining the derivatives of many types of functions. The more basic rules are summarized here.

	Function	*Derivative*
1.	$y = c$	$\frac{dy}{dx} = 0$
2.	$y = a + bx$	$\frac{dy}{dx} = b$
3.	$y = f(x) \pm g(x)$	$\frac{dy}{dx} = \frac{df(x)}{dx} \pm \frac{dg(x)}{dx}$
4.	$y = x^n$	$\frac{dy}{dx} = nx^{n-1}$
5.	$y = [f(x)]^n$	$\frac{dy}{dx} = n[f(x)]^{n-1}\frac{df(x)}{dx}$
6.	$y = e^{f(x)}$	$\frac{dy}{dx} = e^{f(x)}\frac{df(x)}{dx}$
7.	$y = f(x)g(x)$	$\frac{dy}{dx} = f(x)\frac{dg(x)}{dx} + g(x)\frac{df(x)}{dx}$

Applications

1. Rules 1–4 are used in the following problem. If $y = 4 + 6x - x^2$, find the value of x for which y is a maximum.

$$\frac{dy}{dx} = 0 + 6 - 2x$$

Equating dy/dx to 0 we have

$$0 = 0 + 6 - 2x$$

$$2x = 6$$

$$x = 3$$

2. Rule 5 is used in the following problem. If $y = (3x - 1)^2$, find the slope of the tangent line when $x = 1$.

$$\frac{dy}{dx} = 2(3x - 1)(3) = 6(3x - 1)$$

When $x = 1$,

$$\frac{dy}{dx} = 6(3 - 1) = (6)(2) = 12$$

INTEGRAL CALCULUS

Basic Concepts

Integration is the reverse of differentiation. We are interested in finding a function when we know the derivative expressing the rate of change. The resulting function is known as the *integral* of the given function. The symbol $\int$ is used to indicate integration.

In statistics, our principal use of the process of integration will be in the evaluation of a definite integral. For example, we will be interested in such problems as finding the area under a curve or, more specifically, in finding the area bounded by a curve $y = f(x)$, the x axis, and vertical lines at $x = a$ and $x = b$. Symbolically, the area just described would be expressed as

$$\int_a^b f(x)\, dx = F(x)]_a^b = F(b) - F(a)$$

where $F(x)$ is the integral of $f(x)$.

Formulas

We present some of the simpler integration formulas that may be useful to us. For integrals of more complex functions, we suggest consulting the *Handbook of Tables for Probability and Statistics,* edited by W. H. Beyer and published by The Chemical Rubber Company.

1. $$\int_a^b dx = x]_a^b = b - a$$

2. $$\int_a^b c\, dx = cx]_a^b = c(b - a)$$

3. $$\int_a^b kf(x)\, dx = k\int_a^b f(x)\, dx$$

4. $$\int_a^b x^n\, dx = \frac{x^{n+1}}{n+1}\Big]_a^b = \frac{b^{n+1} - a^{n+1}}{n+1}$$

5. $$\int_a^b [f(x) \pm g(x)]\, dx = \int_a^b f(x)\, dx \pm \int_a^b g(x)\, dx$$

6. $$\int_a^b e^{g(x)} \frac{d[g(x)]}{dx}\, dx = e^{g(x)}\Big]_a^b$$

Applications

1. Rules 2 and 4 are used in the following problem. If

$$\begin{aligned} f(x) &= .4 - .08x \qquad && 0 \le x \le 5 \\ &= 0 && \text{elsewhere} \end{aligned}$$

determine the area under the curve between $x = 0$ and $x = 1$.

$$\text{Area} = \int_0^1 f(x)\, dx = \int_0^1 (.4 - .08x)\, dx$$

$$= .4x - \frac{.08x^2}{2}\Big]_0^1 = .4 - .04 = .36$$

See Example 3.3 for more details relating to the preceding problem.

2. Rule 6 is employed in the following problem. If $f(x) = .5e^{-.5t}$, $t \ge 0$, determine the area under the curve between $t = 0$ and $t = 5$.

$$\text{Area} = \int_0^5 .5e^{-.5t}\, dt = -\int_0^5 -.5e^{-.5t}\, dt$$

$$= -e^{-.5t}\Big]_0^5 = -(e^{-2.5} - 1) = -.0821 + 1 = .9179$$

Refer to Example 4.17 for more information on this problem.

APPENDIX B
Tables

TABLE B.1 Selected binomial distributions

$$P(X = x|n, \pi) = C_x^n \pi^x(1 - \pi)^{n-x}$$

Examples: $P(X = 3|10, .3) = .2668$
$P(X = 8|10, .6) = .1209$

$n = 2$

x	.010	.050	.100	.150	π .200	.250	.300	.400	.500	
0	.9801	.9025	.8100	.7225	.6400	.5625	.4900	.3600	.2500	2
1	.0198	.0950	.1800	.2550	.3200	.3750	.4200	.4800	.5000	1
2	.0001	.0025	.0100	.0225	.0400	.0625	.0900	.1600	.2500	0
	.990	.950	.900	.850	.800	.750	.700	.600	.500	x

$n = 3$

x	.010	.050	.100	.150	π .200	.250	.300	.400	.500	
0	.9703	.8574	.7290	.6141	.5120	.4219	.3430	.2160	.1250	3
1	.0294	.1354	.2430	.3251	.3840	.4219	.4410	.4320	.3750	2
2	.0003	.0071	.0270	.0574	.0960	.1406	.1890	.2880	.3750	1
3	.0000	.0001	.0010	.0034	.0080	.0156	.0270	.0640	.1250	0
	.990	.950	.900	.850	.800	.750	.700	.600	.500	x

$n = 4$

x	.010	.050	.100	.150	π .200	.250	.300	.400	.500	
0	.9606	.8145	.6561	.5220	.4096	.3164	.2401	.1296	.0625	4
1	.0388	.1715	.2916	.3685	.4096	.4219	.4116	.3456	.2500	3
2	.0006	.0135	.0486	.0975	.1536	.2109	.2646	.3456	.3750	2
3	.0000	.0005	.0036	.0115	.0256	.0469	.0756	.1536	.2500	1
4	.0000	.0000	.0001	.0005	.0016	.0039	.0081	.0256	.0625	0
	.990	.950	.900	.850	.800	.750	.700	.600	.500	x

$n = 5$

x	.010	.050	.100	.150	π .200	.250	.300	.400	.500	
0	.9510	.7738	.5905	.4437	.3277	.2373	.1681	.0778	.0313	5
1	.0480	.2036	.3280	.3915	.4096	.3955	.3601	.2592	.1562	4
2	.0010	.0214	.0729	.1382	.2048	.2637	.3087	.3456	.3125	3
3	.0000	.0011	.0081	.0244	.0512	.0879	.1323	.2304	.3125	2
4	.0000	.0000	.0005	.0022	.0064	.0146	.0283	.0768	.1562	1
5	.0000	.0000	.0000	.0001	.0003	.0010	.0024	.0102	.0313	0
	.990	.950	.900	.850	.800	.750	.700	.600	.500	x

TABLE B.1 (Continued)

$n = 10$

x	.010	.050	.100	.150	π .200	.250	.300	.400	.500	
0	.9044	.5987	.3487	.1969	.1074	.0563	.0282	.0060	.0010	10
1	.0914	.3151	.3874	.3474	.2684	.1877	.1211	.0403	.0098	9
2	.0042	.0746	.1937	.2759	.3020	.2816	.2335	.1209	.0439	8
3	.0001	.0105	.0574	.1298	.2013	.2503	.2668	.2150	.1172	7
4	.0000	.0010	.0112	.0401	.0881	.1460	.2001	.2508	.2051	6
5	.0000	.0001	.0015	.0085	.0264	.0584	.1029	.2007	.2461	5
6	.0000	.0000	.0001	.0012	.0055	.0162	.0368	.1115	.2051	4
7	.0000	.0000	.0000	.0001	.0008	.0031	.0090	.0425	.1172	3
8	.0000	.0000	.0000	.0000	.0001	.0004	.0014	.0106	.0439	2
9	.0000	.0000	.0000	.0000	.0000	.0000	.0001	.0016	.0098	1
10	.0000	.0000	.0000	.0000	.0000	.0000	.0000	.0001	.0010	0
	.990	.950	.900	.850	.800	.750	.700	.600	.500	x

$n = 15$

x	.010	.050	.100	.150	π .200	.250	.300	.400	.500	
0	.8601	.4633	.2059	.0874	.0352	.0134	.0047	.0005	.0000	15
1	.1303	.3658	.3432	.2312	.1319	.0668	.0305	.0047	.0005	14
2	.0092	.1348	.2669	.2856	.2309	.1559	.0916	.0219	.0032	13
3	.0004	.0307	.1285	.2184	.2501	.2252	.1700	.0634	.0139	12
4	.0000	.0049	.0428	.1156	.1876	.2252	.2186	.1268	.0417	11
5	.0000	.0006	.0105	.0449	.1032	.1651	.2061	.1859	.0916	10
6	.0000	.0000	.0019	.0132	.0430	.0917	.1472	.2066	.1527	9
7	.0000	.0000	.0003	.0030	.0138	.0393	.0811	.1771	.1964	8
8	.0000	.0000	.0000	.0005	.0035	.0131	.0348	.1181	.1964	7
9	.0000	.0000	.0000	.0001	.0007	.0034	.0116	.0612	.1527	6
10	.0000	.0000	.0000	.0000	.0001	.0007	.0030	.0245	.0916	5
11	.0000	.0000	.0000	.0000	.0000	.0001	.0006	.0074	.0417	4
12	.0000	.0000	.0000	.0000	.0000	.0000	.0001	.0016	.0139	3
13	.0000	.0000	.0000	.0000	.0000	.0000	.0000	.0003	.0032	2
14	.0000	.0000	.0000	.0000	.0000	.0000	.0000	.0000	.0005	1
15	.0000	.0000	.0000	.0000	.0000	.0000	.0000	.0000	.0000	0
	.990	.950	.900	.850	.800	.750	.700	.600	.500	x

(*Continued*)

TABLE B.1 (Continued)

$n = 20$

x	.010	.050	.100	.150	π .200	.250	.300	.400	.500	
0	.8179	.3585	.1216	.0388	.0115	.0032	.0008	.0000	.0000	20
1	.1652	.3774	.2702	.1368	.0576	.0211	.0068	.0005	.0000	19
2	.0159	.1887	.2852	.2293	.1369	.0669	.0278	.0031	.0002	18
3	.0010	.0596	.1901	.2428	.2054	.1339	.0716	.0123	.0011	17
4	.0000	.0133	.0898	.1821	.2182	.1897	.1304	.0350	.0046	16
5	.0000	.0022	.0319	.1028	.1746	.2023	.1789	.0746	.0148	15
6	.0000	.0003	.0089	.0454	.1091	.1686	.1916	.1244	.0370	14
7	.0000	.0000	.0020	.0160	.0545	.1124	.1643	.1659	.0739	13
8	.0000	.0000	.0004	.0046	.0222	.0609	.1144	.1797	.1201	12
9	.0000	.0000	.0001	.0011	.0074	.0271	.0654	.1597	.1602	11
10	.0000	.0000	.0000	.0002	.0020	.0099	.0308	.1171	.1762	10
11	.0000	.0000	.0000	.0000	.0005	.0030	.0120	.0710	.1602	9
12	.0000	.0000	.0000	.0000	.0001	.0008	.0039	.0355	.1201	8
13	.0000	.0000	.0000	.0000	.0000	.0002	.0010	.0146	.0739	7
14	.0000	.0000	.0000	.0000	.0000	.0000	.0002	.0049	.0370	6
15	.0000	.0000	.0000	.0000	.0000	.0000	.0000	.0013	.0148	5
16	.0000	.0000	.0000	.0000	.0000	.0000	.0000	.0003	.0046	4
17	.0000	.0000	.0000	.0000	.0000	.0000	.0000	.0000	.0011	3
18	.0000	.0000	.0000	.0000	.0000	.0000	.0000	.0000	.0002	2
19	.0000	.0000	.0000	.0000	.0000	.0000	.0000	.0000	.0000	1
	.990	.950	.900	.850	.800	.750	.700	.600	.500	x

$n = 25$

x	.010	.050	.100	.150	π .200	.250	.300	.400	.500	
0	.7778	.2774	.0718	.0172	.0038	.0008	.0001	.0000	.0000	25
1	.1964	.3650	.1994	.0759	.0236	.0063	.0014	.0000	.0000	24
2	.0238	.2305	.2659	.1607	.0708	.0251	.0074	.0004	.0000	23
3	.0018	.0930	.2265	.2174	.1358	.0641	.0243	.0019	.0001	22
4	.0001	.0269	.1384	.2110	.1867	.1175	.0572	.0071	.0004	21
5	.0000	.0060	.0646	.1564	.1960	.1645	.1030	.0199	.0016	20
6	.0000	.0010	.0239	.0920	.1633	.1828	.1472	.0442	.0053	19
7	.0000	.0001	.0072	.0441	.1108	.1654	.1712	.0800	.0143	18
8	.0000	.0000	.0018	.0175	.0623	.1241	.1651	.1200	.0322	17
9	.0000	.0000	.0004	.0058	.0294	.0781	.1336	.1511	.0609	16
10	.0000	.0000	.0001	.0016	.0118	.0417	.0916	.1612	.0974	15
11	.0000	.0000	.0000	.0004	.0040	.0189	.0536	.1465	.1328	14
12	.0000	.0000	.0000	.0001	.0012	.0074	.0268	.1140	.1550	13
13	.0000	.0000	.0000	.0000	.0003	.0025	.0115	.0760	.1550	12
14	.0000	.0000	.0000	.0000	.0001	.0007	.0042	.0434	.1328	11

TABLE B.1 (Continued)

$n = 25$

x	.010	.050	.100	.150	π .200	.250	.300	.400	.500	
15	.0000	.0000	.0000	.0000	.0000	.0002	.0013	.0212	.0974	10
16	.0000	.0000	.0000	.0000	.0000	.0000	.0004	.0088	.0609	9
17	.0000	.0000	.0000	.0000	.0000	.0000	.0001	.0031	.0322	8
18	.0000	.0000	.0000	.0000	.0000	.0000	.0000	.0009	.0143	7
19	.0000	.0000	.0000	.0000	.0000	.0000	.0000	.0002	.0053	6
20	.0000	.0000	.0000	.0000	.0000	.0000	.0000	.0000	.0016	5
21	.0000	.0000	.0000	.0000	.0000	.0000	.0000	.0000	.0004	4
22	.0000	.0000	.0000	.0000	.0000	.0000	.0000	.0000	.0001	3
23	.0000	.0000	.0000	.0000	.0000	.0000	.0000	.0000	.0000	2
	.990	.950	.900	.850	.800	.750	.700	.600	.500	x

$n = 50$

x	.010	.050	.100	.150	π .200	.250	.300	.400	.500	
0	.6050	.0769	.0052	.0003	.0000	.0000	.0000	.0000	.0000	50
1	.3056	.2025	.0286	.0026	.0002	.0000	.0000	.0000	.0000	49
2	.0756	.2611	.0779	.0113	.0011	.0001	.0000	.0000	.0000	48
3	.0122	.2199	.1386	.0319	.0044	.0004	.0000	.0000	.0000	47
4	.0015	.1360	.1809	.0661	.0128	.0016	.0001	.0000	.0000	46
5	.0001	.0658	.1849	.1072	.0295	.0049	.0006	.0000	.0000	45
6	.0000	.0260	.1541	.1419	.0554	.0123	.0018	.0000	.0000	44
7	.0000	.0086	.1076	.1575	.0870	.0259	.0048	.0000	.0000	43
8	.0000	.0024	.0643	.1493	.1169	.0463	.0110	.0002	.0000	42
9	.0000	.0006	.0333	.1230	.1364	.0721	.0220	.0005	.0000	41
10	.0000	.0001	.0152	.0890	.1398	.0985	.0386	.0014	.0000	40
11	.0000	.0000	.0061	.0571	.1271	.1194	.0602	.0035	.0000	39
12	.0000	.0000	.0022	.0328	.1033	.1294	.0838	.0076	.0001	38
13	.0000	.0000	.0007	.0169	.0755	.1261	.1050	.0147	.0003	37
14	.0000	.0000	.0002	.0079	.0499	.1110	.1189	.0260	.0008	36
15	.0000	.0000	.0001	.0033	.0299	.0888	.1223	.0415	.0020	35
16	.0000	.0000	.0000	.0013	.0164	.0648	.1147	.0606	.0044	34
17	.0000	.0000	.0000	.0005	.0082	.0432	.0983	.0808	.0087	33
18	.0000	.0000	.0000	.0001	.0037	.0264	.0772	.0987	.0160	32
19	.0000	.0000	.0000	.0000	.0016	.0148	.0558	.1109	.0270	31
20	.0000	.0000	.0000	.0000	.0006	.0077	.0370	.1146	.0419	30
21	.0000	.0000	.0000	.0000	.0002	.0036	.0227	.1091	.0598	29
22	.0000	.0000	.0000	.0000	.0001	.0016	.0128	.0959	.0788	28
23	.0000	.0000	.0000	.0000	.0000	.0006	.0067	.0778	.0960	27
24	.0000	.0000	.0000	.0000	.0000	.0002	.0032	.0584	.1080	26

(Continued)

TABLE B.1 (Continued)

$n = 50$

x	.010	.050	.100	.150	π .200	.250	.300	.400	.500	
25	.0000	.0000	.0000	.0000	.0000	.0001	.0014	.0405	.1123	25
26	.0000	.0000	.0000	.0000	.0000	.0000	.0006	.0259	.1080	24
27	.0000	.0000	.0000	.0000	.0000	.0000	.0002	.0154	.0960	23
28	.0000	.0000	.0000	.0000	.0000	.0000	.0001	.0084	.0788	22
29	.0000	.0000	.0000	.0000	.0000	.0000	.0000	.0043	.0598	21
30	.0000	.0000	.0000	.0000	.0000	.0000	.0000	.0020	.0419	20
31	.0000	.0000	.0000	.0000	.0000	.0000	.0000	.0009	.0270	19
32	.0000	.0000	.0000	.0000	.0000	.0000	.0000	.0003	.0160	18
33	.0000	.0000	.0000	.0000	.0000	.0000	.0000	.0001	.0087	17
34	.0000	.0000	.0000	.0000	.0000	.0000	.0000	.0000	.0044	16
35	.0000	.0000	.0000	.0000	.0000	.0000	.0000	.0000	.0020	15
36	.0000	.0000	.0000	.0000	.0000	.0000	.0000	.0000	.0008	14
37	.0000	.0000	.0000	.0000	.0000	.0000	.0000	.0000	.0003	13
38	.0000	.0000	.0000	.0000	.0000	.0000	.0000	.0000	.0001	12
39	.0000	.0000	.0000	.0000	.0000	.0000	.0000	.0000	.0000	11
	.990	.950	.900	.850	.800	.750	.700	.600	.500	x

TABLE B.2 Selected cumulative binomial distributions

$$P(X \le x|n, \pi \le .5) = \sum_{k=0}^{x} C_k^n \pi^k (1 - \pi)^{n-k}$$

or

$$P(X \ge x|n, \pi \ge .5) = \sum_{k=x}^{n} C_k^n \pi^k (1 - \pi)^{n-k}$$

Examples: $P(X \le 3|5, .2) = .9933$; $P(X \ge 6|10, .75) = .9219$

$n = 2$

x	.010	.050	.100	.150	π .200	.250	.300	.400	.500	
0	.9801	.9025	.8100	.7225	.6400	.5625	.4900	.3600	.2500	2
1	.9999	.9975	.9900	.9775	.9600	.9375	.9100	.8400	.7500	1
2	1.0000	1.0000	1.0000	1.0000	1.0000	1.0000	1.0000	1.0000	1.0000	0
	.990	.950	.900	.850	.800	.750	.700	.600	.500	x

$n = 3$

x	.010	.050	.100	.150	π .200	.250	.300	.400	.500	
0	.9703	.8574	.7290	.6141	.5120	.4219	.3430	.2160	.1250	3
1	.9997	.9927	.9720	.9392	.8960	.8437	.7840	.6480	.5000	2
2	1.0000	.9999	.9990	.9966	.9920	.9844	.9730	.9360	.8750	1
3	1.0000	1.0000	1.0000	1.0000	1.0000	1.0000	1.0000	1.0000	1.0000	0
	.990	.950	.900	.850	.800	.750	.700	.600	.500	x

$n = 4$

x	.010	.050	.100	.150	π .200	.250	.300	.400	.500	
0	.9606	.8145	.6561	.5220	.4096	.3164	.2401	.1296	.0625	4
1	.9994	.9860	.9477	.8905	.8192	.7383	.6517	.4752	.3125	3
2	1.0000	.9995	.9963	.9880	.9728	.9492	.9163	.8208	.6875	2
3	1.0000	1.0000	.9999	.9995	.9984	.9961	.9919	.9744	.9375	1
4	1.0000	1.0000	1.0000	1.0000	1.0000	1.0000	1.0000	1.0000	1.0000	0
	.990	.950	.900	.850	.800	.750	.700	.600	.500	x

(*Continued*)

TABLE B.2 (Continued)

$n = 5$

x	.010	.050	.100	.150	π .200	.250	.300	.400	.500	
0	.9510	.7738	.5905	.4437	.3277	.2373	.1681	.0778	.0312	5
1	.9990	.9774	.9185	.8352	.7373	.6328	.5282	.3370	.1875	4
2	1.0000	.9988	.9914	.9734	.9421	.8965	.8369	.6826	.5000	3
3	1.0000	1.0000	.9995	.9978	.9933	.9844	.9692	.9130	.8125	2
4	1.0000	1.0000	1.0000	.9999	.9997	.9990	.9976	.9898	.9687	1
5	1.0000	1.0000	1.0000	1.0000	1.0000	1.0000	1.0000	1.0000	1.0000	0
	.990	.950	.900	.850	.800	.750	.700	.600	.500	x

$n = 10$

x	.010	.050	.100	.150	π .200	.250	.300	.400	.500	
0	.9044	.5987	.3487	.1969	.1074	.0563	.0282	.0060	.0010	10
1	.9957	.9139	.7361	.5443	.3758	.2440	.1493	.0464	.0107	9
2	.9999	.9885	.9298	.8202	.6778	.5256	.3828	.1673	.0547	8
3	1.0000	.9990	.9872	.9500	.8791	.7759	.6496	.3823	.1719	7
4	1.0000	.9999	.9984	.9901	.9672	.9219	.8497	.6331	.3770	6
5	1.0000	1.0000	.9999	.9986	.9936	.9803	.9527	.8338	.6230	5
6	1.0000	1.0000	1.0000	.9999	.9991	.9965	.9894	.9452	.8281	4
7	1.0000	1.0000	1.0000	1.0000	.9999	.9996	.9984	.9877	.9453	3
8	1.0000	1.0000	1.0000	1.0000	1.0000	1.0000	.9999	.9983	.9893	2
9	1.0000	1.0000	1.0000	1.0000	1.0000	1.0000	1.0000	.9999	.9990	1
10	1.0000	1.0000	1.0000	1.0000	1.0000	1.0000	1.0000	1.0000	1.0000	0
	.990	.950	.900	.850	.800	.750	.700	.600	.500	x

$n = 15$

x	.010	.050	.100	.150	π .200	.250	.300	.400	.500	
0	.8601	.4633	.2059	.0874	.0352	.0134	.0047	.0005	.0000	15
1	.9904	.8290	.5490	.3186	.1671	.0802	.0353	.0052	.0005	14
2	.9996	.9638	.8159	.6042	.3980	.2361	.1268	.0271	.0037	13
3	1.0000	.9945	.9444	.8227	.6482	.4613	.2969	.0905	.0176	12
4	1.0000	.9994	.9873	.9383	.8358	.6865	.5155	.2173	.0592	11
5	1.0000	.9999	.9978	.9832	.9389	.8516	.7216	.4032	.1509	10
6	1.0000	1.0000	.9997	.9964	.9819	.9434	.8689	.6098	.3036	9
7	1.0000	1.0000	1.0000	.9994	.9958	.9827	.9500	.7869	.5000	8
8	1.0000	1.0000	1.0000	.9999	.9992	.9958	.9848	.9050	.6964	7
9	1.0000	1.0000	1.0000	1.0000	.9999	.9992	.9963	.9662	.8491	6

TABLE B.2 (Continued)

$n = 15$

x	.010	.050	.100	.150	π .200	.250	.300	.400	.500	
10	1.0000	1.0000	1.0000	1.0000	1.0000	.9999	.9993	.9907	.9408	5
11	1.0000	1.0000	1.0000	1.0000	1.0000	1.0000	.9999	.9981	.9824	4
12	1.0000	1.0000	1.0000	1.0000	1.0000	1.0000	1.0000	.9997	.9963	3
13	1.0000	1.0000	1.0000	1.0000	1.0000	1.0000	1.0000	1.0000	.9995	2
14	1.0000	1.0000	1.0000	1.0000	1.0000	1.0000	1.0000	1.0000	1.0000	1
	.990	.950	.900	.850	.800	.750	.700	.600	.500	x

$n = 20$

x	.010	.050	.100	.150	π .200	.250	.300	.400	.500	
0	.8179	.3585	.1216	.0388	.0115	.0032	.0008	.0000	.0000	20
1	.9831	.7358	.3917	.1756	.0692	.0243	.0076	.0005	.0000	19
2	.9990	.9245	.6769	.4049	.2061	.0913	.0355	.0036	.0002	18
3	1.0000	.9841	.8670	.6477	.4114	.2252	.1071	.0160	.0013	17
4	1.0000	.9974	.9568	.8298	.6296	.4148	.2375	.0510	.0059	16
5	1.0000	.9997	.9887	.9327	.8042	.6172	.4164	.1256	.0207	15
6	1.0000	1.0000	.9976	.9781	.9133	.7858	.6080	.2500	.0577	14
7	1.0000	1.0000	.9996	.9941	.9679	.8982	.7723	.4159	.1316	13
8	1.0000	1.0000	.9999	.9987	.9900	.9591	.8867	.5956	.2517	12
9	1.0000	1.0000	1.0000	.9998	.9974	.9861	.9520	.7553	.4119	11
10	1.0000	1.0000	1.0000	1.0000	.9994	.9961	.9829	.8725	.5881	10
11	1.0000	1.0000	1.0000	1.0000	.9999	.9991	.9949	.9435	.7483	9
12	1.0000	1.0000	1.0000	1.0000	1.0000	.9998	.9987	.9790	.8684	8
13	1.0000	1.0000	1.0000	1.0000	1.0000	1.0000	.9997	.9935	.9423	7
14	1.0000	1.0000	1.0000	1.0000	1.0000	1.0000	1.0000	.9984	.9793	6
15	1.0000	1.0000	1.0000	1.0000	1.0000	1.0000	1.0000	.9997	.9941	5
16	1.0000	1.0000	1.0000	1.0000	1.0000	1.0000	1.0000	1.0000	.9987	4
17	1.0000	1.0000	1.0000	1.0000	1.0000	1.0000	1.0000	1.0000	.9998	3
18	1.0000	1.0000	1.0000	1.0000	1.0000	1.0000	1.0000	1.0000	1.0000	2
	.990	.950	.900	.850	.800	.750	.700	.600	.500	x

$n = 25$

x	.010	.050	.100	.150	π .200	.250	.300	.400	.500	
0	.7778	.2774	.0718	.0172	.0038	.0008	.0001	.0000	.0000	25
1	.9742	.6424	.2712	.0931	.0274	.0070	.0016	.0001	.0000	24
2	.9980	.8729	.5371	.2537	.0982	.0321	.0090	.0004	.0000	23
3	.9999	.9659	.7636	.4711	.2340	.0962	.0332	.0024	.0001	22
4	1.0000	.9928	.9020	.6821	.4207	.2137	.0905	.0095	.0005	21

(Continued)

TABLE B.2 (Continued)

$n = 25$

x	.010	.050	.100	.150	π .200	.250	.300	.400	.500	
5	1.0000	.9988	.9666	.8385	.6167	.3783	.1935	.0294	.0020	20
6	1.0000	.9998	.9905	.9305	.7800	.5611	.3407	.0736	.0073	19
7	1.0000	1.0000	.9977	.9745	.8909	.7265	.5118	.1536	.0216	18
8	1.0000	1.0000	.9995	.9920	.9532	.8506	.6769	.2735	.0539	17
9	1.0000	1.0000	.9999	.9979	.9827	.9287	.8106	.4246	.1148	16
10	1.0000	1.0000	1.0000	.9995	.9944	.9703	.9022	.5858	.2122	15
11	1.0000	1.0000	1.0000	.9999	.9985	.9893	.9558	.7323	.3450	14
12	1.0000	1.0000	1.0000	1.0000	.9996	.9966	.9825	.8462	.5000	13
13	1.0000	1.0000	1.0000	1.0000	.9999	.9991	.9940	.9222	.6550	12
14	1.0000	1.0000	1.0000	1.0000	1.0000	.9998	.9982	.9656	.7878	11
15	1.0000	1.0000	1.0000	1.0000	1.0000	1.0000	.9995	.9868	.8852	10
16	1.0000	1.0000	1.0000	1.0000	1.0000	1.0000	.9999	.9957	.9461	9
17	1.0000	1.0000	1.0000	1.0000	1.0000	1.0000	1.0000	.9988	.9784	8
18	1.0000	1.0000	1.0000	1.0000	1.0000	1.0000	1.0000	.9997	.9927	7
19	1.0000	1.0000	1.0000	1.0000	1.0000	1.0000	1.0000	.9999	.9980	6
20	1.0000	1.0000	1.0000	1.0000	1.0000	1.0000	1.0000	1.0000	.9995	5
21	1.0000	1.0000	1.0000	1.0000	1.0000	1.0000	1.0000	1.0000	.9999	4
22	1.0000	1.0000	1.0000	1.0000	1.0000	1.0000	1.0000	1.0000	1.0000	3
	.990	.950	.900	.850	.800	.750	.700	.600	.500	x

$n = 50$

x	.010	.050	.100	.150	π .200	.250	.300	.400	.500	
0	.6050	.0769	.0052	.0003	.0000	.0000	.0000	.0000	.0000	50
1	.9106	.2794	.0338	.0029	.0002	.0000	.0000	.0000	.0000	49
2	.9862	.5405	.1117	.0142	.0013	.0001	.0000	.0000	.0000	48
3	.9984	.7604	.2503	.0460	.0057	.0005	.0000	.0000	.0000	47
4	.9999	.8964	.4312	.1121	.0185	.0021	.0002	.0000	.0000	46
5	1.0000	.9622	.6161	.2194	.0480	.0070	.0007	.0000	.0000	45
6	1.0000	.9882	.7702	.3613	.1034	.0194	.0025	.0000	.0000	44
7	1.0000	.9968	.8779	.5188	.1904	.0453	.0073	.0001	.0000	43
8	1.0000	.9992	.9421	.6681	.3073	.0916	.0183	.0002	.0000	42
9	1.0000	.9998	.9755	.7911	.4437	.1637	.0402	.0008	.0000	41
10	1.0000	1.0000	.9906	.8801	.5836	.2622	.0789	.0022	.0000	40
11	1.0000	1.0000	.9968	.9372	.7107	.3816	.1390	.0057	.0000	39
12	1.0000	1.0000	.9990	.9699	.8139	.5110	.2229	.0133	.0002	38
13	1.0000	1.0000	.9997	.9868	.8894	.6370	.3279	.0280	.0005	37
14	1.0000	1.0000	.9999	.9947	.9393	.7481	.4468	.0540	.0013	36

TABLE B.2 (Continued)

$n = 50$

x	.010	.050	.100	.150	π .200	.250	.300	.400	.500	
15	1.0000	1.0000	1.0000	.9981	.9692	.8369	.5692	.0955	.0033	35
16	1.0000	1.0000	1.0000	.9993	.9856	.9017	.6839	.1561	.0077	34
17	1.0000	1.0000	1.0000	.9998	.9937	.9449	.7822	.2369	.0164	33
18	1.0000	1.0000	1.0000	.9999	.9975	.9713	.8594	.3356	.0325	32
19	1.0000	1.0000	1.0000	1.0000	.9991	.9861	.9152	.4465	.0595	31
20	1.0000	1.0000	1.0000	1.0000	.9997	.9937	.9522	.5610	.1013	30
21	1.0000	1.0000	1.0000	1.0000	.9999	.9974	.9749	.6701	.1611	29
22	1.0000	1.0000	1.0000	1.0000	1.0000	.9990	.9877	.7660	.2399	28
23	1.0000	1.0000	1.0000	1.0000	1.0000	.9996	.9944	.8438	.3359	27
24	1.0000	1.0000	1.0000	1.0000	1.0000	.9999	.9976	.9022	.4439	26
25	1.0000	1.0000	1.0000	1.0000	1.0000	1.0000	.9991	.9427	.5561	25
26	1.0000	1.0000	1.0000	1.0000	1.0000	1.0000	.9997	.9686	.6641	24
27	1.0000	1.0000	1.0000	1.0000	1.0000	1.0000	.9999	.9840	.7601	23
28	1.0000	1.0000	1.0000	1.0000	1.0000	1.0000	1.0000	.9924	.8389	22
29	1.0000	1.0000	1.0000	1.0000	1.0000	1.0000	1.0000	.9966	.8987	21
30	1.0000	1.0000	1.0000	1.0000	1.0000	1.0000	1.0000	.9986	.9405	20
31	1.0000	1.0000	1.0000	1.0000	1.0000	1.0000	1.0000	.9995	.9675	19
32	1.0000	1.0000	1.0000	1.0000	1.0000	1.0000	1.0000	.9998	.9836	18
33	1.0000	1.0000	1.0000	1.0000	1.0000	1.0000	1.0000	.9999	.9923	17
34	1.0000	1.0000	1.0000	1.0000	1.0000	1.0000	1.0000	1.0000	.9967	16
35	1.0000	1.0000	1.0000	1.0000	1.0000	1.0000	1.0000	1.0000	.9987	15
36	1.0000	1.0000	1.0000	1.0000	1.0000	1.0000	1.0000	1.0000	.9995	14
37	1.0000	1.0000	1.0000	1.0000	1.0000	1.0000	1.0000	1.0000	.9998	13
38	1.0000	1.0000	1.0000	1.0000	1.0000	1.0000	1.0000	1.0000	1.0000	12
	.990	.950	.900	.850	.800	.750	.700	.600	.500	x

TABLE B.3 Selected Poisson distributions

This table is a listing of the probability of exactly $X = x$ successes for selected values of μ, defined by the Poisson probability function

$$P(X = x) = \frac{e^{-\mu}\mu^x}{x!}$$

Examples: If $\mu = 3.5$, then $P(X = 5) = .1322$ and $P(X = 0) = .0302$.

	μ									
x	.1	.2	.3	.4	.5	.6	.7	.8	.9	1.0
0	.9048	.8187	.7408	.6703	.6065	.5488	.4966	.4493	.4066	.3679
1	.0905	.1637	.2222	.2681	.3033	.3293	.3476	.3595	.3659	.3679
2	.0045	.0164	.0333	.0536	.0758	.0988	.1217	.1438	.1647	.1839
3	.0002	.0011	.0033	.0072	.0126	.0198	.0284	.0383	.0494	.0613
4	.0000	.0001	.0003	.0007	.0016	.0030	.0050	.0077	.0111	.0153
5	.0000	.0000	.0000	.0001	.0002	.0004	.0007	.0012	.0020	.0031
6	.0000	.0000	.0000	.0000	.0000	.0000	.0001	.0002	.0003	.0005
7	.0000	.0000	.0000	.0000	.0000	.0000	.0000	.0000	.0000	.0001
8	.0000	.0000	.0000	.0000	.0000	.0000	.0000	.0000	.0000	.0000

	μ									
x	1.1	1.2	1.3	1.4	1.5	1.6	1.7	1.8	1.9	2.0
0	.3329	.3012	.2725	.2466	.2231	.2019	.1827	.1653	.1496	.1353
1	.3662	.3614	.3543	.3452	.3347	.3230	.3106	.2975	.2842	.2707
2	.2014	.2169	.2303	.2417	.2510	.2584	.2640	.2678	.2700	.2707
3	.0738	.0867	.0998	.1128	.1255	.1378	.1496	.1607	.1710	.1804
4	.0203	.0260	.0324	.0395	.0471	.0551	.0636	.0723	.0812	.0902
5	.0045	.0062	.0084	.0111	.0141	.0176	.0216	.0260	.0309	.0361
6	.0008	.0012	.0018	.0026	.0035	.0047	.0061	.0078	.0098	.0120
7	.0001	.0002	.0003	.0005	.0008	.0011	.0015	.0020	.0027	.0034
8	.0000	.0000	.0001	.0001	.0001	.0002	.0003	.0005	.0006	.0009
9	.0000	.0000	.0000	.0000	.0000	.0000	.0001	.0001	.0001	.0002
10	.0000	.0000	.0000	.0000	.0000	.0000	.0000	.0000	.0000	.0000

TABLE B.3 (Continued)

	μ									
x	2.1	2.2	2.3	2.4	2.5	2.6	2.7	2.8	2.9	3.0
0	.1225	.1108	.1003	.0907	.0821	.0743	.0672	.0608	.0550	.0498
1	.2572	.2438	.2306	.2177	.2052	.1931	.1815	.1703	.1596	.1494
2	.2700	.2681	.2652	.2613	.2565	.2510	.2450	.2384	.2314	.2240
3	.1890	.1966	.2033	.2090	.2138	.2176	.2205	.2225	.2237	.2240
4	.0992	.1082	.1169	.1254	.1336	.1414	.1488	.1557	.1622	.1680
5	.0417	.0476	.0538	.0602	.0668	.0735	.0804	.0872	.0940	.1008
6	.0146	.0174	.0206	.0241	.0278	.0319	.0362	.0407	.0455	.0504
7	.0044	.0055	.0068	.0083	.0099	.0118	.0139	.0163	.0188	.0216
8	.0011	.0015	.0019	.0025	.0031	.0038	.0047	.0057	.0068	.0081
9	.0003	.0004	.0005	.0007	.0009	.0011	.0014	.0018	.0022	.0027
10	.0001	.0001	.0001	.0002	.0002	.0003	.0004	.0005	.0006	.0008
11	.0000	.0000	.0000	.0000	.0000	.0001	.0001	.0001	.0002	.0002
12	.0000	.0000	.0000	.0000	.0000	.0000	.0000	.0000	.0000	.0001
13	.0000	.0000	.0000	.0000	.0000	.0000	.0000	.0000	.0000	.0000

	μ									
x	3.1	3.2	3.3	3.4	3.5	3.6	3.7	3.8	3.9	4.0
0	.0450	.0408	.0369	.0334	.0302	.0273	.0247	.0224	.0202	.0183
1	.1397	.1304	.1217	.1135	.1057	.0984	.0915	.0850	.0789	.0733
2	.2165	.2087	.2008	.1929	.1850	.1771	.1692	.1615	.1539	.1465
3	.2237	.2226	.2209	.2186	.2158	.2125	.2087	.2046	.2001	.1954
4	.1733	.1781	.1823	.1858	.1888	.1912	.1931	.1944	.1951	.1954
5	.1075	.1140	.1203	.1264	.1322	.1377	.1429	.1477	.1522	.1563
6	.0555	.0608	.0662	.0716	.0771	.0826	.0881	.0936	.0989	.1042
7	.0246	.0278	.0312	.0348	.0385	.0425	.0466	.0508	.0551	.0595
8	.0095	.0111	.0129	.0148	.0169	.0191	.0215	.0241	.0269	.0298
9	.0033	.0040	.0047	.0056	.0066	.0076	.0089	.0102	.0116	.0132
10	.0010	.0013	.0016	.0019	.0023	.0028	.0033	.0039	.0045	.0053
11	.0003	.0004	.0005	.0006	.0007	.0009	.0011	.0013	.0016	.0019
12	.0001	.0001	.0001	.0002	.0002	.0003	.0003	.0004	.0005	.0006
13	.0000	.0000	.0000	.0000	.0001	.0001	.0001	.0001	.0002	.0002
14	.0000	.0000	.0000	.0000	.0000	.0000	.0000	.0000	.0000	.0001
15	.0000	.0000	.0000	.0000	.0000	.0000	.0000	.0000	.0000	.0000

(Continued)

TABLE B.3 (Continued)

	μ									
x	4.1	4.2	4.3	4.4	4.5	4.6	4.7	4.8	4.9	5.0
0	.0166	.0150	.0136	.0123	.0111	.0101	.0091	.0082	.0074	.0067
1	.0679	.0630	.0583	.0540	.0500	.0462	.0427	.0395	.0365	.0337
2	.1393	.1323	.1254	.1188	.1125	.1063	.1005	.0948	.0894	.0842
3	.1904	.1852	.1798	.1743	.1687	.1631	.1574	.1517	.1460	.1404
4	.1951	.1944	.1933	.1917	.1898	.1875	.1849	.1820	.1789	.1755
5	.1600	.1633	.1662	.1687	.1708	.1725	.1738	.1747	.1753	.1755
6	.1093	.1143	.1191	.1237	.1281	.1323	.1362	.1398	.1432	.1462
7	.0640	.0686	.0732	.0778	.0824	.0869	.0914	.0959	.1002	.1044
8	.0328	.0360	.0393	.0428	.0463	.0500	.0537	.0575	.0614	.0653
9	.0150	.0168	.0188	.0209	.0232	.0255	.0281	.0307	.0334	.0363
10	.0061	.0071	.0081	.0092	.0104	.0118	.0132	.0147	.0164	.0181
11	.0023	.0027	.0032	.0037	.0043	.0049	.0056	.0064	.0073	.0082
12	.0008	.0009	.0011	.0013	.0016	.0019	.0022	.0026	.0030	.0034
13	.0002	.0003	.0004	.0005	.0006	.0007	.0008	.0009	.0011	.0013
14	.0001	.0001	.0001	.0001	.0002	.0002	.0003	.0003	.0004	.0005
15	.0000	.0000	.0000	.0000	.0001	.0001	.0001	.0001	.0001	.0002
16	.0000	.0000	.0000	.0000	.0000	.0000	.0000	.0000	.0000	.0000
17	.0000	.0000	.0000	.0000	.0000	.0000	.0000	.0000	.0000	.0000

	μ									
x	5.1	5.2	5.3	5.4	5.5	5.6	5.7	5.8	5.9	6.0
0	.0061	.0055	.0050	.0045	.0041	.0037	.0033	.0030	.0027	.0025
1	.0311	.0287	.0265	.0244	.0225	.0207	.0191	.0176	.0162	.0149
2	.0793	.0746	.0701	.0659	.0618	.0580	.0544	.0509	.0477	.0446
3	.1348	.1293	.1239	.1185	.1133	.1082	.1033	.0985	.0938	.0892
4	.1719	.1681	.1641	.1600	.1558	.1515	.1472	.1428	.1383	.1339
5	.1753	.1748	.1740	.1728	.1714	.1697	.1678	.1656	.1632	.1606
6	.1490	.1515	.1537	.1555	.1571	.1584	.1594	.1601	.1605	.1606
7	.1086	.1125	.1163	.1200	.1234	.1267	.1298	.1326	.1353	.1377
8	.0692	.0731	.0771	.0810	.0849	.0887	.0925	.0962	.0998	.1033
9	.0392	.0423	.0454	.0486	.0519	.0552	.0586	.0620	.0654	.0688
10	.0200	.0220	.0241	.0262	.0285	.0309	.0334	.0359	.0386	.0413
11	.0093	.0104	.0116	.0129	.0143	.0157	.0173	.0190	.0207	.0225
12	.0039	.0045	.0051	.0058	.0065	.0073	.0082	.0092	.0102	.0113
13	.0015	.0018	.0021	.0024	.0028	.0032	.0036	.0041	.0046	.0052
14	.0006	.0007	.0008	.0009	.0011	.0013	.0015	.0017	.0019	.0022
15	.0002	.0002	.0003	.0003	.0004	.0005	.0006	.0007	.0008	.0009
16	.0001	.0001	.0001	.0001	.0001	.0002	.0002	.0002	.0003	.0003
17	.0000	.0000	.0000	.0000	.0000	.0001	.0001	.0001	.0001	.0001
18	.0000	.0000	.0000	.0000	.0000	.0000	.0000	.0000	.0000	.0000
19	.0000	.0000	.0000	.0000	.0000	.0000	.0000	.0000	.0000	.0000

TABLE B.3 (Continued)

	μ									
x	6.1	6.2	6.3	6.4	6.5	6.6	6.7	6.8	6.9	7.0
0	.0022	.0020	.0018	.0017	.0015	.0014	.0012	.0011	.0010	.0009
1	.0137	.0126	.0116	.0106	.0098	.0090	.0082	.0076	.0070	.0064
2	.0417	.0390	.0364	.0340	.0318	.0296	.0276	.0258	.0240	.0223
3	.0848	.0806	.0765	.0726	.0688	.0652	.0617	.0584	.0552	.0521
4	.1294	.1249	.1205	.1162	.1118	.1076	.1034	.0992	.0952	.0912
5	.1579	.1549	.1519	.1487	.1454	.1420	.1385	.1349	.1314	.1277
6	.1605	.1601	.1595	.1586	.1575	.1562	.1546	.1529	.1511	.1490
7	.1399	.1418	.1435	.1450	.1462	.1472	.1480	.1486	.1489	.1490
8	.1066	.1099	.1130	.1160	.1188	.1215	.1240	.1263	.1284	.1304
9	.0723	.0757	.0791	.0825	.0858	.0891	.0923	.0954	.0985	.1014
10	.0441	.0469	.0498	.0528	.0558	.0588	.0618	.0649	.0679	.0710
11	.0244	.0265	.0285	.0307	.0330	.0353	.0377	.0401	.0426	.0452
12	.0124	.0137	.0150	.0164	.0179	.0194	.0210	.0227	.0245	.0263
13	.0058	.0065	.0073	.0081	.0089	.0099	.0108	.0119	.0130	.0142
14	.0025	.0029	.0033	.0037	.0041	.0046	.0052	.0058	.0064	.0071
15	.0010	.0012	.0014	.0016	.0018	.0020	.0023	.0026	.0029	.0033
16	.0004	.0005	.0005	.0006	.0007	.0008	.0010	.0011	.0013	.0014
17	.0001	.0002	.0002	.0002	.0003	.0003	.0004	.0004	.0005	.0006
18	.0000	.0001	.0001	.0001	.0001	.0001	.0001	.0002	.0002	.0002
19	.0000	.0000	.0000	.0000	.0000	.0000	.0001	.0001	.0001	.0001
20	.0000	.0000	.0000	.0000	.0000	.0000	.0000	.0000	.0000	.0000

	μ									
x	7.1	7.2	7.3	7.4	7.5	7.6	7.7	7.8	7.9	8.0
0	.0008	.0007	.0007	.0006	.0006	.0005	.0005	.0004	.0004	.0003
1	.0059	.0054	.0049	.0045	.0041	.0038	.0035	.0032	.0029	.0027
2	.0208	.0194	.0180	.0167	.0156	.0145	.0134	.0125	.0116	.0107
3	.0492	.0464	.0438	.0413	.0389	.0366	.0345	.0324	.0305	.0286
4	.0874	.0836	.0799	.0764	.0729	.0696	.0663	.0632	.0602	.0573
5	.1241	.1204	.1167	.1130	.1094	.1057	.1021	.0986	.0951	.0916
6	.1468	.1445	.1420	.1394	.1367	.1339	.1311	.1282	.1252	.1221
7	.1489	.1486	.1481	.1474	.1465	.1454	.1442	.1428	.1413	.1396
8	.1321	.1337	.1351	.1363	.1373	.1381	.1388	.1392	.1395	.1396
9	.1042	.1070	.1096	.1121	.1144	.1167	.1187	.1207	.1224	.1241
10	.0740	.0770	.0800	.0829	.0858	.0887	.0914	.0941	.0967	.0993
11	.0478	.0504	.0531	.0558	.0585	.0613	.0640	.0667	.0695	.0722
12	.0283	.0303	.0323	.0344	.0366	.0388	.0411	.0434	.0457	.0481
13	.0154	.0168	.0181	.0196	.0211	.0227	.0243	.0260	.0278	.0296
14	.0078	.0086	.0095	.0104	.0113	.0123	.0134	.0145	.0157	.0169
15	.0037	.0041	.0046	.0051	.0057	.0062	.0069	.0075	.0083	.0090
16	.0016	.0019	.0021	.0024	.0026	.0030	.0033	.0037	.0041	.0045
17	.0007	.0008	.0009	.0010	.0012	.0013	.0015	.0017	.0019	.0021
18	.0003	.0003	.0004	.0004	.0005	.0006	.0006	.0007	.0008	.0009
19	.0001	.0001	.0001	.0002	.0002	.0002	.0003	.0003	.0003	.0004

(*Continued*)

TABLE B.3 (Continued)

x	μ = 7.1	7.2	7.3	7.4	7.5	7.6	7.7	7.8	7.9	8.0
20	.0000	.0000	.0001	.0001	.0001	.0001	.0001	.0001	.0001	.0002
21	.0000	.0000	.0000	.0000	.0000	.0000	.0000	.0000	.0001	.0001
22	.0000	.0000	.0000	.0000	.0000	.0000	.0000	.0000	.0000	.0000

x	μ = 8.1	8.2	8.3	8.4	8.5	8.6	8.7	8.8	8.9	9.0
0	.0003	.0003	.0002	.0002	.0002	.0002	.0002	.0002	.0001	.0001
1	.0025	.0023	.0021	.0019	.0017	.0016	.0014	.0013	.0012	.0011
2	.0100	.0092	.0086	.0079	.0074	.0068	.0063	.0058	.0054	.0050
3	.0269	.0252	.0237	.0222	.0208	.0195	.0183	.0171	.0160	.0150
4	.0544	.0517	.0491	.0466	.0443	.0420	.0398	.0377	.0357	.0337
5	.0882	.0849	.0816	.0784	.0752	.0722	.0692	.0663	.0635	.0607
6	.1191	.1160	.1128	.1097	.1066	.1034	.1003	.0972	.0941	.0911
7	.1378	.1358	.1338	.1317	.1294	.1271	.1247	.1222	.1197	.1171
8	.1395	.1392	.1388	.1382	.1375	.1366	.1356	.1344	.1332	.1318
9	.1255	.1269	.1280	.1290	.1299	.1306	.1311	.1315	.1317	.1318
10	.1017	.1040	.1063	.1084	.1104	.1123	.1140	.1157	.1172	.1186
11	.0749	.0776	.0802	.0828	.0853	.0878	.0902	.0925	.0948	.0970
12	.0505	.0530	.0555	.0579	.0604	.0629	.0654	.0679	.0703	.0728
13	.0315	.0334	.0354	.0374	.0395	.0416	.0438	.0459	.0481	.0504
14	.0182	.0196	.0210	.0225	.0240	.0256	.0272	.0289	.0306	.0324
15	.0098	.0107	.0116	.0126	.0136	.0147	.0158	.0169	.0182	.0194
16	.0050	.0055	.0060	.0066	.0072	.0079	.0086	.0093	.0101	.0109
17	.0024	.0026	.0029	.0033	.0036	.0040	.0044	.0048	.0053	.0058
18	.0011	.0012	.0014	.0015	.0017	.0019	.0021	.0024	.0026	.0029
19	.0005	.0005	.0006	.0007	.0008	.0009	.0010	.0011	.0012	.0014
20	.0002	.0002	.0002	.0003	.0003	.0004	.0004	.0005	.0005	.0006
21	.0001	.0001	.0001	.0001	.0001	.0002	.0002	.0002	.0002	.0003
22	.0000	.0000	.0000	.0000	.0001	.0001	.0001	.0001	.0001	.0001
23	.0000	.0000	.0000	.0000	.0000	.0000	.0000	.0000	.0000	.0000
24	.0000	.0000	.0000	.0000	.0000	.0000	.0000	.0000	.0000	.0000

x	μ = 9.1	9.2	9.3	9.4	9.5	9.6	9.7	9.8	9.9	10.0
0	.0001	.0001	.0001	.0001	.0001	.0001	.0001	.0001	.0001	.0000
1	.0010	.0009	.0009	.0008	.0007	.0007	.0006	.0005	.0005	.0005
2	.0046	.0043	.0040	.0037	.0034	.0031	.0029	.0027	.0025	.0023
3	.0140	.0131	.0123	.0115	.0107	.0100	.0093	.0087	.0081	.0076
4	.0319	.0302	.0285	.0269	.0254	.0240	.0226	.0213	.0201	.0189

TABLE B.3 (Continued)

x	μ 9.1	9.2	9.3	9.4	9.5	9.6	9.7	9.8	9.9	10.0
5	.0581	.0555	.0530	.0506	.0483	.0460	.0439	.0418	.0398	.0378
6	.0881	.0851	.0822	.0793	.0764	.0736	.0709	.0682	.0656	.0631
7	.1145	.1118	.1091	.1064	.1037	.1010	.0982	.0955	.0928	.0901
8	.1302	.1286	.1269	.1251	.1232	.1212	.1191	.1170	.1148	.1126
9	.1317	.1315	.1311	.1306	.1300	.1293	.1284	.1274	.1263	.1251
10	.1198	.1209	.1219	.1228	.1235	.1241	.1245	.1249	.1250	.1251
11	.0991	.1012	.1031	.1049	.1067	.1083	.1098	.1112	.1125	.1137
12	.0752	.0776	.0799	.0822	.0844	.0866	.0888	.0908	.0928	.0948
13	.0526	.0549	.0572	.0594	.0617	.0640	.0662	.0685	.0707	.0729
14	.0342	.0361	.0380	.0399	.0419	.0439	.0459	.0479	.0500	.0521
15	.0208	.0221	.0235	.0250	.0265	.0281	.0297	.0313	.0330	.0347
16	.0118	.0127	.0137	.0147	.0157	.0168	.0180	.0192	.0204	.0217
17	.0063	.0069	.0075	.0081	.0088	.0095	.0103	.0111	.0119	.0128
18	.0032	.0035	.0039	.0042	.0046	.0051	.0055	.0060	.0065	.0071
19	.0015	.0017	.0019	.0021	.0023	.0026	.0028	.0031	.0034	.0037
20	.0007	.0008	.0009	.0010	.0011	.0012	.0014	.0015	.0017	.0019
21	.0003	.0003	.0004	.0004	.0005	.0006	.0006	.0007	.0008	.0009
22	.0001	.0001	.0002	.0002	.0002	.0002	.0003	.0003	.0004	.0004
23	.0000	.0001	.0001	.0001	.0001	.0001	.0001	.0001	.0002	.0002
24	.0000	.0000	.0000	.0000	.0000	.0000	.0000	.0001	.0001	.0001
25	.0000	.0000	.0000	.0000	.0000	.0000	.0000	.0000	.0000	.0000

x	μ 11	12	13	14	15	16	17	18	19	20
0	.0000	.0000	.0000	.0000	.0000	.0000	.0000	.0000	.0000	.0000
1	.0002	.0001	.0000	.0000	.0000	.0000	.0000	.0000	.0000	.0000
2	.0010	.0004	.0002	.0001	.0000	.0000	.0000	.0000	.0000	.0000
3	.0037	.0018	.0008	.0004	.0002	.0001	.0000	.0000	.0000	.0000
4	.0102	.0053	.0027	.0013	.0006	.0003	.0001	.0001	.0000	.0000
5	.0224	.0127	.0070	.0037	.0019	.0010	.0005	.0002	.0001	.0001
6	.0411	.0255	.0152	.0087	.0048	.0026	.0014	.0007	.0004	.0002
7	.0646	.0437	.0281	.0174	.0104	.0060	.0034	.0019	.0010	.0005
8	.0888	.0655	.0457	.0304	.0194	.0120	.0072	.0042	.0024	.0013
9	.1085	.0874	.0661	.0473	.0324	.0213	.0135	.0083	.0050	.0029
10	.1194	.1048	.0859	.0663	.0486	.0341	.0230	.0150	.0095	.0058
11	.1194	.1144	.1015	.0844	.0663	.0496	.0355	.0245	.0164	.0106
12	.1094	.1144	.1099	.0984	.0829	.0661	.0504	.0368	.0259	.0176
13	.0926	.1056	.1099	.1060	.0956	.0814	.0658	.0509	.0378	.0271
14	.0728	.0905	.1021	.1060	.1024	.0930	.0800	.0655	.0514	.0387

(Continued)

TABLE B.3 (Continued)

x	μ = 11	12	13	14	15	16	17	18	19	20
15	.0534	.0724	.0885	.0989	.1024	.0992	.0906	.0786	.0650	.0516
16	.0367	.0543	.0719	.0866	.0960	.0992	.0963	.0884	.0772	.0646
17	.0237	.0383	.0550	.0713	.0847	.0934	.0963	.0936	.0863	.0760
18	.0145	.0255	.0397	.0554	.0706	.0830	.0909	.0936	.0911	.0844
19	.0084	.0161	.0272	.0409	.0557	.0699	.0814	.0887	.0911	.0888
20	.0046	.0097	.0177	.0286	.0418	.0559	.0692	.0798	.0866	.0888
21	.0024	.0055	.0109	.0191	.0299	.0426	.0560	.0684	.0783	.0846
22	.0012	.0030	.0065	.0121	.0204	.0310	.0433	.0560	.0676	.0769
23	.0006	.0016	.0037	.0074	.0133	.0216	.0320	.0438	.0559	.0669
24	.0003	.0008	.0020	.0043	.0083	.0144	.0226	.0328	.0442	.0557
25	.0001	.0004	.0010	.0024	.0050	.0092	.0154	.0237	.0336	.0446
26	.0000	.0002	.0005	.0013	.0029	.0057	.0101	.0164	.0246	.0343
27	.0000	.0001	.0002	.0007	.0016	.0034	.0063	.0109	.0173	.0254
28	.0000	.0000	.0001	.0003	.0009	.0019	.0038	.0070	.0117	.0181
29	.0000	.0000	.0001	.0002	.0004	.0011	.0023	.0044	.0077	.0125
30	.0000	.0000	.0000	.0001	.0002	.0006	.0013	.0026	.0049	.0083
31	.0000	.0000	.0000	.0000	.0001	.0003	.0007	.0015	.0030	.0054
32	.0000	.0000	.0000	.0000	.0001	.0001	.0004	.0009	.0018	.0034
33	.0000	.0000	.0000	.0000	.0000	.0001	.0002	.0005	.0010	.0020
34	.0000	.0000	.0000	.0000	.0000	.0000	.0001	.0002	.0005	.0012
35	.0000	.0000	.0000	.0000	.0000	.0000	.0000	.0001	.0003	.0007
36	.0000	.0000	.0000	.0000	.0000	.0000	.0000	.0001	.0002	.0004
37	.0000	.0000	.0000	.0000	.0000	.0000	.0000	.0000	.0001	.0002
38	.0000	.0000	.0000	.0000	.0000	.0000	.0000	.0000	.0000	.0001
39	.0000	.0000	.0000	.0000	.0000	.0000	.0000	.0000	.0000	.0001
40	.0000	.0000	.0000	.0000	.0000	.0000	.0000	.0000	.0000	.0000

Adapted from Billingsley et al., *Statistical Inference for Management and Economics*, Third Edition.

TABLE B.4 Selected cumulative Poisson distributions

This table is a listing of the probability of at most x—that is, $X \leq x$—successes for selected values of μ, defined by the Poisson cumulative function

$$P(X \leq x) = \sum_{k=0}^{x} \frac{e^{-\mu}\mu^k}{k!}$$

Example: $P(X \leq 5|\mu = 2.8) = .9349$

x	.1	.2	.3	.4	.5	.6	.7	.8	.9	1.0
0	.9048	.8187	.7408	.6703	.6065	.5488	.4966	.4493	.4066	.3679
1	.9953	.9825	.9631	.9384	.9098	.8781	.8442	.8088	.7725	.7358
2	.9998	.9989	.9964	.9921	.9856	.9769	.9659	.9526	.9371	.9197
3	1.0000	.9999	.9997	.9992	.9982	.9966	.9942	.9909	.9865	.9810
4	1.0000	1.0000	1.0000	.9999	.9998	.9996	.9992	.9986	.9977	.9963
5	1.0000	1.0000	1.0000	1.0000	1.0000	1.0000	.9999	.9998	.9997	.9994
6	1.0000	1.0000	1.0000	1.0000	1.0000	1.0000	1.0000	1.0000	1.0000	.9999
7	1.0000	1.0000	1.0000	1.0000	1.0000	1.0000	1.0000	1.0000	1.0000	1.0000
8	1.0000	1.0000	1.0000	1.0000	1.0000	1.0000	1.0000	1.0000	1.0000	1.0000

x	1.1	1.2	1.3	1.4	1.5	1.6	1.7	1.8	1.9	2.0
0	.3329	.3012	.2725	.2466	.2231	.2019	.1827	.1653	.1496	.1353
1	.6990	.6626	.6268	.5918	.5578	.5249	.4932	.4628	.4337	.4060
2	.9004	.8795	.8571	.8335	.8088	.7834	.7572	.7306	.7037	.6767
3	.9743	.9662	.9569	.9463	.9344	.9212	.9068	.8913	.8747	.8571
4	.9946	.9923	.9893	.9857	.9814	.9763	.9704	.9636	.9559	.9473
5	.9990	.9985	.9978	.9968	.9955	.9940	.9920	.9896	.9868	.9834
6	.9999	.9997	.9996	.9994	.9991	.9987	.9981	.9974	.9966	.9955
7	1.0000	1.0000	.9999	.9999	.9998	.9997	.9996	.9994	.9992	.9989
8	1.0000	1.0000	1.0000	1.0000	1.0000	1.0000	.9999	.9999	.9998	.9998
9	1.0000	1.0000	1.0000	1.0000	1.0000	1.0000	1.0000	1.0000	1.0000	1.0000
10	1.0000	1.0000	1.0000	1.0000	1.0000	1.0000	1.0000	1.0000	1.0000	1.0000

x	2.1	2.2	2.3	2.4	2.5	2.6	2.7	2.8	2.9	3.0
0	.1225	.1108	.1003	.0907	.0821	.0743	.0672	.0608	.0550	.0498
1	.3796	.3546	.3309	.3084	.2873	.2674	.2487	.2311	.2146	.1991
2	.6496	.6227	.5960	.5697	.5438	.5184	.4936	.4695	.4460	.4232
3	.8386	.8194	.7993	.7787	.7576	.7360	.7141	.6919	.6696	.6472
4	.9379	.9275	.9162	.9041	.8912	.8774	.8629	.8477	.8318	.8153

(*Continued*)

TABLE B.4 (Continued)

x	μ 2.1	2.2	2.3	2.4	2.5	2.6	2.7	2.8	2.9	3.0
5	.9796	.9751	.9700	.9643	.9580	.9510	.9433	.9349	.9258	.9161
6	.9941	.9925	.9906	.9884	.9858	.9828	.9794	.9756	.9713	.9665
7	.9985	.9980	.9974	.9967	.9958	.9947	.9934	.9919	.9901	.9881
8	.9997	.9995	.9994	.9991	.9989	.9985	.9981	.9976	.9969	.9962
9	.9999	.9999	.9999	.9998	.9997	.9996	.9995	.9993	.9991	.9989
10	1.0000	1.0000	1.0000	1.0000	.9999	.9999	.9999	.9998	.9998	.9997
11	1.0000	1.0000	1.0000	1.0000	1.0000	1.0000	1.0000	1.0000	.9999	.9999
12	1.0000	1.0000	1.0000	1.0000	1.0000	1.0000	1.0000	1.0000	1.0000	1.0000
13	1.0000	1.0000	1.0000	1.0000	1.0000	1.0000	1.0000	1.0000	1.0000	1.0000

x	μ 3.1	3.2	3.3	3.4	3.5	3.6	3.7	3.8	3.9	4.0
0	.0450	.0408	.0369	.0334	.0302	.0273	.0247	.0224	.0202	.0183
1	.1847	.1712	.1586	.1468	.1359	.1257	.1162	.1074	.0992	.0916
2	.4012	.3799	.3594	.3397	.3208	.3027	.2854	.2689	.2531	.2381
3	.6248	.6025	.5803	.5584	.5366	.5152	.4942	.4735	.4532	.4335
4	.7982	.7806	.7626	.7442	.7254	.7064	.6872	.6678	.6484	.6288
5	.9057	.8946	.8829	.8705	.8576	.8441	.8301	.8156	.8006	.7851
6	.9612	.9554	.9490	.9421	.9347	.9267	.9182	.9091	.8995	.8893
7	.9858	.9832	.9802	.9769	.9733	.9692	.9648	.9599	.9546	.9489
8	.9953	.9943	.9931	.9917	.9901	.9883	.9863	.9840	.9815	.9786
9	.9986	.9982	.9978	.9973	.9967	.9960	.9952	.9942	.9931	.9919
10	.9996	.9995	.9994	.9992	.9990	.9987	.9984	.9981	.9977	.9972
11	.9999	.9999	.9998	.9998	.9997	.9996	.9995	.9994	.9993	.9991
12	1.0000	1.0000	1.0000	.9999	.9999	.9999	.9999	.9998	.9998	.9997
13	1.0000	1.0000	1.0000	1.0000	1.0000	1.0000	1.0000	1.0000	.9999	.9999
14	1.0000	1.0000	1.0000	1.0000	1.0000	1.0000	1.0000	1.0000	1.0000	1.0000
15	1.0000	1.0000	1.0000	1.0000	1.0000	1.0000	1.0000	1.0000	1.0000	1.0000

x	μ 4.1	4.2	4.3	4.4	4.5	4.6	4.7	4.8	4.9	5.0
0	.0166	.0150	.0136	.0123	.0111	.0101	.0091	.0082	.0074	.0067
1	.0845	.0780	.0719	.0663	.0611	.0563	.0518	.0477	.0439	.0404
2	.2238	.2102	.1974	.1851	.1736	.1626	.1523	.1425	.1333	.1247
3	.4142	.3954	.3772	.3594	.3423	.3257	.3097	.2942	.2793	.2650
4	.6093	.5898	.5704	.5512	.5321	.5132	.4946	.4763	.4582	.4405
5	.7693	.7531	.7367	.7199	.7029	.6858	.6684	.6510	.6335	.6160
6	.8786	.8675	.8558	.8436	.8311	.8180	.8046	.7908	.7767	.7622
7	.9427	.9361	.9290	.9214	.9134	.9049	.8960	.8867	.8769	.8666
8	.9755	.9721	.9683	.9642	.9597	.9549	.9497	.9442	.9382	.9319
9	.9905	.9889	.9871	.9851	.9829	.9805	.9778	.9749	.9717	.9682

TABLE B.4 (Continued)

x	μ 4.1	4.2	4.3	4.4	4.5	4.6	4.7	4.8	4.9	5.0
10	.9966	.9959	.9952	.9943	.9933	.9922	.9910	.9896	.9880	.9863
11	.9989	.9986	.9983	.9980	.9976	.9971	.9966	.9960	.9953	.9945
12	.9997	.9996	.9995	.9993	.9992	.9990	.9988	.9986	.9983	.9980
13	.9999	.9999	.9998	.9998	.9997	.9997	.9996	.9995	.9994	.9993
14	1.0000	1.0000	1.0000	.9999	.9999	.9999	.9999	.9999	.9998	.9998
15	1.0000	1.0000	1.0000	1.0000	1.0000	1.0000	1.0000	1.0000	.9999	.9999
16	1.0000	1.0000	1.0000	1.0000	1.0000	1.0000	1.0000	1.0000	1.0000	1.0000

x	μ 5.1	5.2	5.3	5.4	5.5	5.6	5.7	5.8	5.9	6.0
0	.0061	.0055	.0050	.0045	.0041	.0037	.0033	.0030	.0027	.0025
1	.0372	.0342	.0314	.0289	.0266	.0244	.0224	.0206	.0189	.0174
2	.1165	.1088	.1016	.0948	.0884	.0824	.0768	.0715	.0666	.0620
3	.2513	.2381	.2254	.2133	.2017	.1906	.1800	.1700	.1604	.1512
4	.4231	.4061	.3895	.3733	.3575	.3422	.3272	.3127	.2987	.2851
5	.5984	.5809	.5635	.5461	.5289	.5119	.4950	.4783	.4619	.4457
6	.7474	.7324	.7171	.7017	.6860	.6703	.6544	.6384	.6224	.6063
7	.8560	.8449	.8335	.8217	.8095	.7970	.7841	.7710	.7576	.7440
8	.9252	.9181	.9106	.9027	.8944	.8857	.8766	.8672	.8574	.8472
9	.9644	.9603	.9559	.9512	.9462	.9409	.9352	.9292	.9228	.9161
10	.9844	.9823	.9800	.9775	.9747	.9718	.9686	.9651	.9614	.9574
11	.9937	.9927	.9916	.9904	.9890	.9875	.9859	.9841	.9821	.9799
12	.9976	.9972	.9967	.9962	.9955	.9949	.9941	.9932	.9922	.9912
13	.9992	.9990	.9988	.9986	.9983	.9980	.9977	.9973	.9969	.9964
14	.9997	.9997	.9996	.9995	.9994	.9993	.9991	.9990	.9988	.9986
15	.9999	.9999	.9999	.9998	.9998	.9998	.9997	.9996	.9996	.9995
16	1.0000	1.0000	1.0000	.9999	.9999	.9999	.9999	.9999	.9999	.9998
17	1.0000	1.0000	1.0000	1.0000	1.0000	1.0000	1.0000	1.0000	1.0000	.9999
18	1.0000	1.0000	1.0000	1.0000	1.0000	1.0000	1.0000	1.0000	1.0000	1.0000

x	μ 6.1	6.2	6.3	6.4	6.5	6.6	6.7	6.8	6.9	7.0
0	.0022	.0020	.0018	.0017	.0015	.0014	.0012	.0011	.0010	.0009
1	.0159	.0146	.0134	.0123	.0113	.0103	.0095	.0087	.0080	.0073
2	.0577	.0536	.0498	.0463	.0430	.0400	.0371	.0344	.0320	.0296
3	.1425	.1342	.1264	.1189	.1118	.1052	.0988	.0928	.0871	.0818
4	.2719	.2592	.2469	.2351	.2237	.2127	.2022	.1920	.1823	.1730
5	.4298	.4141	.3988	.3837	.3690	.3547	.3406	.3270	.3137	.3007
6	.5902	.5742	.5582	.5423	.5265	.5108	.4953	.4799	.4647	.4497
7	.7301	.7160	.7017	.6873	.6728	.6581	.6433	.6285	.6136	.5987
8	.8367	.8259	.8148	.8033	.7916	.7796	.7673	.7548	.7420	.7291
9	.9090	.9016	.8939	.8858	.8774	.8686	.8596	.8502	.8405	.8305

(Continued)

TABLE B.4 (Continued)

x	6.1	6.2	6.3	6.4	6.5	6.6	6.7	6.8	6.9	7.0
10	.9531	.9486	.9437	.9386	.9332	.9274	.9214	.9151	.9084	.9015
11	.9776	.9750	.9723	.9693	.9661	.9627	.9591	.9552	.9510	.9467
12	.9900	.9887	.9873	.9857	.9840	.9821	.9801	.9779	.9755	.9730
13	.9958	.9952	.9945	.9937	.9929	.9920	.9909	.9898	.9885	.9872
14	.9984	.9981	.9978	.9974	.9970	.9966	.9961	.9956	.9950	.9943
15	.9994	.9993	.9992	.9990	.9988	.9986	.9984	.9982	.9979	.9976
16	.9998	.9997	.9997	.9996	.9996	.9995	.9994	.9993	.9992	.9990
17	.9999	.9999	.9999	.9999	.9998	.9998	.9998	.9997	.9997	.9996
18	1.0000	1.0000	1.0000	1.0000	.9999	.9999	.9999	.9999	.9999	.9999
19	1.0000	1.0000	1.0000	1.0000	1.0000	1.0000	1.0000	1.0000	1.0000	1.0000
20	1.0000	1.0000	1.0000	1.0000	1.0000	1.0000	1.0000	1.0000	1.0000	1.0000

x	7.1	7.2	7.3	7.4	7.5	7.6	7.7	7.8	7.9	8.0
0	.0008	.0007	.0007	.0006	.0006	.0005	.0005	.0004	.0004	.0003
1	.0067	.0061	.0056	.0051	.0047	.0043	.0039	.0036	.0033	.0030
2	.0275	.0255	.0236	.0219	.0203	.0188	.0174	.0161	.0149	.0138
3	.0767	.0719	.0674	.0632	.0591	.0554	.0518	.0485	.0453	.0424
4	.1641	.1555	.1473	.1395	.1321	.1249	.1181	.1117	.1055	.0996
5	.2881	.2759	.2640	.2526	.2414	.2307	.2203	.2103	.2006	.1912
6	.4349	.4204	.4060	.3920	.3782	.3646	.3514	.3384	.3257	.3134
7	.5838	.5689	.5541	.5393	.5246	.5100	.4956	.4812	.4670	.4530
8	.7160	.7027	.6892	.6757	.6620	.6482	.6343	.6204	.6065	.5925
9	.8202	.8096	.7988	.7877	.7764	.7649	.7531	.7411	.7290	.7166
10	.8942	.8867	.8788	.8707	.8622	.8535	.8445	.8352	.8257	.8159
11	.9420	.9371	.9319	.9265	.9208	.9148	.9085	.9020	.8952	.8881
12	.9703	.9673	.9642	.9609	.9573	.9536	.9496	.9454	.9409	.9362
13	.9857	.9841	.9824	.9805	.9784	.9762	.9739	.9714	.9687	.9658
14	.9935	.9927	.9918	.9908	.9897	.9886	.9873	.9859	.9844	.9827
15	.9972	.9969	.9964	.9959	.9954	.9948	.9941	.9934	.9926	.9918
16	.9989	.9987	.9985	.9983	.9980	.9978	.9974	.9971	.9967	.9963
17	.9996	.9995	.9994	.9993	.9992	.9991	.9989	.9988	.9986	.9984
18	.9998	.9998	.9998	.9997	.9997	.9996	.9996	.9995	.9994	.9993
19	.9999	.9999	.9999	.9999	.9999	.9999	.9998	.9998	.9998	.9997
20	1.0000	1.0000	1.0000	1.0000	1.0000	1.0000	.9999	.9999	.9999	.9999
21	1.0000	1.0000	1.0000	1.0000	1.0000	1.0000	1.0000	1.0000	1.0000	1.0000
22	1.0000	1.0000	1.0000	1.0000	1.0000	1.0000	1.0000	1.0000	1.0000	1.0000

TABLE B.4 (Continued)

x	μ = 8.1	8.2	8.3	8.4	8.5	8.6	8.7	8.8	8.9	9.0
0	.0003	.0003	.0002	.0002	.0002	.0002	.0002	.0002	.0001	.0001
1	.0028	.0025	.0023	.0021	.0019	.0018	.0016	.0015	.0014	.0012
2	.0127	.0118	.0109	.0100	.0093	.0086	.0079	.0073	.0068	.0062
3	.0396	.0370	.0346	.0323	.0301	.0281	.0262	.0244	.0228	.0212
4	.0940	.0887	.0837	.0789	.0744	.0701	.0660	.0621	.0584	.0550
5	.1822	.1736	.1653	.1573	.1496	.1422	.1352	.1284	.1219	.1157
6	.3013	.2896	.2781	.2670	.2562	.2457	.2355	.2256	.2160	.2068
7	.4391	.4254	.4119	.3987	.3856	.3728	.3602	.3478	.3357	.3239
8	.5786	.5647	.5507	.5369	.5231	.5094	.4958	.4823	.4689	.4557
9	.7041	.6915	.6788	.6659	.6530	.6400	.6269	.6137	.6006	.5874
10	.8058	.7955	.7850	.7743	.7634	.7522	.7409	.7294	.7178	.7060
11	.8807	.8731	.8652	.8571	.8487	.8400	.8311	.8220	.8126	.8030
12	.9313	.9261	.9207	.9150	.9091	.9029	.8965	.8898	.8829	.8758
13	.9628	.9595	.9561	.9524	.9486	.9445	.9403	.9358	.9311	.9261
14	.9810	.9791	.9771	.9749	.9726	.9701	.9675	.9647	.9617	.9585
15	.9908	.9898	.9887	.9875	.9862	.9848	.9832	.9816	.9798	.9780
16	.9958	.9953	.9947	.9941	.9934	.9926	.9918	.9909	.9899	.9889
17	.9982	.9979	.9977	.9973	.9970	.9966	.9962	.9957	.9952	.9947
18	.9992	.9991	.9990	.9989	.9987	.9985	.9983	.9981	.9978	.9976
19	.9997	.9997	.9996	.9995	.9995	.9994	.9993	.9992	.9991	.9989
20	.9999	.9999	.9998	.9998	.9998	.9998	.9997	.9997	.9996	.9996
21	1.0000	1.0000	.9999	.9999	.9999	.9999	.9999	.9999	.9998	.9996
22	1.0000	1.0000	1.0000	1.0000	1.0000	1.0000	1.0000	1.0000	.9999	.9999
23	1.0000	1.0000	1.0000	1.0000	1.0000	1.0000	1.0000	1.0000	1.0000	1.0000

x	μ = 9.1	9.2	9.3	9.4	9.5	9.6	9.7	9.8	9.9	10.0
0	.0001	.0001	.0001	.0001	.0001	.0001	.0001	.0001	.0001	.0000
1	.0011	.0010	.0009	.0009	.0008	.0007	.0007	.0006	.0005	.0005
2	.0058	.0053	.0049	.0045	.0042	.0038	.0035	.0033	.0030	.0028
3	.0198	.0184	.0172	.0160	.0149	.0138	.0129	.0120	.0111	.0103
4	.0517	.0486	.0456	.0429	.0403	.0378	.0355	.0333	.0312	.0293
5	.1098	.1041	.0986	.0935	.0885	.0838	.0793	.0750	.0710	.0671
6	.1978	.1892	.1808	.1727	.1649	.1574	.1502	.1433	.1366	.1301
7	.3123	.3010	.2900	.2792	.2687	.2584	.2485	.2388	.2294	.2202
8	.4426	.4296	.4168	.4042	.3918	.3796	.3676	.3558	.3442	.3328
9	.5742	.5611	.5479	.5349	.5218	.5089	.4960	.4832	.4705	.4579
10	.6941	.6820	.6699	.6576	.6453	.6329	.6205	.6080	.5955	.5830
11	.7932	.7832	.7730	.7626	.7520	.7412	.7303	.7193	.7081	.6968
12	.8684	.8607	.8529	.8448	.8364	.8279	.8191	.8101	.8009	.7916
13	.9210	.9156	.9100	.9042	.8981	.8919	.8853	.8786	.8716	.8645
14	.9552	.9517	.9480	.9441	.9400	.9357	.9312	.9265	.9216	.9165

(*Continued*)

TABLE B.4 (Continued)

x	μ 9.1	9.2	9.3	9.4	9.5	9.6	9.7	9.8	9.9	10.0
15	.9760	.9738	.9715	.9691	.9665	.9638	.9609	.9579	.9546	.9513
16	.9878	.9865	.9852	.9838	.9823	.9806	.9789	.9770	.9751	.9730
17	.9941	.9934	.9927	.9919	.9911	.9902	.9892	.9881	.9870	.9857
18	.9973	.9969	.9966	.9962	.9957	.9952	.9947	.9941	.9935	.9928
19	.9988	.9986	.9985	.9983	.9980	.9978	.9975	.9972	.9969	.9965
20	.9995	.9994	.9993	.9992	.9991	.9990	.9989	.9987	.9986	.9984
21	.9998	.9998	.9997	.9997	.9996	.9996	.9995	.9995	.9994	.9993
22	.9999	.9999	.9999	.9999	.9999	.9998	.9998	.9998	.9997	.9997
23	1.0000	1.0000	1.0000	1.0000	.9999	.9999	.9999	.9999	.9999	.9999
24	1.0000	1.0000	1.0000	1.0000	1.0000	1.0000	1.0000	1.0000	1.0000	1.0000
25	1.0000	1.0000	1.0000	1.0000	1.0000	1.0000	1.0000	1.0000	1.0000	1.0000

x	μ 11	12	13	14	15	16	17	18	19	20
0	.0000	.0000	.0000	.0000	.0000	.0000	.0000	.0000	.0000	.0000
1	.0002	.0001	.0000	.0000	.0000	.0000	.0000	.0000	.0000	.0000
2	.0012	.0005	.0002	.0001	.0000	.0000	.0000	.0000	.0000	.0000
3	.0049	.0023	.0011	.0005	.0002	.0001	.0000	.0000	.0000	.0000
4	.0151	.0076	.0037	.0018	.0009	.0004	.0002	.0001	.0000	.0000
5	.0375	.0203	.0107	.0055	.0028	.0014	.0007	.0003	.0002	.0001
6	.0786	.0458	.0259	.0142	.0076	.0040	.0021	.0010	.0005	.0003
7	.1432	.0895	.0540	.0316	.0180	.0100	.0054	.0029	.0015	.0008
8	.2320	.1550	.0998	.0621	.0374	.0220	.0126	.0071	.0039	.0021
9	.3405	.2424	.1658	.1094	.0699	.0433	.0261	.0154	.0089	.0050
10	.4599	.3472	.2517	.1757	.1185	.0774	.0491	.0304	.0183	.0108
11	.5793	.4616	.3532	.2600	.1848	.1270	.0847	.0549	.0347	.0214
12	.6887	.5760	.4631	.3585	.2676	.1931	.1350	.0917	.0606	.0390
13	.7813	.6815	.5730	.4644	.3632	.2745	.2009	.1426	.0984	.0661
14	.8540	.7720	.6751	.5704	.4657	.3675	.2808	.2081	.1497	.1049

TABLE B.4 (Continued)

	μ									
x	11	12	13	14	15	16	17	18	19	20
15	.9074	.8444	.7636	.6694	.5681	.4667	.3715	.2867	.2148	.1565
16	.9441	.8987	.8355	.7559	.6641	.5660	.4677	.3751	.2920	.2211
17	.9678	.9370	.8905	.8272	.7489	.6593	.5640	.4686	.3784	.2970
18	.9823	.9626	.9302	.8826	.8195	.7423	.6550	.5622	.4695	.3814
19	.9907	.9787	.9573	.9235	.8752	.8122	.7363	.6509	.5606	.4703
20	.9953	.9884	.9750	.9521	.9170	.8682	.8055	.7307	.6472	.5591
21	.9977	.9939	.9859	.9712	.9469	.9108	.8615	.7991	.7255	.6437
22	.9990	.9970	.9924	.9833	.9673	.9418	.9047	.8551	.7931	.7206
23	.9995	.9985	.9960	.9907	.9805	.9633	.9367	.8989	.8490	.7875
24	.9998	.9993	.9980	.9950	.9888	.9777	.9594	.9317	.8933	.8432
25	.9999	.9997	.9990	.9974	.9938	.9869	.9748	.9554	.9269	.8878
26	1.0000	.9999	.9995	.9987	.9967	.9925	.9848	.9718	.9514	.9221
27	1.0000	.9999	.9998	.9994	.9983	.9959	.9912	.9827	.9687	.9475
28	1.0000	1.0000	.9999	.9997	.9991	.9978	.9950	.9897	.9805	.9657
29	1.0000	1.0000	1.0000	.9999	.9996	.9989	.9973	.9941	.9882	.9782
30	1.0000	1.0000	1.0000	.9999	.9998	.9994	.9986	.9967	.9930	.9865
31	1.0000	1.0000	1.0000	1.0000	.9999	.9997	.9993	.9982	.9960	.9919
32	1.0000	1.0000	1.0000	1.0000	1.0000	.9999	.9996	.9990	.9978	.9953
33	1.0000	1.0000	1.0000	1.0000	1.0000	.9999	.9998	.9995	.9988	.9973
34	1.0000	1.0000	1.0000	1.0000	1.0000	1.0000	.9999	.9998	.9994	.9985
35	1.0000	1.0000	1.0000	1.0000	1.0000	1.0000	1.0000	.9999	.9997	.9992
36	1.0000	1.0000	1.0000	1.0000	1.0000	1.0000	1.0000	.9999	.9998	.9996
37	1.0000	1.0000	1.0000	1.0000	1.0000	1.0000	1.0000	1.0000	.9999	.9998
38	1.0000	1.0000	1.0000	1.0000	1.0000	1.0000	1.0000	1.0000	1.0000	.9999
39	1.0000	1.0000	1.0000	1.0000	1.0000	1.0000	1.0000	1.0000	1.0000	.9999
40	1.0000	1.0000	1.0000	1.0000	1.0000	1.0000	1.0000	1.0000	1.0000	1.0000

TABLE B.5 Selected exponential constants for $\mu = c\lambda$

μ	$e^{-\mu}$	μ	$e^{-\mu}$	μ	$e^{-\mu}$
.0	1.0000	4.0	.01832	8.0	.00034
.1	.9048	4.1	.01657	8.1	.00030
.2	.8187	4.2	.01500	8.2	.00027
.3	.7408	4.3	.01357	8.3	.00025
.4	.6703	4.4	.01228	8.4	.00022
.5	.6065	4.5	.01111	8.5	.00020
.6	.5488	4.6	.01005	8.6	.00018
.7	.4966	4.7	.00910	8.7	.00017
.8	.4493	4.8	.00823	8.8	.00015
.9	.4066	4.9	.00745	8.9	.00014
1.0	.3679	5.0	.00674	9.0	.00012
1.1	.3329	5.1	.00610	9.1	.00011
1.2	.3012	5.2	.00552	9.2	.00010
1.3	.2725	5.3	.00499	9.3	.00009
1.4	.2466	5.4	.00452	9.4	.00008
1.5	.2231	5.5	.00409	9.5	.00007
1.6	.2019	5.6	.00370	9.6	.00007
1.7	.1827	5.7	.00335	9.7	.00006
1.8	.1653	5.8	.00303	9.8	.00006
1.9	.1496	5.9	.00274	9.9	.00005
2.0	.1353	6.0	.00248		
2.1	.1225	6.1	.00224		
2.2	.1108	6.2	.00203		
2.3	.1003	6.3	.00184		
2.4	.0907	6.4	.00166		
2.5	.0821	6.5	.00150		
2.6	.0743	6.6	.00136		
2.7	.0672	6.7	.00123		
2.8	.0608	6.8	.00111		
2.9	.0550	6.9	.00101		
3.0	.0498	7.0	.00091		
3.1	.0450	7.1	.00083		
3.2	.0408	7.2	.00075		
3.3	.0369	7.3	.00068		
3.4	.0334	7.4	.00061		
3.5	.0302	7.5	.00055		
3.6	.0273	7.6	.00050		
3.7	.0247	7.7	.00045		
3.8	.0224	7.8	.00041		
3.9	.0202	7.9	.00037		

TABLE B.6 Standard normal distribution

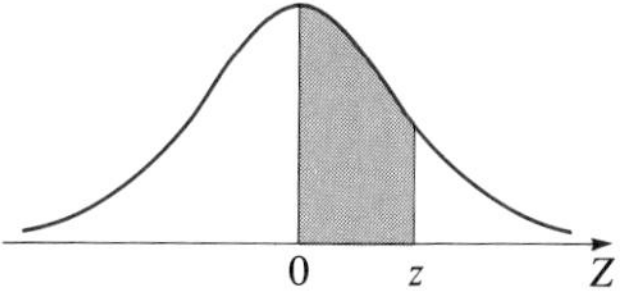

The values in the body of the table are the areas between the mean and the value of z.

z	.00	.01	.02	.03	.04	.05	.06	.07	.08	.09
.00	.0000	.0040	.0080	.0120	.0160	.0199	.0239	.0279	.0319	.0359
.10	.0398	.0438	.0478	.0517	.0557	.0596	.0636	.0675	.0714	.0753
.20	.0793	.0832	.0871	.0910	.0948	.0987	.1026	.1064	.1103	.1141
.30	.1179	.1217	.1255	.1293	.1331	.1368	.1406	.1443	.1480	.1517
.40	.1554	.1591	.1628	.1664	.1700	.1736	.1772	.1808	.1844	.1879
.50	.1915	.1950	.1985	.2019	.2054	.2088	.2123	.2157	.2190	.2224
.60	.2257	.2291	.2324	.2357	.2389	.2422	.2454	.2486	.2517	.2549
.70	.2580	.2611	.2642	.2673	.2703	.2734	.2764	.2793	.2823	.2852
.80	.2881	.2910	.2939	.2967	.2995	.3023	.3051	.3078	.3106	.3133
.90	.3159	.3186	.3212	.3238	.3264	.3289	.3315	.3340	.3365	.3389
1.00	.3413	.3438	.3461	.3485	.3508	.3531	.3554	.3577	.3599	.3621
1.10	.3643	.3665	.3686	.3708	.3729	.3749	.3770	.3790	.3810	.3830
1.20	.3849	.3869	.3888	.3907	.3925	.3944	.3962	.3980	.3997	.4015
1.30	.4032	.4049	.4066	.4082	.4099	.4115	.4131	.4147	.4162	.4177
1.40	.4192	.4207	.4222	.4236	.4251	.4265	.4279	.4292	.4306	.4319
1.50	.4332	.4345	.4357	.4370	.4382	.4394	.4406	.4418	.4429	.4441
1.60	.4452	.4463	.4474	.4484	.4495	.4505	.4515	.4525	.4535	.4545
1.70	.4554	.4564	.4573	.4582	.4591	.4599	.4608	.4616	.4625	.4633
1.80	.4641	.4649	.4656	.4664	.4671	.4678	.4686	.4693	.4699	.4706
1.90	.4713	.4719	.4726	.4732	.4738	.4744	.4750	.4756	.4761	.4767
2.00	.4772	.4778	.4783	.4788	.4793	.4798	.4803	.4808	.4812	.4817
2.10	.4821	.4826	.4830	.4834	.4838	.4842	.4846	.4850	.4854	.4857
2.20	.4861	.4864	.4868	.4871	.4875	.4878	.4881	.4884	.4887	.4890
2.30	.4893	.4896	.4898	.4901	.4904	.4906	.4909	.4911	.4913	.4916
2.40	.4918	.4920	.4922	.4925	.4927	.4929	.4931	.4932	.4934	.4936
2.50	.4938	.4940	.4941	.4943	.4945	.4946	.4948	.4949	.4951	.4952
2.60	.4953	.4955	.4956	.4957	.4959	.4960	.4961	.4962	.4963	.4964
2.70	.4965	.4966	.4967	.4968	.4969	.4970	.4971	.4972	.4973	.4974
2.80	.4974	.4975	.4976	.4977	.4977	.4978	.4979	.4979	.4980	.4981
2.90	.4981	.4982	.4982	.4983	.4984	.4984	.4985	.4985	.4986	.4986
3.00	.4987	.4987	.4987	.4988	.4988	.4989	.4989	.4989	.4990	.4990
3.10	.4990	.4991	.4991	.4991	.4992	.4992	.4992	.4992	.4993	.4993
3.20	.4993	.4993	.4994	.4994	.4994	.4994	.4994	.4995	.4995	.4995
3.30	.4995	.4995	.4995	.4996	.4996	.4996	.4996	.4996	.4996	.4997
3.40	.4997	.4997	.4997	.4997	.4997	.4997	.4997	.4997	.4997	.4998
3.50	.4998	.4998	.4998	.4998	.4998	.4998	.4998	.4998	.4998	.4998
3.60	.4998	.4998	.4999	.4999	.4999	.4999	.4999	.4999	.4999	.4999
3.70	.4999	.4999	.4999	.4999	.4999	.4999	.4999	.4999	.4999	.4999
3.80	.4999	.4999	.4999	.4999	.4999	.4999	.4999	.4999	.4999	.4999
3.90	.5000	.5000	.5000	.5000	.5000	.5000	.5000	.5000	.5000	.5000

Example: If we want to find the area under the standard normal curve between $z = 0$ and $z = 1.96$, we find the $z = 1.90$ row and .06 column (for $z = 1.90 + .06 = 1.96$) and read .4750 at the intersection.

Adapted from Billingsley et al., *Statistical Inference for Management and Economics,* Third Edition.

TABLE B.7 Percentage points of selected *t*-distributions

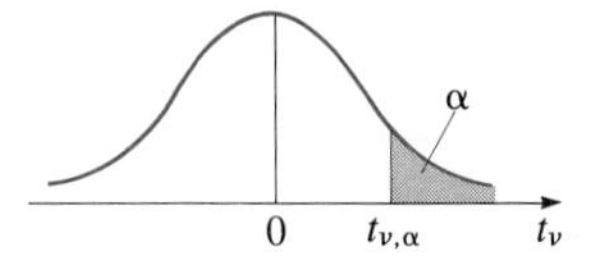

$t_{\nu,\alpha}$ such that $P(t_\nu > t_{\nu,\alpha}) = \alpha$

					α					
ν	.40	.30	.20	.10	.05	.025	.01	.005	.001	.0005
1	.325	.727	1.376	3.078	6.314	12.710	31.820	63.657	318.310	636.620
2	.289	.617	1.061	1.886	2.920	4.303	6.965	9.925	22.330	31.600
3	.277	.584	.978	1.638	2.353	3.182	4.541	5.841	10.214	12.924
4	.271	.569	.941	1.533	2.132	2.776	3.747	4.604	7.173	8.610
5	.267	.559	.920	1.476	2.015	2.571	3.365	4.032	5.893	6.869
6	.265	.553	.906	1.440	1.943	2.447	3.143	3.707	5.208	5.959
7	.263	.549	.896	1.415	1.895	2.365	2.998	3.499	4.785	5.408
8	.262	.546	.889	1.397	1.860	2.306	2.896	3.355	4.501	5.041
9	.261	.543	.883	1.383	1.833	2.262	2.821	3.250	4.297	4.781
10	.260	.542	.879	1.372	1.812	2.228	2.764	3.169	4.144	4.587
11	.260	.540	.876	1.363	1.796	2.201	2.718	3.106	4.025	4.437
12	.259	.539	.873	1.356	1.782	2.179	2.681	3.055	3.930	4.318
13	.259	.538	.870	1.350	1.771	2.160	2.650	3.012	3.852	4.221
14	.258	.537	.868	1.345	1.761	2.145	2.624	2.977	3.787	4.140
15	.258	.536	.866	1.341	1.753	2.131	2.602	2.947	3.733	4.073
16	.258	.535	.865	1.337	1.746	2.120	2.583	2.921	3.686	4.015
17	.257	.534	.863	1.333	1.740	2.110	2.567	2.898	3.646	3.965
18	.257	.534	.862	1.330	1.734	2.101	2.552	2.878	3.610	3.922
19	.257	.533	.861	1.328	1.729	2.093	2.539	2.861	3.579	3.883
20	.257	.533	.860	1.325	1.725	2.086	2.528	2.845	3.552	3.850
21	.257	.532	.859	1.323	1.721	2.080	2.518	2.831	3.527	3.819
22	.256	.532	.858	1.321	1.717	2.074	2.508	2.819	3.505	3.792
23	.256	.532	.858	1.319	1.714	2.069	2.500	2.807	3.485	3.767
24	.256	.531	.857	1.318	1.711	2.064	2.492	2.797	3.467	3.745
25	.256	.531	.856	1.316	1.708	2.060	2.485	2.787	3.450	3.725
26	.256	.531	.856	1.315	1.706	2.056	2.479	2.779	3.435	3.707
27	.256	.531	.855	1.314	1.703	2.052	2.473	2.771	3.421	3.690
28	.256	.530	.855	1.313	1.701	2.048	2.467	2.763	3.408	3.674
29	.256	.530	.854	1.311	1.699	2.045	2.462	2.756	3.396	3.659
30	.256	.530	.854	1.310	1.697	2.042	2.457	2.750	3.385	3.646
40	.255	.529	.851	1.303	1.684	2.021	2.423	2.704	3.307	3.551
60	.254	.527	.848	1.296	1.671	2.000	2.390	2.660	3.232	3.460
120	.254	.526	.845	1.289	1.658	1.980	2.358	2.617	3.160	3.373
∞	.253	.524	.842	1.282	1.645	1.960	2.326	2.576	3.090	3.291

TABLE B.8 Percentage points of selected χ^2 distributions

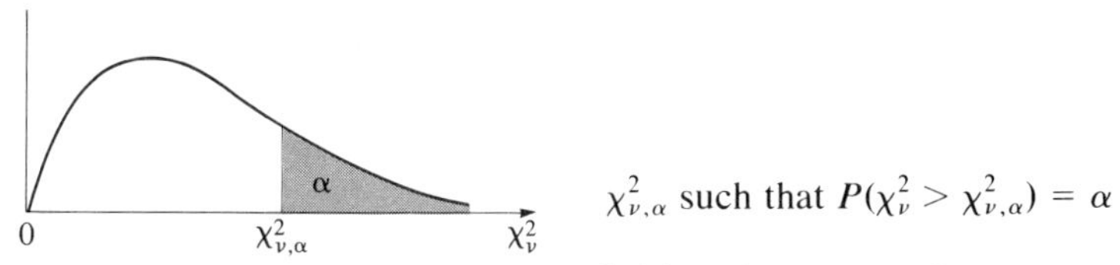

$\chi^2_{\nu,\alpha}$ such that $P(\chi^2_\nu > \chi^2_{\nu,\alpha}) = \alpha$

α = Probability of a Larger Value

ν	.995	.990	.975	.950	.900	.100	.050	.025	.010	.005
1				.004	.016	2.71	3.84	5.02	6.63	7.88
2	.01	.02	.05	.10	.211	4.61	5.99	7.38	9.21	10.60
3	.07	.11	.22	.35	.584	6.25	7.81	9.35	11.34	12.84
4	.21	.30	.48	.71	1.064	7.78	9.49	11.14	13.28	14.86
5	.41	.55	.83	1.15	1.610	9.24	11.07	12.83	15.09	16.75
6	.68	.87	1.24	1.64	2.20	10.64	12.59	14.45	16.81	18.55
7	.99	1.24	1.69	2.17	2.83	12.02	14.07	16.01	18.48	20.28
8	1.34	1.65	2.18	2.73	3.49	13.36	15.51	17.53	20.09	21.96
9	1.73	2.09	2.70	3.33	4.17	14.68	16.92	19.02	21.67	23.59
10	2.16	2.56	3.25	3.94	4.87	15.99	18.31	20.48	23.21	25.19
11	2.60	3.05	3.82	4.57	5.58	17.28	19.68	21.92	24.72	26.76
12	3.07	3.57	4.40	5.23	6.30	18.55	21.03	23.34	26.22	28.30
13	3.57	4.11	5.01	5.89	7.04	19.81	22.36	24.74	27.69	29.82
14	4.07	4.66	5.63	6.57	7.79	21.06	23.68	26.12	29.14	31.32
15	4.60	5.23	6.26	7.26	8.55	22.31	25.00	27.49	30.58	32.80
16	5.14	5.81	6.91	7.96	9.31	23.54	26.30	28.85	32.00	34.27
17	5.70	6.41	7.56	8.67	10.09	24.77	27.59	30.19	33.41	35.72
18	6.26	7.01	8.23	9.39	10.87	25.99	28.87	31.53	34.81	37.16
19	6.84	7.63	8.91	10.12	11.65	27.20	30.14	32.85	36.19	38.58
20	7.43	8.26	9.59	10.85	12.44	28.41	31.41	34.17	37.57	40.00
21	8.03	8.90	10.28	11.59	13.24	29.62	32.67	35.48	38.93	41.40
22	8.64	9.54	10.98	12.34	14.04	30.81	33.92	36.78	40.29	42.80
23	9.26	10.20	11.69	13.09	14.85	32.01	35.17	38.08	41.64	44.18
24	9.89	10.86	12.40	13.85	15.66	33.20	36.41	39.36	42.98	45.56
25	10.52	11.52	13.12	14.61	16.47	34.38	37.65	40.65	44.31	46.93
26	11.16	12.20	13.84	15.38	17.29	35.56	38.89	41.92	45.64	48.29
27	11.81	12.88	14.57	16.15	18.11	36.74	40.11	43.19	46.96	49.64
28	12.46	13.56	15.31	16.93	18.94	37.92	41.34	44.46	48.28	50.99
29	13.12	14.26	16.05	17.71	19.77	39.09	42.56	45.72	49.59	52.34
30	13.79	14.95	16.79	18.49	20.60	40.26	43.77	46.98	50.89	53.67
40	20.71	22.16	24.43	26.51	29.05	51.81	55.76	59.34	63.69	66.77
50	27.99	29.71	32.36	34.76	37.69	63.17	67.50	71.42	76.15	79.49
60	35.53	37.48	40.48	43.19	46.46	74.40	79.08	83.30	88.38	91.95
70	43.28	45.44	48.76	51.74	55.33	85.53	90.53	95.02	100.43	104.22
80	51.17	53.54	57.15	60.39	64.28	96.58	101.88	106.63	112.33	116.32
90	59.20	61.75	65.65	69.13	73.29	107.57	113.14	118.14	124.12	128.30
100	67.33	70.06	74.22	77.93	82.36	118.50	124.34	129.56	135.81	140.17

Abridged from Thompson, Catherine M., "Table of Percentage Points of the χ^2 Distribution," *Biometrika*, Vol. 32 (1942), p. 187, by permission of *Biometrika* Trustees.

TABLE B.9A Percentage points for selected F-distributions

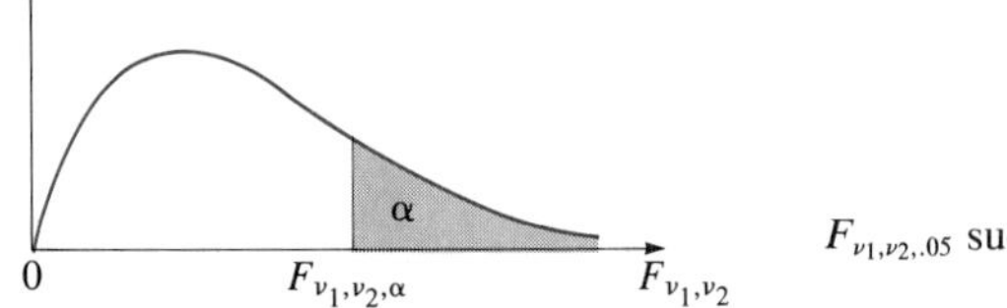

$F_{\nu_1,\nu_2,.05}$ such that $P(F_{\nu_1,\nu_2}\ F_{\nu_1,\nu_2,.05}) = .05$

F-Distribution: .05 Points (α = .05)

ν_2 \ ν_1 (Degrees of Freedom for Denominator \ Degrees of Freedom for Numerator)	1	2	3	4	5	6	7	8	9
1	161.4	199.5	215.7	224.6	230.2	234.0	236.8	238.9	240.5
2	18.51	19.00	19.16	19.25	19.30	19.33	19.35	19.37	19.38
3	10.13	9.55	9.28	9.12	9.01	8.94	8.89	8.85	8.81
4	7.71	6.94	6.59	6.39	6.26	6.16	6.09	6.04	6.00
5	6.61	5.79	5.41	5.19	5.05	4.95	4.88	4.82	4.77
6	5.99	5.14	4.76	4.53	4.39	4.28	4.21	4.15	4.10
7	5.59	4.74	4.35	4.12	3.97	3.87	3.79	3.73	3.68
8	5.32	4.46	4.07	3.84	3.69	3.58	3.50	3.44	3.39
9	5.12	4.26	3.86	3.63	3.48	3.37	3.29	3.23	3.18
10	4.96	4.10	3.71	3.48	3.33	3.22	3.14	3.07	3.02
11	4.84	3.98	3.59	3.36	3.20	3.09	3.01	2.95	2.90
12	4.75	3.89	3.49	3.26	3.11	3.00	2.91	2.85	2.80
13	4.67	3.81	3.41	3.18	3.03	2.92	2.83	2.77	2.71
14	4.60	3.74	3.34	3.11	2.96	2.85	2.76	2.70	2.65
15	4.54	3.68	3.29	3.06	2.90	2.79	2.71	2.64	2.59
16	4.49	3.63	3.24	3.01	2.85	2.74	2.66	2.59	2.54
17	4.45	3.59	3.20	2.96	2.81	2.70	2.61	2.55	2.49
18	4.41	3.55	3.16	2.93	2.77	2.66	2.58	2.51	2.46
19	4.38	3.52	3.13	2.90	2.74	2.63	2.54	2.48	2.42
20	4.35	3.49	3.10	2.87	2.71	2.60	2.51	2.45	2.39
21	4.32	3.47	3.07	2.84	2.68	2.57	2.49	2.42	2.37
22	4.30	3.44	3.05	2.82	2.66	2.55	2.46	2.40	2.34
23	4.28	3.42	3.03	2.80	2.64	2.53	2.44	2.37	2.32
24	4.26	3.40	3.01	2.78	2.62	2.51	2.42	2.36	2.30
25	4.24	3.39	2.99	2.76	2.60	2.49	2.40	2.34	2.28
26	4.23	3.37	2.98	2.74	2.59	2.47	2.39	2.32	2.27
27	4.21	3.35	2.96	2.73	2.57	2.46	2.37	2.31	2.25
28	4.20	3.34	2.95	2.71	2.56	2.45	2.36	2.29	2.24
29	4.18	3.33	2.93	2.70	2.55	2.43	2.35	2.28	2.22
30	4.17	3.32	2.92	2.69	2.53	2.42	2.33	2.27	2.21
40	4.08	3.23	2.84	2.61	2.45	2.34	2.25	2.18	2.12
60	4.00	3.15	2.76	2.53	2.37	2.25	2.17	2.10	2.04
120	3.92	3.07	2.68	2.45	2.29	2.17	2.09	2.02	1.96
∞	3.84	3.00	2.60	2.37	2.21	2.10	2.01	1.94	1.88

TABLE B.9A (Continued)

F-Distribution: .05 Points (α = .05)

ν_2 (Degrees of Freedom for Denominator) \ ν_1	Degrees of Freedom for Numerator									
	10	**12**	**15**	**20**	**24**	**30**	**40**	**60**	**120**	**∞**
1	241.9	243.9	245.9	248.0	249.1	250.1	251.1	252.2	253.3	254.3
2	19.40	19.41	19.43	19.45	19.45	19.46	19.47	19.48	19.49	19.50
3	8.79	8.74	8.70	8.66	8.64	8.62	8.59	8.57	8.55	8.53
4	5.96	5.91	5.86	5.80	5.77	5.75	5.72	5.69	5.66	5.63
5	4.74	4.68	4.62	4.56	4.53	4.50	4.46	4.43	4.40	4.36
6	4.06	4.00	3.94	3.87	3.84	3.81	3.77	3.74	3.70	3.67
7	3.64	3.57	3.51	3.44	3.41	3.38	3.34	3.30	3.27	3.23
8	3.35	3.28	3.22	3.15	3.12	3.08	3.04	3.01	2.97	2.93
9	3.14	3.07	3.01	2.94	2.90	2.86	2.83	2.79	2.75	2.71
10	2.98	2.91	2.85	2.77	2.74	2.70	2.66	2.62	2.58	2.54
11	2.85	2.79	2.72	2.65	2.61	2.57	2.53	2.49	2.45	2.40
12	2.75	2.69	2.62	2.54	2.51	2.47	2.43	2.38	2.34	2.30
13	2.67	2.60	2.53	2.46	2.42	2.38	2.34	2.30	2.25	2.21
14	2.60	2.53	2.46	2.39	2.35	2.31	2.27	2.22	2.18	2.13
15	2.54	2.48	2.40	2.33	2.29	2.25	2.20	2.16	2.11	2.07
16	2.49	2.42	2.35	2.28	2.24	2.19	2.15	2.11	2.06	2.01
17	2.45	2.38	2.31	2.23	2.19	2.15	2.10	2.06	2.01	1.96
18	2.41	2.34	2.27	2.19	2.15	2.11	2.06	2.02	1.97	1.92
19	2.38	2.31	2.23	2.16	2.11	2.07	2.03	1.98	1.93	1.88
20	2.35	2.28	2.20	2.12	2.08	2.04	1.99	1.95	1.90	1.84
21	2.32	2.25	2.18	2.10	2.05	2.01	1.96	1.92	1.87	1.81
22	2.30	2.23	2.15	2.07	2.03	1.98	1.94	1.89	1.84	1.78
23	2.27	2.20	2.13	2.05	2.01	1.96	1.91	1.86	1.81	1.76
24	2.25	2.18	2.11	2.03	1.98	1.94	1.89	1.84	1.79	1.73
25	2.24	2.16	2.09	2.01	1.96	1.92	1.87	1.82	1.77	1.71
26	2.22	2.15	2.07	1.99	1.95	1.90	1.85	1.80	1.75	1.69
27	2.20	2.13	2.06	1.97	1.93	1.88	1.84	1.79	1.73	1.67
28	2.19	2.12	2.04	1.96	1.91	1.87	1.82	1.77	1.71	1.65
29	2.18	2.10	2.03	1.94	1.90	1.85	1.81	1.75	1.70	1.64
30	2.16	2.09	2.01	1.93	1.89	1.84	1.79	1.74	1.68	1.62
40	2.08	2.00	1.92	1.84	1.79	1.74	1.69	1.64	1.58	1.51
60	1.99	1.92	1.84	1.75	1.70	1.65	1.59	1.53	1.47	1.39
120	1.91	1.83	1.75	1.66	1.61	1.55	1.50	1.43	1.35	1.25
∞	1.83	1.75	1.67	1.57	1.52	1.46	1.39	1.32	1.22	1.00

TABLE B.9B .01 percentage points for selected F-distributions

$F_{\nu_1,\nu_2,.01}$ such that $P(F_{\nu_1,\nu_2} > F_{\nu_1,\nu_2,.01}) = .01$

F-Distribution: .01 Points ($\alpha = .01$)

ν_2 (Degrees of Freedom for Denominator) \ ν_1 (Degrees of Freedom for Numerator)	1	2	3	4	5	6	7	8	9
1	4052	4999.5	5403	5625	5764	5859	5928	5981	6022
2	98.50	99.00	99.17	99.25	99.30	99.33	99.36	99.37	99.39
3	34.12	30.82	29.46	28.71	28.24	27.91	27.67	27.49	27.35
4	21.20	18.00	16.69	15.98	15.52	15.21	14.98	14.80	14.66
5	16.26	13.27	12.06	11.39	10.97	10.67	10.46	10.29	10.16
6	13.75	10.92	9.78	9.15	8.75	8.47	8.26	8.10	7.98
7	12.25	9.55	8.45	7.85	7.46	7.19	6.99	6.84	6.72
8	11.26	8.65	7.59	7.01	6.63	6.37	6.18	6.03	5.91
9	10.56	8.02	6.99	6.42	6.06	5.80	5.61	5.47	5.35
10	10.04	7.56	6.55	5.99	5.64	5.39	5.20	5.06	4.94
11	9.65	7.21	6.22	5.67	5.32	5.07	4.89	4.74	4.63
12	9.33	6.93	5.95	5.41	5.06	4.82	4.64	4.50	4.39
13	9.07	6.70	5.74	5.21	4.86	4.62	4.44	4.30	4.19
14	8.86	6.51	5.56	5.04	4.69	4.46	4.28	4.14	4.03
15	8.68	6.36	5.42	4.89	4.56	4.32	4.14	4.00	3.89
16	8.53	6.23	5.29	4.77	4.44	4.20	4.03	3.89	3.78
17	8.40	6.11	5.18	4.67	4.34	4.10	3.93	3.79	3.68
18	8.29	6.01	5.09	4.58	4.25	4.01	3.84	3.71	3.60
19	8.18	5.93	5.01	4.50	4.17	3.94	3.77	3.63	3.52
20	8.10	5.85	4.94	4.43	4.10	3.87	3.70	3.56	3.46
21	8.02	5.78	4.87	4.37	4.04	3.81	3.64	3.51	3.40
22	7.95	5.72	4.82	4.31	3.99	3.76	3.59	3.45	3.35
23	7.88	5.66	4.76	4.26	3.94	3.71	3.54	3.41	3.30
24	7.82	5.61	4.72	4.22	3.90	3.67	3.50	3.36	3.26
25	7.77	5.57	4.68	4.18	3.85	3.63	3.46	3.32	3.22
26	7.72	5.53	4.64	4.14	3.82	3.59	3.42	3.29	3.18
27	7.68	5.49	4.60	4.11	3.78	3.56	3.39	3.26	3.15
28	7.64	5.45	4.57	4.07	3.75	3.53	3.36	3.23	3.12
29	7.60	5.42	4.54	4.04	3.73	3.50	3.33	3.20	3.09
30	7.56	5.39	4.51	4.02	3.70	3.47	3.30	3.17	3.07
40	7.31	5.18	4.31	3.83	3.51	3.29	3.12	2.99	2.89
60	7.08	4.98	4.13	3.65	3.34	3.12	2.95	2.82	2.72
120	6.85	4.79	3.95	3.48	3.17	2.96	2.79	2.66	2.56
∞	6.63	4.61	3.78	3.32	3.02	2.80	2.64	2.51	2.41

TABLE B.9B (Continued)

F-Distribution: .01 Points (α = .01)

Degrees of Freedom for Numerator (ν_1); Degrees of Freedom for Denominator (ν_2)

ν_2 \ ν_1	10	12	15	20	24	30	40	60	120	∞
1	6056	6106	6157	6209	6235	6261	6287	6313	6339	6366
2	99.40	99.42	99.43	99.45	99.46	99.47	99.47	99.48	99.49	99.50
3	27.23	27.05	26.87	26.69	26.60	26.50	26.41	26.32	26.22	26.13
4	14.55	14.37	14.20	14.02	13.93	13.84	13.75	13.65	13.56	13.46
5	10.05	9.89	9.72	9.55	9.47	9.38	9.29	9.20	9.11	9.02
6	7.87	7.72	7.56	7.40	7.31	7.23	7.14	7.06	6.97	6.88
7	6.62	6.47	6.31	6.16	6.07	5.99	5.91	5.82	5.74	5.65
8	5.81	5.67	5.52	5.36	5.28	5.20	5.12	5.03	4.95	4.86
9	5.26	5.11	4.96	4.81	4.73	4.65	4.57	4.48	4.40	4.31
10	4.85	4.71	4.56	4.41	4.33	4.25	4.17	4.08	4.00	3.91
11	4.54	4.40	4.25	4.10	4.02	3.94	3.86	3.78	3.69	3.60
12	4.30	4.16	4.01	3.86	3.78	3.70	3.62	3.54	3.45	3.36
13	4.10	3.96	3.82	3.66	3.59	3.51	3.43	3.34	3.25	3.17
14	3.94	3.80	3.66	3.51	3.43	3.35	3.27	3.18	3.09	3.00
15	3.80	3.67	3.52	3.37	3.29	3.21	3.13	3.05	2.96	2.87
16	3.69	3.55	3.41	3.26	3.18	3.10	3.02	2.93	2.84	2.75
17	3.59	3.46	3.31	3.16	3.08	3.00	2.92	2.83	2.75	2.65
18	3.51	3.37	3.23	3.08	3.00	2.92	2.84	2.75	2.66	2.57
19	3.43	3.30	3.15	3.00	2.92	2.84	2.76	2.67	2.58	2.49
20	3.37	3.23	3.09	2.94	2.86	2.78	2.69	2.61	2.52	2.42
21	3.31	3.17	3.03	2.88	2.80	2.72	2.64	2.55	2.46	2.36
22	3.26	3.12	2.98	2.83	2.75	2.67	2.58	2.50	2.40	2.31
23	3.21	3.07	2.93	2.78	2.70	2.62	2.54	2.45	2.35	2.26
24	3.17	3.03	2.89	2.74	2.66	2.58	2.49	2.40	2.31	2.21
25	3.13	2.99	2.85	2.70	2.62	2.54	2.45	2.36	2.27	2.17
26	3.09	2.96	2.81	2.66	2.58	2.50	2.42	2.33	2.23	2.13
27	3.06	2.93	2.78	2.63	2.55	2.47	2.38	2.29	2.20	2.10
28	3.03	2.90	2.75	2.60	2.52	2.44	2.35	2.26	2.17	2.06
29	3.00	2.87	2.73	2.57	2.49	2.41	2.33	2.23	2.14	2.03
30	2.98	2.84	2.70	2.55	2.47	2.39	2.30	2.21	2.11	2.01
40	2.80	2.66	2.52	2.37	2.29	2.20	2.11	2.02	1.92	1.80
60	2.63	2.50	2.35	2.20	2.12	2.03	1.94	1.84	1.73	1.60
120	2.47	2.34	2.19	2.03	1.95	1.86	1.76	1.66	1.53	1.38
∞	2.32	2.18	2.04	1.88	1.79	1.70	1.59	1.47	1.32	1.00

Adapted from Merrington, Maxine, and Thompson, Catherine M., "Tables of Percentage Points of the Inverted Beta (*F*) Distribution," *Biometrika,* Vol. 33 (1943), by permission of *Biometrika* Trustees.

TABLE B.10 Random numbers

33693	25139	54434	47314	55516	49387	01443	48263	90503	32236
69705	57374	36681	11777	64329	78767	01400	66210	14081	51256
23920	54052	77774	89460	93586	67285	24518	85441	01841	68087
54263	24152	61027	72788	28578	18856	54524	07524	87395	11498
52033	34753	46760	72085	83275	38449	56344	06298	04202	88324
30301	12183	55622	41995	34220	43920	62977	89820	33907	83810
83807	99353	66906	13421	20350	19047	15987	28770	79306	26284
35426	78860	86966	91418	80849	00094	72911	18541	63630	37335
92417	19580	27877	71848	40046	13820	84340	83692	60339	91430
62139	17953	51364	65187	97946	68333	77212	20205	68912	55875
19920	16894	19459	51846	49758	63582	22596	48031	29161	80561
71102	87003	31161	00739	94059	59441	13386	80711	83903	78413
68902	65719	81650	38270	55031	89292	03254	57047	57684	74740
76240	82013	44979	78538	36787	48529	67981	27423	50814	11775
25301	26151	01882	76598	32113	95769	71636	06681	39137	60510
83700	29115	74847	78324	04151	10676	78873	09887	79132	97389
19756	63296	12506	77606	50662	92497	24462	49939	29178	40824
69749	71455	09892	96959	15604	29887	62422	06566	76976	28381
18581	70666	89326	35595	26389	15739	69316	33640	45350	42117
90633	05718	27195	22825	42611	87058	46766	49781	80596	92886
64467	24287	81493	51358	61141	09055	30838	25926	81250	97636
48149	34086	89906	99442	18282	16383	14904	10168	57392	98066
02004	13886	12459	04148	01702	32451	36934	93659	60971	15886
70038	04754	68224	28561	90555	60732	34450	69075	14791	15577
16395	05547	05286	80351	99872	79375	02947	16462	49988	28850
52996	79482	59682	55419	57357	50357	56665	15581	78955	12096
91874	00505	82080	79198	91255	17845	83447	96793	22643	61330
19144	61578	86365	73519	24389	57899	02592	81698	95939	47245
25445	46639	09502	45409	12378	10393	83110	45158	81029	40852
77768	85987	57339	61706	33219	91280	57514	03882	06074	14448
11063	37730	94003	85726	73003	84920	57170	68330	37732	57290
10267	89140	53295	20569	91304	48442	01758	21745	31119	08581
08854	08377	33088	64479	70079	79069	80967	70163	94036	42167
48812	98739	94670	75122	91855	78513	08625	53525	28019	03857
05336	71878	19876	08058	60832	63714	64842	56131	57792	35123
86156	89260	55500	33507	19686	44744	39391	72415	55232	21415
42115	98128	56907	26809	39501	65297	84173	38901	31790	47555
08798	47422	49115	84558	49230	66659	73449	16043	58331	82802
20125	43828	27939	23679	73524	44719	74443	15337	67382	30599
39832	55335	71836	11482	67587	82389	69488	95381	96182	02133
41786	35617	98969	03823	06965	25934	92164	36703	10690	63481
59808	45862	22109	18475	85794	21815	49139	76886	48347	78130
87767	51761	30529	51341	84735	97143	93845	65103	48210	58179
93423	60834	76887	47155	08741	37390	43778	50391	14937	76870
59828	93884	44151	36384	76893	09275	94921	70852	34208	63638
92217	59161	68156	96410	83992	95802	24480	32672	16598	27824
20211	49343	98801	60822	72657	56393	39094	20449	48776	70064
82517	96677	69109	97510	05564	25680	92291	57582	70899	18986
09415	55783	46095	97604	45109	30560	16394	93477	20148	04431
69465	76837	05007	58172	20811	21116	33765	48908	89481	32937

TABLE B.11 Common logarithms

	.00	.01	.02	.03	.04	.05	.06	.07	.08	.09
1.0	.0000	.0043	.0086	.0128	.0170	.0212	.0253	.0294	.0334	.0374
1.1	.0414	.0453	.0492	.0531	.0569	.0607	.0645	.0682	.0719	.0755
1.2	.0792	.0828	.0864	.0899	.0934	.0969	.1004	.1038	.1072	.1106
1.3	.1139	.1173	.1206	.1239	.1271	.1303	.1335	.1367	.1399	.1430
1.4	.1461	.1492	.1523	.1553	.1584	.1614	.1644	.1673	.1703	.1732
1.5	.1761	.1790	.1818	.1847	.1875	.1903	.1931	.1959	.1987	.2014
1.6	.2041	.2068	.2095	.2122	.2148	.2175	.2201	.2227	.2253	.2279
1.7	.2304	.2330	.2355	.2380	.2405	.2430	.2455	.2480	.2504	.2529
1.8	.2553	.2577	.2601	.2625	.2648	.2672	.2695	.2718	.2742	.2765
1.9	.2788	.2810	.2833	.2856	.2878	.2900	.2923	.2945	.2967	.2989
2.0	.3010	.3032	.3054	.3075	.3096	.3118	.3139	.3160	.3181	.3201
2.1	.3222	.3243	.3263	.3284	.3304	.3324	.3345	.3365	.3385	.3404
2.2	.3424	.3444	.3464	.3483	.3502	.3522	.3541	.3560	.3579	.3598
2.3	.3617	.3636	.3655	.3674	.3692	.3711	.3729	.3747	.3766	.3784
2.4	.3802	.3820	.3838	.3856	.3874	.3892	.3909	.3927	.3945	.3962
2.5	.3979	.3997	.4014	.4031	.4048	.4065	.4082	.4099	.4116	.4133
2.6	.4150	.4166	.4183	.4200	.4216	.4232	.4249	.4265	.4281	.4298
2.7	.4314	.4330	.4346	.4362	.4378	.4393	.4409	.4425	.4440	.4456
2.8	.4472	.4487	.4502	.4518	.4533	.4548	.4564	.4579	.4594	.4609
2.9	.4624	.4639	.4654	.4669	.4683	.4698	.4713	.4728	.4742	.4757
3.0	.4771	.4786	.4800	.4814	.4829	.4843	.4857	.4871	.4886	.4900
3.1	.4914	.4928	.4942	.4955	.4969	.4983	.4997	.5011	.5024	.5038
3.2	.5051	.5065	.5079	.5092	.5105	.5119	.5132	.5145	.5159	.5172
3.3	.5185	.5198	.5211	.5224	.5237	.5250	.5263	.5276	.5289	.5302
3.4	.5315	.5328	.5340	.5353	.5366	.5378	.5391	.5403	.5416	.5428
3.5	.5441	.5453	.5465	.5478	.5490	.5502	.5514	.5527	.5539	.5551
3.6	.5563	.5575	.5587	.5599	.5611	.5623	.5635	.5647	.5658	.5670
3.7	.5682	.5694	.5705	.5717	.5729	.5740	.5752	.5763	.5775	.5786
3.8	.5796	.5809	.5821	.5832	.5843	.5855	.5866	.5877	.5888	.5899
3.9	.5911	.5922	.5933	.5944	.5955	.5966	.5977	.5988	.5999	.6010
4.0	.6021	.6031	.6042	.6053	.6064	.6075	.6085	.6096	.6107	.6117
4.1	.6128	.6138	.6149	.6160	.6170	.6180	.6191	.6201	.6212	.6222
4.2	.6232	.6243	.6253	.6263	.6274	.6284	.6294	.6304	.6314	.6325
4.3	.6335	.6345	.6355	.6365	.6375	.6385	.6395	.6405	.6415	.6425
4.4	.6435	.6444	.6454	.6464	.6474	.6484	.6493	.6503	.6513	.6522
4.5	.6532	.6542	.6551	.6561	.6571	.6580	.6590	.6599	.6609	.6618
4.6	.6628	.6637	.6646	.6656	.6665	.6675	.6684	.6693	.6702	.6712
4.7	.6721	.6730	.6739	.6749	.6758	.6767	.6776	.6785	.6794	.6803
4.8	.6812	.6821	.6830	.6839	.6848	.6857	.6866	.6875	.6884	.6893
4.9	.6902	.6911	.6920	.6928	.6937	.6946	.6955	.6964	.6972	.6981
5.0	.6990	.6998	.7007	.7016	.7024	.7033	.7042	.7050	.7059	.7067
5.1	.7076	.7084	.7093	.7101	.7110	.7118	.7126	.7135	.7143	.7152
5.2	.7160	.7168	.7177	.7185	.7193	.7202	.7210	.7218	.7226	.7235
5.3	.7243	.7251	.7259	.7267	.7275	.7284	.7292	.7300	.7308	.7316
5.4	.7324	.7332	.7340	.7348	.7356	.7364	.7372	.7380	.7388	.7396

(*Continued*)

TABLE B.11 (Continued)

	.00	.01	.02	.03	.04	.05	.06	.07	.08	.09
5.5	.7404	.7412	.7419	.7427	.7435	.7443	.7451	.7459	.7466	.7474
5.6	.7482	.7490	.7497	.7505	.7513	.7520	.7528	.7536	.7543	.7551
5.7	.7559	.7566	.7574	.7582	.7589	.7597	.7604	.7612	.7619	.7627
5.8	.7634	.7642	.7649	.7657	.7664	.7672	.7679	.7686	.7694	.7701
5.9	.7709	.7716	.7723	.7731	.7738	.7745	.7752	.7760	.7767	.7774
6.0	.7782	.7789	.7796	.7803	.7810	.7818	.7825	.7832	.7839	.7846
6.1	.7853	.7860	.7868	.7875	.7882	.7889	.7896	.7903	.7910	.7917
6.2	.7924	.7931	.7938	.7945	.7952	.7959	.7966	.7973	.7980	.7987
6.3	.7993	.8000	.8007	.8014	.8021	.8028	.8035	.8041	.8048	.8055
6.4	.8062	.8069	.8075	.8082	.8089	.8096	.8102	.8109	.8116	.8122
6.5	.8129	.8136	.8142	.8149	.8156	.8162	.8169	.8176	.8182	.8189
6.6	.8195	.8202	.8209	.8215	.8222	.8228	.8235	.8241	.8248	.8254
6.7	.8261	.8267	.8274	.8280	.8287	.8293	.8299	.8306	.8312	.8319
6.8	.8325	.8331	.8338	.8344	.8351	.8357	.8363	.8370	.8376	.8382
6.9	.8388	.8395	.8401	.8407	.8414	.8420	.8426	.8432	.8439	.8445
7.0	.8451	.8457	.8463	.8470	.8476	.8482	.8488	.8494	.8500	.8506
7.1	.8513	.8519	.8525	.8531	.8537	.8543	.8549	.8555	.8561	.8567
7.2	.8573	.8579	.8585	.8591	.8597	.8603	.8609	.8615	.8621	.8627
7.3	.8633	.8639	.8645	.8651	.8657	.8663	.8669	.8675	.8681	.8686
7.4	.8692	.8698	.8704	.8710	.8716	.8722	.8727	.8733	.8739	.8745
7.5	.8751	.8756	.8762	.8768	.8774	.8779	.8785	.8791	.8797	.8802
7.6	.8808	.8814	.8820	.8825	.8831	.8837	.8842	.8848	.8854	.8859
7.7	.8865	.8871	.8876	.8882	.8887	.8893	.8899	.8904	.8910	.8915
7.8	.8921	.8927	.8932	.8938	.8943	.8949	.8954	.8960	.8965	.8971
7.9	.8976	.8982	.8987	.8993	.8998	.9004	.9009	.9015	.9020	.9025
8.0	.9031	.9036	.9042	.9047	.9053	.9058	.9063	.9069	.9074	.9079
8.1	.9085	.9090	.9096	.9101	.9106	.9112	.9117	.9122	.9128	.9133
8.2	.9138	.9143	.9149	.9154	.9159	.9165	.9170	.9175	.9180	.9186
8.3	.9191	.9196	.9201	.9206	.9212	.9217	.9222	.9227	.9232	.9238
8.4	.9243	.9248	.9253	.9258	.9263	.9269	.9274	.9279	.9284	.9289
8.5	.9294	.9299	.9304	.9309	.9315	.9320	.9325	.9330	.9335	.9340
8.6	.9345	.9350	.9355	.9360	.9365	.9370	.9375	.9380	.9385	.9390
8.7	.9395	.9400	.9405	.9410	.9415	.9420	.9425	.9430	.9435	.9440
8.8	.9445	.9450	.9455	.9460	.9465	.9469	.9474	.9479	.9484	.9489
8.9	.9494	.9499	.9504	.9509	.9513	.9518	.9523	.9528	.9533	.9538
9.0	.9542	.9547	.9552	.9557	.9562	.9566	.9571	.9576	.9581	.9586
9.1	.9590	.9595	.9600	.9605	.9609	.9614	.9619	.9624	.9628	.9633
9.2	.9638	.9643	.9647	.9652	.9657	.9661	.9666	.9671	.9675	.9680
9.3	.9685	.9689	.9694	.9699	.9703	.9708	.9713	.9717	.9722	.9727
9.4	.9731	.9736	.9741	.9745	.9750	.9754	.9759	.9763	.9768	.9773
9.5	.9777	.9782	.9786	.9791	.9795	.9800	.9805	.9809	.9814	.9818
9.6	.9823	.9827	.9832	.9836	.9841	.9845	.9850	.9854	.9859	.9863
9.7	.9868	.9872	.9877	.9881	.9886	.9890	.9894	.9899	.9903	.9908
9.8	.9912	.9917	.9921	.9926	.9930	.9934	.9939	.9943	.9948	.9952
9.9	.9956	.9961	.9965	.9969	.9974	.9978	.9983	.9987	.9991	.9996

TABLE B.12 Standardized ranges for Duncan's multiple-range test

v − w + 1 = Ordered Pairs $\alpha = .05$

n_2	2	3	4	5	6	7	8	9	10	20	50
1	18.00	18.00	18.00	18.00	18.00	18.00	18.00	18.00	18.00	18.00	18.00
2	6.09	6.09	6.09	6.09	6.09	6.09	6.09	6.09	6.09	6.09	6.09
3	4.50	4.50	4.50	4.50	4.50	4.50	4.50	4.50	4.50	4.50	4.50
4	3.93	4.01	4.02	4.02	4.02	4.02	4.02	4.02	4.02	4.02	4.02
5	3.64	3.74	3.79	3.83	3.83	3.83	3.83	3.83	3.83	3.83	3.83
6	3.46	3.58	3.64	3.68	3.68	3.68	3.68	3.68	3.68	3.68	3.68
7	3.35	3.47	3.54	3.58	3.60	3.61	3.61	3.61	3.61	3.61	3.61
8	3.26	3.39	3.47	3.52	3.55	3.56	3.56	3.56	3.56	3.56	3.56
9	3.20	3.34	3.41	3.47	3.50	3.52	3.52	3.52	3.52	3.52	3.52
10	3.15	3.30	3.37	3.43	3.46	3.47	3.47	3.47	3.47	3.48	3.48
11	3.11	3.27	3.35	3.39	3.43	3.44	3.45	3.46	3.46	3.48	3.48
12	3.08	3.23	3.33	3.36	3.40	3.42	3.44	3.44	3.46	3.48	3.48
13	3.06	3.21	3.30	3.35	3.38	3.41	3.42	3.44	3.45	3.47	3.47
14	3.03	3.18	3.27	3.33	3.37	3.39	3.41	3.42	3.44	3.47	3.47
15	3.01	3.16	3.25	3.31	3.36	3.38	3.40	3.42	3.43	3.47	3.47
16	3.00	3.15	3.23	3.30	3.34	3.37	3.39	3.41	3.43	3.47	3.47
17	2.98	3.13	3.22	3.28	3.33	3.36	3.38	3.40	3.42	3.47	3.47
18	2.97	3.12	3.21	3.27	3.32	3.35	3.37	3.39	3.41	3.47	3.47
19	2.96	3.11	3.19	3.26	3.31	3.35	3.37	3.39	3.41	3.47	3.47
20	2.95	3.10	3.18	3.25	3.30	3.34	3.36	3.38	3.40	3.47	3.47
30	2.89	3.04	3.12	3.20	3.25	3.29	3.32	3.35	3.37	3.47	3.47
40	2.86	3.01	3.10	3.17	3.22	3.27	3.30	3.33	3.35	3.47	3.47
60	2.83	2.98	3.08	3.14	3.20	3.24	3.28	3.31	3.33	3.47	3.47
100	2.80	2.95	3.05	3.12	3.18	3.22	3.26	3.29	3.32	3.47	3.47
∞	2.77	2.92	3.02	3.09	3.15	3.19	3.23	3.26	3.29	3.47	3.47

v − w + 1 = Ordered Pairs $\alpha = .01$

n_2	2	3	4	5	6	7	8	9	10	20	50
1	90.00	90.00	90.00	90.00	90.00	90.00	90.00	90.00	90.00	90.00	90.00
2	14.00	14.00	14.00	14.00	14.00	14.00	14.00	14.00	14.00	14.00	14.00
3	8.26	8.50	8.60	8.70	8.80	8.90	8.90	9.00	9.00	9.30	9.30
4	6.51	6.80	6.90	7.00	7.10	7.10	7.20	7.20	7.30	7.50	7.50
5	5.70	5.96	6.11	6.18	6.26	6.33	6.40	6.44	6.50	6.80	6.80
6	5.24	5.51	5.65	5.73	5.81	5.88	5.95	6.00	6.00	6.30	6.30
7	4.95	5.22	5.37	5.45	5.53	5.61	5.69	5.73	5.80	6.00	6.00
8	4.74	5.00	5.14	5.23	5.32	5.40	5.47	5.51	5.50	5.80	5.80
9	4.60	4.86	4.99	5.08	5.17	5.25	5.32	5.36	5.40	5.70	5.70
10	4.48	4.73	4.88	4.96	5.06	5.13	5.20	5.24	5.28	5.55	5.55
11	4.39	4.63	4.77	4.86	4.94	5.01	5.06	5.12	5.15	5.39	5.39
12	4.32	4.55	4.68	4.76	4.84	4.92	4.96	5.02	5.07	5.26	5.26

(Continued)

TABLE B.12 (Continued) $v - w + 1$ = *Ordered Pairs* $\alpha = .01$

n_2	2	3	4	5	6	7	8	9	10	20	50
13	4.26	4.48	4.62	4.69	4.74	4.84	4.88	4.94	4.98	5.15	5.15
14	4.21	4.42	4.55	4.63	4.70	4.78	4.83	4.87	4.91	5.07	5.07
15	4.17	4.37	4.50	4.58	4.64	4.72	4.77	4.81	4.84	5.00	5.00
16	4.13	4.34	4.45	4.54	4.60	4.67	4.72	4.76	4.79	4.94	4.94
17	4.10	4.30	4.41	4.50	4.56	4.63	4.68	4.73	4.75	4.89	4.89
18	4.07	4.27	4.38	4.46	4.53	4.59	4.64	4.68	4.71	4.85	4.85
19	4.05	4.24	4.35	4.43	4.50	4.56	4.61	4.64	4.67	4.82	4.82
20	4.02	4.22	4.33	4.40	4.47	4.53	4.58	4.61	4.65	4.79	4.79
30	3.89	4.06	4.16	4.22	4.32	4.36	4.41	4.45	4.48	4.65	4.71
40	3.82	3.99	4.10	4.17	4.24	4.30	4.34	4.37	4.41	4.59	4.69
60	3.76	3.92	4.03	4.12	4.17	4.23	4.27	4.31	4.34	4.53	4.66
100	3.71	3.86	3.98	4.06	4.11	4.17	4.21	4.25	4.29	4.48	4.64
∞	3.64	3.80	3.90	3.98	4.04	4.09	4.14	4.17	4.20	4.41	4.60

Reproduced from D. B. Duncan, "Multiple Range and Multiple *F* Tests." *Biometrics*, Vol. 2 (1955), pp. 3–4. With permission from the Biometric Society.

TABLE B.13 Lower-tail critical values for Wilcoxon's signed-rank statistic

$T_{n,\alpha}$ such that $P(T \leq T_{n,\alpha}) \leq \alpha$, $n = n_+ + n_-$

n	*Reference* α .05	.025	.01	.005	n	*Reference* α .05	.025	.01	.005
5	0	—	—	—	23	83	73	62	54
6	2	0	—	—	24	91	81	69	61
7	3	2	0	—	25	100	89	76	68
8	5	3	1	0	26	110	98	84	75
9	8	5	3	1	27	119	107	92	83
10	10	8	5	3	28	130	116	101	91
11	13	10	7	5	29	140	126	110	100
12	17	13	9	7	30	151	137	120	109
13	21	17	12	9	31	163	147	130	118
14	25	21	15	12	32	175	159	140	128
15	30	25	19	15	33	187	170	151	138
16	35	29	23	19	34	200	182	162	148
17	41	34	27	23	35	213	195	173	159
18	47	40	32	27	36	227	208	185	171
19	53	46	37	32	37	241	221	198	182
20	60	52	43	37	38	256	235	211	194
21	67	58	49	42	39	271	249	224	207
22	75	65	55	48	40	286	264	238	220

Reprinted with permission from Beyer, W. H., *Handbook of Tables for Probability and Statistics*, 1966, p. 165. Copyright CRC Press, Inc., Boca Raton, FL.

TABLE B.14 Critical values for the total number of runs (R)

$R_{n_1,n_2,\alpha}$ such that $P(R \leq R_{n_1,n_2,\alpha}) \leq \alpha$, $R_{n_1,n_2,1-\alpha}$ such that $P(R \geq R_{n_1,n_2,1-\alpha}) \leq \alpha$

(n_1,n_2)	$\alpha = .025$	$\alpha = .05$	$\alpha = .10$	$1 - \alpha = .90$	$1 - \alpha = .95$	$1 - \alpha = .975$
(2,3)	—	—	—	5	—	—
(2,4)	—	—	—	—	—	—
(2,5)	—	—	2	—	—	—
(2,6)	—	—	2	—	—	—
(2,7)	—	—	2	—	—	—
(2,8)	—	2	2	—	—	—
(2,9)	—	2	2	—	—	—
(2,10)	—	2	2	—	—	—
(3,3)	—	—	2	6	—	—
(3,4)	—	—	2	7	7	—
(3,5)	—	2	2	7	—	—
(3,6)	2	2	2	—	—	—
(3,7)	2	2	3	—	—	—
(3,8)	2	2	3	—	—	—
(3,9)	2	2	3	—	—	—
(3,10)	2	3	3	—	—	—
(4,4)	—	2	2	8	8	—
(4,5)	2	2	3	8	9	9
(4,6)	2	3	3	9	9	9
(4,7)	2	3	3	9	9	—
(4,8)	3	3	3	9	—	—
(4,9)	3	3	4	9	—	—
(4,10)	3	3	4	—	—	—
(5,5)	2	3	3	9	9	10
(5,6)	3	3	3	9	10	10
(5,7)	3	3	4	10	11	11
(5,8)	3	3	4	10	11	11
(5,9)	3	4	4	10	11	—
(5,10)	3	4	5	11	11	—
(6,6)	3	3	4	10	11	11
(6,7)	3	4	4	11	11	12
(6.8)	3	4	5	11	12	12
(6,9)	4	4	5	11	12	13
(6,10)	4	5	5	12	12	13
(7,7)	4	4	5	11	12	12
(7.8)	4	4	5	12	13	13
(7,9)	4	5	5	12	13	13
(7,10)	5	5	6	13	13	14
(8,8)	4	5	6	12	13	14
(8,9)	5	5	6	13	14	14
(8,10)	5	6	6	13	14	15
(9,9)	5	6	6	14	14	15
(9,10)	5	6	7	14	15	16
(10,10)	6	6	7	15	16	16

TABLE B.15 Lower-tail critical values for the Mann–Whitney U statistic

$U_{n_1,n_2,\alpha}$ such that $P(U \leq U_{n_1,n_2,\alpha}) \leq \alpha$.

$(n_1 \leq n_2)$ (n_1,n_2)	$\alpha = .10$	$\alpha = .05$	$\alpha = .025$	$\alpha = .01$	$\alpha = .005$
(2,3)	0	—	—	—	—
(3,3)	1	0	—	—	—
(2,4)	0	—	—	—	—
(3,4)	1	0	—	—	—
(4,4)	3	1	0	—	—
(2,5)	1	0	—	—	—
(3,5)	2	1	0	—	—
(4,5)	4	2	1	0	—
(5,5)	5	4	2	1	0
(2,6)	1	0	—	—	—
(3,6)	3	2	1	—	—
(4,6)	5	3	2	1	0
(5,6)	7	5	3	2	1
(6,6)	9	7	5	3	2
(2,7)	1	0	—	—	—
(3,7)	4	2	1	0	—
(4,7)	6	4	3	1	0
(5,7)	8	6	5	3	1
(6,7)	11	8	6	4	3
(7,7)	13	11	8	6	4
(2,8)	2	1	0	—	—
(3,8)	5	3	2	0	—
(4,8)	7	5	4	2	1
(5,8)	10	8	6	4	2
(6,8)	13	10	8	6	4
(7,8)	16	13	10	7	6
(8,8)	19	15	13	9	7
(1,9)	0	—	—	—	—
(2,9)	2	1	0	—	—
(3,9)	5	4	2	1	0
(4,9)	9	6	4	3	1
(5,9)	12	9	7	5	3
(6,9)	15	12	10	7	5
(7,9)	18	15	12	9	7
(8,9)	22	18	15	11	9
(9,9)	25	20	17	14	11
(1,10)	0	—	—	—	—
(2,10)	3	1	0	—	—
(3,10)	6	4	3	1	0
(4,10)	10	7	5	3	2
(5,10)	13	11	8	6	4
(6,10)	17	14	11	8	6
(7,10)	21	17	14	11	9
(8,10)	24	20	17	13	11
(9,10)	28	24	20	16	13
(10,10)	32	27	23	19	16

TABLE B.16 Upper-tail critical values for Spearman's rank correlation coefficient

$r_{n,\alpha}$ such that $P(r \geq r_{n,\alpha}) \leq \alpha$

n	$\alpha = .05$	$\alpha = .025$	$\alpha = .01$	$\alpha = .005$
5	.900	—	—	—
6	.829	.886	.943	—
7	.714	.786	.893	—
8	.643	.738	.833	.881
9	.600	.683	.783	.833
10	.564	.648	.745	.794
11	.523	.623	.736	.818
12	.497	.591	.703	.780
13	.475	.566	.673	.745
14	.457	.545	.646	.716
15	.441	.525	.623	.689
16	.425	.507	.601	.666
17	.412	.490	.582	.645
18	.399	.476	.564	.625
19	.388	.462	.549	.608
20	.377	.450	.534	.591
21	.368	.438	.521	.576
22	.359	.428	.508	.562
23	.351	.418	.496	.549
24	.343	.409	.485	.537
25	.336	.400	.475	.526
26	.329	.392	.465	.515
27	.323	.385	.456	.505
28	.317	.377	.448	.496
29	.311	.370	.440	.487
30	.305	.364	.432	.478

TABLE B.17 Critical values for Friedman's statistic

$S_{c,n,\alpha}$ such that $P(S \geq S_{c,n,\alpha}) \leq \alpha$.

(c,n)	$\alpha = .10$	$\alpha = .05$	$\alpha = .01$	$\alpha = .005$
(3,3)	18	18	—	—
(3,4)	24	26	32	—
(3,5)	26	32	42	50
(3,6)	32	42	54	72
(3,7)	38	50	62	86
(3,8)	42	50	72	98
(3,9)	50	56	78	114
(3,10)	50	62	96	122
(3,11)	54	72	104	146
(3,12)	62	74	114	150
(3,13)	62	78	122	168
(3,14)	72	86	126	186
(3,15)	74	96	134	194
(4,2)	20	20	—	—
(4,3)	33	37	45	—
(4,4)	42	52	64	74
(4,5)	53	65	83	105
(4,6)	64	76	102	128
(4,7)	75	91	121	161
(4,8)	84	102	138	184

Reprinted with permission from D. B. Owen, *Handbook of Statistical Tables* (Reading, MA: Addison Wesley Press, Inc., 1962), pp. 407–419.

TABLE B.18 Factors for determining control chart limits for $\overline{X}$ and R charts

Number of Observations in Sample	*Factor for $\overline{X}$ Chart*	*Factor For Estimating σ from $\overline{R}$ ($d_2 = \overline{R}/\sigma$)*	*Factors for R Chart*	
			Lower Control Limit	*Upper Control Limit*
n	A_2	d_2	D_3	D_4
2	1.880	1.128	0	3.267
3	1.023	1.693	0	2.575
4	.729	2.059	0	2.282
5	.577	2.326	0	2.115
6	.483	2.534	0	2.004
7	.419	2.704	.076	1.924
8	.373	2.847	.136	1.864
9	.337	2.970	.184	1.816
10	.308	3.078	.223	1.777
11	.285	3.173	.256	1.744
12	.266	3.258	.284	1.716
13	.249	3.336	.308	1.692
14	.235	3.407	.329	1.671
15	.223	3.472	.348	1.652
16	.212	3.532	.364	1.636
17	.203	3.588	.379	1.621
18	.194	3.640	.392	1.608
19	.187	3.689	.404	1.596
20	.180	3.735	.414	1.586
21	.173	3.778	.425	1.575
22	.167	3.819	.434	1.566
23	.162	3.858	.443	1.557
24	.157	3.895	.452	1.548
25	.153	3.931	.459	1.541

TABLE B.19 Factors for determining a single sampling plan with $\alpha = .05$ and $\beta = .10$

c	$\alpha = .05$ $\pi_1 n$	$\beta = .10$ $\pi_2 n$	$\pi_2 n / \pi_1 n = \pi_2/\pi_1$
0	.051	2.30	45.10
1	.355	3.89	10.96
2	.818	5.32	6.50
3	1.366	6.68	4.89
4	1.970	7.99	4.06
5	2.613	9.28	3.55
6	3.285	10.53	3.21
7	3.981	11.77	2.96
8	4.695	12.99	2.77
9	5.425	14.21	2.62
10	6.169	15.41	2.50
11	6.924	16.60	2.40
12	7.690	17.78	2.31
13	8.464	18.96	2.24
14	9.246	20.13	2.18
15	10.04	21.29	2.12

Adapted with permission from Acheson J. Duncan, *Quality Control and Industrial Statistics,* 5th ed., p. 172. Richard D. Irwin, Inc., © 1986.

TABLE B.20 Sample size code letters: MIL-STD-105D

	Special Inspection Levels				*General Inspection Levels*		
Lot or Batch Size	*S-1*	*S-2*	*S-3*	*S-4*	*I*	*II*	*III*
2 to 8	A	A	A	A	A	A	B
9 to 15	A	A	A	A	A	B	C
16 to 25	A	A	B	B	B	C	D
26 to 50	A	B	B	C	C	D	E
51 to 90	B	B	C	C	C	E	F
91 to 150	B	B	C	D	D	F	G
151 to 280	B	C	D	E	E	G	H
281 to 500	B	C	D	E	F	H	J
501 to 1200	C	C	E	F	G	J	K
1201 to 3200	C	D	E	G	H	K	L
3201 to 10000	C	D	F	G	J	L	M
10001 to 35000	C	D	F	H	K	M	N
35001 to 150000	D	E	G	J	L	N	P
150001 to 500000	D	E	G	J	M	P	Q
500001 and over	D	E	H	K	N	Q	R

TABLE B.21 Single sampling plans for normal inspection: MIL-STD-105D

Acceptable Quality Levels (normal inspection)

Sample Size Code Letter	Sample Size	.010 Ac Re	.015 Ac Re	.025 Ac Re	.040 Ac Re	.065 Ac Re	.10 Ac Re	.15 Ac Re	.25 Ac Re	.40 Ac Re	.65 Ac Re	1.0 Ac Re	1.5 Ac Re
A	2	↓	↓	↓	↓	↓	↓	↓	↓	↓	↓	↓	↓
B	3	↓	↓	↓	↓	↓	↓	↓	↓	↓	↓	↓	↓
C	5	↓	↓	↓	↓	↓	↓	↓	↓	↓	↓	↓	↓
D	8	↓	↓	↓	↓	↓	↓	↓	↓	↓	↓	↓	0 1
E	13	↓	↓	↓	↓	↓	↓	↓	↓	↓	↓	0 1	↑
F	20	↓	↓	↓	↓	↓	↓	↓	↓	↓	0 1	↑	↓
G	32	↓	↓	↓	↓	↓	↓	↓	↓	0 1	↑	↓	1 2
H	50	↓	↓	↓	↓	↓	↓	↓	0 1	↑	↓	1 2	2 3
J	80	↓	↓	↓	↓	↓	↓	0 1	↑	↓	1 2	2 3	3 4
K	125	↓	↓	↓	↓	↓	0 1	↑	↓	1 2	2 3	3 4	5 6
L	200	↓	↓	↓	↓	0 1	↑	↓	1 2	2 3	3 4	5 6	7 8
M	315	↓	↓	↓	0 1	↑	↓	1 2	2 3	3 4	5 6	7 8	10 11
N	500	↓	↓	0 1	↑	↓	1 2	2 3	3 4	5 6	7 8	10 11	14 15
P	800	↓	0 1	↑	↓	1 2	2 3	3 4	5 6	7 8	10 11	14 15	21 22
Q	1250	0 1	↑	↓	1 2	2 3	3 4	5 6	7 8	10 11	14 15	21 22	↑
R	2000	↑	↑	1 2	2 3	3 4	5 6	7 8	10 11	14 15	21 22	↑	↑

Acceptable Quality Levels (normal inspection)

Sample Size Code Letter	Sample Size	2.5 Ac Re	4.0 Ac Re	6.5 Ac Re	10 Ac Re	15 Ac Re	25 Ac Re	40 Ac Re	65 Ac Re	100 Ac Re	150 Ac Re	250 Ac Re	400 Ac Re	650 Ac Re	1000 Ac Re
A	2	↓	↓	0 1	↓	↓	1 2	2 3	3 4	5 6	7 8	10 11	14 15	21 22	30 31
B	3	↓	0 1	↑	↓	1 2	2 3	3 4	5 6	7 8	10 11	14 15	21 22	30 31	44 45
C	5	0 1	↑	↓	1 2	2 3	3 4	5 6	7 8	10 11	14 15	21 22	30 31	44 45	↑
D	8	↑	↓	1 2	2 3	3 4	5 6	7 8	10 11	14 15	21 22	30 31	44 45	↑	↑
E	13	↓	1 2	2 3	3 4	5 6	7 8	10 11	14 15	21 22	30 31	44 45	↑	↑	↑
F	20	1 2	2 3	3 4	5 6	7 8	10 11	14 15	21 22	↑	↑	↑	↑	↑	↑
G	32	2 3	3 4	5 6	7 8	10 11	14 15	21 22	↑	↑	↑	↑	↑	↑	↑
H	50	3 4	5 6	7 8	10 11	14 15	21 22	↑	↑	↑	↑	↑	↑	↑	↑
J	80	5 6	7 8	10 11	14 15	21 22	↑	↑	↑	↑	↑	↑	↑	↑	↑
K	125	7 8	10 11	14 15	21 22	↑	↑	↑	↑	↑	↑	↑	↑	↑	↑
L	200	10 11	14 15	21 22	↑	↑	↑	↑	↑	↑	↑	↑	↑	↑	↑
M	315	14 15	21 22	↑	↑	↑	↑	↑	↑	↑	↑	↑	↑	↑	↑
N	500	21 22	↑	↑	↑	↑	↑	↑	↑	↑	↑	↑	↑	↑	↑
P	800	↑	↑	↑	↑	↑	↑	↑	↑	↑	↑	↑	↑	↑	↑
Q	1250	↑	↑	↑	↑	↑	↑	↑	↑	↑	↑	↑	↑	↑	↑
R	2000	↑	↑	↑	↑	↑	↑	↑	↑	↑	↑	↑	↑	↑	↑

↓ = Use first sampling plan below arrow. If sample size equals, or exceeds, lot or batch size, do 100% inspection.
↑ = Use first sampling plan above arrow.

Ac = Acceptable number.
Re = Rejection number.

TABLE B.22 Sample size code letters: MIL-STD-414

Lot Size	*Inspection Levels*				
	I	*II*	*III*	*IV*	*V*
3 to 8	B	B	B	B	C
9 to 15	B	B	B	B	D
16 to 25	B	B	B	C	E
26 to 40	B	B	B	D	F
41 to 65	B	B	C	E	G
66 to 110	B	B	D	F	H
111 to 180	B	C	E	G	I
181 to 300	B	D	F	H	J
301 to 500	C	E	G	I	K
501 to 800	D	F	H	J	L
801 to 1,300	E	G	I	K	L
1,301 to 3,200	F	H	J	L	M
3,201 to 8,000	G	I	L	M	N
8,001 to 22,000	H	J	M	N	O
22,001 to 110,000	I	K	N	O	P
110,001 to 550,000	I	K	O	P	Q
550,001 and over	I	K	P	Q	Q

TABLE B.23 Normal and tightened inspection sampling plans for single specification limits using sample standard deviation: MIL-STD-414

Sample Size Code Letter	*Sample Size*	*Acceptable Quality Levels (normal inspection)*													
		.04	***.065***	***.10***	***.15***	***.25***	***.40***	***.65***	***1.00***	***1.50***	***2.50***	***4.00***	***6.50***	***10.00***	***15.00***
		k	*k*	*k*	*k*	*k*	*k*	*k*	*k*	*k*	*k*	*k*	*k*	*k*	*k*
B	***3***	↓	↓	↓	↓	↓	↓	↓	▼	▼	1.12	.958	.765	.566	.341
C	***4***	↓	↓	↓	↓	↓	↓	↓	1.45	1.34	1.17	1.01	.814	.617	.393
D	***5***	↓	↓	↓	↓	↓	↓	1.65	1.53	1.40	1.24	1.07	.874	.675	.455
E	***7***	↓	↓	↓	↓	2.00	1.88	1.75	1.62	1.50	1.33	1.15	.955	.755	.536
F	***10***	↓	↓	↓	2.24	2.11	1.98	1.84	1.72	1.58	1.41	1.23	1.03	.828	.611
G	***15***	2.64	2.53	2.42	2.32	2.20	2.06	1.91	1.79	1.65	1.47	1.30	1.09	.886	.664
H	***20***	2.69	2.58	2.47	2.36	2.24	2.11	1.96	1.82	1.69	1.51	1.33	1.12	.917	.695
I	***25***	2.72	2.61	2.50	2.40	2.26	2.14	1.98	1.85	1.72	1.53	1.35	1.14	.936	.712
J	***30***	2.73	2.61	2.51	2.41	2.28	2.15	2.00	1.86	1.73	1.55	1.36	1.15	.946	.723
K	***35***	2.77	2.65	2.54	2.45	2.31	2.18	2.03	1.89	1.76	1.57	1.39	1.18	.969	.745
L	***40***	2.77	2.66	2.55	2.44	2.31	2.18	2.03	1.89	1.76	1.58	1.39	1.18	.971	.746
M	***50***	2.83	2.71	2.60	2.50	2.35	2.22	2.08	1.93	1.80	1.61	1.42	1.21	1.00	.774
N	***75***	2.90	2.77	2.66	2.55	2.41	2.27	2.12	1.98	1.84	1.65	1.46	1.24	1.03	.804
O	***100***	2.92	2.80	2.69	2.58	2.43	2.29	2.14	2.00	1.86	1.67	1.48	1.26	1.05	.819
P	***150***	2.96	2.84	2.73	2.61	2.47	2.33	2.18	2.03	1.89	1.70	1.51	1.29	1.07	.841
Q	***200***	2.97	2.85	2.73	2.62	2.47	2.33	2.18	2.04	1.89	1.70	1.51	1.29	1.07	.845
		.065	***.10***	***.15***	***.25***	***.40***	***.65***	***1.00***	***1.50***	***2.50***	***4.00***	***6.50***	***10.00***	***15.00***	
		Acceptable Quality Levels (tightened inspection)													

All *AQL* values are in percent defective.

↓ Use first samping plan below arrow—that is, both sample size as well as *k* value. When sample size equals or exceeds lot size, every item in the lot must be inspected.

Answers to Exercises

CHAPTER 1

1.1 **(a)** nominal; **(b)** nominal; **(c)** ratio; **(d)** interval; **(e)** nominal; **(f)** nominal; **(g)** interval.

1.2 **(a)** {Able, Baker, Dennis, Ernst}; **(b)** {Dennis}; **(c)** {Able, Cromwell, Fitz}; **(d)** {Baker, Cromwell, Ernst, Fitz}; **(e)** {Cromwell, Fitz}; **(f)** Cromwell and Fitz; **(g)** {Able, Baker, Cromwell, Ernst, Fitz}. Complement of $A \cap B$; **(h)** [a] employees who are in the sales department or are female; [b] employees who are female and in the sales department; [c] employees not in the sales department; [d] employees who are not female; [e] employees who are not in the sales department and who are not female.

1.3 $\varnothing$ is the null set. It contains no elements. {0} is a subset containing the element zero.

1.4 {1, 2, 3}.

1.5 {1, 2, 3, 4, . . . }.

1.6 **(a)** $\{d, e, f, g, h\}$; **(b)** $\{e, f,\}$; **(c)** No, since the universal set is not known.

1.7 **(a)** $(x_1^2 + x_2^2 + x_3^2 + x_4^2)/(y_1 + y_2 + y_3 + y_4)$; **(b)** $[(x_1^2 - a)^2 + (x_2^2 - a)^2 + (x_3^2 - a)^2 + (x_4^2 - a)^2]/c$; **(c)** $x_1y_1 + x_2y_2 + x_3y_3 - 3(x_1 + x_2 + x_3)(y_1 + y_2 + y_3)$; **(d)** $4x_2y_2 + 4x_3y_3 + 4x_4y_4$.

1.8 70.

1.9 29.

1.10 **(a)** $\sum_{i=1}^{4} 4(x_i^2 - y_i)$; **(b)** $\sum_{i=1}^{2} 2(y_i^2 - a)$; **(c)** $\left(\sum_{i=1}^{3} ax_iy_i\right) \Big/ \sum_{i=1}^{3} z_i$.

CHAPTER 2

2.1 **(a)** S = {GG, GS, GP, SG, SS, SP, PG, PS, PP}; **(b)** A = {GG, GS, GP, SG, PG}; **(c)** B = {SS, SP, PS, PP}; **(d)** C = {GG, SS, PP}; **(e)** compound events; **(f)** yes; **(g)** no.

2.2

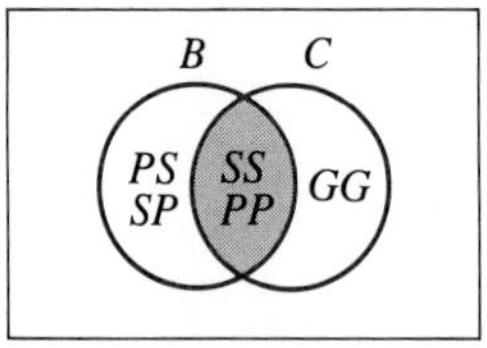

B C
PS SP SS PP GG

2.3

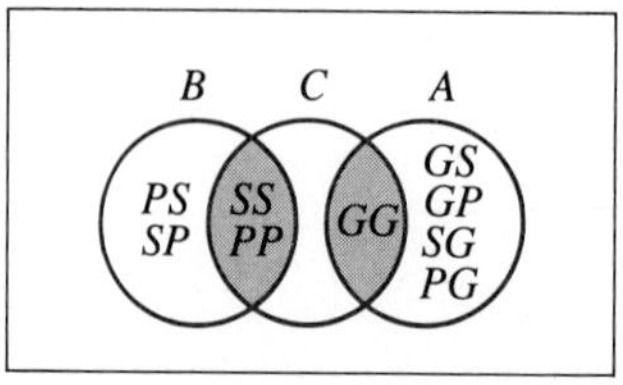

B C A
PS SP SS PP GG GS GP SG PG

2.4 All probabilities between 0 and 1; mutually exclusive and collectively exhaustive; sum of all probabilities adds to 1.
2.5 Not mutually exclusive; $P(C)$ should be greater than .70.
2.6 Subjective.
2.7 $\frac{1}{9}$.
2.8 $\frac{1}{3}$, objective.
2.9 **(a)** $n(H)/5$; **(b)** relative frequency; **(c)** no, too few number of trials.
2.10 8000.
2.11 **(a)** 8; **(b)** $\frac{1}{8}$.
2.12 360.
2.13 6.
2.14 35.
2.15 2520.
2.16 **(a)** $\frac{1}{8}$; **(b)** $\frac{5}{16}$; **(c)** $\frac{1}{3}$.
2.17 $\frac{2}{3}$.
2.18 $\frac{5}{7}$.
2.19 $\frac{4}{15}$.
2.20 **(a)** $\frac{1}{25}$; **(b)** $\frac{1}{5}$; **(c)** days are independent.
2.21 $\frac{1}{625}$.
2.22 $\frac{1}{336}$.
2.23 **(a)** .9; **(b)** .4; **(c)** .9.

2.24

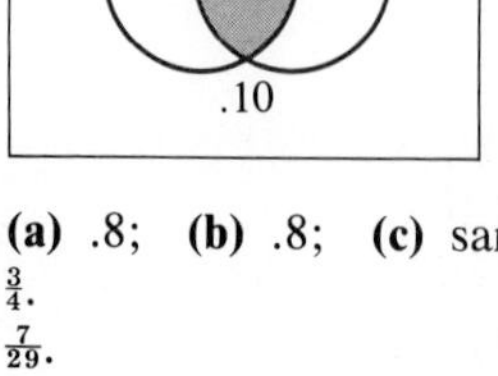

W D
.50 .10 .30
.10

2.25 **(a)** .8; **(b)** .8; **(c)** same problem.
2.26 $\frac{3}{4}$.
2.27 $\frac{7}{29}$.

CHAPTER 3

3.1 **(a)**, **(b)**, **(d)**, and **(f)** would be random variables.

3.2 **(a)**, **(b)**, and **(f)** are discrete.

3.3

X	0	1	2	3
$p(x)$	$\frac{20}{120}$	$\frac{60}{120}$	$\frac{36}{120}$	$\frac{4}{120}$

3.4

Y	0	1	2
$p(y)$	.01	.18	.81

; $P(Y = 2) = p(2) = .81$.

3.5 Since he must have 2 good cakes, $r = 2$, $\pi = .9$.
$P(N = 4) = p(4) = .0243$; $P(N \leq 3) = p(2) + p(3) = .81 + .162 = .972$.

3.6 $P(3 \leq X \leq 4) = \frac{1}{3}(4 - 3) = \frac{1}{3}$; $P(X \leq 4.5) = \frac{1}{3}(4.5 - 2) = \frac{25}{30} = \frac{5}{6}$.

3.7 $P(X \leq 1.5) = \frac{1}{4}$; $P(.5 \leq X \leq 2.5) = \frac{2}{3}$.

3.8 **(a)** At 1, $f(x) = .32 = h$; $P(X \leq 1) = 1 - P(X > 1) = 1 - \frac{1}{2}(.32)(5 - 1) = 1 - .64 = .36$;
(b) At 3, $f(x) = .16 = h$; $P(X \geq 3) = \frac{1}{2}(.16)(5 - 3) = .16$.

3.9 $E[X] = 1.2$ women.

3.10 Eq. 3.8: $(-1.2)^2(20/120) + (-.2)^2(60/120) + (.8)^2(36/120) + (1.8)^2(4/120) = .56$. Eq. 3.10: $E[X^2] = 2$, $\sigma_X^2 = 2 - (1.2)^2 = .56$.

3.11 $\mu_Y = E[Y] = 1.8$ cakes; $\sigma_Y = .18$.

3.12 $\mu_X = 3.5$; $\sigma_X = \sqrt{.75} = .867$.

3.13 $\mu_X = 2.0$; $\sigma_X^2 = .5$; $\sigma_X = .707$.

3.14 $E[B + C] = E[B] + E[C] = 24 + 98 = 122$. Since the Browns and Cavaliers are independent, $\text{Var}[B + C] = \sigma_B^2 + \sigma_C^2 = 7^2 + 9^2 = 130$.

3.15 **(a)** $P(10 \leq B \leq 38) = P(|B - 24| \leq 2(7)) \geq 1 - (\frac{1}{2})^2 = \frac{3}{4}$.
(b) 75% implies $1 - (1/K)^2 = .75$ or $K = 2$. Thus, $2\sigma = 18$, and 98 ± 18 gives between 80 and 116 points.

3.16 **(a)** $K = (30 - 25)/2 = 2.5$; $1 - (1/K)^2 = 1 - (1/2.5)^2 = .84$.
(b) 94% implies $(1/K)^2 = .06$ or $K = 4.08$ and $K\sigma = 4.08(2) = 8.16$. Thus, 94% is expected within 16.84 to 33.16.

CHAPTER 4

4.1 **(a)** $\frac{1}{15}$; **(b)** $\frac{14}{15}$; **(c)** 1.333.

4.2 **(a)** $\frac{21}{3003}$; **(b)** $\frac{504}{3003}$; **(c)** $\mu = 2.333$, $\sigma = .943$; **(d)** $\frac{56}{3003}$.

4.3 **(a)** $\frac{21}{792}$; **(b)** $\frac{672}{792}$; **(c)** $\frac{90}{792}$; **(d)** $\mu = 2.083$, $\sigma = .879$.

(e)

x	0	1	2	3	4	5
$p(x)$	$\frac{1}{792}$	$\frac{35}{792}$	$\frac{210}{792}$	$\frac{350}{792}$	$\frac{175}{792}$	$\frac{21}{792}$

4.4 **(a)** $\frac{15}{330}$; **(b)** $\frac{215}{330}$; **(c)** $\mu = 1.818$.

4.5 **(a)** .0001; **(b)** .7778; **(c)** $\mu = .25$.

4.6 **(a)** .1762; **(b)** .8477; **(c)** .0577; **(d)** graph omitted; **(e)** $\mu = 10$, $\sigma = 2.236$.

4.7 .0459.

4.8 **(a)** .3241; **(b)** .7004; **(c)** $\mu = 126$, $\sigma = 6.148$; **(d)** .0176; **(e)** .0538.

4.9 **(a)** .0613, .0000; **(b)** .6960; **(c)** 3.

4.10 **(a)** .1133; **(b)** .5957; **(c)** .0001, .0362.

4.11 **(a)** .0298; **(b)** .0471; **(c)** .8828; **(d)** $\mu = 1040$, $\sigma = 32.25$.

4.12 **(a)** .1381; **(b)** $\mu = 30.4$; $\sigma = 5.514$; **(c)** .5488.

4.13 .1954; .3712.

4.14 .2240; .8153; .0335.

4.15 **(a)** .99752, **(b)** .1353; **(c)** .5 days, .5 days.

4.16 **(a)** .0821; **(b)** .3496.
4.17 **(a)** .2386; **(b)** .6321.
4.18 **(a)** .4292; **(b)** .9162; **(c)** .6377; **(d)** .3336.
4.19 **(a)** −1.53; **(b)** 1.18; **(c)** ±2.97.
4.20 **(a)** .1515; **(b)** .5704; **(c)** yes, $P(x > 32) = .0007$.
4.21 **(a)** .0062; **(b)** 1.584; 2.416.
4.22 **(a)** .0062; **(b)** .9593.
4.23 .0571.
4.24 .1295; .1131.

CHAPTER 5

5.1 Answers vary with the student. Possible examples are **(a)** market survey from a population of customers in a given area; **(b)** a shipment of items is to be sampled to determine its disposition, with the population being all items in the shipment; **(c)** a sample of employees is to be selected to evaluate employee attitudes about a proposed pension buyout, where the population is all employees; **(d)** a company desires to conduct a quarterly audit of its stores and wants to select a sample of stores to be included in the audit.

5.2 A sample of fuses is to be selected to estimate the average amperage at which the fuses burn out.

5.3 Stratified: a situation in which it is desired to estimate the average income for a population having very high, very low, and middle income segments. Cluster: a retail chain has 150 outlets and it is desired to evaluate customer complaints. It is believed that the variability between outlets is small and not a major consideration but that the time and cost of sampling is.

5.4 Answers may vary slightly with the student. One possibility is the following.

Class Bounds	*Mark*	f_i	$F(B_i)$	*Relative Frequency*
$15 \le X < 19$	17	12	0	.250
$19 \le X < 23$	21	17	12	.354
$23 \le X < 27$	25	13	29	.271
$27 \le X < 31$	29	3	42	.062
$31 \le X < 35$	33	2	45	.042
$35 \le X < 39$	37	1	47	.021
		48	48	1.000

5.5 Graphs vary with students' answers to 5.4.

5.6 Answers vary with answers to 5.4 and 5.5. **(a)** approximately 75%; **(b)** approximately $28.50; **(c)** approximately 64%.

5.7 Array: 23.8, 25.5, 26.7, 26.7, 27.3, 28.4, 29.7, 30.1, 31.6, 34.2. Stem-and-leaf plot:

2	3.8	5.5	6.7	6.7	7.3	8.4	9.7
3	0.1	1.6	4.2				

5.8 **(a)** $\bar{x} = 28.4$ ($1000); **(b)** median = 27.85 ($1000); **(c)** 26.7 ($1000).

5.9 **(a)** Median = $22.50; **(b)** $\bar{x}$ = $22.52; **(c)** mode = $23.95.

5.10 These answers are based on the table in the answer to 5.4. **(a)** $21.82; **(b)** $22.42; **(c)** $24.42. The median from the table is lower than desired, but the mean is close to the true value. Regrouping could be considered.

5.11 **(a)** \$1225; **(b)** \$1916.67.
5.12 Range = \$22.01; variance = 19.87; std. dev. = \$4.46.
5.13 Based on the answer given for 5.4: $s_X^2 = 22.12$; $s_X = 4.70$.
5.14 $s_X^2 = 975{,}378.79$; $s_X = \$987.61$.
5.15 $R = 10.4(\$1000)$, $s_X^2 = 9.447(\$1000)^2$, $s_X = 3.074\ (\$1000)$.
5.16 For the answer given in 5.4, $u_i = \frac{1}{4}(x_i - 21)$.
5.17 $\bar{u} = \frac{17}{48}$; $\bar{x} = 4\bar{u} + 21 = 22.42$; $s_U = 1.1758$; $s_X = 4s_U = 4.70$.
5.18 For $u_i = 10(x_i - 29)$, $\bar{u} = -6.$; $s_U^2 = 944.7$; and $\bar{x} = .1\bar{u} + 29 = 28.4$, $s_x^2 = .01s_U^2 = 9.447$.
5.19 Due to unequal class intervals, we use $u_i = \frac{1}{250}(x_i - 1500)$. Then, $\bar{u} = -1.1$, $s_U^2 = 15.606$, and $s_U = 3.9505$, giving $\bar{x} = 250\bar{u} + 1500 = 1225$, and $s_X = (250)s_U = 987.61$.
5.20 **(a)** $MD = 5.755$; % within $\bar{x} \pm 1\ MD$ is 57.1, which is close to 58; **(b)** $Sk = -.231$; **(c)** $K = 2.939$.
5.21 All are reasonably close to the values expected for samples from normal populations. Thus, it may be reasonable.
5.22 $V_{\text{PE}} = .8821$.
5.23 $V_{\text{ES}} = .1118$. The PE ratios are more variable on a relative basis than the exam scores.

CHAPTER 6

6.1 77.8%; 94.4%; 100% are actual percentages.
68.3%; 95.5%; 99.7% for a normal distribution.
6.2 Answers may vary. One possible frequency table is, for $T = \Sigma X_i$,

Class	f_i
$10 \le T < 14$	2
$14 \le T < 18$	8
$18 \le T < 22$	11
$22 \le T < 26$	9
$26 \le T < 30$	4
$30 \le T < 34$	2

6.3 $\mu_X = 20$; $\sigma_X = 1.2$; $n = 16$.

(a) $$P(19.4 \le \bar{X} \le 20.9) = P\left(\frac{19.4 - 20}{1.2/\sqrt{16}} \le Z \le \frac{20.9 - 20}{1.2/\sqrt{16}}\right)$$
$$= P\left(\frac{-.6}{.3} \le Z \le \frac{.9}{.3}\right) = P(-2.00 \le Z \le 3.00) = .4772 + .4987$$
$$= .9759.$$

(b) For a normal distribution of ages
$$P\left(\frac{19.4 - 20}{1.2} \le Z \le \frac{20.9 - 20}{1.2}\right) = P(-.50 \le Z \le +.75) = .1915 + .2734 = .4649$$

6.4 $P(D \le Q) = .95$; $z_{.05} = +1.645$;

$$\begin{aligned} Q &= 30(28 + 5) + 1.645(11.5)\sqrt{28 + 5} - \text{OH} \\ &= 30(33) + 1.645(11.5)\sqrt{33} - \text{OH} \\ &= 990 + 108.67 - \text{OH or } 1098.67 \text{ tons} - \text{OH} \\ &= 1099 - \text{OH} \end{aligned}$$

6.5 $\mu_X = 80$; $\sigma_X^2 = 576$; $\mu_{\bar{X}} = 80$; $\sigma_{\bar{X}}^2 = \frac{576}{144} = 4$; $\sigma_{\bar{X}} = 2$; $n = 144$.

6.6 $\mu_X = 30$; $US - LS = .24$; **(a)** $P[(-.12/\sigma) \leq Z \leq (+.12/\sigma)] = .95 \Rightarrow (.12/\sigma) = 1.96$ or $\sigma = .0612$ cm. **(b)** $90\% \Rightarrow z = 1.645$; $LL = \mu_X - 1.645(.0612/\sqrt{12}) = 30 - .0291 = 29.9709$; $UL = 30 + 1.645(.017667) = 30 + .0291 = 30.0291$.

6.7 $\mu_X = 16.1$; $\sigma_X = .18$; $n = 24$;

$$P(T > 388) = P\left(Z > \frac{388 - 24(16.1)}{\sqrt{24}\,(.18)}\right) = P(Z \geq 1.81) = .5 - .4649 = .0351$$

No significant effect.

6.8 $\pi = .04$; $n = 2000$; $n\pi = 80$; $n\pi(1 - \pi) = 76.8$;

$$P(X \leq 70.5|2000,\ .04) = P\left(Z \leq \frac{-9.5}{8.76}\right) = P(Z \leq -1.08) = .5 - .3599 = .1401$$

6.9 $\pi = .04$; $\sigma_P = \sqrt{[\pi(1 - \pi)/n]} = .00438$;

$$P(X > 92) = P\left(Z > \frac{92.5 - 80}{8.76}\right) = P(Z > 1.43) = .5 - .4236 = .0764$$

6.10 $P(X \geq 21 \mid 120, \pi = .15) = P\left(Z \geq \frac{20.5 - 18}{3.91}\right)$

$$= P(Z \geq .64) = .5 - .2389 = .2611$$

6.11 $P(p \leq .12) = P(X \leq 24) = P\left(Z \leq \frac{24.5 - 30}{5.05}\right)$

$$P\left(Z \leq \frac{-5.5}{5.05}\right) = P(Z \leq -1.09) = .5 - 3621 = .1379$$

6.12 **(a)** 19.68; **(b)** 13.12; **(c)** 1.34; **(d)** 88.38.

6.13 **(a)** 1.753; **(b)** −2.045; **(c)** 2.764; **(d)** 2.878.

6.14 **(a)** 4.89; **(b)** 2.27; **(c)** .052; **(d)** .094.

CHAPTER 7

7.1 11.992.

7.2 9.7.

7.3 $LL = 72.504$; $UL = 73.096$.

7.4 $LL = 69.839$; $UL = 75.761$; interval is wider.

7.5 $LL = 22.639$; $UL = 26.361$.

7.6 $LL = 22.826$; $UL = 26.174$; interval is smaller.

7.7 $LL = 39920$; $UL = 43280$.

7.8 $LL = .639$; $UL = .661$; We are 90% sure that the interval from .639 to .661 includes the true population mean.

7.9 $LL = 31.521$; $UL = 34.079$.

7.10 $LL = .400$; $UL = .404$.

7.11 $LL = 47.33$; $UL = 49.87$.

7.12 $LL = 47.395$; $UL = 49.805$.

7.13 $LL = .236$; $UL = .484$.

7.14 $LL = .365$; $UL = .435$.

7.15 49.

7.16 445.

7.17 664.
7.18 1309.
7.19 323.
7.20 91.
7.21 230.
7.22 $LL = 2.809$; $UL = 10.010$.
7.23 $LL = .00024$; $UL = .00278$.
7.24 $LL = .186$; $UL = .345$.
7.25 $LL = .015$; $UL = .053$.
7.26 $LL = .00029$; $UL = .00185$.

CHAPTER 8

8.1 Yes, $z = 2.083$.
8.2 $\bar{x}_{LC} = 24.98026$; $\bar{x}_{UC} = 25.01974$.
8.3 The plan is not working, $t = .976$.
8.4 $P(t > .976) = .17391$; decision the same.
8.5 Raise the rental charge, $t = 3.074$.
8.6 Average MPG greater at 55 mpg, $z = 3.35$.
8.7 Friday-Saturday more than $1.50, $z = 1.463$.
8.8 No increase in age, $t = 1.573$.
8.9 Increase in points, $t = 2.084$.
8.10 Expenditure is not greater for winners, $t = 2.185$.
8.11 NEW language is faster, $t = 2.578$.
8.12 Campaign not successful, $z = -5.95$.
8.13 Proportion missed dates decreased, $z = -2.35$.
8.14 Proportions not the same, $z = -3.08$.
8.15 Proportion not more than .2, $z = .712$.
8.16 No reduction, $\chi^2 = 7.00$.
8.17 Yes, there is an increase, $\chi^2 = 44.52$.
8.18 Variability the same, $F = 1.315$.
8.19 Yes, there is a decrease, $F = 6.467$.
8.20 Variability not greater for layer cakes, $F = 1.92$.
8.21 Values for β:

.8739	.7405	.5578	.3613	.1963	.0877
.9185	.8739	.8146	.7405	.6536	.5577

8.22 Values for $1 - \beta$:

.1261	.2595	.4422	.6387	.8037	.9123
.0815	.1261	.1854	.2595	.3464	.4423

8.23 $n = 99$.

CHAPTER 9

9.1 Average lumens/sq ft differs for at least 2 areas, $F = 242.14/5 = 48.43$.
9.2 Average minutes differ for at least 2 manufacturers, $F = 13822.3/217.82 = 63.46$.
9.3 Average weight of kapok differs for at least 2 manufacturers, $F = 142.375/7.442 = 19.132$.

9.4 The average lumens/sq ft is the same for classrooms and laboratories and is greater than hallways and lounges, which are also the same.

9.5 The average length of life differs for all three brands; Cutrite is the longest, Argo the shortest.

9.6 Average weight equal for Mountaineer and Hi-cover and is greater than the Durorest and Restwell, which are also equal.

9.7 No superior method; average production rate is the same for methods 2 and 3 as well as methods 1 and 3.

9.8 Average production does not differ for production lines $F = .409/18.208 = .0225$; it also does not differ for the shifts $F = 42.552/18.208 = 2.337$.

9.9

Source of Variability	*df*
Kinds of wheat	2
Kinds of soil	5
Interaction	10
Error	54
Total	71

9.10 Average stay in hospital does not differ for the hospitals, $F = .67/.236 = 2.84$, but does differ for at least 2 types of birth, $F = 132.54/.236 = 561.02$. No interaction exists, $F = .28/.236 = 1.19$. Average time is longest for Cesarean, shortest for natural.

9.11 Average knots differ for at least 2 boat classes, $F = 3.898/.149 = 26.16$, as well as for at least 2 types of sail designs, $F = 1.997/.149 = 13.40$. Average knots are the same for Meteor and Sunburst; also, Starfish and Viking are the same. Average speed is the same for designs 1, 2, 4, and 5.

9.12 Average number of absences does not differ for the shift, $F = 1.65/3.698 = .446$, or for the length of service, $F = 14.53/3.698 = 3.929$.

9.13 **(a)** $k = 5$;

(b)

Source	*df*	*SS*	*MS*	*F*
Operators	3	59.83	19.94	36.23
Lathes	3	.66	.22	.40
Interaction	9	1.19	.13	.24
Error	64	35.23	.55	
Total	79	96.91		

(c) Average time does not differ by lathe, $F = .22/.55 = .4$ but does differ for at least 2 workers, $F = 19.94/.55 = 36.23$. There is no interaction, $F = 0.13/.55 = .24$. **(d)** average time is the same for Smith and Downs; all other pairs differ.

9.14 Average recognition time differs for at least 2 age categories, $F = 13.3/2.4 = 5.54$, as well as at least 2 product categories, $F = 249.88/2.4 = 104.11$. There is no interaction, $F = 5.134/2.4 = 2.14$. Average time is the same for age categories (21–40) and (over 40). Average time is the same for product categories, cereals, and automobiles.

CHAPTER 10

10.1 Straight line appears best.

10.2 Cubic equation appears best.

10.3 $\hat{Y} = 0 + 10x$.

10.4 $\hat{Y} = 1.54 + .0076x$.

10.5 4.96.
10.6 .076.
10.7 .2394.
10.8 Normality of Y violated. Could have used average points over time for Y.
10.9 $F = 176.39/41.06 = 4.30$; model seems correct.
10.10 $F = 14.59/2.83 = 5.16$; model not correct. $\hat{y} = 14.6$.
10.11 **(a)** $\hat{Y} = -5.091 + .9934x$; **(b)** $F = 174.62$, significant linear relationship; **(c)** $t = 13.214$, significant linear relationship; **(d)** $t^2_{\text{calc}} = (13.214)^2 = F_{\text{calc}} = 174.62$; $t^2_{\text{tab}} = (2.447)^2 = F_{\text{tab}} = 5.99$; **(e)** $UL = 1.1774$; $LL = .8094$.
10.12 **(a)** $\hat{Y} = 2.164 + .1897x$; **(b)** $F_{\text{calc}} = 2.5 < F_{\text{tab}} = 5.32$, fail to reject; **(c)** $t_{\text{calc}} = 1.57 < t_{\text{tab}} = 2.306$, fail to reject; **(d)** $UL = .468$, $LL = -.089$, 0 included, fail to reject.
10.13 $UL = 5.62$; $LL = 4.06$.
10.14 Using $x = 22$, $t = 2.395$, accept claim.
10.15 $UL = 228.08$; $LL = 159.10$.
10.16 Beyond range of data—impossible.
10.19 .967.
10.20 .983.
10.21 $r^2 = .64$; $r = -.8$.
10.22 $r = .486$.
10.23 $t = 1.572$, not significant.

CHAPTER 11

11.3 18.56.
11.5 **(a)** (1) 1 (2) 4 (3) 5; **(b)** No; degrees of freedom for error must be many times larger than of regression.
11.6 **(a)** $\hat{Y} = 4.941 + 0.745x_2 - 0.00879x_3$; **(b)** 7.72; **(d)** $F = 57.44$, the equation is significant; **(e)** $.126 < .632$, no multicollinearity suggested; **(f)** $t_2 = 10.67$ (related); $t_3 = 0.38$ (not related).
11.7 **(a)** $\hat{Y} = 1760.98 - 0.029x_2 + 0.0157x_3 + 0.0938x_4 - 26.245x_5$;
(b) $F = 1499.05$, equation is significant;
(c) all $r > .532$, multicollinearity exists between all independent variables;
(d) $Y = f(x_3)$;
(e) no, because of multicollinearity;
(g) $Y = f(x_2, x_3, x_5)$, minimum standard error $= 13.285$;
(h) multicollinearity;
(i) multicollinearity;
(j) signs and t test invalid with significant multicollinearity.
11.8 $\log \hat{Y} = -0.9189 + 1.2354 \log x$; $\hat{y} = 35.6$.
11.9 **(a)** Straight: 1,052,386; parabola: 1,076,101; cubic: 625,959; ab^x: 6,305,304; ax^b: 1,920,307, **(b)** cubic, provides for decline after 1950; **(c)** Cleveland 1980 census = 573,822.
11.10 $r_{12.34} = .55462$; $r_{12} = .93584$.
11.11 **(a)** x_1 = sq ft; $x_2 = 1$ for colonial, 0 otherwise; $x_3 = 1$ for tudor, 0 otherwise; $x_4 = 1$ for brick, 0 otherwise;
(b) $\hat{Y} = -86{,}764.11 + 80.318\,x_1 + 1063.676\,x_2 + 4369.03\,x_3 + 8257.6\,x_4$.

CHAPTER 12

12.1 6, −2, $\hat{T} = -2 + 6x$, origin = July 1, 1975, x units = 1 year, Y units = thousands of dollars.
12.2 **(a)** $\hat{T} = 11.217 + .437x$, x units = 1 year, Y units = millions of policies; **(b)** $\hat{T} = 13.825 + .452x$; **(c)** $\hat{T} = 11.111 + .452x$; **(d)** 18.20, 18.35, 18.35.
12.3 **(a)** $\hat{T} = 14 + 2x$, Y units = thousands of dollars; **(b)** $\hat{T} = 4 + 4x$; **(c)** 28.
12.4 **(a)** $\hat{T} = 15.862 + .052x$, Y units = millions of employees; **(b)** $\hat{T} = 15.604 + .103x$; **(c)** 16.43.
12.5 **(a)** $\hat{T} = 9 + 2x$; **(b)** $\hat{T} = 3 + 2x$; **(c)** 25.
12.6 **(a)** $\hat{T} = 4.657 + 2.70x + 1.071x^2$, origin = July 1, 1982, x units = 1 year; **(b)** 22.4.
12.7 **(a)** $\hat{T} = (4)(2)^x$, origin = July 1, 1982, x units = 1 year; **(b)** 100%.
12.8 20%.
12.9 10, 13.6, 17.6, 24.4, 31.6, 40.
12.10 7.50, 12.25, 17.00, 20.25.
12.11 12, 12, 11, 12, 14; 14.
12.12 4, 4, 5.2, 7.92, 11.95, 15.17; 18.30.
12.13 91.7, 99.9, 99.6, 108.8.
12.14 110.
12.15 92.4, 100.0, 99.4, 108.2
12.16 95.
12.17 27.3.
12.18 274.6.
12.19 **(a)** 100; **(b)** 110.
12.20

98.0	100.0	99.8	99.6	101.0	97.6	99.4	100.3
102.3	102.8	102.9	99.6	98.3	99.3	98.6	98.9

12.21 1100.
12.22 54.

CHAPTER 13

13.1 125, 150, 100.
13.2 75, 90, 100, 115, 120.
13.3 **(a)** 155.9; **(b)** 122.6.
13.4 98.
13.5 123.9.
13.6 110.
13.7 **(a)** 106; **(b)** 117.9.
13.8 **(a)** 122; **(b)** 86.
13.9 **(a)** 100; **(b)** 105; **(c)** 80; **(d)** 111.1; **(e)** 80.
13.10 **(a)** 106; **(b)** 130.
13.11 **(a)** 90, 120, 111.1, 115, 104.3; **(b)** 144.
13.12 **(a)** 120; **(b)** 200; **(c)** 240; **(d)** 240; answer is the same.
13.13 17% decrease.
13.14 4% increase.

CHAPTER 14

14.1 H_0: $\tilde{X} \geq 1000$, H_a: $\tilde{X} < 1000$. For sign procedure $p = .2122 > .05$. Cannot reject H_0. Conclude: no.

14.2 H_0: $\tilde{X} \leq 12$, H_a: $\tilde{X} > 12$. Since symmetry is doubtful, use the sign procedure. $p = .0592 < .10$. Reject H_0. Conclude: yes.

14.3 H_0: $\tilde{X} \leq 100$, H_a: $\tilde{X} > 100$. Lack of symmetry implies sign procedure. $p = .1316 > .05$. Cannot reject H_0. Conclude: no.

14.4 Requires the assumption of symmetry. $T_- = 76 > 60$. Cannot reject H_0.

14.5 H_0: $\tilde{X} \geq 24$ H_a: $\tilde{X} < 24$. Wilcoxon's procedure. $T_+ = 11.5 < 30$. Reject H_0. Conclude: yes.

14.6 H_0: $\tilde{X} = 12$, H_a: $\tilde{X} \neq 12$. Wilcoxon's procedure, but there are numerous ties that may influence the results. $T = T_- = 29.5 < 30$. Reject H_0. PEs appear to have changed.

14.7 H_0: no pattern, H_a: some pattern. Runs procedure. $R = 4$. Cannot reject H_0 because $3 < 4 < 11$. Conclude: no.

14.8 H_0: random occurrence, H_a: nonrandom. Runs procedure. $R = 9$. Cannot reject because $6 < 9 < 16$. Conclude: yes.

14.9 H_0: no pattern, H_a: cyclical pattern. Runs procedure. Median = 5, $R = 12$. Reject H_0 because $12 \geq 12$. Conclude: yes.

14.10 H_0: $\tilde{X}_F = \tilde{X}_C$, H_a: $\tilde{X}_F \neq \tilde{X}_C$. Nominal data imply sign procedure. $p = .1509 > .025$. Cannot reject H_0. Conclude: no. Matched pairs.

14.11 H_0: $\tilde{X}_{WO} \geq \tilde{X}_W$, H_a: $\tilde{X}_{WO} < \tilde{X}_W$. From the sign procedure, $p = .0193$. Since $.0193 < .05$, reject H_0 and conclude yes.

14.12 H_0: $\tilde{X}_0 = \tilde{X}_L$, H_a: $\tilde{X}_0 \neq \tilde{X}_L$. Nominal data implies sign procedure. Since $.1509 > .025$, H_0 cannot be rejected. Conclude: no difference.

14.13 H_0: $\tilde{X}_{WO} \geq \tilde{X}_W$, H_a: $\tilde{X}_{WO} < \tilde{X}_W$. For $d_i = x_{WO} - x_W$, $T_+ = 4.5$. Since $4.5 \leq 17$, H_0 is rejected. Conclude: yes.

14.14 $\min(T_+, T_-) = T_- = 41.5 > 25$. Conclude: no difference because H_0 cannot be rejected. An ordinal level of measurement is highly questionable; hence Wilcoxon's procedure is not appropriate.

14.15 H_0: $\tilde{X}_F \leq \tilde{X}_A$, H_a: $\tilde{X}_F > \tilde{X}_A$. By Wilcoxon's procedure, $T_- = 13.5 \leq 17$. Thus, reject H_0. Production's claim is supported. Sample of quarters is not random, and symmetry may not be reasonable.

14.16 H_0: $\tilde{X}_A = \tilde{X}_B$, H_a: $\tilde{X}_A \neq \tilde{X}_B$. Cannot reject H_0 because $U = 5.5 > 5.0$. Conclude: difference is not significant.

14.17 H_0: $\tilde{X}_A \leq \tilde{X}_0$, H_a: $\tilde{X}_A > \tilde{X}_0$. Cannot reject H_0 because $U_A = 44 > 22$. Conclude: no.

14.18 H_0: $\tilde{X}_{WO} \geq \tilde{X}_W$, H_a: $\tilde{X}_{WO} < \tilde{X}_W$. $U_W = 34$ and $z = -2.59 < -1.645$. Reject H_0 and conclude that it does retard reaction time. The paired procedure is better because it controls for variability between individuals.

14.19 H_0: $\sigma_A^2 = \sigma_0^2$, H_a: $\sigma_A^2 \neq \sigma_0^2$. $U = 29 > 15$. Thus, we cannot reject H_0. Conclude: no.

14.20 H_0: $\sigma_A^2 = \sigma_B^2$, H_a: $\sigma_A^2 \neq \sigma_B^2$. After adjusting for possible differences in medians, $U = 17.5 > 7$. H_0 cannot be rejected. Conclude: no difference.

14.21 H_0: $\sigma_D^2 \geq \sigma_N^2$, H_a: $\sigma_D^2 < \sigma_N^2$. $U_D = 9 \leq 20$ and H_0 is rejected. Conclude: more variability on night shift.

14.22 The logical relationship would be positive. Thus, H_0: $\rho_S \leq 0$, H_a: $\rho_S > 0$. $r_S = .939 \geq .564$. Reject H_0 and conclude that a positive relationship exists.

14.23 H_0: $\rho_S \leq 0$, H_a: $\rho_S > 0$ because similar tastes would be indicated by similar rankings. $r_S = .776 \geq .497$, and H_0 is rejected. Conclude: they appear to be similar.

14.24 H_0: $\rho_S \geq 0$, H_a: $\rho_S < 0$. $r_S = -.984 \leq -.673$. Thus, reject H_0 and conclude that a negative relationship exists.

CHAPTER 15

15.1 H_0: $\pi_A = .2$, $\pi_B = .3$, $\pi_C = .35$, $\pi_D = .1$, $\pi_{F,W} = .05$, H_a: some $\pi_i \neq \pi_{0i}$. Calculated $\chi^2 = 33.33 > 14.86$. Thus, reject H_0 and conclude no.

15.2 H_0: $\pi_i = .25$ for all i, H_a: some $\pi_i \neq .25$. Calculated $\chi^2 = 9.65 > 7.81$. Reject H_0 and conclude yes.

15.3 H_0: X is Poisson, H_a: X is not Poisson. The last 3 classes must be combined, leaving 4 classes in the table. Calculated $\chi^2 = 98.74 > 9.21$. Reject H_0 and conclude no.

15.4 H_0: $\pi_i = \pi$ all i, H_a: some $\pi_i \neq \pi$. Reject H_0 because calculated $\chi^2 = 9.30 > 7.81$. Conclude: no.

15.5 H_0: $\pi_{1j} = \pi_1$ all j, H_a: some $\pi_{1j} \neq \pi_1$. Reject H_0 because the calculated $\chi^2 = 43.84 > 13.28$. Conclude: yes.

15.6 H_0: ROI independent of industry. H_a: not independent. Reject H_0 because the calculated $\chi^2 = 41.78 > 23.21$. Conclude: ROI is related to industry type.

15.7 H_0: Belief independent of area, H_a: not independent. Reject H_0 because the calculated $\chi^2 = 20.01 > 12.59$. Conclude: customer's belief is contingent on the area.

15.8 H_0: $\tilde{X}_j = \tilde{X}$ for all j, H_a: some $\tilde{X}_j \neq \tilde{X}$. H_0 is rejected because $H = 20.83 > 9.49$. Conclude: yes.

15.9 H_0: $\tilde{X}_j = \tilde{X}$ for all j, H_a: some $\tilde{X}_j \neq \tilde{X}$. $H = 11.34 > 7.81$. Reject H_0 and conclude: yes.

15.10 H_0: $\tilde{X}_1 = \tilde{X}_2 = \tilde{X}_3$, H_a: some $\tilde{X}_i$ differ. $S = 62 \geq 50$. Reject H_0 and conclude that the times differ with the additive.

15.11 H_0: $\tilde{X}_j = \tilde{X}, j = 1, \ldots, 4$, H_a: some $\tilde{X}_j$ differ. $S = 105 > 65$. Reject H_0 and conclude that differences exist among majors.

CHAPTER 16

16.1 **(a)** Payoff table

	Weather		
	Excellent	***Normal***	***Poor***
Till	440	272	104
Rent	340	312	284

(b) Loss table

	Weather		
	Excellent	***Normal***	***Poor***
Till	0	40	180
Rent	100	0	0

(c)

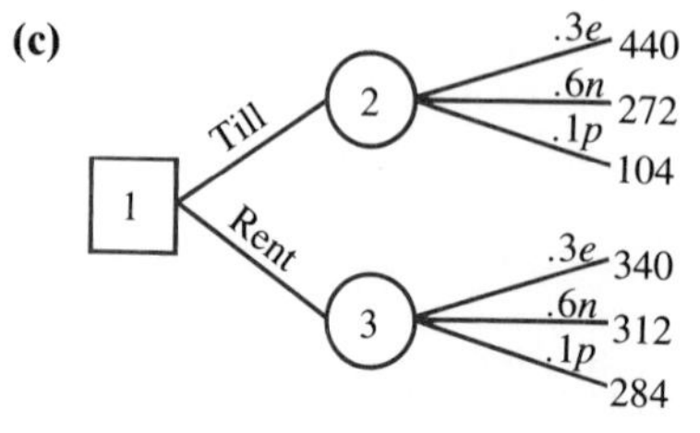

16.2 **(a)** Payoff table

	Demand				
Supply	***25***	***26***	***27***	***28***	***29***
25	2.00	2.00	2.00	2.00	2.00
26	1.88	2.08	2.08	2.08	2.08
27	1.76	1.96	2.16	2.16	2.16
28	1.64	1.84	2.04	2.24	2.24
29	1.52	1.72	1.92	2.12	2.32

(b) Loss table

	Demand				
Supply	***25***	***26***	***27***	***28***	***29***
25	0	.08	.16	.24	.32
26	.12	0	.08	.16	.24
27	.24	.12	0	.08	.16
28	.36	.24	.12	0	.08
29	.48	.36	.24	.12	0

(d) Payoff table

	Demand				
Supply	***25***	***26***	***27***	***28***	***29***
25	2.00	2.00	2.00	2.00	2.00
26	1.90	2.08	2.08	2.08	2.08
27	1.80	1.98	2.16	2.16	2.16
28	1.70	1.88	2.06	2.24	2.24
29	1.60	1.78	1.96	2.14	2.32

16.3 **(a)** a_1; **(b)** a_1; **(c)** a_2; **(d)** a_3; **(e)**

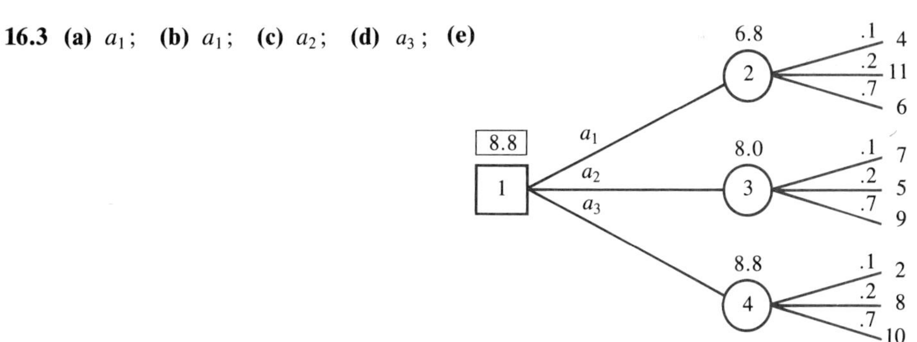

16.4 Maximax-till, maximin-rent, minimax-rent, expected value-rent.
16.5 Buy 27 papers.
16.6 $EVPI = 347.6 - 317.6 = 30 = \min(EOL)$.
16.7 $EVPI = .096$.
16.8 $EMG(\text{buy}) = -19$, don't buy.
16.9 **(a)** $2 < 8$, buy 20 loads; **(b)** $3 < 7$, buy 10 loads.
16.10 $EMG(15) = 4805$, build 15.
16.11 $EVSI = 70 - 32.77 = 37.23$.
16.12 $EVSI = 4 - 2.4 = 1.6$.
16.13 $EVSI = 900 - 731.08 = 168.92$.

CHAPTER 17

17.1 $UCL_{\overline{X}} = 1.04$, $LCL_{\overline{X}} = .97$, process mean in control.
$UCL_R = .11$, $LCL_R = 0$, variation in control.

17.2 $UCL_{\overline{X}} = 11.05$, $LCL_{\overline{X}} = 6.95$, process mean in control.
$UCL_R = 5.15$, $UCL_R = 0$, variation in control.

17.3 **(a)** Out of control, process mean shifted down; **(b)** out of control, cycle, every 4th point high; **(c)** in control, random scatter, normal distribution.

17.4
10.2 X Data appear normally distributed.
10.3 XXXX
10.4 XXXXXX
10.5 XXXXXX
10.6 XXXXX
10.7 XX

17.5 $\overline{X} = 10.467$, $s = 0.134$, $\hat{s} = .267/2.059 = .1295$.

17.6 PC% = 100.5%, $C_{pk} = .829$; the process does not show at least 3-sigma capability. C_{pk} preferred because it does not assume that process mean is centered within specification limits.

17.7 PC% = 77.4%, $C_{pk} = 1.292$; process is 3.876 capable.

17.8 $UCL_p = .1899$, $LCL_p = 0$, point 9 exceeds upper limit.

17.9 $UCL_p = .1311$, $LCL_p = 0$, process not in control.

17.10 $UCL_c = 11.44$, $LCL_c = 0$, process in control.

17.11 $UCL_c = 12.81$, $LCL_c = 0$, process not in control.

17.12

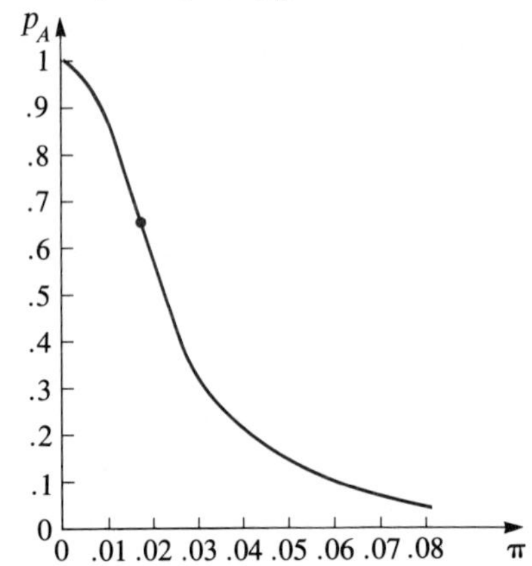

17.13 The OC curve for $n = 100$, $c = 2$ is closer to the ideal OC curve than $n = 50$, $c = 2$ but not as close as $n = 200$, $c = 4$.

17.14 $n = 395$, $c = 7$.

17.15 $n = 125$, $c = 3$.

17.16 $n = 38$, $\bar{x}_k = 19.231$.

17.17 $n = 31$, $k = 1.636$, $\bar{x}_k = 71.82$.

Index

TABLE B.7 Percentage points of selected *t*-distributions

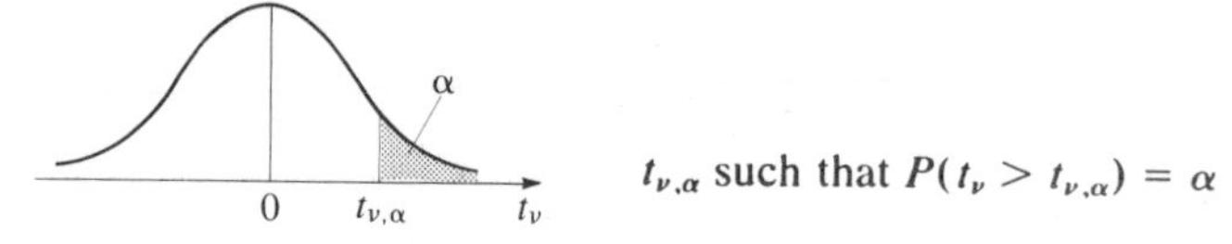

ν	α .40	.30	.20	.10	.05	.025	.01	.005	.001	.0005
1	.325	.727	1.376	3.078	6.314	12.710	31.820	63.657	318.310	636.620
2	.289	.617	1.061	1.886	2.920	4.303	6.965	9.925	22.330	31.600
3	.277	.584	.978	1.638	2.353	3.182	4.541	5.841	10.214	12.924
4	.271	.569	.941	1.533	2.132	2.776	3.747	4.604	7.173	8.610
5	.267	.559	.920	1.476	2.015	2.571	3.365	4.032	5.893	6.869
6	.265	.553	.906	1.440	1.943	2.447	3.143	3.707	5.208	5.959
7	.263	.549	.896	1.415	1.895	2.365	2.998	3.499	4.785	5.408
8	.262	.546	.889	1.397	1.860	2.306	2.896	3.355	4.501	5.041
9	.261	.543	.883	1.383	1.833	2.262	2.821	3.250	4.297	4.781
10	.260	.542	.879	1.372	1.812	2.228	2.764	3.169	4.144	4.587
11	.260	.540	.876	1.363	1.796	2.201	2.718	3.106	4.025	4.437
12	.259	.539	.873	1.356	1.782	2.179	2.681	3.055	3.930	4.318
13	.259	.538	.870	1.350	1.771	2.160	2.650	3.012	3.852	4.221
14	.258	.537	.868	1.345	1.761	2.145	2.624	2.977	3.787	4.140
15	.258	.536	.866	1.341	1.753	2.131	2.602	2.947	3.733	4.073
16	.258	.535	.865	1.337	1.746	2.120	2.583	2.921	3.686	4.015
17	.257	.534	.863	1.333	1.740	2.110	2.567	2.898	3.646	3.965
18	.257	.534	.862	1.330	1.734	2.101	2.552	2.878	3.610	3.922
19	.257	.533	.861	1.328	1.729	2.093	2.539	2.861	3.579	3.883
20	.257	.533	.860	1.325	1.725	2.086	2.528	2.845	3.552	3.850
21	.257	.532	.859	1.323	1.721	2.080	2.518	2.831	3.527	3.819
22	.256	.532	.858	1.321	1.717	2.074	2.508	2.819	3.505	3.792
23	.256	.532	.858	1.319	1.714	2.069	2.500	2.807	3.485	3.767
24	.256	.531	.857	1.318	1.711	2.064	2.492	2.797	3.467	3.745
25	.256	.531	.856	1.316	1.708	2.060	2.485	2.787	3.450	3.725
26	.256	.531	.856	1.315	1.706	2.056	2.479	2.779	3.435	3.707
27	.256	.531	.855	1.314	1.703	2.052	2.473	2.771	3.421	3.690
28	.256	.530	.855	1.313	1.701	2.048	2.467	2.763	3.408	3.674
29	.256	.530	.854	1.311	1.699	2.045	2.462	2.756	3.396	3.659
30	.256	.530	.854	1.310	1.697	2.042	2.457	2.750	3.385	3.646
40	.255	.529	.851	1.303	1.684	2.021	2.423	2.704	3.307	3.551
60	.254	.527	.848	1.296	1.671	2.000	2.390	2.660	3.232	3.460
120	.254	.526	.845	1.289	1.658	1.980	2.358	2.617	3.160	3.373
∞	.253	.524	.842	1.282	1.645	1.960	2.326	2.576	3.090	3.291